U0215793

BIRDS.

THE

ILLUSTRATED

NATURAL HISTORY.

BY THE REV.

J. G. WOOD, M.A. F.L.S.

AUTHOR OF " ANECDOTES OF ANIMAL LIFE," " COMMON OBJECTS OF THE SEA-SHORE AND COUNTRY,"
" MY FEATHERED FRIENDS," ETC. ETC.

WITH NEW DESIGNS BY

WOLF, ZWECKER, WEIR, COLEMAN, WOOD, NEALE, HARVEY, ETC.

ENGRAVED BY THE BROTHERS DALZIEL.

BIRDS.

LONDON:

ROUTLEDGE, WARNE, AND ROUTLEDGE, FARRINGDON STREET.

NEW YORK: 56, WALKER STREET.

1862.

LONDON :
PRINTED BY R. CLAY, SON, AND TAYLOR,
BREAD STREET HILL.

PREFACE.

In this Work I have continued to carry out the same plan which has been employed in the previous volume descriptive of the Mammalia, giving to the body of the work a popular and anecdotal character, and reserving the more scientific portions for the Compendium of Generic Distinctions at the end of the volume. Much pains has been taken with that part of the work, for which I am in a great degree indebted to the invaluable "Genera of Birds" by Dr. Gray, a work which has long established itself as the standard of systematic Ornithology as at present accepted by the learned world.

The system employed, and the names that are given in this work, are those which have been sanctioned by the usage of the national collection in the British Museum; and any one who wishes to study the Birds in a systematic manner can accomplish his object by taking this volume to the Museum, and comparing the specimens with the history of the species in the body of the work, and the characteristic distinctions of the genera at its end. By means of this Table, also, any one can ascertain the approximate position which any bird holds in the system of the present day.

I must here take the opportunity of returning my best thanks to the numerous individuals who have most kindly given their aid to this work, many of whom are even now personally unknown to me.

ERRATA.

Page 5. 10 lines from top, *for* "largest" *read* "longest."

— 13. 3 lines from bottom, *for* "largest" *read* "longest." 10 lines from bottom, *for* "Flesh-bearded" *read* "Flesh-beaked."

— 125. *For* "*Macrodiptex*" *read* "*Macrodipteryx.*"

— 165. After "Barbets" *add* "or Puff-Birds."

— 178. 6 lines from top, *read* "The eggs are four in number, and the young birds resemble," &c.

— 370. 16 lines from bottom, *for* "looped" *read* "lodged." 15 lines from bottom, *for* "conical" *read* "suitable."

— 410. 26 lines from top, *for* "three branches" *read* "thorn branches."

— 511. Transfer title "Scansores," &c. together with first paragraph, to page 507, line 23 from top.

— 652. 25 lines from bottom, *for* "obtained" *read* "attained." 14 lines from bottom, *for* "covering" *read* "wing."

— 675. Transpose titles of birds *thus*, "Bittern, Egret, Heron."

— 695. Transpose titles of Sandpipers *thus*, "Common Sandpiper, Green Sandpiper."

— 705. Transpose titles of Snipes *thus*, "Jack-Snipe, Common Snipe."

— 758. 11 lines from bottom, *for* "entire" *read* "cubic."

GROUP OF VULTURES.

BIRDS.

THE most conspicuous external characteristic by which the BIRDS are distinguished from all other inhabitants of earth, is the feathery robe which invests their bodies, and which serves the double purpose of clothing and progression. For the first of these two objects it is admirably adapted, as the long, slender filaments of the feathers are not only in themselves indifferent conductors of heat, but entangle among their multitudinous fibres a considerable amount of air, which resists the ingress or the egress of external or internal heat, and thus preserves the bird in a moderate temperature through the icy blasts of winter or the burning rays of the summer sun. A similar function is discharged by the furry coats of many mammalia; but the feathers serve another office, which is not possessed by hair or fur. They aid the creature in progression, and enable it to raise and to sustain itself in the atmosphere. Towards the promotion of this latter function the entire structure of the body and limbs is obviously subservient, and even in the comparatively rare instances where the bird—such as the penguin, ostrich, or the kiwi-kiwi—is destitute of flying powers, the general idea of a flying creature is still preserved.

The fuller and more technical description of the Birds runs as follows. They are vertebrate animals, but do not suckle their young, nourishing them in most instances with food which has been partially macerated in their own digestive organs, and which they are able to disgorge at will, after a manner somewhat similar to that of the ruminating quadrupeds. The young are not produced in an actively animated state, but inclosed in the egg, from which they do not emerge until they have been warmed into independent life by the effects of constant warmth. Generally, the eggs are hatched by means of the natural warmth which proceeds from the mother bird ; but in some instances, such as that of the tallegalla of Australia, the eggs are placed in a vast heap of dead leaves and grass, and developed by means of the heat which is exhaled from decaying vegetable substances, and which is generated to such an extent that in some cases, such as a wet haystack, it actually sets the seething mass on fire. Urged by a like instinct, our common English snake deposits its eggs in secret spots, such as dunghills and hotbeds, and there leaves them to be hatched by the constantly generated warmth. An analogous process has long been in vogue among the Egyptians for the hatching of young poultry by artificial heat, and has been, in comparatively recent years, introduced into this country.

When the egg is first produced, the future chicken is merely indicated by a little germ-spot, barely the size of a single oat-grain, and does not attain the power of breathing atmospheric air, and receiving nourishment into its mouth, until a period of many days has elapsed. To watch the gradual development of the young chick is a most interesting experiment, and one which is full of suggestive instruction. There is but little difficulty in the matter, even in the very earliest stages of incubation, for the structure of the egg is so wonderfully balanced, that in order to view the little germ-spot it is only necessary to lay the egg on its side and remove a portion of the shell, when the germ will be seen lying immediately under the aperture. In whatever way an egg may be turned, the germ-spot invariably presents itself at the highest point, provided only that the egg be laid on its side, and that the living principle has not been extinguished, for life, however undeveloped, seems always to aspire. As the chick increases in size, the manipulation becomes easier, but it is always better to immerse the egg in water or other transparent liquid before removing the shell, and to keep it submerged during examination.

There are few objects which will better repay investigation than the young bird in its various stages of development. It is so wonderful to see the manner in which a living creature is gradually evolved from the apparently lifeless substances that are contained within an egg. The being seems to grow under our very gaze, and we arise from the wondrous spectacle with an involuntary feeling that we have been present at a veritable act of creation. To describe fully the beautiful process in which a chick is elaborated out of the germ-spot would occupy very many pages, and cannot be attempted within the compass of the present work. Briefly, however, the order of events is as follows.

When a newly laid egg is opened, it is found to contain a mass of substance which is popularly divided simply into "white" and "yolk," but when examined more closely, by placing it under water and carefully removing the shell, its contents are found to be very elaborately disposed, so as to meet the object for which it was formed. Immediately within the shell lies a semi-transparent and tolerably strong membrane, composed of two distinct layers, pressed closely to each other for the greater portion of its extent, but separated at the widest end of the egg, and containing between the layers a supply of atmospheric air to satisfy the requirements of the young chick. This space gradually increases as the young bird becomes more developed. Within this membrane lies the "white," a liquid, albuminous substance, which is also disposed in two distinct layers, that which is nearest to the shell being rather thin and fluid, while the inner layer is comparatively thick, tenacious, and very transparent. Within the white lies the yolk, surrounded by a slight membrane, which serves to guard it from mixing with the white. In order to prevent the yolk from shifting its place at every change of position in the egg, it is anchored, so to speak, in its proper place by two curious ligaments fastened to the yolk membrane. Upon the yolk, and immediately under the membrane, lies the little germ which in the space of three weeks will be developed into a bird.

After a few hours of warmth, the first idea of the chick is seen in a little whitish streak, barely one-tenth of an inch long, rather wider at one end, and always lying *across* the egg. By degrees, this streak enlarges, and forms a groove between two little ridges, and in a few hours later, a delicate thread is seen lying in the groove, being the first indication of the spinal cord. Presently a number of the tiniest imaginable square white plates make their appearance on each side of the thread, and are the commencement of the vertebræ. It is most curious to see these gradual changes, for the different parts come into view as though they were crystallized from the substance of the egg. By the end of the first day the germ takes a curve, and looks something like a little maggot as it lies in the yolk. The little heart is just perceptible on the second day, and on the third a series of blood-vessels have been formed, and are supplied with blood by a very curious system of arteries and veins. By similar processes the various organs of the body are built up, the feathers beginning to make their appearance about the twelfth day, and on the nineteenth or twentieth day the chick pierces with its beak the air-sac which lies at the blunt end of the egg, and by means of the air which it thus obtains is often able to chirp before it chips the shell.

During this period of its existence the young bird is nourished by the yolk, which is connected with its abdomen, and which is not separated from the body until the chick has broken the shell, and is able to respire freely. When leaving the egg-shell, the chicken pecks in a circle, which nearly corresponds with the shape of the air vesicle, so that when it emerges it walks out of a circular trap-door which it has cut for itself, and which often remains suspended by a hinge formed from an uncut portion of the lining membrane. It is possible that the shell may be softened in this spot by the presence of internal air, and may therefore afford an easier passage to the inclosed chick. In order to enable the tender-billed little creature to penetrate so hard a substance as the egg-shell, the tip of its beak is furnished with a strong horny excrescence, which falls off shortly after the chicken has emerged from the egg, thus carrying out the principle that nature abhors a superfluity.

Having watched the little bird through its life-development, we will now proceed to a short examination of the bird-skeleton, and will take for an example that of the eagle. Even in the mammalia the skeleton presents an appearance very different from that of the living creature, and in many instances the external structure and its bony framework are so unlike each other that an inexperienced observer would probably refer them to different animals. But in the birds the contrast is still more strongly marked, for the skeleton is not only deprived of its fleshy covering, but also of the feathery coat which surrounds the bird so thickly, and which in many cases, such as the owl, entirely masks the general outline of the bird. Taking the skeleton of the eagle as a good example of the bony scaffolding which supports the vital and locomotive organs of birds, we will begin with the head and proceed gradually to the tail.

The chief and most obvious distinctive feature in the skull of a bird and of a mammal lies in the jaw-bones, which in the bird are entirely toothless, and are covered at their extremities with a peculiar horny incrustment, termed the beak or bill. This bill is of very different shape in the various tribes of birds; being in some cases strong, sharp, and curved, as in the birds of prey; in others long, slender, and delicate, as in the creepers and humming-birds ; and in others flat, spoon-like, soft, and sensitive, as in the ducks The movement of a bird's jaw is not precisely similar to that of a mammal, owing to the manner in which a certain little bone, termed from its squared shape the quadrate bone, is articulated to the bones of the skull. On reference to the accompanying illustration, this bone will be seen just at the junction of the lower jaw with the skull.

Passing from the head to the neck, we find a marked distinction from the mammals. In them, the vertebræ of the neck are never more than seven in number ; the long neck of the giraffe and the short one of the elephant being obtained by the prolongation of the seven vertebræ in the former and their compression in the latter. In the birds, however, there are never less than nine vertebræ in the neck, and in some cases the number is considerably greater ; the swan, for example, possessing no less than twenty-three of these bones. The neck is also much longer in the birds, being in many instances longer than the remainder of the body. The vertebræ of the neck are extremely flexible, as is needful for the peculiar habits of birds ; but those of the back are immovably connected with each other, and in many cases are even fused together. The seven or eight short vertebræ which form the tail are movable, and are generally terminated by a single bone of greater length than any of the others.

SKELETON OF EAGLE.

We now proceed to the breast and body. The ribs are chiefly remarkable for a flat appendage, which starts from the lower portion of the bone, and is directed backwards, so that it overlaps each succeeding rib. The breast bone is placed lower than might be supposed from the external aspect of a bird, and is of very great size. Its substance is much flattened, and it possesses a strong ridge or keel of bone, which varies in its depth according to the powers of flight possessed by the particular species to which it belongs. As the eagle is a strong-winged bird, the keel is very prominent, but in such non-flying birds as the ostrich and the apteryx, there is no keel at all. Between the breast bone and the neck lie four clavicles, or collar bones, differing much in size and shape in the various species of birds. One set of them, technically called the *os furculare*, from its forked shape, is sometimes absent, its place being supplied by a ligament ; but the others, termed the *claviculæ coracoidæ*, are invariably present. These two sets of bones are familiar to all who have carved a fowl, under the terms of " merry-thought " and " neck bones."

The limbs now come before our notice, and we cannot but be struck with the curious fact, that in the birds the bipedal mode of walking again makes its appearance, having disappeared through all the mammalia, with the exception of man. There is, however, this analogy between the lower mammals and the birds, namely, that in both instances the anterior limbs are intended for progression, although in the one case these formations belong to earth, and in the other to the air. The bones of the wing present a considerable resemblance to those of a man's arm, as may be seen by comparing the skeleton of the eagle with that of the man in Volume I. The upper arm bone is of various lengths in the different birds, being of wonderful proportions in such long-winged birds as the albatross, but very short in the penguins, the cassowary, and many other birds. The two bones of the fore-arm, technically called the *ulna* and *radius*, are also long in the long-winged birds, and serve to carry a large expanse of feathers. Of these two, the ulna is the larger and more cylindrical. To the end of the ulna and radius are jointed the two little bones of the wrist, which bear a quasi hand, composed of a thumb and two fingers. The thumb is very small, consisting of either one or two bones ; and the fingers, which are only needed for the purpose of bearing feathers, are also small. One of them is composed of either two or three joints, but the other is a very little one, being but one single pointed bone.

The bones of the legs are very similar in their arrangement to those of the mammalian quadrupeds, although they are subject to certain modifications, especially at their extremities. The thigh-bone is tolerably strong and cylindrical, but of no very great length, in proportion to the size of the bird or the length of its limbs. Even in the curious stilt-plover, where the legs are of such extraordinary length, the thigh bone is comparatively short, and not visible outside the feathers. The leg bone, or "tibia," is always the largest bone of the limb, and is accompanied by a very small and undeveloped "fibula," which is only attached to its upper extremity, and tapers gradually to a point. The "instep," as we should term it in a human foot, is merely a single bone, jointed at its upper extremity to the tibia, and its lower to the bones of the toes. In general, birds are furnished with four toes on each foot, but there are several exceptions to this rule, among which the ostrich is the most conspicuous.

Not only do the bones of a bird differ in external form from those of a mammal, but they are also considerably modified in their structure. In the mammals the bones are heavy, solid, and their centre is filled with marrow; but in the birds the bones are of a much lighter make, and many, such as the upper wing bone, the breast bone, and part of the skull, are, moreover, hollow throughout their centres, so as to combine great strength with the least possible weight. These hollow bones communicate with the legs through certain curious appendages called air-sacs, which open into the lungs, and apparently serve as reservoirs of respirable atmosphere, so that the bird is able to force the hot and rarefied air from its lungs into its bones. In some very rare instances even the bones of the feet and toes are hollow, and penetrable with air as far as the insertion of the claws. Some birds, however, especially those of small dimensions, do not possess these hollow bones, and in all cases the cavity is not developed until the creature has attained to maturity. In the apteryx, a non-flying bird, the only hollow bone is that of the lower jaw. So complete is the communication with the lungs through the bones of some birds, that if the bone should be broken they are enabled to breathe through the open extremity, even though the throat be compressed, or the head plunged under water.

THIS slight sketch of the skeleton is necessary as a prelude to the description of the FEATHERS, because several of the most important of these appendages derive their names from the portion of the structure on which they are set.

On a general view of a bird it will be seen that the feathers fall naturally into two orders, namely, those of progression and those of covering. But as in the description of a bird, especially of one that is unknown to science, and of which no figure is extant, it is needful to describe the form and colour of the different portions of the creature with great accuracy, this sweeping division of the feathers into two sets will be quite insufficient for the purpose. On a closer examination, however, it will be seen that the feathers possess a kind of natural arrangement, which, with a few unimportant and obvious additions, is amply sufficient for actual scientific purposes. The best mode of learning the name of the different parts of the plumage is to procure any bird, say a sparrow or pigeon, which may easily be obtained, and to investigate the formation and arrangement of the feathers from actual inspection. It is an interesting little study, and will save much time, as a lesson once so learned will never again be forgotten. We will suppose a dead sparrow to be laid on the table.

Let one of its wings be spread upon the table, and its plumage will be seen to consist of a row of long, flat, and stiff quill feathers, whose insertion is covered by a great number of smaller and softer feathers. The quill feathers are technically termed "principals," and the others are called from their office, "coverts." Before examining the principals, it needs that the coverts be first attacked, because they must be removed before the quill feathers can be properly traced to their sources. Along the upper surface of the wing run two or three rows of these short feathers, which are termed the "greater coverts," and below these a single row of "lesser coverts," the latter of which may be distinguished by their slightly different shape and manner of lying.   The under surface of the wing is clothed with a dense layer of small feathers termed the "under coverts."

Now let all the upper coverts be removed, and the quill feathers will be visible from their insertion to their extremity.   On spreading out the wing it will be seen that ten of these feathers spring from that portion of the wing bone which corresponds to the hand and fingers of man.   As these feathers come first in point of order, beginning at the extremity of the wing, they are termed the "primaries," and indicate, by their shape and development, the mode of flight followed by the bird.   If, for instance, they are comparatively short, rounded, and concave, as is the case with our example, the sparrow, the flight is slow and laborious, accompanied with much beating of the wing and dipping in the air between each stroke.   If they are long, firm, and flat, as seen in the eagles, vultures, and other similar birds, the flight is easy and graceful, though capable of exceeding swiftness when needful.   If they are large, concave, and edged with soft fringes, the flight is quiet and noiseless, as is seen in the owls.   Some birds, such as the ostrich, the cassowary, and other running birds, possess short and pointed primaries, which can hardly be recognised as belonging to so large a bird, and the flight is in consequence reduced to zero.

Next to the primaries come a second set of quills, called for that reason "secondaries." They are often undistinguishable externally from the primaries, into which they imperceptibly merge, but may be at once detected by following them to their roots, which are inserted upon that part of the wing which corresponds to the wrist and elbow of man. They are very variable in number, shape, and size; and although they are in some birds hardly distinguishable from the primaries, are in others very prominent and conspicuous.

Next to the secondaries come the "tertiaries," which take their root in that part of the wing which corresponds to the elbow and shoulder.   In some birds, such as the plovers, the tertiaries are extremely long, giving a very peculiar character to the wing.   In the crane they are developed into long, drooping plumes; but in most birds they are very much shorter than the primaries, and are merged into the little feathers that cover the upper surface of the wing.   Upon the thumb is a little fan-like wing, quite distinct from the remainder of the feathers, and distinguished by the name of "winglet."

A second set of quill feathers is to be found upon the tail, where they assume different shapes and dimensions according to the species of bird, its sex, age, and the nature of its flight.   As these feathers perform the office of a rudder in directing the flight of the bird as it passes through the atmosphere, they are technically termed "rectrices," or directors. The insertion of these quill feathers is concealed above and below by certain little feathers, named from their position the upper and under tail coverts.   Generally, these feathers are of very small dimensions, but in some examples they attain to considerable length, and are very imposing in their appearance.   The magnificent "train" of the peacock is composed, not of the tail quill feathers, which are short, stiff, and used chiefly for the proper displayal of the train, but of the greatly developed upper tail coverts; and from the under tail coverts of the marabout stork are taken those beautiful plumy ornaments that are so well known as articles of feminine decoration.

Lastly, there are some feathers on either side of the head, which shield the orifice of the ear from injury, and are therefore named the ear coverts; and the patch of feathers upon the shoulders is appropriately known by the name of "scapularies."   In the accompanying sketch of the swallow-tailed falcon, the position of the principal groups of feathers is indicated.   A denotes the primaries, or first quill-feathers of the wing; B, the tertiaries; C, D, E, the lesser and greater coverts; F, the scapularies; G, the rectaries; H, the upper tail coverts; I, the ear coverts.

This array of plumage is not obtained until the bird has attained to some amount of development, and the shape and colour of the feathers are so distinct from each other at the different epochs of a bird's life, that in many instances an adult, a half-grown, and a juvenile specimen have been taken for individuals of different species, and noted as such in systematic catalogues.

When the young bird is first hatched its feathers are hardly worthy of the name, being mostly restricted to a kind of soft down.   In the course of a week or two the quill feathers begin to make their appearance, like little yellow or black spikes projecting from the wings, but it is not until after the lapse of some time that they attain sufficient strength to sustain the bird in the air. In a few months after the young bird has gained its first plumage, it loses the feathers with which it has only just been clothed, and by going through the process technically termed "moulting," indues an entirely new plumage, which is often very different from the former in its traits and general aspect. In many cases the bird spends three years of life before it is clothed with the full glory of its adult garments, and during the first and second years the two sexes are so similar as hardly to be distinguished from each other without dissection. The moult takes place annually even in adult birds, and is highly needful as a means of giving them a new set of plumes to replace those which have been worn out by the service of a whole year's wear.

A similar phenomenon is observable in the fur-clad mammalia, who shed the worn and ragged hairs in the autumn, and obtain a new and warm coat in readiness for the colder months. Even in the human race the same principle is observed; but the change

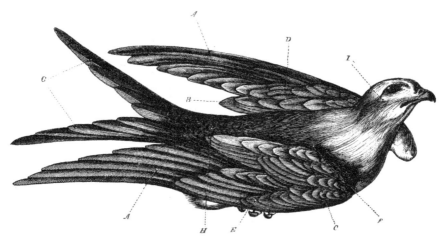

SWALLOW-TAILED FALCON.—(Showing the Feathers.)

of hair is in them so gradual that it is scarcely perceptible, except to those who watch its progress. Indeed, a partial moult can be induced at any time upon a bird, and employed to restore a broken or damaged feather, irrespective of the time of year. If the injured feather be drawn from its socket—an operation which is always attended with some pain and loss of blood—it will soon be replaced by another and a perfect feather, springing from the same socket.

The rapidity of the process is really astonishing, and presents a curious analogy with the phenomenon of the rapid formation of the stag's horns. A remarkable instance occurred lately within my own observation, in the person of a long-tailed Australian parrakeet. The bird contrived to get out of its cage, and in flying along a large room was chased by a man, who made a successful grasp at its tail, but failed in securing the bird, which flew screaming away, leaving its beautiful long tail in the hands of its would-be captor. At last the bird was replaced safely in its cage, but presented a very forlorn aspect in consequence of the loss of its tail. A very few days, however, showed the tips of some new feathers, that had already grown long enough to pass beyond the tail coverts, and in a month or so the long tail was even more beautiful than ever.

There seems, indeed, to be a very marked analogy between the feathers of birds and the tusks or horns of many mammals. Both depend greatly on the sex and age of the animal to which they belong, and their shape and dimensions are unfailing indications of the vigour or feebleness of their owners.

The expanse of the outstretched wings of every flying bird is so very great in comparison with the size of the body, that there is need of very great muscular development in order to give the powerful strokes by which the body of a bird is urged through the atmosphere. It is for this purpose that the breast bone is furnished with the deep keel which has already been mentioned, for its projecting edge and sides afford attachment to muscles of enormous size, which are devoted to the purpose of drawing the wing forcibly downwards. Although in the gallinaceous birds, of which the common barn-door fowl is a familiar example, the pectoral muscle, as it is called, is not so largely developed as in many of the swift-winged birds, it attains to considerable dimensions, as may be seen by every one in carving a common fowl, whether it be boiled or roasted. This muscle forms the solid and delicately flavoured meat which is attached to the wing when removed, and also constitutes the greater part of the " breast."

Strength, however, is not the only requisite in a bird's wing. It is evident that if the stroke were only made upwards and downwards, the bird would never rise in the air, much less make any progress forwards. On gently moving the wing of a dead bird, we shall see how beautifully its opening and closing is managed, so that on the stroke the feathers beat the air with their flat sides, but present their sharp edges as they return for another stroke. This movement is copied by the oarsman as he throws back the blade of his oar for another stroke, and is called "feathering" on account of the source from which it is derived. The means by which this object is attained is through a most perfect and beautiful arrangement of the wing muscles, which are so fashioned as to give the wing a slight and involuntary turn just as it is thrown backwards after making its stroke.

The reader who desires to understand this curious structure, cannot do better than to denude the wing of some bird of its feathers, to remove the skin, and lay bare the muscles. If he then moves the wing as if in flight, he will see, by the play of the different muscles, the part which they take in the general movement, and the wonderful harmony in which every individual muscle works with its fellows. Next let him pass a smooth but blunt edge, such as a small paper-knife, or the flat handle of a scalpel, between the different muscles and separate them throughout their entire length. By pulling each muscle in turn with a pair of forceps, he will see its object, and will be able to form a very good idea of the manner in which all the muscles act while working simultaneously in moving the wing.

In the generality of birds, the senses of touch and taste seem to be but little developed, while those of sight, hearing, and smell are decidedly acute.

The sense of touch can be but very slight in a creature that is covered with feathers over the whole of its body ; whose limbs are either plume-clad or tipped with horn, and whose mouth is defended by a hard, horny beak. There are exceptions in the case of the ducks, and many similar birds, whose beaks are soft and evidently possessed of delicate tactile powers, but in the generality of birds this sense is decidedly dull. Taste, again, can have but little development, as the tongues of most birds are devoid of the soft and sensitive surface which is found in the tongue of man and the mammals in general. At the base of the tongue the nerve-bearing papillæ are found in some genera of birds, but even in them these organs of taste occupy but a small portion of the tongue, and can give but little indication of savour. In many birds, indeed, such as the woodpecker and the humming-bird, the tongue is employed in a manner analogous to the same organ in the ant-eaters, being used to procure food and to draw it into the mouth. This structure will be described more at length when we come to treat of the birds where it is especially developed.

The sight of birds is almost invariably remarkable for its development and its adaptation for near or distant objects. The swallow, for example, when darting through the air with that swiftness which has become proverbial, is capable of accommodating its sight to the insect which it pursues even in the short space of time which is occupied by its swoop at its victim. The same phenomenon may be noticed in the falcon, which is able to perceive a little bird or animal on the ground, and though sweeping downwards with such wonderful rapidity that it looks merely like a dark streak in the air, is able to calculate its distance so exactly, that it just avoids dashing itself to pieces on the ground, and snatches up its prey with the same lightning speed which characterises its descent.

It is very probable that a curious structure, named from its shape the "pecten," or comb, which is found in the interior of the bird's eye, may contribute to this peculiarity of vision. This comb is of a fan-like shape, and is situate upon the spot where the optic nerve enters the eye, projecting obliquely upwards, and evidently playing some very conspicuous part in the economy of the eye. The teeth, or folds of which this fan or comb is composed, are black in colour and very variable in number, being only six or seven in the owls, and twenty or thirty in the sparrow. There is a plentiful supply of blood-vessels in the comb, but no muscular tissues, and it is supposed by several anatomists that its expansion or contraction, caused by the greater or less amount of blood which fills the vessels, may have some effect in the peculiarly delicate adjustment of the eye which has already been mentioned.

From the contact of external substances, as well as for the purpose of excluding unnecessary light, the eye of the bird is furnished with two ordinary eyelids, and a third, or supplementary eyelid, which plays within the others, and is technically called the nictitating membrane. This membrane is elastic, and by its own contractility is kept within the angle of the eye as long as its services are not needed. When, however, the bird wishes to cleanse its eyes from dust or other annoyances, it draws the membrane rapidly over the eye, letting it return to its place by its own powers of contraction. The eye of the bird is further remarkable for a series of bony plates which surround the eye, and are supposed to have a great influence in increasing or lessening the convexity of the eyeball. The number of these plates is nearly as various as the teeth of the comb, but upon an average their number is thirteen or fourteen. There are many other curious and interesting details in the anatomy and general structure of the birds, but as this publication is not intended as a work on comparative anatomy, we must proceed to the histories of the birds themselves.

LAMMERGEYER.—*Gypáëtos barbátus.*

BIRDS OF PREY. VULTURES.

In the arrangement of the various species of living creatures which possess a visible organization, the greater or less perfection of the structure has formed the basis of systematic classification. In a certain sense, however, the development of all animals is equally perfect, inasmuch as it is most perfectly adapted to the necessities of the particular species or individual; so that the term perfection is necessarily rather a conventional one, and the systems of zoological arrangement are as various as their authors. By common consent, however, the VULTURES take the first rank among birds, and in the catalogue of the British Museum, the LAMMERGEYER, or BEARDED VULTURE, stands first upon the list.

This magnificent bird is a native of Southern Europe and Western Asia, and often attains a very great size, the expanse of its wings being sometimes as much as ten feet, and its length nearly four feet.

Before describing this species it may be as well to give a few of the distinguishing points by which the Vultures may be separated from the eagles, hawks, and other diurnal birds of prey. All the birds of prey, called scientifically Raptatores, or Accipitres, are readily known by their compressed and hooked beaks, the powerful talons which arm their toes, and the twelve or fourteen quill feathers of the tail. The Vultures are distinguished by the shape of the beak, which is of moderate size, nearly straight above, curved suddenly and rounded at the tip, and without any "teeth" in the upper mandible. The middle toe of the foot is larger than the others, and the outer toes are connected with them at their base by a small membrane. In the greater number of species the head and upper part of the neck are nearly naked, and the eyes are unshaded by the feathery ridge which overhangs these organs in the eagles. As a general rule, the Vultures feed on dead carrion, and are therefore most beneficial to the countries which they inhabit. When pressed by hunger, however, they will make inroads upon the flocks and herds, and will not disdain to satisfy their wants with rats, mice, small birds, or insects.

The name of Bearded Vulture has been given to the Lammergeyer on account of the tufts of long and stiff bristle-like hairs which take their rise at the nostrils and beneath the bill, and form a very prominent characteristic of the species. The "cere," a soft naked skin which is placed on the base of the beak, is not very large, and the upper mandible is rather higher in front of the cere. The feet are not so large as in many of the birds of prey, and are not very well adapted for seizing or retaining prey. As, however, the Lammergeyer is not a bird of chase, like the eagle and falcon, but obtains its food by striking chamois, goats, and other animals over the precipices near which they are standing, the powerful claws of the eagle would be of little service to it. The claws are therefore comparatively feeble, short, and are covered with feathers down to the toes.

The colour of the Lammergeyer is a grey-brown, curiously dashed with white upon the upper surface, in consequence of a white streak which runs along the centre of each feather. The under surface of the body, together with the neck, are nearly white, tinged with a wash of reddish-brown, which is variable in depth in different individuals. In the earlier stages of its existence, the Lammergeyer is of a much darker hue, and the white dashes upon the back are not so purely white nor so clearly defined. The head and neck are dark brown, and the brown hue of the back is of so deep a tint that the young bird has been classed as a separate species, under the title of *Vultur niger*, or Black Vulture.

Like the true Vultures, the Lammergeyer is invaluable as a scavenger, and if an animal be killed and left exposed to view, the bird is sure to find out the spot in a very short time, and to make its appearance as if called by some magic spell from the empty air. But as there is not a sufficiency of dead animals for the food of this hungry and powerful bird, it makes prey of lambs, kids, hares, and such like animals, nor disdains to feed even on rats, mice, and other small quadrupeds. With the larger animals, such as the chamois, the Lammergeyer cannot successfully cope on level ground, but taking advantage of its wings, it hurls itself suddenly against some devoted animal which is standing heedlessly near a precipice, and by the force of its blow strikes the poor creature into the depths below, whither it is immediately followed by its destroyer. Even mankind is said to be endangered by these sudden attacks of a hungry Lammergeyer, and more than one chamois-hunter is reported as having been killed by an assault from one of these birds.

The Lamb-vulture, as is the import of its name, does not restrict itself solely to the snowy mountains on which it takes up its chief residence, but often makes considerable journeys into the cultivated portion of the country, for the sake of picking up the lambs and other valley-inhabiting animals.

The general aspect of the Lammergeyer is more like that of an eagle than a Vulture, but its carriage and demeanour are devoid of that fearless, regal grandeur which is so characteristic of the eagles of all lands. When flying, however, its appearance is truly magnificent, and on account of its great sweep of wing and powerful flight, the size of the Lammergeyer has been greatly exaggerated. Indeed, it is probable that the celebrated roc which plays so important a part in the adventures of Sinbad the Sailor, and in other portions of the Arabian Nights, is merely the Lammergeyer viewed through the magnifying medium of Oriental exaggeration.

A variety of this bird is found in many parts of Africa, where it is appropriately named Abou-Duch'n, or Daddy-long-beard. It seems to be as audacious as its European and Asiatic relation, and is possessed of even greater boldness. Bruce gives a graphic and amusing narrative of the cool audacity that was displayed by one of these birds. The author, with a number of his attendants, were seated on the summit of a mountain, engaged in cooking their dinner, when a Lammergeyer came slowly sailing over the ground, and boldly alighted close to the dish of boiled meat around which the men were sitting. Undismayed by their shouts of distress, he quietly proceeded to reconnoitre the spot, while the men were running for their spears and shields, and going up to the pot in which some goat's flesh was boiling, he inserted his foot for the purpose of abstracting the meat. Not being prepared for the sudden scalding which ensued, he hastily withdrew his foot, and fastened on a leg and shoulder of goat's flesh which were lying on the dish, carrying them away before he could be intercepted.

The attendants were quite afraid of the bird, and assured Mr. Bruce that it would return in a short time for more meat. Accordingly, in a very few minutes, back came the Lammergeyer, but was evidently rather suspicious at the look of Mr. Bruce, who had taken up his rifle, and was sitting close to the pan of meat. In spite of the shouts of the attendants, the bird, which evidently held in the greatest contempt the warlike capabilities of the natives, and was not prepared for European weapons and hands, settled on the ground about ten yards from the meat, and the next instant was lying dead on the earth, with a rifle-ball through its body. When brought to the scales, the dead bird was found to weigh twenty-two pounds, and the expanse of its wings was eight feet four inches, although it was undergoing its moult at the time.

When the bird was handled, a large amount of yellow dust was shaken from the feathers, and upon the breast was so plentiful that it "flew in full greater quantity than from a hairdresser's powder puff." Mr. Bruce at the time thought that this yellow dust was some extraordinary provision of nature for the purpose of defending the bird against the peculiarly wet climate of the country in which it was found. It is, however, merely a natural deposit of feathery substance, and in many birds, such as the common cockatoo, the heron, and birds of prey, is permanently formed. As this curious powder is produced from the feathers, and is a result of their reproduction, a few lines on the subject can well be spared in the present place.

Each feather is rooted in a socket, which is formed by a fold of the skin, and at the bottom of this tube or socket a peculiar formative fluid is secreted on the commencement of the new feather. By degrees this fluid is inclosed in a little conical vesicle, its closed point being directed outwards, and its open base being held within the cavity of the socket. As it increases in size, the conical point is pushed through the skin, and serves as a wedge by which the feather, which is gradually being developed in its interior, is thrust through the integuments. As the feather gains strength, this conical vesicle is of no service, dries up, and falls off in little plates or scales. In many feathers, however, the development proceeds no farther than the formation of a hollow shaft, the formative fluid drying into powder, and plentifully scattered on the surrounding plumage; this is the yellow dust or powder noticed by Mr. Bruce. The object of its formation is not yet known, but it clearly must serve some important purpose, or it would not be produced in such abundance, as is found in many of the birds where it permanently exists. In the Lammergeyer, for example, it flew from the feathers in clouds, and in the cockatoo is so plentiful, that any one who handles a tame cockatoo for a few minutes will be covered with the particles of this curious production. On examining the feathers of a cockatoo's head and neck, the imperfect and open quills from which the secretion is shaken are many in number, and conspicuous to the sight as the bird bends down its head to receive the caresses of which it is so fond.

The Lammergeyer, like other birds of prey, loves to build upon some elevated spot, and generally places its nest upon the summit of a lofty cliff. The nest is a very rude affair, being chiefly composed of sticks laid inartistically together, and serving merely as a platform, on which the eggs and young may be lifted from actual contact with the rock. Gesner relates an account of a Lammergeyer's nest which was built upon

CONDOR.—*Sarcorhamphus Gryphus.*

three oaks, and was of such dimensions that a wagon might have taken refuge under its shelter. The eggs of this bird are two in number, and their ground colour is a dirty white, washed with irregular brown patches.

ON account of a curious fleshy appendage which decorates the base of the bill and the neighbouring portions of the head, a small group of Vultures has been separated from the remaining species, and gathered into a family under the appropriate title of Sarcorhamphidæ, or Flesh-bearded Vulture. This family is but a small one, comprising the CONDOR, the King Vulture, and the well-known American Vultures, or Zopilotes.

Although not exceeding the lammergeyer in dimensions, the Condor has been long celebrated as a Goliath among birds, the expanse of its wings being set down at eighteen or twenty feet, and its strength exaggerated in the same proportion. In reality, the expanse of a large Condor's wing will very seldom reach eleven feet, and the average extent is from eight to nine feet. In one specimen, where the measurement of the extended wings was only eight feet one inch, the largest quill feather of the wings was two feet two inches in length; the diameter of the body was nine inches, and the total length from the point of the beak to the extremity of the tail, was three feet two inches.

The general colour of the Condor is a greyish-black, variable in depth and glossiness in different individuals. The upper wing coverts are marked with white, which take a greyer tint in the female, and the exterior edges of the secondaries are also white. The adult male bird may easily be distinguished by the amount of white upon the feathers, so that the wings are marked with a large white patch. Around the neck is set a beautifully white downy collar of soft feathers, which does not entirely inclose the neck, but leaves a small naked band in front. This featherless band is, however, so small, that it is not perceptible except by a close examination.

The crest of the male Condor is of considerable size, occupying the top of the head and extending over a fourth of the basal portion of the beak. The nostrils are intersected at the base of the beak, in a space which is created for them by means of the sudden sloping of the crest. Although the crest of the Condor presents an apparent analogy with the wattles of a turkey, it cannot be inflated at pleasure, as is the case with that bird, but is hard of substance and contains but few blood-vessels. As the Condor, when at rest, is in the habit of sinking its head upon its shoulders, and concealing the neck within the collar of white down by which it is surrounded, the aspect of the bird is very curious, as it sits with its large hooked beak and projecting crest lying on the shoulders as if it possessed no neck at all.

There are several curious details in the internal structure of the Condor, among which may be mentioned the remarkable fact that its "gizzard" is furnished with longitudinal rows of horny spikes, which are probably intended to aid the creature in the more rapid comminution and digestion of its food.

The Condor is an inhabitant of the mountain chain of the Andes, and is celebrated not only for its strength and dimensions, but for its love of elevated localities. When enjoying the unrestricted advantages of its native home, it is seldom found lower than the line of perpetual snow, and only seems to seek lower and more temperate regions when driven by hunger to make a raid on the flocks or the wild quadrupeds of its native country. Although preferring carrion to the flesh of recently killed animals, the Condor is a terrible pest to the cattle keeper, for it will frequently make a united attack upon a cow or a bull, and by dint of constant worrying, force the poor beast to succumb to its winged pursuers. Two of these birds will attack a vicugna, a deer, or even the formidable puma, and as they direct their assaults chiefly upon the eyes, they soon succeed in blinding their prey, who rapidly falls under the terrible blows which are delivered by the beaks of its assailants.

The strength of the Condor is really prodigious, a powerful man being no match even for a wounded and tethered bird; and its tenacity of life is such, that a combat of endurance is nearly certain to end in favour of the Condor. Humboldt relates a curious anecdote of a Condor that resisted a series of efforts that were made in order to deprive it of life. In vain was it strangled for many minutes, for as soon as the noose was removed from its neck the bird walked about as composedly as if nothing had happened to it. At last a pistol was brought to bear upon it, and three bullets were fired from a distance of four paces, all fairly entering the body. A fourth ball struck against the leg bone and rebounded without doing much apparent harm. In spite of all the wounds which it had received, this poor bird survived for nearly half an hour.

The Indians are possessed with a strange prejudice against the Condor, and whenever they catch one of these birds alive, they put it to death through the medium of the most cruel tortures. Their mode of capturing so powerful a bird is worthy of notice, as it is based upon the habits of the Condor. A cow or a horse is killed, and the body thrown negligently on one side, so as to be exposed to the open air. In a very short time the Condors begin to assemble, and soon are engaged in feeding voraciously upon the unexpected and welcome banquet. As soon, however, as they have gorged themselves to the full, the Indians dash in among them, armed with their lassos, and make easy captives of the finest birds. When they feel the noose around their necks, the Condors endeavour to eject the huge meal which they have swallowed, but are made hopeless prisoners before they can rid themselves of the enormous mass of food which they have contrived to pack into their interiors.

On account of the exquisitely delicate scent of this bird, the native Mexicans have distinguished it by a name which bears allusion to its keen sense of smell, and has been modified into the more euphonious word Condor.

Although the Condor is not a very social bird, it is generally found in little assemblages of five or six in number, which are seen either seated motionless upon the summits of the rocks, their outlines cutting sharply against the sky, or sailing slowly in circles at an enormous elevation above the ground. The flight of these birds is peculiarly grand and beautiful, and seems to be achieved by the movement of the head and neck rather than by that of the wings. Mr. Darwin gives the following animated description of the flight of the Condor.

"Except when rising from the ground, I do not recollect ever having seen one of these birds flap its wings. Near Lima I watched several for nearly half an hour without once taking off my eyes. They moved in large curves, sweeping in circles, descending and ascending without once flapping. As they glided close over my head I intently watched from an oblique position the outlines of the separate and terminal feathers of the wing : if there had been the least vibratory movement these would have blended together; but they were seen distinct against the blue sky. The head and neck were moved frequently, and apparently with force, and it appeared as if the extended wings formed the fulcrum on which the movements of the neck, body, and tail acted. If the bird wished to descend, the wings were for a moment collapsed, and then when again expanded with an altered inclination, the momentum gained by the rapid descent seemed to urge the bird upwards with the even and steady movements of a paper kite.

In case of any bird soaring, its motion must be sufficiently rapid, so that the action of the inclined surface of its body on the atmosphere may counterbalance its gravity. The force to keep up the momentum of a body moving in a horizontal plane in that fluid (in which there is so little friction) cannot be great, and this force is all that is wanted. The movement of the neck and body of the Condor we must suppose sufficient for this. However this may be, it is truly wonderful and beautiful to see so great a bird, hour after hour, without any apparent exertion, wheeling and gliding over mountain and river."

The Condor deposits its eggs, for it makes no nest whatever, upon a bare shelf of some lofty rock. The eggs are two in number, greyish-white in colour, and are laid about November or December. When the young Condor is hatched it is nearly naked, but is furnished with a scanty covering of down, which in a short time becomes very plentiful, enveloping the body in a complete vestment of soft black plumage. The deep black grey of the adult bird is not attained until a lapse of three years, the colour of the plumage being a yellowish-brown.

The KING VULTURE has gained its regal title from a supposition which is prevalent among the natives of the country which it inhabits, that it wields royal sway over the aura, or zopilote Vultures, and that the latter birds will not venture to touch a dead carcass until the King Vulture has taken his share. There is some truth for this supposition, for the King Vulture will not permit any other bird to begin its meal until his own hunger is satisfied. The same habit may be seen in many other creatures, the more powerful lording it over the weaker, and leaving them only the remains of the feast instead of permitting them to partake of it on equal terms. But if the King Vulture should not happen to be present when the dead animal has reached a state of decomposition which renders it palatable to vulturine tastes, the subject Vultures would pay but little regard to the privileges of their absent monarch, and would leave him but a slight prospect of getting a meal on the remains of the feast.

Waterton, who often mentions this species in his interesting works, gives several curious instances of the sway which the King Vulture exercises over the inferior birds. "When I had done with the carcass of the large snake, it was conveyed into the forest, as I expected that it would attract the king of the Vultures, as soon as time should have rendered it sufficiently savoury. In a few days it sent forth the odour which a carcass should send forth, and about twenty of the common Vultures came and perched on the

KING VULTURE. —*Sarcorhamphus Papa.*

neighbouring trees. The king of the Vultures came too, and I observed that none of the common ones seemed inclined to begin breakfast until his majesty had finished. When he had consumed as much snake as nature informed him would do him good, he retired to the top of a high mora-tree, and then all the common Vultures fell to and made a hearty meal."

The King Vulture is a native of tropical America, and is most common near the equator, though it is found as far as the thirtieth degree of south latitude, and the thirty-second of north latitude. Peru, Brazil, Guiana, Paraguay, and Mexico are the chosen residences of this fine species. It is a forest-loving bird, caring nothing for the lofty home of the condor, but taking up its residence upon the low and heavily-wooded regions, in close proximity to swampy and marshy places, where it is most likely to find abundance of dead and putrefying animal substances. Its nest, or rather the spot on which it deposits its eggs, is within the hollow of some decaying tree. The eggs are two in number.

In its adult state the King Vulture is a most gorgeously decorated bird, though its general aspect and the whole expression of its demeanour are rather repulsive than otherwise. The greater part of the feathers upon the back are of a beautiful satiny white, tinged more or less deeply with fawn, and the abdomen is of a pure white. On account of its colour, the bird is termed the White Crow by the Spaniards of Paraguay. The long pinions of the wing and tail are deep black, and the base of the neck is surrounded with a thick ruff or collar of downy grey feathers.

The most brilliant tints are, however, those of the naked skin of the head and neck. "The throat and back of the neck," says Waterton, "are of a fine lemon colour; both sides of the neck, from the ears downwards, of a rich scarlet; behind the corrugated part there is a white spot. The crown of the head is scarlet, betwixt the lower mandible and the eye, and close by the ear, there is a part which has a fine silvery-blue appearance. Just above the white spot a portion of the skin is blue, and the rest scarlet; the skin which juts out behind the neck, and appears like an oblong caruncle, is blue in part, and part orange. The bill is orange and black, the caruncles on the forehead orange, and the cere orange, the orbits scarlet, and the irides white."

These gorgeous tints belong only to the adult bird of four years old, and in the previous years of its life the colours are very obscure. In the first year, for example, the general colour is deep blue-grey, the abdomen white, and the crest hardly distinguishable either for its colour or its size. In the second year of its age the plumage of the bird is nearly black, diversified with white spots, and the naked portions of the head and neck are violet-black, interspersed with a few dashes of yellow. The third year gives the bird a very near approach to the beautiful satin fawn of the adult plumage, the back being nearly of the same hue as that of the four-year-old bird, but marked with many of the blue-black feathers of the second year. When full grown, the King Vulture is about the size of an ordinary goose.

ALL the Sarcorhamphidæ are natives of America, some of them, such as the condor and the king vulture, being comparatively scarce, while others are so common that they swarm like sparrows in our streets. One of the commonest of these useful but repulsive birds is the BLACK VULTURE, ZOPILOTE, or URUBU, which together with the turkey buzzard and the Californian Vulture are placed in one genus, termed, characteristically of their habits, Catharista, or Cleanser.

The Black Vulture bears so close a resemblance to the turkey buzzard that it has often been confounded with that bird by superficial observers. It may, however, be readily distinguished by the shape of the feathers round its neck, which in the turkey buzzard form a circular ring completely round the throat, while in the Black Vulture they descend from the back of the head towards the throat in a sloping direction. The shape of the bill is more slender, and the nostrils not so rounded as in the turkey buzzard. The general colour of the Black Vulture is a dull black; the primaries are, however, rather white on the inside, and their shafts are also white. The head and part of the neck are devoid of feathers, and covered with a black wrinkled skin sparely furnished with short scattered black hairs in front, and down behind. The throat has a wash of ochreous yellow. The length of the bird is rather more than two feet, and the expanse of its wings is about four feet four inches.

It is a high-flying bird, sweeping through the air with a beautifully easy flight, and often accompanied by the Mississippi kite, which seems to be drawn towards the Zopilote by some common feeling. After the bird has been gorging itself with the putrid meat which it so loves, it gives forth a most horrible stench. But after it has fasted for some time, the unpleasant odour nearly vanishes; and even when the body of the bird is laid open, the only scent which it exhales is a rather strong musky perfume.

The predaceous birds are, like the predaceous beasts, possessed of most powerful appetites, being capable of eating and digesting an amount of food which is perfectly astonishing. As, however, they cannot hope for a constant supply of nourishment, they are gifted with the capability of enduring hunger for a very long time without appearing to suffer very severely from their protracted fast. When in search of food, the Zopilote ascends to a vast height in the air, rising indeed to so great an elevation, that it can hardly be distinguished as a black speck, even when the attention of a spectator

BLACK VULTURE.—*Catharista Iota.*

is drawn towards the bird, and is entirely invisible to those who are not intent upon distinguishing the gnat-like object as it floats about in the upper air.

Every one of these birds is, notwithstanding the enormous height at which it is poised, intently watching the ground in hopes of marking out some dying animal on which it may woop, and hasten its death by the injuries which it inflicts upon the unresisting creature. The movements of the hunters are carefully watched by the Black Vulture, which follows their course with eagerness, knowing how often they may wound an animal which may be able to escape them for a time, but is sure at last to fall a prey to its relentless winged pursuer. Oftentimes the hunters will kill a bison or a deer merely for the sake of the skin, the marrow-bones, or the hump, leaving the remainder on the ground for the benefit of the Zopilotes and the wolves, who soon strip the bones of every particle of the flesh.

According to Don Ulloa, the Zopilotes deserve the gratitude of mankind for the part which they play in destroying the eggs of the alligator, and assisting in keeping down the number of this prolific and dangerous reptile. During the summer, the Zopilote watches the female alligator as she comes to the sandy river-shore for the purpose of depositing her eggs, and permits the reptile to complete her task without any interruption. Scarcely, however, is the alligator fairly out of the way, than the Zopilote issues from its place of concealment, and throwing the sand aside with its bill, feet, and wings, disinters the eggs, breaks the shells, and swallows their contents.

Of the voracity of these birds, Wilson gives the following graphic account :—

"A horse had dropped down in the street in convulsions : and dying, it was dragged out to Hampstead and skinned. The ground for a hundred yards around it was black with carrion crows; many sat on the tops of sheds, fences, and houses within sight; sixty or eighty on the opposite side of a small river. I counted at one time two hundred and thirty-seven, and I believe there were more, besides several in the air, over my head and at a distance. I ventured cautiously within thirty yards of the carcass, where three or four dogs and twenty or thirty Vultures were busy tearing and devouring.

Seeing them take no notice, I ventured nearer, till I was within ten yards, and sat down on the bank. Still they paid little attention to me. The dogs, being sometimes accidentally flapped by the wings of the Vultures, would growl and snap at them, which would occasion them to spring up for a moment, but they immediately gathered in again. I remarked the Vultures frequently attack each other, fighting with their claws or heels, striking, like a cock, with open wings, and fixing their claws into each other's heads. The females, and I believe the males likewise, made a hissing sound with open mouth, exactly resembling that produced by thrusting a red-hot poker into water; and frequently a snuffling, like a dog clearing his nostrils, as I suppose they were theirs. On observing that they did not heed me, I stole so close that my feet were within one yard of the horse's legs, and again sat down. They all slid aloof a few feet; but seeing me quiet, they soon returned as before. As they were often disturbed by the dogs, I ordered the latter home; my voice gave no alarm to the Vultures.

As soon as the dogs departed, the Vultures crowded in such numbers that I counted at one time thirty-seven on and around the carcass, with several within; so that scarcely an inch of it was visible. Sometimes one would come out with a large piece of the intestines, which in a moment was surrounded by several others, who tore it to fragments, and it soon disappeared. They kept up the hissing occasionally. Some of them, having their legs and heads covered with blood, presented a most savage aspect. Still as the dogs advanced, I would order them away, which seemed to gratify the Vultures; and one would pursue another to within a foot or two of the spot where I was sitting. Sometimes I observed them stretching their necks along the ground, as if to press the food downwards."

The Zopilote is rather a familiar bird, and may often be seen marching about the streets in the towns and villages of the Southern States, where it might be easily mistaken for a domestic turkey by a new arriver in the country. By the inhabitants it is popularly called the carrion crow, a confusion of nomenclature which has sometimes led to strange misapprehensions of corvine habits. As the birds, although personally disliked, are so useful to the community, they are protected by common consent, and permitted to roam the streets or prowl among the houses at will.

ANOTHER species of the genus Catharista is the TURKEY BUZZARD, more rightly termed the CARRION VULTURE. Its name of Turkey Buzzard is earned from the strange resemblance which a Carrion Vulture bears to a turkey, as it walks slowly and with a dignified air, stretching its long bare neck, and exhibiting the fleshy appendages which bear some likeness to the wattles of the turkey. Indeed, instances are not wanting, where recent visitors to the country have actually shot these birds, thinking that they had succeeded in killing a veritable edible turkey. This bird is chiefly found in North America, but is also an inhabitant of Jamaica, where it is popularly known as the John crow.

According to Waterton and Darwin, the Turkey Buzzard is not so sociable a bird as the zopilote; for although a little flock of twenty or thirty may be seen together in a corn-field where the refuse stubble has been burned, engaged in feeding on the dead mice, lizards, moles, and other creatures which have perished in the conflagration, each bird comes separately and departs separately, no two individuals having any connexion with each other.

When gorged with food, an event which always takes place whenever there is the least opportunity, the Turkey Buzzard leaves reluctantly the scene of the banquet, and gaining with some difficulty a branch of a neighbouring tree, sits heavy and listless, its head sunk upon its breast, and its wings hanging half open, as if the bird were too lazy even to keep those members closed. The object of this curious attitude seems to be, that the bird may gain as much air as possible, for these feathered creatures are singularly

TURKEY BUZZARD.—*Catharista Aura.*

susceptible to atmospheric influence. It is not improbable that this air-bath may aid the bird in digesting the food which it has so ravenously consumed, as well as to cleanse its feathers from the fetid animal substance which cannot but cling to them after their strong-scented repasts. While engaged in eating they are not at all particular about soiling their feathers, for they will often tear a hole in the skin of a dead animal, and deliberately walk into its interior, for the purpose of getting at some favourite morsel. By this mode of proceeding they soon clear away the softer substance, leaving only the bare ribs standing out, in the midst of which the Vulture continues to move about like a bird in a cage.

Between the Turkey Buzzard and the zopilote there is a certain external resemblance; but these two birds are quite distinct in their movements as well as in their habits. The Turkey Buzzard does not even walk or fly in the same manner as the zopilote. The latter bird, when walking, is very awkward, and hops along in an awkward and lazy manner, while the former moves smoothly forward, even when oppressed with a surfeit of food. In the flight the difference is even more conspicuous: the Turkey Buzzard very seldom flaps its wings, but sails smoothly through the air, its wings being extended

ARABIAN VULTURE.—*Vultur Mónachus.*

almost horizontally; the zopilote, on the contrary, flaps its wings six or seven times in succession, and then sails on for a few hundred yards with its wings raised at a decided angle with the body. The two species never company with each other, nor is the Turkey Buzzard found so familiarly associated with man and his habitation as its darker relation.

The nest of the Turkey Buzzard is a very inartistical affair, consisting merely of some suitable hollow tree or decayed log, in which there may be a depression of sufficient depth to contain the eggs. In this simple cradle the female deposits from two to four eggs, which are of a dull cream-white, blotched with irregular chocolate splashes, which seem to congregate towards the largest end. The young birds are covered with a plentiful supply of white down, and look clean and inviting to the touch. Their motto may, however, be similar to that of the Scotch thistle, "Nemo me impune lacesserit," for at the slightest aggressive touch they will disgorge over the offender the putrid animal substances with which they have been fed, and work sad woe to his hands and garments. May is usually the month in which the young Turkey Buzzards are hatched.

SOCIABLE VULTURE. — *O'togyps auriculáris.*

The adult Turkey Buzzard is rather a large bird, measuring two feet six inches in length, and six feet ten inches across the expanded wings. The weight is about five pounds. The general colour of the plumage is black, mingled with brown, the secondaries being slightly tipped with white, and a few of the coverts edged with the same tint. On the neck, the back, the shoulders, and the scapularies, the black hue is shot with bronze, green, and purple. Beneath the thick plumage is a light coating of soft white down, which apparently serves to preserve the creature at a proper temperature. The bare skin of the neck is not as wrinkled as in the zopilote, and the feathers make a complete ring round the neck. There is but little difference in the plumage of the two sexes, but the bill of the male is pure white.

We now arrive at the true Vultures, the first of which is the common ARABIAN VULTURE, a bird which is spread over a very large portion of the globe, being found in various parts of Europe, Asia, and Africa.

It is a large bird, measuring nearly four feet in length, and the expansion of its wings being proportionately wide. The general colour of this species is a chocolate

PONDICHERRY VULTURE.—*O'togyps calvus.*

brown, the naked portions of the neck and head are of a bluish hue, and it is specially notable for a tuft of long soft feathers which spring from the insertion of the wings. In spite of its large size and great muscular powers, the Arabian Vulture is not a dangerous neighbour even to the farmer, for unless it is pressed by severe hunger, it seems rather to have a dread of living animals, and contents itself with feeding on any carrion which may come in its way. Sometimes, however, after a protracted fast, its fears are overruled by its hunger, and the bird makes a raid upon the sheepfolds or the goat-flocks, in the hope of carrying off a tender lamb or kid. In these illegal excursions the bird often pays the penalty of its transgression with its life, being too hungry to be watchful, and easily shot. Hares and other small animals also fall victims to the starving Vulture, and it is said that even deer are slain by the united efforts of a pair of these birds.

The usual haunts of this species are situated on the mountain tops, and the bird does not descend into the valleys except when pressed by hunger. The specific title of Monachus or Monk has been given to this species on account of the hood-like ruff around its neck, which is thought to bear a fanciful resemblance to the hood of a monk.

THE name of SOCIABLE VULTURE, which has been bestowed upon the bird which is represented in the illustration, is supposed to be founded upon an error of observation.

Le Vaillant, who has given a somewhat detailed description of this species, found several of the nests in close proximity, and supposed from that circumstance that this Vulture was a gregarious bird. It seems, however, from more recent observation, that the proximity of these nests was merely accidental, and that although several nests may have been found near each other, they were not all inhabited simultaneously. It is the habit of many birds to build a new nest close to a deserted one, and such seems to have been the case with the Sociable Vulture. In their character they are anything but social, for it is but seldom that more than three or four of these birds can be seen together, and even in that case, they are drawn together not by any feeling of community, but by the attraction of a dead animal on which they are glad to feed, whether in company or alone.

The Sociable Vulture is a handsome and a large bird. Its length is about four feet, and the measurement of its expanded wings is rather more than ten feet. The general colour of its feathers is black-brown, from which circumstance it is called by the colonists the Black Carrion Bird. The ruff is nearly black, and the feathers of the chest and abdomen are remarkable for their length and narrowness. The naked parts of the head and neck are red, and the skin of the sides of the face droops in folds down the neck. This bird is a native of Southern Africa, and by the Hottentots is called T'Ghaip, the "T" representing one of those strange clicking sounds which play so important a part in the Hottentot language.

A FINE example of the genus Otogyps is also found in the PONDICHERRY VULTURE, a bird which, as its name imports, is an inhabitant of India.

This is not quite so large a bird as the preceding, its length scarcely exceeding three feet. The generic term, Otogyps, which is given to this species and to the sociable Vultures, is of Greek origin, denoting Eared Vulture, and alludes to the folds of skin which arise below the ears and fall for some inches along the sides of the neck. The word "calvus" is Latin, and signifies bald, in allusion to the featherless condition of the flat and broad head of the Pondicherry Vulture. It is a tolerably common bird, but is never seen in great numbers together, as it is not at all sociable in its habits, and associates only in pairs.

The general colour of the plumage is a blackish-brown, the naked portions of the head and neck are flesh-coloured, and the chest is remarkable for a bunch or tuft of white downy feathers, which marks the position of the crop.

The FULVOUS, or GRIFFIN VULTURE, is one of the most familiar of these useful birds, being spread widely over nearly the whole of the Old World, and found in very many portions of Europe, Asia, and Africa.

It is one of the large Vultures, measuring four feet in length, and its expanse of wing being exceedingly wide. Like many of its relations, it is a high-roving bird, loving to rise out of the ken of ordinary eyes, and from that vast elevation to view the panorama which lies beneath its gaze ; not, however, for the purpose of admiring the beauty of the prospect, but for the more sensual object of seeking for food. Whenever it has discovered a dead or dying animal, the Vulture takes its stand on some adjoining tree or rock, and there patiently awaits the time when decomposition shall render the skin sufficiently soft to permit the entrance of the eager beak. As soon as its olfactory organs tell of that desired change, the Vulture descends upon its prey, and will not retire until it is so gorged with food that it can hardly stir. If it be suddenly attacked while in this condition, it can easily be overtaken and killed ; but if a pause of a few minutes only be allowed, the bird ejects by a spasmodic effort the load of food which it has taken into its interior, and is then ready for flight.

A controversy has long raged concerning the manner in which the Vulture obtains knowledge of the presence of food. Some naturalists assert that the wonderful powers of food-finding which are possessed by the Vulture are owing wholly to the eyes, while others as warmly attribute to the nose this curious capability. Others again, desirous of steering a

FULVOUS, OR GRIFFIN VULTURE.—*Gyps fulvus.*

middle course, believe that the eyes and the nostrils give equal aid in this never-ending duty of finding food, and many experiments have been made with a view to extracting the real truth of the matter. The following account has been kindly transmitted to me by Captain Drayson, R.A., who has already contributed much original information to the present work.

"Having shot an ourebi early in the morning, and when about three miles from home, I was not desirous of carrying the animal behind my saddle during the day's shooting, and I therefore sought for some method of concealment by which to preserve the dead quarry from jackals and Vultures. An ant-bear's hole offered a very convenient hiding-place, into which the buck was pushed, and the carcass was covered over with some grass cut for the purpose. As usual in South Africa, there were some Vultures wheeling round at an enormous height above the horizon; these I believed would soon come down and push aside the grass and tear off the most assailable parts of the buck. There was, however, no better means of protection, so I left the animal and rode away. When at about a quarter of a mile from the ant-bear's hole, I thought that it might be interesting to watch how the Vultures would approach and commence operations, so I 'off-saddled,' and kept watch.

After about half an hour, I saw a Vulture coming down from the sky, followed by two or three others. They came down to the spot where the buck had been killed, and flew past this. They then returned, and again overshot the mark. After circling several times within a radius of four hundred yards, they flew away. Other Vultures then came and performed similar manœuvres, but not one appeared to know where the buck was concealed. I then rode off to a greater distance, but the same results occurred.

In the evening I returned for my buck, which, however, was totally useless in consequence of the intense heat of the sun, but which had not been touched by the Vultures."

Whatever may be the general opinion of the scientific world upon this subject, I cannot but think that we shall not discern the true cause of this food-discovering power in the optic or nasal nerve, or indeed in any material structure whatever. It appears to be simply due to that wonderful intuitive teaching which we popularly call instinct, and which, if rightly examined, will most surely prove a key to many mysteries at present unsolved.

The colour of the Fulvous Vulture is a yellowish-brown over the greater part of the body. The quill feathers of the tail and wings are nearly black, the ruff surrounding the base of the neck is composed of long and delicately white slender feathers, and the head and neck are sparingly clothed with short white down.

In its native state the Fulvous Vulture assumes some very curious attitudes, and has the power of altering the contour of its body so completely that it would hardly be recognised as the same bird. At one time it sits upon the branch of a tree in a heavy, indolent manner, its neck hidden in the ruff, and the head just projecting from the feathers. At another time it will be full of life and animation, pacing round the carcass of some animal, tugging furiously at the skin, and snapping fiercely at its companions if they should approach too closely. One remarkable attitude which it is fond of assuming is rather difficult to describe, but has so strange and weird-like an aspect, that it is deserving of mention. While sitting on the ground the Vulture thrusts its legs well to the front, and instead of resting upon the feet, holds them up in the air and sustains the weight of its body upon the tail and ankle-bones. Thus supported, it seems quite at its ease, and reclines with half-spread wings, as if thoroughly enjoying its repose.

Like others of its tribe, the Fulvous Vulture, when satiated with food, will retire to a neighbouring tree-branch, and sit listlessly with hanging wings, as if to rid its feathers of the putrid animal substance on which it has been feeding. It is very probable that the bird may receive great aid from the yellow feather-dust which is so copiously poured from the short and open quills that are found so abundantly upon this and other similar birds, and that by means of quiet repose, aided by the fresh air and a few hearty shakes, the bird may be able to throw off the powder and the putrefaction together.

The ALPINE, or EGYPTIAN VULTURE, is, as its name imports, an inhabitant of Egypt and Southern Europe. It is also found in many parts of Asia, and as it has once been captured on our shores, has been placed among the list of British birds.

The general colour of the adult bird is nearly white, with the exception of the quill feathers of the wing, which are dark brown. The face, bill, and legs are bright yellow, so that the aspect of the bird is sufficiently curious. The sexes are clothed alike when adult. On account of the colour of its plumage, the Egyptian Vulture is popularly termed the WHITE CROW by the Dutch colonists, and AKBOBAS, or White Father, by the Turks. It is also familiarly known by the name of PHARAOH'S CHICKEN, because it is so frequently represented in the hieroglyphical inscriptions of Egypt. When young, the colour of its plumage is a chocolate brown, the neck and shoulders are covered with grey-tipped feathers, and the beak and feet are a very dull ochry yellow. The white plumage of the adult state is not attained until the bird has completed its third year.

As is the case with the Vultures in general, the Egyptian Vulture is protected from injury by the strictest laws, a heavy penalty being laid upon any one who should wilfully destroy one of these useful birds. Secure under its human protection, the bird walks

ALPINE, OR EGYPTIAN VULTURE.—*Néophron percnópterus.*

fearlessly about the streets of its native land, perches upon the houses, and, in common with the pariah dogs, soon clears away any refuse substances that are thrown into the open streets in those evil-smelling and undrained localities. This bird will eat almost anything which is not too hard for its beak, and renders great service to the husbandman by devouring myriads of lizards, rats, and mice, which would render all cultivation useless were not their numbers kept within limits by the exertion of this useful Vulture. It has been also seen to feed on the nara, a rough, water-bearing melon, in common with cats, leopards, mice, ostriches, and many other creatures. The eggs of the ostrich are said to be a favourite food with the Egyptian Vulture, who is unable to break their strong shells with his beak, but attains his object by carrying a great pebble into the air, and letting it drop upon the eggs.

The wings of this species are extremely long in proportion to the size of the bird, and the lofty soaring flight is peculiarly graceful. It is but a small bird in comparison with many of those which have already been mentioned, being not much larger than the common rook of Europe. The nest of the Egyptian Vulture is made upon the shelf or in the cleft of a lofty rock, and the grey-white eggs are three or four in number. It is a curious fact, that during the season of reproduction the male bird slightly changes his aspect, the yellow bill becoming orange, and retaining that tint until the breeding season is over. Like many rapacious animals and birds, the Egyptian Vulture does not disdain to feed on insects, and has been observed in the act of following a ploughman along his furrows, picking up the worms and grubs after the fashion of the common rook.

EAGLES.

NEXT in order to the vultures, are placed the splendid birds which are so familiar to us under the general title of EAGLES, and which form the first group of the great family Falconidæ, which includes the Eagles, falcons, and hawks. In common with the Vulturidæ, the whole of the Falconidæ are diurnal birds, and are therefore classed into one large order, termed Accipitres Diurni. All the Falconidæ possess powerful hooked beaks, not running straight for some distance, and then suddenly curved, as in the vultures, but nearly always bent in a curve from the very base. The head and neck are covered with plumage, and above the eyes the feathers are so thick and projecting, that they form a kind of roof or shade, under which the eye is situated and effectually sheltered from the bright rays of the noontide sun. There is often a tooth-like projection in the upper mandible, and the nostrils are placed within the cere. The females are always larger and more powerful than their mates, and the colour of both sexes is very variable, according to the age of the individual.

The preceding characteristics are common to the entire family of Falconidæ, and the true Eagles may be distinguished by the following additional particulars. The beak is remarkably powerful, and for a short distance from the base is nearly straight; when the mouth is open, the edges of the upper mandible are seen to be slightly wavy, something like the cut edges of an indenture. The tail is of no very great length, but strong and rigid, and the legs are feathered down to the toes. Upwards of forty species have been placed in this genus; but as many of them present characteristics which admit of a further subdivision, they have been grouped together in certain sub-genera, for the purpose of attaining greater perspicuity.

The whole of the Falconidæ are eminently destructive birds, gaining their subsistence chiefly by the chase, seldom feeding on carrion except when pressed by hunger, or when the dead animal has only recently been killed. Herein they form a complete contrast to the vultures, whose usual food is putrefying carrion, and fresh meat the exception. Destructive though they may be, they are by no means cruel, neither do they inflict needless pain on the object of their pursuit. Like the lion and other carnivorous animals, they certainly carry out the great principle for which they were made, and which has already been mentioned in Volume I. They are not cruel birds, for although they deprive many birds and beasts of life, they effect their purpose with a single blow, sweeping down upon the doomed creature with such lightning velocity, and striking it so fiercely with the death-dealing talons, that in the generality of instances the victim must be absolutely unconscious even of danger, and be suddenly killed while busily engaged in its ordinary pursuits, without suffering the terrors of anticipation, or even a single pang of bodily pain. There certainly are some instances where an animal, such as a lamb, has been carried while still living to the Eagle's nest, and there slaughtered. But we must not judge the feelings of such a victim by our own, for the lamb can form no conception of the purpose for which it is conveyed through the air, and doubtlessly feels nothing but astonishment at the strange journey which it is making.

When the Eagle perceives a bird on the wing, the mere shock caused by the stroke of the Eagle's body is almost invariably sufficient to cause death, and the bird, should it be a large one, such as a swan, for example, falls dead upon the earth without even a wound. Smaller birds are carried off in the talons of their pursuers, and

EAGLES AND NEST.

are killed by the grip of their tremendous claws, the Eagle in no case making use of its beak for the purpose of killing its prey. If the bird carries off a lamb or a hare, it grasps the body firmly with its claws, and then by a sudden exertion of its wonderful strength, drives the sharp talons deep into the vitals of its prey, and does not loosen its grasp until the breath of life has fled and all movement has ceased.

The structure by means of which the Eagle is enabled to use its talons with such terrible effect is equally beautiful and simple, and as it is closely connected with many of the habits of birds, deserves separate mention.

Many observant persons have been struck with the curious fact, that a bird can hold its position upon a branch or perch even whilst sleeping, and that in many instances the slumbering bird retains its hold of the perch by a single foot, the other limb being drawn up and buried in the feathers. As this grasp of the perch is clearly an involuntary one, it is evidently independent of the mere will of the bird, and due to some peculiar formation. On removing the skin from the leg of any bird, and separating the muscles from each other, the structure in question is easily seen. The muscles which move the leg and foot, and the tendons which form the attachment of the muscles to the bones, are so arranged, that whenever the bird bends its leg the foot is forcibly closed, and is relaxed as soon as the leg is straightened. A bird is totally unable to keep its foot open when its leg is bent, as may be seen by watching a common fowl as it walks along, closing its toes as it lifts the foot from the ground, and spreading them as they come to the ground again. It will be seen, therefore, that when a bird falls asleep upon a branch, the legs are not only bent but pressed downwards by the weight of the body, so that the claws hold the perch with an involuntary grasp, which is necessarily tightened according to the depth of the bird's slumbers. When, therefore, an Eagle desires to drive his talons into the body of his prey, he needs only to sink downwards with his whole weight, and the forcible bending of the legs will effect his purpose without the necessity for any muscular exertion. Exertion, indeed, is never needlessly used by the Eagle, for it is very chary of exercising its great muscular powers, and unless roused by the sight of prey, or pressed to fly abroad in search of food, will sit upon a tree or a point of rock for hours together, as motionless as a stuffed figure.

Voracious though it be, and capable of gorging itself to the full like any vulture, the Eagle can sustain a prolonged fast from meat or drink; and on one occasion, when wounded, made voluntary abstinence for a fortnight before it would touch the food with which it was liberally provided.

The first, and one of the finest, of these grand birds is the well-known GOLDEN EAGLE. This magnificent bird is spread over a large portion of the world, being found in the British Islands, and in various parts of Europe, Asia, Africa, and America. The colour of this bird is a rich blackish-brown on the greater part of the body, the head and neck being covered with feathers of a rich golden red, which have earned for the bird its popular name. The legs and sides of the thighs are grey-brown, and the tail is a deep grey, diversified with several regular, dark-brown bars. The cere and the feet are yellow. In its immature state the plumage of the Golden Eagle is differently tinged, the whole of the feathers being reddish-brown, the legs and sides of the thighs nearly white, and the tail white for the first three quarters of its length. So different an aspect does the immature bird present, that it has been often reckoned as a separate species, and named accordingly. It is a truly magnificent bird in point of size, for an adult female measures about three feet six inches in length, and the expanse of her wings is nine feet. The male is less by nearly six inches.

In England the Golden Eagle has long been extinct; but it is still found in some plenty in the highlands of Scotland and Ireland, where it is observed to frequent certain favourite haunts, and to breed regularly in the same spot for a long series of years. Their nest is always made upon some elevated spot, generally upon a ledge of rock, and is most inartistically constructed of sticks, which are thrown apparently at random, and rudely arranged for the purpose of containing the eggs and young. A neighbouring ledge of rock is generally reserved for a larder, where the parent Eagles store up the food which they bring from the plains below. The contents of this larder are generally of a most miscellaneous description, consisting of hares, partridges, and game of all kinds, lambs, rabbits, young pigs, fish, and other similar articles of food. An Eagle's nest might therefore be supposed to be an unpleasant neighbour to the farmers, but it is said that the birds

GOLDEN EAGLE. — *A'quila chrysáetos.*

respect the laws of hospitality, and, provided that they are left unmolested, will spare the flocks of their immediate neighbours, and forage for food at a considerable distance.

In hunting for their prey, the Eagle and his mate mutually assist each other. It may here be mentioned, that the Eagles are all monogamous, keeping themselves to a single mate, and living together in perfect harmony through their lives. Should, however, one of them die or be killed, the survivor is not long left in a state of widowhood, but vanishes from the spot for a few days, and then returns with a new mate. As the rabbits and hares are generally under cover during the day, the Eagle is forced to drive them from their place of concealment, and manages the matter in a very clever and sportsman-like manner. One of the Eagles conceals itself near the cover which is to be beaten, and its companion then dashes among the bushes, screaming and making such a disturbance, that the terrified inmates rush out in hopes of escape, and are immediately pounced upon by the watchful confederate.

The prey is immediately taken to the nest, and distributed to the young, if there should be any eaglets in the lofty cradle. It is a rather remarkable fact, that whereas the vultures feed their young by disgorging the food which they have taken into their crops, the Eagles carry the prey to their nests, and there tear it to pieces, and feed the eaglets with the morsels.

When in pursuit of its prey it is a most audacious bird, having been seen to carry off a hare from before the noses of the hounds. It is a keen fisherman, catching and securing salmon and various sea-fish with singular skill. Sometimes it has met with more than its match, and has seized upon a fish that was too heavy for its powers, thus falling a victim to its sporting propensities. Mr. Lloyd mentions several instances where Eagles have been drowned by pouncing upon large pike, which carried their assailants under water, and fairly drowned them. In more than one instance the feet of an Eagle have been seen firmly clenched in the pike's back, the body of the bird having decayed and fallen away.

It is a terrible fighter when wounded or attacked, as may be seen by the following anecdote, which is related by Mr. Watters in his "Natural History of the Birds of Ireland."

"An Eagle was at one time captured in the county Meath, by a gamekeeper, who, surprising the bird sleeping, after a surfeit on a dead sheep in the neighbourhood, conceived the idea of taking him alive, and for that purpose approached noiselessly, and clasped the bird in his arms. The Eagle recovering, and unable to use his wings, clutched with his talons, one of which entered the man's chest, the hind claw meeting the others underneath the flesh. The man, unable to disengage the claw, strangled the bird, but the talons were yet too firmly clutched to open. Taking out his knife, he severed the leg from the body, and walked with the penetrating member to the village dispensary to have it removed."

The same writer was acquainted with a tame Eagle which displayed a great fondness for the flesh of cats, a taste which seems inherent in the Eagle nature, and to have been noticed in every specimen of tame Eagles. In every case, as soon as an irritated cat came within reach, the Eagle would pounce upon it, seize it in his talons, and with one gripe destroy its life so effectually that the poor animal never had time even to cry out. The bird indulged this cat-eating taste to such an extent that he caused sad havoc among the feline tribe, and was forced at last to go into exile.

Many anecdotes of tame Golden Eagles are on record, but as they are already familiar to the public, I shall make no mention of them. The following account, however, has never before been written, and as it displays a curious trait of character in the Eagle, is worthy of insertion.

A Golden Eagle had been captured in Scotland, and being very tame, always accompanied the family to which it belonged in all their journeys. For some time it lived near Clifton, where it passed its existence fastened to a post by a tolerably long chain, that allowed it a reasonable freedom of motion. Like other tame Eagles, she—for it was of the feminine sex—would persist in killing cats if they came within reach, although her ordinary food was fowls, rabbits, and similar articles of diet. On one occasion, a sickly, pining chicken, which seemed in a very bad state of health, was given to the Eagle. The royal bird, however, refused to eat it, but seemed to be struck with pity at its miserable state, and took it under her protection. She even made it sit under her wing, which she extended as a shield, and once when a man unkindly endeavoured to take her *protégée* away, she attacked him fiercely, injuring his leg severely, and drove him fairly off her premises. She several times built a rude nest, but never laid an egg.

There is no doubt but that this beautiful bird might be tamed as readily as the falcon, and trained in a similar manner to fly at game. Indeed, such instances are not wanting, both in ancient and modern times. The old hawking authorities did not place much value on the services of the Eagle, for its weight is so great that it could not be conveyed to and from the field of action without considerable inconvenience. In more modern times the Golden Eagle has been successfully trained to catch game. A gentleman in Huntingdonshire succeeded in taming a Golden Eagle, which he taught to chase hares and rabbits ; and several other examples are on record.

Owing to the expanse of the wings and the great power of the muscles, the flight of this bird is peculiarly bold, striking, and graceful. It sweeps through the air in a succession of spiral curves, rising with every spire, and making no perceptible motion with its wings, until it has attained an altitude at which it is hardly visible. From that post of vantage the Eagle marks the ground below, and sweeps down with lightning rapidity upon bird or beast that may happen to take its fancy. It is not, however, so active at rising from the ground as might be imagined, and can be disabled by a comparatively slight injury on the wing. One of these birds, that was detected by a young shepherd boy in the act of devouring some dead sheep, was disabled by a pebble hurled at it from a sling, and was at last ignominiously stoned to death.

When gorged with food the Eagle dislikes the exertion of flying, and generally runs forward a few paces before taking to flight. The Scotch shepherds have discovered this propensity, and have invented a very ingenious trap, which is made so as to take advantage of this habit.

A circular inclosure is built of stone, about four feet in height, without any roof, and with a small door on one of its sides. A dead sheep is then thrown into the centre of the inclosure, and a noose adjusted round the door. The Eagle soon discerns the sheep, and after making a few circles in the air, alights upon the dead animal, and feeds to his heart's content. After eating until he can eat no more, he thinks of moving, but as he does not choose to take the trouble of flying perpendicularly in so narrow a space, he prefers to walk out through the door, and is straightway strangled by the ready noose.

The Eagle is supposed to be a very long-lived bird, and is thought to compass a century of existence when it is living wild and unrestrained in its native land. Even in captivity it has been known to attain a good old age, one of these birds which lived at Vienna being rather more than a hundred years old when it died.

So splendid and suggestive a bird as the Eagle could not escape the notice of any human inhabitant of the same land, and we accordingly find that in all nations, even the most civilized of the present day, an almost superstitious regard has attached itself to this bird. The Eagles of ancient Rome and of modern monarchies and empires are familiar to all, and it is hardly possible to pay a higher compliment to a poet or a warrior than to liken him to the royal Eagle.

The IMPERIAL EAGLE is an inhabitant of Asia and Southern Europe, and bears a rather close resemblance to the golden Eagle, from which bird, however, it may be readily distinguished by several notable peculiarities.

The head and neck of this species are covered with lancet-shaped feathers of a deep fawn colour, each feather being edged with brown. The back and the whole of the upper parts are black-brown, deeper on the back, and warming towards a chestnut tint on the shoulders. Several of the scapularies are pure white, and the tail is ash-coloured, bordered and tipped with black. The cere and legs are yellow. The surest mark by which the Imperial may be distinguished from the golden Eagle, is the white patch on the scapularies. This is most distinct in the adult bird; for in the plumage of the young, the scapulary feathers are only tipped with white, instead of being wholly of that hue.

The Imperial Eagle is seldom seen sweeping over the plains, as it is a forest-loving bird, preferring the densest woods to the open country. As far as is known, it never builds its nest on the rocks, but always chooses a spreading and lofty tree for that purpose. In habits it resembles the preceding species, and in disposition is fierce and destructive. No specimen of this bird has yet been taken in England, although it is not at all uncommon in the warmer parts of Europe.

AUSTRALIA possesses a fine example of the aquiline birds in the BOLD EAGLE, so called from the extreme audacity which it displayed on first coming in contact with mankind.

This handsome bird is found in the whole of Southern Australia and Van Diemen's Land, but Mr. Gould believes that it does not inhabit the intertropical regions. The colour of the Bold Eagle is a blackish-brown, becoming paler on the edges of the wings. The back of the neck takes a decided reddish hue, which forms a very conspicuous characteristic in the colouring of the plumage. When young, the edge of each feather is

IMPERIAL EAGLE.—*Aquila Mogilnik.*

tinged with red, and the tail is slightly barred. The eyes of this bird are hazel, and the beak is yellowish except at the tip, which is black.

The food of this bird consists naturally of kangaroos, bustards, and other beasts and birds of its own country. Since sheep have been so plentifully bred in Australia, the Bold Eagle has derived considerable advantage from the enterprise of the agriculturists, and has become a perfect pest to the shepherds, from its fondness for mutton. In consequence of its marauding propensities, it is hunted and persecuted in every way by the colonists, but without much apparent result, as the bird is only driven farther inland, and seems in a fair way to hold its own for many years to come. The young cannot be taken, nor the eggs destroyed, as the bird always builds its nest on the summit of some lofty tree, which is inaccessible to any human being except the native Australian. These trees often rise for a hundred feet without a branch, thus presenting an insurmountable obstacle to the efforts of any white man.

It will not disdain to feed upon carrion, a flock of thirty or forty having been observed by Mr. Gould seated round the carcass of an ox, and gorged with food like so many vultures. Like the vultures, it will follow the white kangaroo hunters day

BOLD EAGLE.—*A'quila audax.*

after day, in order to avail itself of the offal which they throw aside. Of the black hunters it takes no heed, knowing well that the black man has no idea of leaving any portion whatever of his prey for any creature except himself, and that if any part of the slain animal should be distasteful to his palate it is handed over to his wives, who wait round him at a respectful distance, receiving humbly any morsels that he may be pleased to throw to them.

A rather amusing account of the discomfiture of a pair of these Eagles is given by Captain Flinders in his "Voyage to Terra Australis." In company with a friend, he had landed on an uninhabited island, and had captured a snake, which he was taking to the ship for the benefit of the naturalist. While so engaged, an Eagle "with fierce aspect and outspread wing came bounding towards us, but stopping short at twenty yards off, he flew up into a tree. Another bird of the same kind discovered himself by making a motion to pounce down upon us as we passed underneath; and it seemed evident they took us for kangaroos, having probably never before seen an upright animal of any other species in the island. These birds sit watching in the trees, and should a kangaroo come out to feed in the daytime, it is seized and torn to pieces by these voracious creatures."

MARTIAL EAGLE.—*Spizaetus bellicosus.*

THERE are many other examples of the genus Aquila, the smallest of which is the BOOTED EAGLE (*Aquila pennáta*).

This little bird is not larger than an ordinary falcon, for which, indeed, it might be mistaken but for the lancet-shaped feathers in the head and neck, which plainly speak of the Eagle. The general colour of this bird is dark brown; a light yellowish-brown stripe runs across the wings; the abdomen is white, and the chest is also white, each feather having a brown dash down its centre. The legs are thickly clothed or "booted" with white feathers. The Booted Eagle is spread over a considerable portion of the world, being found in many parts of Europe, as well as in Asia, which seems to be its natural residence. It has, however, been known to build in Hungary, near the Carpathian mountains, and makes annual migrations. It is not a very destructive bird, its food consisting generally of small birds, rats and mice, bats, insects, and similar articles of diet.

OF the genus Spizaetus, the MARTIAL EAGLE forms an excellent example. This handsome bird is a native of Southern Africa, where it was discovered by Le Vaillant.

The colour of this bird is dark brown, the feathers being paler at their edges; the

LAUGHING FALCON.—*Herpetótheres cachinnans.*

under surface is whitish, the quills being black ; the legs are paler and feathered to the toes. The name Spizaetus signifies "piping Eagle," and has been given to this and several other species on account of their very peculiar cry. Mixed with the rough, barking scream of the ordinary Eagles, there is a piercingly shrill cry which can be heard at a very great distance, even though the bird be out of sight. The nest of the Martial Eagle, or GRIFFARD, as it is sometimes called, is rather peculiar in its structure. being composed of three distinct layers of building materials. The first layer is made with sticks, as is usual among Eagles, and is of considerable dimensions. Upon this foundation is placed a second layer of wood, moss, and roots, to the depth of twenty inches or two feet, and upon this again is laid a quantity of little dry sticks, on which the eggs are laid. The eggs are two in number, white, and very globular.

The Martial Eagle is a bold, powerful, and rapacious bird, feeding mostly upon gazelles and smaller African animals. In the particular locality which it frequents it reigns supreme, and will not permit any other bird of prey to come within a considerable distance of the tree or rock whereon its nest is built. Whilst flying, it permits its legs to hang downward.

THE curious bird which is called from its strange cry the LAUGHING FALCON, is a native of Southern America, where it is found inhabiting the vicinity of marshes and swamps, in which localities it finds the reptile food on which it chiefly subsists. It is also a keen fisher, and haunts rivers and lakes for the sake of the finny prey which they contain. The colour of this bird is nearly white, diversified with a broad band of brown that passes over the back, wings, and the space round the eyes, and is prolonged into a belt that surrounds the neck, so that the bird looks as if it had been wrapped in a brown mantle fastened under the throat. The tail is banded alternately with brown and white. The wings of this species are not very long, and the beak is short. The tarsus is also short, and is covered in part with net-like markings. The head is surmounted with a crest, composed of long, narrow feathers, which pass over the head and droop gracefully until they reach the back of the neck.

JEAN LE BLANC EAGLE.—*Circaëtus Gállicus.*

A SMALL number of the Falconidæ are remarkable for their long tarsi, feathered below the heel, their long, even tail, and the union of the outer claws by a membranous skin. The JEAN LE BLANC EAGLE, so called on account of the generally white colour of its plumage, is a good example of this genus, which includes the bacha, the cheela, and other so-called Eagles.

The colour of the Jean le Blanc Eagle is white, speckled with brown spots, and diversified on the back with brown. The white, however, predominates largely, and even in the back and wings, the bases of the feathers are white. The tail is darker than the rest of the plumage, being of a light grey-brown, barred with dark brown. The long tarsi and toes are blue, and the claws are black. The length of the bird is about thirty inches, but the expanse of its wings is not so proportionally great as in the osprey. As the birds of this genus possess several characteristics of the Eagles, and others of the ospreys, they are supposed with justice to form a connecting link between the genera Aquila and Pandion. The Jean le Blanc is spread over considerable portions of Asia and Europe, but has never yet been discovered in England.

The food of this bird consists chiefly of snakes, frogs, rats, mice, and insects, and it is generally found haunting the low forest lands where such creatures most abound. Its nest is of considerable dimensions, and is generally built on the summit of some lofty tree. The eggs are either two or three in number, and are of a pure, spotless grey.

CRESTED, OR HARPY EAGLE. – *Thrasäetus Harpyia.*

AFTER many attempts to associate the CRESTED, or HARPY EAGLE, with any other bird in some acknowledged genus, systematic zoologists have at last been obliged to consider it a family or single genus in itself, under the appropriate title of Thrasaetus, or Courageous Eagle.

The most obvious external characteristic which serves to distinguish this species is the manner in which the feathers of the head and neck are arranged, so as to form a bold ruff or fan-like crest when erect. As long as this crest lies flatly on the head and neck, the bird might be taken for a common Eagle ; but as soon as the fierce temper is roused, the crest is raised, and the bird assumes an indescribably bold and courageous aspect. The colour of this noble bird is very variable, differing greatly in the several epochs of an Eagle's life. When adult, the general colour is blackish-slate, the head is grey, and the chest and abdomen white, with a band of a darker hue across the chest. The tops of the feathers which compose the crest are black, and the tail is barred alternately with black and grey. The beak and claws are black.

This bird is a native of various parts of Southern America, and prefers the deepest forests to the plains or the rocks as its place of residence.

The Harpy Eagle is a most powerful bird, exceeding even the golden Eagle in the extent of its muscular development. The bones of the Harpy are enormously thick in proportion to the size of the bird, and the claws are nearly twice the size of those which belong to the golden Eagle. The wings, however, are not largely developed, being rather short and rounded, so that the bird is not fitted so much for a swift and active flight as for the power of grasping with considerable force, and using its talons with the greatest effect. This formation is easily accounted for by the fact that the Harpy Eagle is not intended as an aerial hunter, chasing its prey through the air and overcoming it by means of superior activity and strength, but feeds mostly on various mammalia, and is a sad enemy to the sloth. Young deer fall victims to this voracious bird, which also destroys vast quantities of cavies, opossums, and other animals. Even the large parrots and aras are slain and eaten by the Crested Eagle. As soon as the Harpy Eagle pounces upon a sloth, a fawn, or an opossum, the fate of its victim is sealed, for the long curved claws are driven so forcibly into its vitals, that it speedily sinks dead beneath the fatal grasp.

From the thickly wooded nature of the localities in which this Eagle dwells, a lengthened chase would be impossible, as the dense foliage and tangled boughs would enable the intended victim to place itself in security if it were only able to receive intimation of its pursuer's presence. The great object of the Harpy Eagle is, therefore, to steal quietly upon its prey, by gliding on noiseless wing over the tops of the trees, and to swoop suddenly and unexpectedly on the unfortunate sloth or fawn that it may chance to discover. When successful in its chase, and standing exultingly on the body of its quarry, its talons firmly holding the prey which it has gained by its own quickness and strength, and its fierce eyes looking jealously around lest any intruder should endeavour to despoil it of the fruits of its victory, the Harpy Eagle presents a truly magnificent sight. Its crest is raised and continually in motion, and its eyes seem to flame with mingled fury and triumph.

BRAZILIAN EAGLE, OR URUBITINGA.—*Morphnus Urubitinga.*

THERE is rather a curious bird found in Brazil, Cayenne, and various parts of the West Indies, named the BRAZILIAN EAGLE, or URUBITINGA.

This bird is a great contrast in dimensions to that which has just been mentioned, being only about the size of an ordinary raven, whereas the harpy Eagle is among the largest of

OSPREY.—*Pandion halidëtus.*

the diurnal birds of prey. The colour of the Urubitinga is nearly black, diversified with some greyish marks upon the wings, and the white tail-coverts and base of the tail. The beak is powerfully made, and very convex above, and the claws are very sharply pointed. The colour of the legs and cere is yellow, and the tarsus is marked in front with a series of shield-like scales. When young, the plumage of the Urubitinga is very different from that of the adult bird, being largely mixed with yellow and dark brown. It is always to be found near rivers, lakes, and swamps, as it feeds upon the aquatic reptiles which are found so plentifully in such localities, and also upon the smaller mammalia which also inhabit wet and marshy situations.

ONE of the most interesting of the predaceous birds which belong to Great Britain is the celebrated OSPREY, or FISHING HAWK. This fine bird was formerly very common in England, but is now but rarely seen within the confines of the British Isles, although isolated species are now and then seen.

As the bird is a fish-eater, it is generally observed on the sea-coast or on the banks of some large river, but has occasionally been observed in some comparatively waterless situation, where it has probably been driven by stress of weather. In some parts of Scotland the Osprey still holds its own, and breeds year after year on the same spot, generally choosing the summit of an old ruined building or the top of a large tree for that purpose. The nest is a very large one, composed almost wholly of sticks, and contains two or three whitish eggs, largely blotched with reddish brown, the dark patches being collected towards the large end of the egg. As is the case with the Eagles, the Osprey is monogamous; but on the death of either of the pair, the survivor soon finds another mate, and is straightway consoled by a new alliance. From all accounts it is an affectionate and domestic bird, paying the greatest attention to its mate and home, and displaying a constancy which is not to be surpassed by that of the turtle-dove, so celebrated for matrimonial felicity.

Wilson, in his well-known work on the birds of America, gives a very interesting account of the proceedings of a pair of Ospreys. The female had unfortunately lost one of her legs, and was in consequence disabled from catching fish. Her mate, however, redoubled his efforts on her behalf, and, leaving her in the nest, used to set himself to work with such perseverance that he kept the nest well supplied with food by his sole endeavours, so that his mate never was obliged to leave her charge in search of sustenance. Even after the young had been fledged, this model husband continued his efforts, and relieved his wife of the necessity for hunting.

The flight of the Osprey is peculiarly easy and elegant, as might be expected from a bird the length of whose body is only twenty-two inches, and the expanse of wing nearly five feet and a half. Living almost wholly on fish, the Osprey sails in wide undulating circles, hovering over the water and intently watching for its prey. No sooner does a fish come into view than the Osprey shoots through the air like a meteor, descends upon the luckless fish with such force that it drives a shower of spray in every direction, and soon emerging, flies away to its nest, bearing its prey in its grasp. In order to enable it to seize and retain so slippery a creature as a fish, the claws of the Osprey are long, curved, and very sharp, the soles of the feet are rough, and the outer toe is capable of great versatility. When the bird has settled upon its nest, or upon any spot where it intends to eat its prey, it does not relinquish its hold, but, as if fearful that the fish should escape, continues its grasp, and daintily picks away the flesh from between its toes.

Sometimes in making its swoop it arrests itself for a second or two, as if to watch some change of position on the part of its intended prey.

The singular beauty of the Osprey's flight attracted the attention of M. de Quatrefages, who remarked, that the bird was able with outstretched and immovable wings, not only to withstand the power of a " squall" that would have flung a man to the ground, but even to work its way against the wind. How this feat was performed he confesses to be a mystery to him, and that the so-called scientific theories of " acquired velocity" or " tremulous movement" of the wings could not at all account for the phenomenon which he observed.

When unmolested by human foes, the Osprey is a bold bird, as may be seen from the following little anecdote, related by Mr. D'Ewes in his "Sporting in both Hemispheres." " I observed an Osprey, or fishing Eagle, hovering about the river some distance down stream, as if he were regarding my movements with much curiosity. Having caught a small barbel, perhaps a little less than a pound in weight, and extricated the hook with some difficulty, something induced me to throw him back again, as not worth taking, which I did with a sharp jerk, sending him some distance into the middle of the stream. In the space of a few moments, and a hundred yards downwards, I saw the Osprey make a sudden swoop, a dive, and soar aloft with the fish in his mouth—no doubt my identical barbel, which, puzzled with his sudden change of circumstances, and not having regained vigour and instinct sufficient to seek his usual haunts, had floated down stream, and became an easy victim to his destroyer."

In Southern America it is very common, and has been well described by Wilson, Audubon, and other well-known writers, to whom we can but refer for the present. The bird is held in great favour, and protected by common consent, so that any one who shot a fish-hawk would draw down upon himself the anger of the person who constituted himself its protector. The bird is in the habit of building its nest upon the roofs of houses, and is thought to bring good luck to the household which it selects as its protectors.

There is a good reason for the love which the fishermen bear towards the Osprey, as it is the harbinger of their best seasons, and by its headlong sweeps after the basse and other fish intimates that their nets may be successfully employed.

Harmless though the Osprey be—except to the fish—it is a most persecuted bird, being not only annoyed by rooks and crows, but robbed by the more powerful white-headed Eagle. Mr. Thompson records an instance where an Osprey, which had been fishing in Loch Ruthven, was greatly harassed by an impertinent Royston crow, which attacked the nobler bird as soon as it had caught a fish, and, as if knowing that it was incapable of retaliation, actually struck it while on the wing. The Osprey kept quietly on its way, but was so wearied by the repeated attacks of the crow, that when pursued and pursuer had vanished out of sight, the poor Osprey had not been able to commence his repast.

How this species is robbed by the white-headed Eagle, who strikes the Osprey on the wing, and snatches from the poor bird the results of its morning's labours, is well known through the graphic descriptions of Wilson and Audubon. The passages in which this thievish habit is recounted are so familiarly known, and have been so frequently quoted, that I prefer merely to mention them, and to insert in the present pages another account of the same proceedings, written also by an eye-witness. The author is Mr. Webber, well known for his "Wild Scenes and Wild Hunters," and other works of a similar character.

Haliaetus leucocephalus

"The bald Eagle, who is a sort of omnipresent predator wherever the primeval nature holds her own upon the continent, makes his appearance sometimes suddenly on his wide-visiting wings amidst these solitudes, that seem rightly to belong to the fish-hawk alone. His hoarse bark startles the deep silence from afar, and every natural sound is mute. Wheeling grandly amidst the dim blue cliffs, he subsides on slow and royal spread upon some blasted pine beside the lake-river, and with quick short screaming—while he smooths his ruffled plumes—announces to awed nature that its winged monarch has come down to rest. The friendly fish-hawks, in silent consternation, dart hither and yon in vexed uncertain flight, the tiny songsters dive into deep thickets, and the very cricket, underneath dead leaves, pauses for a moment in its cheerful trill, while the shadow of that drear sound passes over all. But now the kingly bird grows quiet, and with many a shift of feet and restless lift of wing—while fierce, far-darting eyes are taking in all the capabilities of his new perch—he sinks into an attitude of deep repose, one yellow-heated eye upturned, watching the evolutions of the startled fish-hawks, whose movement, becoming less and less irregular as they wheel to and fro, gradually subsides into the measured windings of their habitual flight in seeking prey, while the buzz, the hum, the chirp, the chatter, and the carol creep up once again, and nature becomes voiceful in her happy silence.

Now, to witness, as I have done, from the mountain tops, the Osprey sweep down from the dizzy height, almost level with my feet, and hear the faint whirr of arrowy-falling plumes, and see the cloud-spray dimly flash through the blue steep of distance—ah, that was a sight! And then the strong bird's scream of exultation faintly heard, and the far flash of scales glittering as he drags his spoil to sunlight from its dark slumberous home, and on strong vans goes beating up towards the clouds; ah, that too was a sight! But then to see deep down, that couchant tyrant deep down below, 'levelling his neck for flight' (as the 'glorious weaver' has it), his war crest raised, his wings half-spread, pausing for the moment on his stoop, and then one clamorous shriek of confident power, and see him vault away, up, up, with a swift cleave, conquering gravitation, and go lifted on the spell of wings! Wonderful sight—that upward struggle! The fish-hawk has taken warning from the exulting cry of his old enemy, and with yet louder cries, as if for help, goes up and upward, swifter still, with vain beatings that scatter the fleece-forms of cloud above me, and stir them in whirling gyrations. But no; the conqueror with overcoming wings is upon him, with fierce buffetings the stirred chaos cannot hide from me, and the fisher drops its prey with a despairing shriek, while it goes gleaming headlong towards its ravished home. Now but an instant's poise while the sunlight can flash off a ray from steadied plumes, and the Eagle goes, dimmed with swiftness, roaring down to catch the falling prey before it reach the wave.

But the fish-hawk, although the mildest, the most generous and social of all the Falconidæ, still recognises that point beyond which forbearance is no virtue. When the plundering outrages of the bald Eagle have been at length carried to an intolerable extreme in any particular locality, the fish-hawks in the neighbourhood combine in a common assault upon the tyrannical robber. I have frequently witnessed such scenes along the coast of the Gulf of Mexico. They abound in great numbers along the estuaries of its great rivers. I remember particularly to have noted the greatest collection of them at the mouth of the Brazos River, at Texas. Twenty or thirty of them are constantly congregated at this place during the spring months, to feed upon the great shoals of the luscious red fish which then make their appearance here ; though otherwise a barren and uncouth spot, it is constantly enlivened by the aerial gambols of these powerful and graceful flighted birds, and many's the battle between them and the bald Eagle that I have witnessed among the clouds at this place. They seemed to have formed a sort of colony for mutual protection, and the moment their foe, the Eagle, made his appearance among them, the cry of alarm was raised, and the vigilant colonists, hurrying from all quarters, attacked the robber without hesitation, and always succeeded in driving him away.

There was always a desperate battle first before the savage monarch could be routed, and I have seen them gathered about him in such numbers, whirling and tumbling amidst a chaos of floating feathers through the air, that it was impossible for a time to distinguish which was the Eagle, until, having got enough of it amidst such fearful odds, he would fain turn tail, and with most undignified acceleration of flight would dart toward the covert of the heavy forest to hide his baffled royalty and shake off his pertinacious foes amidst the boughs, as do the smaller hawks when teased by the little king-birds. I was told by the residents of Valasco, at the mouth, who from sympathy with the fish-hawks seemed to greatly relish the scenes, that year after year the Eagles made persevering attempts to attain a lodgment in the neighbourhood of this colony, but were always promptly repulsed and finally driven off."

There is but one species of Osprey, although it has been thought that the American bird ought to be reckoned as a different species. The general colour of the Osprey is dark brown, but it is pleasingly variegated with various shades of black, grey, and white. The crown of the head and the nape of the neck are covered with long, grey-white feathers, streaked with dark brown. The under surface of the body is white, with the exception of a light brown band which extends across the chest. The primaries are brown tipped with black, and the tail is barred above with a light and a deep brown, and below with brown and white. The legs, toes, and cere are blue, the eyes golden yellow, and the beak and claws black.

The CINEREOUS, or SEA EAGLE, is by far the most common of the larger British Falconidæ, being much more frequently seen than the golden Eagle. On account of the peculiar white rounded tail the bird is sometimes called the WHITE-TAILED EAGLE.

This species is found in all parts of Europe, but is not known to visit America. As it is a fish-loving bird, and is nearly as great an adept at angling as the osprey, it is generally found on the sea-coast. It possesses, however, a very accommodating appetite, and often makes considerable inland journeys in search of food. Young fawns, lambs, hares, and other animals then fall victims to its hunger, and it is said to watch for disabled or dying deer, and to hasten their end by the injuries which it inflicts upon them. On the shores, the Sea Eagle seems to have regular hunting-grounds, and to make its rounds with perfect regularity, appearing at a certain spot at the same hour daily, keeping an anxious eye on the multitude of sea fowl as they hover about the rock ledges in attendance upon their mates and families.

One of these birds that was domesticated at Oxford for some years, and was very generally known throughout the neighbourhood, contrived, on one occasion, to eat a hedgehog that had strayed too near his quarters. It might naturally have been supposed

CINEREOUS, WHITE-TAILED, OR SEA EAGLE.—*Haliaëtus albicilla.*

that the prickly skin of the animal would have caused some discomfort in the Eagle's interior. Nothing of the kind, however, happened; for the Eagle, as is universal among rapacious birds, ejected the skin and indigestible portions of the hedgehog, and seemed to have felt no inconvenience whatever from the array of prickly spines. The same bird used to spend much of its time in trying to eat a tortoise, a proceeding which the tortoise treated with perfect equanimity. The whole story of this bird is rather a curious one, but would occupy too much space in a work of this character.

It is a fierce and determined bird, having a strange look of lowering self-will in its eyes. When wounded, it fights most fiercely; and even when disabled by a broken wing, it has been known to strike so sharply with the sound wing, that the utmost exertions of two men were required before it could be subdued and bound.

As it is rather an unpleasant neighbour to the farmer, the poultry-keeper, or the sheep-owner, it is much persecuted, and many ingenious traps are constructed for its destruction. In Norway a small conical hut is built, having the roof open, and a piece of stick, to which is attached a bait, laid across the aperture. Inside the little hut sits a man, looking out for the Eagle. As soon as the bird sees the bait, which is generally a rabbit, or some such dead animal, it sweeps down and alights upon the stick. The moment that it settles, it is grasped by the concealed inhabitant of the hut, who jerks it through the opening into the little edifice. Owing to the conical shape of the hut, the bird is unable to use its wings, which are its best weapons, and is, therefore, soon mastered and destroyed.

The nest of this species is constructed after the fashion of the Eagle tribe, and is made of a large mass of sticks, put together in a very inartificial manner. Unlike the generality of the Eagles, it does not return year after year to the same spot, but is of a more roving nature, leaving its young in possession of the dwelling-places, and going farther afield in search of some new hunting-ground. The golden Eagle acts in a precisely opposite manner; for as soon as the young Eagles are able to shift for themselves, their parents drive them from the locality, and will not permit them to come within a considerable distance of the spot where they were hatched.

Although it is not as common in the British Islands as was formerly the case, it still breeds regularly in some parts of Scotland, in Shetland, the Hebrides, and many other localities where it is permitted to spend its life in peace. Even now, it is sometimes observed inland; it is quite recently that a notice appeared in the *Field* newspaper of a Sea Eagle that was shot at Livermere Park, near Bury St. Edmunds. The bird measured three feet in length, and seven feet one inch across the wings. It had been observed for some days hovering about, and apparently taking fish from the water in the park.

The head of the Sea Eagle is covered with long drooping feathers, each feather being ashy brown, and darker at its centre than at the edges. The rest of the body is dark brown, with here and there a lighter spot or streak, the primaries being nearly black. The tail is rounded, and of a pure white colour in the adult Eagle, and brown in the immature bird. The legs, toes, beak, and cere are yellow, and the claws black. The generic name, Haliaëtus, is of Greek origin, and signifies Sea Eagles.

THE noble bird which is represented in the accompanying illustration is celebrated as being the type which has been chosen by the Americans as the emblem of their nation.

The name of BALD, or WHITE-HEADED EAGLE, has been applied to this bird on account of the snowy white colour of the head and neck, a peculiarity which renders it a most conspicuous bird when at large in its native land. The remainder of the body is a deep chocolate brown, inclining to black along the back. The tail and upper tail coverts are of the same white hue as the head and neck. In its earlier stages of existence the creature is of more sombre tints, not obtaining the beautifully white head and tail until it is four full years of age.

The nest of the Bald Eagle is generally made upon some lofty tree, and in the course of years becomes of very great size, as the bird is in the habit of laying her eggs year after year in the same nest, and making additions of fresh building materials at every fresh breeding season. She commences this task at a very early period of the year, depositing her eggs in January, and hatching her young by the middle of February. This statement is made by Wilson, and is corroborated by the following incident, which is narrated in a note to Thompson's Birds of Ireland :—"During a tour made by Richard Langtrey, Esq., of Fort William, near Belfast, through the United States, in 1836, he, in the middle of January, observed a pair of these birds flying about a nest in the top of a gigantic pitch pine, which stood a little remote from other trees, on the bank of the Fish River, Mobile Bay. On the 6th of February he returned to the place, in the hope of procuring a young bird alive. The nest being inaccessible, the tree was cut down, and with it one young bird (unfortunately killed by the fall) came to the ground. The eaglet was covered with down, interspersed with a few feathers. The nest was rather flat, and composed of sticks; it contained the heads and bones of mullet, and two heads of the grey pelican. The parent birds were in great consternation during the felling of the pine, and to the last moment continued flying clamorously about the nest. Mr. Langtrey was told that two or three pair of Bald Eagles build annually about Mobile Bay, and had their nests pointed out to him."

It is always a very affectionate bird, tends its young as long as they are helpless and

BALD, OR WHITE-HEADED EAGLE.—*Haliaëtus Leucocéphalus.*

unfledged, and will not forsake them, even if the tree on which they rest be enveloped in flames.

How the Bald Eagle takes advantage of the fishing talents of the osprey has already been duly related. The Eagle is, in truth, no very great fisher, but is very fond of fish, and finds that the easiest mode of obtaining the desired dainty is to rob them who are better qualified than himself for the sport. He is capable of catching fish, it is true, but he does it in a very awkward manner, wading into the shallows like a heron, and snatching suddenly at any of the finny tribe that may be passing in his direction. This predatory propensity aroused the wrath of Benjamin Franklin, who objected strongly to the employment of the Bald Eagle as the type of the American nation, urging, as his grounds for opposition, that it is "a bird of bad moral character, and does not get his living honestly."

The Bald Eagle is very accommodating in his appetite, and will eat almost anything that has ever possessed animal life. He is by no means averse to carrion, and has been seen seated regally upon a dead horse, keeping at a distance a horde of vultures which were collected round the carcass, and not permitting them to approach until he had gorged himself to the full. Another individual was seen by Wilson in a similar state of things.

He had taken possession of a heap of dead squirrels that had been accidentally drowned, and prevented any other bird, or beast of prey, from approaching his treasure. He is especially fond of lambs, and is more than suspected of aiding the death of many a sickly sheep by the dexterous use of his beak and claws. Sometimes he pays the penalty of his voracity, as was very recently the case. A Bald Eagle had caught a wild duck, and carrying it to a large piece of ice, tore his prey in pieces, and began to eat it. When he had finished his repast, he spread his wings for flight, but found himself unable to stir, his feet having been firmly frozen to the ice. Several persons who witnessed the scene endeavoured to reach the bird, but were unable, owing to the masses of loose ice that intervened between the Eagle and the land. At last the poor bird perished, as was supposed, having been seen to flap his useless wings in vain endeavours to escape until night drew on and darkness hid him from view.

The manner in which the Bald Eagle hunts for, procures, and kills his prey, is so admirably told by Mr. Audubon, that it would be impossible to do justice to the subject without quoting his own words :—

" The Eagle is seen perched, in an erect attitude, on the summit of the tallest tree by the margin of the broad stream. His glistening, but stern eye, looks over the vast expanse. He listens attentively to every sound that comes to his quick ear from afar, glancing every now and then on the earth beneath, lest even the light tread of the fawn may pass unheard. His mate is perched on the opposite side, and should all be tranquil and quiet, warns him, by a cry, to continue patient. At this well-known call he partly opens his broad wings, inclines his body a little downwards, and answers to her voice in tones not unlike the laugh of a maniac. The next moment he resumes his erect attitude, and again all around is silent. Ducks of many species—the teal, the widgeon, the mallard, and others—are seen passing with great rapidity, and following the course of the current, but the Eagle heeds them not; they are at that time beneath his attention.

The next moment, however, the wild, trumpet-like sound of a yet distant, but approaching swan is heard. A shriek from the female Eagle comes across the stream, for she is fully as alert as her mate. The latter suddenly shakes the whole of his body, and, with a few touches of his bill, aided by the action of his cuticular muscles, arranges his plumes in an instant. The snow-white bird is now in sight; her long neck is stretched forward; her eye is on the watch, vigilant as that of her enemy; her large wings seem with difficulty to support the weight of her body, although they flap incessantly; so irksome do her exertions seem, that her very legs are spread beneath her tail to aid her in her flight. She approaches, however. The Eagle has marked her for his prey.

As the swan is passing the dreaded pair, starts from his perch the male bird, in preparation for the chase, with an awful scream, that to the swan's ear brings more terror than the report of the large duck-gun. Now is the moment to witness the display of the Eagle's powers. He glides through the air like a falling star, and, like a flash of lightning, comes upon the timorous quarry, which now, in agony and despair, seeks by various manœuvres to elude the grasp of his cruel talons. It mounts, doubles, and willingly would plunge into the stream, were it not prevented by the Eagle, which, possessed of the knowledge that by such a stratagem the swan might escape him, forces it to remain in the air, by attempting to strike it with his talons from beneath.

The hope of escape is soon given up by the swan. It has already become much weakened, and its strength fails at the sight of the courage and swiftness of its antagonist. Its last gasp is about to escape, when the ferocious Eagle strikes with its talons the under-side of its wing, and, with unresisted power, forces the bird to fall in a slanting direction upon the nearest shore.

It is then that you may see the cruel spirit of this dreaded enemy of the feathered race, whilst exulting over his prey, he for the first time breathes at his ease. He presses down his powerful feet, and drives his sharp claws deep into the heart of the dying swan; he shrieks with delight as he feels the last convulsions of his prey, which has now sunk under his efforts to render death as painful as it possibly can be. The female has watched every movement of her mate, and, if she did not assist him in capturing the swan, it was not from want of will, but merely that she felt full assurance that the power and courage of her lord were quite sufficient for the deed. She now sails to the spot where he eagerly awaits her, and when she has arrived, they together turn the breast of the luckless swan upwards, and gorge themselves with gore."

The Bald Eagle is found throughout the whole of North America, and may be seen haunting the greater part of the sea-coasts, as well as the mouths of the large rivers.

The RED-THROATED FALCON, which affords a good example of the genus Ibycter, is a native of South America.

The birds comprising this genus are remarkable for the convexity of the upper mandible, and the semi-blunt, notched lower mandible. The claws are sharp, and the cheeks, the throat, and part of the crop are naked. This species is a very handsome one, the general tint of the plumage being a very deep blue on the back and the upper surface, and reddish-white below. The neck has a purplish-red hue, from which the bird derives its popular name, the claws are black, the feet and cere yellow, and the beak a deep blue. This bird has been known under a variety of names, such as the Little American Eagle, the Bare-necked Falcon, the White-billed Ibycter, the Bare-necked Polyborus, together with many similar appellations. The habits of the bird are not known.

CLOSELY related to the preceding Bird is the Black CARACARA, also a native of South America. This is a much darker bird than the Red-throated Falcon, the back and upper parts being blue-black, and the rounded tail white only at its base. The feet are yellow, the beak and claws black, the cere a grey-brown, and the space round the eyes devoid of feathers, and flesh-coloured. On account of the short and stout beak, and the large tarsus, this species was formerly placed in a separate genus, with the title of Daptrius.

The CARRION HAWKS, as the CARACARAS are popularly termed, are natives of Southern America, and from their great numbers,

RED-THROATED FALCON.—*Ibycter Americanus*

their boldness, and their unpleasant habits, are sufficiently familiar to any one who has had occasion to travel in the country where they teem. There are several species of Caracaras, which are placed in one genus, and are very similar in their habits to the vultures, but on a smaller scale.

SOUTHERN CARACARA.—*Milvågo Austrális.*

The SOUTHERN CARACARA is not quite so large as some of its brethren, but is quite as useful and as repulsive a bird. Its length is about eighteen inches, and its colour a grey-brown upon the back and upper surface, and paler beneath, diversified with reddish bands. The thighs are of a banded rusty-red, and the tail is yellowish-grey. The cere and feet are yellow, and the bill blue-grey. It is an omnivorous bird, eating vegetable or animal substances with equal willingness, and is said to do damage to the potato crop by digging into the cuttings before they have time to sprout. It may be that the bird is urged by the desire of eating, not the potatoes, but the grubs which have taken up their residence therein, and so confers a favour on the planter instead of doing him an injury. This opinion is strengthened by an observation of Mr. Darwin, who says that he has seen them by scores following the plough, like English rooks, and picking the worms and grubs out of the furrows.

The ordinary food of the Southern Caracara is vermin and putrid meat, and it is believed never to kill either birds or quadrupeds. The flight is very different from that of the vulture, being slow, heavy, and laborious, and the bird is never known to soar in the vulturine fashion, neither does it generally perch on trees, but prefers to seat itself upon stones, walls, and similar resting-places.

The Southern Caracara is a most impudent and mischievous bird, as may be seen from Mr. Darwin's admirable account :—" They actually made an attack on a dog that was lying asleep close to one of the party, and the sportsmen had difficulty in preventing the wounded deer from being seized before their eyes. It is said that several together wait at the mouth of a rabbit-hole, and seize on the animal as it comes out. They were constantly flying on board the vessel when in the harbour, and it was necessary to keep a good look-out to prevent the leather from being torn from the rigging, and the meat and game from the stern. These birds are very mischievous, and most acquisitive ; they will pick up almost anything from the ground ; a large-sized glazed hat was carried nearly a mile, as was a pair of heavy balls (bolas) used in catching cattle. Mr. Wilson experienced during the summer a more severe loss in their stealing a small Kater's compass in a red morocco case, which was never recovered. These birds are, moreover, quarrelsome, and very passionate, tearing up the grass with their bills in rage. They build on the rocky cliffs of the sea-coast, but only in the small islets, and not in the two main lands. This is a singular precaution in so tame and familiar a bird. The dealers say that the flesh of these birds, when cooked, is quite white and very good eating."

The BRAZILIAN KITE, CARRANCHA, or BRAZILIAN CARACARA, is also a native of the southern portions of America, and is found inhabiting the same localities as the Southern Caracara, to which bird it bears a considerable resemblance in general appearance and in its general colour.

The Brazilian Kite is blackish-brown, deepening to dull black from the top of the head, and varied across the neck and shoulders with wavy bands of dark brown on a greyish ground. The tip of the tail is black, and the remainder is greyish-white, traversed by many narrow wavy bands of dusky brown. The bill is tinged with blue at the base, the claws are black, and the legs yellow. This bird is essentially a carrion eater, following the line of road in order to feed on the poor worn-out animals that sink exhausted on the journey, and are left to perish by their hard-hearted drivers. It will watch the course pursued by hunters, and in hopes of obtaining the rejected portions of the slain animals, will follow them in their expeditions with as much perseverance and confidence as is exhibited by the American wolf under the same circumstances. It also frequents the slaughter-houses, and is of great service in devouring the offal, which would otherwise be left to sink into putrefaction, and to taint the air with its deadly odour. Fortunately for the country, there is but little chance of any such catastrophe, as long as there is a Carrancha within a radius of many miles, for the keen sense of

BRAZILIAN KITE.—*Polýborus Braziliensis.*

the bird will enable it to distinguish a feeble animal, or a dead carcass, at a wonderful distance, and its insatiable appetite is never appeased as long as there is a particle of flesh remaining on the bones.

The Carrancha is often seen feeding in company with several closely-allied birds, such as the Chimango Caracara, but is not on friendly terms with them, although the two birds may be seated in close proximity, engaged in their common banquet, and being employed in picking the same bone. Sometimes the Carrancha is subject to a kind of small persecution on the part of its temporary companion, but seems to heed its proceedings with great stolidity. "When the Carrancha," says Mr. Darwin, "is quietly seated on the branch of a tree, or on the ground, the Chimango often continues for a long time flying

E 2

backwards and forwards, up and down, in a semicircle, trying each time at the bottom of the curve to strike its large relative. The Carrancha takes little notice, except by bobbing its head." The cry of the Carrancha is very peculiar, and is thought to resemble the popular name which has consequently been given to the bird. While uttering the strange, rough sounds, the Carrancha gradually raises its head, bending it farther and farther backwards, until at last the top of its head almost touches the back of its neck. This habit is observed in other Caracaras. By sailors, and other imaginative persons, the cry of the Carrancha is thought to resemble that of the English rook, and the bird is in consequence denominated by them the Mexican Crow. The similitude is increased by the dark colour of the plumage and the character of the flight, which bears a considerable resemblance to that of the crows and rooks.

The Carrancha, although persecuted by the Chimango, is in its turn a persecutor, chasing the Zopilote for the same reason that the Bald Eagle chases the Osprey, and forcing it to disgorge the food which it had swallowed. Besides carrion, the Carrancha eats young lambs, many of the smaller mammalia, reptiles, and various insects, and is indeed a very general feeder. Sometimes it will pursue and devour the smaller birds, and has been observed to secure a partridge on the wing, after urging a rather lengthened chase on the ground. It is also in the habit of frequenting the sea-shore, and feeding on the fish, crabs, molluscs, and other edible substances that are found between high and low water, or are flung upon the beach by the waves. Besides all these articles of diet, it feeds much on eggs during the breeding season, taking them from the nests with great audacity and cunning.

Although seen in considerable numbers when attracted by a dead animal, or other prey, the Carrancha is not a gregarious bird, being seen solitary or in pairs in desert places. Under the name of the Mexican Eagle, this bird is embroidered on the banners of the Mexican Government.

THE common BUZZARD is one of our handsomest Falconidæ, and is one which, although banished from the greater part of England, is still found plentifully in many parts of Scotland and Ireland.

The plumage of this bird is looser and more downy than is seen in the generality of the hawk-tribe, and bears a certain resemblance to that of the owl. This peculiarity is explained by the habits of the bird, which will presently be narrated. The average length of a Buzzard is from twenty to twenty-two inches, and the tinting of its plumage is extremely variable, even in adult birds. The usual colouring is as follows. The back and whole of the upper surface is a rich brown, becoming lighter on the head and neck, and diversified with longitudinal streaks of the darker hue. The tail is also dark-brown, but is varied with stripes of a lighter colour, and the primary feathers of the wings are nearly black. The under portions of the body are grey-white, marked on the neck, and chest, and abdomen, with spots and streaks of brown. The claws are black, the bill is a deep blue-black, and the legs, toes, and ears are yellow.

In its habits the Buzzard is a very sluggish bird, never engaging in open chase like the true falcons, but pouncing suddenly and unexpectedly on its prey. The use of the soft downy plumage is now apparent. The Buzzard, in seeking its food, sails slowly over the ground at no great elevation, surveying every spot in search of some living object. At the sight of any eatable being, whether it be rat, mouse, bird, or reptile, down comes the Buzzard, and bears off the doomed creature in its claws, before its victim has taken the least alarm at the presence of its destroyer. The noiseless passage through the air is caused by the down-edged feathers, by which the Buzzard is clothed. Sometimes it sits upon a branch, keeping a vigilant watch, and keenly eyeing every passing creature. Whenever a desirable bird or animal passes within easy reach, the Buzzard darts from its post, and after seizing its prey, returns to the same spot, and recommences its watch. This custom is singularly like the well-known habits of the common fly-catcher.

As, from its comparatively short wings and inactive temperament, the Buzzard is incapable of chasing the swift-winged game birds, it is never trained for the sport of falconry, and among the ancient sportsmen was reviled as a useless and cowardly bird. Like many large birds of prey, it is exceedingly liable to persecution from the vulture, hawks, the rook and crow; and the grey or Royston crow is a notable and constant antagonist. When wounded or disabled from flying, the Buzzard can still maintain a stout fight, and by laying itself on its back, and striking fiercely with its sharp and crooked talons, can drive off an apparently superior foe.

It is easily tamed, and is rather an amusing bird in its new state of domestic life.

One of these birds, that was captured in Ireland, and whose history is related by Mr. Thompson, displayed some very curious peculiarities, and afforded some insight into the Buzzard nature. It was fond of catching mice in a barn, darting at them as they traversed the floor, and striking at them through the straw. In many instances, the bird missed its stroke, but was never discomfited, and was always ready to make a fresh attack. It would also catch and kill rats, but preferred mice, probably because they gave it less trouble. It detested strangers, and used to fly fiercely at them and knock their hats over their ears, or fairly off their heads. A rather remarkable amusement in which this bird indulged, was to jump on its master's feet and untie his shoestrings. It would eat magpies and jackdaws, but did not seem to care very much for such diet, magpies being even more distasteful than jackdaws. On one occasion a jackdaw had been shot, and fell into a mill dam. The Buzzard pounced on the dying bird, and grasping it in his talons, held it beneath the water until it was dead. Whether the act was intentional or not is not certain, but as the bird remained in so awkward a position with its legs wholly immersed in the water until the jackdaw was quite dead, the act does not seem to have been without some definite motive. The same bird was very fond of worms and grubs, and used to attend upon the potato-diggers, for the purpose of eating the subterranean insects and other creatures which are thrown up in the operation.

BUZZARD.—*Búteo vulgáris.*

The nest of the Buzzard is made either in some suitable tree, or upon the rocks, according to the locality, and is generally composed of grass and heather stems, intermingled with long soft roots, and lined with wool, heather leaves, and other substances. One curious instance is known, where the Buzzards took a liking to a nest which had been tenanted by a pair of crows for a series of years, and, after a severe contest, succeeded in ejecting the original inhabitants, and establishing themselves in the ill-gotten premises. The intruders seemed to have been dissatisfied with the internal arrangements of the nest, and relined it with the fur of hares and rabbits. The eggs are from two to five in number, and their colour is greyish-white diversified with a few spots of pale brown. In the stolen crows' nest, four young were hatched, and were taken from their parents just before they were fairly fledged.

The flight of the Buzzard is rather variable. At times the bird seems inspired with the very soul of laziness, and contents itself with pouncing leisurely upon its prey, and returning to the branch on which it has been perched. Sometimes, however, and especially in the breeding season, it rises high in the air, and displays a power of wing and an easy grace of flight which would hardly be anticipated from its formerly sluggish movements.

The Buzzard seems to be a most affectionate mate and parent, attending closely upon its home duties, and watching the safety of its young with anxious care. When this natural living instinct can find no vent in its proper direction, it sometimes exhibits itself in a very curious manner, as was the case with a captive Buzzard whose conduct has been rightly immortalized by Mr. Yarrell, in his history of the British Birds, by the aid of pen and pencil.

"A few years back a female Buzzard, kept in the garden of the Chequers Inn, at Uxbridge, showed an inclination to sit, by collecting and bending all the loose sticks she could obtain possession of. Her owners, noticing these actions, supplied her with materials; she completed her nest and sat on two hen's eggs, which she hatched, and afterwards reared the young. Since then she has hatched and brought up a brood of chicken every year. She intimates her desire to sit, by scratching holes in the ground, and breaking and tearing everything within her reach. One summer, in order to save her the fatigue of sitting, some young chickens, just hatched, were put down to her, but she destroyed the whole. Her family in June, 1831, consisted of nine; the original number was ten, but one had been lost. When flesh was given to her she was very assiduous in tearing and offering it as food to her nurslings, and appeared uneasy if, after taking small portions from her, they turned away to pick up grain."

This curious anecdote is suggestive of the many instances recorded where a predaceous animal has taken to the young of some creature, which would, under other circumstances, have been killed and eaten as soon as it was seen, but which, under the influence of the loving instinct which warms alike the heart of the tiger and the ewe, the hawk and the dove, has been cherished and protected. The interesting anecdote of the protectrix Eagle, which has been already recorded, is another example of the instinctive exhibition of kindly feelings, and finds a parallel in the well-known case of the lion, which, instead of eating a little dog that had been placed in his cage, took it under his care, and would suffer no one to approach his new friend.

The British Islands possess another species of Buzzard, closely allied to the bird which has just been described. This is the ROUGH-LEGGED FALCON, so called from the manner in which its legs are covered with feathers as far as the margin of the toes.

It is rather a larger Bird than the common Buzzard, and the colouring of the feathers is rather different. The beak and upper surface is like that of the Buzzard, but the head and upper part of the neck are of a pale yellow hue, each feather having a streak of the darker colour down its centre. The chin, throat, and breast are of a rusty fawn, and the abdomen nearly of the same tint as the back. The whole of the plumy legs are light fawn, spotted with brown, and the pinions of the wing are brownish-black. The beak and claws are black, and the cere yellow. The habits of this bird are very like those of the common Buzzard, excepting that it is even more sluggish and lazy in its movements. Like the preceding bird, it feeds on various birds and animals, which it seizes as they pass near the spot on which it is standing, or pounces upon them as they sit on the ground. Its flight is very owl-like, and the more so as this species is in the habit of searching for its food by night as well as by day, and especially favours the hours of dusk for its peregrinations. Sometimes it sits upon a tree-branch, after the fashion of

ROUGH-LEGGED FALCON.—*Archibúteo Lágopus.*

the common Buzzard, watching for its prey, seizing it, and retiring to the same spot; but it also makes circling flights at a low elevation from the ground, and darts suddenly upon any bird, animal, or reptile that may take its fancy. All these movements are performed with great deliberation, and the bird is so slow in all its proceedings that Audubon, who had enjoyed many opportunities of watching its habits, says that "the greatest feat he had seen them performing was scrambling at the edge of the water to secure a lethargic frog." He also frequently shot them long after sunset, as they sat patiently waiting for their prey at the edge of a ditch.

Despite of its laziness, the Rough-Legged Falcon is a powerful bird, and is in no wise deficient in strength whenever it chooses to exert itself. When roused by hunger it will not be content merely with frogs and mice, but addresses itself to the capture of larger game, such as wild-ducks and rabbits, the latter of which creatures seem to be a favourite article of diet for this bird, and are almost unfailing in their operation when used to bait a trap.

Although scarcer than the common Buzzard, the Rough-Legged Falcon still holds its place as an inhabitant of the British Isles, and is occasionally taken throughout Great Britain. Several specimens have been killed in Ireland, one of which, recorded by Mr. Thompson, was knocked on the head with a stick, as it sat gorged and sleepy after its meal. In its stomach were found the remains of several birds and of a full-grown rat which had been torn into four pieces. This bird was killed near Dundonald, in the county of Down. Two other specimens were seen in Killinchy, in the same neighbourhood, and one of them was shot. It is also seen near Scarborough, and used to build annually in a rocky dell near Hackness.

The localities in which the Rough-Legged Falcon are most commonly found, are Northern Europe and North America, where it is quite a common bird. Specimens have been seen on the shores of the Mediterranean, and it is also noted as inhabiting many portions of Africa. The nest of the Rough-Legged Falcon is built on lofty trees, and contains from two to four eggs of a brownish-white, covered towards the large end with brown blotches.

In the HONEY BUZZARD we find a singular instance of a predaceous bird, endowed with many capabilities of catching and destroying the ordinary kinds of game, yet preferring to feast upon insect food in preference to the flesh of quadrupeds or birds.

Whenever a Honey Buzzard has been killed, and the stomach opened, it has always been found to contain insects of some kind. In one case, when a Honey Buzzard was shot in Ireland, and examined by Mr. Thompson, the stomach contained some larvæ of small beetles, as well as the perfect insects, which it had evidently obtained by grubbing in cow-dung, as its bill and forehead were covered with that substance in a perfectly fresh state. Some white hairy caterpillars, the pupæ of a butterfly, and three of the common six-spot Burnet moth (*Zygæna filipendula*), were also discovered in the stomach, together with some short lengths of grass stems, which had probably been swallowed together with the pupa-case of the Burnet moth, as that insect always suspends itself upon a stalk of grass when it is about to change into the perfect state.

Another specimen, which was captured in Northumberland, was observed by Mr. Selby, who makes the following remarks.

"The district around Twizel appears to have something attractive to this species, for within these few years several specimens have been procured both in the adult and immature plumage. The bird in question was observed to rise from the situation of a wasp's nest, which it had been attempting to excavate, as, in fact, to a certain extent, it had accomplished; and the large hole which had been scooped showed that a much greater power could be employed, and the bird possessed organs much better fitted to remove the obstacles which generally concealed its prey than a superficial examination of the feet and legs would warrant us in ascribing to it. A few hours afterwards, the task was found to be entirely completed, the comb torn out and cleaned from the immature young; and after dissection proved that at this season (autumn), at least, birds or mammalia formed one part of the food. A steel trap, baited with the comb, secured the aggressor in the course of the next day, when he had returned to view the scene of his previous havoc."

This bird seems to be specially defended by nature against the attacks of the irritated wasps, who would constantly use their stings very freely against the invader of their premises. The only vulnerable parts which they could find in a well-feathered bird, would be the naked skin round the eyes and at the base of the beak. In the genus Pernis, however, this skin is thickly covered with feathers, so that the bird can bid defiance to the poisoned lances of its irritated foes.

The Honey Buzzard does not, however, restrict itself solely to insect food, for it has often been observed to catch and devour birds and various quadrupeds. An instance of its predatorial propensities is given by Mr. Watters, in his "Birds of Ireland." The Honey Buzzard had been seen for several successive summers haunting the same locality, and killing the coots that frequented a piece of water. A coot was therefore shot, poisoned with strychnine, and laid out as a bait for the Honey Buzzard, and on the next day the bird was found dead at some distance from the spot. When in confinement this bird will eat mice, rats, birds, meat, and similar articles of diet.

The nest of the Honey Buzzard is made in some lofty tree, and is generally composed of little twigs as a foundation upon which are laid leaves, wool, and other soft materials. The eggs are generally two in number, and are very variable in colouring, some having a dark red band round the middle, and others being covered with dark red blotches. A curious description of a Honey Buzzard's nest is given by Willoughby. "We saw one that had made use of an old kite's nest to breed in, and that fed its young with nymphæ of wasps, for in the nests were found the combs of wasp's nests, and in the stomachs of the young the limbs and fragments of wasp maggots. There were in the nest but two young ones, covered with a white down, spotted with black. Their feet were of a pale yellow; their bills, between the nostrils and head, white; their craws large, in which were lizards, frogs, &c. In the crop of one of them we found two lizards entire, with their heads lying towards the bird's mouth, as if they sought to creep out."

The colouring of the Honey Buzzard is very variable, but is generally as follows. The back and upper portions are brownish-black, the primaries being chiefly black. The top of the head and back of the neck are yellowish-white marked with brown dashes and streaks; the under portions are yellowish-brown, each feather being marked with a stripe of brown down its middle, and a number of narrow bands run irregularly across the abdomen.

HONEY BUZZARD.—*Pernis Apivorus.*

The whole of the legs are mottled with white and yellowish-brown, and the tail is barred with light and deep brown alternately. The claws and beak are black, and the space between the beak and the eyes is thickly covered with little round feathers. The length of the Bird is twenty-two or twenty-four inches, the female being always the larger.

The KITE may be known, even on the wing, from all other British birds of prey, by its beautifully easy flight, and the long forked tail. Indeed, while flying, the Kite bears no small resemblance to a very large swallow, excepting that the flight is more gliding, and the wings are seldom flapped.

Despite the ill savour into which the name of the Kite has fallen, it is really a magnificent specimen of the falconidæ, and deserves its specific title of "regalis," or regal, quite as much for its own merits as from the fact that it had once the very great honour to be chased by royalty. It seems that the later kings of France were in the habit of marking the Kite as the quarry which was specially suitable to their regal state, and were accustomed to fly their hawks at Kites, instead of herons, as was usually the mode of procedure in the noble sport of falconry. The Kite is therefore termed regal, not on account of any innate royalty in the bird, but simply because royal personages chose to pursue it.

The Kite was in former days one of the commonest of the British birds, swarming in every forest, building its nest near every village, and being the greatest pest of the farmer and poultry-keeper, on account of its voracity, craft, and swiftness. Even the metropolis was filled with these birds, who acted the same part that is played by vultures in more eastern lands, and were accustomed to haunt the streets for the purpose of eating the offal which was so liberally flung out of doors in the good old times, and which, but for the providential instincts of the Kites, would have been permitted to decompose in the open streets of our obtusely-scented ancestors. In consequence of the services which they rendered, the Kites were protected by common consent, and were therefore extremely familiar, not to say importunate, in their habits, settling on the butcher's blocks, and bearing off pieces of meat almost within reach of his hand. In the northern parts of Africa, where they absolutely swarm, the Kite bears the same character for cool audacity, having been often known to sweep suddenly down, snatch a piece of meat from a man's hand, and disappear with its booty before he could recover from his surprise.

In the present day, however, the Kite is comparatively seldom seen in England, and when observed, is of sufficient rarity to be mentioned in the floating records of natural history. A correspondent of the "Zoologist" states that one of these birds was seen flying over London on June 24th, 1859. The bird passed over Piccadilly at a supposed elevation of a hundred yards, and flew with perfect steadiness over the vast expanse of smoke and chimney-pots, which must have presented a strange contrast with the green fields and leafy forests of its country home.

Advancing civilization has done its work with the Kite, as with all other destructive animals, and driven it far away from human habitations. Man chooses to be the only destructive animal within his own domains, and, eagle-like, permits no inferior to poach on his territory. The trap of the farmer, and the ready gun of the gamekeeper, have gradually expelled the Kite from farm and preserve, and it is now to be found only in the wide wooded district where it can remain comparatively free from persecution.

The flight of this bird is peculiarly easy and graceful, as the wings are seldom flapped, and the Kite sails through the air as by the mere power of volition. From the gliding movements of the Kite when on the wing, it has derived the name of Gled, from the old Saxon word *glida*. When in pursuit of prey, the Kite sails in circles at a considerable height from the ground, watching with its penetrating gaze the ground beneath, and sweeping with unerring aim upon any bird, quadruped, or reptile that may take its fancy.

Should it pass over a farm-yard, the whole establishment is in an uproar, quite inexplicable to any one who did not observe a certain little black speck sailing about in the heights of air. As soon as one of the smaller birds sees a Kite, it crouches to the ground and lies there motionless, as if transformed into a stone or a clod of earth. This instinctive movement is of great service, as it affords the only means of escaping the keen eye of the rapacious foe, who hovers above the spot, and is sure to notice any object that gives the least sign of life. Taking advantage of this habit, the fowlers make use of trained Kites or falcons to aid them in securing their prey. When they have marked down a covey of birds, they loose one of their trained hawks, who flies over the spot where the birds are lying, and causes them to crouch to the earth, heedless of anything but the foe above. While their attention is thus occupied, the fowlers come up with their nets, and easily secure the whole covey. Even in ordinary sporting, where the birds are very wild, a common paper kite is employed with great success, and in a very simple manner. The kite is raised in the air, and allowed to take out one or two hundred yards of string; a boy then takes charge of the kite, and walks over the land where the partridges are known to be. The birds mistake the paper kite for some soaring bird of prey, and permit the sportsmen to come within gun-shot before they rise.

Sometimes in making its stoop upon the poultry, it avoids their gaze by making a detour close to the ground, gliding suddenly over the wall or hedge, pouncing upon a chicken, and disappearing almost before its presence has been discovered. These raids, however, are not invariably successful, for the Kite has been often foiled in his stoop by the watchful care of the mother bird, who has seen the enemy coming, and valiantly waged successful battle in defence of her young family.

The food of the Kite is rather general in its nature, consisting of various quadrupeds, young rabbits, hares, rats, mice, and moles, of which latter animals no less than twenty-two were discovered in the nest of a single Kite, showing how rapid and noiseless must be its movements when it can secure so wary and keen-eared an animal as a mole. It does not chase the swift-winged birds through the air, but pounces on many a partridge as it sits on the ground, and is remarkably fond of taking young and unfledged birds from their nests; reptiles of different kinds, such as snakes, frogs, lizards, and newts also form part of its food, and it will not disdain to pick up a bee or a grasshopper when it can find no larger prey. The Kite is also a good fisher, waging nearly as successful war against the finny inhabitants of the rivers or ponds as the osprey itself; sweeping suddenly down upon the fish as they rise to the surface in search of food, or in their accustomed gambollings, and bearing them away to the shore, where it settles down and eats them in peace.

The nest of the Kite is chiefly built with sticks as a foundation, upon which is placed a layer of moss, wool, hair, and other soft and warm articles. The locality which is chosen for the nest is generally in some thick wood, and the bird prefers a strong, forked branch for the resting-place. The eggs are generally two in number, and sometimes three, of a greyish or light brownish-white colour, speckled with reddish chestnut blotches, which, as is the case with so many hawk's eggs, are gathered towards the larger end.

KITE.—*Milvus regális.*

The Kite still breeds in several parts of Great Britain, a recent instance being mentioned by the Hon. G. Berkeley in a communication to the *Field* newspaper, which, as it illustrates something of the disposition of the bird, shall be given in his own words. " I discovered the fact of a Kite's nest in one of the woods in my possession, while at Harrold Hall in Bedfordshire, near which I could not secrete myself sufficiently to witness the return to it of the old bird, because she soared above the wood, and did not consider the cover safe until I had taken my departure. One day I took my keeper with me to the nest, when the Kite, as usual, took to the skies. I then concealed myself, and sent my keeper away. The Kite soared over him, or 'watched' him safe away, and being unable to count even two, she boldly, and without further precaution, came back to her eggs, and I killed her."

A curious incident connected with the Kite is mentioned in the " Journal of a Naturalist." On a winter's evening a heavy fog came on, succeeded by a severe frost, and actually froze the feet of a number of birds to the boughs on which they were roosting. Among them were several Kites, which were thus fastened into the icy stocks, and no less than fifteen of these birds were captured by certain adventurous youths, who climbed the trees in spite of the frosty coating, and made easy prey of the poor prisoners. It is a strange circumstance that so many Kites should have been found together, as the Kite is not a gregarious bird, and associates only in pairs. The same author draws a very neat distinction between the flight of the buzzard and that of the Kite. " Though we see it sometimes in company with the buzzard, it is never to be mistaken for that clumsy bird, which will escape from the limb of some tree, with a confused and hurried flight indicative of fear, while the Kite moves steadily from the summit of the loftiest oak, the scattered crest of the highest poplar, or the most elevated ash—circles round and round, sedate and calm, and then leaves us."

The Kite is possessed of a very docile and agreeable temper, and is easily tamed. Mr. Thompson records an instance, where a pair of these birds were taken from a nest near Loch Awe, in Argyleshire, and were so thoroughly domesticated that they were permitted to fly at liberty every morning. When thrown into the air, they always soared aloft in their graceful circling flights, displaying their wonderful command of wing, and exulting in its exercise, but still so affectionate in their nature that they always returned to the hand of their owner when called. They were generally fed on rats and mice, and were very fond of catching the former animals as they were let loose from a cage. The bird has even been trained for the purpose of falconry, and has been found to perform its task to the satisfaction of its owner. A rather curious "lure" was employed in order to induce the Kite to return to its master after its flight, consisting of a living owl with a fox's brush tied to one of its legs, partly in order to impede its flight, and partly to make it more attractive to the gaze of the wanderer. The extraordinary dislike which is felt for the owl by most of the day-flying birds is well known, and it was probably on account of that aversion that the owl was used as a decoy.

The ordinary length of the common Kite is about twenty inches, but the sexes are rather variable in that respect, the females being always larger than the males. The colouring of the bird is very elegant, although composed of few tints, and is remarkable more on account of the delicate gradations and contrasts of hue than for any peculiar brilliance of the feathers. The general aspect of the Kite is reddish-brown, which on a close inspection is resolved into the following tints. The back and upper portions are dark brown, relieved by a reddish tinge upon the edges of the feathers ; the primaries are black, and the upper tail coverts chestnut. There is a little white upon the edges of the tertiaries, and the head and back of the neck are covered with greyish-white feathers, the centre of each feather being streaked with brown. The forked tail is reddish-brown, barred on the under surface with dark brown stripes, the centre feathers being the darkest. The chin and throat are coloured like the head, and the abdomen and under portions are reddish-brown. The under tail coverts are white, with a slight reddish tinge, and the under surface of the rectrices are also white, but washed with grey.

The ARABIAN KITE still plays the same part in Africa as was formerly taken in England by its European relative.

It is a bold and familiar bird, haunting the habitations of man, and audaciously carrying off its prey, undeterred by human presence. As it will eat garbage of almost any nature, it is a valuable ally to the unclean villagers ; carrying away the offal which is liberally flung out of the houses, and scarcely permitting it to rest on the ground before it is seen and devoured. The bird is strictly protected on account of the services which it continually renders; and so utterly fearless does it become through long experience of the ways of man, that it pays visits to every house in the village, in hopes of finding food of some kind. When Le Vaillant was employed in preparing his dinner at his wagons, the Kites came and boldly carried off the meat, heedless of the shots that were fired and the cries that were raised, and even returned for a second supply as soon as they had disposed of their former booty.

Sometimes the Arabian Kite becomes rather troublesome than useful, for in a dearth of its usual food it will attack the poultry yards, and coolly fly off with the young chickens before the eyes of their owner. On account of this propensity it is popularly termed " Kuyken-dief," chicken-thief, by the Dutch colonists.

This Arabian Kite is also a good fisher, and will dash into a river and carry off its finny prey with a success almost as invariable as that of the osprey. It is also something of a tyrant, and if it sees a smaller and weaker bird, such as a crow, carrying a piece of meat in its mouth, it immediately assaults the unfortunate creature, and, by dint of pecking and buffeting, forces it to relinquish its prize. Whenever it forages the country for food, it follows the habit of many similar birds, and continually haunts the marshy lands, where it is likely to find a plentiful supply of sustenance. The nest of this bird is not unfrequently found in such localities, built in a thick bush well shaded with reeds. Generally, however, it is placed upon a tree or a rock, after the fashion of the common Kite. The eggs are greyish-white, covered irregularly with reddish spots.

The colouring of the Arabian Kite is rather pleasing. The back and upper portions are greyish-brown, the head and throat being marked longitudinally with stripes of black and white. The under parts are reddish-brown, each feather being streaked down its centre with the deeper hue, and the feathers of the thigh are dusky red. The tail is marked with nine or ten transverse bands, and is slightly forked. The length of this species is not quite two feet. As its plumage is so dark, it has been termed the Black Kite by many authors, and its house-haunting propensities have earned for it the specific title of "parasiticus."

THE beautiful bird which is so well known under the appropriate title of the SWALLOW-TAILED KITE, is an inhabitant of various parts of America, though it has occasionally been noticed on the British shores, and in virtue of such casual visits has taken its place as one of the British birds. This species seems to be distributed over a considerable tract of country, according to the observations of many practical ornithologists. Mr. Nuttal has the following remarks on the habitat of the Swallow-tailed Falcon.

"This beautiful bird breeds and passes the summer in the warmer parts of the United States, and is also probably resident in all tropical and temperate America; emigrating into the southern as well as the northern hemisphere. In the former, according to Vieillat, it is found in Peru, and as far as

ARABIAN KITE.—*Milvus Ægyptius.*

Buenos Ayres; and though it is extremely rare to meet with this species as far as the latitude of forty degrees, in the Atlantic States, yet, tempted by the abundance of the fruitful valley of the Mississippi, individuals have been seen along that river as far as the falls of St. Anthony, in the forty-fourth degree of north latitude. They appear in the United States about the close of April or the beginning of May, and are very numerous in the Mississippi territory, twenty or thirty being sometimes visible at the same time. In the month of October they begin to return to the south, at which season Mr. Bateman observed them in great numbers assembled in Florida, soaring steadily at great elevations for several days in succession, and slowly passing to their winter quarters along the Gulf of Mexico."

Audubon remarks that it has never been seen farther eastward than Pennsylvania, and that only a few solitary individuals have been discovered in that locality. Towards the south it becomes more numerous, and in Louisiana and Mississippi it is extremely abundant, arriving in considerable numbers at the beginning of April, as many as a hundred having been counted in the space of a single hour, all passing directly from east to west.

On their first arrival they are so fatigued with their journey that they are easily approached; but owing to their habit of soaring at an immense height, they are tolerably safe even from man at all other seasons.

This falcon bears so strong an external resemblance to the swallow, that it might easily be taken for a common swallow or swift, as it flies circling in the air in search of the insect prey on which it usually feeds. Even the flight is very much of the same character in both birds, and the mode of feeding very similar. The usual food of the Swallow-tailed Kite consists of the larger insects, which it either catches on the wing, or snatches from the leaves as it shoots past the bushes. Various locusts, cicadæ, and other insects, are captured in this manner. It also follows the honey buzzard in its fondness for wasps and their larvæ, and has been noticed to excavate a wasp's nest, and to tear away the comb precisely like that bird. Reptiles, such as small snakes, lizards, and frogs, also form part of the food of this elegant bird. While it is engaged in the pursuit of such prey, or in catching the large insects upon the branches, it may be approached and shot without much difficulty, as it is so intent upon its prey that it fails to notice its human foe.

SWALLOW-TAILED FALCON.—*Élanoides furcatus.*

Audubon found that when he had succeeded in killing one of these birds, he could shoot as many more as he chose, because they have a habit of circling round the body of their slaughtered comrade, and sweep round it as if they were endeavouring to carry it away. Taking advantage of this peculiarity, he was enabled to procure as many specimens as he desired, shooting them as fast as he could reload his gun.

The nest of the Swallow-tailed Hawk is generally found on the very summit of some lofty rock or pine, and is almost invariably in the near vicinity of water. It is composed of small sticks externally, and is lined with grasses, moss, and feathers. The eggs are rather more numerous than is generally the case with the Hawks, being from four to six in number. Their colour is white with a greenish tinge, and they are marked with some dark-brown blotches which are gathered towards the larger end. There is only one brood in the year, and when the young birds are first hatched, they are covered with a uniformly buff-coloured downy coat. The colour of the adult bird is variable, consisting mostly of white and black, but on account of the bold manner in which their hues are contrasted, is remarkably pleasing in its effect. The back, the upper part of the wings, with the exception of the inner webs of the tertiaries, upper tail coverts and rectices, are a deep purple-black, the head, neck, and all other parts of the plumage being pure white. The legs and toes are blue with a green tinge, the cere is blue, and the beak blue-black. The claws are orange-brown. The length of this bird averages twenty inches.

The young of this species very rapidly acquire the tints of the adult bird. When they put off the buff downy mantle of their childhood, they are provided with black and white plumage which bears a close resemblance to the hues of the adult bird, but is devoid of the glossy purple sheen which is so beautiful a characteristic in the colouring. At this period of their existence, the tail is hardly so deeply cleft as that of the common kite of Europe. By the end of the autumn, however, the tail assumes its peculiarly beautiful forked form, and the plumage attains its perfect colouring, so that the bird of a year old can hardly be distinguished from one of six or seven years of age.

The small but brilliant BLACK-WINGED FALCON is a native of Africa, but is found in nearly all the temperate portions of the Old World. It has also been seen in New Zealand and Java.

It is a fierce and daring little

BLACK-WINGED FALCON.—*Elanus melanópterus.*

bird, striking so sharply with beak and claws, that even when wounded it cannot be approached without considerable precaution. The food of the Black-winged Falcon consists chiefly of grasshoppers and various insects, from which it is thought to derive the powerful musky odour which is exhaled from its body, and marks every spot on which it has recently sat. It is generally to be seen perched on the extreme top of some lofty tree, and while looking out for prey or engaged in active pursuit, pours forth a succession of ear-piercing cries, earning thereby from several ornithologists, the specific title of "vocíferus." Besides insects, it also feeds upon snakes and various small reptiles, and will sometimes, though but rarely, kill small birds or mice.

The wings are remarkably long in this species, and the legs short and feathered, a structure which gives clear indication that the bird is strong on the wing, and excels more in swiftness and activity of flight than in strength of beak or clutch of talons. In many of its habits, this species closely resembles the swallow-tailed falcon, and, like that bird, is capable of chasing and capturing insects on the wing. The nest of the Black-winged Falcon is rather large, and is generally built on a convenient forked branch. It is usually lined with moss and feathers, and contains four or five whitish eggs. Although the bird may often be seen darting at the crows, shrikes, and other predaceous birds that may pass near its residence, it has no intention of killing or eating them, but only wishes to drive them away from the vicinity of its home.

The head and neck of the Black-winged Falcon are silver-grey, the centres of its wings are black, and the primaries and secondaries are greyish-brown, with grey edges. The shoulder and the wings, breast, abdomen, and tail are pure white; the cere and toes are yellow, and the bill and claws black. When young, the back is brown, each feather being tipped with white, and the breast is brown spotted with white.

CROOK-BILLED FALCON.—*Cymindis uncinatus.*

The CROOK-BILLED FALCON derives its name from the shape of the beak, which is rather long and narrow, and is curved over at the point so as to form a rather large and sharp hook.

The distinctive characteristics in this bird, which was placed in the genus Cymindis, are the short tarsus armed with net-like markings, and half-clothed with feathers down their front, the wings shorter than the tail, and small narrow nostrils, which are so closely contracted as to resemble a mere cleft in the beak. The word Cymindis is Greek, and ought rather to have been used to designate the nightjar than this Falcon. This species possesses scales as well as reticulation upon the front of the tarsus. Its colour when adult is a leaden-blue, or grey on the upper portions of the body, and paler beneath. The tail is white at the base, and deepens into an orange-grey at the extremity. Its quill feathers are edged with a brownish ash, and the feet and cere are yellow. In its earlier stages of existence, the bird is of an almost uniform brown, relieved by reddish hues on the cap of each feather, a yellow stripe runs beneath the eyes, and little patches of the same colour appear on the cheeks, and the front of the neck is greyish-white. All the species that belong to this family come from America.

There is another allied species called the CAYENNE FALCON (Cymindis Cayanensis), whose specific name announces its habitation. This species is notable for a small tooth or notch at the bend of the beak. The colouring of this species is rather different from that of the former. The general colour of the adult bird is white, with a blue-black mantle, an ashy-white head, and pure white bands on the tail. When young, the mantle is variegated with brown and red, and marked here and there with reddish spots.

MISSISSIPPI KITE.—*Ictinia Mississippensis.*

AMERICA also furnishes us with the genus Ictinia, a member of which is very familiar to ornithologists under the name of MISSISSIPPI KITE.

This fine bird is a native of various parts of America, where it may be seen at a vast elevation in the air, sailing about in strange companionship with the turkey buzzard, and equalling those birds in the power, grace, and readiness of its flight. Why two such dissimilar birds should thus inhabit the same region of air, and delight in each other's society, is a very perplexing question, and requires a much clearer knowledge of the species and its habits before it can be satisfactorily settled. The Mississippi Kite cares not for carrion, and is not absolutely known to make prey of anything bigger than a locust. Yet, as Wilson well observes, the powerful hooked beak and sharp claws seem as if they were intended by nature for the capture of prey much more formidable than grasshoppers, locusts, and butterflies. In its flight, the Mississippi Kite needs not to flap its wings, but sails on its airy course with the same easy grace and apparent absence of exertion that is so characteristic of the flight of the vultures.

The very great proportionate length of its wings may account for this habit; the entire length of the body and tail being only fourteen inches, while the expanse of wing equals three feet. Being possessed of such power of flight, it emulates the swallow-tailed Falcon in many of its evolutions, and in a similar manner is fond of sweeping rapidly past a branch, and snatching from the leaves a choice locust or beetle without checking its progress. Like that bird it also feeds while on the wing, holding its prey in its claws and transferring it to its mouth without needing to settle. In character it seems to be a most fierce and courageous bird, as may be gathered from a short narrative given by Wilson of one of these birds which he had shot.

"This Hawk, though wounded and precipitated from a vast height, exhibited in his distress symptoms of great strength, and an almost unconquerable spirit. I no sooner approached to pick him up, than he instantly gave battle, striking with his claws, wheeling round and round as he lay partly on his rump, and defending himself with great vigilance and dexterity, while his dark-red eye sparkled with rage. Notwithstanding all my caution in seizing him to carry him home, he stuck his hind claw into my hand with such force as to penetrate into the bone. Anxious to preserve his life, I endeavoured gently to disengage it; but this made him only contract it the more powerfully, causing such pain that I had no alternative but that of cutting the sinew of his heel with my penknife.

2.   F

"The whole time he lived with me, he seemed to watch every movement I made, erecting the feathers of his broad head, and eyeing me with savage fierceness ; considering me no doubt as the greater savage of the two. What effect education might have had on this species under the tutorship of some of the old European professors of falconry, I know not ; but if extent of wing and energy of character, and ease and rapidity of flight, could have been any recommendation to royal patronage, this species possesses all these in a very eminent degree."

The attention of Mr. Wilson was greatly taken with these birds, and he on several occasions opened the stomachs of those which he had shot, in order to discover the food on which they had been sustained. On every occasion he found nothing but the legs, wings, and other indigestible portions of beetles, grasshoppers, and other large insects. He suggests that its lofty flight is for the purpose of preying upon those insects which choose the highest region of air for their pleasure trips, and not merely for the better convenience of seizing prey on the ground, as is the case with so many of the more carnivorous hawks.

The colours with which this bird is decorated are, though simple in themselves, exceedingly pleasing in their general effect. The head, neck, and part of the secondaries are a greyish-white, and the whole of the lower parts are whitish-ash. The back and upper portions of the body are ashy-black, and the pinions are deep black, as is its deeply forked tail. The legs are scarlet, and the claws, bill, and cere black.

CLOSELY allied to the Mississippi Kite is the SPOTTED-TAILED HOBBY, or LEADEN ICTINIA, both names being derived from the colouring of the plumage. It is, in common with the preceding bird, a native of America, and resembles it closely in many of its habits and manner of feeding. It is fond of soaring at a very great elevation, and will often remain stationary in a single spot, hanging as it were self-poised in air, much after the manner of the common kestrel, or windhover of England. The back and wings of this species are a slate or leaden blue, and the head and remainder of the plumage of whitish-grey, spotted rather singularly with brown. The eye is bright red. Specimens of this bird have been found both in North and South America.

The true FALCONS are known by their strong, thick, and curved beak, the upper mandible having a projecting tooth near the curve, which fits into a corresponding socket in the under mandible. The talons are strongly curved, sharp pointed, and are either flat or grooved in their under sides. These birds were formerly divided, by authorities in the art of falconry, into noble and ignoble hawks, the former being known by the forma- tion of the wings, in which the second primary feather is the longest, and is supported nearly to its tip by the first primary. In consequence of this structure the flight is extremely quick and powerful, so as to adapt the bird to the peculiar purposes which it was desired to serve. The Falcons all obtain their prey by striking it while on the wing ; and with such terrible force is the attack made, that a Peregrine Falcon has been known to strike the head completely from the shoulders of its quarry, while the mere force of its stroke, without the use of its claws, is sufficient to kill a pigeon or a partridge, and send it dead to the ground.

In striking their prey the Falcons make no use of the beak, reserving that weapon for the purpose of completing the slaughter when they and the wounded quarry are struggling on the ground. Should a small bird be the object of pursuit, and the country be open, a Falcon will sometimes drive full against the object of his pursuit and kill it immediately by the blow from its keeled breast. Generally, however, the bird strikes with its talons, employing chiefly the claw of the hinder foot, on which, from its forma- tion, is concentrated the whole force of the assailant. Sometimes when flying at quails, pigeons, or partridges, birds which generally associate together, a Falcon will flash among them with its lightning swoop, and dash one after another to the ground before he descends to feed upon the product of the chase. Mr. Thompson saw a Peregrine Falcon strike no less than five partridges out of a single covey. The mode of attack differs con- siderably in each species, and will be described in connexion with the habits of the birds themselves.

GROUP OF FALCONS.

Among the true Falcons the JERFALCON is the most conspicuous on account of the superior dimensions of its body and the striking power of its wing.

This splendid bird is a native of Northern Europe, being mostly found in Iceland and Norway, and it also inhabits parts of both Americas. Some naturalists believe that the Norwegian and Icelandic birds ought to be reckoned as different species, but others think that any differences between them are occasioned by age and sex. It is said that of the two birds the Iceland variety is the more powerful, of bolder flight, and greater age, and therefore better adapted for the purpose of falconry. Sometimes it is seen in the northern parts of the British Islands, having evidently flown over the five hundred miles or so of sea that divides Scotland from Iceland; this journey, however, is no difficult task for the Jerfalcon, who is quite capable of paying a morning visit to these islands and returning to its home on the same day. In 1859 one of these birds was shot in Northumberland, and others have been observed in the more southern counties. Towards the south, however, it has seldom if ever been observed.

The power of flight possessed by this bird is wonderfully great, and has been well described by Mr. Mudie in his history of the British Birds. "It pays occasional visits to the northern and western isles, more especially to those places of them that abound with rock doves; and few sights can be finer than that of the Jerfalcon driving through a flock of them. When the Jerfalcon comes within sight of her prey she bounds upwards, every stroke of the wings producing a perpendicular leap, as if she were climbing those giant stairs with which nature moulds the basaltic rocks; and when she has "got the sky" of her prey to a sufficient height for gaining the necessary impetus, her wings shiver for a moment—she works herself into proper command and poise, and to the full extent of her wings. Then, prone she dashes, with so much velocity that the impression of her path remains in the eye in the same manner as that of the shooting meteor or the flashing lightning, and you fancy that there is a torrent of Falcon rushing for fathoms through the air. The stroke is as unerring as the motion is fleet. If it takes effect on the body, the bird is trussed and the hunt is over; but if a wing only is broken, the maimed bird is allowed to flutter to the earth, and another is marked out for the collision of death.

It sometimes happens that the mountain crow comes in for the wounded game, but in order to do so it must proceed stealthily along the ground, for woe betide it if it rises on the wing and meets the glance of the Falcon. The raven himself never scoops out another eye if he rises to attempt that one; and it is by no means improbable that in the early season in those cold northern countries, when the lambs are young and the flock weak, and the crows and ravens prowl about blinding and torturing, the Jerfalcon may be of considerable service to the shepherd."

When at liberty in its native land, it seems to prefer birds to any other kind of prey, and will resolutely attack birds of considerable size, such as the heron or stork. It will also chase hares and rabbits, and in the pursuit of this swift game is so eager, that after knocking over one hare it will leave the maimed animal struggling on the ground while it goes off in chase of another. Although its home is in the chilly wastes of those northern regions, the bird is in no want of food, finding ample supply in the sea birds which swarm around the tall cliffs that jut into the waves, and being able from its great powers of flight to range over a vast extent of country in search of its daily food.

On account of the singular power, swiftness, and courage of this bird, it was in former days held in the highest estimation, and could only be purchased at a most extravagant price. Not only must it be taken at the imminent risk of life, from the almost inaccessible cliffs on which it builds its nest, but it must also be specially brought from Iceland or Norway, and trained after its arrival at its new residence. As the bird is a most unruly and self-willed creature, its instruction was a matter of very great difficulty, and could only be achieved by the most patient of skilful teachers. So highly indeed was this bird valued, that after the hawking season was over, and the ordinary hawks permitted to fly at liberty according to custom, the Jerfalcon was retained by its owner, and kept for the next year. The training of this bird is a long and tedious process, and is managed after the following manner.

It is allowed that all hawks are fierce and untameable in proportion to the latitude in which they reside, those which inhabit the northern and colder parts of the earth being much fiercer and less tameable than those of more southern regions; so that the course of training through which a Jerfalcon is forced to pass, is much more severe than that which suffices to render a Peregrine Falcon subservient to its teacher.

The first object which the trainer bears in mind is, to reduce the strength of the bird by slow degrees, so as to prevent it from injuring itself by the fierce and protracted

JERFALCON.—*Falco Gyrfalco.*

struggles with which it would endeavour to resist any advance on the part of the teacher. This object is obtained by giving the bird only half the usual allowance of food, and by steeping the meat in water before the Falcon is permitted to touch it. A leathern hood, which answers the double purpose of blinding the eyes and keeping the beak closed, is placed on the head, and never removed except at night, so that the bird remains in perpetual darkness for ten days or a fortnight. If the bird attempts to bite when the hood is removed, cold water is splashed in its face, and if it is very savage, it is plunged entirely under water. By the end of fifteen or sixteen days, the Falcon becomes used to the handling to which it is subjected, and will permit the hood to be removed and replaced, without offering any resistance.

The next part of the instruction is to teach the bird to pounce upon any object that may be pointed out by the instructor, whether it be a heron moving in the air, or a hare running on the ground.

The skin of the intended prey is employed for this purpose, and the bird is invariably fed while standing on this skin. When it is accustomed to associate the idea of the hare or heron skin with the pleasure of satisfying its hunger, the skin, if it be that of a hare, is drawn along the ground, and the falcon encouraged to pursue it. As soon as the bird pounces, the teacher looses his hold of the skin, and permits his pupil to feast on the meat which has been previously attached to it. Next day the skin is placed at a distance of several yards before it is started, and the distance is gradually increased, so that the bird learns to search in every direction for its expected prey, as soon as the hood is removed from its eyes. Lastly, the teacher mounts on horseback, and holding in his hand a long string, the other end of which is attached to the skin, he darts off at full gallop, so that the Falcon is forced to put out its best speed before it can overtake the horse or pounce on the skin as it flies leaping and striking along the ground. On the first two or three days, the Falcon is almost quite breathless when it has overtaken the horse, and sits panting, with open beak, upon the skin ; but in a week or so, it becomes much stronger, and is not in the least distressed by its severe chase.

To teach the bird to pounce upon herons, buzzards, kites, or other winged prey, a stuffed skin is employed after much the same fashion that is followed with the hare skin. Instead, however, of being dragged along the ground, the skin is flung into the air, and the bird encouraged to pounce upon it before it reaches the ground. In all cases the attendants keep up a great noise and shouting as soon as the Falcon begins to feed, in order to accustom tho bird to the uproar which is the inevitable concomitant of the chase. Horses and dogs are then brought close to the feeding bird, and the dogs are encouraged to break out in full cry.

When the bird has become sufficiently docile to recognise its keeper and to know his voice, it is then instructed to come to his hand when called. This accomplishment is taught by means of a " lure" and a whistle. The lure is a gaudy apparatus of feathers and leather, on which is placed a small piece of some special dainty. The Falcon is encouraged to jump on the lure and devour the food, the whistle being blown continually while the bird is eating. Next day the teacher stands at a few yards' distance from his pupil, blows the whistle, exhibits the lure, and permits the bird to make its little feast. In a very short time the sound of the whistle attracts the attention of the Falcon, which immediately looks around for the lure and sets upon it at once. When the huntsman takes the field, the lure is attached to a leathern strap, and slung to the side of the horse, so that whenever a flying Falcon is to be recalled, the huntsman whistles sharply in order to attract the bird's attention, and at the same time swings the lure round his head, so as to render it more visible to the bird.

This process of training, of which a very slight and rapid sketch has been given, occupies from six weeks to two months, whereas that of the peregrine, goshawk, or merlin only requires some fifteen or twenty days. Even when the whole series of instructions has been completed, its ultimate success is very dubious, for it sometimes happens that when the bird finds itself wholly at liberty for the first time, it forgets all its teaching, and, heedless of lure or whistle, flies exultingly to its rocky home.

The colour of the adult Jerfalcon is nearly white, being purely white on the under surface and flecked with narrow transverse bars of greyish-brown upon the upper parts. The sharp claws are black, the beak of a bluish tint, and the cere, tarsus, and toes yellow. When young, however, the bird presents a very different aspect, and would hardly be recognised as belonging to the same species. In its earlier stages of life it is almost wholly of a greyish-brown tint, the feathers being slightly marked with a little white upon their edges. As the bird increases in age the white edges become wider, and by degrees the entire feather is of a snowy whiteness. The name Jerfalcon is supposed to be a corruption of " Geyer-falcon," or Vulture Falcon.

LESS powerful, but more graceful than the Jerfalcon, the PEREGRINE FALCON has ever held the first place among the hawks that are trained for the chase.

The temper of the latter bird is incomparably more docile than that of the former, the lessons of the instructor are received with more readiness, occupy far less time, and seem to be more powerfully impressed upon the memory. For training this bird the process is very similar to that which is employed in the instruction of the Jerfalcon, but the system is not nearly so severe, and occupies scarcely one fourth of the time that is needful to render the fierce and fearless Jerfalcon subservient to the dominion of man. The whole process is very simple in its theory, being based on the principle of placing the bird in such situations that it is absolutely unable to disobey the orders which are given by its trainer, and consequently imagines that it is equally bound to obey every order which he may afterwards give. In order to obtain this result two qualities are needful in the instructor, namely, patience and gentleness, for without these traits of character no man can hope to be a successful teacher of hawks, or, indeed, of any other being whatever.

When thoroughly tamed, the Peregrine Falcon displays a very considerable amount of attachment to its owner, and even while flying at perfect liberty will single him out from a large company, fly voluntarily towards him, and perch lovingly on his hand or shoulder. Several of these beautiful birds that had been tamed by Mr. Sinclair were so thoroughly domesticated that they were permitted to range at liberty, and were generally accustomed to perch on a tree near the house. One of these Falcons was permitted to seek her own food whenever she could not find any meat upon the accustomed spot, and would take flights of several miles in extent. Yet she would immediately recognise her master if he were out shooting, and would aid him by striking down the grouse as they rose before his dogs. On one occasion the fearless bird met with an accident which might have proved fatal, but was ultimately found to be of little consequence. Unaware of the presence of his Falcon, her master fired at a grouse, and as the bird was at the same moment making a "stoop" upon the bird, one of the leaden pellets struck the Falcon, and inflicted a slight wound.

The dash and fury with which this hawk makes its stoop is almost incredible. In a little coast town in Yorkshire, a part of a greenhouse had been divided off by wire so as to form an aviary, the roof of the aviary being the glass tiling of the greenhouse. In this edifice were placed a number of small birds, which attracted the attention of a Peregrine Falcon that was passing overhead. Totally unmindful of all obstacles, he shot crashing through the glass without injuring himself in the least, seized one of the terrified birds, and carried it off in safety. Several other birds were found dead, apparently from fright, or perhaps by the shock of the hawk and glass which came flying among their number.

It is said that when the Peregrine Falcon takes up its residence near the moors it is a very mischievous neighbour, slaughtering annually great numbers of grouse. Although very fond of pigeons and similar game, the Peregrine Falcon seems to have sufficient sporting spirit to prefer the grouse to the pigeon, and never to trouble

PEREGRINE FALCON.—*Falco peregrinus.*

itself about the former bird as long as it has a chance of obtaining one of the latter. A correspondent of the *Field* newspaper speaks very strongly on this subject, in answer to those who wished that so noble a bird as the Peregrine Falcon should be spared by the proprietors of moors and other preserves. From observations which have extended over a long series of years, he has come to the conclusion that a single nest of Peregrine Falcons will destroy in a single season nearly three hundred brace of grouse alone. Mr. Thompson relates several anecdotes which bear strongly upon the game-destroying propensities of the Peregrine Falcon.

"Mr. Sinclaire, many years ago, while exercising his dogs on the Belfast mountains, towards the end of July, preparatory to grouse-shooting, saw them point, and coming up, startled a male Peregrine Falcon off a grouse (*Tetrao Scoticus*) just killed by him, and very near the same place came upon the female bird, also upon a grouse. Although my friend lifted both the dead birds, the hawks continued flying about, and on the remainder of the pack which lay near being sprung, either three or four more grouse were struck down by them. Thus, two and a half or three brace were obtained by means of these wild birds, being more than had ever been procured out of a pack of grouse by my friend's trained falcons. The same gentleman has frequently, when out shooting, obtained a single grouse which has been killed by wild Peregrine Falcons, but, except in the above instance, never more than one.

Another friend, walking in Devis mountain, near Belfast, on the first of September, 1832, saw one of these birds pursue a couple of grouse for some distance without success, and subsequently kill a snipe high in the air, after a good chase. A sportsman states that woodcocks shot by him in the south of Ireland have more than once been pounced upon and carried off by wild Peregrine Falcons before they reach the ground." The same writer mentions that the Peregrine Falcons would often follow the sportsmen in spite of the flash and report of their guns, and would boldly carry off the birds that were struck by the shot. The eagles which inhabited the same locality were much less courageous, and used to fly away at the first discharge.

The Peregrine Falcon appears to be very discriminating in its tastes, preferring birds to all other prey, and generally choosing those very species which mankind has acknowledged to be delicacies for the table. Grouse, as has already been seen, are a favourite quarry of the Peregrine, and the bird is also very fond of partridges, snipes, and woodcocks. In the chase of the last-mentioned bird, the Peregrine displays the greatest imaginable command of wing, for it will follow the " cock" into its place of refuge among the branches, and in many cases will carry it off without even touching one of the boughs among which the woodcock shoots and twists with such singular celerity.

Sometimes, however, both the pursuer and the pursued have suffered severe injuries from their too heedless flight, a misfortune of which Mr. Thompson mentions several instances. In one case the woodcock and Falcon shot through the drooping branches of a weeping-ash tree, and, striking against the stem, both fell stunned to the ground. The woodcock was the first to show signs of life, and after waiting for a short time, scrambled to the bank of a neighbouring glen, and was permitted to escape. The Falcon was longer in reviving, and when picked up was bleeding at the mouth ; she, however, shortly recovered. On another occasion the hawk and woodcock came against a large stone, the latter being disabled, but the former suffering but little injury. The force with which the terrified bird flies from its pursuer may be imagined from the fact that one of these birds, when endeavouring to escape from a Falcon, struck against the top of a wall with such violence as to split its breast completely open.

The same author relates some curious incidents relative to the comparative powers of flight of the Falcon and woodcock. "The finest flights are those in which the bird 'climbs the air.' Once, when from fifty to sixty persons were present, a woodcock sprung near Andersson's town in the Falls, climbed the air, and the hawk swept after him until both got out of sight of all persons present except one, who insisted that the quarry was captured : it soon proved so, by the hawk's coming down with its victim. The trial between the birds, which should be highest, was so well contested from the moment the woodcock sprung and went right up, that the issue was most eagerly looked to ; numerous even bets depended on it. Again, at the head of Colin Glen, in the same district, a woodcock, pursued by one of the hawks, climbed until both were lost to view. The death of the woodcock was, however, soon announced by its rapidly falling through the air, until soused in the deep pool of the rocky river, called, from the peculiar sound its waters make, the Rumbling Burn. It was observed on laying hold of the victim, that it had been struck in the back by the hawk, but not laid hold of, which accounted for its coming down singly."

The Peregrine will chase and kill many of the coast birds, such as the dunlin, the gull, and the plover. The curlew is a very favourite prey, and being a strong-winged bird, affords great sport. It is rather remarkable, that the dunlin, together with birds of similar habits, fly instinctively to the sea, lake, or river when attacked by the Peregrine Falcon, as if knowing that the winged hunter is very unwilling to swoop upon any object that is flying upon the surface of the water. The Falcon has been seen to drive a dunlin repeatedly into the sea before it could intercept the poor bird between the dry land and its watery refuge. It will also strike at the grey crows, or at herons, but unless especially trained to the pursuit, will not trouble itself further about them.

Mr. Thompson gives us the following spirited description of a curlew chase, which is valuable as exhibiting the capacities of both curlew and Falcon.

" My friend and his companions were fishing in Loch Ruthven, when a flock of these birds (curlews) appeared. Immediately afterwards a tercel (or male Peregrine) came in sight, bearing down upon them so suddenly as to be hardly seen until he had singled out and swept one of them from a height of about fifty yards into the lake. Here he pounced at it, but without striking, although it did not go below the surface of the water. On the tercel's flying a little way off to take one of his bold circles when the quarry is put down, or at bay, the curlew rose to follow the flock, and had got away about a hundred yards when the tercel again bore down upon it. Refuge was a second time taken upon the lake. This was repeated not less than ten times. The speed of the tercel's flight was considered to be twice that of the curlew's, as when circling about two hundred yards off, he never gave his desired victim leave to get more than about half that distance ahead, until he had it down again.

The curlew, although apparently more fatigued and worn out every time it was put down, was the last time hardly able to rise from the lake, and escaped in consequence of the flock from which it came, or a similar number of birds, appearing in sight, when its persecutor betook himself after them. He very soon had one of this flock also in the water, and enacted just the same part towards it as he had done towards the other. It was put down to the lake at least a dozen times, and along a great extent of its surface, once between the boats of the fishing party, not more than about fifty yards distant from each other. The hawk and the curlew were both several times within about twenty yards of the boats, and once indeed the latter, closely pursued, took the water just before the bow of one of them. Eventually the tercel left off the chase, though, as in the former instance, the curlew was nearly worn out. The poor bird, now seeing two of his species come in sight, joined them, and they all went off swiftly in company." An experienced falconer who witnessed the scene, remarked that the swoops made by the wild Falcon were bolder, and its flight stronger than he had seen in any trained bird.

The full speed of the Peregrine Falcon has been computed at a hundred and fifty miles per hour, and a single chase will often occupy a space of eight or ten miles. Its power of wing is not only useful in enabling it to wage successful pursuit of swift-winged birds, but in giving it sufficient buoyancy to carry off the prey which it has secured. So strong is the Peregrine's wing that it has often been observed to bear in its talons a bird larger than itself, and to carry it to the nest without difficulty. Even a guillemot has been struck and carried off by the Peregrine.

To give a full account of the sport of falconry would demand a separate work, for even the technical terms of that art would require a glossary of many pages. It must suffice to say that the hawks were especially trained to fly at herons, and their mode of taking their prey was by mounting perpendicularly in hopes of gaining a vantage point above the game, from which they might swoop down with lightning-like force, and strike the quarry to the ground. For this sport the female birds were always preferred, as they are not only larger and more powerful than their mates, but are also possessed of more perseverance and daring courage. Technically, the adult males are termed tercels, tiercels, or tarsels, the females only being distinguished by the title of Falcon. The difference between the dimensions of the two sexes is very strongly marked, the male being about fifteen inches in length, while the female measures eighteen inches.

There are various methods of taking these birds for the purpose of the chase, some falconers preferring to remove them from the nest before they are fully fledged, and others caring little for their Falcons unless they are caught while in the vigour of full age. The training which is bestowed on the latter birds is necessarily much longer than that which is needful for treating the young, but as the birds are stronger and fiercer in character, they are well suited for taking the larger game, such as herons and hares. To take the young Peregrine from the nest is a difficult business, and one which needs the possession of a strong arm and a cool head, for the nests are always built on a shelf of some precipitous rock, and the person who desires to take the young has the choice of climbing from below or of being lowered from above by means of ropes. On account of these rock-loving propensities, the bird is known in some parts of England as the Cliff Hawk.

When captured at adult age, it is generally enticed into the toils by means either of a decoy of its own species, or by the aid of a great brown owl, which is trained to flutter in such a manner as to draw the attention of the Falcon. All the hawk tribe seem to be animated with a deadly hate of the night-flying rapacious birds, and never can see an unfortunate owl without longing to attack it. No sooner, therefore, does the Falcon see the owl, than it darts fiercely at its intended victim, and is captured between the meshes of the net which is cunningly set for the purpose. There is another most ingenious mode of taking this Falcon by the aid of the great grey shrike; but as an account of that mode of Falcon-catching will be given in the history of the shrike, it will only be casually mentioned in the present page. One of these birds was caught in a very curious manner. A gentleman heard a number of jays making a very great chattering, as if wrought up to the highest pitch of excitement, and on going to the spot, he found that a Peregrine Falcon had pounced upon a crow, and had contrived to entangle itself in such a manner, that it could neither loosen its grasp of its victim nor carry it away.

The eggs of this bird are generally two or three in number, although a fourth is some times known to be laid in the same nest. The fourth egg is, however, generally addled, but the mother Falcon does not fling it out of her cradle, as is the case with many birds. The colour of the egg is a very pale reddish brown, usually mottled with a darker tint. The young are most voracious creatures, and are kept very constantly supplied with food by the exertions of their parents. A very curious instance of sagacity in this bird was observed by Mr. Sinclair. When a farmer was being lowered by a rope towards the nest, both parents flew from the spot, the female hovering close to her young, and the male circling high in the air with the prey in his mouth. Not desiring to come down to the nest, the male bird dropped from his beak the morsel of food, which was then caught by the female as it fell through the air, and conveyed to her young.

Even in captivity, the greatest regularity of feeding is needful, for if the supply of food be not given with the utmost punctuality, the bird never attains its proper development of colour or size, the primaries of the wing and the quill feathers of the tail being marked with light bands, and the wings being at least an inch shorter than their proper length. Such birds are termed "tainted," and are comparatively useless in falconry, because their wings are too short to permit the full power of flight, and their quills are apt to be soft and weak in the shaft.

The term Peregrine is of Latin origin, and signifies a pilgrim. This title has been given to it because it is in the habit of making very long journeys, and is consequently found spread over a very large extent of country. In its adult state, the Peregrine Falcon is very elegantly coloured. The top of the head, the back of the neck, the primaries and a stripe beneath the eye, are of a deep black-brown; the upper parts of the body are ashy brown, the latter tint becoming fainter in each successive moult, and being always marked with a series of dark bars upon its back, tail, and wing coverts: the breast is white, deepening into a chestnut hue, and being barred transversely with reddish-brown upon the breast, and marked on the front of the throat with longitudinal dashes of very dark brown. The remainder of the under plumage is greyish-white, profusely barred with dark-brown. When young the plumage is altogether of a more ruddy hue, and the birds are termed, in the language of falconry, Red tercels, or Red Falcons, according to their sex.

The true LANNER (*Falco Lanarius*) is a native of Northern Europe, and is not known as a visitor of the British shores. It is a rather large bird, considerably exceeding the Peregrine Falcon in its dimensions, and being little inferior in size to the Jerfalcon itself. This bird was formerly much esteemed for the purpose of falconry, and was specially trained to fly at the kite, a bird which is too strong to afford the ordinary Peregrine Falcon any possible hope of success. The male of this species is considerably smaller than his mate, and is therefore called a Lanneret. The English bird, to which the title of Lanner has often been wrongly applied, is nothing more than the young female Peregrine Falcon.

THE small but exquisitely shaped HOBBY is found spread over the greater part of the old world, specimens having been taken in Northern Africa, and in many portions of Asia, as well as in Europe, which seems to be its chief residence. It was formerly very common in England, but is year by year less seldom seen in our island, as is the case with all its predaceous relations. From all accounts, it seems to be rather a local bird, being partially influenced by the nature of the ground and the quantity of food which it is able to procure.

This bird appears to favour inland and well-wooded lands rather than the sea-shore or the barren rocks; thus presenting a strong contrast to the Peregrine Falcon. We may find an obvious reason for this preference in the fact that a considerable proportion of its food is composed of the larger insects, especially of the fat-bodied beetles, which it seizes on the wing. Chaffers of various kinds are a favourite prey with the Hobby, and in several cases the stomachs of Hobbies that had been shot were found to contain nothing but the shelly portions of the larger dung-chaffer (*geotrupes stercorarius*). As therefore the common cock-chaffer is a leaf-eating insect and frequents forest lands for the purpose of attaining

HOBBY.—*Hypotriorchis subbuteo.*

its food, the Hobby will constantly be found in the same locality for the object of feeding on the cock-chaffer. And as the dung-chaffer swarms wherever cattle are most abundantly nourished, the Hobby is attracted to the same spot for the sake of the plentiful supply of food which it can obtain.

Larks, finches, and various small birds, fall victims to the swift wings and sharp claws of the Hobby; but its predilections for insect-hunting are so great, that even when trained for the purpose of falconry and flown at small birds, it is too apt to neglect the quarry to which its attention was directed, and to turn aside after a passing beetle or grasshopper. Although it is by no means a powerful bird, and seldom of its own free will attacks any prey larger than a lark, it has been successfully trained to fly at pigeons, and has even been known to strike down so comparatively large a quarry as the partridge. When in a state of domestication, its food should consist chiefly of the smaller birds, and it may also be fed upon beef cut into small pieces and very fresh. Its temper is so gentle, and its disposition so mild and docile, that it is easily tamed and taught to obey the instructions of its owner. It is also very hardy in constitution, so that it is well adapted to the purpose of those who wish to possess a trained hawk, but do not care for flying it at the larger game.

The nest of the Hobby is almost invariably built among the branches of a lofty tree, and is never placed upon a rocky ledge except under very peculiar circumstances. The eggs are from two to five in number, these being the usual orange, and some of a greyish-white tint, irregularly speckled over their whole surface with spots of reddish-brown.

The colour of the adult Hobby is of a greyish-black upon the back and upper portions, the tint softening into a blue-grey as the bird increases in years. Each feather is slightly edged with yellowish white, and the primaries and secondaries are black-brown, edged with greyish-white. The tail is grey-black, barred on its upper surface with the lighter hue in all the feathers except the two central rectrices. The lower surface of the tail is barred with black and greyish-white. The chin of the male is pure white, that of the female being of a duller tint, and the breast and abdomen are yellowish-white marked with longitudinal splashes of dark brown. The cere is greenish-yellow, the claws black, and the beak a slaty-brown colour, deepening into black at the tip. In the young bird the upper surface is much darker than in the adult, or the feathers are nearly black in the centre, and are edged with reddish-brown. The general form of the body is elegant and slender, and the tips of the wings project beyond the tail, so that when the bird is hovering in the air, it bears some resemblance to a common swallow or swift.

ALTHOUGH of the smallest of the British Falconidæ, being only from ten to thirteen inches in length, according to the sex of the individual, the MERLIN is one of the most dashing and brilliant of all the hawks which frequent our island.

This beautiful little bird is almost invaluable to the young falconer, as it is so docile in disposition, and so remarkably intelligent in character, that it repays his instructions much sooner than any of the more showy, but less teachable Falcons. Every movement of this admirable little hawk is full of life and vivacity ; its head turns sharply from side to side as it sits on its master's hand, its eyes almost flame with fiery eagerness, and it ever and anon gives vent to its impatience by a volley of ear-piercing shrieks. There is, however, a singular capriciousness in the character of the Merlin, for it seems to be so sensitive to certain influences which are quite imperceptible to human organization, that the same individual which on one day or at one hour is full of fierce energy, chasing large and powerful birds of its own accord, following the erratic course of the snipe with a wing as agile and far more enduring than its own, or shooting suddenly through the tangled branches of the underwood in pursuit of some prey that is fleeing to the leafy abode for refuge, will at another time become listless and inanimate, and even if it be induced to fly at its quarry, will turn suddenly away as if alarmed, and return languidly to its perch.

With all these drawbacks, however, the Merlin is one of the very best little hawks that ever was put into training, for it can be taught to fly at anything that is indicated, and seems to care nothing for disparity of size. As a general rule, the smaller hawks are unable to fly systematically at any prey larger than a lark or a thrush, but the Merlin has no scruples on the point of size, and will freely dash at a snipe, a partridge, or a grouse, at the desire of its owner. One of these birds has been known to make voluntary chase of a magpie, and to follow it up with the greatest perseverance. The great point in the instruction of this bird seems to be that the teacher must never permit his charge to feed upon any bird smaller than a snipe or partridge, nor, if possible, must he allow it to see a sparrow, or even a thrush.

Before the young bird is able to tear to pieces its winged prey, it should always be accustomed to have its food placed upon the stuffed skin of a partridge, and when it has attained sufficient strength, the breast of a real partridge should be cut open, and a small portion of its ordinary food placed within the aperture, so as to encourage the bird to tear away the flesh in order to satiate its hunger. The next step is to substitute an entire partridge for the ordinary diet, and by degrees to teach it to pounce upon the dead bird as it is flung to a daily increasing distance. It is a good pigeon-hunter, and if the owner choose to train it for the smaller game, it is unrivalled as a chaser of thrushes, larks, and similar birds, owing to the pertinacity with which it carries on the pursuit, and the resolutely agile manner with which it will thread the mazes of branch and leaf in chase of a bird which seeks for refuge in the covert.

Even on the wing, the Merlin may be known by its peculiar flight. Sometimes it may be seen skimming over the ground at a swift pace, but at no great elevation, at another it will urge its spiral course upwards in pursuit of some prey which has taken to "climbing the air," while at another time it may be observed following the course of some prey with such singular exactness that the two birds seem to be animated by the same spirit; and the turn is hardly commenced by the fugitive before it is taken up by the pursuer. In striking its prey the Merlin is possessed of a wonderful skill, the quarry falling down almost as soon as it is touched. It seems that the bird is able to strike a vital part with an almost imperceptible touch of its claw or beak.

The Merlin frequently breeds in England, and makes its nest on the ground, generally choosing for that purpose some spot where large stones are tolerably plentiful, and may serve as a protection to the nest, as well as for a perch on which the Merlin, like the

MERLIN.—*Hypotriorchis æsalon.*

harrier, loves to sit and survey the prospect. From this habit of perching on pieces of stone it has derived the name of STONE FALCON, a title which has been applied to this bird in Germany and France as well as in England. Sometimes, but not often, the nest is made on some rocky shelf on a precipice. The eggs are four or five in number, of a light reddish-brown hue, covered with mottlings and splashes of a deeper tint.

The colour of the Merlin is very pleasing, but not very easy to describe, as it is not so conspicuous as in many of the hawks, and moreover is rather different in the two sexes.

The top of the head is a slaty grey, marked with dark streaks running along the line of the head; the beak and upper portions of the body are of a similar slaty grey, but without the dark lines. The shafts of each feather are, however, of a dark brown, and give a very rich and peculiar colouring to those portions of the plumage. The pinions are black, the upper surface of the tail is nearly grey, with the exception of three faint dark bands, the last being the broadest, and the tip white. The chin and throat are white, and the under parts of the body are reddish fawn, thickly marked with patches of a darker colour, and streaks of deep brown. The cere, legs, and toes are yellow, the claws black, and the beak a slaty grey, deepening towards the point, and slightly marked with longitudinal dark lines. Round the neck runs a band of pale reddish-brown, which also extends to the cheeks, and there forms a patch on each side.

This description belongs to the male bird, the colouring of the female being of a rather different nature. The beautiful blue-grey which tints the upper parts of the male bird is in the female of a dark reddish-brown, marked with slender longitudinal streaks covered by the black-brown shafts of each feather. The secondaries and the wing-coverts are of the same hue as the back. The tail is brown, varied with five narrow streaks of dark brown, and the under surface of the body is a very pale brown, marked with longitudinal dashes of a darker hue. The young of both sexes are nearly alike for the first year, after which time the males assume their peculiar colouring, and the females retain the same tints.

THE genus Hypotriorchis is rather rich in interesting birds, among which may be noticed the PIGEON HAWK of America and the Chicquera Falcon of India.

The former of these birds is found not only in America, but also in parts of the West Indies, a specimen from Jamaica being in the collection of the British Museum. Generally, however, it is found in Southern America, where it is rather plentiful, and may be seen hunting for its prey in the proper localities. It is a spirited and swift-winged bird, although not a very large one, measuring barely eleven inches in length, and not quite two feet in the expanse of wing. A rather remarkable peculiarity in its plumage is found in the feathery covering of the legs, which is singularly long, the tips of the feathers reaching nearly to the feet.

The usual prey of the Pigeon Hawk consists of mice, small birds, reptiles, and various insects, and it has a remarkably sharp eye for any unfortunate half-fledged bird that may have strayed from its nest, or crippled itself in its first endeavours to fly. It is a terrible foe to the reed birds, grackles, and other similar members of the winged race; hovering continually about the crowded flocks, and picking off the stragglers or the weakly at leisure. Sometimes, however, the Hawk seems to lose patience, and dashing suddenly into the flock, will bear away an unfortunate bird from the midst of its companions. It has derived the name of Pigeon Hawk, because it is well known as one of the numerous birds of prey that hover around the myriad armies of the Passenger Pigeon, as they make their wonderful migrations which have rendered them so famous.

Further information concerning this bird may be obtained from the pages of Wilson and Audubon.

THE last member of this genus which can be separately noticed in the present work is the CHICQUERA FALCON of India. This bird is often trained by the native sportsmen, and employed for the purpose of chasing the bustard and similar game. It is not good at an aerial flight, and therefore is not used against soaring game; but when employed in the pursuit of the running birds, its peculiar low, skimming flight is admirably adapted to the purpose. In order to keep the bustard from taking to wing, a Hawk of another species is trained to fly above the quarry and beat it down whenever it endeavours to raise itself into the air and escape by flight.

THIS fine bird, which is called the BROWN HAWK, or CREAM-BELLIED FALCON, by civilized men, and the BERIGORA by the natives, is an inhabitant of Van Diemen's Land and New South Wales.

It is a rather sluggish and slow-moving bird, easily obtaining a sufficiency of food, and then settling down upon some neighbouring tree until the calls of hunger urge it to fresh exertions. The principal food of the Brown Hawk consists of insects, although it will also eat carrion, and kills mice, small birds, lizards, and other creatures. The land-holding colonists think it to be a great pest, because it sometimes picks up a young chicken or two ; but in the opinion of Mr. Gould it is in reality one of the farmer's best friends, on account of its services in destroying the insect hosts with which Australia is overrun. Although it is not a gregarious bird, living only in pairs, it may be seen assembled in flocks of a hundred or more, congregated over the localities where the destructive caterpillars most abound.. So plentiful is this bird, and so sluggish is its

character, that they may be seen seated in the tall eucalypti, thirty or forty occupying a single tree, and all so ill-disposed to move that any number of them may be killed without difficulty.

The structure of this bird approaches in many respects to that of the kestrel, and its character is not at all dissimilar. It does not seem to be a bold or dashing bird, and in pursuing its prey either sweeps suddenly upon it from the lofty regions where it is soaring. or snaps up a bird, reptile, or an insect, as it courses over the ground after the manner of the harrier.

The nest of the Cream-bellied Falcon is placed upon the loftiest branches of the eucalypti, or gum-trees as they are popularly called, and to a spectator on the ground bears a strong resemblance to that of the common crow, being built in a similar manner with sticks and twigs. The lining of the nest is made of leaves, stringy bark and similar substances, and the eggs are from two to three in number. Their colour is a yellowish-white, covered irregularly over their whole surface with reddish-brown spots and blotches, very irregularly sown, and very variable in their depth, number, and arrangement. The breeding season is from the end of September to the beginning of December.

In its colour the bird varies greatly, according to its age, the creamy white of the under por-

CREAM-BELLIED FALCON.—*Ieracidea Berigora.*

tions being of a deeper yellow, and the brown portions of the plumage more inclining to black in the young than in the adult bird. The general colour of the adult bird is rusty-brown above and creamy-white below, the darker feathers being marked with a fine black line down their centre. The tail is barred with reddish-chesnut, and there is a very conspicuous black stripe on the sides of the throat, which commences at the corner of the mouth and passes down each cheek. The feet and bill are lead colour, the tip of the beak passing into black. The eyes are very dark brown.

The common KESTREL is one of the most familiar of the British Hawks, being seen in almost every part of the country where a mouse, a lizard, or a beetle may be found.

It may be easily distinguished while on the wing from any other hawk, by the peculiar manner in which it remains poised in air in a single spot, its head invariably pointing towards the wind, its tail spread, and its wings widely extended, almost as if it were a toy kite raised in the air by artificial means, and preserved in the same spot by the trammels of a string. While hanging thus strangely suspended in the air, its head is bent downwards, and its keen eyes glance restlessly in every direction, watching every blade of grass beneath its ken, and shooting down with unerring certainty of aim upon any unhappy field-mouse that may be foolish enough to poke his red face out of his hole while the Kestrel is on the watch. The marvellous powers of the Kestrel's eye may be easily imagined by any one who has any experience of the field-mouse and the extreme difficulty of seeing the little creature while it is creeping among the grass straws. Its ruddy coat blends so well with the mould, and the grass blades bend so slightly under the pressure of its soft fur, that an unpractised eye would fail to detect the mouse even if its precise locality were pointed out.

The number of field-mice consumed by this hawk is very great, for it is hardly possible to open the stomach of a Kestrel without finding the remains of one or more of these destructive little animals. On account of its mouse-eating propensities, the Kestrel is a most useful bird to the farmer, who in his ignorance confounds all hawks together, and shoots the Kestrel because the kite steals his chickens.

Not that the Kestrel is wholly guiltless of chicken-stealing, or even of the greater crime of poaching on the preserves. Like all animals, it occasionally changes its diet, and pounces upon a chicken, a young pheasant, or a partridge. One of these birds has actually been shot with a young pheasant hanging in its claws, and the legs of young game have been found strewed beneath a nest where a Kestrel had reared her young. Young rabbits and hares have sometimes fallen victims to the hunger of the Kestrel, which is, however, unable to carry them away except during their earliest stages of independent existence. Reptiles of different kinds, such as frogs, small moles, and newts, are also a favourite prey of this bird, which has often been known to snatch fish out of the water by a dexterous sweep of its ready claw. When the Kestrel lives among the rocks upon the sea-coast, it haunts the shore at low water in search of food, and makes many a meal on little crabs, shrimps, small fish that have been left in the rock-pools of the receding tide, and many other marine creatures.

In the use of its claws the Kestrel is remarkably quick and ready, and being also a swift-winged bird, it is in the habit of chasing cock-chaffers and other large beetles on the wing, and catching them neatly with its claws as it shoots past their course. Without pausing in its flight, the bird transfers the insect from the foot to the mouth, and eats it without taking the trouble to alight. With such eagerness does it pursue this kind of prey, which we may suppose to be taken as a dessert after a more substantial meal upon mouse-flesh, that it continues its chase far into the evening, and may be seen in hot pursuit of the high-flying beetles long after dusk. Caterpillars and other larvæ are also eaten by the Kestrel, which does not disdain to alight on the ground, and draw the earthworms out of their holes.

Sometimes, but rarely, it attacks the smaller birds, choosing especially to pounce upon them as they are gathered together in little flocks, and are so confounded at the sudden appearance of their enemy, that they fall easy victims to the destroyer. On this account, the Kestrel is often seen in the winter months hovering near the farmyards, in hopes of making a successful dash among a plump of sparrows as they congregate over some fresh straw, or settle among the fowls for the purpose of picking up the grain which was intended for the poultry. Mice, however, are always its favourite diet, and as the multiplication of these little quadrupedal pests is much increased by the abundant food which they find in cultivated grounds, and stacks and barns, the Kestrel has learnt to attach himself to human residences, instead of becoming self-banished, as is the case with almost every other hawk. There is hardly a village where the Kestrel may not be seen hovering with outspread wings, and surveying the fields below.

In general, however, it troubles itself little about feathered prey, unless it can pick up a very young pheasant or partridge, as is indeed seen by the conduct of the sparrows and other small birds. If a sparrow-hawk, merlin, or hobby should appear in sight, the little birds are at once in an uproar, shrieking, chattering, darting from place to place, and expressing their alarm in a thousand ways. But when a Kestrel comes into view, they display hardly any uneasiness, and do not suffer themselves to be disturbed by its presence. Swallows, however, trusting to their speed of wing, are very fond of mobbing the Kestrel, and are so impertinent that even a single swallow has been seen suddenly to turn the tables on a Kestrel which was pursuing it, and to attack its astonished opponent with equal skill and audacity. On one occasion when a Kestrel had caught a sparrow, its cries took the attention of a number of swallows, which made a united attack, and forced the hawk to release its frightened but unhurt victim.

With the aid of a good telescope, every movement of the bird may be discovered as it hangs in the air, and the sight is a very interesting one. Its wings keep up a continual shivering, its widely spread tail is occasionally moved so as to suit the slight changes of the breeze, the spirited little head is in perpetual motion, and the dark-brown eyes gleam with animation as they keep their restless watch. It seems from various observations that each Kestrel has its regular beat or hunting-grounds, and may be observed punctually repairing to the same spot at the same hour, much after the manner of the golden eagle.

The Kestrel is known by various names in different parts of the country. Its most common name is Windhover, in allusion to its peculiar mode of flight. For the same reason it is termed Stannel, Stand-gall, or Stand-gale, and has also obtained the title of Vanner-hawk.

KESTREL.—*Tinnunculus Alaudarius.*

The nest of the Kestrel is generally placed upon the topmost bough of some lofty tree, although it is sometimes found upon a ledge of some precipitous cliff, should the bird have taken up its residence in a mountainous country. Many of these birds have built their nests upon the rocky heights of Dovedale in Derbyshire, and may be seen hovering in mid-air near the spot where their young are nourished. The nest itself is a very simple construction of sticks and moss; and the bird is so averse to trouble that it often takes possession of the deserted nest of the carrion crow. I have several times been greatly surprised in my nest-hunting expeditions, by finding the ruddy eggs of the Kestrel lying in the nest which I thought only to be that of the crow. This bird also deposits its eggs in the crannies of old ruined buildings and lofty towers, but I have never as yet been fortunate enough to find them in such a situation.

A few years ago a pair of Kestrels built for several successive seasons on the top of a tall elm-tree in a field near Oxford, and whether from the altitude of the nest terrifying the bird-nesting boys, or whether its real position was never discovered by them, the birds seemed to enjoy perfect security, and brought up their young without molestation. It was really a pretty sight to see the young essaying their wings round the tree, sometimes attempting longer flights, but always encouraged by the presence and instructions of their parents. The number of eggs is generally three or four, although a fifth has sometimes been detected. The colour is rather variable, but is in all cases sufficiently characteristic to point out the species to an accustomed eye. The ground tint is either pale reddish-brown, or even a ruddy-white, and the entire surface of the egg is blotched and spotted with dark red-brown. The young birds make their appearance at the end of spring or in the first weeks of summer.

The colour of the male Kestrel is very pleasing, and is briefly as follows. The head, cheeks, and back of the neck are ashen grey, marked with narrow longitudinal streaks of deeper grey. The back and upper portions of the body, together with the tertiaries and wing-coverts, are bright ruddy fawn, dotted with little triangular black spots, caused by the extreme tips of the feathers being black. The larger quill feathers of the wing are black-grey, marked with a paler hue; the under portions of the body are pale reddish-fawn, marked with dark streaks on the chest and spotted on the abdomen; the thighs and under tail-coverts are of the same hue as the abdomen, but without the spots. The upper surface of the tail is of the same hue as the head, marked with a single broad band of black near its extremity and tipped with white, while its under surface is grey-white, marked with a number of narrow irregular bars of a darker hue, in addition to the black band and white tip, which are the same as on the upper surface. The legs, toes, cere, and orbits of the eyes are yellow, the claws are black, and the beak is slaty-blue, deepening towards the point.

The females and young males are differently marked, and are altogether of a darker and more ruddy hue. The head and neck are ruddy-fawn, marked with many transverse darker stripes, and the back, upper portions, and tail are red-brown, covered with numerous irregular blue-black bars. The males do not assume their appropriate plumage until they have completed their first year. The length of the male bird is about thirteen inches, and that of the female fifteen inches.

ANOTHER British species, which belongs to the same genus as the kestrel, is the RED-FOOTED FALCON or INGRIAN FALCON (*Tinnúnculus vespertínus*). It is but a rare bird in England, being only a straggler to our coasts, and having its usual residence in Austria, Russia, and Poland. Specimens have also been taken in Athens, Nepâl, and Tunis, so that the species seem to have a very extensive range of country. It goes through considerable changes of tinting before its plumage attains the adult colours, but the full-grown bird may readily be distinguished from the common kestrel by the legs and toes, which are of a reddish flesh tint instead of the yellow hue which is found in the former bird. The claws, too, instead of being black, are yellowish-white, deepening into a greyish-brown on the tips.

Among other members of the same genus, we may notice the LITTLE FALCON (*Tinnúnculus sparvérius*) of America, an interesting account of which bird may be found in the pages of "Wilson's American Ornithology." Its habits are very similar to those of the common kestrel, and like that bird it preys chiefly on mice, lizards, grasshoppers, and the larger insects. It will, however, attack and carry off chickens and the young of other birds during the breeding season. Its nest is always made on some elevated situation, and is generally found on the top of a lofty tree, although the bird sometimes builds upon rocks, in the crevices of towers, or even in the hollows of trees.

THE NOTCHED FALCON is remarkable for the peculiar form of the beak, which exhibits a double notch or tooth on each side, and has therefore been distinguished by the specific title of bidentatus, or "two-toothed."

This species is a native of Southern America, being found most commonly in Brazil and Guinea. In size it is about equal to the common kestrel, its length being thirteen or fourteen inches. The general colour of the Notched Falcon is a slaty-blue or blue-grey upon the upper surface of the body, and the tail is dusky-brown, marked with several transverse bars of greyish-white. The throat and under tail-coverts are white, and the breast and abdomen are rusty-red, marked with undulating streaks of yellowish-white. Very little is known of the habits of this species, but on account of the peculiar form of its beak, it cannot be passed over without notice.

THE members of an allied genus, termed Ierax, also possess a similarly formed beak, but the structure of the wings and arrangement of the feathers are so different as to give reason for placing the bird in a separate genus. One of the most beautiful examples of this genus is the little BENGAL FALCON (*Ierax cœrulescens*), a native of Java, Borneo, and many parts of India. This tiny Falcon is barely six inches in length, and is popularly known in India by the name of "Mooty," a word which signifies a handful, and is given to the bird because when it is flown at game, it is taken in the hand and flung at the quarry as if it were a stone rather than a living missile. It is a most daring little bird, and has been known to strike in succession ten or twelve quails before alighting. The general colour of this species is bluish-black above, and rusty-white below. The plumage of the thighs is long and silken, and the wings are comparatively short.

NOTCHED FALCON.—*Hárpagus bidentátus.*

We now come to a large and important genus of hawks, which is represented in England by the GOSHAWK.

This handsome bird is even larger than the jerfalcon, the length of an adult male being eighteen inches, and that of his mate rather more than two feet. It is not, however, so powerful or so swift-winged a bird as the jerfalcon, and its mode of taking prey is entirely different. The jerfalcon dashes at every flying creature that may take its fancy, and attacks successfully the largest winged game. But the Goshawk, although possessed of the most undaunted courage and of great muscular power, is unable to cope with such opponents, and prefers terrestrial to aerial quarry. Owing to the shape of the wing, and comparative shortness of the feathers, the Goshawk is unable to take long flights, or to urge a lengthened and persevering chase. Moreover, although its courage is of the most determined character, it soon loses heart if often baffled by the same quarry, and in such cases will turn sulky and yield the chase.

When trained, the Goshawk is best employed at hares, rabbits, and other furred game, and in this particular sport is unrivalled. Its mode of hunting is singularly like that of the chetah, which has already been mentioned in the volume on the Mammalia. Like that animal, it is not nearly so swift as its prey, and therefore is obliged to steal upon them, and seize its victim by a sudden and unexpected pounce. When it has once grasped its prey it is rarely found to loose its hold, even by the most violent struggles or the most furious attack. The gripe is so enormously powerful, that a Goshawk has often been observed to pounce upon a large hare, and to maintain its hold even though the animal sprang high into the air, and then rolled upon the ground in the vain hope of shaking off his feathered antagonist. Only the female bird is able to cope with so powerful a creature as a full-grown hare or rabbit, for the male, although more swift of wing, and therefore better adapted for chasing birds than the female, is comparatively feeble.

GOSHAWK.—*Astur palumbarius.*

It never attempts to follow its quarry into cover, as is done by the Peregrine and Merlin, but if its intended prey should seek safety in some place of refuge, the Goshawk perches upon a convenient bough and waits patiently. As the hawk is very endurant of hunger, although sufficiently ravenous when it meets with a supply of food, it "wins, like Fabius, by delay," and pounces upon the unlucky quarry, as it steals out in search of food or water. When it has once seized its prey, it is full of exultation, and being generally rather of a ferocious disposition, is apt to turn savagely upon the hand that attempts to remove it from its victim. Its temper, indeed, is so bad, that if it should happen to escape from its jesses and get among other Falcons, it will almost certainly attack and kill as many of them as it can reach. For the same reason it needs to be kept constantly hooded, and is less to be trusted at liberty than any other Falcon. Its short flights, however, render its recapture a comparatively easy matter, so that there is but little danger of losing it.

Its constitution is very hardy, and as it will feed on almost any animal nourishment, it gives very little trouble to its owner.

This species is found spread over nearly the whole of Europe and Asia, and has also been seen in Northern Africa. The nest of this bird is generally placed on the topmost boughs of some lofty tree, and the eggs are of a uniform spotless blue-white. Their number is from three to four, and the young are hatched about May or the beginning of June.

In colour, the adult birds of both sexes are very similar to each other, the tinting of the plumage being briefly as follows. The top of the head and the entire upper portions of the body and wings are grey-brown, and the under portions of the body, together with a band over the cheeks and the back of the neck, are nearly white, diversified with numerous irregular spots, splashes, and partial bars of black. The cheeks and ear-coverts are dark greyish-brown, the upper surface of the tail is the same hue as the back, and barred with dark brown; the under tail-coverts are white. The cere, legs, and toes are yellow, the claws black, and the beak blue-black. In the female the grey-brown of the back is a more ruddy hue, and in the young the plumage is curiously diversified with reddish-white, buff, and grey.

A VERY beautiful species of this genus, the NEW HOLLAND WHITE EAGLE (*Astur Novæ Hollandiæ*) is found in Australia, and is remarkable for the frequency with which its plumage assumes a snowy-white hue, the ordinary colouring being grey above and white below. The eyes of this bird are very curious, for in some specimens they are of a rich brown, in others of a topaz-yellow, while in others they are ruby-red. The cere, legs, and claws are yellow, and the bill black. The disproportion between the comparative dimensions of the sexes is remarkably great in this species, the male being barely half the size of his mate.

The well-known SPARROW HAWK is almost as familiar to us as the kestrel, the two birds being, indeed, often confounded with each other by those who ought to know better. This fine and active little bird is an inhabitant of many portions of the world, being very common in nearly all parts of Europe, equally so in Egypt and Northern Africa, and being very frequently found in India and other Asiatic countries. The genus Accípiter finds representatives in every quarter of the globe, species being found in North and South America, in Madagascar, in Western and Southern Africa, in Java, and Australia.

Although the Sparrow Hawk inhabits England in great numbers, it is not so often seen as might be imagined, for it is a most wild, shy, and wary bird, and never ventures near human dwellings, or within a considerable distance of human beings, unless urged by hunger or carried away by the ardour of pursuit. As a general rule, to get within ordinary gunshot of a Sparrow Hawk is no easy matter; but if the Hawk be watched as he is hovering about a flock of sparrows or other small birds, he may be approached without much difficulty, his entire attention being engaged on his expected prey. Indeed, while engaged in the chase, the ardour of this bird is so great, that all its faculties seem to be absorbed in the gratification of the ruling passion, and it is evidently unmindful of anything but its flying prey. A Sparrow Hawk has even been known to dash furiously at a man who endeavoured to rescue a small bird which it had attacked.

The courage of the Sparrow Hawk is of the most reckless character, for the bird will fly unhesitatingly at almost any other inhabitant of air, no matter what its size may be. Mr. Thompson relates the following curious instance of the exceeding audacity of this bird.

"Once, at the end of July, when walking along the sides of the river Lagan, near Belfast, I was attracted by the loud screams of herons, which appeared above the trees at the north-west extremity of Belvoir Park. A couple of these giants of the air kept flying above the tops of the trees with tremendous uproar, in consequence of the presence of a single Sparrow Hawk. This bird was circling about, and the herons awkwardly and quite unavailingly endeavouring to strike him. Flying quite at ease, his turns were so short, and at the same time so full of grace, that he seemed to laugh to scorn their heavy, lumbering movements.

The herons' savage cries were apparently (evidently might almost be said) caused by the Hawk's make-believe attempts to carry off their young, as they were particularly violent and vociferous whenever he made a swoop—as I remarked him to do thrice—at the top of a particular tree. It seemed a mere play or bravado on the part of the Hawk, as he could easily, in spite of the herons, have borne off the contents of the nest any time, were the prey not too bulky for his purpose. Mr. R. Langton has not only observed a wild Sparrow Hawk strike his sea-eagles when perching on their sheds, but when his golden eagle was on the wing, has seen one of these birds strike it in passing, and once even witnessed the Hawk's turning back and repeating the impertinence."

The same author also mentions several instances of the extreme audacity of the Sparrow Hawk when urged by hunger. One of these birds actually snatched up a little white pea-chick, selecting it from the rest of the brood, while a lady was engaged in feeding it. A similar circumstance occurred to a gamekeeper who was feeding young pheasants, a Sparrow Hawk suddenly sweeping down upon them and carrying off one of their number. Next day it repeated the attempt, but as the keeper had taken the precaution to bring his gun, the Hawk fell a victim to his own temerity. Again, as some persons were shooting dunlins from a boat, in Belfast Bay, a Sparrow Hawk suddenly shot through the smoke of the discharged gun, and poising itself for an instant, swept a wounded dunlin from the surface of the water with such marvellous dexterity, that it did not wet a feather of its wings.

In consequence of the headlong courage possessed by this handsome little Hawk, it is very valuable to the falconer if properly trained, for it will dash at any quarry which may be pointed out to it. Unfortunately, however, the Sparrow Hawk is one of the most difficult and refractory of pupils, being shy to a singular degree, slow at receiving a lesson and quick at forgetting it. Besides, its temper is of a very crabbed and uncertain nature, and it is so quarrelsome, that if several of these birds should be fastened to the same perch, or placed in the same cage, they will certainly fight each other, and, in all probability, the conqueror will eat his vanquished foe. Such an event has actually occurred, the victrix—for it was a female—killing and devouring her intended spouse.

Few birds are so easily startled as the Sparrow Hawk, for even when it is comparatively tame, the presence of a stranger, or even the shadow of a passing bird in the air, will throw it into a paroxysm of excitement, during which it seems to lose all consciousness of external objects. This curious trait of character a practical falconer, in a communication addressed to the *Field* newspaper, describes most graphically in the following terms. " The young falconer will naturally be disappointed to find the bird which came so well to hand yesterday, now on the first day of its being carried, stare wildly with its mad eyes, and bate violently. It will probably hang down at the end of the jesses and swivel, and dart off again the moment it is quietly replaced. More than this, the very power of standing will appear to have left it ; the claws will be clenched and distorted ; the whole creature will be changed ; instead of a tolerably bold and very handsome bird, the transition of a few minutes will present you with a terrified, crouching, vicious, abject wretch ; a horrible mixture of fright and feathers.

Some people think that the helpless look of the feet and legs arises only from temper, and that it is a sham. It may arise from temper, but it is not a sham. It appears to me that this bird's brain is overcharged with electricity or something fearfully subtle ; and that on the smallest provocation, these fluids shoot through the whole frame, over-turning and decaying everything that is healthy and regular. The Sparrow Hawk's legs are, during these fits of fright and passion, in a temporary paralysis. Still, they are of short duration, and when the bird is trained, they pass away altogether." The same writer sums up the character of the Sparrow Hawk as a pupil in the following energetic language: " The Sparrow Hawk is, in my opinion, the wildest, in some sense the most intractable, the most ungrateful, the most provoking and temper-trying of all birds or beasts that ever were taken under the care of man from the beginning of the world."

With this writer's opinion my own experience to a very great measure coincides, though as I never attempted to train a Sparrow Hawk to falconry, I cannot answer for some of its deficiencies.

One of these birds afforded an excellent example of the shyness and timidity above mentioned. Although he was most kindly treated and liberally fed, he used to scream in the most ear-piercing manner when approached, even by the person who generally carried his food. The only companion whose presence he would tolerate, was a little Skye terrier, named Rosy, and the two strangely matched comrades used to execute the most singular gambols together, the dog generally taking the initiative, and persecuting the Hawk until she forced him to fly. The great object of the dog was to catch the Hawk by the wing, while the bird gave his attention to flying at the dog's throat, hanging on by his claws and boxing her ears with his wings until she was fain to shake him off. Once, Rosy caught the Hawk by his tail, and having the game all her own way, careered round the yard in great exultation, dragging after her the unfortunate Hawk, who could not possibly resist or retaliate, and was reduced to scream abjectly for succour.

Another Sparrow Hawk which I procured for some time was, curiously enough, a most arrant coward, and so far from chasing the little birds, as was his duty, and keeping them from eating the pease and fruit, he allowed them to bully him shamefully, and would run away from a wagtail. The little birds soon learned his incapacity, and the blue titmice used to watch the time when he was fed, and run off with the meat before his eyes. The bird was not a young one when it came into my possession, and had probably been broken in spirit by cruel treatment.

The credit of the race was, however, better kept up by a Sparrow Hawk that belonged to a lady friend, but it was not taught any artificial accomplishments. The bird took a great fancy to its mistress, and would perch on her shoulder or eat from her hand. But it would permit no other person to touch it, neither

SPARROW HAWK.—*Accipiter Nisus.*

would it allow any one to approach its mistress while it was at hand. In such cases it would fly savagely at the fancied foe, and was so determined in its attack upon the ankles, that any one who attempted to cross its path was obliged to fend it off with an umbrella, which it would fight and scold as it was being pushed away.

It was a terrible thief, and crafty to a degree. Once, having made itself acquainted with the fact that a partridge was hanging in the larder, it hung about the spot until it saw a servant approaching the spot. As soon as she opened the door, the Hawk shot noiselessly over her head, and sat quietly until she had retired. It then proceeded to demolish the partridge. Between this servant and the Hawk there was a deadly feud, owing to a depredation committed by the bird and resented by the servant. A chicken had just been plucked and was lying on the kitchen table, when the Hawk glided softly through the door, and perching on the chicken, had devoured its breast before the theft was discovered. The servant struck it with a broom, when the bird flew at her head, and pushing its claws into her hair, it buffeted her face with its wings, and could not be removed until it had torn out no small quantity of hair. After a while the bird disappeared, perhaps stolen, but very probably killed by its foe.

The propensity of the Sparrow Hawk to attack larger birds of prey has already been mentioned, and the creature only suffers poetical justice in being made the subject of similar attacks. The swallows and other swift-winged birds are wonderfully fond of mobbing the Sparrow Hawk, although in many instances they pay dearly for their audacity. I once saw a Sparrow Hawk that was being mobbed by a number of sand martins, and was flying about in a seemingly purposeless and bewildered manner, suddenly turn on its pursuers, seize one of them in its terrible grasp, and instantly sail away bearing its screaming victim in its talons. The Hawk was almost within reach when this circumstance occurred.

The general colour of the adult male is dark brown upon the upper surface of the head, body, and wings, softening into grey as the bird increases in years. The entire under surface is rusty brown, marked with narrow bands of a darker hue; the long and slender legs and toes are yellow, as is the cere, but with a tinge of green; the long, sharp, armed claws are black, and the beak is a slate-blue, darkening towards the point. The length of the male bird is about one foot. The female, which is about fifteen inches in length, is coloured differently from her mate, the upper parts of the body and wings being hardly so rich a tint as in the male bird, and covered with numerous little white spots, caused by the white hue which is found on the base of each feather. The primaries and rectrices are of a lighter brown, and coloured with transverse dark bars, and the under surface of the body is grey-white, also barred transversely. These hues are also found in the young male, who has in addition a reddish edging to the feathers of the back.

The nest of the Sparrow Hawk is placed in some elevated spot, and contains three or four eggs, rather variable in their marking, but always possessing a certain unmistakeable character. The ground tint of the egg is a greyish white, slightly tinged with blue, and a number of bold blotches of a very dark brown are placed upon the surface, sometimes scattered rather irregularly, but generally forming a broad ring round the larger end. The bird seldom troubles itself to build a new nest, but takes possession of the deserted tenement of a crow or rook.

As a general rule, the voices of all the rapacious birds are notable for the rough, strident dissonance of the larger species, or for the piercing shrieks of the smaller birds. There is, however, an exception to this rule, which is supposed at present to be quite unique, in the person of the CHANTING FALCON of Africa.

In a certain sense, even the scream of the eagle and the shriek of the Falcon possess a sort of wild music, which is sufficiently appropriate to the localities in which they dwell, but is singularly out of place when the bird is seated on a perch or immured within the confines of wiry walls. The Chanting Falcon, however, possesses a really musical voice, its very peculiar notes having been compared to the thrilling sounds of musical glasses. Le Vaillant tells us that it sings in the morning and evening, and that its song lasts for about one minute, being very frequently repeated in the course of an hour, and with very short intervals. It is at all other times of the day a very shy and suspicious bird, but while singing is so occupied with its task, that it can be approached, and, if desired, shot without much difficulty.

It is rather a large and powerful bird, being nearly two feet in length, and somewhat resembling the jerfalcon in proportion. Its prey consists of hares, rabbits, and similar quadrupeds, and it also wages successful war against the larger birds, such as the bustard or "pauw." The general colour of this curious bird is greyish on the upper parts of the body, and white on the lower parts, barred with brown streaks.

THE very remarkable SECRETARY BIRD derives its name from the curious feathery plumes which project from each side of its head, and bear a fanciful resemblance to pens carried behind the ear by human secretaries. In allusion to the same peculiarity, the Arabs term the bird Selazza Izn, or Thirty-ears.

The Secretary Bird has long been a standing perplexity to systematic zoologists, having been placed by some writers among the wading birds on account of its long legs, while others consider its proper place to be among the hawks and other birds of prey. It is an inhabitant of Southern Africa, and is most invaluable in destroying the serpent race, on which creatures it almost exclusively feeds. Undaunted by the deadly teeth of the cobra, the Secretary Bird comes boldly to the attack, and in spite of all the efforts of the infuriated and desperate reptile, is sure to come off victorious. Many other creatures fall victims to the ravenous appetite of the Secretary, and in the stomach of one of these birds which was found by Le Vaillant, were discovered eleven rather large lizards, eleven small tortoises, a great number of insects nearly entire, and three snakes as thick as a man's arm. The following description of the habits and mode of hunting which is employed by this bird has been kindly forwarded to me by Captain Drayson, to whom I have already been indebted for much curious information respecting the quadrupeds of Southern Africa.

"The Secretary Bird is not very common in any part of South Africa, still one or two are frequently seen during a day's ride on the plains. Sometimes two or three of these birds may be seen stalking over the ground, with a bold, military, and jaunty stride, which is quite in character with the nature of the reptile-eating bird, but more frequently a solitary individual pursues his investigations of newly-burnt grass, or likely and deserted ant-heaps.

CHANTING FALCON.—*Melierax músicus.*

Frogs and toads appear the more favoured repast of the Secretary Bird, but a snake of even three or four feet in length is easily disposed of.

On one or two occasions I have seen a Secretary Bird busily engaged with a snake, and it appeared that the bird by means of activity escaped from the deadly fangs of its prey. A Secretary Bird might be seen sailing slowly along at about a hundred yards from the ground; suddenly he would stop and descend, attracted evidently by some prey, towards which he would stalk. The bird would then appear very busy, now striking with his wings and pecking, as though engaged in thrust and parry; then, when his adversary

SECRETARY BIRD.—*Serpentárius Secretárius*

made a fierce attack, the bird would rise with a spring in the air, and descend some twenty feet from his foe. Advancing again to the attack, he seldom failed to dispose of his enemy by eating him at once, or he would carry him off wriggling vainly in attempts to escape.

The Secretary has a curious habit of occasionally breaking from his staid military step, and running in a fussy excited way for about ten or twelve paces, with apparently no object, and again resuming his march. This bird is very wary, and rarely allows of a nearer approach than one hundred yards. He is rarely if ever shot, for a sportsman values the bird for its deeds, and there is a fine in the Cape Colony imposed upon those who are known to have shot one."

In these combats the wing of the bird is its most important weapon, and answers equally all the purposes of a shield and a club. As the serpent rises to strike, the Secretary presents the front of its wing as a buckler, and almost immediately dashes the snake to the ground by a blow from the same member. It also kicks with considerable force, and almost invariably concludes the combat by a violent blow on the head from its beak, which lays the skull of the enemy completely open. Sometimes, when the serpent is very full of vitality, the Secretary makes a sudden snatch at its neck, soars to a great height in the air, carrying the struggling foe with it, and then drops it upon the hard ground, a process which effectually expels the last remnant of life. When domesticated it has been known to go through a similar series of manœuvres, by way of gamesomeness; the snake being represented for the nonce by a straw or a twig.

One of the most notable peculiarities of this strange bird, is the manner in which it runs and walks. While young its mode of walking is ungraceful in the extreme, and can but be characterised as a hobble. When it has attained adult age, however, its gait, although rather odd, and like a person walking on stilts, is yet easy and unconstrained, but when the bird puts forth its speed, it runs with a swiftness so remarkable that the Arab has given it the name of Ferras Seytan, or devil's horse. This astonishing speed is probably useful in carrying out the great business of its life, and in attacking or avoiding the onset of its poison-bearing enemies. The nest of the Secretary is built on the summit of a lofty tree, and contains two or three large white eggs.

The ordinary length of the adult Secretary bird is about three feet, and its colour is almost wholly a slaty-grey. The peculiar feathers which form the crest are black, as are the primaries and the feathers of the thigh. There is a lighter patch towards the abdomen. The tail is black with the exception of the two central rectrices, which are grey with a white tip and a broad black bar towards their extremities.

WE now arrive at the Harriers, probably so called because they "harry" and persecute the game. Several species of this genus are found in England, the most common of which is the HEN HARRIER.

The Harrier may be readily distinguished from the other hawks by the manner in which the feathers radiate around the eyes, forming a kind of funnel-shaped depression, somewhat similar to but not so perfect as that of the owl. This structure is thought to be serviceable to the bird in giving it a wide range of vision in its hunting excursions. The flight of the Harrier is very low, seldom being more than a few yards above the ground, and as the bird flies along it beats

HEN HARRIER.—*Circus cyáneus.*

every bush, and pries into every little covert in search of prey. There are few of the smaller animals that do not fall victims to the Hen Harrier, which is always ready to pick up a field-mouse, a lizard, a small snake, a newt, or a bird, and will even pounce upon so large a bird as a partridge or pheasant. Sometimes it sits on a stone or small hillock, and from that post keeps up a vigilant watch on the surrounding country, sweeping off as soon as it observes indications of any creature on which it may feed.

The flight of the Hen Harrier, although it is not remarkable for its power, is yet very swift, easy, and gliding, and as the bird quarters the ground after its prey, is remarkably graceful. The Harriers prefer to live on moors and similar localities, where they can pursue their rather peculiar mode of hunting, and where they may find a secluded spot for a secure home. Like the Kestrel, the Hen Harrier appears to have regular hunting-grounds, and is very punctual in its visits. The nest of this bird is generally placed under the shadow of some convenient furze-bush, and is composed of a few sticks thrown loosely together, in which are deposited four or five very pale blue eggs. The young are hatched about the middle of June.

MOOR BUZZARD.—*Circus ceruginosus.*

The two sexes differ very greatly in colour, and until comparatively recent times were recorded as distinct species. The general colour of the adult male is ashen grey from the beak and upper parts, the only exception being the primaries, which are black. The throat and chin are nearly of the same hue as the beak, but the chest and abdomen are white, with a slight blue tinge, which is lost upon the plumage of the thigh. On the under surface of the tail are several indistinct dark bars, and the hair-like feathers between the eye and the base of the beak are black. The legs, toes, and cere are yellow, the claws black, and the beak nearly black, with a bluish tinge. The length of the male bird is about eighteen inches.

The female is a much darker bird, the head being mottled brown, and the back and upper portions of a deep dusky brown, the primaries being but a little darker than the plumage of the back. The feathers of the under parts are lighter brown, with pale margins, so as to present a kind of mottled buff and chestnut aspect; the upper surface of the tail is marked with partial dark bands, and its under surface is very distinctly bound with broad bands of black and greyish-white. The funnel-shaped depression round the eyes, technically called the concha, or shell, is brown towards the base of the feathers, but merges into a white eyebrow above, reaching to the cere, and in a white streak below, edged with brown. The length of the female is about two inches more than that of the male, and her spread of wing is about three feet six inches.

ANOTHER British example of this genus is to be found in the MOOR BUZZARD as the bird has very wrongly been termed, or the MARSH HARRIER, as it ought more properly to be named. The bird is also known as the Duck Hawk and Harpy.

This handsome bird is considerably larger than the preceding species, the female being about two feet in length, and the male about three inches shorter. It is not a very uncommon bird, being found most extensively upon marshy ground, where it can obtain abundance of food. It generally preys on water birds, mice, water rats, various reptiles, frogs, rats, and fish. It is rather partial to young game, and is apt to be a dangerous neighbour to a preserve, snatching the young partridges and pheasants from their parents. Sometimes it is sufficiently bold to enter the precincts of the farm, and to carry away a young chicken or a duckling. Rabbits also, both young and old, fall victims to this rapacious bird, which sweeps on noiseless wing over the common, carefully choosing the morning and evening, when the rabbits are almost sure to be out of their burrows.

The Marsh Harrier appears never to take up its residence in dry localities, but always to prefer the fenny district, whether of the coast or inland. The bird may be found plentifully in Cambridgeshire, as well as in Scotland, Ireland, and parts of Wales. The nest of this species is placed on the ground, and is composed of twigs or stems of coarse grass, and is sheltered from observation by an overhanging bush, or by a tuft of rushes, fern, or long grass. The eggs are white, and about three or four in number.

Like the hen harrier, the male Marsh Harrier is of a much greyer tint than his mate, the grey hues being not fully assumed until the bird has completed his third year, and spreading more widely on each successive year.

THE ASH-COLOURED FALCON, sometimes called MONTAGUE'S HARRIER, is also a British resident, although it is frequently found in Nepâl and other parts of Asia.

This species is more slender in its form than either of the two preceding birds, being hardly more than two-thirds of the weight of the hen harrier, although its length is nearly the same. In its habits

JARDINES HARRIER.—*Circus Jardinii.*—(See p. 94.)

it is very like the hen harrier, skimming over the ground in much the same manner, but with a more rapid flight. Its food consists chiefly of small birds and reptiles, to the latter of which creatures the Ash-coloured Falcon appears to be especially partial, no less than five lizards having been found in the stomach of one of these birds. It is not very uncommon in England, being found most plentifully in Cambridge, Lincoln, and other fenny districts, and being also met with in Devonshire and Cornwall. Specimens of this bird have also been obtained in Nepâl and other parts of India, and it is also said to be an inhabitant of Southern Africa.

The colour of the adult male is bluish-grey on the upper parts of the body, the secondaries being marked with three bars of dark-grey brown, and the primaries black. The upper surface of the tail is bluish-grey upon the central feathers, and white upon those at the side, marked with several bars of orange-red; their under surface is greyish-white, with several transverse bars of greyish-brown. The under parts of the body are nearly white, barred with numerous transverse streaks of orange-red, like those on the tail. The legs and toes are yellow, the cere is yellowish-green, and the beak is almost black. The length of the bird is about seventeen inches.

The very remarkable bird which is now known as the JARDINES HARRIER is one of the myriad strange creatures which are produced by Australia, that land of wonders.

According to Gould, it is generally found in plains, and specially frequents the wide and luxuriant grass flats that intervene between the mountain ranges. Like all the birds of the same genus, it is never seen to soar, but sweeps over the surface of the ground at a low elevation, seeking after the mice, reptiles, small birds, and other creatures on which it feeds. It is very fond of small snakes and frogs, and in order to obtain them may be seen hovering over the marshes, or beating the wet ground after the fashion of the hen harrier. It is seldom known to perch on trees, preferring to take its stand on some large stone or elevated hillock from which it may survey the surrounding land. The nest of this bird is supposed to be built on the ground, overshadowed by some bush or tuft of grass, like that of other harriers, and placed upon the top of one of the numerous "scrub" hills.

The colouring of this bird is quite unique, and would attract attention even if it were not an anomaly among birds of this genus. The head, cheeks, and ear-coverts are dark streaked chestnut, the streaky appearance being given by a deep black line down the centre of each feather. A grey collar or band passes round the neck and the back of the head, the primaries are buff towards their base, and black for the latter two-thirds of their length. The tail is barred alternately with dark brown and grey, the extremity being brown. The back and scapularies are dark-grey sprinkled with a number of little white dots, and the entire under surface is a bright ruddy chestnut, covered profusely with nearly circular white spots of considerable size. The legs are yellow, and the bill dark slaty-blue, becoming black at the extremity.

OWLS.

THERE are few groups of birds which are so decidedly marked as the OWLS, and so easy of recognition. The round, puffy head, the little hooked beak just appearing from the downy plumage with which it is surrounded, the large, soft, blinking eyes, and the curious disk of feathers which radiate from the eye and form a funnel-shaped depression, are such characteristic distinctions, that an Owl, even of the least Owl-like aspect, can at once be detected and referred to its proper place in the animal kingdom. There is a singular resemblance between the face of an Owl and that of a cat, which is the more notable as both these creatures have much the same kind of habits, live on the same prey, and are evidently representatives of the same idea in their different classes. The Owl, in fact, is a winged cat, just as the cat is a furred Owl.

These birds are, almost without an exception, nocturnal in their habits, and are fitted for their peculiar life by a most wonderfully adapted form and structure. The eyes are made so as to take in every ray of light, and are so sensitive to its influence, that they are unable to endure the glare of daylight, being formed expressly for the dim light of evening or earliest dawn. An ordinary Owl of almost any species, when brought into the full light of day, becomes quite bewildered with the unwonted glare, and sits blinking uncomfortably, in a pitiable manner, seemingly as distressed as a human being on whose undefended eyes the meridian sun is shining.   The nictitating membrane, or inner eyelid, with which the Owl, in common with many other birds and animals, is furnished, stands it in good stead under such circumstances, and by repeatedly drawing its thin membranous substance over the aching eyeball, the Owl obtains some relief from the pain which it is suffering.

The eyes of Owls are very curiously formed, as are their ears and plumage, and their structure will be briefly described in the course of the next few pages.

THE transition from the falcons to the Owls is evidently through the harriers, as may be seen by comparing the engraving of any harrier with that of the CANADA OWL, or HAWK OWL, as it is often termed.   In the harriers we find the commencement of the peculiar facial disk, and in the Hawk Owl this disk, or "concha," is not nearly so large or so well defined as in the other members of the same group. The eyes, too, are rather differently formed, as the bird is able to follow its prey by day as well as in the dark, and therefore requires a character of eye which will not be injured or half blinded by ordinary daylight. The plumage is closer than that of the generality of owls, whose feathers are fringed with delicate downy filaments, for the purpose of enabling them to float noiselessly through the air, for the Hawk Owl is a swift-winged bird, and obtains its prey by fair chase.

CANADA OWL.—*Surnia úlula.*

The food of the Canada Owl consists chiefly of rats, mice, and insects, during the summer months ; but in the winter, while rats and mice keep within their homes, and the insects are as yet in their pupa state, the Canada Owl turns its attention to birds, and will even chase and kill so powerful a prey as the ptarmigan.   It is a very bold bird, and has been known to pounce upon and carry away wounded game that has fallen before the sportsman's gun. While chasing the ptarmigan it follows the course of their migration, hanging about the flocks and making sad havoc in their numbers.

The Hawk Owl is an inhabitant of the more polar regions, being most commonly seen in the extreme north of Asia and America, though it sometimes pays a visit to Northern Europe. Richardson tells us that it seldom travels farther north than Pennsylvania, but very few specimens having been noticed in that locality, and those only when the winter has been more than usually severe.

Although so bold and so successful a hunter, the Hawk Owl is by no means a large bird, being only from fifteen to seventeen inches in length, and therefore not equalling the common hen harrier in dimensions.   Its nest is generally made on the summit of a tree, contrary to the usual habit of Owls, which usually take possession of a hollow in some dead branch and lay their eggs on the soft decaying wood, or make their home in a convenient crevice of some old building.   The male Hawk Owl is rather less than the female, as is the case with most predaceous birds.

SNOWY OWL.—*Nyctea ivea.*

The general colour of this bird is dark spotted brown above and striped white below, arranged briefly as follows. The top of the head and back is brown, covered with white spots, the spots disappearing at the insertion of the wings, where a large patch of very dark brown is placed. The outer edge of the concha is jetty black, and its inner surface is greyish-white. The throat is also white, and the chest and abdomen are of the same tint, marked with a number of irregular stripes of ashen-brown. The tail is brown, covered with a few narrow intercepted black bands. The legs are feathered as far as the claws, and the bill is yellow with a few spots of black. These colours are slightly variable in individuals, owing most probably to the difference of age, and in the female they are not so bright as in her mate.

THE SNOWY OWL is one of the handsomest of this group, not so much on account of its dimensions, which are not very considerable, but by reason of the beautiful white mantle with which it is clothed, and the large orange eyeballs that shine with a lustre as of a living topaz set among the snowy plumage.

This bird is properly a native of the north of Europe and America, but has also a few domains in the more northern parts of England, being constantly seen, though rather a scarce bird, in the Shetland and Orkney Islands, where it builds and rears its young. Like the Hawk Owl, it is a day-flying bird, and is a terrible foe to the smaller mammalia, and to various birds. Mr. Yarrell, in his well-known History of the British Birds, remarks that "one wounded on the Isle of Balta disgorged a young rabbit whole ; and that one in my possession had in its stomach a young sandpiper with its plumage entire." It is rather remarkable that the bird should have thus been swallowed whole, as I have always remarked that when an Owl devours a little bird, he tears it to pieces before eating it, though he always swallows a mouse entire.

In proportion to its size the Snowy Owl is a mighty hunter, having been detected in chasing the American hare, and carrying off wounded grouse before the sportsman can secure his prey. According to Yarrell, the Swedish name of Harfang, which has been given to this bird, is derived from its habit of feeding on hares. It is also a good

COQUIMBO, OR BURROWING OWL.—*Athéne cunicularia.*

fisherman, posting itself on some convenient spot overhanging the water, and securing its finny prey with a lightning-like grasp of the claw as it passes beneath the white-clad fisher. Sometimes it will sail over the surface of the stream, and snatch the fish as they rise for food, but its general mode of angling is that which has just been mentioned. It is also a great eater of lemmings; and in the destruction of these quadrupedal pests, does infinite service to the agriculturist and the population in general.

The large round eyes of this bird are very beautiful, and even by daylight are remarkable for their gem-like sheen, but in the evening they are still more attractive, and glow like two balls of living fire. There is an amusing anecdote respecting one of these Owls, which settled on the rigging of a ship by night to rest itself after a long journey. The bird was quietly seated on one of the yards, when it was suddenly roused by a sailor who was sent aloft upon some nautical duty. The man, terrified at the two glowing eyes that suddenly opened upon him, descended precipitately from the rigging, declaring that "Davy Jones" was sitting on the main yard. Several instances are known where Snowy Owls have made use of a ship as a temporary resting-place. On one such occasion, the ship was visited by no less than sixty of these birds, which were so fatigued that they permitted themselves to be captured by the crew.

The colour of an old Snowy Owl is pure white without any markings whatever; but in the earlier years of its life, its plumage is covered with numerous dark-brown spots and bars, caused by a dark tip to each feather. Upon the breast and abdomen, these markings form short abrupt curves, but on the back and upper surface they are nearly straight. The beak and claws are black. The length of the male Snowy Owl is about twenty-two inches, and that of the female twenty-six or twenty-seven.

THE quaint, long-legged little Owl which is represented in the accompanying illustration is a native of many parts of America, where it inhabits the same locality with the prairie dog. The description of that curious marmot and its peculiar burrow may be found in Volume I.

The prairie dogs and Burrowing Owls live together very harmoniously; and this strange society is said also to be augmented by a third member, namely, the rattlesnake. It is now, however, ascertained with tolerable accuracy that the rattlesnake is nothing but a

2. <center>H</center>

very unwelcome intruder upon the marmot, and, as has been shown by the Hon. G. F. Berkeley's experiments, is liable to be attacked and destroyed by the legal owner of the burrow. If all had their rights, it would seem that the Owl is nearly as much an intruder as the snake, and that it only takes possession of the burrow excavated by the prairie dog in order to save itself the trouble of making a subterranean abode for itself. Indeed, there are some parts of the country where the Owl is perforce obliged to be its own workman, and in default of convenient "dog" burrows, is fain to employ its claws and bill in excavating a home for itself.

The tunnel which is made by the Owl is not nearly so deep or so neatly constructed as that which is dug by the marmot, being only eighteen inches or two feet in depth, and very rough in the interior. At the bottom of this burrow is placed a tolerably-sized heap of dried grass, moss, leaves, and other soft substances, upon which are deposited its white-shelled eggs.

Some persons have supposed that the Coquimbo Owl is attracted to the habitation of the prairie dog by the charms of the young and tender marmots, which would furnish a delicate and easily obtained meal whenever the bird might happen to be hungry. As, however, the stomach of the Burrowing Owl has only been found to contain the wing-cases and other indigestible parts of beetles and various large insects, the bird may be pronounced guiltless on this charge. Those specimens, however, that inhabit the plains of Buenos Ayres are proved, on the authority of Mr. Darwin, to feed on mice, small reptiles, and even on the little crabs of the sea-shore.

The Coquimbo Owl is by no means a nocturnal bird, facing the glare of the midday sun without inconvenience, and standing at all times in the day or evening on the little heaps of earth which are thrown up at the entrance of the burrows. It is a lively little bird, moving about among the burrows with considerable vivacity, rising on the wing if suddenly disturbed, and making a short undulating aerial journey before it again settles upon the ground. When it has alighted from one of these little flights, it turns round and earnestly regards the pursuer. Sometimes it will dive into one of the burrows, heedless of prior occupants, and thus it is that marmot, owl, and snake come to be found in the same burrow.

Lizards and other reptiles have also been found in the burrows of the prairie dog. While sitting on the little earth mounds, or moving among the burrows, the Coquimbo Owl presents a very curious likeness to the prairie dog itself, and at a little distance might easily be taken for the little marmot as it sits erect at the mouth of its domicile.

The colour of the Burrowing Owl is a rather rich brown upon the upper parts of the body, diversified with a number of small grey-white spots, and altogether darker upon the upper surface of the wings. The under parts are greyish-white. The length of the bird is not quite eleven inches. The cry of this curious bird is unlike that of any other Owl, and bears a very great resemblance to the short, sharp bark of the prairie dog.

The genus ATHÉNE is a very large one, and contains many curious and interesting birds which cannot be described at length except in a monograph of the Owl tribe. Among these may be mentioned two remarkable birds, the BOOBOOK OWL (*Athéne Boo-book*) and the WINKING OWL (*Athéne connivens*), both natives of Australia.

The former of these birds is popularly called the Australian Cuckoo by the colonists on account of its cry, which bears no small resemblance to the well-known song of "Spring's harbinger." As the bird, after the manner of Owls, utters its cry by night, it is often noted as an instance of the perversity of the Australian climate, which reverses the usual operations of nature, and forces the cuckoo to take the place of the nightingale, and pour forth its song at night. This species is diurnal in its habits, and remarkably swift and agile on the wing, being able to chase successfully the quick-winged insects that are so numerous in Australia. When roused, it is said by Mr. Gould to resemble the wood-cock in the manner of its flight, and to further carry out the likeness in its habit of rising out of gun-shot, and diving rapidly into the nearest covert, where it lies safely housed until its enemies have withdrawn from the neighbourhood. The quaint title of Boobook is the name by which it is known among the natives.

THE WINKING OWL is also a day flier, strong and powerful on the wing, though with flight nearly as noiseless as that of the common barn Owl of England. It is a large and powerful bird, delighting to capture the young koala, or native bear, together with other prey of equal strength and magnitude. Berries have been found in the stomach of one of these Owls ; but Mr. Gould thinks that they have probably come from the crop of some unfortunate bird which had fallen a prey to the Winking Owl. The cry of this species is remarkably resonant, and is said to resemble the lowing of an ox. If wounded it becomes a very dangerous opponent, flinging itself on its back, striking fierce and rapid blows with its well-armed feet, and seeking to seize its foe in the terrible clutch of its curved talons.

The general colour of this species is a dark clove-brown, diversified by many bars and stripes.

THIS genus also finds a British representative, in the person of the LITTLE OWL (*Athéne passerina*), many specimens of which bird have been captured in England, and even the nest and young occasionally discovered.

The name of Little Owl is very appropriate, for it is only eight inches in length including the plumage, and when stripped of its feathers appears hardly so large as a common starling. It is properly a native of Germany, Holland, France, and Austria, and has sometimes been called the Austrian Rufous Owlet, or the Sparrow Owl. Although so small a creature, its food is the same as that of any of its larger relatives, consisting of small birds, bats, mice, and various insects. The general colour of this curious little Owl is clove-brown, banded and marked with yellowish-brown, grey, and white. It may easily be distinguished from the British Owls by the legs, which are very long in proportion to the dimensions of the bird, and instead of being feathered down to the toes, are covered with very short hair-like plumage, becoming very scanty over the toes. It is easily domesticated, and in a tamed state is so voracious that, according to Bechstein, it can swallow five mice at a single meal.

ANOTHER curious little Owl is sometimes found in England, and has therefore gained a place among the British birds. This is the TENGMALM'S OWL or DEATH BIRD, the latter name having been given to it on account of a common superstition that reigns among several of the North-American Indian tribes. When an Indian hears one of these birds uttering its melancholy cry, he whistles towards the spot from whence the sound proceeded, and if the bird does not answer him, he looks for a speedy death.

This species is at first sight not unlike the Little Owl, but may be at once distinguished from that bird by the structure of its legs and toes, and the thick feathery coating with which they are clad. It is a very common bird over the whole of the

TENGMALM'S OWL.—*Nyctale funerea.*

inhabited portions of North America, but is frequently found in Norway, Sweden, Russia, and even in Northern France and Italy. It is a nocturnal bird, seldom wandering from its home during the hours of daylight, as it is almost blinded by the unaccustomed glare, and may be easily captured by hand while thus bewildered. The nest of the Tengmalm's

H 2

Owl is generally made of grass, and is placed about half-way up some convenient pine-tree. The eggs are seldom more than two in number, are pure white in colour, and not quite so globular as is the case with the generality of Owls' eggs.

The colour of this bird is more rich and better defined than that of the Little Owl. The whole of the upper parts of the body are a rich chocolate-brown, dotted and splashed with many white markings, which are very minute upon the top of the head, and larger upon the back and wings, some indeed being arranged on the lower portions of the wings so as to form irregular stripes. Similar white spots are placed on the tail, which is usually of a dark brown. The eye disk is greyish-white, excepting a bold black-brown ring just round the eye. The under portions of the body are greyish-white, covered with numerous brown bars and spots, and the plumage of the legs and toes is also grey-white sprinkled with brown spots. The size of the Tengmalm's Owl is nearly the same as that of the Little Owl.

SCOPS EARED OWL.—*Ephialtes Scops.*

WE now arrive at a large group of Owls which are remarkable for two tufts of feathers which rise from the head, and occupy nearly the same relative position as the ears of quadrupeds. These "ears," as they are called, have, however, nothing to do with the organs of hearing, but are simply tufts of feathers, which can be raised or depressed at the will of the bird, and give a most singular expression to the countenance.

The first of these birds is the SCOPS EARED OWL, a most singular little creature, which is sometimes, though rarely, taken in England, and has therefore been placed in the catalogue of British Birds.

The geographical range of this species is very great, specimens now in the British Museum having been taken in Germany and several parts of Europe, India, Malacca, China, Gambia, and the Cape of Good Hope. It is by no means an uncommon bird in Southern Europe, and is said even to have bred several times in England. A very good description is given of the habits of the Scops Eared Owl by Mr. Spence.

"This Owl, which in summer is very common in Italy, is remarkable for the constancy and regularity with which it utters its peculiar note or cry. It does not merely ' to the moon complain,' but keeps repeating its plaintive and monotonous cry of *Kew! kew!* (whence its Florentine name of *Chiù*, pronounced almost exactly like the English letter Q) in the regular intervals of about two seconds the livelong night, and until one is used to it, nothing can well be more wearisome. Towards the end of April, last year, 1830, one of these Owls established itself in the large *Jardin Anglais*, behind the house where we resided at Florence, and until our departure for Switzerland in the beginning of June, I recollect but one or two instances in which it was not constantly to be heard, as if in spite to the nightingales, who abounded there from nightfall to midnight (and probably much later), whenever I chanced to be in the back part of the house, or took a friend to listen to it, and always with precisely the same unwearied cry, and the intervals between each as regular as the tickings of a pendulum.

This species of Owl, according to Professor Savi's excellent *Ornitologia Toscana*, Vol. I. p. 74, is the only Italian species which migrates ; passing the winter in Africa and

Southern Asia, and the summer in the south of France. It feeds wholly upon beetles, grasshoppers, and other insects."

The length of this tiny Owl is only seven inches and a half, the female being a little longer than her mate. The nest is generally placed in a hollow tree or the cleft of a rock, and contains from two to four white eggs. It is a pretty little bird, the general colouring being much as follows. The head is light brown, marked with several narrow dark-brown streaks ; the back is variegated brown and chestnut, marked with dark bands and grey mottlings. The wing is brown, speckled largely with white and grey, and the tail is similarly barred and dashed with black and pale brown. The facial disk **is** greyish-white, thickly covered with small brown spots, and the two feather-tufts of the head are similarly tinted. The under portions of the body are greyish-white, with several streaks and dashes of dark brown, and the legs are covered as far as the toes with short speckled feathers. The claws are nearly white at their base, declining to blackish-brown at the tip ; the toes are brown and the beak black.

GREAT OWL.—*Bubo máximus.*

ALTHOUGH seemingly exceeding the golden eagle in dimensions, the GREAT OWL is in reality a very much smaller bird, owing its apparent magnitude to its feathers and not to its body. In weight it hardly exceeds one quarter of that of the eagle, but in power of muscle it is little inferior even to that royal bird itself.

The Great Owl, or EAGLE OWL as it is often called, inhabits the northern parts of Europe, being especially common in Sweden, Norway, Switzerland, and Russia, and being found even in some parts of Italy and Turkey. It is a very rare bird in our country, and is only an occasional visitant to our shores. When captured, however, the Eagle Owl is easily reconciled to its habitation, and has frequently been known to hatch and bring up its young while in captivity. In its wild state it makes a very rude nest upon some convenient ledge of rock or other similar locality, and lays two or three pure white, rather globular eggs. The young, when in their first few days of independent life, would hardly be recognised as Owls at all, being mere shapeless lumps of grey woolly down. The parent birds take great care of their young, and are so fond of their offspring, that when an Eagle Owl's nest has been harried, and the young birds removed, the parents have been known to supply them with food for a period of fourteen days, laying dead partridges and other prey before the bars of the cage in which the young birds were confined.

The food of this Owl consists generally of grouse, partridges, hares, and other similar game, and the bird is so powerful that it will successfully chase even larger prey. Mr. Lloyd, in his well-known " Field Sports," gives the following description of the Eagle Owl as it appears in the Scandinavian forests.

" These Owls, Dr. Mellerborg assured me, will sometimes destroy dogs. Indeed, he himself once knew an instance of the kind. He states another circumstance, showing the ferocity of these birds, which came under his notice. Two men were in the forest for the purpose of getting berries, when one of them happening to approach near to the nest of the Owl, she pounced on him while he was in the act of stooping, and fixing her talons in his back, wounded him very severely. His companion, however, was fortunately near at hand, who, catching up a stick, lost no time in destroying the furious bird.

Mr. Nilsson states that these Owls not unfrequently engage in combat with the eagle himself, and that they often come off victorious. These powerful and voracious birds, that gentleman remarks, occasionally kill the fawns of the stag, roebuck, and reindeer. The largest of the birds common to the Scandinavian forests, such as the capercali, often become their prey. The hooting of these Owls may often be heard during the night-time in the northern forests ; the sound, which is a most melancholy one, and which has given rise to many superstitions, is audible at a long distance." This bird is also most invaluable in destroying the lemmings.

The cry of the Eagle Owl is a very deep and doleful note, sounding most lugubriously in the depth of the lonely forests during the hours of night. When angry, the bird utters a sharp hiss, not unlike the sound which is produced by the common brown Owl of England when irritated. It is generally seen during the twilight hours, although it some-times continues to search after prey during the entire night. While engaged in hunting, it flies low over the ground, and displays great quickness of eye and wing in discerning and pouncing upon everything that has life and can be eaten. This bird has been employed for the purpose of decoying falcons towards the snare, by being fastened in a convenient spot where it can be seen by any passing bird. The falcon cannot possibly resist the pleasure of dashing at the great, solemn, winking bird, and is accordingly captured in the fowler's ready net as he swoops after the Owl, which runs for refuge precisely in the direction of the toils.

The length of this fine bird is rather more than two feet, and the aspect of its outstretched wings is wonderfully magnificent. The general colour of the Eagle Owl is brown, mingled with a yellow tinge, and covered on the upper surface with bars, dashes, and streaks of blackish-brown. The facial disc is pale brown, decorated with many small spots of black, and the under surface of the body is nearly yellow, traversed by longitudinal stripes of black upon the chest, and barred transversely with many bars of dark brown on the abdomen. The legs and thighs are pale brown, with many narrow bars of a darker hue. The long armed claws are black, and the beak is also nearly black. The eyes are of a bright radiant orange, and have a very fierce appearance when the bird looks the spectator in the face. The female is rather larger than the male, the difference in length being about four inches.

The VIRGINIAN EARED OWL holds the same place in America as the eagle Owl in Europe, and is even now a familiar bird, though it has been extirpated from many localities where it once reigned supreme.

It is a very large bird, nearly equalling the great Owl in magnitude, and being in no way its inferior in strength or courage. This species is found spread over the greatest portion of North America, and in former days did great damage among the poultry of the agriculturists, being a bold as well as a voracious bird. Now, however, the ever-ready rifle of the farmer has thinned its numbers greatly, and has inspired the survivors with such awe, that they mostly keep clear of cultivated lands, and confine themselves to seeking after their legitimate prey.

The Virginian Eared or Horned Owl is a terrible destroyer of game, snatching up grouse, partridges, hares, ducks, sparrows, squirrels, and many other furred and feathered creatures, and not unfrequently striving after larger quarry. The wild turkey is a favourite article of diet with this Owl; but on account of the extreme wariness of the turkey nature, the depredator finds an unseen approach to be no easy matter. The usual mode in which the Owl catches the turkey is, to find out a spot where its intended prey is quietly sleeping at night, and then to swoop down suddenly upon the slumbering bird before it awakes. Sometimes, however, the Owl is baffled in a very curious manner. When the turkey happens to be roused by the rush of the winged foe, it instinctively ducks its head and spreads its tail flatly over its back. The Owl, impinging upon the slippery plane of stiff tail feathers, finds no hold for its claws, and glides off the back of its intended victim, which immediately dives into the brushwood before the Owl can recover from the surprise of its unexpected failure.

The following admirable description of the Virginian Eared Owl, as it used to be in the earlier days of cultivation, is given by Audubon in his well-known History of the Birds of America.

"It is during the placid serenity of a beautiful night, when the current of the waters moves silently along, reflecting from its smooth surface the silver radiance of the moon, and when all else of animated nature seems sunk in repose, that the great Horned Owl, one of the Nimrods of the feathered tribes of our forests, may be seen sailing silently and yet rapidly on, intent on the destruction of the object destined to form its food.

The lone steersman of the descending boat observes the nocturnal hunter gliding on extended pinions across the river, sailing over one hill and then another, or suddenly sweeping downwards and again rising in the air, like a moving shadow, now distinctly seen, and again, mingling with the sombre shades of the surrounding woods, fading into obscurity.

The bird has now floated to some distance, and is opposite the newly-cleared patch of ground, the result of a squatter's first attempt at cultivation in a place lately shaded by the trees of the forest. The moon shines brightly on his hut, his light fence, the newly planted orchard, and a tree which, spared by the axe, serves as a roosting-place for the scanty stock of poultry which the new-comer has procured from some liberal neighbour. Amongst them rests a turkey-hen, covering her offspring with extended wings.

The great Owl, with eyes keen as those of any falcon, is now seen hovering above the place. He has already espied the quarry, and is sailing in wide circles, meditating his plan of attack. The turkey-hen, which at another time might be sound asleep, is now,

VIRGINIAN EARED OWL.—*Bubo Virginianus.*

however, so intent upon the care of her young brood, that she rises on her legs, and purrs so loudly as she opens her wings and spreads her tail, that she rouses her neighbours, the hens, together with their protector. The cacklings which they at first emit soon become a general clamour.

The squatter hears the uproar, and is on his feet in an instant, rifle in hand; the priming examined, he gently pushes open the half-closed door and peeps out cautiously, to ascertain the cause by which his repose has been disturbed. He observes the murderous Owl just alighting on the dead branch of a tall tree, when, raising his never-failing rifle, he takes aim, touches the trigger, and the next instant sees the foe falling dead to the ground. The bird is unworthy of his further attention, and is left a prey to some prowling opossum or other carnivorous quadruped. In this manner falls many a Horned Owl on our frontier, where the species abounds."

The flight of this bird is remarkably powerful, easy, and graceful, as may be gathered from the enormous expanse of wing, in comparison with the weight and dimensions of the body. Its voice is of a hollow and weird-like character, and when heard by night from some spot on which the Owl has silently settled, is apt to cause many a manly but superstitious cheek to pale. As Wilson well observes, the loud and sudden cry of Waugh O ! Waugh O ! is sufficient to alarm a whole garrison of soldiers. Probably on account of the peculiar sounds which are uttered by this bird, the Cree Indians know it by the name of Otowuck-oho !

The Virginian Horned Owl takes up its residence in the deep swampy forests, where it remains hidden during the day, and comes out at night and morning, heralding its approach with its loud, unearthly cries, as of an unquiet, wandering spirit. Sometimes, according to Wilson, "he has other nocturnal solos, one of which very strikingly resembles the half-suppressed screams of a person suffocating or throttled."

Sir W. Jardine, in his notes to his well-known edition of Wilson's American Ornithology, gives the following account of a captive Owl, which affords an excellent idea of the peculiar sounds that can proceed from an Owl's throat.

"An Eagle Owl in my possession remains quiet during the day, unless he is shown some prey, when he becomes eager to possess it, and when it is put within his reach, at once clutches it, and retires to a corner to devour it at leisure. During night he is extremely active, and sometimes keeps up an incessant bark. It is so similar to that of a cur or terrier, as to annoy a large Labrador house-dog, who expresses his dissatisfaction by replying to him, and disturbing the inmates nightly. I at first mistook the cry also for that of a dog, and, without any recollection of the Owl, sallied forth to destroy the disturber of our repose ; and it was not until tracing the sound to the cage, that I became satisfied of the author of the annoyance. I have remarked that he barks more incessantly during a clear winter night than at any other time, and the thin air at that season makes the cry very distinctly heard to a considerable distance. This bird also shows a great antipathy to dogs, and will perceive one at a considerable distance, nor is it possible to distract his attention so long as the animal remains in sight. When first perceived, the feathers are raised, and the wings lowered as when feeding, and the head moved round, following the object while in sight. If food is thrown, it will be struck with the foot and held, but no further attention paid to it."

The nest of this bird is extremely large, and consists of a large bundle of sticks, grass, leaves, and feathers, placed in the fork of some large bough, and containing three or four white eggs. The colour of the Virginian Eared Owl is reddish brown upon the upper surface, mottled with various splashes of black, and covered with regular bands of the same hue. The facial disc is brown, edged with black. The under surface is of a light reddish-brown colour, covered with numerous transverse bars of dusky brown, with a few white lines and dashes among them. The throat is pure white, the beak and claws are black, and the eyes are of a bright orange, gleaming out strangely even by day and burning with double radiance in the twilight.

The common BROWN OWL, or TAWNY OWL as it is often named, is, with the exception of the Barn Owl, one of the best known of the British Owls.

Although rather a small bird, being only about fifteen inches in total length, it is possessed of a powerful pounce and audacious spirit, and when roused to anger or urged by despair, is a remarkably unchancy antagonist.

In the *Field* newspaper there is a curious account of the conduct of a pair of Brown Owls, who built a nest in the attic of an untenanted house. The writer proceeds to say, "I should have been a little afraid of molesting them, so ferocious did the old gentleman look when his wife and children were approached. One morning the cat was missing, and I found, on inquiry, that some strange sounds had been heard the evening before in the room where the Owls were. On going up that evening I found poor puss quite dead, one of her eyes actually picked out, and her antagonist, also killed, lying on the side of the nest. The mamma Owl was absent, probably in search of food, but she may have been present and have assisted at the death. I have see a cat on another occasion cowed by an old Owl that came down the chimney into the dining-room."

In the same paper is recorded an anecdote of a pair of Brown Owls that were kept in confinement, and which, when approached by any stranger, would fly at him and fasten their talons into his head with such angry violence that they could but be removed by direct force.

The food of this Owl is of a very varied nature, consisting of all the smaller mammalia, many reptiles, some birds, fishes when it can get them, and insects. It seems to be a good fisherman, and catches its finny prey by waiting on the stones that project a little above the water, and adroitly snatching the fish from the stream by a rapid movement of the foot. Sometimes it flies at much higher game, especially when it has a young family to maintain, and will then attack birds and quadrupeds of very great size when compared with its own dimensions. In a single nest of this bird have been found, according to a writer in the *Field*, three young Owls, five leverets, four young rabbits, three thrushes, and one trout weighing nearly half a pound. All these achievements, however, sink into insignificance in comparison with a feat which is described by Mr. Carr.

"In 1844 a pair of Tawny Owls reared and ushered into the world their hopeful young, after having fed them assiduously upon the trees for many weeks after they had left the nest. The food must often have consisted in great part of worms, snails, and slugs, for the old birds brought it every minute from the ground in the immediate vicinity of the trees where the young were perched. This, however, might only be considered as a whet to their appetites before dinner, for the parents made repeated and persevering attacks upon three or four magpies' nests, sometimes during half an hour at a time. As the defence was gallant and spirited, they were often repulsed, but finally I found the remains of young magpies under the favourite perch of the Owls, and one morning the bloody head and feathers of an old magpie, conspicuous for its size and the want of any cerous skin about the beak. This, then, I thought, must have been taken while roosting.

BROWN OWL.—*Surnium Aluco.*

In 1845 the Owls alone were seen, and they passed the summer in sedate retirement, and seemed to rest from the labours of propagation, neither did they molest the magpies. But in 1846 they began to be very active early in the spring, and by the beginning of May again had their young Owlets out upon the branches. Walking about nine o'clock one evening, I heard a pertinacious attack going on against a pair of magpies that had had their nest in the top of a very tall sycamore. At last, instead of the frantic chattering of the poor magpies, one of them began to shriek in agony like a hare when caught in a noose, and it was evident that the Owl was trying to drag it—the mother bird—by the head from the entrance of the nest.

I ran down in time to separate the combatants by striking against the stem of the tree with a stick. Before the next morning, the young of an only pair of rooks had disappeared from the nest in a retreat in which none but the Owls could have imagined them. This was too bad; a decree went forth against their young Owls, and they paid the penalty of their voracious appetites."

The voice of the Brown Owl is a loud monotonous hoot, that may be often heard in the evening in localities where the bird has made its home. The nest is usually placed in the hollow of a tree, and contains several white eggs. The colour of the Brown or Tawny Owl is an ashen-grey upon the upper parts of the body, variegated with chocolate and wood-brown. Several whitish-grey bars are seen upon the primaries, and there are several rows of whitish spots upon the wings and scapularies. The facial disc is nearly white, edged with brown, and the under surface of the body is of the same hue, covered with longitudinal mottlings of variously tinted brown. The claws are nearly white at their base, darkening towards their extremities, and the beak is nearly of the same colour. The eyes are of a very dark black-blue.

This species is found in many parts of Europe, and is said to be one of the indigenous birds of Japan.

LONG-EARED OWL.—*Otus vulgáris.*

THE fine bird which is known by the name of the URAL OWL (*Syrnium Uralense*) belongs to the same genus with the barn Owl. This bird is nearly two feet in length, and preys on hares, rabbits, grouse, and other large quarry, after the manner of the eagle Owl. It is a native of the colder regions of Europe and Asia, being found in Lapland, Norway, Northern Russia, and in similar localities.

WE now come to a familiar example of the British Owls, a bird that has attracted great notice on account of its singular aspect. This is the LONG-EARED OWL, its popular name being derived from the great length of the "ears," or feather-tufts which are placed upon the head, and erect themselves whenever the bird is excited.

The Long-eared Owl is found in almost all parts of England, and also inhabits portions of Asia, Africa, and America, so that it possesses a very large geographical range. It is not a very large bird, being only fourteen or fifteen inches in length, but is a most rapacious being, preying upon all the smaller mammalia, and capturing the finches and other small birds with as much success as if it were a hawk rather than an Owl. Even moles fall victims to the Long-eared Owl, and in the "castings" of this species have been found the remains of mice, rats, and various birds.

While the young are still in the nest, the parent birds display a singular assiduity in collecting food for their infant charge, and make sad havoc among the half-fledged nestlings of the neighbourhood. The nest of this species seems seldom to be built by the bird itself, as the Owl prefers to take to the deserted nest of some other bird, and to fit up the premises for its own use. According to Mr. Yarrell, this Owl has been known to take possession of the nest of a squirrel, and therein to rear its young. The eggs of the Long-eared Owl are generally four or five in number, and white, as is the case with nearly all Owls' eggs.

It is a decidedly nocturnal bird, seldom being seen in the light of day, and being always greatly disturbed if it should chance to issue from its concealment while the sun is above the horizon. When it can take its choice of locality, it seems always to prefer some spot where the foliage is thick, dark, and heavy, and if possible will build its nest in the shade of some large evergreen.

The colour of this bird is very handsome, but so complicated that it is not easy to describe. The colour of the back and upper feathers of the body is pale brown, diversified on the neck and shoulders by sundry longitudinal streaks of black-brown. The upper surface of the wings is variously splashed with black, fawn, and brown, and the primaries are light chestnut barred and spotted with dark brown. The facial disc is curiously marked with several shades of brown and white, and the "ears" are composed each of seven or eight blackish-brown feathers. The under surface of the body is greyish-white intermixed with fawn and various longitudinal brown streaks, and the legs are covered up to the claws with pale-brown plumage. The sharp curved claws are black, as is the bill, and the eyes are of a light orange.

As the facial disc is very conspicuous in this species, I shall take the opportunity of inserting a few remarks upon that portion of the Owl's structure which have already appeared in "My Feathered Friends."

"It is said that the use of this circle is to collect the rays of light and throw them upon the eye, a provision necessary in dark nights. This principle is apparently carried out in the case of the Barn Owl, where the feathery circle, being of a whitish hue, may be supposed to act as a reflector of the light. But it must be remembered that in the Brown Owls this circle is also brown, and therefore would rather absorb than reflect the light. Besides, objects are seen by the light reflected from *them* to the eyes, while light reflected upon the eyes from the sky would rather distract than aid the vision. When, on a bright day, we put our hands to our eyes in order to view a distant object, we do so not to collect scattered rays and to force them to converge upon the pupil, but rather to keep these scattered rays from interfering with those that proceed directly from the object of vision. The same thing may be observed when people look at a picture through a tube.

In my own opinion the radiating feathery circle is very simple in its operation, being only a kind of circular splay window cut through the thick mass of plumage in which the head of the bird is enveloped, in order to give it a wider sphere of vision, just as architects cut a splay window in the thick wall of a fort so as to permit a musket-barrel to be pointed in any direction. And the radiating formation of the feathers is preserved because the natural elasticity of their stems presses aside the softer downy plumage of the head, and preserves the circular form complete. If examined, they will be found to be formed in a very peculiar manner, and quite distinct from those on which their extremities press."

For the following interesting account of the habits of a tame Long-eared Owl, I am indebted to the kindness of a correspondent.

"The Horned Owlet has a peculiarly cat-like expression of face, and this I think was the chief attraction possessed by a downy greyish-white ball, that was thrust into my lap by one of my boy friends, who at the same time announced its name and nature.

With great delight I proceeded to introduce him to my other bird pets, but the intense excitement caused by his appearance compelled me to remove him with all speed. The small birds were all afraid of him, but the jackdaw and magpie both charged poor 'Blinker' at once. It then struck me that the cat-like face and nocturnal mousing habits of the creature indicated the deep secret of its nature, and if so, that it would have more sympathy with the feline establishment than with that of the birds.

Acting upon this impression, I at once conveyed him to pussy's closet, and introduced him to its occupants, namely, Mrs. Fanny and her blind kitten. Pussy regarded him at first with very suspicious looks ; but the poor bird, feeling pleased with the dim light and pussy's soft warm coat, soon nestled up to her.   This act of confidence on Blinker's part appeared to affect Fanny favourably, and she at once purred him a welcome. From this time they were fast friends, and many mice did she good-naturedly provide Blinker with in common with her own kitten. When he grew large enough, he used to sit on the side of her basket, and would never settle quietly for the night until the two cats were asleep in their bed.

It was quite beautiful to observe the warm affection which grew up between the Owlet and the kitten. The only cause of discord that we ever noticed between the two, was when the kitten would play with a living mouse. This evidently hurt Blinker's feelings, for he would always pounce down and seize the mouse by the back of its neck, and kill it in a moment. Still, he had a sense of justice in his nature ; for when the mouse was dead, he would drop it down to its rightful owner.

I had him for a year, and was much attached to him ; but he fell ill, and went the way of all pets."

SHORT-EARED OWL.—*Otus brachyótus.*

ANOTHER species of Eared OWL is also found in England, and is nearly as common as the preceding bird. This is the SHORT-EARED OWL, or SHORT-HORNED OWL as it is sometimes called.

This species is remarkable for the very small size of the head, which is even smaller than the neck, and gives a very un-Owl-like look to its aspect.   In its habits the Short-eared Owl is very unlike its British relations, as it flies much by day, and haunts the heaths and open lands in preference to the woods.   The eggs of this bird are laid upon the bare earth, which is scraped away by the parent bird until a small hollow is made, and is undefended by the slightest lining.   Richardson, however, says that in North America its nest is formed of dried grasses and moss laid on the ground.   Sometimes the eggs are laid on a depression among sedges or heather, and there are instances where Owls' eggs, supposed to be of this species, have been found in rabbit burrows.   The eggs are seldom more than three in number.   On account of its small head and day-hunting propensities, it is sometimes known by the local name of Hawk Owl.

It does not seem to be very powerful on the wing, seldom flying for more than a hundred yards, and at a very low elevation.   If observed, it seeks the nearest covert, and dives so deeply among the brushwood that it is not easily seen, and cannot be driven out if the covert should be of any great extent.   Its food consists chiefly of mice and birds ; and Mr. Yarrell mentions that he has discovered in the stomach of a Short-eared Owl, the remains of a bat and a half-grown rat.

This bird has a very large geographical range, being found spread over the whole of Europe and in many parts of Asia and Africa.   In North America it is a very common bird, and is thought to be also an inhabitant of Chili.   Specimens have been brought from the Sandwich Isles, Brazil, and the Straits of Magellan.

In colour it is a pretty bird ; the upper surface of the body, together with the head and neck, are fawn, covered with dark-brown patches, the darker tint being placed on the centre of each feather, and the same round the edges.   A few spots of yellow are seen on the wing-coverts, and the ruddy-brown primaries are bound with dark brown and tipped with ashen-grey.   Several very decided bars of dark brown cross the tail, and the facial disc is dark towards the eye, becoming lighter towards the circumference and edged with a white line.   The feathers of the disc are long, and almost conceal the basal portions of the beak.   With the exception of the white skin, the whole of the under surface of the body is light buff, thickly crossed with longitudinal dashes of dark brown upon the breast, and with a few long streaks of the same colour upon the abdomen.   The legs and thighs are pale buff, the claws and beak are black, and the eyes golden yellow.   The length of the bird is about fifteen inches, the female being longer than the male.

> " In the hollow tree, in the old grey tower
>   The spectral Owl doth dwell ;
> Dull, hated, despised in the sunshine hour,
>   But at dusk he's abroad and well !
> Not a bird of the forest e'er mates with him,
>   All mock him outright by day ;
> But at night, when the woods grow still and dim,
>   The boldest will shrink away.
>   O ! when the night falls, and roosts the fowl,
>   Then, then is the reign of the Horned Owl.
>
> And the Owl hath a bride who is fond and bold,
>   And loveth the woods' deep gloom ;
> And with eyes like the shine of the moon-stone cold,
>   She awaiteth her ghastly groom :
> Not a feather she moves, not a carol she sings,
>   As she waits in her tree so still,
> But when her heart heareth his flapping wings,
>   She hoots out her welcome shrill !
>   Oh ! when the moon shines and dogs do howl,
>   Then, then is the reign of the Horned Owl.
>
> Mourn not for the Owl, nor his gloomy plight !
>   The Owl hath his share of good.
> If a prisoner he be in the broad daylight,
>   He is lord in the dark greenwood !
> Nor lonely the bird nor his ghastly mate ;
>   They are each unto each a pride ;
> Thrice fonder, perhaps, since a strange dark fate
>   Hath rent them from all beside !
>   So when the night falls and dogs do howl,
>   Sing Ho ! for the reign of the Horned Owl !
>   We know not alway
>   Who are kings of day,
>   But the king of the night is the bold brown Owl ! "

BARRY CORNWALL.

THE best known of the British Owls is the WHITE, BARN, OR SCREECH OWL, by either of which appellations the bird is familiarly known over the whole of England.

This delicately coloured and soft-plumed bird is always found near human habitations, and is generally in the vicinity of farmyards, where it loves to dwell, not for the sake of devouring the young poultry, but of eating the various mice which make such havoc in the ricks, fields, and barns.   The " feathered cat," as this bird has happily been termed, is a terrible foe to mice, especially to the common field-mouse, great numbers of which are killed daily by a single pair of Owls when they are bringing up their young family.   In the evening dusk, when the mice begin to stir abroad in search of a mole, the Owl starts in search of the mice, and with noiseless flight quarters the ground in a sportsmanlike and systematic manner, watching with its great round eyes every movement of a grass-blade, and catching with its sensitive ears every sound that issues from behind.   Never a field-mouse can come within ken of the bird's eye, or make the least rustling among the leaves within hearing of the Owl's ear, that is not detected and captured.   The claws are the instruments by which the Owl seizes its victim, and it does not employ the beak until it desires to devour the prey.

It is curious that the Owl should have two modes of eating, which, as far as my own experience goes, are invariably followed.   If the bird has caught a mouse and is going to eat it, the mouse is first bitten smartly across the back so as to destroy all life, and when it hangs motionless from the bird's beak, it is thrown up into the air in a most adroit manner, so as to fall with its head downwards.   The Owl then catches the little quadruped in such a manner that its head falls into the bird's mouth, where it hangs for a few seconds.   A sharp toss of the head then sends the whole of the mouse down the Owl's throat with the exception of the tail, which hangs out of one side of the beak, generally the left side, and is then rolled about just as a boy rolls a stick of sweetmeats between his lips.   After carrying on this process for two or three minutes, the Owl again jerks its head, and the mouse vanishes wholly from sight.   But when the Owl has to deal with a bird, it eats it after the manner of the hawks, partially plucking it, and tearing it to pieces with its beak before swallowing it.

A cat with which I am well acquainted always follows the example of the Owl in its method of eating prey.   If it catches a mouse, she disposes of it without ceremony, beginning at the head and gradually eating towards the tail ; but if she has captured a bird, she places her feet upon its body, and with her teeth seizes the feathers and deliberately pulls them out before she will attempt to eat the carcase.   It may be that while the Owl is twisting and turning the mouse in its mouth, it may be lubricating its skin in order to admit of its easier passage down the throat.   The feathers of birds are too stiff and absorbent to admit of this process, and are therefore removed by the Owl before it swallows its prey.

Some doubts have been raised respecting the bird-killing propensities of the Barn Owl, many writers having asserted that it never kills adult birds, and that at the worst it only takes a young finch or so out of the nest.   Now, as my own Owl was always delighted with a full-grown bird, and proceeded to eat it in a very business-like manner, it seemed to me as if the process were by no means a new one, and these suspicions were confirmed by some "castings" of a Barn Owl which were sent to me, and which contained, among other matters, the bones of birds and an entire skull of a full-grown sparrow.   Since that time, I have seen several accounts of similar objects being discovered in the "castings," and it is rather a curious fact that the skull is always unbroken.   Generally, the "castings" are composed of the bones and skins of mice, together with the hard portions of various beetles, mostly in the specimens which I have examined belonging to the genera Carabus, Abax, Agonum, and Steropus.

Sometimes the Owl has been detected in robbing the pigeons' nests of their young; but such conduct seems to be very exceptional, as there are many instances on record where the Owl has actually inhabited the same cote with the pigeons without touching their young or disturbing the peace of the parents.   This Owl is also an experienced fisher, and has been seen to drop quietly upon the water, and return to its nest bearing in its claws a perch which it had captured.

This bird is easily tamed when taken young, and is a very amusing pet.   If properly treated, and fed with appropriate diet, it will live for a considerable time without requiring very close attendance.   Even if it be set at liberty, and its wings permitted to reach their full growth, it will voluntarily remain with its owner, whom it recognises with evident pleasure, evincing its dislike of strangers by a sharp hiss and an impatient snap of the bill.   One of these Owls, belonging to a friend, was, although a sufficiently amusing bird to its owner, so incorrigibly mischievous and spiteful, that it was at last doomed to death.

It seemed to fear nothing, and to care for nothing with one curious exception, in the person of a free but tame skylark, which was accustomed to sleep in a cage with the door open, and to forage for food on its own account when it was not satisfied with the quantity or quality of the diet that was daily furnished. With this lark the Owl contracted a firm alliance, permitting its little friend to sit upon its back and bury itself among the mass of soft plumage with which it was clad. This Owl always welcomed the approach of its friend, and when it perched upon its back, seemed as pleased as a horse when his favourite cat comes to bear him company. No other bird was so honoured, and a pair of goldfinches that were kept in a cage were constantly persecuted by the Owl, which could never understand that they were not to be killed, and was in the habit of pushing his feet through the bars, in vain attempts to secure the inmates.

It was a confirmed murderer of bats, and small birds as well as mice, and was accustomed to push its prey into a hole in an old wall that had been occasioned by the fall of a brick. In this odd larder were constantly found a strange variety of slaughtered

WHITE, OR BARN OWL.—*Strix flammea.*

game. Six to eight small birds were often counted when the hole was explored in the early morning, and once the Owl had poked fourteen bats into the aperture. On several occasions, the bird had contrived to pack a moderately sized eel into its storehouse, having always killed the eel by a bite across the back of the neck. The Owl was always attracted by bright and glittering objects, and once was seen to pounce upon a knitting-needle that lay glistening in the moonshine, and to carry it away to its usual receptacle.

This bird was remarkably fond of half-cooked chicken, and was wonderfully delighted if its meal were seasoned with a very slight sprinkling of sugar and salt, a fact which is rather remarkable, because, as a general rule, the predaceous birds do not care for sugar.

The hunting hour of this Owl varied much according to the time of year, and was about six P.M. in April and May, and eight in June and July. It was a spiteful bird, and very much given to attacking strange men and beasts. His last escapade was of such a serious nature, that he was summarily handed over to the executioner. He dashed at a pony which was coming towards the house, and fastening on its nose with its claws, battered the poor beast with his wings to such an extent that it became quite frantic, and by a powerful toss of the head flung its assailant violently on the ground and broke one of his legs.

Nothing daunted by this mishap, the Owl returned to the attack, and, grasping the pony's nose with the sound foot, struck his curved beak into its face and recommenced his buffeting. He was at last torn away by main force, and paid the penalty of his mischief with his life.

This species is generally considered to be the typical example of the Owl tribe, as it exhibits in great perfection the different characteristics of the Owls, namely, the thick coat of downy plumage, the peculiar disc round the eye, the large eye-balls, and the heavily feathered legs and toes. The feathers are so thickly set upon this bird, that it appears to be of much greater dimensions than is really the case. When standing on its feet, or while flying over the fields like a huge bunch of thistle-down blown violently by the night breeze, the Barn Owl appears to be rather a large bird; but when the creature is lying on the bird-stuffer's table, after its skin and feathers have been removed, the transformation is really astonishing. The great round head shrinks into the shape and size of that of a small hawk, the body is hardly larger than that of a pigeon, and but for the evident power of the firm muscles and their glistening tendinous sheaths, the bird would appear absolutely insignificant.

Although so small, it is a terrible bird to fight, and when it flings itself defiantly on its back, ire glancing from its eyes, and its sharp claws drawn up to its breast ready to strike as soon as its antagonist shall come within their range, it is really a formidable foe, and will test the nerves of a man to some extent before he can secure the fierce little bird, as I can assert from experience, having had my hands somewhat torn in such an encounter. So fiercely does this bird strike, that I knew an instance where a dog was blinded by the stroke of a Barn Owl's claws. The Owl was a tame one, and the dog—a stranger—went up to inspect the bird. As the dog approached the Owl, the bird rolled quietly over on its back, and when the dog put his head to the prostrate bird, it struck so sharply with its claws that it destroyed both the eyes of the poor animal, which had to be killed on account of the injury. While its young are helpless, the White Owl watches over their safety with great vigilance, and if any living thing, such as a man or a dog, should approach too closely to the domicile, the Owl will dash fiercely at them, regardless of the consequence to itself.

The nest of this species is placed either in a hollow tree, or in a crevice of some old building, where it deposits its white, rough-surfaced eggs upon a soft layer of dried "castings." These nests have a most ill-conditioned and penetrating odour, which taints the hand which is introduced, and cannot be removed without considerable care and several lavations. The young are curious little puffs of white down, and the Barn Owl is so prolific that it has been known to be sitting on one brood of eggs while it is feeding the young of a previous hatching.

As may be supposed from its popular title of White Owl, this species is very light in its colouring. The general colour of this bird is buff of different tints, with grey, white, and black variegations. The head and neck are light buff, speckled slightly with black and white spots, and the back and wings are of a deeper buff spotted with grey, black, and white. The tail is also buff, with several broad bars of grey. The facial disc is nearly white, becoming rusty-brown towards the eye, and a deeper brown round the edge. The under surface of the male bird is beautifully white, the claws are brown, the beak nearly white, and the eyes blue-black. The sexes are very similar in their colouring, but the females and young males may be distinguished by the under surface of the body, which is fawn instead of white.

TRINIDAD GOAT-SUCKER.—*Steatornis Caripensis.*

GOAT-SUCKERS.

WITH the owls closes the history of those birds which are called predaceous, although to a considerable extent nearly all birds are somewhat predaceous, even if they prey upon smaller victims than do the vultures, eagles, falcons, or owls. Next to the Accipitres come the Passeres, distinguished by their cere-less and pointed beak, their legs feathered as far the heel, their tarsus covered in front with shield-like scales, and their slightly curved and sharply pointed claws. This order is a very large one, and embraces a vast variety of birds.

First among the Passerine birds are placed the Fissirostres or cleft-beaked birds, so called from the enormous gape of the mouth, a structure which is intended to aid them in the capture of the agile prey on which they feed. Some of the birds, such as the Goat-suckers, swallows, bee-eaters, and others, prey upon insects, which they take upon the wing; while some, such as the kingfishers, feed upon fish, which they snatch from the water and bear to their homes in spite of their hard, slippery, scale-covered armour, or the watery element in which they dwell.

The GOAT-SUCKERS, as they are familiarly termed, from a stupid notion that was formerly in great vogue among farmers, and is not even yet quite extinct, that these birds were in the habit of sucking the wild goats, cows, and sheep, are placed first among the Fissirostres on account of the wonderfully perfect manner in which their structure is adapted to the chasing and securing of the swift-winged insects on which they feed. The colour of all these birds is sombre; black, brown, and grey being the prevailing tints. The gape of the mouth is so large that when the bird opens its beak to its fullest extent, it seems to have been severely wounded across the mouth, and the plumage is lax and soft like that of the owl.

2. I

THE singular bird which is known by the name of the TRINIDAD GOAT-SUCKER or GUACHARO is remarkable for the peculiarity from which it derives its name of Steatornis or Fat-Bird.

The Guacharos congregate in vast multitudes within the shelter of certain dark caverns, the greater portions of which remain unexplored owing to the superstitious fear of the natives, who fancy that the spirits of their ancestors hold their gloomy state in the innermost recesses of the caverns where the Guacharos reside. So strongly is this idea imprinted in their minds, that to "join the Guacharos" is, in the native language, an expression which signifies the death of the person of whom it is said. In these caverns the Guacharos build their nests, choosing for that purpose certain holes which exist in the roof, some forty or fifty feet in height.

During the breeding season, the Indians arm themselves with long bamboo poles, and entering the caves as far as they dare to penetrate, strike the young birds from their nests in spite of the outcries of the afflicted parents, which dash to and fro over their heads, emitting their strange hoarse cries. When a sufficient number of young birds has been felled, they are collected and taken to little temporary huts near the entrance of the cave. The young birds are then stripped of their fat, which lies in extraordinary volume on the abdominal regions, and the fat is melted down in clay pots, whence a peculiarly soft and limpid oil is produced. This oil has the useful property of remaining pure for a very long time, and even after the lapse of a year is found to be inodorous and free from rancidity. These birds are killed annually in such vast numbers that they would have long since been extirpated but for the religious scruples of the Indians, who are afraid to venture for any great distance into the caves, and can therefore only injure those unfortunate novices which have made their nests near the entrance.

The Guacharos remain in the caverns during the entire day, and issue forth at night in search of food. Contrary to the general habit of the Goat-suckers, the Guacharo feeds largely on fruits of various kinds, especially those of a hard and dry nature. Some persons suppose that the bird is strictly frugivorous in its habits, but this opinion has not been sufficiently strengthened to bear much value. It is certainly true that in the crop of the young birds which are annually slain, nothing is found except hard and dry fruits, which are carefully preserved by the natives and employed medicinally as a remedy against fever. The noise which these birds make in the recesses of their caverns, however, is quite deafening, and from the strange unearthly character of the cries uttered by the Guacharo, there is good reason for the superstitious terror of the Indians, who, like all heathens, have a very doleful idea of the spiritual condition, and when they hear by night a loud wailing cry, set it down at once as proceeding from some wretched wraith longing to resume the body in which it had been clothed, and piteously lamenting its sad doom.

Among other peculiarities of this bird may be reckoned the structure of the vocal organs, by means of which it is enabled to utter the unearthly sounds which have given to its residence so evil a reputation. It must first be mentioned that whereas in the mammals the chief part which the trachea or windpipe plays in the production of sound, is that it conveys the air towards the larynx where the voice is produced; in birds, however, the voice is produced at some distance down the trachea; in our vocal organ it is situated just at the spot where the trachea divides into the two bronchi which supply the lungs of each side with air. On dissecting the throat of the Guacharo, it is found that the "inferior larynx," as this curious organ is rather wrongly termed, is double, each of the bronchi being furnished with one of these structures at some distance below the division.

These birds have a very unpleasant smell, of a powerful and fish-like nature, compared by some persons to the odour of the cockroach. This smell, however, disappears when the bird is cooked, and seems to aid in conveying to the palate a peculiarly agreeable flavour. They are very awkward subjects for the taxidermist, as the fat is so plentiful and so adherent to the skin, that it liquifies under the touch and runs over the feathers, ruining their beautiful colours, and can hardly be separated from the skin without causing some damage to its delicate texture.

The colour of this curious bird is a ruddy fawn, mottled with dark brown, and

TAWNY-SHOULDERED PODARGUS.—*Podargus numeralis.*

spotted here and there with square white marks, the squares being mostly set with one of the angles upwards, in lozenge fashion. The head is very hawk-like, owing to the large hooked beak and the manner in which the feathers are set on the top of the head. The Guacharo was discovered by Humboldt.

THE members of the genus Podargus are chiefly remarkable for the great width of their beaks, which at their base are broder than the forehead of the bird. The tip of the beak is hooked, and the upper mandible overlaps the lower at its edges. They are all handsome birds, and many of them are of considerable dimensions.

The TAWNY-SHOULDERED PODARGUS is a native of Australia, and an admirable account of the bird, as well as good figures, may be found in Gould's magnificent work on the Birds of Australia.

This bird is one of the drowsiest of creatures, being less easily roused by day than any other slumberer of night. All the day long it sits sleeping upon a branch, its body crouched closely to the bough, its head buried amid the masses of soft feathers upon the neck and shoulders, and its whole form as motionless as if it were carved out of the branch on which it reposes. It is worthy of notice that the Podargi always sit across the branch on which they rest, whereas the generality of Goat-suckers recline longitudinally upon the bough. This bird, however, is so quiet, and its sombre colour harmonizes so well with the bark of the branch to which it clings, that even by day it needs a quick and practised eye to discern its form.

These birds almost invariably sit close together in pairs, and they are so incorrigibly drowsy, that if one of the pair be shot, its mate will not be disturbed even by the report

I 2

of the gun or the fall of its companion, but sit quietly in its place, and may either be knocked down with a stick, shot with a second charge, or taken by hand as is most convenient. If pushed off the branch by a long rod, the Podargus can barely summon sufficient energy to save itself from falling to the ground, and flapping its wings languidly to the nearest bough, settles, and is almost immediately wrapped in sleep, thus practically carrying out the complaint of Dr. Watts' sluggard, " You have waked me too soon, I must slumber again." Sometimes, however, it is known to fly by day without being thus violently disturbed, but such instances of diurnal liveliness are extremely rare, and may be considered exceptional.

As soon, however, as the daylight fades out of the sky, the Podargus awakes from its stupor, and after a few shakings and plumings of feathers, becomes a most lively and animated bird, quick, light, and active in all its movements, and scarcely to be recognised as the same being which but a few hours ago seemed hardly able to move a head or limb without difficulty. At the earliest approach of nightfall the Tawny-Shouldered Podargus sets off on its travels in search of food, and chases the insects on which it feeds with great agility and perseverance. Sometimes it runs stealthily along a bough, and picks off the locusts and cicadæ, as they cling sleeping to the bark, or traces them into the hollow eucalyptus branches, where they pass the night, and pecks them out with such haste, that it swallows no small quantity of the decayed wood together with the insects.

It is not particularly good on the wing, its flights being but short and abrupt, owing to the comparative small size and concave form of the wings. This structure is found in all the birds belonging to this genus, and causes them to trust for their food rather to their power of detecting and capturing sleeping insects, than to their capabilities of pursuing their flying prey through the air. In this particular they differ greatly from the Goat-sucker of England, which is remarkable for its great powers and agility of wing, and taking almost the whole of its prey while flying. The stomach of this bird is found, when dissected, to be lined with hair, in a manner very similar to that of the English cuckoo. The voice is loud, hoarse, and rather startling when heard for the first time, although not so stridulous as that of the Guacharo.

The nest of the Tawny-Shouldered Podargus is of very slight construction, being made of little sticks laid upon the horizontal part of some convenient tree, usually the eucalyptus, or gum-tree as it is more commonly termed, and is nearly flat in shape. The eggs are white, and generally two in number; their length is nearly two inches. It is proved that the male takes his turn in sitting upon the eggs, and that while one bird is engaged in incubation, its mate always seats itself on a neighbouring bough and bears it company.

In its tinting this bird is extremely variable, some species being much gayer than others. A rich tawny hue is dashed liberally over the sombre brown which forms the ground tint of its body and wings, and is especially conspicuous on the shoulders, where it warms into a ruddy chesnut. The species may be easily distinguished from its relatives by the white spots which are scattered upon the top of its head.

ANOTHER very curious species of Podargus is that bird which is popularly known to the settlers under the title of " More Pork ! " because its curious cry forms a very excellent imitation of those words. Its more scientific name is CUVIER'S PODARGUS, or *Podargus Cuvieri.*

This is a smaller bird than the preceding, but resembles it in many of its habits, sleepiness among the rest. While slumbering, the Cuvier's Podargus throws its head backward, so that the top of the beak points upwards, and it sleeps so soundly that even the report of a gun will not disturb it from its repose. It is a very familiar bird, and is fond of approaching human habitations and feeding in the verandahs. As it is also in the habit of frequenting burial grounds and sitting on the tomb-stones, it is reckoned a bird of ill omen, and its visits are not at all encouraged. Like the preceding bird, it is extremely variable in the colours of its plumage.

A MUCH smaller but very beautiful example of the Podargi is the MOTH PLUMED PODARGUS, (*Podargus phalænoïdes*,) so called on account of the exquisitely soft tinting of its feathers, which bear a very close resemblance to the velvet plumage of the large moths. The moth-like hues are most perceptible upon the wing-coverts, which are thickly sown with markings that closely resemble the "eyes" found upon the wings of so many large moths, and that add so much to their beauty. These "eyes" are very like those which decorate the wings of the well-known Emperor moth of England, but are devoid of the brilliant colours of that insect, and are composed of different shades of grey and brown.

Lastly, the PLUMED PODARGUS deserves notice on account of the singular development of the feathers of the nostrils, which rise to a considerable height, and are richly spotted with black, brown, and white, forming a kind of plume over the forehead. It is a large and handsome bird, and is generally found in the deepest bushes upon the banks of rivers of New South Wales. Its scientific title is *Podargus plumíferus*.

The NEW HOLLAND GOAT-SUCKER is a very fine and beautifully marked bird, its plumage being richly mottled with black and brown upon its upper surface, while the under surface is rusty grey, curiously variegated with buff. The tail is barred with darker bands.

This bird is very owl-like in its appearance and many of its habits, and has therefrom received the popular title of OWLET NIGHT-JAR. In the shape of its head, and the steady upright carriage, it bears a great resemblance to the Coquimbo Owl, a likeness which is further carried out by the sharp, angry hiss which it emits when irritated. Like the owls, it also possesses the habit of twisting its head so that the beak is brought on a level with the spine.

NEW HOLLAND GOAT-SUCKER.—*Ægothéles Novœ Hollandiœ.*

The New Holland Goat-sucker resides in the hollow branches of the eucalypti, technically called "spouts" by the colonists. When the sportsman wishes to know whether a "spout" is occupied by one of these birds, he has nothing more to do than to administer a sharp tap to the branch with a stick or axe. Should the bird be at home, it runs quickly to the entrance, pops out its head, and, after surveying the intruder for a moment, retires into the seclusion of its domicile. It will repeat this process several times, but at last loses patience at the frequent interruptions, and fairly takes to flight. In these "spouts" the eggs are laid, being placed simply upon the soft decaying wood. The eggs are white in colour, and from three to five in number. There are generally two broods of young in the year. The bird is nocturnal, and its principal food consists of insects, mostly of the coleopterous order.

A CLOSELY allied species is worthy of notice on account of the very singular arrangement of colour upon its head.

This is the White-bellied Nightjar (*Ægothéles leucogaster*), which may be distinguished from all its relatives by the white hue of the under surface of its body, and the three broad dark bands which surround its head and the upper part of its neck. One of these bands is short, and is placed just above the beak, while the other two sweep in bold curves, being very wide upon the top of the head, and narrowing rapidly to the corner of the eye,

GREAT-EARED GOAT-SUCKER.—*Batrachóstomus auritus.*

where they cease. It is extremely shy, but withal a dull bird, and when alarmed flies leisurely to the nearest tree, and, perching upon a branch, turns round to reconnoitre the cause of its alarm. When it sits in this manner it has very much the aspect of a common hawk.

On account of a difference in the arrangement of the quill feathers of the wing, the birds which form the genus Batrachóstomus have been separated from the Podargi. The generic name is of Greek origin, and is very appropriate, as it signifies " frog-mouthed," and is analogous to the French name of Crapaud-volant, or Flying Toad, which is given to all the Goat-suckers. The birds of this genus are not equal in dimensions to the Podargi, and they are all inhabitants of the Indian Archipelago, instead of being found in Australasia.

The GREAT-EARED GOAT-SUCKER is chiefly remarkable for the extreme length of certain feathers which start horizontally from the upper part of the head, and are evidently analogous to similar structures in the eared or horned owls. The gape of the bill is wonderfully wide in this bird, and with its soft plumage, great round eyes, and large head, with its tufts of feathers at each side, it has a peculiarly owl-like aspect. The colour of its plumage is black, grey, buff, and brown, all curiously mottled and intermixed with each other in a manner almost similar to the Goat-sucker, and which cannot be described without a needless expenditure of time and space. It is a nocturnal bird, and seems to be very shy in disposition.

One of the largest examples of the Goat-suckers is the GRAND GOAT-SUCKER (*Nyctibius Grandis*), sometimes called the GREAT IBIGAU

This fine bird is nearly thirteen inches in length, and in its habits resembles closely the common Goat-sucker of Europe. It is a nocturnal bird, as may be gathered from its generic title of Nyctibius, a word derived from the Greek, and signifying Night-liver. It is stronger on the wing than the preceding birds, and is capable of chasing its insect prey as they fly through the air in the dusk of evening, or just before the dawn of morning. In its outward form there is little to attract notice, and its colouring is the same as that of its allies, excepting that some dark bars extend across the head, neck, and lower parts. It is a native of South America, and has been taken in Cayenne.

THERE are many well-known proverbs relating to the power of calumny, and the readiness with which an evil report is received and retained, notwithstanding that it has been repeatedly proved to be false and libellous. The common GOAT-SUCKER is a good instance of the truth of this remark, for it was called Aigothéles or Goat-sucker by Aristotle in the days of old, and has been religiously supposed to have sucked goats ever afterwards. The Latin word caprimulgus bears the same signification. It was even supposed that after the bird had succeeded in sucking some unfortunate goat, the fount of nature was immediately dried up, and the poor beast also lost its sight. Starting from

EUROPEAN GOAT-SUCKER.—*Caprimulgus Europœus.*

this report, all kinds of strange rumours flew about the world, and the poor Goat-sucker, or NIGHTJAR, as it ought more rightly to be called, has been invariably hated as a bird of ill omen to man and beast.

As usual, mankind reviles its best benefactors, for there are very few creatures which do such service to mankind as the Nightjar. Arriving in this country in the month of May or June, it reaches our shores just in time to catch the cockchaffers, as they fly about during the night in search of their food, and does not leave us until it has done its best to eat every chaffer that comes across its path.

The damage which is done by these brown-backed, white-ribbed, hook-tailed beetles is almost incredible, for they are not only extremely destructive in their larval states, but are scarcely less voracious when they have assumed their perfect form. Passing a life of three years or so below the level of the ground, the larvæ of the cockchaffer shear away the grass-roots and other subterranean vegetation with their scissor-like jaws, and are constantly busy in satiating the hunger of their huge stomachs, which occupy nearly the whole of the body of the grub. When they have passed through their earlier changes of form, the cockchaffers rise from the ground, and, taking to flight, settle upon the trees and devour the foliage just as they had previously fed upon the roots. Sometimes a whole series of trees may be seen, which have been entirely stripped of their leaves by the chaffers. I well remember seeing a row of trees that extended along a country road near Dieppe, that had been totally despoiled of their foliage, and which stretched their naked branches abroad as if they had been blasted by the destroying breath of the Simoom.

The Nightjar also feeds on moths of various kinds, and catches them by sweeping quickly and silently among the branches of the trees near which the moth tribes most love to congregate. While engaged in their sport, they will occasionally settle on a bank, a wall, a post, or other convenient perch, crouch downward until they bring their head almost on a level with their feet, and utter the peculiar churning note which has earned for them the name of Churn-Owls, Jar-Owls, and Spinners. Their cry has been rather well compared to that sound which is produced by the larger beetles of the night, but of course much louder, and with the addition of the characteristic "chur-r-r!—chur-r-r!" Sometimes, although but seldom, the Nightjar utters its cry while on the wing. When it settles, it always seats itself along a branch, and almost invariably with its head pointing towards the trunk of the tree.

There is also a strange squeaking sound which is emitted by the Nightjar while playing round the trees at night, and which is supposed to be a cry of playfulness, or a call to its mate.

Although rather a shy bird, and avoiding the presence of mankind, it is bold enough on occasion, and when it finds an abundance of food, or when it desires to defend its young, it cares little for any strange form, whether of man or beast. On one occasion, while I was travelling to Paris by railway, a Nightjar accompanied us for a considerable distance, hunting after the flies that are so plentifully attracted by a moving train of railway carriages. Should an intruder come too near its nest, the Nightjar will sweep repeatedly over his head, producing a sharp ruffling sound, intended to terrify him, and formed, as some supposed, by striking the wings smartly together over the back.

Unlike the Falconidæ, the Goat-sucker catches its prey, not with its claws, but with its mouth, and is aided in retaining them in that very wide receptacle, by the glutinous secretion with which it is lined, and the "vibrissæ" or hair-like feathers which surround its margin. On an examination of the foot of this bird, the claw of the middle toe is seen to be serrated like the teeth of a comb, a structure which has never yet been satisfactorily explained, notwithstanding the various theories which have been put forward concerning its use. The hind toe of each foot is very mobile, and can be brought round to the remaining toes, so that all the claws take their hold in the same direction. Apparently, this structure is intended to enable them to run along the branches of trees, in their nocturnal chase after beetles and other insects.

This bird is spread over Europe, and has been captured in Africa, whither it retires in order to pass the winter. Specimens have been taken in Ireland, and I once saw one of these birds which had been shot close to Oxford.

The Nightjar makes no nest, but choosing some sheltered hollow under the shade of a grass tuft, a bunch of fern, bramble, or other defence, there lays two eggs on the bare ground. The colour of the egg is greyish-white, plentifully mottled with pale buff and grey. The young are very similar to those of the cuckoo. The plumage of the Nightjar is very rich in its colouring, the tints of buff, grey, black, white, brown, and chestnut, being arranged in pleasing but most intricate patterns, and easier to be understood from a pencil illustration than a description of the pen. The sexes are very similar in their plumage, but the male may be distinguished from his mate by a number of oval white spots which are found on the inner side of the first three quill feathers of the wings, and upon the outside quill feathers of the tail. The length of the adult bird is about ten inches.

A VERY remarkable form of plumage is seen in the LYRE-TAILED GOAT-SUCKER.
This beautiful bird is a native of Columbia, and is notable for the extraordinary development of the outer tail feathers. Although the bird itself is by no means large, very

little exceeding the common English Nightjar in dimensions, the total length of an adult male Lyre-tailed Goat-sucker is nearly three feet. Indeed, the general contour of the body and plumage remind the observer strongly of the resplendent Trogon, a bird which will very shortly be described and figured. The general colour of this species is the mottled dark and light brown which is universal among the Goat-suckers, but is diversified by a band round the neck of rich chestnut. The primaries are nearly black, with the exception of a few chestnut spots scattered irregularly upon their necks. The extremely elongated tail-feathers are deep brown-black, edged with a warm band of pale brown upon the inner web. The outer web is hardly a quarter of an inch wide, while the inner is almost an inch and a half in width. Several feathers of the tail project for some distance, and lie upon the base of the elongated feathers.

The CAROLINA GOAT-SUCKER is more popularly known under the title of CHUCK-WILL'S-WIDOW, a name which it has earned in consequence of its repeated utterance of a cry that exactly resembles those words.

This pretty and interesting bird resides in the deepest ravines, swamps, and pine ridges, where it can not only obtain shelter and a convenient nesting-place, but is also sure of finding a plentiful supply of insect prey. It prefers to roost in the hollows of decayed trees, or other retired spots, and is not unfrequently found tenanting the same habitation together with a large company of bats.

The nest is as open and undefended as is the case with most of the Goat-suckers, and the eggs and young would probably be exposed to considerable danger, were it not for the wonderful care and ingenuity displayed by the parents when their offspring are in danger. The following account of the behaviour of the bird when it fears that its nesting-place has been discovered, is given by Audubon.

"When the Chuck-Will's-Widow, either male or female, for each sits alternately, has discovered that its eggs

LYRE-TAILED GOAT-SUCKER.—*Caprimulgus Lyra.*

have been touched, it ruffles its feathers and appears extremely dejected for a minute or two, after which it emits a low murmuring cry, scarcely audible to me as I have lain concealed at a distance of eighteen or twenty yards. At this time I have seen the other parent reach the spot, flying so low over the ground that I thought its little feet must have touched it as it skimmed along. After a few low notes and some gesticulations, I have witnessed each take an egg into its large mouth, and both fly off together, skimming closely over the ground, until they disappeared among the branches and trees. But to what distance they remove their eggs I have never been able to ascertain, nor have I ever had an opportunity of witnessing the removal of the young."

M. Audubon proceeds to say that the birds do not carry away the eggs unless they have been touched, and that if the parent bird be merely frightened from her nest by the sudden shout of a stranger, she will return to her eggs as soon as the intruder has left the spot.

It is a nocturnal bird, seeming to lose all its energies during the hours of daylight, and to recover its alertness at the first coming of dusk. The full light of day appears to be exceedingly painful to its eyes, and when suddenly taken from the dark resting-places in

CAROLINA GOAT-SUCKER.—*Caprimulgus Carolinensis.*

which it loves to dwell, into the glare of daylight, it seems to be quite confused, and is unable to escape by flight, contenting itself with snapping its bill, hissing and puffing out its feathers after the manner of the owls. It is a migratory bird, arriving in the United States in March or April, according to the state of the weather and the latitude of the locality, and remaining in these parts until the end of August, when it takes flight for Mexico and other warm regions, and there passes the winter.

The plumage of this pretty bird is like that of the ordinary Goat-suckers, both in colour and texture, but is notable for the bright motley yellowish-red with which the dark-brown feathers of its head and back are plentifully but minutely sprinkled, as well as for the bars of deep sooty-brown and yellowish-red which extend across the wings and tail. The three outer quill feathers of the tail are white upon their inner webs, a band of greyish-white extends across the front of the neck, and the whole of the under parts are sooty-black profusely covered with little yellowish-red markings.

THE SCISSOR-TAILED NIGHTJAR is also worthy of a short notice, on account of the peculiar structure of the rectrices, from which it derives its name. In the male bird, the

LONG-TAILED GOAT-SUCKER. — *Scotornis Climacurus.*

two central quill feathers of the tail are greatly elongated, so as to give the tail a deeply forked appearance. As the bird flies it has a curious habit of closing and opening these feathers alternately, an action which bears a potent resemblance to the shutting and opening of a pair of scissors, and has given rise to the name by which it is popularly known. The plumage of this species is in colour like that of other Goat-suckers, with the exception of a rather wide band of sooty-red which extends partly round the neck. The tail of the female is comparatively small, and hardly exceeds the ordinary length. It is a native of tropical America.

The LONG-TAILED GOAT-SUCKER is one of the most conspicuous of this group of birds; the long and slightly curved feathers of its tail giving it some resemblance in outline to the European Cuckoo. The body of this species is by no means large, but the bird appears to be considerably above its real dimensions on account of the great length of its tail.

In the colour of its plumage it is rather a handsomer bird than the generality of Goat-suckers, owing to the quantity of white which is laid in bold markings on several parts of its feathers. The chin is white, as is also a streak that passes from the corner of the mouth. A broad band of white passes across the extremities of the lesser wing coverts, and there is a smaller band of cream colour upon the tips of the greater coverts. Another beautifully white band is drawn across the middle of the first six primary feathers, and the remaining primaries have a spot of white on their tips. The rest of the plumage is variegated with black and brown, warmed here and there with a more ruddy hue. The tail is also white in several parts, and has a number of very narrow dark bars across the middle pair of feathers. The Long-tailed Goat-sucker is an inhabitant of Western Africa.

THE LONG-WINGED GOAT-SUCKER is a scarcely less wonderful bird than the Lyre Goat-sucker which has already been mentioned, the extraordinary development of feather being in the present case transferred from the tail to the wing.

The colour of the Long-winged Goat-sucker is generally of the usual tints of chestnut and brown, but is diversified by a broad greyish-white irregular band, which passes across

LONG-WINGED GOAT-SUCKER. — *Caprimulgus vexillarius.*

the centre of the secondaries, and part of the base of the primaries. From the white band, a dark-brown stripe runs towards the back, the feathers composing it being tipped with white. The elongated feathers of the wing increase the length of the bird to two feet or even more, and their colour is very dark brown on the outer web, and greyish-white on the inner. The Long-winged Goat-sucker is an inhabitant of Western Africa.

The LEONA NIGHTJAR affords another example of the singular form which plumage so often takes without any apparent object.

In the male bird, a pair of very long and very elastic feather shafts rise from the middle of the wing-coverts, and extend to a length of eight-and-twenty inches, according to the individual. These shafts are totally destitute of barbs, except at the extremity, where they suddenly give out a broad web of four or five inches in length. The transition from the bare shaft to the broad web is so abrupt that the bird looks as if it had originally possessed a pair of very long perfect feathers, which had been stripped with the exception of a few inches at their extremities. The shafts are very slight indeed, and as the webbed ends are easily acted upon by the wind, they are continually moving, and float about in the breeze in a most graceful manner. The inner web of these curious feathers is nearly two inches in width at its broadest part, while the outer web is barely one fourth of that measurement.

The object of these curious appendages is not known. They are only found in the male bird, and evidently bear an analogy to the train of the peacock and the long tail-feathers of the pheasant among the birds, as well as to the beards, horns, tusks, manes, and

LEONA GOAT-SUCKER.—*Macrodiptex longipennis.*

similar masculine appendages of male quadrupeds. The plumage of the Leona Nightjar is very prettily marked with spots and bars of rusty-red and black upon the usual brown ground. Every primary feather possesses nine rusty-red spots, and as many of a black hue, and there are many other spots and bars scattered over the body and wings. There is a considerable amount of creamy white upon the scapularies, a few white mottlings upon the throat of the male, and a reddish-white stripe down the outer web of the two exterior tail-feathers.

The beak of this species is not so powerful as in many of its relatives, but the vibrissæ are long and well developed. The wings are long, overpassing the tip of the tail while the bird is at rest, and showing that the powers of flight are considerable. The bird is not a large one, measuring only eight or ten inches in total length. It is a native of Western Africa.

The VIRGINIAN GOAT-SUCKER, MOSQUITO HAWK, or NIGHT-HAWK, inhabits the northern parts of the American continent, and in the summer months is seen even in the Arctic regions.

It is not so exclusively nocturnal a bird as most of the Goat-suckers, but will voluntarily leave its home on a cloudy day, and commence its task of hunting after flies, moths, beetles, and other insects on which it feeds. It is a bird of vigorous and active wing, and follows its insect prey even into the loftier regions of air, where it seems as much at its ease as the swift or swallow. While chasing the insects, the Night-hawk constantly utters a shrill squeaking kind of cry, reminding the spectator of the screaming cries of the common swift. It also has a curious habit of hovering over its mate as she sits on her eggs, darting down upon her from a considerable elevation, and then suddenly sweeping up again with a loud booming sound, occasioned either by the wings or by the vocal organs. This strange manœuvre is constantly repeated, and appears to be performed for the purpose of showing a delicate attention to the sitting bird, and amusing her during her long and tedious task.

The eggs of this bird are placed on the bare ground, and when a stranger happens to approach the spot where they are lying, the parent bird immediately flings herself in the way of the intruder, and by tumbling about in front of him, as if she had broken a wing or otherwise disabled herself, endeavours to induce him to leave the sacred spot and give chase. If she succeeds in decoying him from the locality, she darts into cover, and takes the earliest opportunity of returning quietly to her nest. Many birds pursue this curious contrivance, the common peewit or lapwing being a very familiar instance among ourselves.

The eggs of this species are generally two in number, greyish-white in colour, covered with a number of streaks and dashes of brown. The young are odd little creatures,

VIRGINIAN GOAT-SUCKER.—*Chordeiles Virginianus.*

clothed with a quantity of fine brownish-grey down, and of a very indeterminate shape. The tail of this bird is forked, and the long wings overpass the tail when they are closed. The bill is rather small. The colour of the Night-hawk is rather different from that of the ordinary caprimulgidæ, being notable for a greenish gloss upon the dark red-brown of the general plumage. A number of yellowish spots occur upon the head, neck, and wing-coverts, there is a well-defined white band across the middle primaries, together with a white patch on the throat, and an irregular stripe above the eyes. The total length of this bird is between nine and ten inches.

The WHIP-POOR-WILL also belongs to this group of birds, and is familiarly known by the peculiar melancholy cry, which very much resembles the other odd names by which it is called.

THE birds which belong to the genus Podargus have less of the peculiar Goat-sucker aspect than any of their relatives, owing to the comparative paucity and scantiness of the vibrissæ, the naked legs, the shortness of the tail, and the comparatively small gape of the mouth. The wings of these birds are extremely long, and the powers of flight are very considerable.

The NACUNDA GOAT-SUCKER, one of the best examples of this genus, is a diurnal bird, like the Virginian Goat-sucker, and excepting on very bright days, may be seen abroad even at noon, chasing the insects at a great elevation, and wheeling and diving after them with the activity of the swallow. It also descends close to the ground, and pursues the gnats and other aquatic flies as they rise from the surface of the water, or attempt to settle for the purpose of depositing their eggs. It is a handsome bird, possessing a brilliantly variegated plumage. On the upper parts of the body, the feathers are generally of a

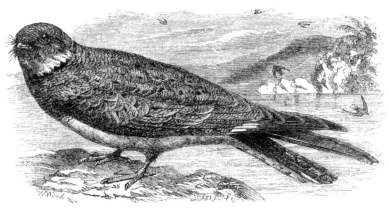

NACUNDA GOAT-SUCKER.—*Podager nacunda.*

greyish-brown variegated with large spots of black-brown, edged with rusty-red, and mottled here and there with the same tint. The under parts of the body are nearly white, the chin being tinted with cream-yellow, and covered with a few greyish bars, and the breast white, striped with grey-brown horizontal lines. The tail is beautifully mottled with yellow and brownish-black, and crossed by a number of black bars, sprinkled with dots of brown. The Nacunda Goat-sucker is a native of Brazil and Paraguay.

SWALLOWS.

THE close-set plumage of the SWALLOW TRIBE, their long sickle-like wings, their stiff, firm tail, forked in most of the species, and their slight legs and toes, are characteristics which mark them out as birds which spend the greater part of their existence in the air, and exercise their wings far more than their feet.

They all feed upon insects, and capture their prey in the air, ascending at one time to such a height that they are hardly perceptible to the naked eye, and look merely like tiny dots moving upon the sky, while at other seasons they skim the earth and play for hours together over the surface of the water, in chase of the gnats that emerge in myriads from the streams, during the time and season when they assume the perfect form. The gape of the mouth is therefore exceedingly great in these birds, reaching as far as a point below the eyes. The bill itself is very short, flattened, pointed, slightly curved downwards, and broad at the base.

GROUP OF SWALLOWS.

The group, which is scientifically termed the Hirundínidæ, is a very large one, and is divided into two lesser groups, the members of one being classed together under the title of Swifts, while the others are known by the name of Swallows. With the former birds we have first to deal.

THE Swifts, technically called the " Cypselinæ," or Cypseline birds, are readily distinguished from the Swallows by the very great comparative length of the two first primary feathers of the wing, which are either equal to each other, or have the second feather longer than the first. The secondaries are remarkably small, being nearly concealed under the coverts. There are ten primaries in the wing, and the same number of quill feathers in the tail.

The AUSTRALIAN SPINE-TAILED, NEEDLE-TAILED SWALLOW, or SWIFT, as it ought rather to be called, is the largest of all the Swallow tribe, measuring no less than twenty-eight inches in total length when the wings are closed, and twenty inches to the extremity of the tail.

The title of NEEDLE-TAIL has been given to this bird on account of the curious formation of the tail-feathers, which are short and even, and have their extremities devoid of web, so as to form a row of sharp, short points, as if a number of needles had been thrust through the shaft of each feather. This species is a native of Australia, and as may be seen from the following extract from Mr. Gould's work on the Birds of Australia, is very interesting and curious in its habits, and worthy of especial attention from any naturalist who may chance to have an opportunity of watching it.

"The keel or breast-bone of this species is more than ordinarily deep, and the pectoral muscles more developed than in any other bird of its weight with which I am acquainted. Its whole form is especially and beautifully adapted for aërial progression, and, as its lengthened wings would lead us to imagine, its power of flight, both for rapidity and extension, is truly amazing. Hence it readily passes from one part of the country to another, and if so disposed may be engaged in hunting for flies on the continent of Australia at one minute, and in half-an-hour be similarly employed in Van Diemen's Land.

AUSTRALIAN NEEDLE-TAILED SWALLOW.—*Acánthylis caudacúta.*

So exclusively is this bird a tenant of the air, that I never in any instance saw it perch, and but rarely sufficiently near the earth to admit of a successful shot; it is only late in the evening and during lowery weather that such an object can be accomplished. With the exception of the crane, it is certainly the most lofty as well as the most vigorous flier of the Australian birds. I have frequently observed in the middle of the hottest days, while lying prostrate on the ground with my eyes directed upwards, the cloudless blue sky peopled at an immense elevation by hundreds of these birds, performing extensive curves and sweeping flights, doubtless attracted thither by the insects that soar aloft during serene weather. Hence, as I have before stated, few birds are more difficult to obtain, particularly on the continent of Australia, where long droughts are so prevalent; on the contrary, the flocks that visit the more humid climate of Van Diemen's Land, where they necessarily seek their food nearer the earth, are often greatly diminished by the gun during their stay.

I regret that I could ascertain no particulars whatever respecting the nidification of this fine bird, but we may naturally conclude that both rocks and holes in the larger trees are selected as sites for this purpose, as well as for a roosting-place during the night. Before retiring to roost, which it does immediately after the sun has gone down, the Spine-tailed Swallows may frequently be seen either singly or in pairs, sweeping up the gullies or flying with immense rapidity just above the top of the trees, their never-tiring wings enabling them to perform their evolutions in the capture of insects, and of sustaining themselves in the air during the entire day without cessation."

The Australian Needle-tailed Swallow is a most beautiful bird in its colouring, as well as handsome in size and elegance of shape. The general colour of this bird is olive-brown, exceedingly dark in the chest and abdomen, and washed with a dark green tinge upon the back of the head, the wings, and the tail. Before the eyes there is a velvet-black patch, and a large portion of the under parts of the body is white, including the chin, throat, under tail-coverts, and the inner web of the secondaries. There is also a white band extending across the forehead.

A CLOSELY allied species is the Aculeated Swallow (*Acanthylis Pelasgia*).

This bird is a native of many parts of America, being found in Louisiana, Carolina, and even in Pennsylvania. It is chiefly remarkable for the ingenuity which it exhibits in the construction of its singular nest. Choosing some convenient locality, such as a rocky crevice, or the unused chimney of a house, the bird commences its labours by putting together a slight platform of dry twigs, which it cements together with certain vegetable gums. So large is this platform, that it sometimes causes considerable inconvenience to the inhabitants of the house where the Swallow has taken up its residence, as it completely stops up the orifice of the chimney. Upon this platform is formed a kind of cradle nest, also composed of small twigs, which are woven into a kind of rude basket, and also cemented together.

The eggs of this species are very large in proportion to the dimensions of the parent bird; their colour is greyish-white, streaked and spotted with black and brown towards the large end. The general colour of the plumage is brown, and the throat is whitish-grey.

THE birds which belong to the genus Macropteryx, or Long-winged Swifts, possess wings of very great comparative length, owing to the development of the two first primary feathers of the wings, which are nearly equal to each other in length, the second being slightly the longer of the two. They may, however, be easily distinguished from the members of the genus Acanthylis, by the formation of the tail, which, instead of being composed of feathers of nearly equal length, and tipped with sharp points, is forked after the manner usually observed in the smaller tribes. The feet are furnished with four toes, the hinder toe being directed backward.

The KLECHO SWALLOW, or LARGE-WINGED SWIFT, as it is more rightly termed, as it belongs to the ranks of the Cypseline birds, is considered of great value by ornithologists as supplying a link in the chain that connects the Swifts with the Swallows. Like the Swifts in general, its feet are well adapted for climbing, and supplied with firm curved claws; and like the Swallows, its hinder toe is directed backwards, and cannot be brought round in a line with the remaining toes. It is a very handsome bird, in its colouring nearly equalling the Needle-tailed Swift which has just been described. The colour of the upper parts of the body is deep brown, through which runs a strong tinge of green that gives a beautiful glossy aspect to that part of the plumage. The throat and breast, together with the under tail-coverts, are very light grey, and the abdomen, part of the scapularies, and a well-defined streak over the eye, are white.

The bird is a native of many parts of India, and has been taken in Java and Malacca.

The true SWIFTS, of which England affords two examples, one very familiarly known, and the other a very rare and almost unnoticed species, are remarkable for the feathered tarsus, the long wings, and the peculiar form of the feet. In this member, all the toes are directed forward, a structure which is admirably adapted to the purpose which it fulfils. The Swifts build their nests, or rather lay their eggs, for the nest is hardly worthy of the name, in holes under the eaves of houses, or in similar localities, and would find themselves greatly inconvenienced when seeking admission into their domiciles, but for the shape of the feet, which enables them to cling to the slightest projection, and to clamber up a perpendicular surface with perfect ease and safety. In one species, the White Collared Swift (*Cypselus Cayanensis*), the feet are clothed with feathers to the base of the claws.

The WHITE-BELLIED, or ALPINE SWIFT is the largest of our British Hirundinidæ, being rather more than eight inches in total length. It is but rarely found in the British Isles, but is common enough on the continent of Europe, and in many parts of Africa and Asia.

Several specimens have been taken in England, and in Mr. Thompson's work on the Natural History of Ireland, three examples are noticed that had been shot in that country, one near Dublin, another in the county Cork, and the third a few miles from land off Cape Clear. Unlike the common Swift of England, which is possessed of a loud and stridulous note, the Alpine Swift is sweet of voice; its cry, although loud, being musical in its intonation. The popular name of this bird is given to it on account of the white hue of the under portions of its body, the only exception being a broad

KLECHO SWALLOW.—*Macrópteryx Klecho*

dusky bar across the breast. The toes are brown with an orange tint, and the black beak is longer than that of the common Swift. The general colour of its plumage is brown. The nest of this bird is made in crevices of lofty cliffs or buildings, and is composed of straw, hay, moss, and other substances, connected firmly together with a glutinous secretion furnished by certain glands, and rendered very hard and firm when the cement is dry. The eggs are four or five in number, white, and very long in proportion to the breadth.

K 2

The following interesting account of the habits of this bird is given by Mr. Thompson :—

" The first place I met with the Alpine Swift was almost ten miles to the north of Naples, on the 12th of August, 1826, when a great number were observed associated together in flight, at a high elevation. Their evolutions in the air were similar to those of a common Swift. Independently of their superior size, which at once distinguishes them from that bird, the white colour of a portion of the under plumage, from which they have received the name of White-bellied Swift, is conspicuous, even when the bird is at a considerable altitude.

When on the continent in 1841 with my friend Professor E. Forbes, this species was first seen by us on the 9th of April, as we descended the Rhone, from Lyons to Avignon. About half way between these cities, several appeared flying over the river, and a few at all suitable places thence to Avignon. On the morning of the 28th of April, as we entered the splendid bay of Navarino, great numbers appeared careering high overhead. When walking through the pretty town of the same name, later in the day, Alpine

WHITE-BELLIED SWIFT.—*Cypselus Melba.*

Swifts were observed flying very low over the streets and houses, though the weather was delightfully warm and fine. On my visiting the island of Sphacteria, the western boundary of the bay, on the 29th, these birds were very abundant. The attraction here was a range of noble precipitous cliffs rising directly above the sea, at the western side of the island. These Swifts inhabited the cliffs, which are similar to those tenanted by the common species in the north of Scotland.

Although the day was as fine and as warm as our northern summers ever are, these birds, as I walked along the top of the cliffs, swept about low and in numbers, occasionally within a few yards of my head. This remark is made from the circumstance of the common Swift being generally high in the air in fine weather; we do, however, occasionally observe it sweeping near the earth at such times. Though larger, they in general appearance and flight strongly resemble the common Swift: they are very noisy, almost constantly uttering a loud twitter, beside which, they occasionally give a brief scream, nowise resembling the long drawn and shrill cry of the common species. Towards the end of May, I saw a few Alpine Swifts at Constantinople, wheeling about the heights of Pera, and near the high tower of Galata, in which they probably build. In the month of June, I met with this species at the island of Paros, and about the Acropolis of Athens.

Throughout this town, the common Swift was more frequently seen than the *Cypselus Alpinus*, and at one locality only did they both appear—this was at Constantinople, where the former species was abundant, and a few of the latter were observed. This seemed rather remarkable, as in no scene did I meet with the one species, in which the other would not have appeared equally at home. The only difference in their habits which struck me, was, that the Alpine Swift is apparently more partial to cliffs than buildings, the common Swift more partial to artificial structures than to rocks."

THE WHITE-COLLARED SWIFT (*Cypselus Cayanensis*), to which bird a passing reference has already been made, is a native of the Brazils, and is easily to be distinguished by the peculiarity of colouring from which it derives its name. The general tint of the plumage is the deepest violet-blue, so deep, indeed, that except in certain lights it appears to be velvet-black. Round the neck runs a band or collar of the purest white, the two contrasting tints having a remarkably fine effect.

The nest of this species is very singular in its form, being a short truncated cone, the bottom being about five inches in diameter, and the middle about three inches. The material of which it is built is dogs-bane, and the young are defended from the air by a quantity of the soft woolly down that grows on that plant, and it is pressed into the cavity so as to form a sort of plug. The nest is usually made within houses, after the common fashion of many swallows.

DEVOID of all pretensions to the brilliantly-tinted plumage which decorates so many of its relations, and clad only in sober black and grey, the SWIFT is, nevertheless, one of the most pleasing and interesting of the British birds ; resting its claims to favourable notice upon its graceful form, and its unrivalled powers of wing.

There are very few birds which are so essentially inhabitants of air as our common Swift, which cuts the atmosphere with its sabre-like wings with such marvellous ease and rapidity, that at times its form is hardly discernible as it shoots along, and it leaves the impression of a dark black streak upon the eyes of the observer. The plumage of this bird is constructed especially with a view to securing great speed, as may be seen by an inspection of the closely set and firmly webbed feathers with which the entire body and limbs are clad. The muscles which move the wings are enormously developed, and in consequence the breast-bone is furnished with a remarkably strong and deep "keel."

The flight of the Swift is quite peculiar to the bird, and cannot be mistaken even for that of the swallow by any one who has a practical acquaintance with the habits of the two species. The Swift does not flap its wings so often as the swallow, and has a curious mode of shooting through the air as if hurled from some invisible bow, and guiding itself in its headlong course by means of its wings and tail. While flying, the Swift makes very great use of its tail, a habit which has been admirably described by Mr. Thompson :

"It was highly interesting to watch their motions as they flew noiselessly a few yards above my head. The tail would at one moment be drawn to a point, the root appear square at the end ; would then present a 'tender fork,' and the root its full formation ; again, it would be expanded to the uttermost, with the feathers simply touching at their margins, and the whole tail appearing so membraneous, that the light shone through it ; lastly, it would be thrown into the form of an arch, which had a singular effect, and generally when thus exhibited, the whole body was like a well-strung bow, an appearance which was several times observed with very high interest. Within a few seconds of time all these appearances were assumed by the same bird."

After making some further remarks on the subject, the same writer proceeds to observe that a similar habit is also found in the common swallow, as seen by himself when looking down upon the birds from the summit of Mount Pagus. "The swallows, as they gently floated on the bosom of the air a few yards beneath, exhibited the tail expanded to such a degree, that the beautiful white portion towards its base was quite conspicuous ; presenting in this respect so great a difference from its ordinary appearance, that I did not feel certain at the moment of their being our own common bird."

There are few hours of the day when this ever active bird may not be seen on the wing, employed either in sport with its companions, or in pursuit of the insects on which it feeds, and of which it carries such numbers to its young. Several authors have said that the Swift prefers the morning and evening for its aerial evolutions, remaining quietly in the dusky recesses where it has built its nest during the fierce heat of the summer's noon. The bird is also reported to retire to the same retreat while rain falls and wind blows, and to rest at home until the weather changes for the better. As far as my own personal observations go, both these assertions are too sweeping, if not entirely erroneous. The Swift has a special love for the bright heat of a July noon, but it must be sought, not near its usual haunts, but far up in the sky, where it may be seen like a little black mote against the blue heavens, and hardly visible except to experienced observers.

As to the alleged habit of keeping under cover during a storm, it has been decidedly contradicted by Mr. Thompson, who has observed the Swifts engaged in the pursuit of their prey during stormy as well as in fine weather. I am able to contradict this assertion from the events of the day on which I write—July 23d, 1860. The whole of the day has been most stormy, the rain falling heavily and without cessation, and the wind howling furiously in intermittent gusts. While standing at the window at Margate, and watching the black clouds come sweeping over the sky, I saw numbers of Swifts dashing through the air at a very low elevation, seldom rising above the roofs of the adjoining houses, and especially affecting some small gardens and the fruit trees therein planted. The same fact was observable on the two preceding days, but as an occasional respite from the rain was enjoyed on those days, the presence of Swifts was not so remarkable.

This indefatigable bird is an early riser, and very late in returning to rest, later indeed than any of the diurnal birds. Though engaged in flight during the live-long day, the Swift appears to be proof against fatigue, and will, during the long summer days, remain upon the wing until after nine in the evening. As the days become shorter, the Swift is found to retire earlier, but during its stay in this country, it is almost invariably later than other birds, sometimes being on the wing together with the owl. Indeed, the air seems to the Swift even a more familiar element than the earth, and the bird is able to pass the whole of its life, and to perform all the bodily functions except those of sleep and repose, while upborne on the untiring pinions with which it is furnished. The Swift that has a nest to take care of is forced to descend at intervals for the purpose of supplying its family with food, but except when urged by such considerations, it is able to remain in the air for many successive hours without needing to rest.

The Swifts may generally be found near buildings, rocks, and cliffs, for in such localities they build their nests, and from their homes they seldom wander to any great distance, as long as they remain in the country. These birds appear to be singularly susceptible to home influence, and will return year after year to the same nest, attracted by some subtle but most powerful influence, which guides them across sea and land to the spot in which they had first settled themselves, and cherished their young families. One of these birds was marked in order to ascertain its powers of returning to the same spot, and was observed to make its appearance regularly for seven successive years.

In general, the Swift loves to build its nest in a hole under a roof, whether slated, tiled or thatched, preferring, however, the warm, thick straw-thatch to the tile or slate. Sometimes it makes a hole in the thatch, through which it gains access to the nest, but in most instances it makes use of some already existing crevice for that purpose. In all cases, the nest is placed above the entrance, and generally may be found about eighteen inches or two feet from the orifice. Even by the touch, the eggs of the Swift may be discerned from those of any other bird, as their length is singularly disproportionate to their width.

The activity with which the Swifts enter their holes is really remarkable, and is well worthy of observation. The bird stoops suddenly from its aerial flight, and with a loud scream shoots under the eaves of the house in which it has fixed its residence. Turning quickly aside, it glides towards the orifice of the tunnel, and settling for a moment with closed wings, runs nimbly into the hole, like a rat or mouse.

When the Swifts have become accustomed to human beings, they become wonderfully indifferent to their presence, and will permit their movements to be watched without displaying any signs of fear. I well remember a certain street which was copiously favoured by the Swifts, who congregated in such great numbers, that they became a positive nuisance on account of the continual screaming which they kept up. The houses were mostly of a very ancient fashion, and their eaves were so low, that a man could introduce his hand into the Swifts' tunnels merely by standing on a chair. Yet the birds cared nothing for their apparent danger, even though their nests were several times robbed of their contents. At one time, the small boys, who abounded in the neighbourhood, took a fancy to manufacture bows and arrows, with which they kept up a persevering fire upon the Swifts, as they went to and fro upon their avocations, or visited and returned from their nests. The birds, however, looked upon these weapons with supreme contempt, and never troubled themselves in the least about them.

The sound which these birds utter is of the most piercing description, and can be heard at a very great distance, thus betraying them when they are hawking after the high-flying insects at such an altitude that their forms are hardly perceptible to the unassisted eye. Whether the Swift uttered this cry as a call or serenade to its mate, was

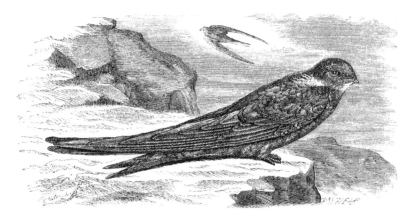

SWIFT.—*Cypselus apus.*

once a mooted point, but is now clearly settled. The bird certainly uses its cry when it is far away from its mate, but it also employs its voice in giving encouragement to its mate as she sits upon her eggs in the dark recesses of her home. Darting closely by the orifice of the hole, the Swift gives forth a loud and piercing scream, as a signal of his presence, and is answered by a soft chattering twitter from the female bird, in acknowledgment of his courtesy. While thus employed, the agility with which it sweeps along by the loved spot is truly marvellous, and the manner in which it shoots round any projecting angle is almost incredible to one who has not observed this bird while performing this feat.

The nest is a very firmly made but yet rude and inartificial structure. The materials of which it is made are generally straw, hay, and feathers, pieces of rag, or any soft and warm substance which the bird may find in its rambles, and when woven into a kind of nest, are firmly cemented together with a kind of glutinous substance secreted by certain glands. In Norway and Sweden the Swift builds in hollow trees. The eggs are from two to five in number, not often, however, exceeding three, and in colour they are pure white. The shell is very fragile, and the inexperienced collector will often break the shell in attempting to remove the contents.

The young of the Swift are rather later in appearing than most young birds, seldom being hatched until the end of June, and often delaying their advent until the beginning of July. While in their juvenile plumage, they may be distinguished from the older birds by their white chins and the yellowish white spots which appear on various parts of the body. Owing in all probability to the lateness of the hatching time, there is only one brood in the year.

In this country the Swift pays but a very short visit, as the bird evidently requires a very high temperature, and is forced to depart as soon as the weather becomes chilly. Africa seems to be the true home of this species, and in various parts of that continent the Swifts may be found throughout three-fourths of the year, forming a curious link between countries so far removed from each other. Generally the Swifts leave England by the end of August, but there are often instances where a solitary bird has delayed its voyage for some good reason. A pair of Swifts have been known to remain in England until the beginning of October, having overstayed their associates by a period of seven weeks, for the purpose of remaining with a late brood of young, and acting as guides in their long journey to Africa. Sometimes, however, the migrating instinct has conquered the parental feelings, and the old birds have taken flight in company with their neighbours, leaving their unfortunate young to perish miserably in their nests.

It appears that the cause of a brood being delayed until so late in season, may be found in the fact, that an accident had occurred to the former brood, and that the reproductive instinct of the birds forced them to fulfil their destiny, and to rear a pair of living young, in spite of the bitterness of the season. Such, at all events, was the case with the birds, whose strange behaviour was so well recorded by Gilbert White. Even in this instance the male bird yielded to the migrating impulse, and flew away with or after his companions, leaving his mate to the hard task of bringing up her young without his aid :

"Our Swifts in general withdrew this year about the first day of August, all save one pair, which in two or three days was reduced to a single bird. The perseverance of this individual made me suspect that the strongest of motives, that of an attachment to her young, could alone occasion so late a stay. I watched, therefore, until the twenty-fourth of August, and then discovered that under the eaves of the church, she attended upon two young, which were fledged, and now put out their white chins from a crevice. These remained till the twenty-seventh, looking more alert every day, and seeming to long to be on the wing.

After this day they were missing at once, nor could I ever observe them with their dam, coursing round the church in the act of learning to fly, as the first broods generally do. On the thirty-first I caused the eaves to be searched, but we found only two callow dead Swifts, on which a second nest had been formed."

As a general fact, Swifts and Swallows hold little communion with each other, though they inhabit the same localities, and pursue the same description of prey. When, as is often the case, they make their residence in the same street, it has been observed that the two species occupied different sides of the street, the Swifts taking the north, and the Swallows preferring the south. Even when upon the wing, the Swallows and Swifts seem to have nothing in common with each other, but hold aloof in little parties of the same species.

The colour of the Swift is remarkably sombre, more so, indeed, than that of almost any British bird. The whole of the plumage is dark sooty-black, devoid of the rich green and purple gloss which is seen upon the rook and other dark-feathered birds, and only relieved from its dull monotony by a grey patch below the chin. The beak is black in colour, and very small, but the gape of the mouth is remarkably wide. The legs, toes, and claws, are of the same dull black as the beak, and the eyes are brown. The entire length of a full-grown Swift is about seven inches, the end of the wing reaching more than an inch beyond the tip of the tail. The second primary is the longest feather in the wing.

AMONG the many "travellers' tales" which called forth such repudiation and ridicule from the sceptical readers of the earlier voyagers, the accounts of the Chinese cuisine were held to be amongst the most extravagant.

That civilized beings should condescend to eat dogs and rats specially fattened for the table, was an idea from which their own better sense revolted; that the same nation should reckon sharks' fins and sea-slugs among their delicacies, was clearly an invention of the writer; but that the Chinese should make soup out of birds' nests, was an absurdity so self-evident, that it destroyed all possibility of faith in the writers' previous assertions. Very witty remarks were made on the subject, and many jokes made on the manner of cooking a birds' nest, so as to convert it into soup, the humourist having no conception of the possibility that a birds' nest could be made of anything but sticks, moss, feathers, and mud. Yet it is now a well-known fact, that certain birds have the faculty of producing or discovering a curious substance with which they make these very singular nests, and which is perfectly capable of being cooked and eaten.

The birds that make these remarkable nests belong to several species, four of which have been acknowledged. There are the ESCULENT SWALLOW, the Linchi, (*Collocalia fuciphaga*), the White-backed Swallow, (*Collocalia troglodytes*), and the Grey-backed Swallow, (*Collocalia Francica*).

These nests could hardly be recognised as specimens of bird architecture by any one who had not previously seen them, as they look much more like a set of sponges, corals, or fungi, than nests of birds. They are most irregular in shape, are adherent to each other, and are so rudely made, that the hollow in which the eggs and young are intended to live, is barely perceptible. They are always placed against the face of a perpendicular rock, generally upon the side of one of the tremendous caverns in Java and other places where these strange birds love to dwell. The men who procure the nests are lowered by ropes from above, and their occupation is always considered as perilous in the extreme.

ESCULENT SWALLOW.—*Collocalia nidifica.*

While adherent to the rocks, or when gathered into baskets, the nests are not at all attractive in their aspect, and it is not until they have been carefully washed and cleansed, that they begin to show their semi-fibrous structure, shining through its partially transparent substance. The nests are of very different value, those which have been used in rearing a brood of young being comparatively low in price, while those which are quite new and nearly white, are held in such esteem, that they are worth their weight in silver. When placed in water and allowed to remain in soak, the nests, being made of a partially gelatinous substance, begin to soften and swell, and when thoroughly dressed, are said to bear some resemblance to rather stiff turtle fat. To European palates, however, they appear very insipid, and not worthy of the great value which is set upon them by the Chinese.

In the British Museum may be seen a very fine specimen of the nest of the Esculent Swallow, comprehending a mass of the nests still adhering to the rock. It is rather remarkable that the birds have a habit of building these curious nests in horizontal layers.

The substance of which these nests are composed is evidently of an animal nature to some extent, although certain vegetable matters, such as the gelatinous fuci or sea-weeds, may be admixed with it.   Whatever. may be the basis of the nutriment that forms these nests, it is clear that a very large portion of it is furnished by certain glands, which pour out a viscid secretion.

The trade in these strange articles of diet is a very large one, and still holds its ground, the annual weight of nests that is obtained from the various caves in which the Esculent Swallows take up their residence being upwards of fifty thousand pounds, and the value of the goods more than two hundred thousand pounds.   In the Philippine Islands, the bird is called the Salangana.   The tribe of the genus *Collocalia* approach very nearly to the true Swallows in their structures, but have many points which are identical with the Swifts.   The second primary feather is the largest, the first three toes point forward, and the fourth toe backward, and the tail is moderate in length.

The nests are harried about three times in every year, and it is said that the natives who are employed in procuring them are careful to destroy the old and deep-coloured nests, in order to force the birds to build new habitations, which command a high price in the market.   The construction of a single nest is a work of considerable time, occupying nearly two months, and the structure of these wonderful habitations seems to show that the bird forms them by procuring out of its mouth a viscid secretion, and which hardens into adhesive threads as it comes in contact with the air.   A close examination of the nest shows that it is composed of a great many layers of irregular network, the

WHITE-BREASTED SWALLOW.—*Atticora leucosternon.*

meshes of which connect them in every direction.   Whatever it may be, it certainly possesses great strengthening and restorative powers when cooked, and is said to be an excellent specific in cases of indigestion.

The Esculent Swallow is a small bird, and its colour is brown on the upper parts of the body, and white beneath.   The extremity of the tail is greyish-white.   The British Museum possesses specimens of all the Swallows which are known to make these curious edible nests, and have for that reason been somewhat wrongly termed edible Swallows.   In point of fact, a Swallow is not at all an edible bird, possessing a most nauseatingly sweet flavour, as I can testify from personal experience.

The elegant little WHITE-BREASTED SWALLOW is the Australian representative of the genus Atticora.

It is a very small bird, the total length being only about three inches and a half. The colour is chiefly of two sorts, white and black-brown of different depths, according to the individual, and the position of the feathers.   The crown of the head is light brown, dotted with darker spots, and succeeded by a white ring.   A black band passes from the corner of the mouth round the back of the head, embracing the eyes in its course.   The chin, throat, and chest are pure white, and the remainder of the plumage is deep black-brown, the line of demarcation between the two tints being very strongly drawn.

All that is known of the habits of this pretty little bird is contained in the notes of Mr. Gilbert, quoted in Gould's "Birds of Australia :"

" I only observed this bird in the interior, and as far as I can learn, it has not been seen to the westward of York : I am told it is merely a summer visitor. It is a very wandering species, never very numerous, and is generally seen in small flocks of from ten to twenty in number, flying about, sometimes in company with the other Swallows for about ten minutes, and then flying right away. I noticed this singular habit every time I had an opportunity of observing the species. It usually flies high, a circumstance which renders it difficult to procure specimens. Its flight more nearly resembles that of the Swift than the Swallow; its cry also at times very much resembles that of the former. Its food principally consists of minute black flies.

This bird chooses for its nest the deserted hole of either the Dalgyte (*Perameles lagotis*), or the Boodee (a species of Bettongia), in the sides of which it burrows for about seven or nine inches in an horizontal direction, making no nest, but merely laying its eggs on the bare sand.

The White-breasted Swallow is termed by the colonists the Black-and-white Swallow, in allusion to the bold contrasts of the colours with which its plumage is decorated, and the natives know it under the title of Boo-de-boo-de."

SWALLOW.—*Hirundo Rústica.*

THE elegantly shaped and beautifully coloured SWALLOW may be readily distinguished from any of its British relations by the very great elongation of the feathers which edge its tail, and which form nearly two-thirds of the bird's entire length.

It is the most familiar of all the Hirundinidæ of England, and from its great familiarity with man, and the trustfulness with which it fixes its domicile under the shelter of human habitations, is generally held as an almost sacred bird, in common with the robin and the wren. In eastern countries, the protection of man is extended towards this beautiful little bird even more extensively than in England, where too often it is killed or wounded by the unfeeling possessor or hirer of a gun merely by way of practice in "shooting flying."

Independently of any question of humanity or the barbarity of a disposition which can find amusement in the death and cruel maiming of beings full of life and enjoyment, it is a matter of very bad policy to shoot a swallow. There are some birds which afford some excuse to their destroyers by reason of their fondness for grain and fruit, but the Swallow is exclusively an insect-eating bird, and plays a most important part in preserving the balance between the various departments of the animal kingdom. There are many noxious insects which are most valuable in themselves, and which, until the conditions which cause their existence be removed or altered, are actual visible providences towards mankind. But these very creatures are necessarily so extremely prolific, that their increase outgrows their task, and they would themselves become nuisances, did not the Swallow and other similar birds keep down their numbers by day, and the goat-suckers and bats by night.

To ascertain the truth of this observation, nothing more is necessary than to open the mouth of a Swallow that has been shot while flying, and to turn out the mass of small flies which will be found collected there, and which the bird was intending to take home to its hungry little family. The extraordinary amount of flies and other insects which a Swallow can thus pack into its mouth is almost incredible, for when relieved by the constant pressure to which it is subjected, the black heap begins to swell and enlarge, until it attains nearly double its former size.

The Swallow wages a never-ceasing war against many species of insects, and seems to be as capricious in its feeding as are the roach and other river fish. At one time it will feed almost exclusively upon gnats and other small flies, and will destroy many thousands of these obnoxious flies in a single day. At another time it will prefer beetles, chasing the Geotrupidæ rather than those of any other order. On another occasion it will confine itself to May-flies, catching them as they emerge from their pupal envelopes and flutter soft, fat, and languid on the river bank. Sometimes the Swallow flies at larger prey, and frequenting the neighbourhood of bee-hives, swoops with unerring aim upon their inmates as they enter or leave their straw-built houses. It is a very remarkable fact that the working bee is generally unharmed by the Swallow, which directs its attack chiefly upon the comparatively useless drone. Perhaps the bird may possess an instinctive knowledge of the poisoned weapon with which the worker is armed, and may therefore prefer to attack the large but stingless drone.

Owing, in all probability, to this insect diet, the flesh of the Swallow is quite unfit for the table, and possesses a very disagreeable flavour. Out of curiosity I once cooked and tried to eat some Swallows that had been shot, and was effectually deterred from the attempt by the peculiar and nauseous character of the flesh, which has some resemblance to a sweet potato in its flavour. Like the generality of predaceous birds, the Swallow ejects the legs, wings, and other indigestible portions of its insect prey in little pellets, or "castings."

The flight of this bird is very rapid and graceful, and is readily distinguished from that of the Swift by certain peculiarities which are not easy to be described, but can be recognised without difficulty. Unlike the Swift, which never settles except on some elevated spot, the Swallow is fond of resting a while on the ground, and may often be seen dusting itself after the fashion of the common sparrow. I have often seen it settle on the patches of sand that are left among the rocks at low water, and from the busy activity which it displayed on such occasions, imagine it to have been engaged in chasing the sand-flies, or perhaps even the sand-hoppers that swarm so abundantly in such localities.

When taken young, the Swallow is easily tamed, and after having passed the season of emigration, becomes reconciled to its enforced home and is a very docile and loveable little pet. The poor bird must suffer greatly during this period when its brothers are voyaging to warmer climates, for the organization of all birds is sensitive to a high degree, and especially so in the case of birds of passage. The extreme delicacy of the bird's nature was well shown in the time of cholera. In the town of Verviers, while that fell disease was carrying away twenty inhabitants per diem out of a population of two thousand, the Swallows and all the singing birds left the spot, and did not return until the cholera had passed away.

The voice of the Swallow is vastly more agreeable than the shrill scream of the Swift, and is, although weak and twittering, very musical in its tone and pleasing to the ear.

The nest of the Swallow is always placed in some locality where it is effectually sheltered from wind and rain. Generally it is constructed under the eaves of houses, but as it is frequently built within disused chimneys, it has given to the species the popular title of Chimney Swallow. The bird is probably attracted to the chimney by the warmth of some neighbour fire.

The nest is composed externally of mud or clay, which is brought by the bird in small lumps, and stuck in irregular rows so as to build up the sides of its little edifice. There is an attempt at smoothing the surface of the nest, but each lump of clay is easily distinguishable upon the spot where it has been stuck. While engaged at the commencement of its labours, the Swallow clings perpendicularly to the wall of the house or chimney, clinging with its sharp little claws to any small projection, and sticking itself by the pressure of its tail against the wall. The interior of the nest is lined with grasses and other soft substances, and after it has been inhabited by a young brood, becomes very offensive to the nostrils and unpleasant to the touch, in consequence of the large parasitic ticks which are peculiar to the birds of this tribe, and which swarm in the nest.

Persons who value the cleanly exterior of their houses more than the pleasure of affording shelter to these beautiful and graceful little birds, and of watching them through the interesting period of building, hatching, and rearing, have tried many methods of preventing the birds from building, and have found none so effectual as soap or oil laid on the wall with a brush, a substance which will not suffer the clay to hold to the wall's perpendicular surface.

Sometimes the Swallow is seized with a fit of eccentricity, and builds its nest in very odd localities. One of these birds actually made its home in the outspread wings of an owl which had been nailed against a barn door, and it is not at all unusual to see the nests of the Swallow built in the shaft of an old mine or wall. Various other localities are recorded by zoological observers, such as a half-open drawer, an old cap hung upon a peg, and in one curious instance, which is mentioned and figured by Mr. Yarrell, upon the forked branch of a sycamore tree which hung over a pond. A brood of young was hatched in this nest, and a second batch of eggs was laid, but came to nothing.

There are sometimes two broods in the year, and when the second brood has been hatched at a very late period of the year, the young are frequently deserted and left to starve by their parents, who are unable to resist the innate impulse that urges them to seek a warmer climate. It has occasionally, but very rarely, happened that the parents have remained for some time in order to bring up their young brood. When fully fledged, and before they are forced to migrate, the young birds generally roost for the night in osiers and other water-loving trees.

Except in confinement, the Swallow knows not the existence of frost, nor the extreme of heat, passing from Europe to Africa as soon as the cold weather begins to draw in, and migrating again to the cooler climes as soon as the temperature of its second home becomes inconvenient to its sensitive existence. The time of its arrival in England is various, and depends almost entirely on the state of the weather. Solitary individuals are now and then seen in very early months, but as a general fact, the Swallow does not arrive until the second week in April; the time of its departure is generally about the middle of September, although some few lingerers remain in the country for more than a month after the departure of their fellows.

Before the time of migration they may be seen assembled in great numbers, chattering away eagerly, and appearing to be holding a great parliament for the settlement of affairs before starting on their long journey. The dome of the Radcliffe Library at Oxford is a favourite assembling place of these birds, all the lines of its architecture being studded with Swallows, whose white bellies look like pearl beads strung upon the dark surface of the leaden dome. They do not, however, migrate in flocks, but pass in little families of two or three in number across the vast space that separates them from the end of their journey. Although such powerful and swift fliers, they become fatigued in passing the sea, and will flock in great numbers to rest upon the rigging of some ship that may happen to pass their course. It is rather curious that the birds almost invariably fly in a line directly north and south, influenced doubtlessly by the magnetic current that flows everlastingly in that direction.

Sometimes the poor birds are so utterly worn out with fatigue, that when they have perched upon the side of a boat, they are unable to take again to the wing, and if disturbed, can hardly fly from one end of the boat to the other. They have been even seen to settle upon the surface of the waves, and to lie with outspread wings until they were able to resume their journey.

Guided by some wondrous instinct, the Swallow always finds its way back to the nest which it had made, or in which it had been reared, as has frequently been proved by affixing certain marks to individual birds, and watching for their return. Sometimes it happens that the house on which they had built has been taken down during their absence, and in that case the distress of the poor birds is quite pitiable. They fly to and fro over the spot in vain search after their lost homes, and fill the air with the mournful cries that tell of their sorrows.

The Swallow is widely spread over various parts of the world, being familiarly known throughout the whole of Europe, not excepting Norway, Sweden, and the northern portions of the continent. It is also seen in Western Africa, and Mr. Yarrell mentions an instance where it was observed in the island of St. Thomas, which is situated upon the equator. The martin and the swift were seen at the same place.

The colour of the Swallow is very beautiful. Upon the forehead the feathers are of a light chesnut, which gives place to deep glossy steel-blue upon the upper portions of the body and wings. The primaries and secondaries are black, as are the tail feathers, with the exception of a few white patches. The throat is chesnut, and a very dark-blue band crosses the upper part of the chest. The under parts are white, and the beak, legs, and toes black. The female is distinguished by the smaller chesnut patch on the forehead, the smaller tint of the feathers, and the narrowness of the dark band across the chest.

Many examples of white Swallows are on record, and specimens may be seen in almost every collection of British birds.

RUFOUS-BELLIED SWALLOW.—*Hirundo erythrogaster.*

The two Swallows which next come before our notice, are natives of America, and are high in favour among the lower inhabitants of the land, one species taking up its abode with civilized men, and the other preferring, at all events at present, the habitations of the indigenous savage tribes.

The RUFOUS-BELLIED SWALLOW is plentifully found in the United States, and is fond of building its nest in outhouses and barns, and is frequently furnished by the kindness of the proprietor with convenient boxes fastened to poles or nailed on trees. It is never known to build in chimneys, like our own Chimney Swallow.

The nest of this bird is rather peculiar in form, being according to Wilson, " in the form of an inverted cone, with a perpendicular section cut off on that side by which it adheres to the rafters. At the top it has an extension of the nest, or an off-set, for the male or female to sit on occasionally: the upper direction is about six inches by five, the height externally seven inches. This shell is formed with mud mixed with fine hay, as plasterers do mortar with hair to make it adhere the better; the mud seems to be placed in regular layers from side to side: the shell is about an inch in thickness, and the hollow of the cone is filled in with fine hay well stuffed in, and above that is laid a handful of downy feathers." The nest of the Pinc-pinc (*Cisticola tectrix*) is also remarkable for possessing a supplementary erection on which one of the birds sits while the other is engaged in hatching the eggs.

As the nest is rather complicated in its structure, it occupies some time in preparation, a week generally passing before it is fit to receive the eggs. When the Rufous-bellied Swallow builds in barns or out-houses, it is very gregarious in its habits, twenty or thirty nests being often placed in close proximity to each other, and generally within an inch or two ; yet there is no quarrelling among the birds, and the whole society is remarkably harmonious. While the female is sitting on the eggs, the male often places himself on the mud perch, and pours forth his complacent little twitter of a song for her consolation. There are generally two broods in each season.

In size the Rufous-bellied Swallow is not quite equal to the common Swallow of England, being only about seven inches in length. Its colour is not unlike that of the Swallow, excepting that the under portions of the body are of a ruddy chestnut in the male, and of a rusty white in the female.

The RUFOUS-NECKED SWALLOW is easily to be distinguished from the preceding bird by its comparatively small dimensions, its entire length being hardly six inches, and by the form of its tail, which is without the usual fork.

It is also known by the name of the Republican Swallow, on account of its habit of building its nests in close proximity to each other, and in very great numbers. The nest is of a singular shape, being formed something like a Florence flask with its neck shortened and widened. Mud is the chief substance with which it is built, and the lining is composed of dried grasses and other soft substances. These birds are rapidly beginning to attach themselves to civilized habitations, and may be often found building their nests upon walls of barns and out-houses. When they prefer to live in the open country, they choose a convenient wall of rock for the purpose, and place their nests in some spot where they are sheltered by overhanging heaps of stone. The nest is very rapidly built, being ready for occupation within five or six. days from the time when it was begun. The eggs are generally four in number, and their colour is white, covered with many brown spots.

The plumage of this pretty little bird is a deep glossy violet, almost black upon the back, head, and wings ; the under portions of the body are greyish-white excepting the breast, which is a ruddy-ashen tint. A band of pale rusty-red passes over the forehead, and the upper tail coverts are of the same hue. Above the bill a narrow black band is drawn as far as the eye.

AMONG the most ingenious of bird architects, the FAIRY MARTIN holds a very high place in virtue of the singular nest which it constructs.

The nest of the Fairy Martin has a very close resemblance to a common oil flask, and reminds the observer of the flask-shaped nests which are constructed by the Pensile Oriole and similar birds, although made of harder material. The Fairy Martin builds its curious house of mud and clay, which it kneads thoroughly in its beak before bringing it to the spot where it will be required. Six or seven birds work amicably at each nest, one remaining in the interior enacting the part of chief architect, while others act as hodsmen, and bring material as fast as it is required. Except upon wet days, this bird only works in the evening and early morning, as the heat of mid-day seems to dry the mud so rapidly that it cannot be rightly kneaded together. The mouths, or " spouts " of these nests vary from eight to ten inches in length, and point indifferently in all directions. The diameter of the widest portion of the nest is very variable, and ranges between four and seven inches.

The exterior of the nest is as rough as that of the common swallow of England, but the interior is comparatively smooth, and is lined with feathers and fine grass. The eggs are generally four or five in number, and the bird rears two broods in the course of the year.

The Fairy Martin is very capricious in its choice of locality. Sometimes it will take a fancy to a house, and will build its nests in regular rows under the shelter of the eaves. Sometimes it prefers the perpendicular face of a rock, and in that case will build several

FAIRY MARTIN.—*Hirundo Ariel.*

hundreds of nests in close proximity to each other, but without the slightest attempt at regularity or order, and with the parts sticking out in all directions. Now and then, the nest of this bird is found within the hollow of some decayed tree. In every case, the nest is built in some place where water is in the near vicinity, but it is a very remarkable fact that it has never been seen within twenty miles of the sea. The Fairy Martin is spread over the whole of Southern Australia, arriving in August, and departing in September.

THE WIRE-TAILED SWALLOW is chiefly remarkable on account of the peculiarity from which it derives its name.

The external feathers of the tail are singularly elongated, and for the greater part of their length are devoid of web, resembling in some degree the filamentary appendages of the Bird of Paradise. The general colour of this bird is a rich steel-blue, the head being chesnut, and the under portions of the body white, with the exception of a large black patch upon the back of the thigh. The wiry portion of the tail feathers is black, and the same tint runs across the edge of the webbed portions, which in the centre are white like the abdomen. Specimens of this bird in the British Museum have been brought from Madras and Abyssinia.

THE handsome PURPLE SWALLOW is a native of the United States of America, where it is one of the most familiar, and at the same time one of the most generally beloved of the indigenous birds.

It instinctively resorts to human habitations, and even finds favour in the eyes of the American Indian, a being who is little given to mercy, and who makes the possession of a head but a theme for self laudation. Yet even the copper-skinned native respects the Purple Martin, and takes care to prepare a convenient resting place for the little bird, by hanging on a neighbouring tree an empty gourd in which a hole has been roughly cut. In this receptacle the Martin makes its inartificial nest, and cheers the heart of its host by its monotonous though sweet-toned song. The more civilized inhabitants of farms provide for the roosting of this bird by fastening nest-boxes against the wall, and some persons even build regular cotes, of which the sociable birds soon take possession.

Sometimes the Purple Martins become rather presuming in their familiarity, and actually turn the pigeons out of their own nest-boxes when they do not find sufficient accommodation for themselves. The negro, too, takes pleasure in domesticating this most trustful of birds, and provides for its accommodation by fastening hollow calabashes to the tops of long bamboo canes, which are stuck in the ground for the purpose.

Like the swallow of England, the Purple Martin exhibits a great predilection for the spot where it has once built its nest, and will return year after year to the beloved locality.

As is the case with many familiar birds, such as the robin, the sparrow, and the swallow, the Purple Martin is a most fearless and withal quarrelsome bird, greatly delighting in annoying any other bird that may happen to be larger than itself, and trusting to its great command of wing for impunity. Hawks of all kinds, crows, jays, and similar birds live in constant terror of the Purple Martin, which no sooner sees the hateful form of a hawk or crow in the distance than it flies at him savagely, and makes such rapid and vicious pounces, that the wretched victim is fain to escape as he best can from the attacks of his small but determined foe. Even the eagle enjoys no immunity from the persecution of the Purple Martin, which dashes at the regal bird with as much assurance as if it were only chasing a pigeon. It is rather remarkable that although the Purple Martin will generally fly at the king-bird, it will make common cause with that bird against the eagle, and unite in a temporary alliance until the common enemy is driven off.

The Purple Martin feeds mostly upon the larger insects, such as wasps, bees, and beetles, caring little for the gnats, flies, and other small insects which form the food of the generality of swallows. The flight of this species is wonderfully rapid and active, the little bird dashing to and fro with lightning speed, and wheeling with such remarkable suddenness that it really has nothing to fear from the larger but less active claws of the eagle or falcon.

When this bird builds in a crevice or other spot which has not been prepared by the hand of man, the nest is found to consist of a rather large mass of dried grass, leaves, moss, feathers, and other similar substances, and contains from four to six white eggs. When several birds are building in proximity to each other, they make an extraordinary noise at the break

WIRE-TAILED SWALLOW.—*Hirundo filifera.*

of day, which, although very useful in awaking the farmers and their men in time for their daily work, is by no means agreeable to those whose tastes do not incline them to early rising. There are generally two broods in each year, and both parents take their fair turn in sitting upon the eggs.

As might be gathered from the popular name of this bird, the colour of this species is a rich, deep purple, of a very glossy kind. This purple hue is peculiar to the male, and extends over the greater part of the body, with the exception of the wings and tail, which

2. L

PURPLE SWALLOW.—*Progne purpurea.*

are of a deep blackish-brown. The female and young male birds are brown, with a strong blue tinge upon the upper parts of the body, and only a greyish-white below.

Several examples of the Purple Martin have been taken in Great Britain, and Mr. Yarrell has therefore admitted it into his list of British birds. He is of opinion that as two specimens were shot within a single week in the same locality, the brood might have been reared in this country. Another specimen was killed near Kingston.

The pretty little SAND MARTIN is, in spite of its sober plumage and diminutive form, a very interesting bird, and one which adds much to the liveliness of any spot where it may take up its abode.

In size it is less than any of the other British Hirundinidæ, being less than five inches in total length. The colour of this bird is very simple, the general tint of the entire upper surface of the head and body being a soft brown, relieved from too great uniformity by the sooty-black quill feathers of the wings and tail. The under surface is pure white, with the exception of a band of brown across the upper part of the chest. The young bird possesses a lighter plumage than the adult, owing to the yellowish-white tips of the back, tertiaries, and upper coverts. The beak is dark brown, and the eyes hazel.

Although its little beak and slender claws would seem at first sight to be utterly inadequate for the performance of miner's work, the Sand Martin is as good a tunnel-driver as the mole or the rat, and can manage to dig a burrow of considerable depth in soil that would severely try the more powerful limbs of the quadrupedal excavators. The soil which it most loves is light sandstone, because the labour which is expended in the tunnelling is very little more than that which would be required for softer soils, and the sides of its burrow are sufficiently firm to escape the likelihood of breaking down. In default of sandstone, this bird will drive its burrow through the bank of a railway-cutting or of a river, even though the soil should be of softer and less-enduring material. It has even been known to make its nest in a belt of soft clay above the basaltic cliffs of the Giant's Causeway.

The depth of the burrow is extremely variable, some tunnels being only eighteen inches or two feet deep, while others run to a length of nearly five feet. During some five years' experience and constant watching of these birds in Derbyshire, I generally found that the hand could reach to the end of the burrows, and remove the eggs, provided that the birds had not been forced to change the direction of the tunnel by the intervention of a stone or a piece of rock too hard for their bills to penetrate. The approach to the nests and the mode of taking them was hazardous in the extreme, as it was necessary to clamber along the perpendicular face of a loose, crumbling sandstone rock, and to cling by the feet and one hand, after the fashion of the Sand Martin itself, while the other hand was groping for the eggs at the extremity of the burrow.

In excavating its domicile the Sand Martin displays wonderful activity and ingenuity, and abandons itself to its work with a thorough recklessness of enjoyment.

Clinging to the face of the rock, it delivers thereon a firm sharp blow with its closed beak, as if to test the quality of the material, and then nimbly runs or flutters to another spot, where it repeats the same process, until it has fixed upon a suitable locality. It then sets fairly to work, and by dint of repeated blows in the same spot, loosens a considerable piece of soil, which comes tumbling to the ground. The bird then cuts a circular funnel-shaped depression, by running nimbly round the circumference, and working from the centre outwards, and in wonderfully short time succeeds in forming a well-defined circular hole. Having made so much progress, it rests for a short space, and then redoubles its ardour, chipping away the sandstone at a great rate with repeated blows of its sharp, conical little bill, and clearing the fallen material with its claws. While it works, it clings to the sides or the roof of its burrow with equal facility, and traverses the tunnel with singular ease and rapidity.

This bird is not very readily satisfied with a locality, and being in no wise sparing of its labour, will often dig three or four holes before it will make a final settlement. As has been already mentioned, the burrows are generally straight unless turned out of their course by some impediment, but in all cases they are slightly globular at the extremity where the nest is deposited, and slope gently upwards, so that the eggs and young cannot be inundated by rain. The Sand Martin is very gregarious in its habits, and crowds its burrows

SAND MARTIN.—*Cótile ripária.*

closely together, so that a cliff is often absolutely honey-combed by these persevering little diggers. Perhaps the quality of the soil may have some influence on this association, as it is quite common to see one part of a bank crowded with nests, while the remainder is left deserted.

As is generally the case with burrowing birds, the Sand Martin takes very little trouble about the construction of its nest, but contents itself with laying down a small handful of various soft substances, such as moss, hay, and feathers. The eggs are very small and fragile, and are not easily removed from the burrow without being fractured. Their colour is, when freshly laid, a delicate semi-transparent pink, which darkens to a dull opaque grey when incubation has proceeded to some extent, and changes to a beautiful white when the contents are removed from the shell. Their number is from four to six.

The voice of the Sand Martin is a weak twitter, soft and musical, and when the birds congregate in numbers, can be heard at a considerable distance. At times, however, when irritated by the presence of a bird of prey, or when engaged in quarrelling with one of its own species, a weakness to which this bird is especially prone, it pours out from its little throat a succession of harsh screaming cries that may express either rage or fear according to circumstances. I have often seen the Sand Martins "mob" larger birds; and on one occasion, while they were engaged in pursuing a hawk that was passing near their habitation, they suffered for their temerity by the loss of one of their companions, who was carried off screaming in the hawk's talons, and whose sad fate at once dispersed the noisy assembly. I fancy that in this instance the hawk intentionally provoked the attack, as his flight was very unhawklike, and he seemed to stagger, so to speak, while fluttering amid the crowd of Sand Martins.

The food of this bird is composed of insects, and, in spite of the small dimensions of the little creature, it will pursue, capture, and eat insects of considerable dimensions and strength of wing, such as wasps and dragon-flies. Gnats and similar insects, however, form the staple of its diet.

The Sand Martin is not much of a wanderer, always hanging about the vicinity of its abode, and satiating itself with the insects of the locality. Generally the nest is placed near water, such as a stream or a lake, so that the bird is sure to find plenty of food among the innumerable insect tribes that frequent such localities. These birds have been noticed to perch by hundreds on reeds and patches of sand, where they are able to rest their wings, and to pursue a terrestrial chase after their insect food.

This bird generally makes its appearance in England about the beginning of April, and has even been noticed before the end of March, so that its arrival is earlier than that of the swallow or Martin. It departs about the beginning of September, and like other British Hirundinidæ, makes its way to Africa, where it remains until the succeeding year.

RESEMBLING the common swallow in habits and general appearance, the HOUSE MARTIN may easily be distinguished from that bird by the large white patch upon the upper tail-coverts, a peculiarity which is even more notable when the bird is engaged in flight than when it is seated on the ground or clinging to its nest. In the dusk of evening the Martins may often be seen flying about at so late an hour, that their bodies are almost invisible in the dim and fading twilight, and their presence is only indicated by the white patches upon their backs, which reflect every fading ray, and bear a singular resemblance to white moths or butterflies darting through the air.

This beautiful little bird is found in all parts of England, and is equally familiar with the swallow and sand Martin. It places its clay-built nest principally under the shelter afforded by human habitations, and becomes so trustful and fearless that it will often fix its nest close to a window, and will rear its young without being dismayed at the near presence of human beings.

It is rather a curious fact that the Martin should be so apparently capricious in its architectural taste, as has frequently been observed. The birds will often take a great fancy to one side of a house, and will place whole rows of their nests under the eaves, totally neglecting the remaining sides of the house, even though they offer equal or superior accommodation. Generally the Martins avoid the south side, apparently from a well-grounded fear that the heat of the midday sun might crack and loosen the mud walls of their domiciles. A north-eastern aspect is in great favour with the Martins, and I lately observed a very great number of their nests affixed to the eastern walls of a row of houses, together with several isolated cottages, and, on a careful examination, could not see a single nest upon any other part of the buildings.

The nest of this species is extremely variable in shape and size, no two being precisely similar in both respects. Generally the edifice is cup-shaped, with the rim closely pressed against the eaves of some friendly house, and having a small semicircular aperture cut out of the edge, in order to permit the ingress and egress of the birds. Sometimes, however, the nest is supported on a kind of solid pedestal, composed also of mud, and often containing nearly as much material as would have made an ordinary nest. These pedestals are generally constructed in spots where the Martin finds that her nest does not find adequate support from the wall.

The material of which the nests are built is said to consist principally of the finely pulverized mould which is swallowed by earthworms as they feed, and is ejected at the surface of the ground in the well-known "casts" that often disfigure our lawns, and excite the wrath of the gardener. This substance is evidently well moistened and kneaded before it is applied, and it is very probable that the bird may supply some viscid secretion which renders it more tenacious. The exterior of the nest is very rough, but has a picturesque appearance by reason of that very roughness.

When once they have attached themselves to any locality, the Martins are thorough conservatives in their feelings, and set their faces against any alteration or improvement. One of my friends, on finding that these little birds were beginning to build their nests under the eaves of his house, was desirous of attracting them to his residence and affording them the best hospitality in his power. He therefore ordered a kind of verandah to be erected along the side of the house, so that the Martins might find a better shelter than was afforded by the shallow eaves. The birds, however, took a different view of the matter; deserted the nests which had already been built, and never came back again.

In all cases the Martins exhibit a strong dislike to smooth walls. Stucco they detest, and only tolerate new brick when they can find no other resting-place. But their chief delight is in walls that are covered with rough cast, or that are built of roughened stone. They also take advantage of any projection, such as a spout or a piece of sculpture, and employ it as a foundation on which they may rest their domiciles.

Not only is the Martin capricious in choosing certain points of the compass, but also in fixing upon a locality wherein to build its habitation, it exhibits no small fancifulness. Generally it affects human dwellings, and rests safely under the protection of their inmates; but it will often fly far from the presence of man, and build its nest in uninhabited spots. Precipitous rocks of various kinds, whether limestone, sandstone, or chalk, are frequently studded with the nests of the Martin. The basaltic rocks of the Giant's Causeway are in great favour with this bird, which has even been known to plant its nests thickly in the arches of a bridge.

HOUSE MARTIN.—*Chelidon urbica.*

In Northern Europe the Martin is held in very high estimation, as it seems to keep down the numbers of the mosquito and those winged pests which swarm in those cold regions as profusely as under a tropical sky, and use their poisoned weapons with quite as much severity. Its presence is therefore courted, and the inhabitants strive to attract the Martins to themselves by preparing certain attractive boxes, in which they are able to build their nests without expending a tithe of the labour that is required for fixing a nest upon a perpendicular wall.

In this place it may be as well to mention the well-known assertion that a sparrow which had usurped the nest of a Martin, and refused to relinquish his ill-gotten home, was summarily punished by the entire efforts of the society, who gathered a large supply of clay, and fairly immured him in a prison and a tomb. This story appears to be very improbable, because it would seem that a very few blows of a sparrow's beak would break through the frail walls of a Martin's nest, and release the prisoner. But in Mr. Thompson's work on the Birds of Ireland the strange tale is confirmed in the following passage:—

"When the House Martins returned in that year to a thatched cottage, belonging to Mr. John Clements, where they had annually built for a long period—and which then displayed fourteen of their nests—a pair found that sparrows had taken possession of their domicile. On perceiving this, they kept up 'such a chattering about the nest' as to attract the attention of the owner of the house. After its continuance for some time, apparently until they were convinced that the sparrow was determined to retain possession, they flew away, and did not return for a considerable time, when they reappeared with about twenty of their kindred. They now immediately commenced 'claying up the entrance to the nest,' which was done in the course of the day; next morning, the construction of a new nest was commenced against the side of the old one, and in it they reared their brood undisturbed. After some time, the proprietor of the cottage, who had never heard of any similar case, pulled down both nests, and in that occupied by the sparrow found its 'rotten corpse,' together with several eggs. A particular note of the entire proceeding, as related by Mr. Clements, was made by my brother soon after the occurrence; but to make 'assurance doubly sure,' before publishing the account in 1842, I inquired of Mr. Clements whether he remembered such a circumstance, and he repeated it just as narrated nine years before. Some other persons, too, of our mutual acquaintance were witness to the chief part of the proceeding, and saw the sparrow, together with the eggs, in the sealed-up nest.

What appears to me the most singular feature in this case is, that the sparrow should remain in the nest, and allow itself to be entombed alive. But this bird was sitting on the full complement of eggs, probably in the last stage of incubation, at which period we know that some birds leave the nest only to procure such a scanty morsel as will barely support life. Occasionally, at such times they allow themselves to be lifted off their eggs, and when placed on again, continue to sit as intently as if they had not been disturbed. The filling up of the aperture is not in itself a singular proceeding on the part of the Martins, but on this occasion, when the assistance of their neighbours was called in, would almost seem to be intended as an act of retributive justice on the sparrow. Their building against the side of the old nest is quite a common occurrence."

There are generally several broods in the course of the year, two being the usual number, and three or even four being sometimes noticed. In such cases, however, the young birds seldom reach maturity, for they are hatched at such a late period of the year that the parents are unable to withstand the instinct that leads them to migrate, and in obeying the promptings of this principle, leave their unfortunate family to perish miserably of hunger. The parents do not seem to grieve over their dead children, and when they return to the nest in the succeeding season, they unconcernedly pull the dried and shrivelled bodies out of the nest, and rearrange it in readiness for the next brood. It has been well suggested that a change in the nature of the bird takes place at every season of migration, and that the "storgè," or love of offspring, is suddenly quenched when the creature is called upon to undergo its long journey.

Although they are by no means brave birds, excepting in familiarity with man, they display great courage in defending their homes from the attacks of foes, and will oppose, to the best of their power, any adversary, whether man, beast, or bird. If their nests are broken down, they hover round the destroyer, and dash at him with all their might, uttering hoarse screams of rage and sorrow.

The habits of the Martin are very like those of the swallow, and in the first year of their existence the two species are frequently companions. Their friendship is, however, but of brief duration, for as soon as the birds have obtained their full development of form, the stronger wings and more enduring powers of the swallow deter the Martin from continuing the companionship.

In their migration, the Martin differs slightly from the other Hirundinidæ, being rather later than the swallow and sand Martin. It is seldom seen in England before the middle of April, whereas the swallow is often noticed by the beginning of that month, although a solitary specimen may occasionally be noticed, which has acted as pioneer to its companions, and arrived simultaneously with the swallow. It leaves this country about the middle of October, seldom staying beyond the thirteenth or fourteenth of that month.

It migrates in considerable bodies, collecting for several days before starting, and seeming to arrange the order in which the migration shall be accomplished.

The general colouring of this bird is composed of rich blue-black and white, arranged in bold masses, so as to present a fine contrast of two very opposite tints. The head and upper portions of the body are of a very deep glossy blue, with the exception of the quill feathers of the wings and tail, which are sooty black, and the upper tail-coverts, which are snowy white. The chin, breast, and abdomen are of the same pure white as the upper tail-coverts, except in the young birds, which are greyish-white beneath. The female bird is rather grey on the under portions of the body. A number of tiny white feathers are spread over the legs and toes, and the beak is black and the eyes brown. The total length of the Martin is rather more than five inches.

ROLLERS.

THE ROLLERS evidently form one of the connecting links between the swallows and the bee-eaters, as may be seen by the shape of their feet, which have the two hinder toes partially joined together, while those of the bee-eaters are wholly connected, or, as it were, soldered together. The Rollers, as is evident from their long pointed wings, stiff tail, and comparatively feeble legs and feet, are to a great extent feeders on the wing, although they do not depend wholly on their powers of flight for subsistence, but take many insects, worms, and grubs from the ground.

The birds of the genus Eurystomus, or wide-mouthed birds, may be known by the peculiarity from which they derive their generic name. The beak is remarkably wide at its base, and the gape of the mouth is very large. The point of the beak is flattened, and rather abruptly curved. There are some bristle-like feathers at the angle of the mouth, and the wings are extremely long, the second primary feather being the longest. The tail is moderate in extent and not forked.

ORIENTAL ROLLER.—*Eurystomus Orientalis.*

The ORIENTAL ROLLER is found spread over a large expanse of country, being a native of many parts of India, Java, and the Polynesian Islands. It is a very handsome bird, the greater part of its plumage gleaming with the most brilliant green, which has been compared, though not very happily, to the peculiar sea-green of the aquamarine. A brilliant azure colours the throat and the points of the wings, and the primary feathers of the wing are black, diversified with a white bar. The tail is deep black. In its habits it is quick, active, and vigorous, chasing its insect prey through the air, and displaying great command of wing and powers of endurance.

ANOTHER closely allied species is the AUSTRALIAN ROLLER (*Eurystomus Pacificus*). This bird is popularly known to the Australian colonists by the title of DOLLAR BIRD, on account of a circular white spot upon the inside of each wing, which is very conspicuous when the bird is flying overhead. The flight of the Australian Roller is heavy and laboured, and the bird does not appear to chase insects with the activity and perseverance of the preceding species. Generally it is fond of sitting on some convenient bank overhanging the water, and from that post of vantage pouncing upon a passing insect, much after the fashion of the harriers. While thus engaged, it frequently utters a peculiar chattering cry. Its most active seasons are sunrise and sunset; at other parts of the day it is but sluggish in its movements.

It is, however, a very bold and fearless bird, and will attack man, beast, or bird that approaches within a certain distance of the spot where its cradle lies. True nest there is none, as the bird contents itself with a hole in a decaying tree-trunk, and depositing its eggs upon the soft wood. The eggs are from three to four in number.

GARRULOUS ROLLER.—*Coracias Garrula.*

ALTHOUGH tolerably common on several parts of the Continent, the GARRULOUS ROLLER is at the present time a very rare visitant to this country. There seems, however, to be reason to believe that in former days, when England was less cultivated and more covered with pathless woods, the Roller was frequently seen in the ancient forests, and that it probably built its nest in the hollows of trees, as it does in the German forests at the present day.

Africa is the legitimate home of the Roller, which passes from that land in the early spring, and makes its way to Europe, *viâ* Malta and the Mediterranean Islands, which afford it resting-places during its long journey. Accordingly, in those islands the Rollers are found in great plenty, and as they are considered a great delicacy when fat and in good condition, they are killed in considerable numbers, and exposed for sale like pigeons, whose flesh they are said greatly to resemble. Even in its flight it possesses something of the pigeon character, having often been observed while flying at a considerable elevation to "tumble" after the manner of the well-known tumbler pigeons. It is rather curious that throughout Asia Minor the Rollers and magpies were always found in close proximity to each other.

Mr. Thompson records a very curious anecdote of this bird, a specimen of which was seen flying across the grounds of the Duke of Leinster, in September, 1831, and was pursued by a large number of rooks, who were mobbing it after their usual custom whenever they meet with a strange bird. The Roller did not seem to be in the least disconcerted; but, instead of endeavouring to escape, darted repeatedly among its foes, as if for the purpose of increasing their irritation. The bird was not killed for the sake of ascertaining its species; but its peculiar mode of flight, its size, and its gorgeous plumage were so characteristic, that the observer, Mr. Bell, was quite satisfied of its identity.

The food of the Roller is almost wholly of an insect nature, but is diversified with a few berries and other vegetable productions. It has even been known to become carnivorous in its habits, for, according to Temminck, it sometimes feeds on the smaller mammalia. Worms, slugs, millipedes, and similar creatures also fall victims to its voracity.

The position and structure of the nest is remarkably variable. Generally it is placed in the hollows of decaying trees, but it is often found deposited in the extremity of a hole which has been burrowed in a river-bank, like that of the common kingfisher of England. The eggs are of a beautiful shining white in colour, and from five to seven in number. Nests of the Roller have been found in many portions of Europe and Africa, as well as in Malta and Japan, so that the species possesses a very extensive range of country.

In the colouring of its plumage it is truly a gorgeous bird. In its size and general shape, the Roller bears a considerable semblance to the rooks and crows, and like these noisy birds it is gifted with a harsh and loud voice, which it is very fond of raising, and which often leads to its detection as it sits hidden among the deep foliage of the oak and birch forests, which it best loves to inhabit. In allusion to its fondness for the birch-tree, it is known in Germany by the title of Birkhäter, or Birch-jay. In its habits it is fond of seclusion, shy, wary, and restless, so that it is not easily approached by the sportsman.

One of my friends, while residing in Worcestershire, saw frequently some "tumbling" birds, which, from the description, I cannot but think to have been Rollers. They always tumbled when on the ascent, and never while descending. The colours described were precisely those of the Roller—and the size and general mode of flight accorded with that bird—almost equalling the kingfisher in the brilliancy of its hues. The general tint of the head, neck, breast, and abdomen is that peculiar green-blue termed "verditer" by artists, changing into pale green in certain lights, and deepening into deep rich azure upon the shoulders. The back is

MADAGASCAR BRACHYPTERACIAS.—*Brachypteracias leptosomus.*

a warm chestnut-brown, changing to purple upon the upper tail-coverts. The tail is of the same verditer hue as the head and neck, with the exception of the exterior feathers, which are furnished with black tips. The quill feathers of the wings are of a dark blue-black, becoming lighter at their edges, and the legs are covered with chestnut-brown feathers like those of the back. These gorgeous hues are not attained until the bird has passed through the moult of its second year. Both male and female are nearly equally decorated, the latter being slightly less brilliant than her mate. It is not a very large bird, scarcely exceeding a foot in total length.

THERE are many examples of the group which is gathered together under the general title of Rollers, the last of which is the BRACHYPTERACIAS, a bird which is remarkable, as its name imports, for the shortness of its wings.

In colour it is rather a handsome bird, although it suffers somewhat from the proximity of its more brilliant relatives. The upper parts are a warm chestnut-brown, with a green gloss upon the shoulder. The wings are brown also, glossed with sheeny green, and marked with a number of black spots edged with white. The under parts are greyish white, splashed on the throat with chestnut, and transversely barred upon the abdomen with the same tint, leaving a white band across the chest. As its name implies, it is a native of Madagascar.

THE curious little birds which are termed TODIES bear a considerable resemblance to the kingfisher, from which they may be easily distinguished by the flattened bill. The gape of the mouth is very wide, and a number of vibrissæ are set around its margin. The wings and tail are short and rounded, and the outer toes are connected as far as the last joint. The Todies are natives of tropical America, and are very conspicuous among the brilliant plumaged and strangely shaped birds of that part of the world.

The GREEN TODY is a very small bird, being hardly larger than the common wren of England, but yet very conspicuous on account of the brilliant hues with which its plumage is decorated. The whole of the upper surface is a light green, the flanks are rose coloured,

GREEN TODY.—*Todus viridis.*

deepening into scarlet upon the throat and fading into a pale yellow upon the abdomen and under tail-coverts. The under surface of the wings is bare. These tints may be easily examined, even during the life of the bird, for the Green Tody is a sluggish creature, and so disinclined to move, that it may be approached quite closely, and watched as it sits with its head sunk beneath its shoulders, and its bill projecting, as if without life or sensation.

It has but little power of wing, flying always near the ground, and never venturing on a long aërial journey. From this habit of remaining near the earth, it is popularly known by the name of Ground Parrot. The food of the Green Tody is chiefly of an insect nature, and the bird is able to secure its prey as they crawl about in the muddy banks of ponds or rivers. It also searches the grass and herbage for insects, and catches them with much adroitness. The nest of this bird is placed on the ground, generally in some hole in a river's bank, but often in a depression made for the purpose, and is built of dried grasses, moss, cotton, feathers, and similar substances. The eggs are four or five in number, of a bluish grey, diversified with bright yellow spots. The length of this bird is barely four inches. There is another species of Tody (*Todus Mexicánus*), inhabiting the same country.

The JAVAN TODY is a truly remarkable bird, and is so curiously formed that its proper position in the kingdom of birds has long been uncertain.

The extraordinary beak of this bird is shorter than the head, and at its base is wider than the portion of the head to which it is attached. The centre toes are connected together as far as the second joint. This bird is a native of Java and Sumatra, and in many of its habits resembles the green Tody. It feeds mostly on aquatic insects, worms, and larvæ, which it obtains from the banks of the rivers near which it loves to dwell. It does not keep so closely to the earth as the green Tody, but builds a pendent nest, hanging to the slender bough of some tree that grows near the water. Although not a very rare bird, it is but seldom seen, owing to its habit of withdrawing itself to the most inaccessible wooded lands of its native country, and there taking up its residence near the swampy grounds that are often found within the precincts of vast forests.

It is rather variable in its plumage; some specimens having a black bar across the chest. In all cases it is a striking bird, owing to the forcible contrast between the deep velvet purple of the back and the bright golden yellow with which it is relieved. Another species, belonging to the same genus, the Hooded Eurylaïmus (*Eurylaïmus ochromalus*), is still more beautiful, on account of the delicate rose hue with which its throat is tinted, and the bold black, white, and yellow marking of the remainder of the plumage.

Although a very handsome bird, it does not equal the preceding species in the brilliancy of its plumage. The general colour is a deep rich purple, diversified by yellow, black, and brown. The yellow is chiefly seen in the coverts and edges of the wing, and the tips of the tail-coverts. The forehead, tail-coverts, and tail feathers are black, with the exception of a white mark upon the quill feathers of the tail.

JAVAN TODY.—*Eurylaïmus Javanicus.*

The GREAT-BILLED TODY has been placed by some authors in the genus Eurylaïmus, together with the Javan Tody, but in the catalogue of the British Museum it is separated into a new genus under the title of Cymbirhynchus, or Boat-billed, in allusion to the singular form and shape of its beak. The specific title of Macrorhynchus is also given in allusion to the same peculiarity, and signifies Long-billed.

It is rather a thickly made bird, possessing a stout, heavy-looking body, which harmonises well with the great boat-shaped beak. The curiously-shaped bill is very large, thick, and strong, very wide at its base, well arched above, and hooked at the point. Both mandibles are about the same length, and the colour is blue. The bird is an inhabitant of the Indian Archipelago, and is most numerously found in the interior of Sumatra, where it may be seen haunting the banks of rivers and searching for its food, which consists chiefly of insects, worms, and various aquatic creatures. Its nest is ingeniously constructed of slender twigs woven into a nearly globular form, and

GREAT-BILLED TODY.—*Cymbirhynchus macrorhynchus.*

is fastened to the extremity of some convenient branch which overhangs the water, so that the young and eggs are safe from the attacks of the many enemies which assail them in these regions. The eggs are from two to four in number, and of a pale blue tint.

The colouring of its plumage is rather handsome, although quaint and peculiar. The general tint of the upper parts of the body is dead black, and that of the abdomen and lower parts a dark red. Around the throat runs a broad belt of stiff, wiry feathers of a red hue, which point upward on each side, and are probably intended for the purpose of defending the eyes. At each side of the bill there are several similar stiff bristly hairs, which also point upward. The scapularies are long and sharp in form, and their colour is a beautifully pure white, contrasting strongly with the deep black of the upper part of the body. At the upper angle of each wing there is a well-defined orange line and a white spot on the inside. The wedge-shaped tail is black, the thigh is blackish-brown, and the legs are brown. The colour of the eyes is blue, which changes to green soon after death and then fades wholly into dulness.

By the natives the bird is called Burong-palano, or Tam-palano. Several other species of Eurylaïmus are known to science, all of them being handsome and remarkable birds. The Great Eurylaïmus (*Eurylaïmus Córydon*), for example, is notable for the great width of the beak, its bright rosy hue, its hooked form, and the very wide gape of the mouth. The plumage too is coloured in a very bold and striking manner. The general hue is jet black, relieved by a large white mark on the middle of the wing, another at the extremity of the tail, and a small scarlet patch of elongated feathers in the centre of the back. As a general rule, the birds of this group adhere to the above-mentioned colours, but there is a curious and notable exception in the person of the Dalhousie's Eurylaïmus (*Psarisómus Dalhousiæ*), whose plumage is tinted with blue, green, and yellow, after the manner of many paroquets. Indeed, the general aspect of the bird irresistibly reminds the observer of a paroquet, and the semblance

BRAZILIAN MOTMOT.—*Mómotus Braziliensis.*

is farther increased by its long azure tail feathers.

The MOTMOTS, so called from their monotonous cry, which is thought to resemble the syllables mot-mot, continually repeated, are inhabitants of tropical America and the adjacent parts of the world. There are several species of these curious and beautiful birds; but as their habits and form are very similar, they can be sufficiently represented by a single example. The Motmots are among the number of those creatures which have perplexed the systematic naturalist, and their position in the kingdom of birds is even yet subject to doubt. On account of their large and deeply serrated mandibles, their long bearded tongue, and the similarity of some of their habits, they have been placed close to the toucans, to which birds they bear no small resemblance. Their feet, however, are of entirely different construction; and instead of congregating in flocks like the toucans, they lead solitary lives in the forest depths. In these birds the tail is wedge-shaped, and in several of the species the two central feathers are remarkable for a naked space before their termination.

The BRAZILIAN MOTMOT is, like the other species of the same genus, a very solitary bird, being seldom seen except by those who care to penetrate into the deepest recesses of the tropical forests. In its habits, it is not at all unlike the common fly-catcher of England, delighting to sit motionless upon a branch that overlooks one of the open spaces that are found in all forests, or that commands a view of a path made by man or beast. On its perch it remains as still as if carved in wood, and sits apparently without thought or sensation until a tempting insect flies within easy reach. It then launches itself upon its prey, catches the insect in its bill, and returning to its perch, settles down again into its former state of languid tranquillity. The Motmot is not formed for long or active flight, as its wings are short and rounded, and the plumage, especially about the head, very loosely set.

Some writers say that the Motmots do not confine themselves to such small prey as insects, but that they steal young birds out of their nests, and are also in the habit of eating eggs.

All the Motmots are about the size of the common magpie, and are remarkably handsome birds, their plumage being tinted with green, blue, scarlet, and other brilliant hues. The Brazilian Motmot is bright green on the upper parts of the body, excepting a spot of velvety-black upon the head, edged with green behind. The primary feathers are blue, and the under portions are green, "shot" with crimson, and a black spot is found on the breast.

TROGONS.

FOR our systematic knowledge of the magnificent tribe of the TROGONS we are almost wholly indebted to Mr. Gould, who by the most persevering labour and the most careful investigations has reduced to order this most perplexing group of birds, and brought into one volume a mass of information that is rarely found in similar compass. There are few groups of birds which are more attractive to the eye than the Trogons, with all their glowing hues of carmine, orange, green, and gold; and few there are which presented greater difficulties to the ornithologist until their various characteristics were thoroughly sifted and compared together. The two sexes are so different from each other, both in the colour and shape of the feathers, that they would hardly be recognisable as belonging to a single species, and even the young bird is very differently coloured from his older relatives.

These beautiful birds are found in the Old and the New Worlds, those which inhabit the latter locality being easily distinguishable by their deeply barred tails. Those of the Old World are generally found in Ceylon, Sumatra, Java, and Borneo, while only a single species, the Narina Trogon, is as yet known to inhabit Africa. The Trogons are mostly insect-eaters, taking their prey easily by means of their widely-opening mouths, and making no use of their slender feet and claws in the capture of their active enemy. Although gifted with such brilliant colouring, they are but seldom seen, for they prefer the deep forest to the more open grounds, and remain seated among the dense foliage of some chosen tree as long as the sun remains above the horizon.

MASSENA'S TROGON.—*Trogon Masséna*   MEXICAN TROGON.—*Trogon Mexicánus.*

The Trogons are mostly silent birds, the only cry used being that of the male during the season of pairing. It is not a very agreeable sound, being of a sombre and melancholy cast, and thought to resemble the word "couroucourou," a continuation of syllables which has therefore been applied to the entire tribe. The Trogons have been separated into five genera, each of which will find an example in the following pages.

THE splendid bird which has been called MASSENA'S TROGON, in complimentary allusion to the celebrated prince of that name, is an inhabitant of Central America, specimens from Honduras and Mexico being in the collection of the British Museum. In size it is rather large, measuring fourteen inches in total length.

In the tinting of the plumage the two sexes are very different from each other, and are coloured briefly as follows. In the male, the crown of the head, the back and chest are a deep rich green, contrasting well with the jetty, glossy black of the ear-coverts and throat; the breast and abdomen are of a rich scarlet. The ground tint of the centre of the wings is a soft grey, pencilled with exquisitely delicate lines of jetty black. The quill feathers of the wing are jetty black, each feather being edged with pure white; and the quill feathers of the tail are also black, with the exception of the two central feathers, which are imbued with changeable hues of dark green and purple throughout the greater part of their length, and are tipped with a black patch at the extremity. The bill is light yellow.

The female bird possesses a more sober plumage than her mate. The upper parts of the body, instead of being richly coloured with deep green, are of a dark bluish-grey; and the wings, instead of being finely pencilled with black upon grey, are powdered with the same tints. The abdomen and breast are scarlet, and the bill is rather curiously coloured, the upper mandible being black and the lower yellow.

This diversity of colouring in the sexes, which holds throughout the entire group, is productive of very great trouble to the systematic naturalist, as the two sexes are in many cases so very unlike each other that there is hardly any criterion for settling the species to which they belong, except by patient and careful observation of their habits when at liberty in their native haunts. When, moreover, the birds are shy and retiring in their habits, as is the case with many of the Trogons, the amount of labour which is entailed upon the observer is more than doubled, and the value of such a work as Mr. Gould's monograph is proportionately increased.

The MEXICAN TROGON is, as its name implies, an inhabitant of that country whose name it bears, being generally found in the northern districts of Mexico. It is worthy of remark that the Trogons of America are all similar in their colouring, the upper parts of the adult males being green, and the under portions either scarlet or yellow. The young male and the female birds are not so brilliant in hue. In most instances the outer quill feathers of the tail are barred with black and white. The beak of these species is marked with notches along the tip of the mandibles and pointed with bristly hairs at the base.

In the Mexican Trogon, the females and young males are so different from the adult bird in their plumage, that each must be described separately. The adult male, when he has obtained the full glory of his feathery adornment, is coloured as follows. The entire upper surface of the body, together with the chest, are of a beautifully rich glossy green, and the whole of the under surface, with the exception of the chest, is bright scarlet. The throat and ear-coverts are black, and a crescent-shaped band of white surrounds the throat. The wings are black, finely dotted with grey, except the primaries, which are wholly black. The tail is curiously diversified with black, white, and green. The two central feathers are green, tipped with black, the two next are black, and the remainder black tipped with white. The head is bright yellow.

The female is not coloured so gorgeously as the male, but is sufficiently brilliant in hue to be considered a very handsome bird were not her colours overpowered by the superior beauty of her mate. The rich green feathers of the male are in her of dark olive-brown, taking a slight green hue on the back, and chestnut on the upper part of the breast. The under parts are scarlet, as in the male, and there is a grey streak across the chest. The wings are black, splashed with brown and white, the latter hue being found mostly on the edges of the primaries. The tail is scarlet, similar to that of the male, except that the green feathers of the centre are replaced with brown, tipped with black, the two next entirely black, and the other feathers are barred with black and white throughout. The bill is yellow, tinged with brown.

The young male possesses the black and white barred tail, and white fringed primary feathers of the female bird, but may be distinguished from her by the green hue of its upper surface and chest. The wings are not so black as in the adult bird, but take a decided tinge of brown on the secondaries, and are spotted rather profusely with grey. The total length of the Mexican Trogon is about one foot, of which the tail occupies very nearly eight inches.

The NARINA TROGON is an inhabitant of Africa, and is generally found in the densest forests of the southern portions of that continent. Its rather graceful name was given to it by Le Vaillant, in remembrance of Narina, a young Gonaqua Hottentot girl, whose dusky charms and savage graces made an instantaneous and most powerful impression on the heart of the susceptible Frenchman.

This species has many of the habits of the flycatcher, with the exception of its partially nocturnal mode of life. During the daytime it seeks the darkest recesses of its native forests, and selecting a dead branch as its perch, sits dull and motionless until the evening. It then sallies from its place of refuge, and settling upon a convenient bough, sweeps upon every insect that may pass within a convenient distance, and carrying its prey back to the perch, devours it at leisure. It also is fond of chasing various beetles as they run upon the ground or along the branches of trees, and feeds largely upon the caterpillars and other larvæ which abound in such localities. The early morning is also

NARINA TROGON.—*Apaloderma Narina.*

chosen by the Narina Trogon for a feeding season; but when the sun has risen high above the horizon, the bird ceases its labours, and betakes itself to its resting-place.

The general colour of this beautiful bird is emerald green, shining with an almost metallic lustre. This hue is spread over the whole of the upper surface, except the wings and tail, and also tinges the throat and chest. The abdomen and remainder of the under surface are bright red. The wings are brown covered with minute dots of grey upon the secondaries and greater coverts, and the tail is coloured with several shades of green above, diversified by the pure white of the three outer feathers on each side. The beak is yellowish blue. The female bird is differently coloured, her plumage being of the following hues :—The green of the upper surface and throat is not quite so resplendent as in her mate, and a rusty brown tint is spread over the throat and round the eyes, warming into a delicate rose upon the chest. The total length of the bird is nearly one foot.

As a general rule it is a very silent bird, but during the pairing season the male constantly utters a loud and rather doleful cry, so that on the whole it is thought to be rather a sad and melancholy bird. The nest is made in the hole of a tree, and the eggs are four in number, rather globular, of a rosy-white hue, caused most probably by the semi-transparency of the shell. The young are said to be able to follow their mother as soon as hatched.

In describing these birds, it is not easy to avoid a considerable sameness in the language, on account of the frequent recurrence of the same gorgeous tinting. The words green, scarlet, black, and white necessarily occur so frequently, that the aid of colour is almost needed to enable the reader to realize the full vividness of the plumage which decorates these wonderful birds, which receives but scant justice from the plain black and white of a wood engraving. From the feathers of the Trogons, the ancient Mexicans were accustomed to make their justly famed feather pictures and mantles. For this purpose thousands of these and similar birds were kept in confinement, a whole army of attendants being maintained for the purpose of attending to them and securing their valuable plumage.

CUBA TROGON.—*Prióteltus temnúrus.*　　　MALABAR TROGON.—*Harpactes Malabáricus.*

THE very rare and curiously formed CUBA TROGON is a native of the country from which it derives its name.

According to Gould, it bears a singular resemblance to the woodpeckers, both in its habits and in the general formation of its plumage. Like those birds, it runs about the trunks and branches of trees, peers into the hollows, and dislodges from under the bark the insects on which it feeds. The most striking peculiarity in its form is the shape of the tail feathers, which have the web extending beyond the shaft in such a manner that they seem to have been trimmed with scissors. The specific name "temnurus" signifies "clipped wing," and is given to the bird on account of this remarkable formation.

The back and upper tail-coverts of the Cuba Trogon are bright grassy green, and the head and ear-coverts are steel-blue. The wings are beautifully barred with white, green, and black; and the tail is blue-green in the centre, the feathers being green with blue edges, those of the exterior are white, and the rest barred with white and green. The total length of this bird is about eleven inches. In the illustration the left-hand figure represents the Cuba Trogon.

The MALABAR TROGON is a very local bird, and is thought never to be seen in any locality except that from which it derives its name.

It is a somewhat nocturnal bird, and is so totally different in its habits according to the time of day, that it would hardly be recognised for the same creature. During the day the Malabar Trogons sit in pairs on the topmost branch of some tree, and seldom stir from their post until evening. Sometimes they rouse themselves sufficiently to pounce upon a passing insect, but immediately return to the perch, and resume their position.

2.　　　　　　　　　　　　　M

BEAUTIFUL TROGON.—*Calúrus antisiínus.*

But when the dusk approaches, the Malabar Trogon shakes off its drowsiness, and becomes one of the most spirited and active of birds, flitting from branch to branch, and tree to tree, or traversing the boughs in search of its prey, with wonderful adroitness, and almost meteoric rapidity.

The head and neck of the adult male bird are deep sooty black, and the back and upper surface are brownish yellow. A white crescent-shaped stripe runs round the chest, and separates the black hue of the throat from the brilliant scarlet of the breast and remainder of the under surface. The primary quill feathers of the wings are black edged with white, and the centre of the wings is pencilled with very delicate white lines on a blacker ground. The tail is boldly marked with chestnut and black, and is decorated with white tips at the extremity of the feathers. The bill and the space round the eyes are light blue. The female is easily distinguished from her mate by the duller hue of her plumage, and the absence of the beautiful scarlet which decorates the abdomen of the male bird.

SEVERAL of the Trogons are distinguished from their relatives by the length and downy looseness of many of the feathers, more especially the lance-shaped feathers of the shoulders, and the elongated upper tail-coverts. On account of this structure of the plumage, they are gathered into a separate genus under the appropriate title of Calurus, or Beautiful-tailed Trogons.

The first of these birds, the BEAUTIFUL TROGON, is a native of South America, and well deserves its name, as it is not only richly gorgeous in the colours of its plumage, but is also elegant in form. On account of the looseness of its feathers it is not able to chase insects in the air with as much adroitness as is exhibited by the firmer feathered Trogons, and is found to make its diet chiefly upon berries, fruits, and the insects which it can pick off the branches without being forced to pursue them on the wing. While engaged in the search after food, it is a sufficiently active bird, running about the boughs with great agility, and clinging with its powerful feet in every attitude, seeming to care little whether it be sitting on a branch, after the custom of most birds, or hanging with its head downward, like the parrots.

Although so brilliant in colouring, it is not so easily seen as might be supposed, for its colour harmonizes well with the foliage and bark of the trees among which it dwells, and even the rich carmine of its under surface is not very conspicuous in that land of flowers.

As may be seen from the engraving, the head is decorated with a curiously-shaped tuft of slight and elastic feathers, which spring from the forehead, and by their peculiar curve overshadow the nostrils and a considerable portion of the beak. This crest, together with the head, the throat, the back, wing-coverts, and upper tail-coverts, are of the richest imaginable green glazed with gold, glowing with a changeable sheen as the breeze plays with the delicate fibres of the plumage. The quill feathers of the wing are black, as are the six central feathers of the tail. The upper tail-coverts are very long, exceeding the tail by two inches, flowing gracefully over the stiffer feathers by which they are supported, and contrasting beautifully with their glossy black. The abdomen and remainder of the under surface is rich carmine. The total length of an adult bird is about fourteen inches.

BEFORE entering into any detailed description of the RESPLENDENT TROGON, we must explain that in order to bring it within the limits of our pages, it has been drawn in smaller proportions than any of the other Trogons. In size the Resplendent Trogon is larger than the species which has just been described, so that if it had been drawn to the same scale of proportion the engraving would

RESPLENDENT TROGON.—*Calurus resplendens*

M 2

have been rather more than sixteen inches in length, being precisely double the length of the present pages.

Of all the birds of the air there is hardly any which excites so much admiration as the Resplendent Trogon.   Many, such as the humming-bird, are gifted with greater brilliancy of colour; but for gorgeousness of hue, exquisite blending of tints, elegance of contour, and flowing grace of plumage, there is no worthy rival in all the feathered tribe. This magnificent bird is a native of Central America, and was in former days one of the most honoured by the ancient Mexican monarchs, who assumed the sole right of wearing the long plumes, and permitted none but the members of the royal family to decorate themselves with the flowing feathers of this beautiful bird.

In all the Trogons the skin is very delicate, and the feathers are so loosely attached that they are always liable to be lost when the bird is handled; but in the Resplendent Trogon the skin is so singularly thin that it has been not inaptly compared to wet blotting-paper, and the plumage has so slight a hold upon the skin, that when the bird is shot, the feathers are plentifully struck from their sockets by its fall and the blows which it receives from the branches as it comes to the ground.   These peculiarities render the preservation of the skin no easy task; and the difficulty of removing the skin without injury is so well known to the natives, that they almost invariably dry the body without attempting any further preservation.

This species is fond of inhabiting the densest forests of Southern Mexico, and generally haunts the topmost branches of the loftiest trees, where it clings to the boughs like a parrot, and traverses their ramifications with much address.   It does not seem to expend much time on the wing, and to all appearance feeds more on vegetable diet than is the case with its relatives.

The colour of the adult male bird is generally of a rich golden green on the upper parts of the body, including the graceful rounded crest, the head, neck, throat, chest, and long lancet-shaped plumes of the shoulders.   The breast and under parts are brilliant scarlet, the central feathers of the tail are black, and the exterior white with black bars. The wonderful plumes which hang over the tail are generally about three feet in length, and in particularly fine specimens have been known to exceed that measurement by four inches, so that the entire length of the bird may be reckoned at four feet.   The bill is light yellow.

As is often the case with birds, where the male is remarkable for the beauty of his plumage, the female is altogether an ordinary and comparatively insignificant bird, at least to human eyes, although beautiful enough in those of her mate.   She possesses only the rudiments of a crest or elongated plumes, as may be seen by reference to the engraving, where both sexes are represented.   The colour of the upper surface is nearly the same as that of the male, although hardly so vivid, and the head, throat, and chest are of a decidedly dull green.   The breast and abdomen are greyish brown, and the under tail-coverts are scarlet.   The elongated feathers of the shoulders are not so long as in the male, nor so sharply pointed, nor so vividly coloured.   The central feathers of the tail are black, and the exterior are white marked with black bars ; the bill is black.   The young of the first year, whether male or female, assume this dress, and do not put on their full glory of apparel until they have passed through the moult of the second year.

In reviewing this group of birds, the thoughtful observer cannot fail to be struck with wonder and admiration at the extreme beauty of their forms, and the indescribably gorgeous hues with which their plumage is interpenetrated.   The mode in which these marvellous colours are produced, and the reason for which their existence is necessary, are two of the many mysteries which abound in all nature, and which excite to the highest degree the minds of those who care to look below the surface, and who take more interest in causes than in effects.

What strange vital chemistry is that, which, in addition to supplying the ordinary substances of the body, extracts from dead insects and gathered fruits the glowing hues which bedeck the plumage of these resplendent birds, and lays every tint in a true and just gradation which sets at defiance the brush and pencil of the most accomplished artist ?   What is the reason—for we may be assured that the Creator does nothing without some very deep purpose—that necessitates each colour ; its depth, brilliancy, and tone, and regulates the position of every hue ?

There is, doubtlessly, some grand and simple law that governs all these apparently capricious variations of colour, tint, and intensity ; and which, if it should ever be discovered, will be found to rule with the most rigid and undeviating precision every shade of colour and depth of tone throughout the three kingdoms of material nature, painting the plumage of the bird, the leaves and corolla of the flowers, and the translucent glories of the gem, in obedience to one and the same universal principle. Even in the human race the rule of the colour-producing principle is most powerful, and exercises a potent influence upon the individual being and those with whom it comes in contact.

The various shades of black that tinge the skins of many African and even more northern tribes, the coppery hue of the North American savage, the yellow tint of the Mongol, and the pure white and red of the Caucasian, are not to be considered as mere casual differences, or as the results of food, climate, and habit only, but as the external indications of some vast but hidden principle that can only exhibit itself by its effects. Even in a single race, such as that of our own island, the effect of this chromic law as manifested in the colour of the eyes and hair is strongly marked, and to a certain degree tells its own story. By a kind of natural instinct, teaching us to read intuitively the hieroglyphs of nature, we deduce the character of an individual as much from the tone and clearness of the colour of his eyes, as from the shape of the features, or that subtle and almost spell-like power which is known by the term " expression."

There are some eyes dull and emotionless, behind which a curtain seems to be drawn, leaving them as devoid of meaning as the glass optics of a waxen image ; while others gleam with gem-like translucency, beaming from the spirit-light of the ingenuous human soul. Other eyes there are which resemble in their profound mystery the unfathomable depths of ocean, rich with pearls and hidden treasures, and ever disclosing greater and still greater stores of intellectual wealth. We are quite wrong whenever we undervalue the importance of any variety of external form or colour, for we may rest assured that whatever God found reason to make, we may find reason to observe with reverence and read with awe. It is only when we remain satisfied with a mere knowledge of external form, without inquiring into its cause or its influence, that our efforts are wasted, or our time expended in vain.

These questions are not insignificant, for even our modern discoveries have shown that the apparently merest trifle may be the key to hidden treasures of knowledge ; neither ought they to be heedlessly disregarded so long as we are told that we may learn deep lessons from the lilies of the field and the birds of the air.

BARBETS.

THE BARBETS evidently form a connecting link between the trogons and the kingfishers, possessing several of the peculiarities of the former birds, together with some characteristics of the latter.

In shape they bear a close resemblance to the kingfishers, and none of them are of any great size. Their food consists chiefly of insects, which they chase much after the manner of the woodpeckers, prying into the hollows of trees, and striking away the bark in their endeavours to secure the concealed prey. They can cling to the perpendicular trunk of a tree, and support themselves by the pressure of their short stiff tails against the bark. They also possess some of the habits which belong to the flycatchers, and taking their perch upon a twig, will wait patiently until an unfortunate insect passes within a short distance, when they will launch themselves on the devoted creature, and return to the identical twig from which they started.

COLLARED BARBET.—*Bucco colláris.*

To all appearance the Barbets are dull and heavy birds, seeming to pass a very unenviable kind of existence; chained as it were to a single spot, and apparently feeling every movement a source of trouble. But to the Barbet itself, this kind of inactive life constitutes its best happiness; and we should be as wrong to attribute sadness and melancholy to it, as was Buffon when he spoke in such forcible and eloquent terms of the miserable existence passed by the woodpeckers. While sitting upon the twig which it has chosen for its perch, the Barbet has a curious habit of puffing out its plumage, so as to transform itself into an almost cylindrical ball of feathers, and has, on account of this odd custom, been termed the Puff Bird.

There are many species of Barbet, one of which, the COLLARED BARBET, a native of South America, is an example of the typical genus Bucco. It is rather a pretty little bird, the head and neck being of a chestnut fawn, the chest white, and the under parts of the body the same hue as the head, but of a lighter tint. A well-defined black collar or band runs across the chest, and extends over the shoulders, where it merges into the chestnut brown of the back. The wings and back are darker than the head, and covered with a number of small black bars. The tail is chestnut and barred with black.

The WHITE-FACED BARBET is not so handsome a bird, being more sombre in its clothing than the collared Barbet. The general colour of this bird is black, and the forehead and face are white, together with the chin. In size it is about equal to our

WHITE-BACKED BARBET.—*Cœlidóptua tenebrósa.*

WHITE-FACED BARBET.—*Monasa leucops.*

common starling. It is also a native of Southern America. This bird has been chosen as a representative of the genus Monasa, a small group of birds which has been separated from the other Barbets on account of the form of the beak and the structure of the wing.

The last example of these curious birds is the WHITE-BACKED BARBET, which serves to represent the genus Chelidoptera. This is a much smaller bird than either of the preceding examples, but is notable on account of the curious manner in which its plumage is diversified with black and white. The general tint of the body is sooty black, but upon the back there is a conspicuous patch of white, and a considerable amount of white is scattered over the middle of the wings, and upon the under tail-coverts. It is also a native of Southern America.

So highly gifted are these birds with that quality which is called " adhesiveness " by phrenologists, that when they have once selected a twig as a resting-place, they will remain faithful to their choice, and for month after month may be seen sitting on the identical perch, lethargic and happy. They are solitary birds, never being seen in flocks, and very seldom in pairs, residing always in the murkiest recesses of the deep forests, in preference to the open country, and sitting on their low perch in spots which the foot of man seldom penetrates.

KINGFISHERS.

THE KINGFISHERS form a tolerably well marked group of birds, all of which are remarkable for the length of the bills and the comparative shortness of their bodies, which gives them a peculiar bearing that is not to be mistaken.

The bills of these birds are all long and sharp, and in most cases are straight. Their front toes are always joined together more or less, and the number of the toes is **very**

GROUP OF KINGFISHERS.

variable in form and arrangement; some species possessing them in pairs, like those of the parrots, others having them arranged three in front and one behind, as is usually the case with birds, while a few species have only three toes altogether, two in front and one behind. The wings are rounded. As may be gathered from their popular name, they mostly feed upon fish, which they capture by pouncing upon the finny prey; although in

LAUGHING JACKASS.—*Dacélo gigas.*

some instances, such as that with which we shall commence our history, they make the greatest part of their diet of insects and crustaceans. In colour they are very variable, some being comparatively dull in tint, possessing no colours but black, brown, and white, while others are decorated with the most brilliant plumage, which nearly equals that of the trogons in gorgeousness of hue, although the colours are not distributed in such large masses, nor are the feathers so exquisitely soft and downy. They are to be found in nearly all parts of the world, and our own island boasts of one of the handsomest, although one of the smallest, species.

Our first example of the Kingfishers is the Laughing Jackass, or Giant Kingfisher, its former title being derived from the strange character of its cry.

This bird is an inhabitant of Australia, being found chiefly in the south-eastern district of that country, and in New South Wales. In Van Diemen's Land Mr. Gould believes that it does not exist. In no place is it found in any great numbers; for although it is sufficiently common, it is but thinly dispersed over the country. It is rather a large bird, being eighteen inches in total length, and is powerful in proportion, being able to wage successful war against creatures of considerable size.

Although one of the true Kingfishers, it so far departs from the habits of the family as to be comparatively careless about catching fish, and, indeed, often resides in the vast arid plains where it can find no streams sufficiently large to harbour fish.in their waters. Crabs of various kinds are a favourite food with this bird, which also eats insects, small mammalia, and reptiles. Mr. Gould mentions an instance where he shot one of these birds for the sake of possessing a rare and valuable species of rat which it was carrying off in its bill. It is also known to eat snakes, catching them with great dexterity by the tail, and crushing their heads with its powerful beak. Sometimes it is known to pounce upon fish, but it usually adheres to the above-mentioned diet.

The cry of this bird is a singular, dissonant, abrupt laugh, even more startling than that of the hyæna, and raising strange panics in the heart of the novice, who first hears it while bivouacking in the "bush." Being of a mightily inquisitive nature, the Laughing Jackass seems to find great attraction in the glare of a fire, and in the evening is apt to glide silently through the branches towards the blaze, and, perching upon a neighbouring bough, to pour forth its loud yelling cry. The "old hands" are in nowise disconcerted at the sudden disturbance, but shoot the intruder on the spot, and in a very few minutes convert him into a savoury broil over the fire which he had come to inspect.

At the rising and the setting of the sun the Laughing Jackass becomes very lively, and is the first to welcome the approach of dawn, and to chant its strange exulting pæans at the return of darkness. From this peculiarity, it has been called the Settler's Clock. In allusion to the cry of this bird, which has been compared by Sturt to the yelling chorus of unquiet demons, the natives call it by the name of Gogobera.

We evidently have in this bird another example of the frequency with which one idea runs through and intersects the various divisions of the animal kingdom, mystically uniting by undefinable bonds the various departments and innumerable groups of living beings. Several of these remarkable facts have already been mentioned, where the question was of form ; and we have in the Laughing Jackass, and its resemblance in that respect to the laughing hyæna, a similarity of voice in two very opposite beings. In the same manner, the voice of the harmless ostrich is a roar so precisely resembling that of the fierce and carnivorous lion, that even the Hottentots have been unable to discriminate between the bird and the quadruped. As a general rule, colour is but little developed in the mammalian forms, and very greatly so among the birds. Yet we have several instances among the mammals—such as the mandril and several other quadrumana—where the vivid colouring of the skin is but little inferior to that which paints the plumage of the tropical birds.

There is, again, one characteristic which is even more universal in its occurrence than either of the preceding, namely, the development of odour. Probably on account of the forcible manner in which it strikes the senses, those creatures which exhale an unpleasant scent are more readily known than those which emit a pleasant, or at all events a non-offensive emanation. As the most prominent types of this principle, we have among the quadrupeds the skunk, among the birds the vulture, and among the reptiles the crocodile with its intolerable musky scent. Among insects the skunk principle is very prominent, and is found in many of the most lovely and exquisite forms of insect life as well as in those creatures which are repulsive to the eye as well as to the nostrils.

The Laughing Jackass is in no way fastidious in choosing a locality, as it may be found in equal plenty in the bush, the forest, or the open plain. While at rest upon a branch, it sits in a rather dull and "lumpy" attitude, its chin resting upon its breast, and arousing itself at intervals to utter its discordant laugh. It is readily tamed, and bears the climate of England with tolerable hardihood.

The home of the Laughing Jackass is usually made in the hole of a gum-tree (*eucalyptus*), where it makes no sort of nest, but simply lays its eggs upon the soft decaying wood. The eggs are pearly white, and the bird keeps a vigilant watch over the burrow which holds its treasures, fiercely combating any creature that may approach the entrance, and aiming the most desperate blows with its long pointed and powerful beak.

It is a really handsome bird, and although not possessing such an array of brilliant plumage as falls to the lot of many Kingfishers, is yet very richly coloured. The bird is

BUFF DACELO.—*Dacélo cervina.*

decorated with a dark brown crest, and the general tint of the back and upper surface is olive brown. The wings are brown black, a few of the feathers being slightly tipped with verditer, and the breast and under portions are white, washed with pale brown, which forms a series of faint bars across the breast. The tail is rather long, and rounded at the extremity, and is of a rich chestnut colour, banded with deep black and tipped with white.

SEVERAL species belonging to the same genus are worthy of a passing notice, among which we may mention LEACH'S DACELO (*Dacelo Leachii*), and the BUFF DACELO (*Dacelo cervina*). The former of these birds is a remarkably handsome creature, and inhabits the north-eastern parts of Australia, where, according to Mr. Gould, it takes the place of the laughing jackass. It is a little smaller than that bird, and resembles it greatly in its form and general habits. Its head and crest are dark brown, and the abdomen is covered with numerous narrow wavy brown bars. In the male bird the wings and tail are richly coloured with deep orange; but in the female the tail is chestnut, barred with a bluish black.

The Buff Dacelo inhabits the thickly wooded portions of the northern and north-western districts of Australia, where it may be seen and heard sitting on the topmost branches of the loftiest trees, taking observations of the surrounding country, and yelling in a most unmusical manner. When three or four pairs of these birds get together upon a single tree, they become quite excited by mutual noise, and make such a horrid uproar that nothing can be heard except their deafening outcries. It is a very shy bird, and not easily approached within range of shot.

The general colour of this bird is a pale fawn marked with brown, and with a considerable amount of rich blue in the wings and tail of the male, the tail feathers being largely tipped with white. The tail of the female is chestnut, boldly barred transversely with deep black, and tipped with buff.

AUSTRALIAN KINGFISHER.—*Halcyon Sancta.*

Of the genus Halcyon, the AUSTRALIAN KINGFISHER affords a good example.

This bird is a resident in New South Wales from August to December or January, and then passes to a warmer climate. Like the preceding birds, it cares little for the presence of water, making its subsistence chiefly on large insects, such as locusts, caterpillars, grasshoppers, and cicadæ, which it seizes in its bill, and beats violently against the ground before eating them. It is also very fond of small crabs and other crustaceans. Mr. Gould mentions that the stomachs of Australian Kingfishers that had been shot were found crammed with these creatures. To obtain them, it is in the habit of frequenting the sea-shore, and pouncing upon the crabs, shrimps, prawns, and various other creatures as they are thrown on the strand by the retiring tide, or forced to take refuge in shallow rock-pools, whence they can easily be extracted by the long bill of this voracious bird.

On the banks of the Hunter River this Kingfisher resorts to a very curious method of obtaining food. There is a kind of ant which builds a mud nest upon the dead branches and stems of the gum-trees, and by the unpractised eye would be taken for fungi or natural excrescences. The Kingfisher, however, knows better, and speedily demolishes the walls with his powerful beak, for the purpose of feeding upon the ants and their larvæ.

Like the preceding bird, the Australian Kingfisher is a most noisy creature, and remarkably fond of exercising its loud startling cry, which is said to resemble the shriek of a human being in distress, sharp, short, urgent, and frequently repeated. There is hardly any real nest of this species, which chooses a convenient hollow branch or "spout" as its domicile, and there lays its eggs. They are generally from three to five in number, and are of a pure white.

It is rather a fine bird, being nearly the same size as the laughing jackass. The top of its head and the back of the scapularies are tinged with a dull green, and the throat, neck, and abdomen are buff, abundantly flecked with brown spots. The wings and the tail are of a rather peculiar greenish blue, in which the latter hue prevails, and the ear-coverts and a line round the back of the head are blackish green.

THE genus Tanysíptera is well illustrated by the well-known though somewhat scarce TERNATE KINGFISHER, a bird which may be easily recognised by the peculiar form of the tail. The generic name is of Greek origin, and signifies Long-winged, and is rather longer than needful, the simpler form of the word being Tanýptera, or more properly Tanypteryx. But when once a systematic naturalist begins to indulge in so-called classical nomenclature, he seems to be irresistibly attracted by the words in proportion to their length and abstruseness.

Thus it happens that the pages of our scientific works are disfigured by vast quantities of cacophonic combinations of syllables, many of them entirely needless, and the greater number of such barbarous construction, that neither Roman nor Greek would acknowledge them as belonging to his own language. The general common sense of those who love Nature for her own sake, and not for the sake of the harsh vocabulary which has been appended to science, has now begun to revolt against the cumbersome phraseology which has been so needlessly employed, and which serves in many cases to deter real lovers and observers of Nature from entering into the details of science at the cost of so great a task to the memory as is now needed to gain the character of a scientific man.

The unnecessary multiplication of genera has now come to such a pitch, that, according to a well-known writer on this subject, "the study of ornithology (we may say zoology) is merging into a study of barbarous nomenclature; we shall soon have a genus for every species; and this is called science!"

The Ternate Kingfisher is one of those species which are decorated with richly coloured plumage, and is a truly handsome and striking bird. The head is of a bright ultramarine blue, and the upper parts of the body are of a deeper tint of the same colour, being of a "Prussian" blue, that is almost black in its intensity. The wing-coverts are of the same ultramarine as the head, as are also the edges of the quill feathers

TERNATE KINGFISHER.—*Tanysíptera Dea*

of the tail. The two central tail feathers are much prolonged, considerably exceeding the others in length, and are very curiously shaped, being webbed at their bases, bare for nearly the whole of their length, and again webbed at the extremities. Their colour throughout is blue, the tips being white, as are the remaining feathers of the tail with the exception of their blue edges. The whole of the under parts are white.

The Ternate Kingfisher is a native of New Guinea, and from thence its skin has often been sent to Europe, but in a mutilated state, the natives being in the habit of depriving it of its legs and wings before parting with the skin. The Paradise birds were long treated in the same manner, until the sportsmen learned that they could sell the entire bird at a better price than when it had been mutilated.

In the birds which form the genus Ceÿx there are only three toes, and one of them very strong, the tail is very short, and the bill straight, like that of the common Kingfisher of Europe. The reader must note that the word Ceÿx is dissyllabic.

TRIDACTYLOUS KINGFISHER.—*Ceÿx Tridáctyla.*

The TRIDACTYLOUS KINGFISHER is a native of Java, Borneo, and the whole of the Indian Archipelago, and is said to have been discovered even upon the continent itself. Although a very little bird, it is one of the most brilliantly coloured of the entire group, and hardly yields even to the gem-like humming-bird in the metallic and glittering colour of its plumage. Even the united aid of pencil and brush can give but little idea of the extreme beauty of the colouring of this bird, for the glowing richness of the tints as they flash and glitter with every movement of the bird and vary momentarily in hue and tone, is far beyond the power of art, and sets at nought the colours of the most skilful painter.

The head of the Tridactylous Kingfisher, as well as the whole upper surface of the body, is a deep rich lilac, and the wings are stained with a most beautiful and singular mixture of deep blue and ultramarine, the centre of each feather being of the former tint, and the edges of the latter hue. The whole of the under surface is pure white, the feet are red, and the bill is a pale carmine. In its dimensions it is exceedingly small, being one-third less than the common Kingfisher of Europe.

THE interesting birds which are gathered into the genus Ceryle may be known by the thick, compressed, and sharply pointed beak, the comparatively long and rounded tail, and the length of the front inner toe. To this genus may be referred all the American species of this group, one of which, the BELTED KINGFISHER, forms the subject of the illustration.

The Belted Kingfisher is an inhabitant of many parts of America, and as it is in the habit of migrating northward or southward according to the season of year and the state of the temperature, it is a very familiar bird throughout the greater part of America, from Mexico to Hudson's Bay. So common is it in these regions that, according to Wilson, "mill-dams are periodically visited by this feathered fisher, and the sound of his pipe is as well known to the miller as the sound of his own hopper."

The sight of the Belted Kingfisher is singularly keen, and even when passing with its

BELTED KINGFISHER.—*Cérylc A'lcyon.*

meteor-like flight over the country, it will suddenly check itself in mid career, hovering over the spot for a short time, watching the finny inhabitants of the brook as they swim to and fro, and then with a curious spiral kind of plunge will dart into the water, driving up the spray in every direction, and after a brief struggle will emerge with a small fish in its mouth, which it bears to some convenient resting-place, and after battering its prey with a few hearty thumps against a stump or a stone, swallows it, and returns for another victim. Waterfalls, rapids, or "lashers" are the favoured haunts of the Belted Kingfisher, whose piercing eye is able to discern the prey even through the turmoil of dirty water, and whose unerring aim fails not to seize and secure the unsuspecting victims, in spite of their active fins and slippery scale-covered bodies.

"Rapid streams," says Wilson, "with high perpendicular banks, particularly if they be of a hard, clayey, or sandy mixture, are also the favourite places of resort for this bird, not only because in such places the small fish are more exposed to view, but because those steep and dry banks are the chosen situation of his nest."

In these banks the Belted Kingfisher digs a tunnel, which often extends to the length

SPOTTED KINGFISHER.—*Céryle guttáta.*

of four or five feet, employing both beak and claws in the work. The nest is of a very simple nature, being composed of a few small twigs and feathers, on which are laid the four or five pure white eggs. The birds seem to be much attached to their homes, and the same pair will frequent the same hole for many successive years, and rear many broods within the same habitation. The extremity of the burrow where the eggs are placed is always rather larger than the tunnel itself, and is expanded into a globular chamber for the purpose of affording a sufficiency of space for the parents and their young. It is said that when a supposed enemy approaches the nest, the parent birds employ various artifices to draw his attention away from the sacred spot, and by fluttering about as if wounded or disabled, will often succeed in their endeavours. When the young are hatched, the parents are remarkably attentive to them, as might be supposed from the reckless manner in which they expose themselves to danger for the sake of their offspring.

The colouring of this spirited little bird is rather complicated and not very easy of description. The head is furnished with a crest of long pointed feathers, which can be raised or depressed at will, and the whole upper surface of the body is light blue, marked with a great number of narrow dark streaks caused by the black-blue shaft of each feather. The wings are blackish-brown, bound with white upon the primaries, and diversified with blue upon the exterior web of the secondaries. The sides are covered with blue mottlings, a belt of the same bright hue crosses the chest, and a broad white band encircles the neck, throat, and chin. The tail is black-brown barred with white, with the exception of the two central feathers, which are blue. The length of this bird is about one foot.

GREAT AFRICAN KINGFISHER.—*Céryle máxima.*

The voice of the Belted Kingfisher is loud, dissonant, and startling, and has been compared by Wilson to the sound produced by twirling a watchman's rattle.

Mr. Thompson records two instances where the Belted Kingfisher has been seen and shot in Ireland. One specimen was placed in the Museum of Trinity College, Dublin.

ASIA presents us with a remarkably fine representative of the same genus in the handsome SPOTTED KINGFISHER, several specimens of which bird are to be found in the British Museum. This magnificent bird, which is appropriately called by the natives Muchee-bag, *i.e.* Fish-Tiger, is an inhabitant of India, where it seems to be confined to the Himalayan district.

In size it is but little inferior to the Great African Kingfisher, being one foot three inches in length, and bearing a bill three inches long. The chest and sides of the neck of the Fish-Tiger are of a beautiful greyish-white, which slightly deepens into a very pale fawn on the abdomen and the under tail-coverts. The remainder of the body is covered with jetty-black plumage, relieved by numerous spots of pure white, and the head is decorated with a large and noble-looking crest, composed of elongated feathers of the same boldly contrasting hues. A few black spots form a curved line between the bill

2. N

and the shoulder, and also are scattered in an uninterrupted band across the chest. The food of this bird consists mostly of fish, although it feeds also on aquatic insects. Its nest is made among large stones, and is not composed only of fish-bones, as is the case with the common European Kingfisher, but is rather elaborately constructed of mud lined with grasses, and adheres to the stones after the fashion of the well-known swallows' nest. The eggs are four in number, and resemble the parent bird in their markings.

AFRICA affords an instance of another species belonging to the same genus, Ceryle. This is the GREAT AFRICAN KINGFISHER, an inhabitant of Senegal, Congo, and many parts of Western Africa; a bird which well deserves its specific title of "máxima," or greatest, as it is the largest species of the genus to which it belongs, measuring nearly fifteen inches from the point of the very long beak to the extremity of the rather short tail.

It is by no means a gaudy bird, possessing none of the metallic green and blue plumage which is so conspicuous in many species belonging to the same group, but is nevertheless adorned with bold and striking marks of black, white, and deep ashen-grey. The top of the head is black, with the exception of a white streak that runs from the eye towards the nostrils, and is furnished with a feathery tuft. The whole of the upper surface is of a dark ashen-grey, approaching to brown. The same hue extends over the wings, but is diversified with a great number of short white bars and dashes, which are also found, but very small in size and very thinly scattered, upon the back. The tail is also ashen-grey, and marked with seven distinct white bars on each feather, and on a close examination, the middle of the feathers will be found to be black. Across the whole of the chest runs a very broad reddish-brown band, obscurely spotted with a darker hue. The chin is grey, and the cheeks are pure white marked with a number of black "tears," which arrange themselves in several lines. The abdomen and under surface are greyish-white, marked on the sides with dark-brown spots, and towards the tail with short irregular bands of the same hue.

ANOTHER species of Ceryle, the BLACK AND WHITE KINGFISHER (*Céryle rudis*), is remarkable for the peculiarity of tinting from which it derives its name; which affords a rather remarkable contrast to the brilliant hues of blue and green which decorate the majority of these birds. It is a native of many parts of the Old World, being spread over various portions of Asia and Africa.

THE common KINGFISHER is by far the most gorgeously decorated of all our indigenous birds, and can bear comparison with many of the gaily decorated inhabitants of tropical climates.

It is a sufficiently common bird, although distributed very thinly over the whole country, and considering the great number of eggs which it lays, and the large proportion of young which it rears, is probably more plentiful than is generally supposed to be the case. The straight, glancing flight of the Kingfisher, as it shoots along the river-bank, its azure back gleaming in the sunlight with meteoric splendour, is a sight familiar to all those who have been accustomed to wander by the sides of rivers, whether for the purpose of angling, or merely to study the beauties of nature. So swift is the flight of this bird, and with such wonderful rapidity does it move its short wings, that its shape is hardly perceptible as it passes through the air, and it leaves upon the eye of the observer the impression of a blue streak of light. This straight, arrow-like course is that which is generally adopted by the bird, but on some occasions the Kingfishers will become very playful, and sport with each other in the air, turning and wheeling with much adroitness as they mutually chase or avoid each other in their game.

The food of this bird consists chiefly, though not exclusively, of fish, which it takes, kills, and eats in the following manner :—

Seated upon a convenient bough or rail that overhangs a stream where the smaller fish love to pass, the Kingfisher waits very patiently until he sees an unsuspecting minnow or stickleback pass below his perch, and then, with a rapid movement, drops into the water like a stone and secures his prey. Should it be a small fish, he swallows it at once; but if it should be of rather large dimensions, he carries it to a stone or stump,

KINGFISHER.—*Alcédo Ispida.*

beats it two or three times against the hard substance, and then swallows it without any trouble.

The Kingfisher is sometimes given to hoarding, and having caught more fish than he can eat, will take them to his secret storehouse, and there hide them until he is able to eat them. In one such case, of which an account has been forwarded to me, the bird seemed to employ its storehouse for the reception of fish which it had caught, and which were too large for it to swallow. The treasury chosen by this individual was a crevice formed by the roots of a willow-tree that grew close to the water's edge, and it usually contained one or two fish. Sometimes there would be five or six fishes lying in the hole, and some of them so large that they have been removed and cooked. The bird must sometimes have found a great difficulty in getting its prey fairly ashore, as the tracks were evident on the soft mud of the bank where the fish had been dragged, and the bird's feet had trodden. Young trout were the general occupants of the storehouse, and in every case the fish had been killed by a bite across the back of the head or neck.

Sometimes the bird has been known to meet with a deadly retribution on the part of his prey, and to fall a victim to his voracity. One such example I have seen. A Kingfisher had caught a common bull-head, or miller's thumb, a well-known large-headed fish, and on attempting to swallow it had been baffled by the large head, which refused to pass through the gullet, and accordingly choked the bird. The Kingfisher must have been extremely hungry when it attempted to eat so large a morsel, as the fish was evidently of a size that could not possibly have been accommodated in the bird's interior. Several similar examples are known; but one, which is recorded by Mr. Quekett, is of so remarkable a kind, that it is worthy of notice. The bird had caught and actually attempted to swallow a young dabchick, and, as might be supposed, had miserably failed in the attempt.

The most complete instance of poetical justice befalling a Kingfisher, is one which occurred in Gloucestershire, and was related to me by an eye-witness. The narrator was sitting on the bank of a favourite river and watching the birds, fish, and insects that disport themselves upon and in its waters, when some strange blue object was seen floating down the stream, and splashing the water with great vehemence. On a nearer approach it was seen to be a Kingfisher, from whose mouth protruded the tail and part of the body of a fish. The struggles of the choking bird became more and more faint, and had well-nigh ceased, when a pike protruded his broad nose from the water, seized both Kingfisher and fish, and disappeared with them into the regions below.

The same person who related this tragical story, and who has always felt an ardent love for birds, beasts, and things that enjoy the blessings of life, and to whom I have been indebted for much curious information and many valuable specimens of natural history, also tells me that Kingfishers are greatly susceptible of music, provided that it be played in a slow and solemn strain like the old ecclesiastic chants. There was an organ in the house placed in a room that looked toward the stream which the Kingfishers frequented, and it was observed by the household that whenever music of such a character was played upon the organ, the Kingfishers would soon make their appearance at the bottom of the garden, and sit as if enchanted with the strains. Quick and lively airs seemed rather to disconcert the birds, a fact which was not discovered until after many experiments and the consumption of much time.

With the fish it generally feeds its young, being able to disgorge at will the semi-digested food which it has swallowed, after the manner of most birds of prey. Fish, however, do not constitute its sole nourishment, as it is known to eat various insects, such as dragon-flies and water-beetles, and will often in cold weather pay a visit to the sea-shore for the purpose of feeding upon the little crabs, shrimps, and sandhoppers that are found upon the edge of the tide. Still, by far the greatest portion of its diet is composed of fish ; and I have never yet found any fragments of insect anatomy among the _débris_ which forms the nest.

The nest of the Kingfisher is always made in some convenient bank, at the extremity of a hole which has previously been occupied and deserted by the water-rat or other mining quadrupeds, and been enlarged and adapted for use by the Kingfisher. Now and then the nest of this bird has been found built in the deserted hole of a rabbit-warren. It is always found that the tunnel slopes gently upward, and that the bird has shaped the extremity into a globular form in order to contain the parent bird, the nest, and eggs. Sometimes the nest is placed in the natural crevices formed by the roots of trees growing on the water's edge. In many cases it is easily detected, for the birds are very careless about the concealment of their nest even before the eggs are hatched, and after the young have made their appearance in the world, they are so clamorous for food and so insatiable in their appetite that their noisy voices can be heard for some distance, and indicate with great precision the direction of their home.

Some writers say that the interior of the burrow is kept so scrupulously clean that it is free from all evil scents. My own experience, however, contradicts this assertion, for after introducing the hand into a Kingfisher's nest, I have always found it imbued with so offensive an odour that I was fain to wash it repeatedly in the nearest stream. As the Kingfisher is so piscatorial in its habits, it would naturally be imagined that the nest would be placed in close connexion with the stream from which the parent birds obtained their daily food. I have, however, several times seen a Kingfisher's nest, and obtained the eggs, in spots that were not within half a mile of a fish-inhabited stream. The bird is greatly attached to the burrow in which it has once made its nest, and will make use of the same spot year after year, even though the nest be plundered and the eggs stolen.

The eggs are from six to eight in number, rather globular in form, and of an exquisitely delicate pink in colour while fresh, changing to a pearly white when the contents are removed. As soon as the young are able to exert themselves, they perch on a neighbouring twig or other convenient resting-place, and squall incessantly for food. In a very short time they assume their yearling plumage, which is very nearly the same as that of the adult bird, and soon learn to fish on their own account.

The nest of the Kingfisher has long been known to consist of the bones, scales, and other indigestible portions of the food, which are ejected from the mouth in " castings," like those of the hawk or owl ; but until Mr. Gould recently procured a perfect Kingfisher's nest, its shape and the manner of construction was entirely unknown. His account of its discovery, and the ingenious manner in which it was procured, is so interesting that it must be given in his own words.

" Ornithologists are divided in opinion as to whether the fish-bones found in the cavity in which the Kingfisher deposits its eggs are to be considered in the light of a nest, or as merely the castings from the bird during the period of incubation. Some are disposed to consider these bones as entirely the castings and fæces of the young brood of the year before they quit the nest, and that the same hole being frequented for a succession of years, a great mass is at length formed ; while others believe that they are deposited by the parents as a platform for the eggs, constituting, in fact, a nest ; in which latter view I fully concur, and the following are my reasons for so doing :—

On the 18th of the past month of April, during one of my fishing excursions on the Thames, I saw a hole in a precipitous bank, which I felt assured was a nesting-place of the Kingfisher, and on passing a spare top of my fly-rod to the extremity of the hole, a distance of nearly three feet, I brought out some freshly-cast bones of fish, convincing me that I was right in my surmise. On a subsequent day, the 9th of May, I again visited the spot with a spade, and after removing nearly two feet square of the turf, dug down to the nest without disturbing the entrance-hole or the passage which led to it. Here I found four eggs placed on the usual layer of fish-bones ; all of these I removed with care, and then filled up the hole, beating the earth down as hard as the bank itself, and replacing the sod on the top in order that barge-horses passing to and fro might not put a foot in the hole. A fortnight afterwards the bird was seen to leave the hole again, and my suspicion was awakened that she had taken to her old breeding-quarters a second time.

The first opportunity I had of again visiting this place, which was exactly twenty-one days from the date of my former exploration and taking the eggs, I again passed the top of my fly-rod up the hole, and found not only that the hole was of the former length, but that the female was within. I then took a large mass of cotton wool from my collecting-box, and stuffed it to the extremity of the hole, in order to preserve the eggs and nest from damage during my again laying it open from above. On removing the sod and digging down as before, I came upon the cotton wool, and beneath it a well-formed nest of fish-bones, the size of a small saucer, the walls of which were fully half an inch thick, together with eight beautiful eggs and the old female herself. This nest and eggs I removed with the greatest care, and I now have the pleasure of exhibiting it to the Society, before its transmission to the British Museum ; the proper resting-place of so interesting a bird's nest. This mass of bones then, weighing 700 grains, had been cast up and deposited by the bird, or the bird and its mate, besides the unusual number of eight eggs, in the short space of twenty-one days.

To gain anything like an approximate idea of the number of fish that had been taken to form this mass, the skeleton of a minnow, their usual food, must be carefully made and weighed, and this I may probably do upon some future occasion. I think we may now conclude, from what I have adduced, that the bird purposely deposits these bones as a nest ; and nothing can be better adapted, as a platform, to defend the eggs from the damp earth."

The Kingfisher, if unmolested, soon learns to be familiar with man, and has no hesitation in carrying on the daily affairs of its life without heeding the near presence of a human observer. I have known a Kingfisher to sit upon a projecting stone that over-hung a stream running at the foot of a garden, and to permit the owners of the garden to watch its proceedings without exhibiting any alarm. If managed properly, this interesting bird will so far extend its confidence to man as to become partially domesticated, speedily rivalling the robin or the sparrow in the bold familiarity of its manners. One such bird, that was tamed by a friend, owed its domestication to the loss of its parents.

Three young Kingfishers were seen sitting in a row upon a branch of a tree close to a stream, and drew the attention of their future guardian by their constant wailing after food. Various kinds of food were accordingly procured for the poor desolate birds ; but as the right sort of diet was not obtained for some days, two of the young birds died. The third, however, survived, and lived for a considerable time, coming regularly for his food, and receiving it at the hands of his protector, but never venturing into the house. In process of time he met with a mate and founded a family after the usual Kingfisher fashion. But he soon discovered that it was easier to supply his family with food by resorting to his kind friends and asking them for fish, than by spending time and trouble in capturing fish for himself.

One of these birds became self-tamed, if such an expression may be used, and was remarkably familiar with the person to whom it owed its self-acknowledged allegiance. The association began as follows :—

A young man was fishing in a preserved stream, and had caught, as is sometimes the case, a very little fish instead of the trout which he was endeavouring to capture. He took the insignificant prey off the hook and flung it towards the river, intending to return it to the water. His aim, however, was not a good one, and the fish fell upon the bank instead of reaching the stream, and was immediately pounced upon by a Kingfisher that shot unexpectedly through the air. Being rather amused at so bold a proceeding, the angler threw the next little fish on the grass, and had the pleasure of seeing the bird come and seize it as before. By degrees the bird became more and more familiar, until, encouraged by impunity, it would snatch up a fish within a yard of the angler, and after swallowing it or carrying it home, would perch on a neighbouring bough and wait for more.

After a while the angler bethought himself of accompanying each fish with a peculiar cry, and in a very short time the bird understood the call, and would come whenever it was uttered. This strange friendship endured for upwards of three years, but the ultimate fate of the bird I have not been able to discover. It never would take the fish out of the hand, but was in every other way so exceedingly tame that the keepers were utterly astounded, the possibility of taming a wild bird never having entered their dull heads.

Many young Kingfishers have been successfully reared from the nest, and seem to be hardy little birds, provided that they are furnished with a very large supply of minnows and other small fish. The number of fish which a single bird will consume is almost incredible, for its voracity is so great and its digestion so rapid that it causes no small trouble to the owner in insuring a proper supply of food.

The Kingfisher is a very solitary bird, never assembling in flocks, and seldom being seen except when single. Sometimes it has been observed to engage in aërial sports with a companion, and it frequently happens that the two parent birds are seen in company, or that the whole family sits amicably upon the same branch. With these exceptions, however, the Kingfisher is remarkably eremitical in its habits, and appears to suffer no rival establishment near the spot where it has fixed its home. It is, however, a very affectionate bird to its family, and a very remarkable instance of its loving nature has been recorded by a correspondent of the *Field* newspaper, who saw a female Kingfisher fly to the aid of her mate when shot, and picking him up by his bill, endeavour to carry him to a place of safety. The poor bird was unable to carry her wounded mate more than a few yards, but the affectionate unselfishness of the act is as praiseworthy as if she had succeeded in her attempt.

The voice of the Kingfisher is a peculiarly shrill and piping cry, that can be heard at some distance, and is not easily mistaken for any other sound.

The colour of this bird is very gorgeous, and rather complicated in its arrangement. The top of the head and back of the neck are dark green, flecked with many spots of verditer blue upon the tips of the feathers. The upper part of the back is also dark green, and the lower part is light violet or blue, gleaming vividly under a strong light, and being very conspicuous as the bird is on the wing. The tail is deep indigo, and the quill feathers of the wing are dark blackish-green, lightened by a brighter hue of green on the outer webs, and set off by the verditer-blue spots of the tertiaries. A white patch or streak passes from the eye to the back of the neck, and a dark green streak is drawn immediately under the white patch. The throat and chin are yellowish-white, and the whole of the under surface is chestnut. The eyes are crimson, and the bill is black, with the exception of the orange-tinted base of the lower mandible. The total length of the bird is about seven inches.

Brilliant as is the plumage of this bird, and unmistakable as the varied colouring seems to be, it loses all its gorgeousness when viewed against freshly fallen snow, and is not a whit less dull than that of the dipper. This curious fact is noticed by Mr. Thompson, whose acute and practised eye was unable to recognise the species of a small brown bird that continually flitted across his path as he was engaged in shooting woodcocks, and who was so perplexed by its curious style of flight, that he at last shot it, and to his regret found that he had killed a Kingfisher.

So conspicuous a bird as the Kingfisher takes, as may be imagined, a conspicuous part in legends of antiquity, and the traditions of the present day.

The classical scholar is familiar with the expression " Halcyon days," which is so frequently employed to denote a season of special security and joyousness, and is derived from an old fable, that the Halcyon or Kingfisher made its nest on the surface of the seas and possessed some innate power of charming the waves and winds to rest during the time of its incubation. Fourteen days of calm weather were in the power of the Kingfisher, or Alcyonë, who was fabled to be the daughter of Æolus, wearing a feathered form in token of grief for the loss of Ceÿx, her husband, and to have derived her authority from her father, the lord of winds.

In many parts of England at the present day there is a singular idea concerning the Kingfisher, which seems to have its origin in the same mythical history. Those who are familiar with cottage life in the rural districts will often have noticed a Kingfisher suspended by the point of the beak from the beams of the ceiling, and if they have asked the object of the custom, will learn that the bird always turns its breast towards the quarter from which the wind is blowing. Some writers in mentioning this custom have said that the bird is so suspended as to point with its beak towards the wind; but in every case where I have seen this curious wind-vane, it has been hung by the very extremity of the beak, so as to rotate freely in every direction. The bird is not stuffed, but various spices are placed in its interior after the viscera have been removed, and the body is then dried by exposure to the sun.

TINY KINGFISHER.—*Alcyone pusilla.*

The Tiny Kingfisher is found, according to Gould, in Northern Australia and New Guinea, and is a remarkably beautiful little creature.

In its habits it is very shy, and seems to prefer the deepest thicket as its place of residence, so that it cannot easily be approached without taking the alarm, and, indeed, is but seldom seen at all, even by those who give their attention wholly to the search after objects of natural history. Its voice, however, will often betray its presence, as it is fond of hearing itself talk, and frequently utters a shrill piping note, which can be heard at a considerable distance, and cannot be mistaken for the voice of any other bird. Although it is able to fly with considerable swiftness, it is not very powerful on the wing, its flight being strangely unsteady.

In its habits it resembles the European Kingfisher, catching and feeding on fish in much the same manner.

The general colour of the Tiny Kingfisher is a most intense blue, which, with few exceptions, is spread over the whole of the upper surface. Upon the eyes and below the ear-coverts there is a rather large white patch, the primary feathers of the wings are blackish brown, and the secondaries blue, edged with shining green. The throat, chest, and abdomen are of a beautifully pure white, contrasting boldly with the deep blue of the upper parts of the body.

Another species, the Azure Kingfisher (*Alcyone azurea*), is closely allied to the tiny Kingfisher, and is also a native of Australia, but inhabits a different locality, being found in New South Wales and Southern Australia.

The nest of this bird is made in holes in the banks, and is simply composed of the disgorged bones, scales, and other indigestible portions of the fishes which have been rejected after the manner of most carnivorous birds. The number of eggs is rather large, being from five to seven. The young are remarkably noisy, and whenever the parent birds pass the entrance of the hole the young Kingfishers immediately think themselves hungry, and set up a clamorous appeal for food. It is a very remarkable fact that the young birds assume the plumage of the adult at their first moult, and being always rather precocious, soon manage to get their own living.

The food of this bird consists chiefly of fish and aquatic insects. It is solitary in its habits, being never seen assembled in numbers, and appearing to exercise a watchful jurisdiction over a certain amount of land which it chooses to consider as its own property. The intrusion of a stranger is instantly resented, and as the temper of the bird is naturally quarrelsome, it is no uncommon event to see a pair of them engaged in conflict, dashing to and fro like angry meteors, and whirling through the air in transports of rage. The general colour of the Azure Kingfisher is bright ultramarine blue above, buff upon the neck, chest and abdomen, and pure white upon the chin.

JACAMARS.

The curious birds which are popularly known by the name of JACAMARS are all natives of the New World, and, as might be imagined from the metallic brilliancy of their plumage, are denizens of the tropical regions of their native land.

In all these birds the bill is straight, long, rather compressed, pointed, with a decided keel on the upper mandible, and with the corners of the mouth defended by some bristle-like hairs. As will be seen, the toes are varied in number, some species possessing only three toes, while the remainder are furnished with four toes as usual. The two front toes are united only as far as the claws, and the thumb or "hallux" is either very short or altogether absent. They are insect-eaters, and greatly resemble the trogons in many of their habits.

The PARADISE JACAMAR is a striking little bird, on account of the beautiful colours with which its plumage is decorated, its graceful form, and the long forked tail. It is but a small bird, being not as large as an ordinary thrush, but its plumage is so beautiful in its colouring and so graceful in the arrangement of its feathers that the spectator entirely forgets its size in admiration of its beauty. The neck of this species is rather long and mobile, enabling the bird to dart its long straight bill in every direction with great rapidity. The tail is rather curiously formed, the feathers being so graduated that the central pair extend far beyond the others, and form a kind of fork, alterable at the will of the bird. As the Jacamars bear a very close resemblance to the kingfishers, they were formerly supposed to belong to that group of birds, and the Paradise Jacamar was termed the Fork-tailed Kingfisher.

The head of the Paradise Jacamar is brown tinged strongly with violet, and the throat, the neck, and some of the wing-coverts are pure white. The back, wings, and remainder

PARADISE JACAMAR.—*Gálbula Paradisea.*   GREEN JACAMAR.—*Gálbula viridis.*

of the body is a rich golden green, and the bill and feet are black. The feet are feathered nearly as far as the toes.

In its habits the Paradise Jacamar is not unlike the trogons and flycatchers, seldom troubling itself to chase its prey through the air, but preferring to sit upon a bough and catch the butterflies as they pass unconsciously near the feet of their destroyer, and then pounce suddenly upon them and secure them in its long bill. So persevering are they in their watchfulness, and so strong is their attachment to the spot where they have taken up their residence, that the locality where they feed can readily be discovered on account of the wings, legs, and other uneatable portions of their prey, which they twist off and throw away before endeavouring to swallow their victim. The Paradise Jacamar is a native of Surinam.

The GREEN JACAMAR, receives its popular name from the slight preponderance which green holds above the other hues in the colouring of its plumage.

Nearly all the Jacamars present a very similar arrangement of colours, which is by no means easy to describe, as the feathers are tinted with glowing hues of green, azure, gold, and metallic red, all of which seem to have been scattered at random over the plumage, and to have become so intermixed that the eye fails to separate them, or to assign any particular locality to any particular colour. Indeed, the plumage of the Jacamars is a very Turkey-carpet of tints, all the colours being very bright but without any definite arrangement; so that, although clad in gorgeous raiment which nearly equals the plumage of the humming-birds in its bright effulgence when examined feather by feather, the Jacamars are by no means conspicuous birds, and at a little distance do not appear nearly so handsome as our English starling.

The colour which is most conspicuous in this and among other Jacamars is a bright metallic coppery-red, which continually changes to a purplish hue, and irresistibly reminds the observer of a copper tea-kettle that has been subjected to the action of fire. The top of the head is green, and the breast is marked with the same hue plentifully mixed with the peculiar coppery tint which has just been mentioned. The chin is greyish white marked with a few brown spots, the chest is dark green and copper, and the wings are also coppery-green, but possess a large admixture of blue. The breast is green with a little copper, and the abdomen chocolate, marked with a few dark longitudinal dashes. The upper surface of the tail is dark shining green, and its under surface is nearly of the same colour as the abdomen. The bird is quite a little one, not so large as our English kingfisher.

THREE-TOED JACAMAR.—*Jacamaralcyon tridactyla.*

Of the genus Jacamaralcyon we have a good example in the THREE-TOED JACAMAR.

This little bird, which is even smaller than the preceding species, possesses none of the brilliant hues which decorate the majority of the group, but is clad in colours even more sombre than those of the sparrow. The whole of the plumage, with very few exceptions, is of a dark, dull, lustreless, sooty-black, beside which the blackbird of England would look quite brilliant. On a closer inspection a dark olive-green reflection is visible on the upper surface of the body and tail. The top of the head is marked with two or three chocolate streaks, and there is another stripe of the same colour drawn from the corner of the mouth towards the back of the neck. The flanks are of the same sooty black as the back, but without the green reflection, and the white with a slight rusty-red tinge. The under surface of the tail is a grey brown.

The GREAT JACAMAR, or BROAD-BILLED LAMPROTILA as it is sometimes called, is so like the kingfishers in form and general outline of contour, that it might easily be mistaken for one of those birds by one who had not studied the characteristics of the group with some attention.

In this bird, which evidently forms a link of transition between the Jacamars and the Bee-eaters, and whose generic name of Jacamarops has been given to it in allusion to that fact, the beak is extremely broad when compared with the compressed bills of the other Jacamars, and the dilated ridge on the upper mandible is distinctly curved. The tail is broad and moderately long, and the feathers of the head form a partial crest. The short neck, rounded wings, and long bill of this bird give it a great resemblance to the kingfishers, and in its attitudes it has a great air of those birds. Like them, it poises itself upon a branch and darts down to secure its active prey in its bill, but differs from them in the fact that it feeds almost exclusively upon insects, and knows not how to snatch from the stream the scaly inhabitants of the waters.

In its colouring this bird very closely resembles the green Jacamar, which has already been described, but does not possess quite so much cf the green hue.

GREAT JACAMAR. — *Jacamerops grandis.*

BEE-EATERS.

THE BEE-EATERS may at once be distinguished from the jacamars by the shape of the bill, which, although somewhat similar in general shape to the beak of those birds, is curved instead of straight, and by the formation of the wings, which, instead of being short and rounded, are long and pointed, and give to their owners a wonderful command of the air, while engaged in chasing their winged prey. Some short bristles overhang the nostrils, and the long and broad tail has generally the two outer feathers longer than the others. Their plumage is remarkably handsome; being painted with rich and at the same time with extremely delicate hues of many colours. Green predominates throughout the group, a verditer-blue seeming to be generally mixed with the green. Some species, such as the Nubian Bee-eater (*Merops Nubius*), are clothed in bright red; while others, such as the Rose-breasted Night-feeder (*Nyctiornis Amicta*), are decorated with a rich rose tint upon the face and breast.

THE common BEE-EATER of Europe is very frequently found in many parts of the Continent, and has been several times taken in England. It is, however, a scarce bird in Great Britain, and is of sufficient rarity to excite some curiosity whenever it is found within the confines of our shores.

In Mr. Yarrell's well-known work on the British birds, there is a most elaborate enumeration of the specimens which had been shot in various parts of England; getting shot being a fate that inevitably befalls any rare or strange bird that may happen to visit this country. Even a stray parrot runs a great risk if it should escape into the rural districts, as can be unwillingly testified by many a sorrowing and bereaved parrot-owner. So far does this cruel and abominable custom go, that I have known a paroquet to be

BEE-EATER.—*Merops apiaster.*

shot by a farmer, although it had escaped from a house in the same little village. The destructive propensity is truly developed to a wonderful extent in some persons, who quite justify the sarcastic foreigner in his remark, that a heavenly day always inspires an Englishman with a desire to go out and kill something.

The food of the Bee-eater consists wholly of insects, the bees and others of the hymenopterous order being the favourite article of diet. In chasing these insects, which are for the most part very active of flight, the Bee-eater displays very great command of wing, and while urging its pursuit, can twist and turn in the air with as much ease and skill as is exhibited by the swallow or the roller.

Undaunted by the poisoned weapons of the wasp, hornet, or bee, the bird makes many a meal upon these insects, contriving to swallow them without suffering any inconvenience from their stings. It is probable that there may be some peculiarity in the structure of this and several other birds, that renders them indifferent to the poisonous influence of the sting, for it is difficult to account for their immunity on any other theory. Mr. Yarrell imagines that the Bee-eater renders its prey harmless by much pinching and biting, and that by "repeated compression, particularly in the abdomen, the sting is either squeezed out, or its muscular attachments so deranged, that the sting itself is harmless."

I cannot coincide in this opinion, for the sting cannot be entirely squeezed out of the abdomen by any amount of pressure, and its poisonous properties are quite as rife after it has been separated from the muscular attachments as during its connexion with them. I speak from experience, having suffered rather severely from the effect of a sting which I had received from a common honey-bee, and which was carelessly suffered to pierce the skin of my thumb. In the case of a wasp, too, we may also have noticed that after the abdomen has been completely crushed, the sting appears to possess a separate vitality of its own, and moves about as if still guided by the will of its dead owner.

To the apiarian, who resides in the same country with the Bee-eater, the bird is a terrible foe, as it has an insatiable appetite for the honey-making insects, and haunts every spot where it is likely to meet them. The hives are constantly visited by the Bee-eaters, who are ingenious enough to resort to the turpentine pines for the sake of catching the bees that come to carry away the exudations for the purpose of converting them into "propolis," or that substance with which they harden the edges of their cells, caulk the crevices of the hives, and perform many other useful tasks. It does not, however, confine itself to the hymenopterous insects, but is fond of beetles, cicadæ, grasshoppers, and similar creatures. These it eats entire, ejecting the wing-cases or "elytra," the wings, legs, and other indigestible portions, after the manner of the hawk and owl.

Taking advantage of its insect-eating propensities, the boys of the Greek Archipelago, where the Bee-eater is very common, are in the habit of capturing it by means of a hook and line, in a kind of aërial angling, in which the atmosphere takes the place of water, and the victim is hauled struggling downwards, instead of being drawn struggling upwards. A hook is attached to the end of a strong but slender line, and fastened to a cicada or other insect, in such a manner as to cause no impediment to its flight, and the cicada is then allowed to fly about at will. The Bee-eater soon perceives its fluttering prey, and darting upon it with open beak, is caught by the hook and made prisoner. Swallows are often captured in a somewhat similar manner; and many a fisherman has been disappointed of his cast by finding that a swallow had taken his fly while it was passing through the air, in the backward sweep preparatory to its fall upon the water.

While engaged in the pursuit of its prey, the Bee-eater flies at various heights, according to the weather and the species of insect which it is engaged in eating. Sometimes it may be seen careering high in air at so great an elevation that its beautiful colours cannot be distinguished, but attracting great admiration on account of its great command of wing and easy gliding movements. At another time it sweeps over the very surface of the ground, snapping up the bees, wasps, and other insects that are not in the habit of ascending to any great height.

In allusion to its insect-eating propensities, the colonists of the Cape of Good Hope, where the Bee-eater is found in plenty, call it by the name of "gnat-snapper," and the Hottentots find it a very useful bird, as it often directs them to the spot where the wild bees have made their nests. In Egypt it is called by a name which signifies "bees' enemy."

The Bee-eater is a gregarious bird, being generally seen in flocks that vary from ten to forty or even fifty in number, and presenting a very handsome appearance as they wheel round in the air, displaying the metallic effulgence of their beautiful plumage, and occasionally perching upon the branches of a tree and resting awhile from their toil. While flying after insects the Bee-eater continually utters its curious cry, which is a rich-toned warbling chirrup. In many places where the Bee-eater is common, it is killed and eaten by the natives, who are not deterred by any æsthetical prejudices from destroying so lovely a bird for the sake of roasting, but estimate the creature in proportion to its gastronomical value. It will be remembered that the equally beautiful roller meets with a similar fate in the lands where it is most frequently found.

The nest of the Bee-eater is not unlike that of the kingfisher, being placed at the extremity of a burrow made in some convenient bank. The burrow is excavated by the bird itself, and it often happens that the Bee-eaters are as gregarious in their nesting as in their flight, honeycombing the clay banks in a manner very similar to that of the sand-martin. The burrows do not run to any great depth, seldom exceeding six or eight inches in length. The nest is composed of moss, and contains about five or six beautifully white and pearly eggs.

Of the habits of this bird, Mr. Thompson has given so graphic and elegant a description in his work on the birds of Ireland, that his account must be subjoined in his own words.

"I have had the gratification of seeing the Bee-eater in scenes with which its brilliant plumage was more in harmony than with any in the British Isles. It first excited my admiration in August, 1826, when visiting the celebrated grotto of Egeria, near Rome. On approaching the classic spot, several of these birds, in rapid swift-like flight, swept closely past and around us, uttering their peculiar call, and with their graceful form and brilliant colours proved irresistibly attractive. My companion, who, as well as myself, beheld them for the first time, was so greatly struck with the beauty of their plumage and their bold sweeping flight, as to term them the presiding deities over Egeria's grotto.

Rich as was the spot in historical and poetical associations, it was not less so in pictorial charms. All was in admirable keeping;—the picturesque grotto with its ivy-mantled entrance and gushing spring; the gracefully reclining though headless white marble statue of the nymph ; the sides of the grotto covered with the exquisitely beautiful maiden-hair fern in the richest luxuriance ; the wilderness of wild flowers around the exterior attracting the bees, on which the Merops was feeding; and over all the deep blue sky of Rome, completing the picture.

On the 26th of April, 1841, three Bee-eaters, coming from the south, flew close past H.M.S. *Beacon*, sailing from Malta to the Morea, but did not alight. We were then about ninety miles from Zante (the nearest land), and a hundred and thirty from Navarino. On the morning of the next day, when forty-five miles from Zante, and sixty west of the Morea, a Bee-eater, coming from the south-west, alighted for a moment on the vessel, and then flew towards Zante in a south-eastern direction. Soon afterwards, a flock consisting of fifteen came from the same quarter, lurked about the lee side of the vessel for a short time, and then proceeded north-east. One hour after their departure (ten o'clock) a flock of eight appeared, and alighting on a rope astern the ship, remained there for nearly an hour. They were perched so closely together and so low down on the rope, that by its motion the lowest one was more than once ducked in the water, but nevertheless did not let go its hold or change its position for a drier one.

These birds were but a few yards from the cabin windows, and looked so extremely beautiful, that they were compared by some of the spectators to paroquets, and not very inaptly, on account of their gaudy plumage. After these left us, others were seen through-out the day, but generally singly : they rarely alighted : all flew in the same course.

When not very far to the westward of Cape Matafan, on the first of May, a flock of twenty-nine of the *Merops apiaster* flew close past the ship towards the Morea."

The Bee-eater is very common in Southern Russia, about the Don and the Volga, and is a familiar inhabitant of Turkey, Greece, Egypt, and Asia Minor. Africa seems to be its ordinary residence, from which country it migrates over the Mediterranean and pours into the various districts of Europe. In Spain, France, Germany, and Switzerland the Bee-eater is often observed, and is very common in Malta, Sardinia, and Sicily.

The colours of the adult male bird are extremely varied and very beautiful. The top of the head is rich chestnut brown, extending to the neck, back, and wing-coverts. Over the rump the chestnut changes to light reddish yellow. The primaries and secondaries of the wing are bright blue-green tipped with black, and their shafts painted with the same colour, and the tertiaries are green throughout their entire length. The upper tail-coverts are of the same hue as the wings, and the tail is likewise green, tinted with a darker hue, graphically called by Mr. Yarrell "duck-green." The chin and throat are reddish yellow, and around the throat runs a band of deep blue-black. The under part of the body is green with a blue tinge, and the under surface of the wings and tail is greyish brown. The ear-coverts are black, and the eye is light scarlet, which contrast beautifully with the chestnut, black, and yellow of the head and neck.

In the young birds the tints are not nearly so brilliant, and they are different in hue and arrangement; the rich saffron-yellow of the whole having a greenish tinge in the second year's bird, and the chestnut hue of the head extending only to the neck. A first year's bird, described by Mr. Yarrell, had the top of its head green, no red colour on the back, and no black collar round the neck. The tail feathers were all of the same length, whereas in the adult bird the narrow ends of the central pair of feathers extend beyond the others.

The female may be distinguished from the male by the paler hue of the reddish yellow on the throat, and the reddish tinge that runs throughout the green of the body and wings. In size the Bee-eater is nearly equal to the English starling.

The CHESTNUT BEE-EATER is most commonly found in the different islands of the Indian Archipelago, but has also been taken in Malacca and the South of France.

It is not a very large bird, but is remarkably slim and elegant in its form, appearing as if it were possessed of very great activity. The colours of its plumage, although not so brilliant as those of other Bee-eaters, are yet soft and fine, and very pleasing in their general effect. The general colour of this bird is a very dark green upon the upper surface, so deep as to approach nearly to black in a dim light, and the upper surface of the tail is a deep blue-green. Under each eye is a small black patch. The chin is white, and the under surface of the tail and the tail-coverts are greyish brown.

CHESTNUT BEE-EATER.—*Merops bádius.*

THE truly magnificent AZURE-THROATED BEE-EATER is an inhabitant of India, and is found, although very rarely, in the interior of that country.

It is a very rare bird, perhaps not so much on account of the actual paucity of its numbers, as from its extreme shyness, and the nature of the localities where it makes its residence. The home of this bird is always in the deepest recesses of the vast Indian forests, and in spite of its glowing colours and noisy tongue, it is so wary and fearful of man that it is seldom seen. When fairly discovered, however, it often falls an easy prey to the native hunter on account of the extreme nervousness of its nature. The report of a gun in close proximity will have such an effect upon its nervous system as to afflict it with a momentary paralysis, and it sometimes happens that in the great hunting expeditions of the native chiefs, this Bee-eater is so stupified by the unwonted turmoil, and repeated explosion of fire-arms, that it lies helplessly on the branch, and permits itself to be taken by hand.

The manners of the Azure-throated Bee-eater are particularly quiet, and during the daytime it is seldom to be seen in motion. At the approach of night, however, it becomes very active, and utters its peculiar short grating cry in rapid succession. The generic name, Nyctiornis, or Night-bird, has been given to this Bee-eater because it is supposed to feed mostly by night. Mr. Gould, however, doubts the accuracy of this theory, and the appropriateness of the title. The food of this species consists of various insects, such as bees, wasps, and other similar creatures. It seems, however, to feed mostly upon beetles, preferring those which belong to the Geotrupidæ, such as the chaffers and scarabæi, probably on account of their slow wheeling flight, which renders them an easy capture, and their solid fat bodies, which insure a plentiful meal attended with very little trouble.

It is but a solitary bird, seldom being seen even in company with its mate, and never associating in flocks like the Bee-eaters of Europe.

In plumage the Azure-throated Bee-eater is a really splendid bird, and is chiefly remarkable for the long soft azure feathers which hang from the throat and neck, like the nuchal mane of many antelopes. The top of its head is bright scarlet and blue, and the whole of the upper surface a brilliant green. The pendent feathers of the throat are verditer-blue, and those of the neck are bluish-green, edged with the same verditer-blue

AZURE-THROATED BEE-EATER.—*Nyctiornis Athertóni.*

as the plumes of the throat. The general hue of the under surface is buff dashed with green, and the under side of the wings and tail is dark buff without the green tinge. The eyes are reddish-brown, and the beak is blue-grey.

The sexes are very similar in colour, but the female is to be distinguished from her mate by her inferior dimensions. The young birds are not so brilliant in their colours, and they are devoid of the beautiful pendent plumage of the neck and throat.

There is a very beautiful and closely-allied species to the above bird, namely, the Rose-breasted Nyctiornis, or Red-faced Night-feeder (*Nyctiornis amicta*), of which a passing mention has already been made. This beautiful bird is a native of India, and is supposed to feed chiefly by night, although the fact is not very clearly ascertained. In the beauty and delicacy of the tints which stain its plumage, it may challenge comparison even with the trogons themselves. The crown of the head is a fine lilac, and the face, part of the throat, and the upper part of the breast, are a bright rose-carmine. The remainder of the plumage is golden green. The total length rather exceeds one foot.

The very handsome bird which is indifferently known by the name of Bullock's Bee-eater, and the Blue-bellied Bee-eater, is an inhabitant of Western Africa, and has been taken in Senegal. It is a truly beautiful creature, elegant in form and very brilliant in colour. It is not a very large bird, being considerably smaller than the common Bee-eater of Europe, and measuring only seven inches in total length, of which the bill occupies one inch.

BULLOCK'S BEE-EATER.—*Melittóphagus Bullockii.*

The upper portions of this species are light green with the exception of the upper parts of the neck, which is coloured by a reddish crimson hue, the two tints merging gradually into each other without any definite line of demarcation. From the gape of the mouth a black stripe runs towards the back of the neck, enveloping the eye in its progress, and a small spot of the same jetty hue is seen upon the tip of the chin. The throat is of the most brilliant scarlet, and the breast and upper parts of the abdomen are crimson like the neck. The lower part of the abdomen is clear ultramarine blue, a peculiarity from which the bird derives its popular name. The wings are green like the back, but the secondaries and tertiaries are tipped with velvety black, about half an inch in depth. The feathers of the tail are of equal length, and the bill and legs are black.

THE last example of the Bee-eaters which can be mentioned in these pages is the RED-THROATED BEE-EATER (*Melittóphagus guláris*). This bird is remarkable for the singular colouring of its plumage, and the vivid contrasts presented by a few spots of bright colour upon a dark ground.

The general colour of this species is a deep velvety black, through which a green hue shines in certain lights. Upon the forehead and over the eyes are scattered a few tiny but most brilliant verditer-green feathers gleaming with a metallic lustre, and a patch of long plumy feathers of the same brilliant hue occurs on the end of the back. The throat is decorated with a patch of light chestnut-red, and the remainder of the breast and abdomen are of the same deep velvety-black, over which a number of isolated verditer feathers are very thinly scattered.

SLENDER-BILLED BIRDS.

UPÚPIDÆ, OR HOOPOES.

THE large group of birds which are termed TENUIROSTRAL, or Slender-billed, always possess a long and slender beak, sometimes curved, as in the creepers, hoopoes, and many humming birds, and sometimes straight, as in the nuthatch and other humming birds. The feet are furnished with lengthened toes, and the outer toe is generally connected at the base with the middle toe.

The first family of the Tenuirostres is called after the hoopoe, and termed Upupidæ. In all these birds the bill is curved throughout its entire length, long, slender, and sharply pointed. The wings are rounded, showing that the birds are not intended for aërial feats, and the tail is rather long. The legs are short, and the claws strong and decidedly curved. As several of the families embrace a great number of species, it has been thought advisable to separate them into sub-families, for greater convenience of reference and more precision of arrangement.

The first sub-family is that of the Plume Birds, or Epimachinæ, containing some very beautiful species, all of exotic birth, and inhabiting Australia, New Holland, New Guinea, and the neighbouring islands. In these birds the long and slender bill is cloven as far as the eyes, the nostrils are placed at its base, and covered with soft silken plumes, and the thumb-toe or "hallux" is of considerable length and very strong, evidently for the purpose of aiding the birds in the pursuit of their prey. The fourth quill-feather of the wing is generally the longest.

The PTILORIS, or RIFLE-BIRD, is, according to Gould, the most gorgeous of all the Australian birds, although the full beauty of the creature is not at first sight so striking as that of the parrots or other gaudy-plumaged birds, and needs to be seen by a favourable light before the full glory of the colouring can be made out.

In size the Rifle-Bird is equal to a large pigeon, and in spite of its beauty it is not very often seen, as it is retiring in its habits, and seems to be confined to a very limited range of country. As far as is at present known, it is found only in the thick "bush" of the south-eastern portions of Australia, and even there appears to be a very local bird. It is no wanderer, never flying to any great distance from its home, and procuring its food in the near vicinity of its nest. For lengthened flight, indeed, it is singularly incapacitated by the shortness and rounded form of the wings, which is a never-failing characteristic of weakness in the flight and want of sustaining power. While in its native woods it seems never to make more use of its wings than is needful for the purpose of conveying it from one tree to another.

The habits of this bird are very like those of the common creeper of England, for it is generally seen upon the trunks and large branches of trees, running nimbly round them in a spiral course, and extracting the insects on which it feeds from the crevices and recesses of the bark.

Although in many instances, some of which have already been mentioned, the two sexes are clothed in very different plumage, there are few species where the distinction is so great as is the case with the Rifle-Bird. In the male bird, the upper part of the body is deep velvet-black, with a tinge of purple in a cross light, and the breast, abdomen, and under parts are of the same velvety hue, but diversified with a fine olive-green, which stains the edges of each feather. The crown of the head and the throat are covered with a multitude of remarkably little patches of the most brilliant emerald-green, glancing with a lustrous metallic sheen that equals the well-known emerald feathers of the humming-bird, and is in vivid contrast with the velvet-black of the body. The tail is black, with the exception of the two central feathers, which are of a rich metallic green, nearly as gorgeous as those of the head and neck. The bill is black.

RIFLE BIRD.—*Ptilóris Paradíseus.*

The plumage of the female and young male bird is strongly contrasted with these vivid colours, being as brown and homely as that of the English thrush, and giving no indication of the gorgeous hues that dye the feathers of the adult male. The upper parts of the female and undeveloped male are an obscure rusty brown, the wings and tail being edged with a reddish hue. A whitish-buff streak runs through the eye, and the under surface is half- covered with many spear-headed black marks, something like the arrow-headed characters of Nineveh, caused by the black hue which tips each feather, and very partially stains their edges.

Van der Hoeven, in his "Handbook of Zoology," places this bird in the genus Epimachus, together with the two beautiful species which will be next described. It is separated from them by other systematic zoologists, on account of the formation of the tail, which is comparatively short, and the feathers of nearly equal length.

THE very remarkable bird which is depicted in the accompanying illustration has been very appropriately named Neomorpha, or New-form, as it exhibits a peculiarity of formation which, so far as is at present known, is wholly unique.

When this bird was first discovered, Mr. Gould very naturally considered the specimen with the straight beak to be of a different species from that which has the curved bill, and accordingly set them down in his list under different titles. In process of time, however, he discovered the real state of the matter, as will shortly be seen by his own account. This very curious anomaly in form is of considerable value to systematic zoologists, against over-estimating the importance of form in a single limb or organ. Any one would be justified in considering so decided a difference of beak as a mark of distinction between two separate species ; but it must not be forgotten that there are many *genera*, not only of birds, but of every class of living beings, which have been established upon a far slighter foundation than is afforded by the straight and curved beak of these birds, which have been found to be nothing more than mere sexual distinctions of the same species.

The locality and habits of the Neomorpha are briefly but graphically described by Mr. Gould, in the following passage, which is taken from his "Birds of Australia":—
" These birds, which the natives call E. Elia, are confined to the hills in the neighbourhood

GOULD'S NEOMORPHA.—*Neomorpha Gouldii.*

of Port Nicholson, whence the feathers of the tail, which are in great request among the natives, are sent as presents to all parts of the island. The natives regard the bird with the straight and stout beak as the male, and the other as the female. In three specimens which I shot this was the case, and both birds are always together.

These fine birds can only be obtained with the help of a native, who calls them with a shrill and long-continued whistle, resembling the sound of the native name of the species. After an extensive journey in the hilly forest in search of them, I had at last the pleasure of seeing four alight on the lower branches of the tree near which the native accompanying me stood. They came quick as lightning, descending from branch to branch, spreading out the tail and throwing up the wings. Anxious to obtain them, I fired; but they generally come so near, that the natives kill them with sticks. Their food consists of seeds and insects; of their mode of nidification, the natives could give no information. The species are apparently becoming scarce, and will probably be soon exterminated."

In the colouring of its plumage, it is, although rather dark, a really handsome bird when closely inspected in a good light. The general hue of the feathers is a very dark green, appearing to be black in some lights, and having a bright glossy surface. Upon each side of the neck is a fleshy protuberance, or "wattle," analagous to the wattle of the common turkey, and of a rich orange colour during the life of the bird. After death, however, the bright colour rapidly fades, and the full, round, fleshy form quickly contracts, so that after a while the only remnants of the wattle are to be found in two flat, shrivelled, dusky projections, which give no idea of their former shape and beauty, and look as if they had been cut out of old parchment.

The same unfortunate result is to be found in every stuffed or dried skin, whenever the skin itself is not concealed by fur, scales, or feathers. It is very much to be regretted that some plan cannot be discovered for preserving such portions of the creatures in their original form and colour, as in many cases they are extremely important in affording distinctive marks of species, and in all are so characteristic in their appearance, that their total absence, or any change in their shape and hue, entirely alters the whole aspect of the creature. At present, the only mode of getting over the difficulty is to model the organ in wax, but this is at best but a kind of charlatanry; and as it depends entirely on the skill of hand and faculty of observation possessed by the individual taxidermist, is not sufficiently reliable to be of much value in a museum.

Moreover, the greater number of rare and new species are obtained when there are no means of obtaining the wax and other appliances which are needful for this mode of proceeding, and even if it were otherwise, the skins are seldom set up before they reach their final destination, on account of the space which they would occupy, and the great risk of injury they would run.

Until some method has been discovered by which these naked parts can be restored to their original shape and brilliancy, they will always present that repulsive shrivelled appearance which is too familiar to all who have compared a stuffed skin with the living, or even with the dead creature before decay has fairly set in. Mr. Waterton's method is the best that has as yet been put forward, but it is too tedious to be of much service even in the closet, and in the field or forest would occupy so much time that the collector would find his days taken up with the never-ending labour of preserving the skin, and could give no time to the observation of habits, or the procuring of specimens. Perhaps some mode of injection might be discovered which would answer the purpose of preserving the form and colour of these appendages, as well as it serves the purpose of the anatomist in preserving the form and colour of the veins and arteries, and which would not require any cumbrous apparatus beyond the usual outfit of a hunter-naturalist.

To proceed with the description of the Neomorpha. The tail is of the same deep black-green as the rest of the body, but the uniform monotony of the tint is pleasingly interrupted by a broad band of pure white which is drawn round its edges. The bill is of a rather dark-brown colour, and is rather lighter towards the extremity than at the base. This bird is a native of New Zealand.

WE now come to the true Plume Birds, which have been placed in various positions by the different zoologists who have written upon this subject; some classing them with the bird of paradise, to which they certainly bear a great external resemblance, both on account of the luxuriancy and the peculiar brilliant hues of the plumage, while others have considered them as nearer allied to the honey-suckers, and have, in consequence, placed them in close proximity to those beautiful birds.

The SUPERB EPIMACHUS is a native of New Guinea, and is one of the most lovely creatures that inhabit the face of the earth. Although in the size of the body it is by no means large, its plumage is so wonderfully developed, that the bird measures nearly four feet from the point of the bill to the extremity of the tail.

"To add to the singularity of this bird," says Lesson, whose description is too vivid and life-like to be neglected, "nature has placed above and below its wings feathers of an extraordinary form, and such as one does not see in other birds; she seems, moreover, to have pleased herself in painting this being, already so singular, with her most brilliant colours. The head, the neck, and the belly are glittering green; the feathers which cover these parts possess the lustre and softness of velvet to the eye and touch; the back is changeable violet; the wings are of the same colour, and appear, according to the lights in which they are held, blue, violet, or deep black; always, however, imitating velvet. The tail is composed of twelve feathers; the two middle feathers are the longest, and the lateral feathers gradually diminish; it is violet or changeable blue above, and black beneath. The feathers which compose it are as wide in proportion as they are long, and shine both above and below with the brilliancy of polished metal.

Above the wings the scapularies are very long and singularly formed; their points

SUPERB PLUME BIRD. — *Epimachus magnus.*

being very short on one side, and very long on the other. These feathers are of the colour of polished steel, changing into blue, terminated by a large spot of brilliant green, and forming a species of tuft or appendage at the margin of the wings. Below the wings spring long curved feathers, directed upwards; these are black on the inside, and brilliant green on the outside. The bill and feet are black."

The same author, in referring to the brilliant metallic hues of this and other birds, takes occasion to notice the iridescent effect which is produced by the different angle at which light falls on the feathers. The emerald green, for instance, will often fling out rays of its two constituent primary colours, at one time being blue-green, at another gold-green, while in certain lights all colour vanishes, and a velvet-black is presented to the eye. The ruby feathers of several birds become orange under certain lights, and darken to a crimson-black at other times. This change of hue is analogous to the well-known iridescent changeableness of the nacre which lines various shells, and is owing to the structure of its surface refracting the light in different rays according to the angle at which it falls upon the feathers.

THE adult male TWELVE-THREAD PLUME BIRD presents so strong a resemblance to the birds of paradise that it might easily be mistaken for one of those gorgeous creatures, than which, indeed, it is scarcely less splendid. Not only does its plumage glow with all the resplendency of brilliant emerald-green and velvety violet-black, but the bird is also provided with a number of long thread-like plumes, which are very similar to those of the paradise birds. Like the preceding species, it is a native of New Guinea, and is, if possible, even a more beautiful creature; the white floating plumes compensating for the absence of the extremely lengthened tail.

The general colour of the Twelve-thread Plume Bird is rich violet, so intense as to become black in some lights, and having always a velvet-like depth of tone. Around the neck is placed a collar of glowing emerald-green feathers, which stand boldly from the neck, and present a most brilliant contrast with the deep

TWELVE-THREAD EPIMACHUS.--*Epimachus albus.*

violet of the back and wings. The tail is short in comparison with the dimensions of the bird. From the back and the rump spring a number of long silken plumes of a snowy white colour, and a loose downy structure that causes them to wave gracefully in the air at the slightest breeze. Six of these lower plumes at each side are furnished with long, black, thread-like prolongations of the shaft, a peculiarity which has earned for the bird its title of Twelve-thread.

Albino specimens of this bird have been found, in which the entire plumage was of the same snowy white as the downy plumes.

In attempting to describe these gorgeously-decorated creatures, it is impossible to avoid a feeling of dissatisfaction when mentally comparing the wondrous beauty of the beings under consideration and the imperfect words in which the writer has endeavoured to portray their beauties. Even with the assistance of colour, any idea that can be given of these birds would necessarily be very imperfect, and the most admirable illustrations that ever were drawn, rich in ultramarine, carmine, and gold, would "pale their ineffectual fires" even before the stiff and distorted form of the stuffed bird. Yet that very stuffed semblance of the living creature fails egregiously in reproducing the bird as it was during life, as every one must have observed who has visited a museum.

Putting aside the inevitable shrinking and darkening of the soft parts about the head, legs, and claws, which change from their natural forms into dry and shrivelled pieces of dull, black parchment, the feathers always present an unsightly staring appearance; and there is no taxidermist whose hand, be it ever so skilful, can give to the stuffed creature the exquisite swell and rounding of the various parts, and that air and carriage of the body which is so indicative of the character. Not only is this the case with the stuffed bird, but immediately after death the plumage loses half its beauty, for during its lifetime the bird is able, by smoothing or ruffling its plumage, to give to its form a vast variety of expressions, which sink in death to one listless aspect, which tells that life has fled. The very respiration of the bird keeps the feathers in continual motion, causing them to change their tints with every breath. Such being the case, even with the recently slain bird or the preserved skin, it may well be imagined that no artist is sufficiently skilful to delineate, no artificial colour sufficiently brilliant to reproduce, and no pen sufficiently accomplished to describe, the glowing tints with any degree of success, when the drawings and the descriptions are compared with the living originals.

In the Plume Birds the nostrils are partly covered with a number of velvet-like plumes, but in the Hoopoes they are protected by a membranous scale. The bill is long, curved, pointed at the tip, and keeled at the base. The crown of the head is surmounted by a tuft of feathers which can be raised or depressed at will. The wings are rather long, the first quill being short, and the fourth the longest, and the tail is composed of ten feathers of nearly equal length.

The common Hoopoe enjoys a very wide range of country, being found in Northern Africa, where its principal home is generally stationed, in several parts of Asia, and nearly the whole of Europe. On account of its very striking and remarkable form, it has attracted much notice, and has been the subject of innumerable legends and strange tales, nearly all of which relate to its feathery crest. One of the Oriental legends is worthy of notice inasmuch as it contains a moral exclusive of the interest of the story.

It is related that Solomon was once journeying across the desert and was fainting with heat, when a large flock of Hoopoes came to his assistance, and by flying between the sun and the monarch, formed an impenetrable cloud with their wings and bodies. Grateful for their ready help, Solomon asked the birds what reward they would choose in return for their services. After some consultation among themselves, the Hoopoes answered that they would like each bird to be decorated with a golden crown; and in spite of Solomon's advice, they persisted in their request, and received their crowns accordingly. For a few days they were justly proud of their golden decoration, and strutted among the less favoured birds with great exultation, and repaired to every stream or puddle in order to admire the reflection of their crowns in the water.

But before very long, a fowler happened to see one of the promoted birds, and on taking it in his net, discovered the value of its crown. Immediately the whole country was in an uproar, and from that moment the Hoopoes had no rest. Every fowler spread his nets for them, every archer lay in wait for them, and every little boy set his springle or laid his rude trap in hope of catching one of these valuable birds. At last they were so wearied with persecution, that they sent one or two of the survivors to Solomon, full of repentance at their rejection of his advice, and begging him to rescind the gift which they had so unwisely demanded. Solomon granted their request, and removed the golden crown from their heads; but being unwilling that the birds should be left without a mark by which they might be distinguished from their fellows, he substituted a crown of feathers for that of gold, and dismissed them rejoicing.

The Turks call the Hoopoe Tir-Chaous, or Courier-Bird, because its feathery crown bears some resemblance to the plume of feathers which the chaous or courier wears as a token of his office. The Swedes are rather fearful of the Hoopoe, and dread its presence, which is rare in their country, as a presage of war, considering the plume as analogous to a helmet. Even in our own country the uneducated rustics think it an unlucky bird, most probably on account of some old legend which, although forgotten, has not entirely lost its powers of exciting prejudice.

The food of the Hoopoe is almost entirely of an insect nature, although the bird will frequently vary its diet with tadpoles and other small creatures. Beetles and their larvæ, caterpillars and grubs of all kinds, are a favourite food with the Hoopoe, which displays much ingenuity in digging them out of the decayed wood in which they are often found. The jet-ant (*Formica fuliginósa*), which greatly haunts the centre of decaying trees, is also eaten by this bird.

HOOPOE.—*Upupa Epops.*

The nest is made in hollow trees, and consists of dried grass stems, feathers, and other soft substances. The eggs are of a light grey colour, and in number vary from four to seven. They are laid in May, and the young make their appearance in June. It is worthy of notice that the beak of the young Hoopoe is short and quite straight, not attaining its long curved form until the bird has attained its full growth. The nest of the Hoopoe has a very pungent and disgusting odour, which was long thought to be caused by putrid food brought by the parent birds to their young, and the Hoopoe was therefore supposed to enact a part analogous to that of the vulture, and to perform the office of a scavenger. But as the reader will doubtlessly have observed, the food of the Hoopoe consists chiefly of living insects, and could have no such ill effects. The real reason of the evil odour is that the tail-glands of these birds secrete a substance that is extremely offensive to human nostrils, although it is unheeded by the birds themselves.

The name Hoopoe is doubly appropriate to this bird, as it may be either derived from the crest (*huppe*), or from the peculiar sound which the bird is fond of uttering, and which resembles the syllable *hoop! hoop!* which, as Mr. Yarrell observes, "is breathed out so softly, yet rapidly, as to remind the hearer of the note of the dove." The pace of the Hoopoe is a tripping kind of walk, which is at times very quick and vivacious, and sometimes is slow and stately as if the bird were mightily proud of its crested head. When at liberty it is generally found in sequestered spots, preferring low, marshy grounds, and the vicinity of woods, because in these places it is certain to find plenty of food.

The Hoopoe is a hardy bird in captivity, and from all accounts seems to be very interesting in its habits. From the many histories of caged or domesticated Hoopoes, I select the following, as they give a very good idea of the bird and its peculiarities. The first account is written by Mr. Blyth, in a contribution to the Magazine of Natural History.

"On beholding six of these birds confined in a very roomy cage, I was particularly struck with their vivacity and quick and expressive physiognomy; and a scene not a little amusing was exhibited on holding to them a morsel of meat. In a moment they all crowded eagerly to seize it, uttering a wheezing cry, and following my hand with rapidity about the cage, one or two of them sometimes clinging to the wires; and when at length two or three pieces were given to them, the scramble, though they could not have been very hungry, and the subsequent struggle for possession, was maintained with a pertinacity that was truly surprising. Two might be seen tugging with might and main at the same morsel, till wearied with repeated efforts they would give over for a while, still retaining, however, their hold, to resume the contest after an interval of rest; and it was not unusual on such occasions for a third individual, generally a smaller and weaker bird, to quietly watch the issue of the contest, when it would endeavour to deprive the victor of its prize. Certainly, I never saw birds struggle so vigorously before, nor pull with such determined force and energy, tumbling over not unfrequently from the violence of their efforts."

This pugnacious disposition appears to be universal among the Hoopoes, for M. Necker remarks that in his own country they fight in the most desperate manner and leave the scene of their combat covered with feathers that had been torn off in the struggle. Another history of tame Hoopoes was communicated to Bechstein, the well-known author of " Cage Birds," by the owner of the Hoopoes. Of little birds, which he had taken from nests on the top of an oak-tree, he says, " They followed me everywhere, and when they heard me at a distance, showed their joy by a peculiar chirping, jumped into the air, or, as soon as I was seated, climbed upon my clothes, particularly when giving them their food from a pan of milk, the cream of which they swallowed greedily. They climbed higher and higher, till at last they perched on my shoulders, and sometimes on my head, caressing me very affectionately ; notwithstanding this, I had only to speak a word to rid myself of their company ; they would then immediately retire to the stove. Generally they would observe my eyes to discover what my temper might be, that they might act accordingly.

I fed them like the nightingales, with the universal paste, to which I sometimes added insects ; they would never touch earthworms, but were very fond of beetles and May-bugs (cockchaffers) ; these they first killed, and then beat them with their beak into a kind of oblong ball. When this was done they threw it into the air, that they might catch it and swallow it lengthwise ; if it fell across the throat, they were obliged to begin again.

I took them one day into a neighbouring field, that they might catch insects for them-selves, and had then an opportunity of remarking their innate fear of birds of prey, and their instinct under it. As soon as they perceived a raven, or even a pigeon, they were on their bellies in the twinkling of an eye, their wings stretched out by the side of their head so that the large quill-feathers touched ; they were thus surrounded by a coat of armour formed by the feathers of the tail and wings, the head leaning on the back with the bill pointed upwards ; in this curious position they might be taken for old rags. As soon as the bird which frightened them was gone, they jumped up immediately, uttering cries of joy. They were very fond of lying in the sun ; they showed their content by repeating in a quivering tone, *vec! vec! vec!* When angry, their notes are harsh ; and the male, which is known by its colour being redder, cries *hoop! hoop!*

The female had the trick of dragging its food about the room ; by this means it was covered with small feathers and other rubbish, which by degrees formed into an indigestible ball in its stomach about the size of a nut, of which it died. The male lived through the winter ; but not quitting the heated stove, its beak became so dry that the two parts separated and remained more than an inch apart ; thus it died miserably."

One of these birds which was seen in captivity by Mr. Yarrell was in the habit of concealing any superabundant food, and resorting to his hidden stores whenever he felt hungry. It was mostly fed upon meal worms, which it always killed before eating, by repeated bites from the end of the bill, and by a succession of pecks as they lay on the ground after being disabled by the bites. On account of this habit, and from the fact that the horny portions of the bill are very much longer than the bony structures, some zoologists have considered the Hoopoe to be related to the Hornbills, and have accordingly placed them next to those remarkable birds.

The general colours of the Hoopoe are white, buff, and black, distributed in the following manner. The plumes of the crest, which is composed of a double row of feathers, are of a reddish-buff, each feather being tipped with black. The remainder of the head, neck, and breast is purplish buff, and the upper part of the beak purple-grey. Three semicircular black bands are drawn across the back, and the quill-feathers of the wings are marked with broad bands of black and white. The tail is also black, with the exception of a sharply defined white semicircular band that runs across its centre. The under portions of the body are pale yellowish buff, and the under tail-coverts are white. In their colours the two sexes are rather different from each other, the male being of a more ruddy hue than his mate, and having a larger crest. The total length of the adult Hoopoe is not quite thirteen inches.

As a general fact the Hoopoe is but a rare visitant to England, and has little inducement to fix its habitation in so inhospitable a country ; for the persecution to which the poor bird is subjected is nearly as severe at the present day as that which was suffered by the Hoopoes in the old times when they wore their golden crowns. It seldom is found in the northern parts of England, but in the southern and eastern counties is not unfrequently seen. Many notices of these birds have been sent to the *Field* newspaper and other periodicals which treat of natural history ; and it seldom happens that a year passes by without several such notices. In Cornwall it seems to be more plentiful than in any other part of England, and to be quite a familiar bird.

FIERY-TAILED SUN-BIRD.—*Nectarinia ignicauda.*

SUN-BIRDS.

THE beautiful and glittering SUN-BIRDS evidently represent in the Old World the humming-birds of the New. In their dimensions, colour, general form, and habits, they are very similar to their brilliant representatives in the western hemisphere, although not quite so gorgeous in plumage, nor so powerful and enduring of wing. They are termed Sun-birds, because the hues with which their feathers are so lavishly embellished gleam out with peculiar brilliancy in the sunlight. Our common sun-beetles of England, that are so familiar to us as they run about the ground in the hot weather, their glittering surface flashing rainbow-tinted light in every direction, have earned their popular and expressive name in a similar manner.

These exquisite little birds feed on the juice of flowers and the minute insects that are found in their interior, but are not in the habit of feeding while on the wing, hovering over a flower and sweeping up its nectar with the tongue, as is the case among the humming-birds. The Sun-Birds generally, if not always, perch before they attempt to feed, and flit restlessly from flower to flower, picking the blossoms in rapid succession, and uttering continually a sharp, eager cry, that indicates the earnestness of their occupation. In accordance with their peculiar habits, the feet and legs are very much stronger than those of the humming-birds; their wings are shorter, rounder, and less powerful, and their plumage is not so closely set. Moreover, the feathers, although bedecked with the most brilliant of hues, lack, except in certain spots, such as the crown of the head and the throat, the scintillating radiance of the humming-bird, and do not possess in an equal degree the property of changing their hues with every movement.

The brilliant colours of the Sun-birds belong, as a general rule, only to the male sex, the female being comparatively sober in her plumage, possessing neither the beauty of form nor colour which is so conspicuous in the other sex. Even in the male bird, the gorgeous plumage has but a temporary existence, becoming developed at the commencement of the breeding season, and being lost at the moult which always follows the rearing of the young. At all other seasons of the year, the male birds are nearly as simply clothed as their mates, and even the glittering, scaly feathers of the head and throat are replaced by a dull brown plumage, hardly distinguishable, except by difference of structure, from the surrounding feathers of the neck. The change of colour and form is so great in these birds, that many zoologists have described the immature male, the adult male, and the female as three distinct species, and have consequently wrought great confusion among their ranks.

The young male birds are not unlike the female, but may be known by one or two feather structures, which will be presently mentioned; and it is a rather curious fact that the adult male always returns after the breeding season to the plumage of immaturity. Some writers have questioned the truth of this statement, but without sufficient reason. As soon as the time arrives when the birds begin to choose their mates, and the brilliant feathers have fully developed themselves, the male Sun-bird becomes very animated, and makes the most of his gorgeous plumage, puffing up the feathers of the neck and head, so as to make them flash in the sun's rays, as if conscious of the fascination which his brilliant costume must exert upon the susceptible hearts of the gentler sex.

The Sun-birds usually make their nests in the hollows of decaying trees, or within the centre of thick brushwood. In many cases the nest is concealed with great care; and in some instances is constructed with consummate art. The material of the nest is generally composed of very fine fibres, interwoven and lined with the soft cottony down that is found in the seed-vessels of many plants, and ingeniously set round with various lichens, so as to give it a close resemblance to the tree in which it is placed. One species has even been known to make a thick spider's web the foundation of its nest, and to cover it so completely with little bits of moss, lichens, paper, cloth, and all kinds of miscellaneous substances, as to destroy its nest-like appearance, and make it look like a chance bundle of scraps entangled in the branches.

When taken young, the Sun-birds are very susceptible to human influence, rapidly becoming tame, and learning to fly about the room and take their food from the hand of their owners with charming familiarity. It has already been mentioned that the Sun-bird utters a shrill, sharp whistle, while engaged in seeking food. This, however, is not their only cry, as many of them possess considerable musical powers, their cry, although feeble, being sweet and agreeably undulated. It is thought by many observers that the Sun-birds, while flitting from flower to flower, aid in the work which is so efficiently carried out by bees and similar insects, and help to carry the fructifying pollen from one blossom to another.

AMONG these birds the FIERY-TAILED SUN-BIRD, although not the largest, is yet one of the most striking and beautiful in form and colour.

This most lovely little creature is an inhabitant of India, being found near the foot of the Himalaya mountains, and most plentifully near Nepâl. In dimensions it is extremely small, owing to the great difference which exists, even in adult males, in the length of the central feathers of the tail, the disparity often amounting to two inches, so that the length of the bird may be from three to five and a half inches. The forehead and the top of the head are brilliant steel-blue, and the neck, the back, and the upper tail-coverts are the most beautiful scarlet vermilion, diversified by a broad patch of bright yellow upon the bend of the back. The two long central feathers of the tail are also bright vermilion, and the side feathers are brown edged with the same brilliant hue. The upper surface of the wings is olive-brown, each feather being brown and edged with olive ; the under surface of the wing is greyish-white, worked here and there with very pale brown. The breast is beautiful gold yellow, with a wash of crimson in the centre ; and the abdomen, and remainder of the under parts, are rather pale olive-green. The bill is blackish-brown.

Before and after the breeding season the Fiery-tailed Sun-bird assumes a more sober plumage, the general colour being olive, with a slight mark of pale scarlet upon the back. The crimson patch on the breast vanishes, and the tail-feathers are all of equal length. The female is olive-green above and greenish yellow below, and there is a slight mark of red upon the base of the tail.

The COLLARED SUN-BIRD is an inhabitant of many parts of Africa, stretching from the northern portions of that continent as far as the western coasts. It is extremely plentiful in the larger forests of the Cape and the interior, but there is very little information concerning its habits, saving that they resemble those of its relations. The nidification of this species differs according to the locality, for it places its nest in the interior of hollow trees wherein it resides in the forests, and is content with the shelter of a thick bough when there are no decaying trees within reach.

The male Collared Sun-bird is a most beautiful little creature, bedecked with glowing tints of wonderful intensity. The general colour of the upper parts of the body and breast is a rich golden green, the upper surface of the wings and tail being blackish brown with green reflections. Across the breast are drawn several coloured bands, which have earned for the bird its popular and expressive name, as all names should be. A narrow band of bright steel-blue runs across the upper part of the breast, being rather wide in the centre and narrowing rapidly towards the sides of the neck. Below this blue band runs a broad belt of rich carmine, and immediately below the carmine is a third narrow band of bright golden yellow. From the sides of the breast proceed several small feathery plumes of the same golden hue. The remainder of the abdomen is greyish brown, and the upper tail-coverts are violet-purple.

The female is rather less in dimensions than her mate, and is very sober in her attire, wearing a suit of uniform olive-brown, darker upon the wings and tail, and very pale behind. The total length of this species is rather more than four and a half inches.

THERE is another species of Sun-bird which closely resembles the last-mentioned bird in its colouring, and is often mistaken for it. This is the GREATER COLLARED SUN-BIRD (*Nectarinia Afra*), a rather larger bird, measuring at least one inch more in total length than the preceding species. It is also an inhabitant of Africa, but is seldom seen in the extreme south of that country, preferring the deep forests of the interior, and rarely descending to the plains. Its nest is made in some hollow tree, and the eggs are four or five in number and white in colour, plentifully variegated with tawny markings. It may be distinguished from the common Collared Sun-bird by the greater amount of the bronze-green hue, and by the shortness of the blue collar.

The JAVANESE SUN-BIRD is a native of the country from which it derives its name. It is a very pretty little creature, although its colours are not so resplendent as in several of the species. The upper parts of the body are shining steely-purple, and the under surface is olive-yellow. The throat is chestnut, and a bright violet streak runs from the angle of the mouth to the breast.

The GOALPORAH SUN-BIRD (*Nectarinia Goalpariensis*) is also worthy of a passing notice.

This beautiful species is an inhabitant of several parts of Asia, and is rather plentiful in and about Nepâl.   In dimensions it is equal to the preceding species, the adult male measuring about five and a half inches in length.   The nest is beautifully constructed, and is of the pendulous order.   The food of this bird consists chiefly of minute insects, spiders, and various larvæ, chiefly those of flies.   It lives mostly in the depths of the densest forests, where it may be found in tolerable numbers by those who choose to take the trouble to search after it.

In the plumage of this pretty bird, red is the prevailing colour. The crown of the head is rich golden green, and the nape of the neck, the breast, and scapulæ are of a dazzlingly brilliant scarlet. The long central tail-feathers are of a rich green, and when closed, as is the case while the bird is at rest, completely conceal the bright yellow tint of the feathers below. The remainder of the tail is brownish black, and all the plumage of the lower part of the back is loose and downy in its structure.

Some of the tribes of the Sun-birds, and their behaviour when in captivity, are well recorded in the following description of some tame Sun-birds, by Captain Boys, quoted in Gould's "Birds of Asia." The species which is described is another Indian species, the Asiatic Sun-bird (*Nectarinia Asiàtica*), called by the natives "Shukur-khor," or sugar-eater :—

"In 1829 I slightly wounded a male in the bastard wing, secured and brought it home.   By some neglect it was unthought of for four days, when, on looking into the bag in which it had been placed, I found that it was not only alive, but that the wing had completely cicatrised : I should observe, however, that the broken part of the wing had been taken off with a pair of scissors immediately after the bird was brought home.

COLLARED SUN-BIRD.—*Nectarinia chalybeïa.*
JAVANESE SUN-BIRD.—*Nectarinia Javànica.*

I placed it in a cage, and succeeded in keeping it alive for several weeks by feeding it on sugar and water, of which it took great quantities, but, owing perhaps to a want of variety in its food, it became thinner and thinner until it died.   During its captivity it was very sprightly, and from the first day readily fed itself by dipping its tongue into the dish of syrup with which it was supplied."

BLUE-HEADED HONEYSUCKER.—*Nectarinia cyanocéphala.*

It is probable that the poor little bird died, as was supposed, from the effects of its diet. A similar story is told of some of the humming-birds, by Webber, where the little creatures pined after long feeding upon syrup alone, but, on being permitted to fly at liberty, immediately set to work upon the little garden spiders, and soon recovered their health and brilliancy. This need of animal food seemed to be periodical and irresistible.

In the same account, a portion of which has just been extracted, Captain Boys asserts that the nest of the Asiatic Sun-bird is very rudely made, whereas Mr. Layard tells us that it is constructed in a remarkably neat manner, and that it is often suspended from a twig in such a manner that the spiders cover it with their webs, and make it almost invisible.

THIS beautiful bird, which is represented in the accompanying illustration, is a good specimen of the genus Nectarinia.

The BLUE-HEADED HONEYSUCKER is an inhabitant of Brazil, where it is extremely common, and by the bright gorgeousness of its plumage, and the restless activity of its movements, adds much to the beauty of the wondrous scenery among which it dwells. It is found spread over the whole of Brazil, and may always be found haunting the blossoming trees and plants, dashing to and fro with its glancing flight, hovering with tremulous wing over the flowers while indetermined in its choice, and plunging its long beak eagerly into their newly-opened blossoms, where it finds its food. It is not known to feed while on the wing, as is the case with the humming-birds, but perches near or upon the flower, and clings with its strong little feet while taking its meal.

The Blue-headed Honeysucker derives its name from the azure-blue which decorates its head, and which is very changeable in different lights. The throat, the back, the tail, and the wings are black, except that the quill-feathers are edged with blue. The female bird does not possess the beautiful tints of her mate, the greater part of her plumage being green, tinged with blue upon the head and the scapularies; the throat is grey. This bird is known by several other titles, such as the Cayenne Warbler, the Blue-headed Warbler, and the Blue-headed Creeper.

LARGEST of all the group, the MALACHITE SUN-BIRD has long attracted the attention of ornithologists, on account of its great comparative size and its beautiful plumage.

It is one of the African species, being an inhabitant of the Cape of Good Hope, where it remains throughout the entire year, and is in the habit of frequenting the gardens, and soon becomes familiar with the proprietors, provided that it be not disturbed. Sometimes the Malachite Sun-birds take a violent fancy to some particular shrub or tree, and may be seen in flocks of forty or fifty in number congregating upon its branches and amusing themselves among its blossoms. Day after day these birds may be seen in the same spot, attracted by some irresistible though obscure charm resident in the tree which they favour. The nest of this species is composed of very tiny twigs covered with moss, and contains four or five green eggs.

MALACHITE SUN-BIRD.—*Nectarinia famosa.*

The title of Malachite Sun-bird has been given to this creature on account of the brilliant malachite-green of its plumage.

The male bird when dressed in full nuptial costume is a remarkably handsome bird, and is nearly double the length of any other species, often exceeding nine inches in total length The whole of the upper surface is rich golden green marked with a reddish bronze. The feathers of the throat and forehead are of the same hue, but of so deep a tone that they appear to be velvety-black at first sight, and are so constructed that they have a velvet-like feel to the touch as well as to the sight. Whenever the bird moves, even by the act of respiration, waves of bright hues seem to ripple upon its surface, caused by the peculiar colouring of the feathers, which are black at their bases and coloured at their extremities. The wings and tail are black, and the secondaries and wing-coverts are edged with green and violet. There is a tuft of bright yellow feathers under each shoulder.

The female is much smaller than her mate, and is of a dull olive-brown, except the exterior feathers of the tail, which are edged with white.

Among other long-tailed Sun-birds may be mentioned *Nectarinia pulchella*, which may be known by its green-edged black tail-feathers and the bright double collar of carmine and golden-yellow that runs across the chest. Another species, also long-tailed, *Nectarinia platúra*, is remarkable for the brilliant golden-yellow of the breast and abdomen, and the rich violet-purple of the upper tail-coverts.

THE beautiful little DICÆUM, although very common throughout the whole of Australia, and a remarkably interesting little bird, was, when Mr. Gould wrote his animated description, so little known among the colonists that there was no popular name for the bright little creature.

This tiny bird is fond of inhabiting the extreme summits of the tallest trees, and habitually dwells at so great an elevation that its minute form is hardly perceptible, and not even the bright scarlet hue of the throat and breast can betray its position to the unaccustomed eye of a passenger below. The song of the Dicæum, although very sweet and flowing, is very soft and faint, and seems to be an inward warbling rather than the brilliant melody which is flung so energetically from the vocal organs of many singing-birds. The little bird, however, is possessed of considerable endurance, for its strain, although weak, is long continued. The Dicæum is mostly found among the thick foliage of the Casuarinæ, and Mr. Gould relates that he frequently saw it flitting about the branches of a remarkably beautiful parasitic plant termed scientifically the Loranthus, which it seems to visit either to eat the soft viscid berries, or for the purpose of preying upon the little insects that come to feed on the flowers. Mr. Gould prefers the latter supposition.

AUSTRALIAN DICÆUM.—*Dicæum hirundinaceum.*

The flight of the Dicæum is very quick and darting, and it makes more use of its wings and less of its feet than any of the insect-hunting birds. The nest is remarkably pretty, being woven as it were out of white cotton cloth, and suspended from a branch as if the twigs had been pushed through its substance. The peculiar purse-like shape of the nest is shown in the illustration. The material of which it is woven is the soft cottony down which is found in the seed-vessels of many plants. The eggs are four or five in number, and their colour is a dull greyish-white profusely covered with minute speckles of brown.

The two sexes differ considerably in the colouring of their plumage, the male bird being much more brilliant than his mate. The head, back, and upper parts of the adult male are deep black with a beautiful steely-blue gloss, the sides are brownish-grey, and the throat, breast, and under tail-coverts are a bright glaring scarlet. The abdomen is snowy-white, with the exception of a tolerably large black patch on its centre. The female is more sombre in her apparel, the head and back being of a dull sooty-black, and the steel-blue reflection only appearing on the upper surface of the wings and tail. The throat and centre of the abdomen are buff, the sides are pale greyish brown, and the under tail-coverts scarlet, of a less brilliant hue than in the male. In its dimensions the Dicæum is hardly so large as our common wren.

ANOTHER species belonging to the same genus, but an inhabitant of a different part of the world, is equally remarkable for its minute form and the bold richness of its colours. This is the RED-BACKED DICÆUM (*Dicæum cruentátum*) of Asia.

This beautiful wee bird is plentiful in India, extending over a wide range of country, and being found in the vast tracts which reach from Calcutta to Assam on the east, and as far as Malacca on the north. Like the Australian Dicæum, it resides on the summits of the loftiest trees, and on account of its very small size is not very often seen, and even

2. P

if seen is so hard to shoot that it is but seldom killed except by those who make it their business to collect specimens. The male bird is remarkable for a broad line of the brightest scarlet which extends from the top of the head along the back, and reaches nearly to the extremity of the tail. The remainder of the upper surface is black, marked with green upon the wing-coverts, and the lower parts are of a light buff.

Nearly allied to the preceding species is the FIRE-BREASTED MYZANTHE (*Myzanthe ignipectus*), a bird which is remarkable as being the smallest bird of India. So very small is this beautiful little bird, that an adult specimen is hardly two and a half inches in total length, and weighs only three and a half drachms. In its habits it is very like the Dicæum, frequenting the tops of trees, and keeping itself well out of sight. The general colour of this bird is a dark glossy green above and buff below. Upon the chest there is a bold patch of bright crimson, and immediately below is a curious little black mark something like a frond of seaweed in shape. It is an inhabitant of Nepâl.

AZURE CÆREBA.—*Cœreba cyanea.*

THE two following species of birds are of small size, but are remarkable for the extreme beauty of the plumage, which glows with the most brilliant hues, but is not endowed with the peculiar changing tint of the Epimachi and humming-birds.

The first of these creatures, the AZURE CÆREBA, is a most glowing little bird, its feathers being deeply and gorgeously dyed with azure, verditer, and velvet-black, arranged in the following bold and striking manner. The crest is of a brilliant verditer-blue, possessing a metallic splendour, and almost flashing with emerald rays when placed in a strong light. A black velvet-like patch of feathers is placed on the back of the head and neck, affording the most decided contrast with the light plumage of the crest. Another but larger patch of the same deep hue occurs upon the shoulder, the wings are also black, and a black streak is drawn from the angle of the mouth towards the back of the neck. With these exceptions, the whole of the body is a bright azure.

This species is an inhabitant of Cayenne, Guiana, and the neighbouring localities. It is a little bird, hardly larger than a common sparrow.

The Azure Cæreba may generally be found upon the various flowering trees and shrubs of its native land, where it occupies its time in a perpetual search after the tiny insects that conceal themselves within the newly-opened blossoms.

In its nesting and in other parts of its economy it is a most singular bird. The nest is of the pensile order, being neatly woven upon the extremity of some slender twig, which sways to and fro even with the trifling weight of the mother and her tiny brood, and will in nowise bear the heavy bodies of the various snakes and lizards that abound among the branches of the trees, and keep up a relentless persecution of young nestlings and eggs. The shape of the nest is not unlike that of a large "jargonelle" pear, the lower extremity being produced into a long tube with the mouth below, and the eggs placed in the large rounded portion of the nest. No predaceous reptile could venture itself into so formidable a stronghold, and any noxious insect that might make its way through the tunnel would soon be snapped up by the watchful parent. The substance of the nest is composed of very slender grasses and fibres, and the entire structure is put together with a delicate firmness that human fingers would strive in vain to imitate.

It is said, upon very good authority, that the young of the Azure Cæreba are blind when they emerge from the eggshell, and that they do not attain the full power of vision until they are able to fly and to get their own living.

SCARLET DREPANIS.—*Drepanis coccinea.*

The SCARLET DREPANIS is well worthy of notice, not only on account of the position which it holds in the present system of ornithology, but by reason of the extreme value which is set upon it, and upon other species of the same genus, by the natives of the country where it dwells.

The colour of this bird is, as its name implies, scarlet upon the greater part of its plumage, the wings and tail being black, so that the two contrasting tints have a remarkably good effect. It is an inhabitant of the Sandwich Islands, and is in very great favour with the natives, who employ its plumage in the manufacture of those wonderful feather mantles and helmets which cannot but excite the wonder of all who ponder upon the singular amount of mechanical skill, dogged perseverance, and true artistic taste that has been employed in their manufacture.

The mantles, some fine specimens of which are in the British Museum, are made with the greatest care, the precious feathers being so judiciously disposed that none are wasted, while, at the same time, they cannot be discomposed by any movement of the wearer so as to betray the groundwork on which they are woven. Their colours, too, are arranged with great artistic feeling, and produce a very brilliant effect without offending the eye, or appearing to be needlessly gaudy. The helmets, which are in like manner decorated with the glowing feathers of these beautiful birds, are even more wonderful than the mantles, as they are not only skilfully constructed, but their form is absolutely classic in its graceful simplicity, and recalls to the spectator the best efforts of Greek art.

These mantles are so extremely beautiful in the soft flowing grace of their folds, are so light to the wearer, and so exquisitely brilliant in colour, that they would soon be in great request in the world of fashion, were they once introduced by one of the leading votaries of that capricious deity. The feather head-dress, too, would be so soft, light and brilliant, that it would soon vanquish all other costumes, and reign supreme.

The birds of this genus are very gregarious, delighting to associate in large flocks, and haunting the flower-bearing plants for the purpose of feeding upon the sweet juices and tiny insects which are found within the blossoms. In feeding they thrust their long bill and tongue to the very bottom of the flowers, and greatly resemble the bees in that respect. The natives take advantage of their flower-loving and gregarious habits, and by setting snares in the spot which they love best to haunt, contrive to immolate them in considerable numbers. As the Scarlet Drepanis is but a small bird, being hardly larger than the Cæreba, and as neither the tail nor wing appear to be employed in the structure of the mantles and helmets, it is evident that a vast number of these beautiful little creatures must perish before one chief can be gratified with the completion of a single mantle or the adorning of a single helmet.

HONEY-EATERS.

THE true HONEY-EATERS form a very numerous group of birds, all of which are graceful in their forms and pleasing in the colour of their plumage, while in some instances the hues with which they are decorated are so bright as to afford ground for classing them among the really beautiful birds. They all feed on similar substances, which, as indicated by their name, consist chiefly of honey and the sweet juices of flowers, although they also vary their diet by insects and other small living beings.

The NEW HOLLAND HONEY-EATER is a remarkably pretty bird, the whole of its body being covered with black, white, and yellow markings, which stand out in bold contrast to each other. The top of the head is black, and a number of little white feathers are gathered on the forehead. The sides of the head and neck are marked very conspicuously with three streaks of pure white, one of which is drawn over each eye, as if it were intended to stand in the place of an eyebrow, another passes from the nostrils towards the back of the neck, like a moustache, and the third is seen on the side of the neck, so that its whole aspect presents a sufficiently curious appearance. The body and upper part of the wings are deep brown-black, diversified with a narrow line of pale yellow upon the outer edge of each quill-feather, and a slight edging of white around their extremities. The tail is of the same brown-black as the body, edged with yellow, and tipped with white on the under surface. The lower parts and abdomen are greyish white, profusely covered with dashes of black.

This bird is admirably described and figured in Gould's "Birds of Australia," from which the following account of its habits is taken. This bird, according to Gould, is "one of the most abundant and familiar birds inhabiting the colonies of New South Wales, Van Diemen's Land, and South Australia; all the gardens of the settlers are visited by it, and among their shrubs and flowering plants it annually breeds. It is not a migratory species, but occasionally deserts some district for others whose flowering plants offer it a more tempting *locale*, and furnish it with a more abundant supply of food. The belts of Banksias, growing on sterile, sandy soil, afford it so congenial an asylum, that I am certainly not wrong in stating that they are never deserted by it, or that the one is a certain accompaniment of the other.

The range enjoyed by this species appears to be confined to the south-eastern portions of Australia; it is abundant on the sandy districts of South Australia, wherever the Banksias abound; but to the westward of this part of the country I have not traced it. At the Swan, and the other parts of the western coast, it certainly is never found. In Van Diemen's Land it is much more numerous on the northern than the southern portions of the island; it is also most abundantly dispersed over all the islands in Bass's Straits, whose sterile, sandy soil favours the growth of the Banksias. It is equally common over many parts of the colony of New South Wales; which may, in fact, be regarded as the

NEW HOLLAND HONEY-EATER.—*Meliphaga Novæ Hollándiæ.*

great stronghold of the species. At the same time, I must not fail to observe that the districts bordering the sea-coast are most favourable to the growth of their favourite tree; hence, while it is there most numerous, in the interior of the country it is seldom to be seen.

It evinces a more decided preference for shrubs and low trees than for those of a larger growth; consequently it is a species particularly subject to the notice of man, while it flits from bush to bush. Nor is it the least attractive of the Australian fauna; the strikingly contrasted markings of its plumage, and the beautiful appearance of its golden-edged wings, when passing with its quick, devious, and jumping flight from shrub to shrub, rendering it a conspicuous and pleasing object.

It has a loud, shrill, liquid, though monotonous, note. Its food, which consists of the pollen and juices of flowers, is procured while clinging and creeping among them in every variety of position; it also feeds on fruit and insects.

It usually rears two or three broods during the course of the season, which lasts from August to January: the nest is very easily found, being placed, in the forest, in any low open bush; and, in the gardens, among the shrubs and flowers. One of the nests in my collection was taken from a row of peas in the kitchen-garden of the Government House at Sydney. It is usually placed at about eighteen inches or two feet from the ground, and is a somewhat compact structure, composed of small wiry sticks, coarse grasses, and broad and narrow strips of bark; the inside is lined with the soft woolly portions of the blossoms of small ground plants.

It usually lays two, but occasionally three, eggs, which are of a pale buff, thinly spotted and freckled with deep chestnut-brown, particularly at the larger end, where they not unfrequently assume the form of a zone."

The WHITE-PINIONED HONEY-EATER is found, according to Mr. Gould, upon the north coast of Australia, and is very plentiful, especially near the settlement at Port Essington.

In its habits it is partly gregarious, being seen in little flocks, perhaps families, of six or seven in number, flitting about the tops of lofty trees and ever in active motion. Partly on account of the great elevation at which it loves to dwell, and partly because of

WHITE-PINIONED HONEY-EATER.—*Entomyza albipennis.*

the extreme shyness and wariness of its disposition, the gunner finds considerable difficulty in approaching within gunshot, so that although the bird is so common, it is not very often shot. In its flight it is strong and steady, not contenting itself with mere flittings from tree to tree, but on occasion launching boldly into the air, and shaping its course for some distant point. In taking these aerial journeys it always commences by rising perpendicularly to a very great height, and then, after having settled the direction in which it intends to go, it shoots off with a swift and steady flight.

This bird may lay claim to the title of Australian clock, which has been given to the laughing jackass, for no clock can be more invariable than the White-pinioned Honey-eater in giving its warning note of the approach of day. Precisely one half-hour before the sun rises, this bird, urged by some strange instinct, awakes from its sleep and pours forth a succession of long-drawn plaintive notes that have been compared to the word *Peet! peet!* uttered in a wailing kind of tone. But as soon as the sun has fairly risen, the melancholy cry changes into a quick, harsh, squeaking sort of cry, entirely unlike the previous note, and remarkably unpleasant and grating to the ear. The bird is very fond of exercising its voice, and continually utters its rough cry while it is on the wing.

The White-pinioned Honey-eater is boldly coloured, and both sexes, when adult, possess much the same kind of plumage.

The crown of the head and nape of the neck are jetty black, and the skin around the eye is of a lovely azure. In the young bird, this skin is rich orange, and forms a conspicuous mark whereby the young and old birds may be distinguished from each other. On each side of the nape is a crescent-shaped mark of snowy white, contrasting finely with the black and azure of the surrounding parts. The under parts are most curiously coloured. The abdomen is beautifully white, and the same tint runs towards the base of the bill, but is interrupted by a black patch which commences under the chin as a narrow line, and rapidly widening embraces nearly the whole of the chest, leaving a white stripe at either side. The primary quill-feathers of the wing are black, with the exception of the half nearest their bases, which are pure white in the inner webs, the outer web being black like the remainder of the feather. The upper surface and the wings are olive-green, changing in certain lights to ochreous yellow. In size it equals a large starling. The native name for it is Wur-ra-luh.

Another species belonging to the same genus, the BLUE-FACED HONEY-EATER (*Entomyza cyanótis*), is worthy of a passing notice. This bird, although so like the white-pinioned Honey-eater as to be easily mistaken for it by a hasty observer, is readily to be distinguished from that bird by its greater size and the absence of the white patches upon the pinions.

This species is a native of New South Wales, and is one of the most familiar birds of that country, remaining in the same locality throughout the entire year, and caring not to emigrate according to the change of season. It is usually found upon the eucalypti when the flowers are in blossom, and feeds upon nectar and insects, after the fashion of all the Honey-eaters. It has also been observed to employ its hard-tipped tongue in licking something from the clefts in the bark, whence had issued some of the gummy secretion from which the tree derives its familiar name of gum-tree. Whether it was eating the gum itself, or whether it was merely engaged in capturing the little insect that had rashly ventured upon the adhesive surface and become affixed thereto, is a question which is not satisfactorily settled. The latter supposition, however, appears to be the correct one.

Perhaps the bird may have resorted to the gum as a medicine, for I knew of a jackdaw that often used to eat the gum that exuded from plum-trees, and always did so when it was unwell. In connexion with this subject, it may as well be mentioned that a careful observer would find himself repaid by watching the modes of cure employed by sick or wounded creatures. We all know that the dog and cat resort to grass when they feel out of health, and hares to a species of moss. I was also told, on the authority of the eye-witness, that a goldfinch, which had been struck by a hawk and wounded, made its way to a dry puff-ball, tore it open with its beak, and dusted the wounded shoulder with the spores, thereby stopping the effusion of blood. The spectator was greatly surprised by this incident, and being induced to try the effect of the same remedy upon a wounded finger, found that the experiment was perfectly successful.

It is a most vivacious and active bird, ever on the move, and running about the boughs with the most surprising activity; caring nothing for the attitude which it may assume, and even hanging quietly by the grasp of a single claw while it employs itself in securing its insect prey with its long tongue. It is mightily pugnacious in its disposition, quarrelling incessantly even with its own species in default of better amusement, and waging ceaseless combat with any other kind of bird that may choose to come to the same tree. While employed in its search after food it frequently utters its cry, which is not very pleasing to human ears, being loud and monotonous in its character.

One of the most singular circumstances connected with this bird, is the manner in which it makes its nest.

There is a certain bird called the Pomatorhinus, not very unlike a rather dull-feathered bee-eater, which builds a very large, dome-shaped nest, similar in shape and size to the well-known edifice of our common magpie. The Blue-faced Honey-eater is in the habit of taking possession of the deserted nests of the pomatorhinus, and of saving itself the trouble of building by making a small depression in the domed roof of the deserted domicile, and therein laying its eggs. It might be expected that the bird would prefer to avail itself of the protection offered by the peculiar form of the usurped nest, and would take possession of its interior. Such, however, is not the case; and Mr. Gould states that he has seen many of the females in the nest and always found that the eggs were placed upon the roof. Whether the bird is forced to depend upon the pomatorhinus for a locality wherein to build, or whether it is capable of constructing a nest on its own account, is not at present satisfactorily ascertained. Mr. Gould thinks that the bird must be often forced to build her own nest, although he never found any such edifice.

The native term for this species is Batikin.

ANOTHER of these birds, the GARRULOUS HONEY-EATER, so named on account of its singularly talkative propensities, is a native of Van Diemen's Land and New South Wales, in both of which localities it is very common.

It enjoys, however, but a very limited range, being contained within certain boundaries with such remarkable strictness, that in some cases it is found in great numbers on one

GARRULOUS HONEY-EATER.—*Myzantha górrula*.

side of a river, while on the other side not a single bird can be seen. Those which inhabit Van Diemen's Land are rather larger than those of New South Wales, the greater size being probably caused by a greater profusion and more nourishing properties of the food. The Garrulous Honey-eater generally takes up its habitation among the thick forests of eucalypti that are found upon the plains and the hills of low elevation, and there passes a very lively existence. Its food consists of the sweet nectar of flowers, which it procures after the manner of Honey-eaters in general, by plunging its long tongue into the depths of the flowers, and licking up their luscious store. It also feeds upon various insects, being always ready to eat those minute creatures which inhabit the flowers, and delighting also in chasing the beetles and larger insects as they run upon the ground at the foot of the eucalypti.

In its habits the Garrulous Honey-eater is very amusing, although it often is the cause of no small annoyance to the traveller or the sportsman, as will be seen by the following remarks made by Gould in his "Birds of Australia." The Garrulous Honey-eater, he says, "is not gregarious, but moves about in small flocks of from five to ten in number. In disposition it is unlike any other bird I ever met with, for if its haunts be in the least intruded upon, it becomes the most restless and inquisitive creature possible, and withal so bold and noisy that it is regarded as a nuisance rather than an object of interest.

No sooner does the hunter come within the precincts of its abode, than the whole troop assemble round him, and perform the most grotesque actions, spreading out their wings and tails, hanging from the branches in every possible variety of position, and sometimes suspended by one leg; keeping up all the time one incessant babbling note. Were this only momentary, or for a short time, their droll attitudes and singular notes would be rather amusing than otherwise; but when they follow you through the entire forest, leaping and flying from branch to branch, and almost buffeting the dogs, they become very troublesome and annoying, awakening, as they do, the suspicions of the other animals of which you are in pursuit."

The nest of the Garrulous Honey-eater is a rather large edifice when the dimensions of the bird are taken into consideration, but is very neatly put together. The materials of which it is composed are very slender twigs and grass as a framework, and the lining is made of wool, hair, and any other soft and warm substance that the bird may be able to obtain. It is generally placed among the upright branches of some small tree, so that in spite of its size it is not so easily detected as might be imagined. The eggs are a bluish-white, covered uniformly with small spots of reddish-brown.

The colouring of this bird is sober, but pleasing in its general effect. The top of the head is black, and the face and remainder of the head grey, with the exception of a broad dash of black which reaches from the ear to the angle of the mouth, and looks exactly as if it had been laid on with a single sweep of a brush dipped in ivory black. The whole of the upper surface is a light greyish-brown, each feather on the neck being tipped with a beautiful silver grey. The wings are dark brown, diversified with longitudinal streaks of yellow, produced by the bright yellow colour which stains the outer web of each quill-feather. The tail is greyish-brown, streaked with narrow dark-brown lines in consequence of the shafts of the feathers being coloured by that hue. The under surface is grey, with the exception of a dark-brown patch which is found upon the chin. The eyes are hazel.

The size of the Garrulous Honey-eater equals that of a large thrush.

ANOTHER very curious species of Honey-eater is placed in the same genus, and attracts admiration, not so much on account of its plumage or its interesting habits, as on account of its voice, which is so bell-like in its tone that the colonists know it by the popular name of BELL BIRD. This species must not, however, be confounded with the Bell Bird, or Arapunga, of tropical America, which belongs to a totally different tribe. Moreover, the voice of the two birds is very different ; that of the Arapunga resembling the slow solemn tolling of a church-bell, while that of the Australian Bell Bird is wonderfully similar to the sharp merry tinkle of the sheep-bell. The scientific name for the Australian Bell Bird is *Myzantha melanophrys.*

In his "Gleanings of a Naturalist," Dr. Bennett speaks as follows of this curious little bird :—

"Among the dense forest trees skirting the margins of the rivers, the note of the Bell Bird is almost incessantly heard; it is sometimes uttered by a solitary bird, and at others by many congregated together: this I observed on the banks of the Nepean river, in October, when I saw them in greater numbers than usual. The Bell Bird is named *Gilbulla* by the blacks of the Murrumbidgee district. The peculiar tinkling sound made by this little bird is heard with delight by the wearied and thirsty traveller, as an indication of water near at hand. I have also heard these birds utter loud garrulous notes. At the Nepean they sported among the branches of the trees in search of insects, and I remember that the tinkling note was uttered while they were quietly perched upon a branch, but the garrulous notes were used only when they were seen flitting in sportive gaiety amid the branches of the trees."

AMONG this group of birds the POË BIRD, or TUE, or PARSON BIRD, is one of the most conspicuous, being nearly as remarkable for its peculiar colouring as the rifle bird itself, although the hues of its feathers are not quite so resplendently brilliant as in that creature.

The Poë Bird is a native of New Zealand, where it is far from uncommon, and is captured by the natives for the purposes of sale. Many individuals are brought over to Sydney, where, according to Dr. Bennett, they are kept in cages, and are very amusing in their habits, being easily domesticated and becoming very familiar with those who belong to the household. Independently of its handsome and rather peculiar colour, which make it very effective in a room, it possesses several other qualifications which render it a very desirable inhabitant of an aviary. Its native notes are very fine, the bird being considered a remarkably fine songster, and it also possesses the power of mimicking in a degree surpassing that of the common magpie or raven, and hardly yielding even to the famous mocking-bird himself. It learns to speak with great accuracy and fluency, and readily imitates any sound that may reach its ear, being especially successful in its reproduction of the song of other birds.

While at liberty in its native land it is remarkable for its quick, restless activity, as it flits rapidly about the branches, pecking here and there at a stray insect, diving into the recesses of a newly opened flower, and continually uttering its shrill sharp whistle. Although one of the large group of Meliphagidæ or Honey-eaters, the Poë Bird feeds less upon honey than upon insects, which it discovers with great sharpness of vision and catches in a particularly adroit manner. It will also feed upon worms, and sometimes varies its diet by fruits.

In New Zealand it is often killed for the sake of its flesh, which is said to be very delicate and well flavoured, its beautiful feathers and interesting character affording no safeguard against the voracity of hungry New Zealanders; although, to borrow the expression of Dr. Bennett when speaking of the nautilus, such delinquency is enough to put any scientific naturalist into a fever.

The general colour of the Poë Bird is a very deep metallic green, becoming black in certain lights, and having a decided bronze reflection in others. The back is deep brown, also with a bronze reflection, and upon the shoulders there is a patch of pure white. On the back of the neck the feathers are long and lancet-shaped, each feather having a very narrow white streak along its centre. From each side of the neck depends a tuft of snowy curling downy feathers, spreading in fan-like fashion from their bases. This creature is called the Parson Bird because these white tufts are thought to bear some resemblance to the absurd parallelograms of white lawn that are denominated "bands," and which flutter beneath the chins of ecclesiastics in their official costume. The brown tint of the lower part of the back changes to steely-blue, and the tail is brown "shot" with the same beautiful tint. The quill-feathers of the wings are brown edged with blue, and the whole under surface is reddish-brown.

POE BIRD.—*Prosthemadéra Novæ Zeelandiæ.*

In size the Poë Bird equals a large blackbird or a small pigeon, being about twelve inches in total length. The long generic title of Prosthemadéra is formed from two Greek words—the former signifying an appendage, and the latter the neck—and is given to the bird in allusion to the white tufts of feathers which depend from the neck.

THE very quaint, and rather grotesque bird which is represented in the accompanying illustration is an inhabitant of Australia, and is very common in the southern parts of that continent, although at present it has not been seen in Van Diemen's Land.

By the colonists it is known by a variety of names, some relating to its aspect and others to its voice. Thus, it is named the FRIAR BIRD because the bare, oddly shaped head, with its projecting knob upon the forehead, is considered as resembling the bare shaven poll of the ancient friar. Another analogous name is the MONK BIRD. Another name is LEATHER-HEAD, a title which refers to the dark leathery aspect of the whole head, which is as rigid in outline and as dark in colour during life as after death. On account of its peculiar voice, it is also known by the names of "PIMLICO," "POOR-SOLDIER," or "FOUR-O'CLOCK," as its cry is said to resemble these words. The resemblance, however, cannot be very close, as neither of the words which it is supposed to utter could be mistaken for the other, so that the Friar Bird cannot be very remarkable for the distinctness of its articulation. By the natives it is called Coldong.

The Friar Bird is possessed of unextinguishable loquacity, delighting to get upon the topmost branch of some lofty tree, and there chatter by the hour together at the top of its loud and peculiar voice, as if it were desirous of attracting attention to its powers of elocution. Among the branches it is extremely active, traversing them in all directions with great ease, and clinging to their rough bark by the grasp of its powerful toes and curved claws. So strong is the grip of the foot, that the bird may often be seen hanging from a branch suspended only by a single foot, while it is engaged in peering into the recesses of the bark in search of the little insects that may be concealed under its rough surface.

FRIAR BIRD.—*Tropidorhynchus corniculatus.*

Like all the honey-birds, it is fond of feeding upon the nectar and pollen of flowers, generally preferring those of the eucalyptus, or gum-tree as it is popularly termed, and also delights in fishing out the little insects that are to be found in the depths of all honey-bearing flowers. It does not, however, disdain to feed upon the larger beetles and other insects that take up their residence under the flakes of bark, and is also known to eat various kinds of berries.

The wings of the Friar Bird are rather short in proportion to its size, so that its flight is undulating in its character, the body rising and dropping alternately as the bird flies from one tree to another. It is, however, a strong-winged bird, and can maintain a flight of considerable length. Although apparently unprovided with weapons of offence, this bird is a most courageous and determined fighter, and when wounded and unable to escape from its captor, can strike so fiercely with the feet, that the sharp curved claws will make very painful wounds in the hands of any one who attempts to seize it without proper precaution.

Indeed, the disposition of this bird is decidedly of the pugnacious order, and when it becomes a parent it is the terror of all large birds that may happen to pass near the spot where it has made its nest. Every hawk, no matter how powerful it may be, and every crow, no matter how long and strong of bill, is immediately attacked by this valiant bird, who launches itself at the marauder with such reckless courage, that it always forces the enemy to take to flight, and drives it far from the sacred spot over which it exercises such watchful vigilance.

The nest certainly stands in need of a protector, for it is very large and clumsy in appearance, and is placed on the low branch of a tree, with such little care for concealment that it is visible from afar, and if left unprotected would soon be pillaged of eggs or young by the crows, and other predaceous birds, who are always on the watch for eggs or unfledged birds, and make desolate many a happy nesthold. The nest is composed externally of fine twigs and strips of "stringy bark," and is lined with various grasses, fine roots, and wool. The eggs are generally two or three in number, and their colour is pale salmon, variegated with small spots of dark red. The breeding season is in the month of November. In Gould's "Birds of Australia" may be seen some admirable figures of this bird, together with a full description of its habits and plumage.

In its colouring the Friar Bird is rather sombre, although the uniform dulness of its hue is relieved by a sparing admixture of white. The entire head is devoid of feathers, being covered with a dull black, leather-like skin, sufficiently singular in itself, but deriving an additional shade of grotesqueness from the large round black knob upon the base of the bill. The whole of the upper surface is a greyish-brown, and the tail is of the same hue with the exception of the extremity, which is tipped with pure white. A number of long lancet-shaped feathers of the same beautiful white hang from the breast, each feather being marked with a line of rather dark brown down its centre. With the exception of these pendent breast-feathers, the whole of the under surface is light brown marked with a rusty red. The colour of the eye is red.

This is not the only species of Friar Birds, several other examples of this curious genus being known. There is, for example, the SILVERY-CROWNED FRIAR BIRD (*Tropido-rhynchus argénticeps*), an inhabitant of the north-western coast of Australia, and the YELLOW-THROATED FRIAR BIRD (*Tropidorhynchus citreoguláris*), which takes its place in the interior of Southern Australia. The generic name Tropidorhynchus is of Greek origin, and signifies keel-beaked.

ANOTHER common and rather striking example of the Honey-eaters is the WATTLED HONEY-EATER, or BRUSH WATTLE BIRD of Australia.

This pretty bird is spread over the whole of Southern Australia, and is one of the best known of the birds belonging to that country. It may generally be found upon lofty trees, and, like others of the same group, especially haunts the eucalypti for the purpose of feeding upon the juices of the flowers. It always chooses the most recently opened blossoms, as they are not so likely to be rifled of their sweet stores as those which have been exposed to the attacks of the honey-eating insects and birds. The method of feeding is the same as that which is pursued by the other Honey-eaters, viz. by plunging the long bill and slender tongue into the very depths of the blossoms, and brushing out their contents. It also has a great affection for the flowers of the Banksia, and is sure to be found wherever these plants are in blossom, thereby doing good service to the intending purchaser of land; for the Banksia always grows upon poor soil, so that, according to

Mr. Gould, the harsh cry of the Wattle Bird is a trusty indication to the wary settler that the land on which it is heard is not worth purchasing.

The Banksia especially loves scrubby and sandy soils, and in such localities may be found in the greatest profusion. A celebrated example of this peculiarity may be found in the immense belts of Banksias which border the coast of the delta formed by the embouchure of the Great Murray River, and called in consequence the "Great Murray Scrub."

It is extremely active and quick of foot among the branches, running about the boughs in any position, and seeming to care nothing whether its back or head be downwards or upwards. It is a lively, restless creature, ever on the move, tripping over the branches with a quick, easy step, examining every flower, and diving its long tongue into its recesses and flying quickly from tree to tree as fancy may dictate, or whenever it sees a flower which it thinks likely to contain its liquid treasure. The wings being short and the tail rather long, the flight is of an undulating character, and is said to resemble that of the European magpie.

The Wattled Honey-eater can hardly be termed gregarious, although many specimens may be seen on a single tree, as it lives in pairs, and the two may be generally seen together. The males are very combative in their nature, and being very liable to take offence, are often engaged in single conflict without any apparent reason. It seems to be rather a shy bird, withdrawing itself from human presence. The voice is harsh, loud, rough, and screaming, and of a nature that is not easy to describe, but is said to resemble the peculiar sounds produced by a person who is suffering from the effects of an emetic.

BRUSH WATTLE BIRD.—*Anthochœra carunculáta*

The native name Goo-gwar-ruck is tolerably descriptive of its character.

The nest of this species is rather large and rude of construction, and is simply laid upon the fork of some horizontal branch. The materials of which it is composed are mostly slender twigs and soft dried grasses. The eggs are two or three in number, and their colour is a reddish yellow with a slight brown tinge, thickly covered with deep chestnut and brown spots, impinged with others of a dark grey. The breeding season is in the months of September and October.

Although not decorated with very brilliant plumage, the Wattled Honey-eater is a very pretty bird, deriving great character from the curious appendages from which it has received its popular name. These wattles are of moderate length, and of a blood-red colour, producing a most singular effect as they hang down the sides of the neck. The general colour of the upper surface is light brown, relieved by a white streak along the centre of each feather; the wings are also brown, excepting that the primary quill-feathers are white at their extremities, and the secondaries are tipped with grey. Under each eye there is a patch of white. The throat and breast are of a light greyish brown, covered with multitudinous little longitudinal white streaks, and the centre of the abdomen is yellow. The tail feathers are tipped with white on the under surface, and the tail is black.

ANOTHER species belonging to the same genus is worthy of a passing notice, although it possesses some of the habits of the brush wattle bird. This is the YELLOW WATTLE HONEY-EATER (*Anthochœra inauris*), a bird which may readily be distinguished from the preceding species by the golden yellow colour of its wattles, and the greater length and more handsome appearance of the tail, which is covered on its under surface with bold bands of black and white.

Instead of the semi-solitary life led by the brush wattle bird, this species is partially gregarious in its character, assembling in large flocks of thirty or forty in number upon one tree, and traversing its branches with amazing celerity. The voice of this bird is loud, harsh, and screaming, and it is very fond of exercising its vocal powers, so that the proximity of a flock of the noisy creatures is not very agreeable to a person endowed with sensitive ears. In the winter months it thrives wonderfully, and becomes a perfect ball of fat, the accumulation of this substance being so great that half a tea-cup full of pure oil is often extracted from a single bird. This oil is peculiarly soft and limpid, and is very useful for lamps, as it gives a better light than can be obtained from candles. This state of obesity does not, however, last for any long continuance, and after the breeding season it gradually disappears until the bird becomes as remarkable for its emaciated condition as it was formerly for its extreme corpulence.

There are many other species of Honey-eaters; but the habits of all are so similar, that the examples already given must suffice in a work of such dimensions.

HUMMING-BIRDS OR TROCHĬLIDÆ.

"Bright Humming-bird of gem-like plumeletage,
By western Indians 'Living-Sunbeam' named."—BAILEY, *Mystic.*

THE wonderful little HUMMING-BIRDS are only found in America and the adjacent islands, where they take the place of the sun-birds of the Old World. It is rather remarkable that, as yet, no Humming-birds have been discovered in Australia.

These little winged gems are most capricious in their choice of locality, some being spread over a vast range of country, while others are confined within the limit of a narrow belt of earth hardly more than a few hundred yards in width, and some refuse to roam beyond the narrow precincts of a single mountain. Some of these birds are furnished with comparatively short and feeble wings, and, in consequence, are obliged to remain in the same land throughout the year, while others are strong of flight, and migrate over numerous tracts of country. They gather most thickly in Mexico and about the equator, the number of species diminishing rapidly as they recede from the equatorial line.

The name of Humming-birds is given to them on account of the humming or buzzing sound which they produce with their wings, especially while they are hovering in their curious fashion over a tempting blossom, and feeding on its contents while suspended in air. This name is so appropriate that it holds good in other languages, and expressive

titles have been given to these birds which are either descriptive of the sound or endeavour to imitate it. So characteristic is this humming sound, that it is not precisely the same in any two species, and in many instances is so very decided in its tone, that a practised and observant ear can often detect the species of a Humming-bird by the sound which it produces in flight. For example, Mr. Gosse records that the Black-capped Humming-bird produces a noise exactly like the whizzing of a wheel driven by machinery, while that of another species is very like the droning hum of a large bee.

The number of species of these birds is truly wonderful, as more than three hundred are known and have been described, while new species are being continually discovered. It is evident to any one who has examined these exquisite little birds, and studied the inexhaustible variety of form and colour which they exhibit, that many forms are yet wanting as links needed to complete the chain of species, and that in all probability there are in existence Humming-birds which possess forms quite as strange and colours quite as glowing as any of those which have found a place in our collections.

The legs of these birds are remarkably weak and delicate, and the wings are proportionately strong, a combination which shows that the creatures are intended to pass more of their time in the air than on foot. Even when feeding they very seldom trouble themselves to perch, but suspend themselves in the air before the flower on which they desire to operate, and with their long slender tongues are able to feed at ease without alighting. In the skeleton, especially in the shape of the breast-bone and wings, as well as in the comparative small size of the feet, the Humming-birds bear some analogy to the swifts, and, like those birds, never lay more than two eggs.

The flight of these birds is inconceivably rapid, so rapid indeed that the eye cannot follow it when the bird puts forth its full speed; and with such wonderful rapidity do the little sharp-cut wings beat the air, that their form is quite lost, and while the bird is hovering near a single spot, the wings look like two filmy grey fans attached to the sides. While darting from one flower to another the bird can hardly be seen at all, and it seems to come suddenly into existence at some spot, and as suddenly to vanish from sight. Some Humming-birds are fond of towering to a great height in the air, and descending from thence to their nests or to feed, while others keep near the ground, and are seldom seen at an elevation of many yards.

The food of the Humming-birds is much the same as that of the honeysuckers, except, perhaps, that they consume more honey and fewer flies. Still, they are extremely fond of small insects, and if kept away from this kind of diet soon pine away, in spite of unlimited supplies of syrup and other sweet food.

In Webber's "Wild Scenes and Song Birds," there is an interesting description of some ruby-throated Humming-birds, in which their necessity for insect food is well shown. He had several times succeeded in capturing and taming specimens of these lovely little birds, but always found that they began to pine away and look doleful until they were set at liberty. As soon as they were free, they darted away into the air, but soon returned to their old quarters, attracted by the sweet repast which was plentifully prepared for them. They had evidently been greatly benefited by their short absence, for they resumed their accustomed vivacity, and continued in good health for a fortnight, at the expiration of which time they again drooped, and again needed a short period of freedom. Anxious to discover their proceedings during their absence, Mr. Webber and his sister watched them carefully when they were next set at liberty, and at last were fortunate enough to succeed in their endeavours.

"When we opened the cage this time, it was a bright summer's morning, just after sunrise. What was our surprise to see the ruby-throat, instead of darting away as usual, remain with the young ones, which had immediately sought sprays, as if feeling a little uncertain what to do with themselves. Scarlet flew round and round them; then he would dart off to a little distance in the garden, and suspend himself on the wing for an instant, before what I at first could not perceive to be anything more than two bare twigs; then he would return and fly around them again, as if to show them how easy it was.

The little bold fellows did not require long persuasion, but were soon launched in the air again. They too commenced the same manœuvres among the shrubbery, and as there were no flowers there, we were sadly puzzled to think what it was they were dipping at so eagerly, to the utter neglect of any of the many flowers, not one of which they appeared to notice. We moved closer to watch them to better advantage, and in doing so, changed our relative position to the sun.

At once the thing was revealed to me. I caught friend Ruby in the very act of abstracting a small spider, with the point of his long beak, from the centre of one of those beautiful circular webs of the garden spider, that so abound throughout the South. The thing was done so daintily, that he did not stir the dew drops, which, now glittering in the golden sun, crowded the gossamer tracery all diamond strung.

'Hah! we've got your scent, my friends! Hah! ha! hah!' And we clapped and danced in triumph.

Our presence did not disturb them in the least, and we watched them catching spiders for half an hour. They frequently came within ten feet of our faces, and we could distinctly see them pluck the little spider from the centre of the wheel where it lies, and swallow it entire. After this, we let them out daily, and although we watched them closely, and with the most patient care, we never could see them touch the spiders again until the usual interval of about a fortnight had elapsed, when they attacked them again as vigorously as ever; but the foray of one morning seemed to suffice. We observed them carefully, to ascertain whether they ate any other insect than these spiders; but, although we brought them every variety of the smallest and most tender insects that we could find, they did not entice them at all. But if we could shut them up past the time, until they began to look drooping, and then bring one of those little spiders along with other small insects, they would snap up the spider soon enough, but pay no attention to the others."

The writer then proceeds to remark, as the result of his experiments, that the chief part of their diet consists of nectar, but that they require a feed of insects at certain intervals in order to preserve them in health. He furthermore observes that the birds could not live upon an exclusive diet of insects or of honey, and thinks that they do not eat any creatures except small spiders. Here, however, he seems to be too hasty in his conclusions, as Mr. Davison has discovered the wings of small flies in the stomach of King's Humming-bird, floating in a yellow fluid. In many of the birds which he shot and opened, the stomach was nearly filled with a black mass of insects.

There are some Humming-birds which are very open in their manner of capturing insects, and are fond of perching upon a twig, darting at a passing insect and returning to the same perch, as has been related of some of the trogons, and various other birds. They seem to inherit a singular affection for the twig which they have chosen, and may be found day after day in precisely the same spot.

In order to enable the Humming-bird to extract the various substances on which it feeds from the interior of the flowers, the beak is always long and delicate, and in shape is extremely variable, probably on account of the particular flower on which the bird feeds. In some instances the bill is nearly straight, in others it takes a sharp sickle-like downward curve, while in some it possesses a double curve. The general form of the beak is, however, a very gentle downward curve, and in all instances it is pointed at its extremity. At the base the upper mandible is wider than the lower, which is received into its hollow. Their nostrils are placed at the base of the beak, and defended by a little scale-like shield.

The plumage is very closely set on the body, and is possessed of a metallic brilliancy in every species, the males being always more gorgeously decorated than their mâtes. The tail is composed of ten feathers, although in several species some of the feathers are so slightly developed that they can hardly be seen under the larger rectrices, and, in consequence, their owner has been set down as possessing only six feathers in its tail.

The tongue is a very curious structure, being extremely long, filamentous, and double nearly to its base. At the throat it is taken up by that curious forked bony structure, called the hyoid bone, the forks of which are enormously elongated, and passing under the throat and round the head, are terminated upon the forehead. By means of this structure, the Humming-bird is enabled to project the tongue to a great distance from the bill, and to probe the inmost recesses of the largest flowers. If cut transversely, the filaments of the tongue look as if they had been made of flat, horny parchment, and partially rolled up. The sense of taste or touch, or both, must be extremely delicate in the Humming-birds, for when they are feeding they cannot by any possibility see into

2.   Q

the flowers which they are ransacking, and are perforce obliged to trust implicitly to the tongue. The common woodpecker has a very similar description of tongue, and employs it in a similar manner.

In their habits the Humming-birds are mostly diurnal, although many species are only seen at dawn and just after sunset. Many, indeed, live in such dense recesses of their tropical woods, that the beams of the sun never fairly penetrate into their gloomy depths, and the Humming-bird dwells in a permanent twilight beneath the foliage. It is worthy of notice that the name Trochilidæ is not a very apt one, as the Trochilus was evidently a bird which had nothing in common with the Humming-bird, and was most probably the zic-zac of Egypt.

There are many other peculiarities of habit, structure, and form in these interesting birds, which will be mentioned at length when we come to treat of individual species.

Before commencing the history of these birds, I must acknowledge with pleasure the great assistance which has been derived from Mr. Gould's magnificent monograph of the Trochilidæ.

ONE of the most peculiar forms among these exquisite little creatures is the RUBY-THROATED HUMMING-BIRD, so called on account of the glowing metallic feathers that blaze with ruby lustre upon its throat, and gleam in the sunshine like plumes of living fire. This beautiful species is found in Northern America, and is one of the migrating kind. Passing over a large range of country, it arrives in Pennsylvania about the end of April, and is found during the summer months of the year in different parts of North America, even venturing into the lands owned by the Hudson's Bay Company.

It is a most lovely little creature ; the general colour of its upper surface and the two central tail-feathers is light shining green glazed with gold. The under parts of the body are greyish white intermingled with green, and the throat is of the most gorgeous ruby-carmine. When placed under a moderate magnifier, the feathers of the throat are seen to be constituted in a different manner from those of the other parts of the body, the wonderful refulgent property being due to certain minute furrows which are traced upon the surface, and are analogous in their mode of action to the delicate lines which give to nacre its peculiar iridescent splendour. As is generally the case with Humming-birds, the wings, as well as many of the tail-feathers, are of a purplish-brown hue.

In consequence of the peculiar structure of the throat-feathers, they change their tints with every variation of light, or even with the quick respiration of the little fiery creatures, and fling out at one moment the most dazzling rays of ruby and carmine, and on the instant change to the deepest velvety-black.

Of the Ruby-throat, Audubon speaks in the following terms :—

" I have seen these birds in Louisiana as early as the 10th of March. Their appearance in that State varies, however, as much as in any other, it being sometimes a fortnight later, or, although rarely, a few days earlier. In the middle district they seldom arrive before the 15th of April, more usually the beginning of May. I have not been able to assure myself whether they migrate during the day or by night, but am inclined to think the latter the case, as they seem to be leisurely feeding at all times of the day, which would not be the case had they long flights to perform at that period.

They pass through the air in long undulations, raising themselves for some distance at an angle of about forty degrees, then falling in a curve ; but the smallness of their size precludes the possibility of following them with the eye farther than fifty or sixty yards without great difficulty, even with a good glass. A person standing in a garden by the side of a common Althæa in bloom, will be as surprised to hear the humming of their wings, and then see the birds themselves within a few feet of him, as he will be astonished at the rapidity with which the little creatures rise into the air, and are out of sight and hearing the next moment."

Trusting in its matchless power of wing, the Ruby-throated Humming-bird cares nothing for eagle, hawk, or owl; and though only three inches or so in length, thinks nothing of assaulting any bird of prey that may happen to come within too close a proximity of its home. The tiny creature is in fact a shocking tyrant, jealous to an extreme of its own territories, launching itself furiously at any bird that may seem to be an intruder. It has even been seen to attack the royal eagle itself, and to perch itself upon the head of its gigantic enemy, pecking away with hearty good will, and scattering the eagle's feathers in a stream as the affrighted bird dashed screaming through the air, vainly attempting to rid itself of its puny foe.

The Ruby-throat is very easily tamed, and is a most loving and trustful little creature. Mr. Webber, in the work to which I have already made allusion, has given a most interesting account of a number of Ruby-throats which he succeeded in taming. On several occasions he had enticed the living meteors into his room by placing vases of tempting flowers on the table, and adroitly closing the sash as soon as they were engaged with the flowers, but he had always lost them through their dashing at the window and striking themselves against the glass. At last, however, his attempts were crowned with success, and "this time I succeeded in securing an un-injured captive, which, to my inexpressible delight, proved to be one of the Ruby-throated species, the most splendid and diminutive that comes north of Florida. It immediately suggested itself to me that a mixture of two parts refined loaf-sugar, with one of fine honey, in ten of water, would make about the nearest approach to the nectar of flowers.

RUBY-THROATED HUMMING-BIRD.—*Tróchilus colubris.*
AVOCET HUMMING-BIRD.—*Avocetta Recurvirostris.*

While my sister ran to prepare it, I gradually opened my hand to look at my prisoner, and saw, to my no little amusement as well as suspicion, that it was actually 'playing 'possum'—feigning to be dead most skilfully. It lay on my open palm motionless for some minutes, during which I watched it in breathless curiosity. I saw it gradually open its bright little eyes to peep whether the way was clear, and then close them slowly as it caught my eye upon it. But when the manufactured nectar came, and a drop was touched upon the point of its bill, it came to life very suddenly; and in a moment was on its legs, drinking with eager gusto of the refreshing draught from a silver tea-spoon. When sated, it refused to take any more, and sat perched with the coolest self-composure on my finger, and plumed itself quite as artistically as if on its favourite spray. I was enchanted with the bold innocent confidence with which it turned up its keen black eyes to survey us, as much as to say, 'Well, good folks! who are you?'

Q 2

Thus in less than an hour this apparently tameless rider of the winds was perched pleasantly clinging to my finger, and received its food with edifying eagerness from my sister's hand. It seemed completely domesticated from the moment that a taste of its natural food reassured it, and left no room to doubt our being friends. By the next day it would come from any part of either room, alight upon the side of a white china cup containing the mixture, and drink eagerly, with its long bill thrust into the very base, after the manner of pigeons. It would alight on our fingers, and seem to talk with us endearingly in its soft chirps. Indeed, I never saw any creature so thoroughly tamed in so short a while."

The writer then proceeds to remark, that after he had kept the bright little thing for three weeks, it began to droop daily, so that he was obliged to let it fly at liberty. As soon as the window was opened, the bird darted out like a ruby meteor, and vanished immediately from sight. In hopes of attracting him back again, the two enthusiasts prepared a fresh cup of nectar, hung the cage with flowers, and placing the cage and cup invitingly in the window, retired to a distance and waited patiently. After watching in vain for a whole hour, they were just about to give up the point in despair, when they saw their pet Ruby-throat hovering before the window.

"The little fellow was darting to and fro in front of his cage, as if confused for a moment by the flower dressing; but the white cup seemed to overcome his doubts very quickly, and with fluttering hearts we saw him settle upon the cup as of old, and while he drank we rushed lightly forward on tiptoe to secure him. We were quite rebuked for our want of faith, threw open the door again, and let him have the rest of the day to himself."

One of the most curious circumstances connected with a family of Ruby-throated Humming-birds which Mr. Webber succeeded in domesticating was, that after they had left the country at the ordinary migrating season, they retained the memory of their kind entertainers, and, on the return of spring, flew straight to the well-known window. As soon as the white cup was placed in the room, the birds, as if they had been only waiting for its appearance, dashed through the casement, and plunging their beaks into the syrup, drank long draughts of its welcome nectar. By degrees they found mates, and brought their companions to partake of the same hospitality, until at last there was quite a company of these exquisite little creatures, who brought their friends and families to the familiar feast.

Mr. Webber also discovered a curious habit connected with their nesting. He had frequently observed, while watching for their nest, that the Ruby-throats, after leaving his station, shot suddenly and perpendicularly in the air, until they became invisible. At last he had the great satisfaction of seeing the female bird fall, like a fiery aerolite from the sky, upon the spot where she had built her nest; so that this curious habit of ascending and descending must have been instinctively taught to the birds for the purpose of concealing the precise position of the nest.

As to the nest itself, an admirable description is given by Audubon. Here I must pause for a moment, to explain the reason why there will be so many quotations in the history of the Humming-birds. These little creatures exist only in exotic lands, and on that account are not very accessible to the English naturalist, who is debarred from a personal acquaintance with these most exquisite of birds. He must, therefore, rely entirely on the accounts of those who have seen them and studied their habits, and whose personal narratives are so far superior to any abstract or paraphrase, that, in justice to the author, they ought to be given in his own words.

"The nest of this Humming-bird," says Audubon, "is of the most delicate nature, the external parts being formed of a light grey lichen found on the branches of trees or on decayed fence-rails, and so neatly arranged round the whole nest, as well as to some distance from the spot where it is attached, as to seem part of the branch or stem itself. These little pieces of lichen are glued together by the saliva of the bird.

The nest-coating consists of cottony substances, and the innermost of silky fibres obtained from various plants, all extremely delicate and soft. On this comfortable bed, as if in contradiction to the axiom that the smaller the species the greater the number of eggs, the female lays only two, which are purely white and nearly oval. Ten days are required for their hatching, and the birds raise two broods in a season. In one week the young are ready to fly, but are fed by the parents for nearly another week. They receive their food directly from the bill of their parents, who disgorge it in the manner of canaries and pigeons. It is my belief that no sooner are the young able to provide for themselves, than they associate with other broods and perform their migrations apart from the old birds, as I have observed twenty or thirty young Humming-birds resort to a group of trumpet flowers, when not a single old male was to be seen.

They do not receive the full brilliancy of their colours until the succeeding spring, although the throat of the male bird is strongly imbued with the ruby tints before they leave us in autumn."

The reader will doubtless remember that Mr. Webber mentions the fact that one of his captured Ruby-throats "played 'possum" when taken, simulating death in a very perfect manner. In Wilson's well-known work on the birds of America, there is an account of a somewhat similar performance on the part of a captive Ruby-throat, which seemed to simulate death, or, at all events, to fall into a state of semi-torpidity from the effects of cold.

"This little bird is extremely susceptible of cold, and if long deprived of the animating influence of the sunbeam, soon droops and dies. A very beautiful male was brought to me this season (1809), which I put into a wire cage, and placed in a retired shaded part of the room. After fluttering about for some time, the weather being uncommonly cold, it clung to the wires and hung in a seemingly torpid state for a whole forenoon. The motion of the lungs could not be perceived on the closest inspection, though at other times this is remarkably observable : the eyes were shut, and when touched by the finger it gave no signs of life or motion. I carried it out to the open air and placed it directly in the rays of the sun in a sheltered situation. In a few seconds respiration became very apparent, the bird breathed faster and faster, opened its eyes, and began to look about with as much seeming vivacity as ever. After it had completely recovered, I restored it to liberty, and it flew off to the withered top of a pear-tree, where it sat for some time dressing its disordered plumage, and then shot off like a meteor."

It has also been related that Humming-birds are so delicate of constitution, that when caught in a gauze net they die at once from fear. I have no doubt but that in fact they often simulate death, and do it so well as to make their escape when their captor's attention is withdrawn from them.

ANOTHER species belonging to this genus is well worthy of notice, on account of its beauty and interesting habits. This is the long-tailed Humming-bird of Jamaica (*Tróchilus Polytmus*), one of the species which do not migrate, but remain in one locality throughout the year. A very good figure of it may be seen in the engraving on page 223. It is the perched figure at the upper part of the group of Humming-birds, and may be easily noted by means of the two long tail-feathers that hang nearly to the bottom of the engraving.

The upper parts of this beautiful bird are green, glossed with gold, the wings are purple-brown, and the tail black, with a steel-blue reflection. The long streaming feathers of the tail are the pair next to the exterior feathers, and when the bird is in a state of repose they cross each other as is represented in the engraving. The throat, breast, and whole of the lower parts are glowing emerald green, except the under tail-coverts, which are purple-black. The top of the head and nape of the neck are velvet black, and the feathers of the head are rather long, and form a kind of loose plume. The whole length of a male bird is rather more than ten inches, the long tail-feathers being between seven and eight inches in length.

The female is not possessed of the beautiful tail which distinguishes her mate ; the under parts are white, covered with green spots caused by the green tips of the feathers, the top of the head is dirty brown, and her entire length is little more than four inches. Mr. Gosse, in his well-known " Birds of Jamaica," has given some admirable descriptions of this pretty bird and its habits.

" It loves to frequent the margins of woods and roadsides, where it sucks the blossoms of the trees, occasionally descending to the low shrubs. There is one locality where it is abundant,—the summit of that range of mountains just below Bluefields, and which is known as the Bluefields ridge. . . . Not a tree, from the thickness of one's wrist up to the giant magnitudes of the hoary figs and cotton trees, but is clothed with fantastic parasites; begonias with waxen flowers, and ferns with hirsute stems, climb up the trunks; enormous bromelias spring from the greater forks and fringe the horizontal limbs; curious orchideæ, with matted roots and grotesque blossoms, droop from every bough, and long lianes, like the cordage of a ship, depend from the loftiest branches or stretch from tree to tree. Elegant tree-ferns and towering palms are numerous; here and there the wild plantain, or heliconia, waves its long ivy-like leaves from amidst the humbler bushes, and in the most obscure corners, over some decaying body, rises the nobler spike of a magnificent limodarum. The smaller wood consists largely of the plant called glass-eye berry, the blossoms of which, though presenting little beauty in form or hue, are preeminently attractive to the Long-tailed Humming-bird.

And here at any time we may, with tolerable certainty, calculate on finding these very lovely birds. But it is in March, April, and May that they abound. I suppose I have sometimes seen not fewer than a hundred come successively to rifle the blossoms within the space of half as many yards, in the course of a forenoon. They are, however, in no respect gregarious; though three or four may at one moment be hovering round the blossoms of the same bed, there is no association; each is governed by his individual preference, and each attends to his own affairs.

It is worthy of remark, that males uniformly form the greater portion of the individuals observed at this elevation. I do not know why it should be so, but we see very few females there, whereas, in the lowlands, this sex outnumbers the other. In March, a large number are found to be clad in the livery of the adult male, but without long tail-feathers; others have the characteristic feathers lengthened, but in various degrees. These are, I have no doubt, males of the preceding season.

It is also quite common to find one of the long tail-feathers much shorter than the other, which I account for by concluding that the shorter is replacing one that had been accidentally lost. In their aerial encounters with each other a tail-feather is sometimes displaced. One day, several of these 'young bloods' being together, a regular tumult ensued, somewhat similar to a sparrow-fight; such twittering, and fluttering, and dartings hither and thither. I could not exactly make out the matter, but suspected that it was mainly an attack—surely an ungallant one—made by them upon two females of the same species that were sucking at the same bud. These were certainly in the skirmish, but the evolutions were too rapid to be certain how the battle went.

The whirring made by the vibrating wings of the male Polytmus is a shriller sound than that produced by the female, and indicates its proximity before the eye has detected it. The male almost constantly utters a monotonous quiet chirp, both while resting on a twig or while circling from flower to flower. They do not invariably probe flowers on the wing; one very frequently observes them thus engaged when alighted and sitting with closed wings; and often they partially sustain themselves by clinging by the feet to a leaf while sucking, the wings being expanded and vibrating."

Several of these beautiful birds were captured and tamed by Mr. Gosse, who, however, found the task to be one of no ordinary difficulty. It was easy enough to catch them in a gauze net, for they were so inquisitive that they would hover over the net and peep into its recesses; but when they were caught they would generally die within a few hours. Several of the Long-tailed Humming-birds were at last taken from the nest, and were soon tamed. They were fed chiefly upon syrup, but were also supplied with little insects, in imitation of their ordinary diet in a wild state. They were especially pleased with a very small species of ant, which used to get into the vessel of syrup and fairly cover its surface with their bodies.

There is a long and very interesting description of these birds, which resembles, in many respects, the amusing account given by Mr. Webber of his own winged pets. One peculiarity deserves notice. Each bird, as soon as it was introduced into the room in which it lived, made choice of separate perches for roosting, alighting after flight, and for resting-places, and, when it had once settled itself, it would not permit any of its companions to usurp its dominions. Even if their owner endeavoured to make them change their perches, they were quite uneasy, hovered about the spot, and did all in their power to reassume their positions.

The nesting of this beautiful species is very remarkable, as the nests are wonderfully constructed, and are placed in very curious localities. One of these nests was found upon the sea-shore, fastened to a slender twig of wild vine, and actually overhanging the waves. It seems that the bird is in the habit of removing its eggs or young when it has been disturbed, although the mode by which this feat is accomplished has not yet been discovered. The nest is beautifully made of silky cotton threads, intermixed with the web of certain spiders, and is often studded profusely with lichens. Mr. Gosse was fortunate enough to see the bird in the act of making her nest, and describes her movements in the following words :—

"Suddenly I heard the whirr of a Humming-bird, and, looking up, saw a female Polytmus hovering opposite the nest with a mass of silk-cotton in her beak. Deterred by the sight of me, she presently retired to a twig a few paces distant, on which she sat. I immediately sank down among the rocks as gently as possible, and remained perfectly still. In a few seconds she came again, and, after hovering a moment, disappeared behind one of the projections, whence in a few seconds she emerged again and flew off. I then examined the place, and found, to my delight, a new nest. . . .

I again sat down on the stones in front, where I could see the nest, not concealing myself, but remaining motionless, waiting for the bird's re-appearance. I had not to wait long : a loud whirr, and there she was, suspended in the air before her nest. She soon espied me, and came within a foot of my eyes, hovering just in front of my face. I remained still, however, when I heard the whirring of another just above me, perhaps the mate, but I durst not look towards him, lest the turning of my head should frighten the female. In a minute or two the other was gone, and she alighted again on the twig, where she sat some little time preening her feathers, and apparently clearing her mouth from the cotton fibres, for she now and then swiftly projected the tongue an inch and a half from the beak, continuing the same curve as that of the beak. When she arose, it was to perform a very interesting action, for she flew to the face of the rock, which was thickly clothed with soft dry moss, and, hovering on the wing as if before a flower, began to pluck the moss until she had a large bunch of it in her beak. Then I saw her fly to the nest, and having seated herself in it, proceed to place the new materials, pressing and arranging and interweaving the whole with her beak, while she fashioned the cup-like form of the interior by the pressure of her white breast, moving round and round as she sat. My presence appeared to be no hindrance to her proceedings, although only a few feet distant ; at length she left the place, and I left also. On the 8th of April I visited the cave again, and found the nest perfected and containing two eggs, which were not hatched on the 1st of May."

In the same work are contained many interesting descriptions of this exquisite bird and its habits, and to its pages the reader is referred for further information.

The SWORD-BILL HUMMING-BIRD derives its name from the singular shape and size of its beak, which is very nearly as long as the rest of the body.

This curious species is rather large, as it measures about eight inches in length. It inhabits Santa Fé de Bogóta, the Carracas and Quito, and is generally found at considerable elevations, having been often seen at a height of twelve thousand feet above the level of the sea. The inordinately long bill is given to this bird in order to enable it to obtain its food from the very long pendent corollas of the Brugmansiæ, and, while probing the flowers with its beak, it suspends itself in the air with a tremulous movement of the wings. Its movements are singularly elegant, and while engaged in feeding it performs the most graceful manœuvres as it probes the pendent blossoms, searching to their inmost depths. The nest of this species is hung to the end of a twig, to which it is woven with marvellous skill, and its whole construction is very beautiful.

The adult male bird is coloured as follows. The head and the upper part of the body are green, glossed with gold in some parts and with bronze in others, the tints changing according to the light. The wings are dark black-brown with a purple gloss, and the tail is dark black, bronzed on the upper surface. Behind each eye is a small but

SLENDER SHEAR-TAIL HUMMING-BIRD.
Thaumastura enicura.
SWORD-BILL HUMMING-BIRD.—*Docimaster ensiferus.*

conspicuous white spot slightly elongated, and there is a broad crescent-shaped mark of light green on each side of the neck. The under parts are of a bronze-green, and the under tail-coverts are flecked with a little white. The female is of much the same colour as the male upon the upper parts of the body, except that there is a little white upon the lower part of the back and a narrow white line behind the eye. The throat is brown, each feather being slightly edged with grey, and there is a very faint indication of emerald-green on part of the throat. The young male is much like the female, but is more coppery in his hues. The throat is white speckled with brown, because each feather is white with a brown tip. At each side of the throat there is a large patch of green inter-mingled with white.

The SLENDER SHEAR-TAIL is an inhabitant of Central America, and appears to be rather a local bird. It is supposed not to be found south of the Isthmus of Panama, nor to extend more than eighteen degrees northwards. As its wings are rather short, and not remarkable for strength, it is conjectured to be a non-migratory bird. The country where it is seen in the greatest plenty is Guatemala.

The sexes of this creature are very different in their form and colour of their plumage, and could hardly be recognised as belonging to the same species. In the adult male bird, the upper parts of the body are a deep shining green, becoming brown on the head, and changing into bronze on the back and wing-coverts. The wings are purple-brown. The long and deeply-forked tail is black, with the exception of a little brown upon the inner web of the two outermost feathers. The chin is black glossed with green, the throat is deep metallic purple, and upon the upper part of the chest is placed a large crescent-shaped mark of buff. The abdomen is bronze, with a grey spot in its centre; and there is a buff spot on each flank. The under tail-coverts are of a greenish hue.

The female does not possess the long tail, and her colours are golden-green above and reddish-buff below. The tail is very curiously marked. The central feathers are entirely gold-green; the exterior feathers are rusty red at their base, black for a considerable portion of their length, and tipped with white.

Another example of this genus is the well-known CORA'S SHEAR-TAIL, a remarkably pretty bird, and specially notable for the peculiarity from which it derives its popular name. It inhabits Peru, and is found very plentifully between Callao and Lima. The valley of the Andes is also a favourite residence of this bird.

In the male, the head and upper parts of the body are golden-green, with the exception of the wings, which are purple-brown. The throat is violet, changing into metallic crimson, and the under parts are greyish-white. The tail is rather curiously shaped. The two central feathers are double the length of the next pair, and the remaining feathers are regularly gradated, the exterior being the shortest. This long tail is only found in the male bird, the tail of the female being of the ordinary length.

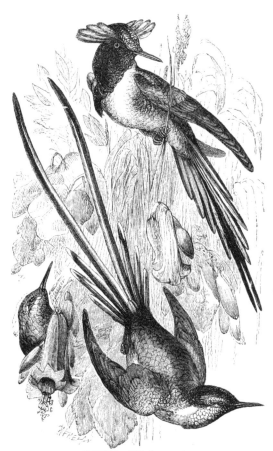

SEVERAL of the Humming-birds are remarkable for a tuft of pure white downy feathers which envelop each leg, and which has obtained for them the popular title of Puff-legs, because the white tufts bear some resemblance to a powder-puff. The generic name Eriocnemis is given to the bird in allusion to this peculiarity, and is formed of two Greek words, the former signifying wool or cotton, and the other the thigh. Owing to the very curious effect of these tufts, the Puff-legs are in great demand among the dealers, as they look remarkably well in a case of stuffed birds.

The COPPER-BELLIED PUFF-LEG is an inhabitant of Santa Fé de Bogóta, and is a very common bird in that locality. It may easily be found, as it is a remarkably local bird, being confined to a narrow strip or belt

SUN-GEM.—*Heliactin cornuta.*
CORA'S SHEAR-TAIL.—*Thaumastúra Coræ.*

of land, which possesses the requisite characteristics of temperature and vegetation.

It must here be remarked that in the mountainous districts where this and many other species of Humming-birds are found, every degree of temperature may be obtained within the compass of a few miles by merely ascending or descending the lofty mountains which form the greater part of the country. A few hours' journey will bring the traveller through every shade of climate, from the perpetual snow and ice at the summit, to the moderate temperatures of the middle regions, and the tropical heat of the mountain's foot. This circumstance must be borne in mind, as we shall find, on examining the habits of many of these birds, that the conditions requisite for their maintenance are very capricious, and that a belt of land a few yards in width will often suffice to separate the habitation of one local species from that of another, neither venturing to trespass into the dominions of its neighbour.

The Copper-bellied Puff-leg is always found in a narrow belt of land varying from six thousand to nine thousand feet above the level of the sea, being, therefore, practically confined to a strip of land barely a thousand yards in width. In all probability the reason of this restricted range may be found in the vegetation of the locality, which supplies the food on which this species lives.

COPPER-BELLIED PUFF-LEG HUMMING-BIRD.
Eriocnémis cupreiventris.

It is a very beautiful little bird, and both the sexes are very similar in their colour and general appearance, except that in the female the puffs of white down are not so large nor so conspicuous as in her mate. In the adult male, the top of the head, the sides of the neck, and the back are green, washed with a decided tint of bronze, except upon the upper tail-coverts, where the green is very pure and of a metallic brilliancy. As is generally the case with Humming-birds, the fine and sharply-cut wings are brown washed with purple. The tail is black, with a purple gloss in a side light. The throat is of a beautiful shining metallic green, and the general colour of the breast and under portions of the body is green glossed with gold, with the exception of the abdomen, where the green takes a coppery hue, from which the bird has received its popular name. The "puffs" are of a snowy whiteness, and look like refined swans'-down.

The female is very similar in colour, except that the hues of the throat are not possessed of so metallic a brilliancy, and, as has already been stated, the leg-tufts are comparatively small.

In the opinion of many observers, the Topaz Humming-birds are the most resplendent and beautiful of all their tribe, the palm of beauty being almost equally divided between the two birds which will be described in the following lines.

The FIERY TOPAZ inhabits the country through which passes the Rio Negro, a tributary of the Upper Amazon. It is a most gorgeous creature, and attracts peculiar attention on account of the very considerable dimensions to which it reaches. Its nest is a very remarkable structure, looking much as if it were made from leather, and woven so adroitly to the bough upon which it is placed that it can hardly be distinguished from the natural bark or from some of the numerous fungi that grow upon trees. Its surface is quite smooth, and the colour is a reddish-dun. The substance of which it is composed is a kind of fungus, of the same order as the well-known Boletus of which German tinder is made. The eggs are two in number, and beautifully white.

The colour of this splendid bird is mostly a blazing scarlet, contrasting boldly with the deep velvet-black of the head and part of the neck. The throat is emerald-green, with a patch of delicate crimson in the centre. The lower part of the back and the upper tail-coverts are beautiful green with orange gloss; and the wings and tail are purple-black, with the exception of the two elongated feathers of the tail, which are purplish-green, and cross each other near the base. The under tail-coverts are green. So vivid are the tints, and so beautiful the form of this bird, that it well deserves the honourable title accorded to it by Prince Lucien Bonaparte of being "*inter Trochilides pulcherrimus.*" The female is without the elongated tail-feathers, and she is of a green-gold colour on the upper parts of the body. This species is very like the following bird, but may be distinguished from it by the purple of its tail-feathers and the fiery effulgence of its body. The entire length of this bird is about eight inches.

The CRIMSON TOPAZ, or ARA HUMMING-BIRD (*Topaza Pella*), closely resembles the fiery topaz, except that the hues of its body are more of a deep crimson than of the flaming scarlet which denotes the preceding species. The tail is reddish-buff, with the exception of the two central feathers which have the same purple-green as in *T. Pyra*. It inhabits Cayenne, Trinidad, and Surinam, and among the natives is known by the name of Karabamiti. It is a shy and retiring bird, living near rivers, and shrouding its beauties in the deepest forests. It is a semi-nocturnal bird, resembling the nightjars in many of its habits, and being most active in the early dawn and the beginning of the evening. Only at those hours does it venture from the deep recesses of its home, and display its flashing colours as it darts along the glades or over the streams in search of its insect prey.

FIERY TOPAZ HUMMING-BIRD.—*Topáza Pyra.*

WE have in the Racket-tailed Humming-birds one of those singular forms which are so often found among these strange little birds.

The RACKET-TAIL HUMMING-BIRD (*Discúra longicauda*) is a native of Cayenne, Surinam, and Demerara, and is also found in several portions of Northern Brazil. It is chiefly remarkable for the curious formation from which it derives its popular and appropriate name.

In the male bird, the face, throat, and part of the neck are light verditer-green, becoming more luminous towards the chest. Under the chin there is a little velvet-black spot, which is very conspicuous against the light green of the surrounding feathers. The upper parts are bronze-green, and a buff-white band crosses the lower end of the back. The very curious tail is deeply forked, the two exterior feathers being twice the length of the second pair, and the others decreasing in length in rapid progression. The general colour of the tail is purple-black, the purple being especially visible on the "rackets." The female possesses no rackets on the tail, nor green on her head or throat. The velvet-black spot on the chin, however, retains its place.

WHITE-BOOTED RACKET-TAIL.—*Spathura Underwoodii.*
Male and Female.

The WHITE-BOOTED RACKET-TAIL inhabits the Columbian Andes, and is very common near Santa Fé de Bogóta. It is a hill-loving bird, being generally found at an elevation of five or ten thousand feet above the level of the sea. It is thought to be confined within the third and tenth degrees of north latitude. This bird is remarkably swift of wing, its darting flight reminding the spectator of the passage of an arrow through the air. At one time it will hover close to the ground, hanging over some favourite flower and extracting the sweet contents of the blossoms; and at the next moment it will shoot to the very summit of some lofty tree, as if impelled from a bow, and leave but the impression of an emerald-green line of light upon the observer's eye. While hovering over the flowers, the long racket-shaped feathers of the tail are in constant motion, waving gently in the air, crossing each other, opening and closing in the most graceful manner. But when the bird darts off with its peculiar arrowy flight, the tail-feathers lie straight behind it.

The male of this species is bronze-green upon the greater part of the body, the green taking a richer and redder hue upon the upper tail-coverts. The throat and breast are brilliant emerald-green. The wings are purple-brown, and the tail is brown, with the exception of the rackets, which are black "shot" with green. The feet are yellow, and upon the legs are placed two beautiful white puffs. The whole length of the bird is rather more than three inches. The female bird does not possess the racket-shaped tail-feathers, and is of a bronze-green upon the upper surface. The tail is brown, with the exception of the two middle feathers, which are bronze-green like the body. The two exterior feathers are tipped with white, and the others with bronze-green. The under surface is white, diversified with bronze-green spots on the breast and flanks. The puffs are smaller than in the male.

There are several species belonging to this genus, among which may be mentioned the PERUVIAN RACKET-TAIL, a bird which may be distinguished by the rusty-red colour of the leg-muffs.

The COLUMBIAN THORNBILL is an inhabitant of Santa Fé de Bogóta, and is remarkable for its adherence to the more temperate regions of that locality. It never seems to ascend to the hills, but prefers remaining in the plains or in some of the valleys where it can find the greatest abundance of food. It appears not to frequent the tops of trees, as is often the case with Humming-birds, but contents itself with the low flowering shrubs of the plains and valleys. The colour of this bird is golden-green on the upper parts, changing into a warmer hue on the upper tail-coverts. Below, it is dull green, with the exception of a remarkable tuft or beard which hangs from the chin, and which is light green towards its base and purple-red towards its extremity. The wings are purple-brown, the tail brown with a bronze gloss, and the under tail-coverts brown-yellow. The female resembles the male, but has not the flame-like mark on the throat. The total length of this species is between five and six inches.

There are several species of Thornbills, among which may be noticed HERRAN'S THORNBILL (*Rhamphomicron Herráni*), a bird which is remarkable for its broad purple tail and the snowy white tips of the three exterior feathers. All the Thornbills possess the curious beard-like appendage to the chin.

COLUMBIAN THORNBILL.—*Rhamphomicron heteropógon.*

At the right-hand lower corner of the engraving on page 223, will be seen a curiously formed bird, remarkable for its long slender crest and the elongated feathers of its tail. This is the POPELAIRE'S THORNTAIL (*Gouldia Popelairii*). This beautiful little creature inhabits Peru and Columbia, and is found in the most elevated regions of that locality. It is a very quick flier, but there is little known of its habits.

In the adult male, the crown of the head, the shorter feathers of the crest, together with the face and throat, are light golden green, and the long slender feathers are black.

The wings are purple brown. The back is gold-green, with the exception of a bold bar of pure white, which crosses the back and forms a patch on each flank. The middle of the abdomen is black, the flanks are brown, and the under tail-coverts are greyish white. The upper surface of the tail is blue, the shafts of the middle feathers are white, and the remaining feathers are white at their bases and brown for the rest of their length. The under surface of the tail is a bright steel-blue, and the shafts are white throughout their length.

The female possesses no crest and no elongated tail-feathers, and bears a very curious resemblance to the well-known insect termed the Humming-bird Moth.

CONVERS' THORNTAIL.—*Goùldia Conversi.*
Male and Female.

ANOTHER curious example of the same genus may be found in CONVERS' THORNTAIL, a native of Santa Fé de Bogóta. This species is very beautiful both in shape and colouring, and, as in the case of the preceding bird, the two sexes differ greatly in appearance. In the male of this bird the general colour is green, a white bar running across the lower end of the back, and the tail-feathers being very long, narrow and pointed. Their colour is shining black, the shafts being white. In the female the general colour of the plumage resembles that of the male, except that the colours are not so brilliant, and the throat is greyish white, covered with brown-green spots. The tail is very short, and is composed of a series of rounded feathers of a dusky hue, and white at the tip. Both the species are swift fliers, and are said to resemble the swallow when on the wing.

ON the extreme left of the engraving on page 223, and about half-way from the top, the reader may observe a very small Humming-bird, remarkable for its curious spiky tail and rich feathery gorget. This is the LITTLE FLAME-BEARER (*Selásphorus scintilla*), one of several species which possess the fiery tuft of feathers from which they derive their name of Flame-bearers.

This species inhabits the inner side of the extinct volcano Chiriqui, in Veragua, at an elevation of nine thousand feet above the level of the sea. It is a tiny bird, measuring only two and a half inches in length, and as it darts about the singular habitation in which it lives, its fiery gorget gleams with such a flaming crimson, that, as Mr. Gould happily remarks, it seems to have caught the last spark from the volcano before it was extinguished.

In the male, the upper surface is of a bronze green. The gorget is of a fiery red, and as the feathers on each side are longer than those in the centre, it necessarily projects from the neck. Below the gorget is a band of white marked with buff, and the wings are purple brown. The central feathers of the tail are brownish black edged with red, and the remaining feathers are brownish black on their outer webs, and reddish rust on the inner webs. The under surface of the tail is a rusty red. The female is duller in her colouring, and the gorget is shorter and of a whitish grey spotted with brown. The flanks are buff, and the tail-feathers are not so pointed as in the male.

There are several species of Flame-bearers, among which may be mentioned the RUFUS FLAME-BEARER (*Selásphorus rufus*), a bird which was originally discovered by Captain Cook. It is an inhabitant of Mexico, and is also found on the Pacific side of Northern America in the summer time, returning to Mexico in the winter. This species is well described by Mr. Nuttal, whose account is quoted by Audubon.

"We began to meet with this species near the Blue Mountains of the Columbia River in the autumn, as we proceeded to the coast. These were all young birds, and were not very easily distinguished from those of the common species of the same age.

We now for the first time (April 16) saw the males in numbers, darting, buzzing, and squeaking in the usual manner of their tribe : but when engaged in collecting its accustomed sweets in all the energy of life, it seemed like a breathing gem, a magic carbuncle of glowing fire, stretching out its glorious ruff as if to emulate the sun itself in splendour. Towards the close of May the females were sitting, at which time the males were uncommonly quarrelsome and vigilant, darting out at once as I approached the tree, probably near the nest, looking like an angry coal of brilliant fire, passing within very little of my face, returning several times to the attack, sailing and darting with the utmost velocity, at the same time uttering a curious reverberating sharp bleat, somewhat similar to the quivering twang of a dead twig, yet also so much like the real bleat of some small quadruped, that for some time I searched the ground instead of the air for the actor in the scene.

At other times the males were seen darting high up in the air, and whirling about each other in great anger and with much velocity. After these manœuvres, the aggressor returned to the same dead twig, where for days he resolutely took his station, displaying the utmost courage and angry vigilance. The angry hissing or bleating note seems something like *wht' t' t' t' sh vee*, tremulously uttered as it whirls and sweeps through the air, accompanied also by something like the whirr of the night hawk. On the 29th of May I found a nest in a forked branch of the Nootka bramble (*Rubus Nutkanus*). The female was sitting upon two eggs of the same shape and colour as those of the common species, *Trochilus colubris*. The nest also was similar, but somewhat deeper. As I approached, the female came hovering round the nest, and soon after, when all was still, she resumed her place contentedly."

The nest of this bird measures, according to Audubon's description, two inches and a quarter in height and an inch and three-quarters in breadth at the upper part, and is composed of mosses, lichens, and feathers, woven together with delicate vegetable fibres. The lining is very soft cotton. Another observer, Dr. Townsend, compares the curious note of this bird to the sound which is produced by the rubbing together of two branches during a high wind.

THE birds which compose the genus Phaëthornis are remarkable for the very long and beautifully graduated tail, all the feathers being long and pointed, and the two central far exceeding the rest. The two sexes are mostly alike, both in the colour and shape of their plumage and in size. These birds inhabit Venezuela and the Carracas, being generally found in the richest district of those localities, where the flowers blossom most abundantly. All the Hermits build a very curious and beautiful nest, of a long funnel-like form tapering to a slender point, and woven with the greatest neatness to some delicate twig or pendent leaf by means of certain spiders' webs. The material of which it is made is silky cotton fibre, intermixed with a woolly kind of furze, and bound together with spider-web. Our present example is SALLÉ'S HERMIT.

Very little is known of its habits, but, like the generality of Humming-birds, it does not possess any great power of voice. Indeed, even in the few instances where one of these birds is gifted with vocal powers, its song is of a feeble and uncertain character, and in England would attract little attention. The best songster of all the Humming-birds appears to be the Vervain Humming-bird (*Mellisuga minima*), which, according to Mr. Bullock, can sing, although not very perfectly.

"He had taken his station on the twig of a tamarind-tree which was close to the barn and overspread part of the yard: there, perfectly indifferent to the number of persons constantly passing within a few yards, he spent most of the day. There were few blossoms on the tree, and it was not the breeding season, yet he most pertinaciously kept absolute possession of his domain; for the moment any other bird, though ten times as large as himself, approached near his tree, he attacked it most furiously and drove it off,

SALLE'S HERMIT HUMMING-BIRD.—*Phaëthornis Augusti.*

always returning to the same twig he had before occupied, and which he had worn quite bare for three or four inches by constantly feeding on it. I often approached within a few feet with pleasure, observing his tiny operations of cleaning and pluming, and listening to his weak, simple, and oft-repeated note. I could easily have caught him, but was unwilling to destroy so interesting a little visitant, who had afforded me so much pleasure.

In my excursions round Kingston I procured many of the same species, as well as the long-tailed black and a few others, as well as the one I have mentioned as the smallest yet described, but which has the finest voice of any. I spent some agreeable hours in the place that had been the Botanical Garden of Jamaica; and on the various trees, now growing to a luxuriant size, met with many curious birds, among which this specimen was perched upon the bread-fruit or cabbage tree. He poured forth his slight querulous note among a most numerous assemblage of the indigenous and exotic plants and trees of the island, on a spot once the pride of Jamaica, but now a desolate wilderness." This beautiful Humming-bird will be described at length in a future page.

To return to SALLÉ'S HERMIT. The upper parts of its body are green-bronze, excepting the upper tail-coverts, which are rusty red. The wings are purple-brown. The central tail-feathers are bronze, largely tipped with white, and the remaining feathers are white, with the exception of a broad black band, drawn obliquely across them near the base. Above and below the eye there is a white streak, and the colour of the under parts of the body is sober grey.

If the reader will again turn to the engraving on page 223, and refer to the bird immediately below Princess Helena's Coquette, he will see a rather large species of Humming-bird, which is remarkable for the manner in which the rounded tail-feathers are arranged, and the very long upper tail-coverts. This bird is the JACOBIN HUMMING-BIRD (*Florisuga mellivora*), a beautifully coloured species, glowing with boldly contrasted hues of white, blue, green and black

It inhabits Cayenne, Guiana, Trinidad, and seems to have rather an extensive range, being found from Cayenne to Peru. It is a very curious species, inhabiting broad and fluviatile districts not more than two or three hundred feet from the level of the sea. The colour is very variable, but is generally a light blue upon the head and throat, with a large white crescentic patch passing over the back of the neck. The back, the very long upper tail-coverts, and a line extending to each side of the neck, are golden green, and the wings are purple-black, edged on the shoulders with golden green. The tail is tipped with a narrow band of black. Some individuals have a green mark upon the blue of the head, and others are curiously mottled with white and brown.

There are several species of this genus, among which may be mentioned the GREAT JACOBIN (*Florisuga flabellifera*), a truly beautiful bird, and much larger than the preceding species. It is found in Tobago, in the Orinocos, and other neighbouring localities. It lives mostly in low marshy situations, chiefly upon planta-tions abroad, and generally feeds while on the wing. Another curious species is the PIED JACOBIN (*Mellisuga Atra*), a bird which is much blacker than either of the preceding. It inhabits the ex-treme parts of Brazil, from Per-nambuco on the north to Rio Janeiro on the south. Like the preceding species, it is very variable in colouring.

LINDEN'S HELMET-CREST, OR BLACK WARRIOR.
Oxypógon Lindénii.

The Helmet-crests are very curious birds, and are at once known by the singular pointed plume which crowns the top of the head, and the long beard-like appendage to the chin. They all live at a very considerable elevation, inhabiting localities of such extreme inclemency that few persons would think of looking for a Humming-bird in such frozen regions. There are several species of Helmet-crest, and their habits are well described

2.   R

by Mr. Linden, the discoverer of LINDEN'S HELMET-CREST, in a letter written to Mr. Gould, and published in his monograph of the Humming-birds.

"I met with this species for the first time in August, 1842, while ascending the Sierra Nevada de Merida, the crests of which are the most elevated of the eastern part of the Cordilleras of Columbia. It inhabits the regions immediately beneath the line of perpetual congelation, at an elevation of from 12,000 to 13,000 feet above the level of the sea. Messrs. Funck and Schlim found it equally abundant in the Paramos, near the Sierra Nevada, at the comparatively low elevation of 9,000 feet. It appears to be confined to the regions between the eighth and ninth degrees of north latitude.

It occasionally feeds upon the thinly-scattered shrubs of this icy region, such as the hypericum, myrtus, daphne, arborescent espeletias, and towards the lower limit on bejarias, but most frequently upon the projecting ledges of rocks near to the snow. Its flight is swift, but very short; when it leaves the spot upon which it has been perched, it launches itself obliquely downwards, uttering at the same time a plaintive whistling sound, which is also occasionally uttered while perched; as well as I can recollect. I have never heard it produce the humming sound made by several other members of the same group, nor does it partake of their joyous spirit or perpetual activity. Neither myself nor Messrs. Funck and Schlim were able to discover its nest, although we all made a most diligent search.

Its food appears principally to consist of minute insects, all the specimens we procured having their stomachs filled with small flies."

ANGELA STAR-THROAT.—*Heliomaster A'ngelæ.*

The head and neck of the adult male are black, a line of white running along the centre. The long plumes of the throat are white. Round the neck and the back of the head runs a broad white band. The upper surface of the body and the two central tail-feathers are bronze-green, and the other feathers are a warm reddish bronze, having the basal half of their shafts white. The under surface is a dim brownish bronze. The length of the male bird is about five and a quarter inches. The female is coppery brown upon the head and upper surface of the body, and there is no helmet-like plume on the head nor beard-like tuft on the chin. The throat is coppery brown covered with white mottlings, and the flanks are coppery brown washed with green. The length of the female is about one inch less than that of her mate.

Another species, the WARRIOR of dealers, and the GUERIN'S HELMET-CREST of naturalists (*Oxypógon Guerenii*), is an inhabitant of the higher parts of the Columbian Andes, where it is tolerably common. It is easily to be distinguished from the preceding species by a bright green line which passes down the centre of the beard, and of which only a very faint indication is perceptible in the Black Warrior. There is also much more white upon the tail.

SICKLE-BILL HUMMING-BIRD.—*Eutóxeres A'quíla.*

THERE are several species of the STAR-THROATED HUMMING-BIRDS, all of which are known by the bright metallic gleam of the feathers on the throat.

The ANGELA STAR-THROAT inhabits Buenos Ayres and many parts of Brazil. It seems not to be a very common bird, or at all events it is rarely found in collections. It is chiefly remarkable for the singular shape of its bill, which is evidently formed for the purpose of enabling the creature to penetrate to the bottom of the curiously-shaped blossoms on which it finds its sustenance. It feeds, apparently, upon the long-blossomed flowers of the llianas, which are very plentiful in the regions inhabited by the Star-throat, and whose cups are always filled with minute insects. The generic name, Heliomaster, is very appropriate, signifying Sun-star.

In the male, the crown of the head is metallic green, "shot" with ultramarine blue and gold, and the upper surface of the body is golden green, with more gold upon the lower part of the back. The wings are purple-brown, and the tail purple-black with dark green gloss. Behind each eye there is a white spot, and a grey streak is drawn through the cheeks. The centre of the throat is a brilliant crimson, shining effulgently as if made of living fire, and edged with long feathers of a deep blue. The under surface is dark green, changing to rich blue in the centre, and on each side of the flanks there is a tuft of white feathers. The under tail-coverts are green, fringed with white.

The female is gold bronze on the upper part of the body, and the crown of her head is greyish. There is no crimson or blue on the throat; it is simply grey, covered with pale brown spots.

THE very remarkable bird whose portrait is seen in the accompanying illustration affords another example of the wonderful adaptation of means to ends which is often found among these birds. In the Sword-bill Humming-bird, sketched on page 232, the beak is enormously lengthened, in order to enable it to feed on the long bell-like flowers wherein it finds its sustenance, and a similar modification of structure may be seen in the Star-throats. In the SICKLE-BILLS, however, which feed on the short curved flowers of those regions, the bill is also short and very sharply curved, in order to suit the peculiar shape of the flowers. This

R 2

SICKLE-BILL is a very rare bird, and is found sparingly in Bogota and Veragua. The plumage is not very brilliant in its hues, but the various tints with which it is coloured are pleasing in their arrangement, and give to the bird a very pretty aspect.

The crown of its head and the little crest are blackish brown, and each feather has one small spot of buff on its tip. The upper parts of the body are of a dark shining green, with a slight buffy wash, and on the tips of several of the secondaries there is a little white spot. The two central feathers of the tail are a dark glossy green with small white tips, and the others are of the same hue in their outer webs, greenish brown on the inner, and largely tipped with white. The under surface is brownish black, diversified with some dark buff streaks upon the throat and breast, and with white streaks upon the abdomen and flanks; the under tail-coverts are brown fringed with buff. The total length of the bird is about four and a half inches.

MARS SUN-ANGEL.—*Heliángelus Mavors.*

Another species belonging to the same genus, CONDAMINE'S SICKLE-BILL (*Eutoxeres Condamini*), is remarkable for its propensity to inhabit high ground. It is a very rare bird, and whenever it is discovered, it is seen feeding among the orchidaceous plants, at an elevation of ten thousand feet above the level of the sea.

THE little group of Humming-birds called the Sun-angels are all remarkable for the exceeding lustre of the feathers which decorate their throats, and the general beauty of their plumage. In nearly every species there is a white or buff crescentic mark immediately below the gorget, and they are all inhabitants of the Andes. Concerning the MARS SUN-ANGEL and its habits Mr. Gould speaks in the following words :—

" Of all the species of the Andean Humming-birds belonging to the genus Heliangelos, I regard this as the most beautiful and interesting; it has all the charms of novelty to recommend it, and it stands alone, too, among its congeners, no other member of the genus similarly coloured having been discovered up to the present time. The throat vies with the radiant topaz, while the band on the forehead rivals in brilliancy the frontlet of every other species. . . . The country in which this rare bird flies is the elevated region of Northern Columbia, particularly the flat Paramos of Portachuela and Zambador, where Messrs. Funck and Schlim found it, at an elevation of from seven thousand to nine thousand feet; they also met with it in the Paramos of Los Conejos at a similar elevation. In those districts there doubtless exist other fine species at present unknown to us, for we can scarcely imagine that these travellers procured examples of all the species of the genera which dwell therein, and which we may reasonably expect to be as rich in the feathered tribes as it is in another department of Nature's wonderful works, Botany."

In the male bird there is a narrow mark upon the forehead of a deep fiery red, and the crown of the head and the upper surface of the body are bronze-green. Behind each eye there is a very small white spot, and a jetty black cross-streak is drawn from the angle of the mouth towards the neck. The throat is decorated with a gorget of deep fiery red, below which is a crescent-shaped band of whitish buff, and the abdomen is deeper buff, changing to green upon the flanks. The two central feathers of the tail are bronze-green, and the remainder bronze-brown. The female has no red upon the throat or forehead. There are several species of Sun-angel, among which may be mentioned CLARISSA'S SUN-ANGEL (*Heliangelus Clarissæ*), a bird which is remarkable for the deep ruby-crimson with which its gorget is dyed. Thousands of these birds are killed annually by means of the deadly blowpipe, and their skins forwarded to Europe, where they are largely employed for various decorative purposes, such as being mounted in ornamental cases of stuffed birds for drawing-rooms, feather fans and fire-screens, or for head-dresses of more than ordinary brilliancy. Two thousand of these birds have been sold at Paris at a single time, merely for the manufacture of head-dresses.

SNOW-CAP HUMMING-BIRD.—*Microchæra albocoronáta.*   SPANGLED COQUETTE.—*Lophornis Reginæ.*

THE two little birds which are represented in the accompanying illustration are remarkable for the manner in which their heads are decorated. One of them is seen to be a dark little creature, with the exception of a snowy white crown to its head, and a bold streak of white upon its tail. This is the SNOW-CAP HUMMING-BIRD, one of the most curious and the most rare of all the Trochilidæ. Its habits and the localities in which it lives are well described in the words of its discoverer, as quoted by Gould:—

"It was in the autumn of 1852, while stationed in the district of Belen, Veraquas, New Granada, that I obtained several specimens of this diminutive variety of the Humming-bird family.

The first one I saw was perched on a twig, pluming its feathers. I was doubtful for a few moments whether so small an object could be a bird, but on close examination I convinced myself of the fact and secured it. Another I encountered while bathing, and for a time I watched its movements before shooting it. The little creature would poise itself about three feet or so above the surface of the water, and then as quick as thought dart downwards, so as to dip its miniature head in the placid pool; then up again to its original position, quite as quickly as it had descended. These movements of darting up and down it would repeat in rapid succession, which produced not a moderate disturbance of the surface of the water for such a diminutive creature. After a considerable number of dippings, it alighted on a twig near at hand, and commenced pluming its feathers."

The colours of this little bird are so dark, that it appears to be uniformly brown, until it is examined more closely, when it is seen to be of a coppery hue, on which a purple reflection is visible in extreme lights, the copper hue taking a warmer tint towards the tail. The crown of the head is dazzlingly white, and the tips of all the tail-feathers, and the bases of all except the two central, are also white.

ON the same drawing may be seen another remarkable little bird, possessed of a most beautiful and graceful crest. This is the SPANGLED COQUETTE, an excellent example of the very remarkable genus to which it belongs. All the Coquettes possess a well-defined crest upon the head, and a series of projecting feathers from the neck, some being especially notable for the one ornament, and others for the other.

TUFTED COQUETTE.—*Lophornis ornátus.*

The Spangled Coquette is a native of several parts of Columbia, and was first brought to England in 1847. The singular crest is capable of being raised or depressed at the will of the bird, and produces a great effect in changing the whole expression of the creature. When raised to its fullest extent it spreads itself like the tail of the peacock, and much resembles the crest of the king tody, a bird which will be described in a future page. When depressed, it lies flat upon the bird, and is so large that it projects on either side, barely allowing the little black eyes to gleam from under its shade.

The crown of the head and the crest are light ruddy chestnut, each feather having a ball-like spot of dark bronze-green at the tip. The throat and face are shining metallic green, below which is a small tuft of pointed white feathers that have a very curious effect as they protrude from beneath the gorget. The upper parts are bronze-green as far as the lower part of the back, where a band crosses from side to side, and the rest of the plumage is dark ruddy chestnut as far as the tail. The tail is also chestnut-brown, with a slight wash of metallic green. The female has no crest nor green gorget.

THE TUFTED COQUETTE is one of the rare species of this beautiful genus.

It seems to be entirely a Continental bird, not being found in any of the West Indian Islands, and its principal residence seems to be in Northern Brazil and along the course of the Amazon as far as Peru. It may be readily known from the other species of Coquettes by the colours of its head, crest, and neck-plumes. The crest and top of the head are a rich ruddy chestnut, and the upper surface of the body is bronze-green, excepting the wings, which are purple-black, and a broad band of white which crosses the lower part of the back.   From the white band to the insertion of the tail is bright chestnut.   The tail is also chestnut, except the two central feathers, which are green at the latter half of their length.   The forehead and throat are emerald green, and the neck-plumes are snowy white tipped with resplendent metallic green.

The female has no crest nor neck-plumes, and the band of white across the back is very narrow.   The total length of the bird is about two inches and a half.

ANOTHER species of this remarkable genus is depicted upon the engraving on page 223, where it occupies the bottom left-hand corner.   This is GOULD'S COQUETTE (*Lophornis Gouldii*), a species which is remarkable for the beautiful pure white of the neck-tufts, and their green tips.   When the crest and tufts of this bird are depressed they lie closely upon the other feathers, the crest coming to a sharp point upon the back of the neck, and the neck-tufts also coming to a point upon the shoulders.   This species seems to be exclusively continental, and not to be found on any of the West Indian islands.   At present it is supposed to inhabit the country from the embouchure of the Amazon to its sources in Peru.   It is a very rare bird.

The crest of the male is rich chestnut-red, the upper surface is bronze-green, and a band of white crosses the lower part of the back.   The forehead and throat are emerald-green.   The female is comparatively a dull bird, having no crest nor neck-plumes.   The length of this species is about two and a half inches.

BUT the most singular of all the genus, if not the most unique and remarkable of all the Humming-birds, is the PRINCESS HELENA'S COQUETTE (*Lophornis Hélenæ*), which is figured on the top right-hand corner of the engraving on page 223.   This wonderful bird is a native of Vera Paz in Guatemala.

The curious forked crest and face are green, and the throat is of a metallic effulgent emerald in the centre, and surrounded with a series of long narrow white feathers, those which start from the neck being longest and generally edged with blue-black, while the others are much shorter and of a jetty black.   From the back of the head start six long hair-like feathers, three on each side.   The upper surface of the body is coppery-bronze, and a buff band crosses the lower end of the back.   The female is quite an ordinary little bird, without crest, neck-plumes, or long hair-feathers, and is generally of a dull bronze-green colour, and greyish-white below, sprinkled with green.

THERE are many species of VIOLET-EARED Humming-birds, all of which are easily recognised by means of the patch of violet feathers which is placed on each side of the face.

The BOLIVIA VIOLET-EAR inhabits the country from which it derives its popular title, and is one of the migratory birds, passing over a considerable tract of territory in the course of its travels.   The localities which it most prefers are the valleys and low grounds where maize is cultivated, and in such situations it is very plentiful.   The character of the species is eminently pugnacious, and it will not permit any other bird to approach its dominions.   It is a very pretty bird : the general colour of the upper parts of the body is golden green, a tint which extends to the two central feathers of the tail ; the remainder of the tail is deep blue-green.   The throat and breast are shining green, and the chin, abdomen, and a patch above the eye, are deep blue.   The length of the bird is rather more than three inches.

THE lovely little SPARKLING-TAIL is an inhabitant of Mexico, and is found very plentifully in Guatemala, where it is remarkably familiar and visits the habitations of mankind without any reluctance, haunting every garden wherein are blooming flowers, and altogether displaying a wonderful amount of confidence.   The nest of this species is very tiny, rounded and beautifully woven from various delicate fibres, cottony down, and spiders' webs, and is covered externally with lichens applied in a very artistic manner.   In this nest are laid two eggs, hardly bigger than peas, of a delicate semi-transparent pearly white, and reminding the observer of the eggs of the common snail.   The nest is always stuck upon

BOLIVIA VIOLET-EAR.—*Petasóphora iolata.*　　　SPARKLING-TAIL HUMMING-BIRD.—*Tryphœna Dupontii.*

a leaf or some slight twig by means of spiders' webs, so that instead of the great spider catching and eating the Humming-bird, as Madame Merian supposed, the Humming-bird is the real depredator, and robs the spider.

In colouring and form the two sexes are quite dissimilar.

The male is bronze-green above, with the exception of the bold crescent-shaped white feathers on the lower part of the back. The throat is rich metallic blue, becoming velvety black in certain lights, because each feather is black at the base and blue at the tip. The wings are of a rich dark purple-brown. Round the neck runs a broad snowy-white crescentic band, and the whole under surface is bronze-green, except the under tail-coverts, across which runs a band of white. The tail is very curious, exhibiting very many tints, and not very easy to describe. The two central feathers are rich shining green; the next green marked with bronze; the next dark brown, with two triangular white spots on the inner web, one near the middle and the other at the tip; the two central feathers are dark brown for the first half of their length, then comes a broad band of deep rusty red, then a broad white band, then a brown band, and the tip is white. The whole length is about four inches.

The female is of a rich bronze-green on the upper surface of the body, and the two crescentic marks on the lower part of the back are buff instead of white, as in her mate. Her tail is short, and of a purple-black bronzed at the base; all the feathers except the two central ones are tipped with white and ringed with buff. The under surface is rusty red, becoming darker on the under tail-coverts. The length is not quite three inches.

WE now come to one of the most imposing of all the Humming-birds, namely the SAPPHO COMET, or the BAR-TAILED HUMMING-BIRD as it is often called. It is a native of Bolivia, and is a migratory species, generally going to Eastern Peru in the winter. It is a remarkably familiar bird, haunting the gardens and orchards while the trees are blossoming, especially while the apple-trees are in flower. The males are extremely fierce and pugnacious, chasing each other through the air with surprising perseverance and acrimony. Of these birds Mr. Bonelli gives a very spirited description :—

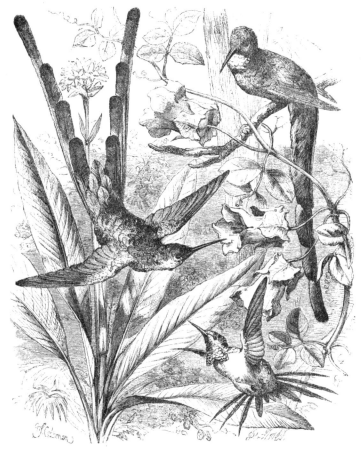

SAPPHO COMET.—*Cométes sparganúrus.*   YARRELL'S WOODSTAR.—*Calothórax Yarréllii.*

"It arrives in the environs of Chuquesaqua in the months of September and October, and takes up its residence in the shrubberies of the city and the gardens of the Indian cottages; the hill-side of the neighbouring country, clothed with indigenous trees and shrubs, also affords it a fit place of abode, whence it descends several times a day to the cultivated plains below, particularly to the fields of maize, pulse, and other leguminous plants; the rich flowers of the large cacti are also frequently visited, as they afford it a constant and abundant supply of insect food.

Soon after their arrival the task of incubation is commenced; and when the summer is over, both the old and the young, actuated as it were by the same impulse, wend their way southward, to return again when the spring has once more gladdened the earth.

The nest is a somewhat loose structure, outwardly composed of interlaced vegetable fibres, slight twigs, moss, &c., and frequently lined with soft hairs like those of the viscacha, with the lower portion prolonged considerably below the bottom of the cup-shaped interior, which is about an inch and a half in diameter and an inch in depth; the total length of the nest averaging from two and a half to three inches. The nest is placed in situations similar to those selected for the like purpose by the spotted flycatcher, namely, against the sides of the walls, supported or entirely sustained by any hanging root or twig that may be best adapted to afford it security; the part of the nest next the wall is much thicker, but of a looser texture, than the similar portion of the true structure. The eggs are two in number, oblong in form, of a pure white, and about half an inch in length by about five-sixteenths of an inch in breadth.

The difficulty of shooting these birds is inconceivably great, from the extraordinary turns and evolutions they make when on the wing; at one instant darting headlong into a flower, at the next describing a circle in the air with such rapidity, that the eye, unable to follow the movement, loses sight of it until it again returns to the flower which at first attracted its attention."

Magnificent as are these gorgeous birds when mounted as specimens, they lose much of their beauty in the needful handling, and give but a faint image of their real effulgent beauty. Many specimens are injured by being left too long before the skins are removed from the bodies, and in the lands where the Sappho lives, putrefaction takes place with such rapidity, that a delay of a few hours is fatal to the perfection of the skin. It has been found that these delicate creatures were much injured by the shot or other missiles employed in their capture, so a very ingenious trap was made for their especial benefit. It was noticed that the birds were accustomed to dash into the long pendent blossoms of certain flowers, so the ingenious collector put a little birdlime into the bottom of every blossom, and thus secured many an unwary Humming-bird as it came to feed.

In the male bird the head, neck, the upper part of the back, the face, the sides of the neck, and great part of the under surface, are light green, bronzed on the sides of the neck and face. The lower part of the back is a deep crimson red. The wings are purple-brown, and the throat metallic green. The tail is ruddy brown at its base, and the remainder of its length is a fiery red, tipped with a velvety black band. The female is smaller than the male, but is not possessed of his length of tail nor brilliancy of hue, her throat being white spotted with green, and the only piece of crimson being a patch on her back.

THERE is another species belonging to the same genus, which has been called the PHAON COMET (*Cométes Phaon*) in allusion to the classical name of the preceding species. This is equally magnificent with the former, and is altogether a larger bird. It may be known from the Sappho by the colour of the tail, which is wholly of a crimson red. It inhabits Peru and Bolivia.

ON the same engraving will be seen a much smaller bird, the YARRELL'S WOODSTAR.

This pretty creature is very rare, and inhabits the eastern parts of Peru and Bolivia. Mr. Gould thinks it is limited to the localities between the mountainous ranges and the sea. In the male, the crown of the head, the whole upper surface of the body, and the four central tail-feathers, are of a light yellow-green, and the chest, the middle of the abdomen, and the under tail-coverts are greyish white. Upon the throat there is a gorget of purple-blue, "shot" with lilac in some lights, and of a brighter blue in the centre. The wings are greyish-brown, and the lateral feathers of the tail are dark brown. The total length of the bird is about two and a half inches.

Among other species belonging to the same genus may be mentioned the SHORT-TAILED WOODSTAR (*Calothórax micrúrus*). This odd little bird is remarkable for the extreme shortness of its tail-feathers, which when closed are, with the exception of the two central feathers, hidden under the upper tail-coverts. It is generally seen in the mimosas, and hums very loudly when flying.

THE RUBY AND TOPAZ HUMMING-BIRD derives its name from the colouring of its head and throat, the former being of a deep ruby tint, and the latter of a resplendent topaz. Sometimes it is called the Ruby-headed Humming-bird, and it is also known under the name of the Aurora. It is very common in Bahia, the Guianas, Trinidad and the Caraccas, and as it is in great request for the dealers, is killed by thousands annually. There is no species so common in ornamental cases of Humming-birds as the Ruby and Topaz. It makes a very beautiful nest, round, cup-like, and delicately woven of cotton and various fibres, and covered externally with little leaves and bits of lichen.

The plumage of this species is extremely variable, but may be described briefly as follows. The forehead, the crown, and the nape of the neck are metallic ruby red, and the chin, throat, and chest are effulgent topaz. The upper parts of the body are velvety bronze-brown, and the wings are purple-brown. The tail is rich chestnut red, tipped with black, and the abdomen is a dark olive-brown. The female has none of the ruby patches on the head, but retains a little of the topaz on the throat.

OCCUPYING the centre of the illustration on page 223, may be seen a very striking and remarkable bird, whose long forked tail extends completely across the engraving. This is the BLUE-TAILED SYLPH (*Cynanthus cyanurus*), one of the most beautiful of the birds which are called by the name of Sylphs, in allusion to their beautiful form and graceful movements.

RUBY AND TOPAZ HUMMING-BIRD.
Chrysolampis moschitus.

This beautiful bird is found in the temperate regions of the Andes, its range extending as far as Panama. It also inhabits the sides of the Cordilleras, at an elevation of five or ten thousand feet above the level of the sea, as the vegetation of those regions is most luxuriant in spite of the coldness of the climate, and affords plentiful nourishment for the Humming-birds. The flower on which it usually feeds is the *Sedum Quitense*, and its flight is extremely rapid as it darts from one flower to another.

In the male bird the crown of the head is of a metallic golden green, and the general colour of the body is a bronze-green, becoming browner on the under parts. Upon the throat is a gorget of the most intense purple-blue, and the wings are purple-brown. The two central feathers of the magnificent tail are shining metallic green, the two next are black at their base and rich blue towards their extremities, and tipped and edged with bright metallic green, "shot" with blue. The outer feathers are black for the first half of their length, and the remaining portion is metallic steely blue. Some white feathers are scattered across the end of the abdomen, the under tail-coverts are green, and there is a little white dab above the eye and another behind it.

The female is something like the male, but not so bright in colouring. The throat is greyish-white covered with green spots, and the abdomen is rusty red, changing into bronze-green upon the flanks. The young male is duller in colour than the adult, the tail is shorter, and there is no blue part upon the throat. Like many other of the Humming-birds, this species is extremely variable in its colouring, especially among the young males. The length of this species is about seven inches. It is not, however, so remarkable for the length of its tail as the TRAINBEARER (*Lesbia Amaryllis*), a native of Quito. This bird, although a small creature, possesses a long and very straight tail, something like that of the Polytmus or the Sappho, but much larger in proportion, the length of the elongated feathers being nearly six inches.

ONE of the most striking forms among the Humming-birds is that which is exhibited by DE LALANDE'S PLOVER-CREST.

This singular bird is remarkable from the fact that the elevated plume which is placed upon the head is terminated by a single feather, instead of being double, as is usually the case with crested birds. This species inhabits the southern parts of Brazil. The nest which it builds is very pretty, and is ingeniously woven into a tuft of leaves or twigs at the extremity of some very slender branch, so that the whole structure droops downwards, and may be reckoned among the pensile nests. Its form is much elongated, and the materials of which it is composed are delicate pieces of roots, mosses and lichens, and spiders' webs.

The male bird is the sole possessor of the beautiful crest, the female being without that decoration. The crest and the top of the head are bright green, with the exception of the long single feather, which is jetty black. The upper surface is green washed with bronze, and the breast and abdomen are of an intense shining violet. Behind the eye there is a small white streak.

DE LALANDE'S PLOVER-CREST.
Cephalepis Delalandii.

IF the reader will again refer to the engraving on page 223, he will see that immediately under the crossed tail-feathers of the Polytmus and the Cynanthus, a Humming-bird is represented with outspread wings and a rather curiously formed tail. This is the CAYENNE FAIRY (*Heliothrix auritus*), an inhabitant of Guiana, Cayenne, and the forests near the mouth of the Amazon.

In the male, the general hue of the upper surface is glossy golden green, very light on the forehead. The four middle feathers of the tail are blue-black, and the three exterior feathers are white. A jetty black line is drawn across each side of the face, including the eye in its breadth, and terminated with a small tuft of violet blue; below the black runs a luminous green line. The under parts of the body are snowy white.

THE very beautiful bird which has been appropriately named the SUN-GEM inhabits the mountain ranges of Brazil, especially those of Minas Geraes, where it seems, as Mr. Gould poetically remarks, to be a veritable gem sprung out of the mountain and suddenly gifted with life. It may generally be found in the open country in nursery grounds, where it finds abundance of the small flowers on which it feeds. During the rainy season it resorts to the forest, and finds its nourishment in the orchidaceous plants which flourish there in such rich profusion.

The male bird is remarkable for a very conspicuous crest, which starts from either side of the head, and has gained for the bird the title of Double-crested Humming-bird. The name "cornuta," or horned, refers to this peculiarity. The forehead and crown of the head are azure blue, and the neck fiery crimson at its base, changing to green near the centre, and taking a rich golden radiance at the extremity. The throat, the sides of the face, and the pendent tuft of feathers from the chin are velvety-black. The back is bronze-green, and the wings purple-brown. The two central feathers of the tail are olive-brown and the rest white, edged with the same tint. The breast, sides of the neck, and the upper part of the abdomen are greyish-white, and the remainder of the under surface is bronze-green. The female has nearly as long a tail as the male, but her throat is buffy white, and she has no crest on the head nor beard hanging from the chin.

A drawing of this species may be seen on page 233, where it occupies the space immediately above the Cora's Shear-tail.

On reference to the engraving on page 227, the reader will see a figure of the AVOCET HUMMING-BIRD, placed immediately below the drawing of the Ruby-throated Humming-bird.

This singular species is remarkable for the curious manner in which the bill is curved upwards at the extremity, after running nearly straight for the greater part of its length. As this formation of beak bears some resemblance to that which is found in the well-known Avocet, the present species has been named the Avocet Humming-bird. When the first specimen of this bird was brought to Europe, the peculiar shape of the beak was thought to be accidental, and owing to pressure against the side of the box in which the bird had been packed ; but it is now clear that the structure is intentional, and that, in all probability, it subserves some very important purpose. Some persons have suggested, with some show of reason, that the beak is recurved in order to enable the bird to feed upon the nectar and insects which reside in the deepest recesses of certain tubular flowers.

It is a pretty bird, but is not gifted with the gorgeous and dazzling hues which are so lavishly shed upon the plumage of many Humming-birds. The head and the whole of the upper parts of the body are shining golden-green, and the throat is bright emerald. The under parts are also gold-green, with the exception of a black streak that runs from the breast through the centre of the abdomen. The wings are purple-brown, and the lateral feathers of the tail are topaz.

Our ideas of Humming-birds are naturally associated with the tropical climate and burning sun of the regions which they inhabit, and few persons would think of looking for any species of Humming-bird in a locality where the temperature seldom rises above that of an ordinary English winter. Yet the CHIMBORAZIAN HILL-STAR is never found except upon the elevated portions of the lofty mountains from which it derives its name, and inhabits exclusively the very edges of the line of perpetual snow.

This bird is never seen on any spot that is less than twelve thousand feet above the level of the sea, and is most commonly seen at a much greater elevation, specimens having been obtained on spots that are at least sixteen thousand feet above the ocean. Beyond this height the creature cannot live, as the line of perpetual snow commences at that point, and places an effectual barrier against the growth of the plants on which the bird finds its subsistence. The two sexes are seldom seen near each other, the males preferring to haunt the extremities of the loftiest branches, while the females hover near the ground. Partly owing to this peculiarity, and partly on account of her sober tinting, the female generally escapes observation. The plant on which the Chimborazian Hill-star is usually found is the *Chuquiraqua insignis*, a flowering alpine shrub, with large pale yellow blossoms, and the bird is so closely attached to this shrub, that it is never found at any great distance from its golden flowers.

The nest of this species is made of lichens, and is fastened to the side of a rock in some situation where it is protected by an overhanging ledge of rock.

Except upon the head and throat, the Chimborazian Hill-star is not so brilliantly clothed as many of its compeers, but upon those parts the creature shines with rainbow lightness. The general colour of the upper parts of the body is pale dusky olive-green, with the exception of the wings, which have the purple-brown tint usual among Humming-birds. The under parts are white, deepening into dusky black upon the under tail-coverts, and there is a line of black down the centre of the abdomen. The head and throat are of the brightest and most resplendent blue, with the exception of an emerald-green patch in the centre of the throat. This patch is triangular in shape, and has one of the angles pointing upwards. Round the neck runs a broad collar of deep velvety black, abruptly dividing the brilliant hue of the head and throat from the plain black and white of the chest and abdomen, and giving the bird an appearance as if the head and throat of some brightly coloured bird had been joined to the neck and body of a plainly clad individual of another species. The two central feathers of the tail are nearly of the same hue as that of the back, the two exterior feathers are white for the first third of their length, and greenish-black for the remaining two-thirds, while the other feathers are white, edged with greenish-black.

The female is a very soberly clad bird, being olive-green upon the head, white spotted with green upon the throat, and the remainder of the body olive-green, white, and brownish black.

THERE are several species of Hill-stars, among which the PI-CHINCHIAN HILL-STAR is the most remarkable.

This bird is very local, inhabiting the volcanic mountain of Pichinca, in the republic of Ecuador, and being only found in a zone of five or six hundred feet in width, at an elevation of about eleven hundred feet above the level of the sea. It is a very remarkable fact, that although both these species inhabit volcanic mountains within thirty miles of each other, and are found at nearly the same elevation, the Pichinchian Hill-star is never seen upon Chimborazo, nor the Chimborazian Hill-star upon Pichinca. This species is very like the preceding, but may be easily distinguished by the absence of the triangular green spot upon the throat.

CHIMBORAZIA HILL-STAR.—*Oreotróchilus Chimbo ázo.*

UPON the engraving on page 223, and in the upper left-hand corner, the reader will observe a very elegantly shaped little bird, sitting close to its nest. This is the GILT-CRESTED HUMMING-BIRD (*Orthorhynchus exilis*), an inhabitant of Martinique and other West Indian islands, where it is always found upon the low-lying grounds.

In colour it is very pretty, the general hue of its body being green with bronze reflections, and its crest glowing with golden-green and emerald; the emerald hue being most

VERVAIN HUMMING-BIRD.—*Mellisuga minima*

conspicuous towards its point, and the gold towards the base. The nest is a very pretty compact little structure, beautifully rounded, and composed chiefly of cotton fibres, inter-mixed with the dried involucres of certain composite plants, and bound together with spiders' web.

THE beautiful little VERVAIN HUMMING-BIRD is one of the minutest examples of feathered life that are at present known to zoologists. In total length this bird does not measure three inches ; while, as the tail occupies nearly an inch and the head half an inch, the actual length of the body will be seen to be not quite an inch and a half. It is a native of Jamaica, and has been admirably described by Mr. Gosse, while treating of the birds which inhabit that island.

The name of Vervain Humming-bird has been given to this tiny creature, because it is in the habit of feeding on the blossoms of the West Indian Vervain, but it is also known under a variety of other titles, and has been described by many scientific writers under different names. Speaking of this bird, Mr. Gosse says :—

"The West Indian Vervain (*Stachytarpheta*) is one of the most common weeds in neglected pastures, shooting up everywhere its slender columns set around with blue flowers to the height of a foot. About these our little Humming-bird is abundant during the summer months, pecking the azure blossoms a few inches from the ground. It visits the spikes in succession, flitting from one to another exactly in the manner of the honey-bee, and with the same business-like application and industry. In the winter, the abun-dance of other flowers, and the paucity of vervain blossoms, induce its attentions to the hedge-rows and woods.

I have sometimes watched with much delight the evolutions of this little species at the moringa tree. When only one is present, he pursues the round of the blossoms soberly enough. But if two are at the tree, one will fly off and suspend himself in the air a few yards distant, the other presently starts off to him, and then, without touching each other, they mount upward with strong rushing wings, perhaps for five hundred feet. They then separate, and each starts diagonally towards the ground like a ball from a rifle, and wheeling round comes up to the blossoms again, and sucks as if it had not moved away at all.

Frequently one alone will mount in this manner, or dart on invisible wing diagonally upwards, looking exactly like a humble-bee. Indeed, the figure of the smaller Humming-birds on the wing, their rapidity, their wavering course, and their whole manner of flight, are entirely those of an insect, and any one who has watched the flight of a large beetle or bee will have a very good idea of these tropic gems painted against the sky."

The Vervain Humming-bird appears to be less susceptible of human influence than the Ruby-throated and the Long-tailed Humming-birds, for although Mr. Gosse succeeded in capturing several specimens of these beautiful little creatures, and confined them in a room, they were so hopelessly timid that nothing could be done with them. More than once he secured the female as she was sitting upon her eggs, and removed her, together with her nest, into a gauze-covered cage, hoping that she would continue her labours while in captivity, and produce a brood of young that would be familiar with mankind from their earliest birth. Maternal love, however, was not proof against the power of fear, and in every case the poor little bird forsook her eggs, fluttered about the cage aimlessly, and died within twenty-four hours.

The general colour of this beautiful little bird is a brilliant metallic green, the wings being, as usual, purple-brown, and the tail deep black. The throat and chin are white, sprinkled profusely with little black spots, and the breast is pure white. The abdomen is also white, but diversified with a slight green tip to each feather, and the flanks are bright metallic green nearly as resplendent as upon the back. The under tail-coverts are white, with a few very pale green spots. The colours of the female are rather more dull than those of her mate, the green being tinged with yellow, and the under parts without the green spots. The first half of the tail is yellowish-green, and all the feathers of the tail, with the exception of the two central feathers, are furnished with white tips.

The nest of the Vervain Humming-bird is very small, in accordance with the dimensions of the architect, is round and cup-like in shape, and beautifully constructed of cotton fibres and other soft and warm substances. As is the case with the nests of almost all the species of Trochilidæ, the rim is so made as to curve slightly inwards, and is, in all probability, constructed for the purpose of preventing the eggs from rolling out of the nest when the "procreant cradle" is rocked by the tempestuous winds of the tropics. A somewhat similar structure may be seen in the nests of many of our British birds, and I well remember seeing the nest of a goldfinch which had been built at the extremity of a long and slender horizontal spray of oak, and which was completely turned inwards at the rim. No ordinary wind could have shaken the eggs out of the nest, as even when the branch was seized and drawn towards the ground, the eggs still remained in their places.

In one species, which was watched by Captain Lyon, R.N., the nest was not completed until the young birds were nearly fledged.

The mother bird built a small and rather shallow nest, in which she laid two eggs, and began to sit as usual. As soon, however, as the young were hatched, she set to work again at her nest, and added fresh material round the edges, so as to raise the sides and prevent her offspring from tumbling out of their house upon the ground. In proportion to the growth of the young, the mother increased the height of her walls, so that by the time the young were ready to fly, the nest had been transformed from a shallow, saucer-like structure, into a round and deeply-hollowed cup. It has been suggested in explanation of this curious behaviour on the part of the mother bird, that her completed nest had been destroyed, and that she was forced to run up a hasty kind of hut for the reception of her young, and could only finish it when she was relieved from the constant duty of sitting on her eggs.

In the nesting of the Humming-birds, there is one peculiarity that is worthy of a passing notice. In almost every case where a nest has hitherto been discovered, the materials of which it is composed are thick, soft and woollen, and in all instances are arranged in such a manner as to shield the eggs even from the effects of rain or atmospheric influences, as long as the mother bird is seated upon them. Mr. R. Hill, who has paid close attention to the nests of the Humming-birds, has ingeniously hit upon a connexion between their structure and the electrical conditions of the atmosphere.

The injurious effect of a sudden increase of electricity is very strongly marked upon the young of all animals, the hurtful influence being in proportion to the growth of the victim. Eggs are peculiarly susceptible to the influence of electricity, and even when the chick is partially matured, are often killed by a passing thunderstorm. In climates where thunderstorms are so frequent and violent as in the lands which are inhabited by these birds, it is needful that the eggs should be protected from the deadly influence, and we accordingly find that the nests are oval or rounded in shape, and are made of substances which are bad conductors of electricity. In accordance with this principle, Mr. Hill proceeds to remark that "in tropical climates, there are a greater number of birds that build close nests than in the temperate climates of Europe. In the West Indian Islands, with the exception of the pigeon tribes and the Humming-birds (which latter build deep, thick, cottony nests), the nests are almost uniformly circular coverings of dried grass, varied by intermingled cotton, moss and feathers, with an opening from below, or an entrance at the side.

The Banana bird weaves a hammock of fibres, sometimes of horse-hair, deep and purse-like, and loosely netted ; the *Muscicapa olivacea* (a fly-catcher), a hanging cot of withered leaves, straw, moss, fibrous thread, and spiders' webs fitted together ; and the mocking-bird builds in the midst of a mass of wicker-work a neat nest of straw lined with hair. The woodpecker and the parrots take to hollow trees, but I hardly know any arboreal beside which constructs any nest that is not wholly covered or domed over.

Very many insects that are exposed to the air during their metamorphoses weave coverings of silk and cotton, in which they lie shrouded, at once impenetrable to moisture, and uninfluenced by the disturbances of the atmosphere. It would seem that the object, whatever it be, is the same in both. It is not for warmth that the insects spin these webs, for they form their coverings of silk and cotton in the hottest period of the year ; and I find that whilst all our birds that build open nests (the Humming-Birds build in May, June, and later) breed early, those that construct the domed and spherical ones nestle in the season between the spring and autumnal rains, when the air is saturated with electricity, and is in a state of constant change."

THE semi-musical voice of the Vervain Humming-bird has already been mentioned. With this exception, the Humming-birds seem to be without any melodious song, and even in this case the song appears to be little more than a pleasing twitter, without much melody, combination of notes, or musical force. Indeed, as a general rule, it is found that the most brilliant songsters among the birds are attired in the plainest garb ; and it may safely be predicted of any peculiarly gorgeous bird, that power, quality, and sweetness of voice, are in inverse ratio to its beauty of plumage.

To this rule there are some exceptions, but these are more in appearance than in reality. For example, the well-known blue-bird of America is very beautiful in plumage, and yet possesses a song whose "low æolian twitterings greet the morn." The canary-bird again possesses, together with its brilliant voice, some degree of beauty in the colouring of its plumage. But it must be remembered that the song of the Blue-bird is, although sweet and pleasing, by no means remarkable for brilliancy of power, and that the canary has little right to be ranked among the bright-plumaged birds, as a very great proportion of the best songsters are of a dull olive green, hardly more showy than the common linnet, and less so than the goldfinch or yellow-hammer. The goldfinch again possesses a low sweet song, but has very little vocal force, requiring to be crossed with the canary before it attains any strength of voice.

Reference is, however, especially made to those birds whose plumage is of a peculiarly resplendent character, absorbing, transmuting, and reproducing all the colours of which we are cognizant, and almost dazzling the eyes of the spectator with their exceeding splendour. Such plumage is to be found among the Humming-birds more than among any other feathered creatures ; although there are examples of redundant brilliancy of attire in many other birds, such as the poe bird, the promerops, some of the sun-birds, and our own common kingfisher. In all these creatures, the male possesses no real song, the glorious beauty of the feathers compensating him and his mate for the absence of poetic utterance. Why this should be the case is a problem which has long attracted the attention of observant men, and it seems to me that a key may be found to its solution

2. S

in the now acknowledged fact that sound and colour run in parallel lines through creation, and closely correspond with each other in their several relations. Some of my readers may be aware that a chromatic piano has been made by some ingenious inventor, although at present the popular mind has not been sufficiently interested in this abstruse subject, to have bestowed upon so admirable an invention the notice which it deserves.

It is also well known that certain sounds are suggestive of certain colours, and *vice versâ*, as has been forcibly exemplified in the well-known experiment upon a congenitally blind person, whose power of vision was developed by means of a surgical operation, and who said that the colour scarlet reminded him of a trumpet blast. It may be, therefore, that on the one side the bird which is possessed of a good voice and a plain dress, pours forth his love and manifests his sympathetic emotions in gushing strains, which are addressed to the ears of his mate ; again, the bright-plumaged bird utters his voiceless song by the vivid hues that flash from his glittering attire, the eye being the only medium through which his partner, whose ears are not attuned to melody, could realize the fulness of his emotional utterance. The one showers his musical notes like vocal rainbows, and the other scatters his scintillating coruscations of many-coloured light in fiercely flashing or softly blending tints of most living hues, and whether through sweet song or glittering vesture, the creature utters the love and sympathy of its nature.

At regularly recurrent intervals, an effluence of Divine Love is poured out upon all organic nature, and finds its level and outburst in graceful form, in sweet sound, in brilliant colouring or in odorous scent, and beyond and above all in these acts of maternal and self-sacrificing love which permeate the entire universe, and are more tinged with divinity than any of the preceding forms because more noiseless and less obtrusive.

Even in flowers the same principle is manifested, for they only put forward their fragrant perfume and gleaming corollas while their offspring is being developed, and as soon as the seed is perfected they lose both scent and petals until the next season of love.

The female needs no song or glittering plumage for the expression of her love, for she performs in loving acts the sympathies which her mate expresses in colour, form, or sound, and while imparting the divine element of love to her callow young, she utters and incarnates her song of praise through the coming generations. These elements of the being, which in the male bird are modified into glorious plumage or vocal melody, are from her transmitted to her descendants, and pass away from her individual self, to be universally and eternally reproduced and multiplied in the persons of the future offspring. He sings the song, but she performs it, and manifests her love in a melody more fruitful than that of her mate, because so many beings are evolved from it. She cherishes and nourishes, he protects and provides.

The mother freely gives her whole being for the welfare of her young, and finds the most exquisite delight in utter abnegation of self. She often robs herself of her own covering, tearing away her warmest feathers to make a soft nest for her young, taming the ever restless bird-nature to absolute quietness, and chaining herself for successive weeks to the one single spot on which all her hopes are concentrated. Her very life itself is valueless in her eyes as long as she holds in her charge the lives of her young ; and she who would but a few weeks before have fled in terror from the meanest foe, now with her feeble powers of offence boldly faces the hawk, the stoat, the snake, and even man himself, and her fear being conquered by her love, boldly attacks with dauntless heart the would-be destroyer of her domestic peace.

When by age or injury the reproductive element cannot be carried out in the usual manner, it manifests itself in outward form, and thus we often see the aged hen assume the warrior armature and plumage of Chanticleer himself. And when the season of love is over, and the young are launched upon the world, the female is released from her home cares, shakes off the chains which have trammelled her so long, and exultingly disports herself once more in the open fields and forests ; the male lays aside his splendid raiment and ceases his melodious song, needing no longer to cheer the spirit of his mate and nestlings by his exertions, and even the very flowers doff their raiment and quench their fragrance until a fresh infusion of the love element shall so intensify and vivify their being that they burst forth with a renewed and redundant beauty, and repeat the same great lesson to those who have eyes to see and ears to hear.

It is evident that song, form, colour and odour are conductors of some subtle and most powerful influence, which is inhaled into the inmost nature of the young and tender beings now being moulded into their matured development, and imbues and inter-penetrates their dual nature with the same element of love. For although there may be many modes of affecting our senses, through touch, or taste, or scent, or sight, or hearing, all these senses must be reduced to one principal faculty of perception, which in human beings is reached through various channels of communication with the outer world. It is not the eye that sees nor the ear that hears, for they exist as perfectly in the lifeless corpse as in the animated and active human form ; but the eye and the ear are only the window and the door, through which the inmost sensual faculty inherent in the soul is enabled to communicate with things that are of a grosser and less ethereal nature than itself.

At times we are all so exalted that we become dimly conscious that there are other ways of affecting our perceptive nature than through those means which we commonly call the five senses. Who has not conversed without a word spoken on either side ? Who, by the mystical workings of their inmost nature, has not perceived the presence of a loving and sympathising friend, or a deadly and vengeful foe, before sight or hearing has given any premonition of his approach ? Sometimes the senses perform each other's duty, and we are hardly conscious to which of them we owe the impression that is being made upon our mind. If for example we read a book, we hear the words as if uttered in sound ; while, if we listen to conversation, the articulate sounds arrange themselves in alphabetical form before our mental eye, ready to be placed on paper if needed.

WE now arrive at the large family of the CERTHIDÆ, or Creepers ; a family which includes many birds of very different forms, and which can only be known to belong to it by their anatomical structure. In fact, the Creepers may be considered as analogous among birds to the antelopes among mammalia, and be considered as a " refuge for the destitute," formed for the purpose of receiving all the slender-billed birds which cannot find accommodation in any other more definite family.

The Certhidæ are mostly small birds, but there is one notable exception in the person of the celebrated lyre-bird of Australia. Many of them are good songsters, and they all feed chiefly upon insects, which they pick out of the bark of trees or unearth from the soil. The beak is rather long and slender, except perhaps in the nuthatch, which, although comparatively long, is possessed of great strength ; and there is always a curve more or less marked. The beak is always sharp at its extremity, and the nostrils are placed in a little groove at the base of the bill, and defended by a membranous scale. The feet are, although slender in several species, possessed of remarkable strength, and furnished with sharp round claws, in order to enable the birds to cling to the tree-trunks in which they find their food.

The OVEN-BIRDS derive their name from the peculiar form of their nests.

The edifice, for it fully deserves that name, is of considerable dimensions when compared with the small size of its architect, and is built in the shape of a dome, the entrance being on one side, so as to present a decided resemblance to an ordinary oven. The walls of the nest are fully an inch in thickness, and the materials of which the structure is composed are clay, grass, and various kinds of vegetable substances, which are woven and plastered together in so workmanlike a manner, that the nest is quite hard and firm when the clay has been dried in the sun. The bird seems to be conscious of the security of its nest, for it takes no pains to conceal its habitation, but builds openly upon some exposed spot, such as the large leafless branch of a tree, the top of palings, or even the interior of houses or barns.

The Oven-bird is not content with barely building this curious domed structure, but adds to its security by separating it into two parts, by means of a partition reaching nearly to the roof, the eggs being placed in the inner chamber. The bed on which the eggs are placed consists mostly of feathers and soft grasses. The number of the eggs is generally about four.

s 2

The Oven-bird is a bold little creature, caring nothing, as has already been implied, for the close proximity of man, and attacking fiercely any other bird that might happen to approach too closely to its residence, screeching defiantly the while. It is a quick, active bird, tripping over the ground with great rapidity while searching after its prey, and is almost invariably found in company with its mate. The flight of the Oven-bird is not at all strong, and it seldom indulges in any aerial excursions beyond a short flight from one bush to another. Both sexes take part in the construction of the nest, each going alternately for supplies of clay, straw, and grass-stems, working them well together, and then flying for a fresh load as soon as its mate has arrived. It feeds principally on insects, having a special liking for those of the beetle kind.

OVEN-BIRD.—*Furnárius fuliginósus.*

There is an allied genus of Oven-birds, termed CINCLODES, the members of which are found upon the western coasts of South America, and generally frequent the sea-shore, where they feed upon the smaller crustaceans and molluscs. They are rather daring little birds, and will seek their prey at some distance from shore, perching upon the fronds of floating sea-weed, and pecking out the various marine creatures that are always to be found in such localities. Like the true Oven-birds, they are careless of the presence of man, and are so fearless that they can almost be taken by hand. Indeed, one voyager relates that he killed ten of these little birds with a stick without any difficulty, and hardly having to change his position.

ANOTHER small group of the Certhidæ is known to zoologists by the title of SYNAL-LAXINE BIRDS, and distinguished by the greater length of the outer toe, and its juncture to the middle toe nearly as far as the first joint. The hinder toe is long and rather powerful, and all the claws are sharply curved, pointed, and strong. The tail is rather long, and is almost always pointed, like that of the common creeper of England.

The Synallaxine birds are inhabitants of tropical America, and, like the oven-birds, are notable for the very curious nest which they construct. Although these birds are of small dimensions, they all build nests which might easily be attributed to the labours of some hawk or crow. The nest of one species is often from three to four feet in length, and is placed very openly in some low bush, where it escapes notice on account of its resemblance to a bunch of loose sticks thrown carelessly together by the wind. In its interior, however, the edifice is very carefully made, and, like the nest of the oven-birds, is divided into two recesses, the eggs being laid in the inner apartment, upon a bed of soft feathers.

The Synallaxine birds are generally found upon the trees, which they traverse with great rapidity in search of the various insects on which they feed, and may often be seen running about upon the ground, peering anxiously into every little hole and cranny, and dragging slugs, snails, worms, and beetles from the recesses in which they are accustomed to conceal themselves during the hours of daylight.

ANOTHER very small group of the Creepers is represented by the CURVED-BILLED CREEPER, a bird about the size of an English blackbird, which is found in the forests of Brazil.

It is chiefly remarkable from the curiously-formed bill, which is very long in proportion to the size of the bird, and is curved in a manner that can best be understood by reference to the engraving. The bill, although so much elongated, is possessed of considerable strength, and is evidently employed for the purpose of drawing the insects on which the creature feeds from the crevices of the bark in which they dwell. As is indicated by the stiff and sharply-pointed feathers of the tail, the Curved-billed Creeper is in the habit of traversing the trunks of trees, and is able to support itself in a perpendicular position by hooking its long curved claws into the inequalities of the bark, and resting the weight of its body upon the stiff tail-feathers. The general colour of this bird is brown, with a wash of cinnamon upon the greater part of the surface. The head and neck are of a greyer brown, and spotted with white.

WE now arrive at the true Creepers, of which birds our well-known ENGLISH CREEPER is an excellent example.

This little bird is one of the prettiest and most interesting of the feathered tribes that are found in this country. It is a very small bird, hardly so large as a sparrow, and beautifully slender in shape. The bill is rather long, pointed, and curved, and the tail-feathers are stiff and pointed at their extremities. The food of the Creeper consists chiefly of insects,

CURVED-BILLED CREEPER.—*Dendrocolaptes procurvus.*

although the bird will sometimes vary its diet by seeds and other vegetable substances. The insects on which it feeds live principally under the bark of various rough-skinned trees, and when it is engaged in running after its food, it runs spirally up the trunk with wonderful ease and celerity, probing every crevice with ready adroitness, its whole frame instinct with sparkling eagerness, and its little black eyes glancing with the exuberance of its delight. While running on the side of the tree which is nearest to the spectator, it presents a very curious appearance, as its dark-brown back and quick tripping movements give it a great resemblance to a mouse, and ever and anon, as it comes again into sight from the opposite side of the trunk, its beautifully white breast gleams suddenly in contrast with the sombre-coloured bark. Its eyes are wonderfully keen, as it will discern insects of so minute a form that the human eye can hardly perceive them, and it seems to possess some instinctive mode of detecting the presence of its insect prey beneath moss or lichens, and will perseveringly bore through the substance in which they are hidden, never failing to secure them at last.

The Creeper is a very timid bird, and if it is alarmed at the sight of a human being, it will either fly off to a distant tree, or will quietly slip round the trunk of the tree on which it is running, and keep itself carefully out of sight. It soon, however, gains confidence, and, provided that the spectator remains perfectly quiet, the little head and white breast may soon be seen peering anxiously round the trunk, and in a few minutes the bird will resume its progress upon the tree, and run cheerily up the bark, accompanying itself with its faint trilling song. It seldom attempts a long flight, seeming to content itself with flitting from tree to tree.

COMMON TREE CREEPER.—*Cérthia familiáris.*

Although so timid a bird, the Creeper soon becomes familiarized with those whom it is accustomed to see, provided that they treat it kindly, and will even come to receive food from their hands. In one instance that has come to my knowledge, the little birds were seen to frequent the patches of gum that exude from the bark of several trees, and in one spot where a number of small branches united, so as to form a kind of cup or hollow, a little heap of gum was found, which seemed to have been placed there by the Creepers, as they were constantly seen haunting the spot. Feeling sure that the birds fed upon the gum, the spectators used to supply their larder not only with gum, but with crumbs of bread, different seeds, and little morsels of raw meat, which disappeared as regularly as they were provided.

Some persons have supposed that in climbing the trees it uses its beak, after the manner of the parrots and other climbing birds. This, however, is not the case, as the beak is only employed for the purpose of probing the bark, and the whole progression is achieved by means of the long, curved, and sharply-pointed claws, which retain their hold so firmly, that I have seen a Creeper hang by its claws after it had been shot, and remain firmly fixed to the bark long after life had fled.

The Creeper is a very nervous bird, and may be temporarily paralysed by a smart blow given to the tree or branch on which it is running. Expert bird-catching boys will often secure this bird by flinging a stone or heavy stick at the tree, and then pouncing on the bird before it has recovered from its alarm. It can even be struck from its hold by suddenly running round the tree, and delivering a sharp blow upon the part of the trunk on which it is clinging. It also takes advantage of the uniform brown tint of its back to conceal itself from a real or fancied foe, by clinging closely to the tree, and pressing itself so flatly into some crevice, that a human eye can scarcely distinguish it from the bark. The Creeper does not confine itself exclusively to trees, but has often been seen running up old walls, and seeking for the insects that are always to be found in such localities.

The nest of the Creeper is usually made in the hollow of some decaying tree, and is made of grasses, leaves, and vegetable fibres, and lined with feathers. The eggs are very small, about seven or eight in number, and of an ashen-grey colour, sprinkled with little grey-brown spots. Sometimes it builds in the hole of an old wall, and has been known to make its nest in a disused spout.

The WALL CREEPER is a native of central and southern Europe, and is found plentifully in all suitable localities. It is called the Wall Creeper because it frequents walls and perpendicular rocks in preference to tree trunks.

In its movements it does not resemble the common Creeper; for, instead of running over the walls with a quick and even step, it flies from point to point with little jerking movements of the wing, and when it has explored the spot on which it has alighted, takes flight for another. The food of this bird is similar to that of the common Creeper, but it is especially fond of spiders and their eggs, finding them plentifully in the localities which it frequents. Old ruined castles are favourite places of resort for this bird, as are also the precipitous faces of rugged rocks.

The nest of the Wall Creeper is made in the cleft of some lofty rock, or in one of the many holes which are so plentifully found in the old ruined edifices which it so loves.

In colour the Wall Creeper is a very pretty bird, the general colour of the plumage being light grey, relieved by a patch of bright crimson upon the shoulders, the larger wing-coverts, and the inner webs of the secondaries. The remainder of the quill-feathers of the wing are black, and the tail is black tipped with white. It is a much larger bird than the Creeper of England, measuring about six inches in total length.

WALL CREEPER.—*Tichódroma murária.*

THERE is a curious genus of the Creeping-bird, known by the name of CLIMACTERIS. All the members of this genus are inhabitants of Australia, and notices of the individual species may be seen in Mr. Gould's well-known work on the birds of that country. They are generally found upon the tall gum-trees, traversing their rugged bark with great rapidity, and probing the crevices in search of insects, after the manner of the common English Creeper. They do not confine themselves to the bark, but may often be seen running into the " spouts," or hollow branches, which are so often found in the gum-trees, and hunting out the various nocturnal insects which take refuge in these dark recesses during the hours of daylight.

The Nuthatches form another group of the Certhidæ, and are represented in England by the common NUTHATCH of our woods. They are all remarkable for their peculiarly stout and sturdy build, their strong, pointed, cylindrical beaks, and their very short tails.

The Nuthatch, although by no means a rare bird, is seldom seen except by those who are acquainted with its haunts, on account of its shy and retiring habits. As it feeds mostly on nuts, it is seldom seen except in woods or their immediate vicinity, although it will sometimes become rather bold, and frequent gardens and orchards where nuts are grown. The bird also feeds upon insects, which it procures from under the bark after the manner of the creepers, and it is not unlikely that many of the nuts which are eaten by the Nuthatch have been inhabited by the grub of the nut weevil. It will also feed upon the seeds of different plants, especially preferring those which it pecks out of the fir-cones. Beech mast also seems grateful to its palate, and it will occasionally take to eating fruit.

In order to extract the kernel of the nut, the bird fixes the fruit securely in some convenient crevice, and, by dint of repeated hammerings with its beak, breaks a large ragged hole in the shell, through which the kernel is readily extracted. The blows are not merely given by the stroke of the beak, but the bird grasps firmly with its strong claws, and swinging its whole body upon its feet, delivers its stroke with the full weight and sway of the body.

NUTHATCH.—*Sitta Europæa.*

The beak, by means of which this feat is accomplished, is remarkably strong and powerful, and can be used with a vigour and endurance that is quite astonishing. Many instances of its powers have been narrated, among which we may mention that one of these birds which had been captured in a common brick trap, and had remained in its dark cell for some hours, was found when released to have been deprived of one-third of its beak, which had evidently been ground away by the continual pecking which had been kept up at the bricks. The person who caught the bird and who narrated the tale is the Reverend Mr. Bree. Another of these birds that had been put into a cage, immediately began to hammer at the wooden supports of its prison, and although severely wounded in the wing, refused to cease from its exertions except to eat and drink, both of which operations it performed with the greatest coolness. For two days the poor bird continued to peck unceasingly at his cage, and at the close of the second day, sank under its extraordinary exertions.

The Nuthatch is a capital climber of tree-trunks, even surpassing the creeper in the agility with which it ascends and descends the perpendicular surface, clinging firmly with its strong claws, and running equally well whether its head be upwards or downwards.

Even the creeper does not attempt to run down a tree with its head towards the ground. It is a very hardy bird, continuing to pick up an abundant supply of food even in the depths of winter, always appearing plump and lively.

The nest of the Nuthatch is placed in the hollow of a decaying tree, and the bird always chooses some hole to which there is but a small entrance. Should the orifice be too large to please its taste, it ingeniously builds up the orifice with clay and mud, probably to prevent the intrusion of any other bird. If any foe should venture too near the nest, the mother bird becomes exceeding valiant, and dashing boldly at her enemy, bites and pecks so vigorously with her powerful beak, hissing and scolding the while, that she mostly succeeds in driving away the assailant. The nest is a very inartificial structure, made chiefly of dried leaves laid loosely upon the decaying wood, and rudely scraped into the form of a nest.

In its colour the Nuthatch is rather a pretty bird, of pleasing though not of brilliantly tinted plumage. The general colour of the upper parts is a delicate bluish-grey, the throat is white, and the abdomen and under parts are reddish-brown, warming into rich chestnut on the flanks. From the angle of the mouth a narrow black band passes towards the back of the neck, enveloping the eye in its course and terminating suddenly before it reaches the shoulders. The tail is black on the base and grey towards the tip, except the two outer tail-feathers, which have each a black spot near the extremity. The shafts are also black.

We now arrive at the family of the Wrens, in which group we find two birds so dissimilar in outward appearance as apparently to belong to different orders, the one being the common Wren of England, and the other the celebrated LYRE-BIRD of Australia.

This bird, which also goes under the name of NATIVE PHEASANT among the colonists, and is generally called BULLEN-BULLEN by the natives, on account of its peculiar cry, would, if it had been known to the ancients, have been consecrated to Apollo, its lyre-shaped tail and flexible voice giving it a double claim to such honours. The extraordinary tail of this bird is often upwards of ten feet in length, and consists of sixteen feathers, formed and arranged in a very curious and graceful manner. The two outer feathers are broadly webbed, and, as may be seen in the illustration, are curved in a manner that gives to the widely-spread tail the appearance of an ancient lyre. When the tail is merely held erect and not spread, the two lyre-shaped feathers cross each other, and produce an entirely different outline. The two central tail-feathers are narrowly webbed, and all the others are modified with long slender shafts, bearded by alternate feathery filaments, and well representing the strings of the lyre.

The tail is seen in its greatest beauty between the months of June and September, after which time it is shed, to make its first reappearance in the ensuing February or March. The habits of this bird are very curious, and are so well and graphically related by Mr. Gould, that they must be given in his own words :—

The great stronghold of the Lyre-bird is the colony of New South Wales, and from what I could learn, its range does not extend so far to the eastward as Moreton Bay, neither have I been able to trace it to the westward of Port Phillip on the southern coast; but further research can only determine these points. It inhabits equally the bushes on the coast and those that clothe the sides of the mountains in the interior. On the coast it is especially abundant at the Western Port and Illawarra; in the interior, the cedar brushes of the Liverpool range, and according to Mr. G. Bennett, the mountains of the Tumat country, are among the places of which it is the denizen.

Of all the birds I have ever met with, the Menura is far the most shy and difficult to procure. While among the mountains I have been surrounded by these birds, pouring forth their loud and liquid calls for days together, without being able to get a sight of them, and it was only by the most determined perseverance and extreme caution that I was enabled to effect this desirable object, which was rendered more difficult by their often frequenting the almost inaccessible and precipitous sides of gullies and ravines, covered with tangled masses of creepers and umbrageous trees : the cracking of a stick, the rolling down of a small stone, or any other noise, however slight, is sufficient to alarm

LYRE BIRD.—*Menura superba.*

it; and none but those who have traversed these rugged, hot, and suffocating bushes, can fully understand the anxious labour attendant on the pursuit of the Menura.

Independently of climbing over rocks and fallen trunks of trees, the sportsman has to creep and crawl beneath and among the branches with the utmost caution, taking care only to advance while the bird's attention is occupied in singing, or in scratching up the leaves in search of food: to watch its action it is necessary to remain perfectly motionless, not venturing to move even in the slightest degree, or it vanishes from sight as if by magic. Although I have said so much on the cautiousness of the Menura, it is not always so alert; in some of the more accessible bushes through which roads have been cut, it may frequently be seen, and on horseback even closely approached, the bird evincing less fear of those animals than of man.

At Illawarra it is sometimes successfully pursued by dogs trained to rush suddenly upon it, when it immediately leaps upon the branch of a tree, and its attention being attracted by the dog below barking, it is easily approached and shot. Another successful mode of procuring specimens is by wearing the tail of a full-plumaged male in the hat, keeping it constantly in motion, and concealing the person among the bushes, when, the attention of the bird being arrested by the apparent intrusion of another of its own sex, it will be attracted within the range of the gun. If the bird be hidden from view by surrounding objects, any unusual sound, such as a shrill whistle, will generally induce him to show himself for an instant, by causing him to leap with a gay and sprightly air upon some neighbouring branch to ascertain the cause of the disturbance; advantage must be taken of this circumstance immediately, or the next moment it may be half-way down the gully.

The Menura seldom, if ever, attempts to escape by flight, but easily eludes pursuit by its extraordinary powers of running. None are so efficient in obtaining specimens as the naked black, whose noiseless and gliding steps enable him to steal upon it unheard or unperceived, and with a gun in his hand he rarely allows it to escape, and in many instances he will even kill it with his own weapons.

The Lyre-bird is of a wandering disposition, and although it probably keeps to the same bush, it is constantly engaged in traversing it from one end to the other, from the mountain base to the top of the gullies, whose steep and rugged sides present no obstacle to its long legs and powerful muscular thighs. It is also capable of performing extraordinary leaps, and I have heard it stated that it will spring ten feet perpendicularly from the ground. Among its many curious habits, the only one at all approaching to those of the Gallinaceæ is that of forming small round hillocks, which are constantly visited during the day, and upon which the male is continually tramping, at the same time erecting and spreading out its tail in the most graceful manner, and uttering its various cries, sometimes pouring forth its natural notes, at others imitating those of other birds, and even the howling of the native dog (Dingo). The early morning and the evening are the periods when it is most animated and active.

Although upon one occasion I forced this bird to take wing, it was merely for the purpose of descending a gulf, and I am led to believe that it seldom exerts this power unless under similar circumstances. It is peculiarly partial to traversing the trunks of fallen trees, and frequently attains a considerable altitude by leaping from branch to branch. Independently of a loud full note, which may be heard reverberating over the gullies for at least a quarter of a mile, it has also an inward warbling song, the lower notes of which can only be heard within about fifteen yards. It remains stationary while singing, fully occupied in pouring forth its animated strain. This it frequently discontinues abruptly, and again commences with a low inward, snapping noise, ending with an imitation of the loud and full note of the satin-bird, and always accompanied by a tremulous motion of the tail.

The food of the Menura appears to consist principally of insects, particularly of centipedes and coleoptera. I also found the remains of shelled snails in the gizzard, which is very strong and muscular."

The same writer, in a recent communication to the Zoological Society, mentions the discovery of a nest of the Lyre-bird, containing a nearly adult young bird of sixteen inches in height. In spite of its large size, it was a most helpless creature, and seemed to be quite incapable of escape, even with the assistance of its mother, who with devoted courage tried to withdraw it from its enemies, and laid down her life in the attempt.

The young Lyre-bird displayed no fear of its captors, but was easily induced to follow any one when allured by the imitation of the mother's cry. It fed well, its chief articles of diet being worms, ants, and their larvæ. For water it seemed to care but little, and seldom, if ever, was seen engaged in drinking. Although tended with great care, and its wants well supplied, this interesting captive died within eight days after it was taken. Dr. Bennett remarks of the young Lyre-bird, that when it is able to leave the nest it is very swift of foot, and as it instinctively conceals itself under rocks and among the densest thickets, its capture is a difficult task even for a native.

Of the adult bird, the same writer speaks as follows:—

"I first saw these birds in the mountain range of the Tumat country; lately they have been very abundant among the Blue Mountain ranges bordering on the Nepean River, above Emeu Plains (about thirty-five miles from Sydney). They are remarkably shy, very difficult of approach, frequenting the most inaccessible rocks and gullies, and on the slightest disturbance they dart off with surprising swiftness through the brakes, carrying their tail horizontally; but this appears to be for facilitating their passage through the bushes, for when they leap or spring from branch to branch as they ascend or descend a tree, the tail approaches to the perpendicular. On watching them from an elevated position, playing in a gully below, they are seen to form little hillocks or mounds, by scratching up the ground around them, trampling and running flightily about, uttering their loud shrill calls, and imitating the notes of various birds."

The nest of the Lyre-bird is a large, loosely-built, domed structure, composed of small sticks, roots and leaves, and of an oven-like shape, the entrance being in front. The lining is warm and soft, being composed of downy feathers.

The egg of this singular bird is quite as curious as its general form, and presents the curious anomaly of an egg as large as that of a common fowl, possessing all the characteristics of the insessorial egg. The general colour of the egg is a deep chocolate tint, marked with purple more or less deep in different specimens, and its surface is covered with a number of stains and blotches of a darker hue, which are gathered towards the larger end, as is usual in spotted eggs.

ANOTHER species of Lyre-bird has been discovered, which is called ALBERT'S LYRE-BIRD (*Menura Alberti*), in compliment to the Prince Consort. This species may be known by the comparative shortness of the lyre-shaped tail-feathers, and the absence of dark bars upon the web.

Dr. Stephenson, in speaking of this bird, says: "The locality it frequents consists of mountain ridges, not very densely covered with brush; it passes most of its time on the ground, feeding and strutting about with the tail reflected over the back to within an inch or two of the head, and with the wings drooping on the ground. Each bird forms for itself three or four 'corroboring places,' as the sawyers call them; they consist of holes scratched in the sandy ground, about two and a half feet in diameter, by sixteen, eighteen, or twenty inches in depth, and about three or four hundred yards apart, or even more.

Whenever you get sight of the bird, which can only be done with the greatest caution and by taking advantage of intervening objects to shelter yourself from its observation, you will find it in one or other of these holes, into which it frequently jumps and seems to be feeding; then ascends again and struts round and round the place, imitating with its powerful musical voice any bird that it may chance to hear around it. The notes of the *Dacelo gigantea*, or laughing jackass, it imitates to perfection; its own whistle is exceedingly beautiful and varied. No sooner does it perceive an intruder, than it flies up into the nearest tree, first alighting on the basement branches, and then ascending by a succession of jumps until it reaches the top, when it instantly darts off to another of its playgrounds.

The stomachs of those I dissected invariably contained insects, with scarcely a trace of any other material. Now, collectors of insects know that gravel-pits and sandy holes afford them great treats, and it appears to me that one, if not the principal use of the excavations made by this bird, is to act as a trap for unwary coleopteras and other insects, which falling in, cannot be again rescued, and are therefore easily secured."

The nest of Albert's Lyre-bird is like that of the preceding species in general shape, but is almost wholly composed of long and slender twigs, and presents a most curious appearance. Specimens of this structure may be seen in the British Museum. The nest resembles nothing so much as a large round mass of loose sticks, into which some giant had thrust his foot and left the impression of his shoe. The hollow of the nest is, in fact, a kind of cave on a small scale, domed over by the sticks as they lie crossing each other in all directions.

WE are all familiar with the WREN, "the king of all birds," as he is termed in ancient rhyme, his title to royalty resting on his defeat of the eagle in upward flight. The story runs that the birds assembled to choose a king, and that the election should fall on the bird who soared the highest. Up sprang all the birds into the sky, but highest of all towered the eagle, who, after mounting until his wearied wings could beat no more, proclaimed himself the sovereign of the birds. But all unperceived, the little Wren had been quietly perching between his shoulders, and as soon as the eagle ceased to mount, the Wren sprang into the air, and, rising on tiny pinion far above the wearied eagle, twittered forth the victory of wit and intellect over bulk and physical strength.

The long and harsh name of Troglodytes, which has been given to this bird, signifies a diver into caves, and has been attributed to the Wren on account of its shy and retiring habits, and its custom of hiding its nest in some hollow or crevice where it may escape observation. The Wren is seldom to be seen in the open country, and does not venture upon any lengthened flight, but confines itself to the hedge-rows and brushwood, where it may often be observed hopping and skipping like a tiny feathered mouse among the branches. It especially haunts the hedges which are flanked by ditches, as it can easily hide itself in such localities, and can also obtain a plentiful supply of food. By remaining perfectly quiet, the observer can readily watch its movements, and it is really an interesting sight to see the little creature flitting about the brushwood, flirting its saucily expressive tail, and uttering its quick and cheering note.

The voice of the Wren is very sweet and melodious, and of a more powerful character than would be imagined from the dimensions of the bird. The Wren is a merry little creature, and chants its gay song on the slightest encouragement of weather. Even in winter there needs but the gleam of a few stray sunbeams to set the Wren a-singing, and the cold Christmas season is often cheered with its happy notes. While skipping among the branches, the Wren utters a continuous little twitter, which, although not worthy of being reckoned as a song, is yet very soft and pleasing.

The nest of the Wren is rather an ambitious structure, being a completely domed edifice, and built in a singularly ingenious manner. If, however, the bird can find a suitable spot, such as the hole of a decaying tree, the gnarled and knotted branches of old ivy, or the overhanging eaves of a deserted building, where a natural dome is formed, it is sure to seize upon the opportunity and to make a dome of very slight workmanship. The dome, however, always exists, and is composed of non-conducting materials, so that the

WREN.—*Troglodytes vulgaris.*

bird always contrives to insulate itself and its young from electrical influences. It is a very singular fact, that a Wren will often commence and partly build three or four nests in different localities before it settles finally upon one spot. Some persons have supposed that these supplementary nests are built by the parent bird as houses for its young after they have grown too large to be contained within the house where they were born, while others have suggested that they are experimental nests made by the inexperienced young while trying their 'prentice beak in the art of bird architecture.

For my own part, I believe that these partial nests have been made by Wrens when building for the first time, and consequently inexperienced in the world and its difficulties. They seem to fix too hastily upon a locality, and then to find, after they have made some progress with their house, that danger lurks near, perhaps in the form of a weasel, a shrike, or a snake. In one such instance of desertion the cause was sufficiently evident, for the head of a snake was seen protruding from the opening of the nest.

The materials of which the nest is composed are always leaves, moss, grass and lichens, and it is almost always so neatly built that it can hardly be seen by one who was not previously aware of its position. The opening of the nest is always at the side, so that the eggs are securely shielded from the effects of weather.

As to the locality and position in which the nest is placed, no definite rule is observed, for the Wren is more capricious than the generality of birds in fixing upon a house for her young. Wrens' nests have been found in branches, hedges, hayricks, waterspouts, hollow trees, barns and outhouses. Sometimes the Wren becomes absolutely eccentric in its choice, and builds its nest in spots which no one would conjecture that a bird would select. A Wren has been known to make its nest in the body of a dead hawk, which had been killed and nailed to the side of a barn. Another Wren chose to make her house in the throat of a dead calf, which had been hung upon a tree, and another of these curious little birds was seen to build in the interior of a pump, gaining access to her eggs and young through the spout.

The eggs of the Wren are very small, and are generally from six to eight in number.

During the winter, the Wren generally shelters itself from the weather in the same nest which it had inhabited during the breeding season, and in very cold seasons it is not an uncommon event to find six or seven Wrens all huddled into a heap for the sake of warmth, and presenting to the eye or hand of the spectator nothing but a shapeless mass of soft brown feathers. It is probable that these little gatherings may be composed of members of the same family, an opinion which is strengthened by the following account which was sent by Mr. Ogilby to Mr. Thompson, and quoted in his work on the natural history of Ireland :—

"These little birds associate in small families of from four or five to a dozen or more, and take refuge in holes, or under the eaves of thatched houses, during the severity of winter nights. I have often, when a boy, watched the little party thus taking up their lodgings for the night, and have on more than one occasion captured and driven them from their retreat. They make a prodigious chattering and bustle upon finally settling for the night, as if contending which shall get into the warmest and most comfortable place, and frequently come to the mouth of the hole to see that they are unobserved. I presume that these little parties are composed of the nestlings of the previous year, with, perhaps, the parent birds, but I have no proof beyond its probability. If such be the case, however, it would show that the bond of social union between the parent birds and their young continues unbroken during the year, and is severed only when the new season prompts the young brood to become parents in their turn."

Sometimes the Wrens have rather odd modes of bivouacking. A lady who was accustomed to attract great numbers of the feathered tribes to her garden by supplying them abundantly with crumbs, seeds, and other dainties, told me that when the weather became cold, the Wrens used to gather themselves upon a moderately large branch of a tree, about four inches above which grew another branch. In the evening the Wrens assembled upon their resting-place, and packed themselves very comfortably for the night, piled three or four deep, apparently for the sake of warmth, the topmost bird always having his back pressed against the upper branch, as if to keep all steady. Pitying their forlorn condition, their benefactress provided a bedroom for them, being a square box lined with flannel, and with a very small round hole by way of door. This was fixed on the branch, and the birds soon took advantage of it, their numbers seeming to increase nightly, until at last upwards of forty Wrens would crowd into a box which did not seem capable of containing half that number. When asleep, they were so drowsy that they would permit the lid of the box to be lifted, and themselves to be handled, without attempting to move. All these Wrens were supposed to come from a number of nests which had been made in the gnarled roots of old hawthorn bushes which grew at the side of a narrow but deep brook running at the end of the garden.

The same lady noticed that Wrens were much attracted by colour, especially by scarlet. At the beginning of her acquaintance she had laid some food for them, but they were so shy that they dared not approach, and the saucy sparrows invariably ate all the egg and crumbs that were intended for the Wrens. One morning, a flower of a scarlet geranium fell upon the saucer of food, and immediately seemed to influence the Wrens, who came hopping and flying nearer and nearer, until they were bold enough to taste the food. For several days the geranium blossoms were used as a decoy, but the little birds soon became so familiar that they needed no such allurement, and would crowd round the saucer as soon as it was placed on the ground.

In the pages of the above-mentioned writer a curious anecdote is related, where a pair of flycatchers who had made a nest and laid three eggs were ousted by a party of young Wrens just able to fly. The little birds had probably been ejected from home for the first time, and seeing so comfortable a nest, had taken possession of it. I have often observed the same conduct in many young canaries, for whenever a family of the newly-fledged birds is turned out of the nest, they generally wend their way to the home of some other female, and instal themselves in possession of her nest and eggs before she is aware of their intentions.

The title of "Kitty Wren," which is often given to this bird, is owing to the peculiar little twittering sound of *Chit! chit!* which it utters while engaged in the pursuit of food.

The colour of the Wren is a rich reddish-brown, paling considerably on the under surface of the body, and darkening into dusky brown upon the quill-feathers of the wings and tail. The outer webs of the former are sprinkled with reddish-brown spots, and the short tail-feathers are barred with the same hue. The bill is slender, and rather long in proportion to the general dimensions of the bird. The total length of the Wren is rather more than four inches. White and pied varieties are not uncommon.

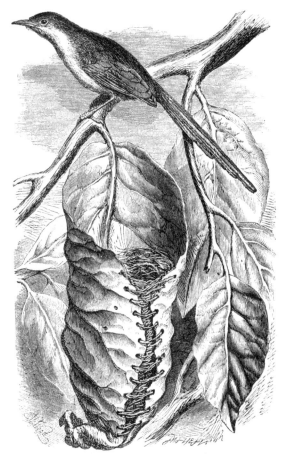

TAILOR-BIRD.—*Orthótomus longicaudus.*

WE now arrive at the very large family of the WARBLERS, a group in which the genera, when taken together, number more members than those of any other family. The first sub-family is that which is known by the name of the *Malurínœ*, or soft-tailed Warblers.

THE first example of the soft-tailed Warblers is the celebrated TAILOR-BIRD of India and the Indian Archipelago. There are many species belonging to the genus Orthotomus, and as they all possess similar habits, there is no need of describing more than the example which has been shown. They are peculiar birds, haunting cultivated grounds,

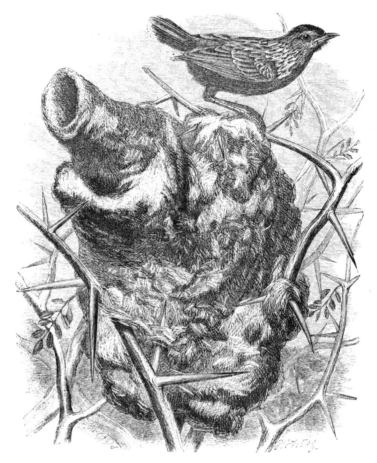

PINC-PINC.—*Drymoïca textrix.*

and being generally seen in pairs on fields and in gardens. They dislike lofty elevations, and may generally be seen near the ground, hopping about the lower branches of trees and shrubs in their search after insects, and occasionally seeking their prey on the ground. Their flight is rapid but undulating, after the manner of many short-winged and long-tailed birds.

The Tailor-bird is a sober little creature, not more conspicuous than a common sparrow, and is chiefly remarkable for its curious nest, which is made in a singular and most ingenious manner. Taking two leaves at the extremity of a slender twig, the bird literally sews them together at their edges, its bill taking the place of the needle and vegetable fibres constituting the thread. A quantity of soft cottony down is then pushed between the leaves, and a convenient hollow scraped out in which the eggs may lie and the young birds may rest at their ease. Sometimes, if the leaf be large enough, its two edges are drawn together, but in general a pair of leaves are needed. A few feathers are sometimes mixed with the down.

This curious nest is evidently hung at the very extremity of the twigs, in order to keep it out of the way of the monkeys, snakes, and other enemies which might otherwise attack and devour mother and young together.

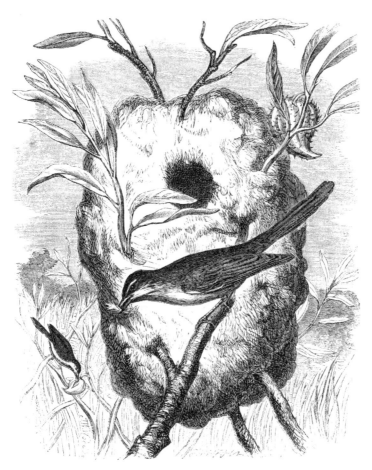

CAPOCIER.—*Drymóica maculósa.*

THE large genus Drymoica, which numbers nearly seventy species, is quite as remarkable as the preceding genus. Some species inhabit Africa, others Southern Europe, some are found in India, and many in Australia. They are always found in open plains where the grass is long and thick, or among the rich rank herbage that grows in marshy localities. They mostly feed on the ground insects, and are well fitted by their very great speed of foot for the chase and capture of their swift insect prey. Being but indifferent in flight, they seldom take to wing, and always try to escape from their foes by means of running among the thick herbage in which they live. While running, they generally lift their tails and hold them over the back. Their song is loud, but by no means agreeable.

The nest of the PINC-PINC is of considerable dimensions, being often more than a foot in circumference and of a most singular shape and structure. The materials of which it is composed are vegetable fibres, beaten, twisted, and woven into a fine felt-like substance, and strongly fastened to the branches among which it is situated. It is of a rough, gourd-like shape, and is always entered by means of a neck, or spout, so that the bird is able from the interior to present its sharply pointed bill to any assailant, and to prevent its entrance. Near the mouth of the nest there are generally one or two projections,

2.

which serve as perches for the bird to rest upon before it enters the nest, and may probably be used by the male as a seat whereon to recline while his mate is sitting upon the eggs within. The colour of the nest varies according to the substance of which it is composed, sometimes being of a snowy whiteness, and at other times of a dingy brown.

The peculiar form of this and other similar nests is evidently for the purpose of enabling the parent bird to defend its home against the intrusions of the many foes to bird life with which these regions swarm. The eggs are generally from six to eight in number.

ANOTHER species of the same genus, the CAPOCIER, builds a nest which, although of coarser texture, is quite as beautiful as that of the pinc-pinc.

The materials are much the same as those employed by the pinc-pinc, but they are only closely felted together in the interior of the nest, and are left to flow loosely on its exterior. It is a very large edifice in proportion to the dimensions of the architect, and the cottony down, the flaxen fibres, and fine moss are woven together in so skilful a manner as to excite our highest admiration. Le Vaillant, who watched a pair of Capociers hard at work upon their nest, says that they were occupied for a whole week in their task. The interior of the nest is of an oval shape, rather roughly corresponding with the external outline, and it is entered through an opening in the side. One of these nests will sometimes reach nine inches in height. The Capocier is a native of Africa.

THE genus which has been chosen as a type of this family is MALURUS, the members of which are only found on the continent of Australia. They possess the habits of Drymoica, and even resemble those birds in the peculiar fashion of tilting up their tails as they run over the ground.

EMEU WREN.—*Stipiturus malachurus.*

OUR last, and pernaps the most curious example of the Malurine birds is the beautiful little EMEU WREN of Australia.

This pretty bird is remarkable for the development of the tail-feathers, which are extremely lengthened, and are nothing more than bare shafts slightly fringed on each side. The bird never perches on high trees, and very seldom takes to wing, but runs over the grass with very great rapidity, holding its tail erect over its back in a singularly pert manner. It is generally found among long grass, and according to Dr. Bennett it congregated some years since in the Sydney Domain, near the Botanic Gardens, but has not since appeared in that locality. The colour of this little bird is mottled brown above, and very light fawn below, deepening into chestnut on the flanks. The throat of the male is tinged with blue, and his tail-feathers are larger than those of his mate.

The nest of the Emeu Wren is very large in comparison with the size of its inhabitant, and is placed on the ground, where it looks like a large ball of grass with a hole in the side. The interior is snugly lined with soft feathers, and there are generally three eggs.

GOLDEN-CRESTED WREN.—*Régulus cristátus.*   FIRE-CRESTED WREN.—*Régulus ignicapillus.*

THE tiny GOLDEN-CRESTED WREN, as it is popularly called, is very common throughout England, and may be seen hopping and flitting merrily among the branches in copses, orchards, and plantations. Although from its diminutive size it has gained the title of Wren, it has no claim to that designation, and is more rightly termed the Kinglet or Regulus.

This "shadow of a bird," as it is happily called by White, in his "Natural History of Selborne," is a remarkably hardy little creature, braving the severest frosts of winter, and mostly disdaining to avail itself of the shelter of human habitations. On account of its minute proportions and its retiring habits, it is a very unobtrusive bird, and is often thought to be extremely rare in localities where it may be found plentifully by those who know where to look for it. In Derbyshire, for instance, it was held to be extremely scarce, but I could always procure specimens at will by a judicious disposition of a little birdlime, and I have frequently discovered the admirably hidden and beautifully constructed nests of these interesting birds.

The Golden-crested Wren is notable for the crest of golden-coloured feathers which is placed upon the crown of its head, which it can raise or depress at pleasure, and which gives so pert and changeful an expression to the little creature. But for this golden crest which is not at all conspicuous when the feathers are lowered, the bird might easily be mistaken for a tree-creeper as it runs up and down the branches, searching into the crevices of the bark for the little insects on which it feeds. The first specimen that I ever saw was traversing the branches of a fine "Blenheim Orange" apple-tree in an adjoining garden, and by my inexperienced eyes was at first taken for a very young creeper. Like the creeper it can even run up a perpendicular wall, peering into every little crevice, and stocking up the moss and lichens for the purpose of obtaining the insects and their eggs that are lying concealed. It will also eat the chrysalides that are found so abundantly upon the walls.

All the movements of the Golden-crested Wren are full of spring and fiery activity, and the manner in which it will launch itself from one tree to another, and then, without a pause, commence traversing the branches, is a sight well worth seeing. Perhaps it is seen to best advantage among the fir-trees, where it finds great scope for its active habits. Up one branch it scuds, down another, then whisks itself through the air to a fresh tree, and then flings itself back again to its former perch. Along the twigs it runs with astonishing rapidity, sometimes clinging with its head downward, sometimes running round and round them spirally, always twisting its pert little head in every direction, and probing each hole and crevice with its sharp slender little bill. The roughest-barked trees are its favourite resort, because in such localities it finds its best supply of insect food.

Mr. Thompson, in the History of the Birds of Ireland, rather controverts the hardy nature of the Gold Crest. "The Gold-crested Regulus seems not to be the hardy bird that authors generally imagine. In the north of Ireland it has frequently been found dead about the hedges, not only in severe winters, but after slight frosts. In the greenhouses and hothouses in the garden of a relative near Belfast, these birds resorted so regularly in the mild winter of 1831—1832, that some were captured weekly throughout the season, and taken to one of our bird-preservers : on the rear wall of the house is a range of sheds accessible to birds, and dense plantations of trees and evergreen shrubs are quite contiguous. They were occasionally caught at all seasons, as were common wrens and titmice—many of both—together with robins, sparrows and chaffinches. . . . Early in the winter of 1835, three of these birds, which had been captured by a cat in a small garden in a very populous part of Belfast, were brought to me, and on the preceding day four or five had, in the same place, shared a similar fate. In the middle of December, 1846, after a few days of frost and snow, I observed a Regulus fly from a plantation at the roadside several times, and alight at the base of the demesne wall bounding the footway on which I walked."

Although these remarks would tend to contradict the statement that the Kinglet is a hardy bird, it must be remembered that the climates of England and of Northern Ireland are very different, and that in the latter locality the constant damp, frequent rain, and bleak winds are sufficiently trying to drive any creature under shelter.

The nest of this beautiful little bird is exquisitely woven of various soft substances, and is generally suspended to a trunk where it is well sheltered from the weather. I have often found their nests, and in every instance have noticed that they are shaded by leaves, the projecting portion of a branch, or some such protection. In one case the nest that was suspended to a fir-branch was almost invisible beneath a heavy bunch of large cones that drooped over it, and forced the bird to gain admission by creeping along the branch to which the nest was suspended. The edifice is usually supported by three branches, one above and one at either side. The nest is usually lined with feathers, and contains a considerable number of eggs, generally from six to ten. These eggs are hardly bigger than peas, and, as may be supposed, their shells are so delicately thin, that to extract the interior without damaging them is a very difficult matter. It may here be noticed that the surest mode of emptying such delicate eggs is to cover them with silver paper saturated with thick gum, and waiting till it is dry before attempting to bore a hole through the shell. After the egg has been cleaned, it should be placed in warm water, when the gum is dissolved and the paper can be washed away.

The following interesting account of some Golden-crested Wrens and their habits was kindly forwarded to me by a lady :—

"I had often questioned and wondered why some birds possess crests and crowns, while the majority are destitute of these feathery ornaments. But in the winter of 1853 I became personally acquainted with some Golden-crested Wrens, who revealed the mystery to me. I now feel sure that crowns and crests typify and are a sort of natural safety-valve to a nature which is imbued with a spirit of empire. At all events it is so in the case of the Golden-crested Wren, for he is running over with the governing spirit, and his cool audacity, fiery courage, and fierce domination beggar description.

That winter we had a family of six tame, but uncaged birds ; they were a strange-looking group, but, nevertheless, a very happy one. There was a jackdaw, a magpie, two skylarks, a goldfinch and a robin, and they lived when at home in a large and well-thatched aviary, which was placed in a very sheltered position. In this abode they kept open house, for there was always a good supply of food kept therein, and the door was never shut save at night, when we closed it to keep out stray cats, rats, and other enemies of the feathered race. After partaking of a good breakfast, they would daily leave their comfortable home on expeditions of pleasure or business, and return regularly to their supper and perch.

During the very cold weather of 1853, they brought home nightly a party of hungry wild birds to share with themselves the hospitality of the aviary; sometimes their guests would number nearly two hundred, and it was really quite astonishing to see the quantity of bread, barley, and fat meat, that these little beings would dispose of. In this assembly most of our small native birds had representatives, and amongst them were the two before-mentioned Golden-crested Wrens, who were the first to reach and the last to quit their good quarters.

Whilst they honoured us with their company, they ruled the whole bird community, and what they could not achieve by force they would accomplish by stratagem. For instance, if one of these tiny creatures took a fancy to a piece of meat to which Mr. Jackdaw had helped himself, and which he was holding firmly down with one foot whilst he pecked away at it after the dawish fashion, this mite of a bird would jump upon the jackdaw's head, and attack the eye that was situated on the side of the occupied foot. The poor jackdaw, not quite understanding what had gone wrong with him, would lift his foot to scratch his tiny tormentor off his head, when in an instant the coveted morsel was seized by the daring thief. If the daw were unwise enough to follow, vainly hoping to recover his property, the wicked wee thing would get upon his back, where he knew himself to be safe, and the poor jackdaw was forced to content himself with other fare.

If, after selecting a piece of meat, the Wrens were left to themselves, they would leave the larger birds and retire to a quiet corner, where they would both peck amicably at the same piece, and if the meat happened to be tough, one of the wise little things would hold it fast in its bill, while the other would pull a morsel off; and then the one that had eaten would perform the same kind office for his friend. Before the winter was over, there was not a bird in the aviary that did not give way to the two little Kinglets, and they always went to roost upon the backs of some other birds; I thought that they did this to warm their feet.

They remained with us until May, and then, when all nature appeared to be bursting with life and beauty, and the huge forest trees and even the old earth herself had donned their glorious resurrection robes, the Golden-crests forsook us. I suppose they found the aviary dull, and we ourselves soon sought another home; hence we never renewed our friendly intercourse with these interesting little birds."

The entire length of this bird is about three inches and a half, and its general colour is brownish above marked with olive-green, and flanked with white on the wing-coverts. The under surface is yellowish-grey, the beak is black, and the eye hazel-brown. The forehead is marked with greyish-white; the crest is brilliant yellow tipped with orange, and on each side of the crest runs a black line. The female is not so brilliant in her colouring, and the crest is wholly of a pale yellow.

The FIRE-CRESTED WREN is very similar to the preceding species, but may be distinguished from it by the ruddy hue of the forehead, the fiery orange of the crest, and the decidedly yellow hue of the sides of the neck. It is an inhabitant of England, but is a much rarer bird than the Golden-crest. Owing to the great resemblance between the two species, they have often been mistaken for each other, and it is only within a comparatively recent period that their diversity was established. Another species of the same genus, the DALMATIAN REGULUS (*Regulus modestus*) has, although very rarely indeed, been found in England. The general colour of the plumage is greenish-yellow, and the crown of the head is marked by a narrow streak of paler yellow.

GROUP OF BRITISH WARBLERS.

The *Luscininæ*, or SONG-WARBLERS, are, as their name denotes, remarkable for their sweet song, to which accomplishment may be added the beauties of elegant shape and graceful movement. In their colour they are very inconspicuous, brown, grey, and olive-green being the hues with which they are generally tinted, in accordance with the principle which has been lately laid down while treating of the Humming-birds. In all these birds the beak is strong, straight, sharply-pointed, and with a notch upon the upper mandible near its extremity.   The feet have three toes in front and one behind, and the claw of the hinder toe is the largest and most sharply curved.   The nostrils are placed at the base of the beak, are pierced through a rather large membrane, and are unprotected by feathers.

THE common WHITETHROAT is abundant in various parts of England, and is, perhaps, the best known of all its tribe.

It is a lively, brilliant little bird, and is remarkable for the curious movement which it makes when singing, and which seldom fails to attract the attention of the observer. Generally starting from some low bush, the Whitethroat begins its sweet quiet song, and then, springing suddenly into the air, wings its way perpendicularly upwards, as if it were about to rival the lark in its lofty flight. But after attaining a height of four or five feet, it slowly sinks upon the spot from which it had started, and again rises into the air with a fresh burst of music. While singing, it seems full of life and eagerness, and the white feathers of its throat, together with those of the crest, vibrate powerfully as the sweet notes are poured forth. The song is not a lengthened one, but is frequently repeated within a short space of time.

The Whitethroat possesses a strong spirit of rivalry, and will match itself against almost any songster that begins to utter his strain, having even been known to challenge the nightingale to a friendly trial of vocal powers. Sometimes it becomes a mimic, and imitates the songs of the other birds by which it is surrounded. In captivity it sings readily, and has been known to start into song only three days after its capture.

WHITETHROAT.—*Sylvia undata.*
GARDEN WHITETHROAT, OR GREATER PETTICHAPS.
Sylvia hórtensis.

The haunts of this bird are generally among low bushes, brambles, underwood and copses, among which it flits with restless activity. Owing to its habit of traversing the low and nettle-overgrown underwood, it has received the popular name of Nettle-creeper in many parts of England.   The food of the Whitethroat consists chiefly of insects, and it is indefatigable in its attacks upon various caterpillars and flies, thereby doing great service to the gardener, who, however, generally aims at its destruction, because in the autumn it repays itself by a few of the fruits which it has saved from the caterpillar and the grub. The caterpillar of the cabbage butterfly is a favourite article of diet with this bird.

THE CHIFF-CHAFF, OR LESSER PETTICHAPS.

The Whitethroat arrives in England towards the end of April, the male always making his appearance before his mate, and immediately upon their arrival they set to work in searching after a fit spot on which to build their nest. For this purpose they generally choose some thick bush, and are often very indifferent about concealing it, placing it openly on the top of a stunted thorn bush, within reach of every boy's hand. It is seldom placed at any great height from the ground, and I have generally found them situated at an elevation of two feet. The nest is not very elaborate in its structure, and is chiefly formed of various grasses twisted into shape, and slightly woven into the branches. The entire nest can, however, be easily removed, without the necessity of cutting the twigs to which it is fastened. The complete number of eggs is five, but their number is usually four. Their colour is ashen-grey with a green wash, and they are boldly covered with ashen-brown and ashen-green spots and blotches.

The general colour of the Whitethroat is reddish-brown on the upper parts of the body and brownish-white below, with the exception of the throat, which is of a beautifully pure white. Its total length is about five inches and a half.

THERE is another species of Whitethroat which is not so often seen as the last-named species. This is the GARDEN WHITETHROAT, sometimes called by the name of GREATER PETTICHAPS. It is also known by the name of GARDEN WARBLER.

This is an active and lively bird, restless to a degree, and constantly flitting from place to place and from branch to branch on unwearied wing, and seldom coming out into view. The song of this bird is much finer than that of the common Whitethroat, being more lengthened, fuller, and more melodious. Some persons have compared the rich mellow notes which it occasionally utters, to those of the blackbird. Although it chiefly frequents shrubberies, copses, and plantations, it will often take a predilection to some garden, and if left undisturbed, soon learns confidence and becomes quite a familiar bird, permitting those with whose persons it is acquainted to approach within very close proximity before it takes alarm. But for its very retiring habits it would be a well-known bird, and many persons have been charmed by its melody who have no idea of the bird from whose small throat it proceeds.

The food of the Garden Whitethroat is mostly of an insect nature, but the bird is not averse to various fruits, and in the autumn often becomes rather obnoxious to the gardener.

The nest is not unlike that of the common Whitethroat, and is placed in similar localities; a low thick bush being the favourite spot, although on occasions the bird will build its house among thick herbage, or even among heavy ivy leaves. The nest is made quite as loosely as that of the preceding species, and is composed of grasses, fine fibrous roots, moss, hair, and lichens, and the interior is softly lined with hair and other similar substances. The eggs are greenish-white, covered with many spots and blotches of light brown and ashen-green.

By its colour the Garden Whitethroat can be distinguished from its relative, as the upper surface is of a more delicate brown, and the abdomen of a purer white than the throat, which is tinged with a pale brown. The under wing-coverts are pale buff, the beak is dark brown, and the eyes are beautiful hazel. The total length of this species nearly reaches six inches.

THE tiny CHIFF-CHAFF, one of the smallest of the British birds, is the first Warbler that makes its appearance in these islands, and that cheers us with its pretty little song and its light, lively actions.

The curious name of this bird has been derived from its cry, which bears some resemblance to the words "Chiff-chaff! Chery-churry!" often repeated. This little song is sometimes uttered while the bird is on the wing, but generally when it is perched on some convenient bough of a lofty tree. The localities which it most frequents are woods and hedgerows, and so lively is it in temper, that its pleasant little voice is often heard before the trees have put forth their verdure. It is a very useful bird, as it feeds almost wholly on insects, and on its first arrival saves many a grand oak-tree from destruction by devouring the caterpillars of the well-known green oak moth, which roll up the leaves in so curious a manner, and come tumbling out of their green houses at the slightest alarm. Gnats and other small flies are a diet much in favour with the Chiff-chaff; and one of these birds that had been captured and tamed was accustomed to dash to the ceiling of the room in which it was kept, and to snatch from thence the flies as they settled after their fashion on the white surface.

This little bird has been seen in England as early as the twelfth of March, and it remains in this country as late as the middle of October, so that it is the first to arrive and the last to depart of all the British Warblers.

The nest of the Chiff-chaff is something like that of the common wren, being a rounded structure with a hole in the side, through which the bird obtains admission into the interior. It is seldom placed at any great elevation from the ground, and is often built upon the soil itself at the foot of some overshadowing bush. The materials of which it is composed are generally leaves, grasses, and moss, and the interior is lined with a warmer bed of soft feathers. The eggs are five or six in number, and their colour is whitish grey, speckled with a few spots of dark red.

The general colour of this bird is ashen brown upon the upper parts of the body, the quill-feathers being, as is usual, of a darker hue. The whole under surface is white, washed with yellowish-brown, and the under wing-coverts are of a fine soft yellow hue. The beak and eyes are brown. The entire length of this pretty little bird is rather less than five inches.

CHIFF-CHAFF, OR LESSER PETTICHAPS.—*Sylvia rufa.*   WILLOW WARBLER.—*Sylvia trochilus.*

ANOTHER interesting member of this large genus is the WILLOW WARBLER, WILLOW WREN, or YELLOW WREN, its various names being derived from the localities which it frequents and the colour of its feathers.

The habits of this bird are very like those of the Whitethroat, and it feeds on much the same kind of food, preferring insects to any other diet, and seldom if ever invading the fruit trees. It generally arrives in England about the middle of April, when its cheery song may be heard enlivening the hedgerows and copses, sometimes being poured forth while the bird is on the wing, but generally from some elevated branch. The nest of the Willow Wren is like that of the chiff-chaff, and is generally placed upon the ground. The eggs are from five to seven in number, and their colour is white-grey, dotted with numerous spots of pinky-red. The young birds are hatched in May or June. In Mr. Yarrell's work on the British birds, there is a very interesting account of the attachment displayed by the Willow Wren to its nest. The narrator is a lady, and the anecdote appeared originally in the "Field Naturalist."

"In the spring of 1832, walking through an orchard, I was attracted by something on the ground in the form of a large ball, and composed of dried grass. I took it up in my hands, and upon examination found that it was a domed nest of the Willow Wren. Concerned at my precipitation, I put it down again as near the same place as I could suppose, but with very little hope that the architect would ever claim it again after such an attack. I was, however, agreeably surprised to find next day that the little occupier was proceeding with its work. The feathers inside were increased, as I could perceive by the alteration in colour.

In a few days two eggs were laid, and I thought my little *protégée* safe from harm, when a flock of ducks, that had strayed from the poultry-yard, with their usual curiosity went straight to the nest, which was very conspicuous as the grass was not high enough to conceal it, and with their bills spread it quite open, displaced the eggs and made the nest a complete ruin. I now despaired, but immediately on driving the authors of the mischief away, I tried to restore the nest to something like its proper form, and placed the eggs inside. That same day I was astonished to find an addition of another egg, and in about a week four more. The bird sat, and ultimately brought out seven young ones, but I cannot help supposing it a singular instance of attachment and confidence after being twice so rudely disturbed."

The general colour of this bird is dull olive green on the upper parts of the body, the quill-feathers of the wing being brown roped with green, and those of the tail dark brown. The chin, throat, and breast are yellowish-white, and the abdomen is nearly pure white. The total length of the bird is about five inches.

The LESSER WHITETHROAT or BRAKE WARBLER is another of our British Warblers, arriving in this country towards the end of April.

It is not nearly so pleasing or so brilliant a songster as the species which have already been mentioned, but it is nevertheless very liberal in the exercise of its vocal powers, and chatters merrily as it flutters about the hedgerows, dives into the branches and reappears on the branch of some convenient tree or bush. It is not so often seen as the common Whitethroat, being more shy, and venturing less frequently from its hiding-place. On account of this habit of chattering, it has been termed the Chatterer by several writers. The nest is something like that of the common Whitethroat, being an open, saucer-like structure, placed among brushwood, and generally found upon some thick bramble or stunted bush. The materials of which it is composed are grasses, roots, and fibres, and it is generally lined with hair.

The colour of this species is dark grey upon the upper surface, and the quill-feathers are dark brown except the two exterior feathers of the tail, which are greyish-white. The under surface is beautifully white, with a slight but perceptible tinge of pale carmine on the abdomen. The total length of the bird rather exceeds five inches.

FROM the willow Warbler, the WOOD WARBLER is distinguished by the green hue of the upper part of the body, the pure white of the under surface, and the light yellow streak over the eye. There are also other distinctions which are of no very great importance, as the difference between the two species, which are in other respects very similar, is readily decided by the above-mentioned characteristics.

This bird is rather late in arriving in our country, seldom making its appearance before the end of April, and towards the northern parts of England being a week or ten days later. It remains with us until September, when it takes its departure for a warmer clime. Immediately on its arrival it commences its lively, though not much varied song, and perching upon a branch pours forth the trilling notes in rapid succession and with the greatest earnestness. Three or four of these birds will often sing against each other, their throats vibrating, their wings shivering, and their whole bodies panting with eager rivalry. The song of the Wood Warbler resembles the word "twee-ee" very much lengthened, and as it closes its song, it changes the last note into a peculiar hissing or whistling sound, drooping its wings at the same time and agitating them in accordance with its notes. Sometimes the bird utters its trilling notes while flying from one tree to another.

The Wood Warbler is elegant in shape, being light and active and possessed of much command of wing, a qualification which is necessary for the procuring of its food. This bird does not feed on fruits or berries, but restricts itself to insects, especially when they are in the caterpillar state. The leaf-rolling caterpillars are its principal food, and of these insatiable devourers it destroys vast quantities before they can do much harm. In obtaining them it flits round the trees, and is able to snap up the caterpillars as they hang by the slender thread to which they always attach themselves when alarmed. Sometimes the Wood Warbler chases the insects on the wing, and in this manner destroys great numbers of the green oak moth that have escaped from its beak while they were in the caterpillar condition.

The nest of this species is placed on the ground under the shelter of thick herbage or an overhanging bush, and a domed structure composed principally of long dried grass, leaves, fibres, and moss. The entrance is by a hole at the side, and the interior is lined with hair and fine fibres, feathers not appearing to be employed for this purpose. The number of eggs is from four to six, and their colour is greyish-white, profusely sprinkled with dark red and ash-coloured spots, gathered most thickly into a belt round the larger end. The general colour of this species is soft green-grey on the upper parts of the body, and pure white below, the latter characteristic having earned for the bird the local name of "linty-white." A streak of bright yellow passes over the eyes, and reaches as far as the chin and the sides of the neck. In total length the Wood Warbler rather exceeds five inches.

LESSER WHITETHROAT.—*Sylvia curruca.*
WOOD WARBLER.—*Sylvia sibilátrix.*

WITH the exception of the nightingale, the BLACKCAP WARBLER is the sweetest and richest of all the British song-birds, and in many points the voice of the Blackcap is even superior to that of the far-famed Philomel.

The Blackcap derives its name from the tuft of dark feathers which crown the head, and which in the males are coal black, but in the females are deep reddish-brown. It is rather late in arriving, seldom being seen or heard until the end of April, and it remains with us until the middle of September. As several specimens of this pretty bird have been noticed in England in the months of December and January, it is probable that some individuals may not migrate at all, but remain in this country throughout the entire winter. Should it do so, it might easily escape notice, as it would not be likely to sing much during the cold months, and owing to its retiring habits it is at all times more likely to be heard than seen.

While singing, the Blackcap chooses some spot where it can conceal itself if alarmed, and there pours forth his melodious notes in security. Sometimes he will sing while perched upon an open branch, but he is very jealous of spectators, and if he fancies himself visible, immediately drops among the foliage and is lost to sight. The song of this bird is well described by Mr. Mudie in the following words :—

" Its song is generally given from a high perch or an elevated branch, on the top twig if the tree be not very lofty. While it sings, the axis of the body is very oblique by the elevation of the head, and the throat is much inflated. While the bird is trilling, in

BLACKCAP WARBLER.—*Sylvia atricapilla.*

which it excels every songster of the grove in rapidity and clearness, and in the swells and cadences which it gives to the same trill, the throat has a very convulsive motion, and the whole bird appears to be worked into a high state of excitement. It has indeed the mildest and most witching notes of all our warblers; it has not certainly, the volume and variety of the nightingale, neither has it the ineffably sweet chant of the garden warbler; but its notes take one by surprise, and the changes and especially the trills are finer than those of any other bird.

The song, when the bird is at rest, appears to be by turns like those of several birds; but it transposes them into a lower, or rather a minor key, and finishes off with variations of its own; and, as is the case with the works of some of the more impassioned musical composers, the very genius (so to speak) of the bird interferes with the melody, and a sort of indescribable wildness is the character of the whole."

The Blackcap has often been known to become a mimic of other birds, and will frequently spoil its own exquisite notes by introducing imitations of the surrounding songsters.

The food of the Blackcap consists chiefly of insects, but it also pays attention to the ripe fruit in the autumn, being especially fond of raspberries. Perhaps it may choose this fruit on account of the little white maggots that are so often found in the centre of the over-ripe raspberry. When in captivity it can be easily tamed, and sings well; but the capture and domestication of a free singing bird always appears to me to be so heartless a business, that I can never recommend any one to act in such a manner.

As to the canary and birds of that class, which have never known liberty, and would be quite bewildered if they were to escape into the open fields, not knowing where to obtain food or where to roost, it would be a cruelty not to give them the welcome shelter of their accustomed home, provided that the cage be roomy enough for them to exercise their wings, and they be well supplied with food and water. But to imprison the restless nature of the free wild bird in the midst of its happiness, to take away the power of flight, to remove it from its well-loved woods and fields, to take it away from its expectant mate, and to imprison it within the narrow precincts of a wire cage merely for the purpose of gratifying our ears with its song, is really so refined a piece of selfish barbarity, that I can but enter my strongest protest against it.

The nest of the Blackcap is generally placed only a foot or so above the ground, within the shelter of a dense bush or tuft of rank herbage, and is composed of vegetable fibres and hairs rather loosely put together. The eggs are four or five in number, and are of a pale reddish-brown dappled with a deeper hue of brown. The general colour of the Blackcap is grey, with a wash of dark green upon the upper surface and ashen grey upon the lower surface. The total length of the bird is not quite six inches, its extent of wings nearly nine inches, and its weight not quite half an ounce.

The well-known and far-famed NIGHTINGALE is, happily for us, an inhabitant of England, visiting us about the middle of April and remaining until the breeding season is over.

It seems to be rather a local bird, some parts of England appearing to be quite unsuited to its habits. The northern counties are seldom visited by this bird, and in Ireland and Scotland it is almost unknown. Attempts have been made to introduce the Nightingale into different parts of England by substituting its eggs for those of robins and other small birds, but although the young were regularly hatched and fledged, they all retired at the usual season and never came back again. Perhaps, however, it really inhabits localities in which it has been said not to exist, and is not discovered simply because no one has taken the trouble to observe it. For example: Devonshire and certain parts of Wales are said to be among the localities which are unvisited by the Nightingale, but several letters have appeared in the *Field* newspaper, wherein the constant presence of the bird in both these places is distinctly proved. It is at all times a hazardous matter to make an unconditional assertion in the negative.

It is very probable that the quality and drainage of the soil and the character of the cultivation may have some influence upon the Nightingale, for it is well known that certain singing birds which had previously been excluded from the northern parts of England have advanced northward together with cultivation, and the Nightingale may probably follow their example.

The food of the Nightingale consists principally of various insects, and it is so powerfully attracted by the common mealworm, that one of these creatures employed as a bait is sure to attract the bird to its destruction. It appears to make great havoc among the caterpillars, which come out to feed at night, and are to be seen so abundantly on damp warm evenings. In the autumn it is somewhat of a fruit-eater, and has been seen in the act of eating "black-heart" cherries, plucking them from the tree and carrying them to its young. In captivity it is best fed upon mealworms, raw beef scraped with a knife and given very fresh, hard-boiled egg and water, all mixed into a kind of paste. The idea, however, of caging a Nightingale, seems so barbarous, that I shall say nothing more on that subject.

As is well known, the song of the Nightingale is almost wholly uttered in the evening, but the bird may sometimes be heard in full song throughout the day. Towards the end of June, when the young birds are hatched, the song changes into a kind of rough croaking

NIGHTINGALE.—*Luscinia Philomela.*

sound, which is uttered by way of warning, and accompanied with a sharp snapping sound of the beak. The time when the Nightingales sing loudest and most constantly is during the week or two after their arrival, for they are then engaged in attracting their mates, and sing in fierce rivalry of each other, hoping to fascinate their brides by the splendour of their voices. When once the bird has procured a partner, he becomes deeply attached to her, and if he should be captured, soon pines away and dies, full of sorrowful remembrances. The bird dealers are therefore anxious to catch the Nightingale before the first week has elapsed, as they can then, by dint of care and attention, preserve the bird in full song to a very late period. Mr. Yarrell mentions an instance where a caged Nightingale sang upon an hundred and fourteen successive days.

The nest of the Nightingale is always placed upon or very near the ground, and is generally carefully hidden beneath heavy foliage. One such nest that I discovered in Wiltshire was placed among the knotted and gnarled roots of an old ivy-covered thorn stump that still maintained its place within a yard of a footpath. The nest is made of grass and leaves, and is of exceedingly slight construction, so slight, indeed, that to remove it without damage is a very difficult process, and requires the careful use of the hands. The eggs are generally four and sometimes five in number, and are of a peculiar smooth olive-brown, that distinguishes them at once from the egg of any other British bird of the same size.

The colour of the Nightingale is a rich hair-brown upon the upper parts of the body, and greyish-white below, the throat being of a lighter hue than the breast and abdomen. The entire length of the bird rather exceeds six inches.

THE little GRASSHOPPER WARBLER has earned its name by its very peculiar song, which bears a singular resemblance to the cry of the grasshopper or the field cricket. It arrives in England some time in April, according to the weather, and leaves us in September.

Speaking of this bird, Mr. White, the naturalist of Selborne, says: "Nothing can be more amusing than the whisper of this little bird, which seems to be close by, though at a hundred yards' distance; and when close at your ear is scarce louder than when a great way off. Had I not been a little acquainted with insects, and known that the grasshopper kind is not yet hatched, I should have hardly believed but that it had been a *locusta*, whispering in the bushes. The country people laugh when you tell them that it is the note of a bird. It is a most artful creature, skulking in the thickest part of a bush, and will sing at a yard's distance, provided it be concealed. I was obliged to get a person to go on the other side of the hedge where it haunted; and then it would run, creeping like a mouse before us for a hundred yards together, through the bottom of the thorns; yet it would not come into fair sight; but in a morning early, and when undisturbed, it sings on the top of a twig, gaping and shivering with its wings."

I can corroborate this account by personal experience of the bird, and generally found that the country people entirely denied that the strange hissing whistle was that of a bird, and attributed it to the field mouse. The ventriloquial power (if it may so be termed) is as remarkable as in the case of the common grasshopper, for it is almost impossible to ascertain from the sound the distance or even the direction of the creature which utters it.

GRASSHOPPER WARBLER.—*Calamodyta locustella.*

The nest of the Grasshopper Warbler is cup-shaped, and made of various kinds of grasses, the coarser being woven round the circumference, and the finer placed in the centre. It is so admirably hidden that it is discovered less frequently than that of any other warbler. In all my bird-hunting days, I was never fortunate enough to secure an egg of the Grasshopper Warbler, although the bird was far from uncommon. A large patch of furze is a favourite locality for the nest, and the bird hides it so ingeniously among the thick roots of the prickly shelter, that even when the bird is watched to its home, its discovery is a matter of very great difficulty. The eggs are from five to seven in number, and their colour is reddish-white, speckled with dark red spots.

The general colour of the Grasshopper Warbler is greenish-brown, each feather being brown in the centre and green at the edges, so that its whole aspect presents rather a spotty or mottled appearance. The under surface is pale brown, diversified with some dark spots on the neck and breast. The total length of the bird is about five inches and a half.

THE generic title of Calamodyta, which has been given to the grasshopper Warbler and the SEDGE WARBLER, signifies a diver into reeds, and has been attributed to these birds in consequence of their habit of diving abruptly among the herbage whenever they are alarmed.

The Sedge Warbler arrives in this country about the same time as the last-mentioned species, and immediately repairs to the low-lying spots where it can find that peculiar herbage that grows near water. Sedges, reeds, rushes and willows are its favourite resorts, and upon the branches of the last-mentioned tree this Warbler may be observed, on the rare occasions when it deigns to present itself in full view. In such localities it conceals itself most effectually, and although it pours forth its pleasant song with great fluency, prefers to remain secluded in the thick foliage of its home. On one or two occasions, while sitting in a boat drawn among the thick reeds that are found in rivers, I have both seen and heard this interesting little creature, and noticed that it seldom shows itself within six or seven inches of the reed tops. By remaining perfectly quiet, a careful observer may note the peculiar fitful movements of the Sedge Warbler, as it dives among the reeds, and ever and anon shows itself in some small open space. only to disappear the next moment.

The song of the Sedge Warbler is not powerful, but is very constantly uttered. It may be heard to the best advantage in the early morning and the dusk of the evening,

SEDGE WARBLER.—*Calamodyta phragmitis.*

and, like that of the nightingale, is often prolonged far into the hours of darkness. The strain is quick, and has a peculiar guttural sound that is quite indescribable in words.

The structure of the nest and its position are extremely variable, according to the locality in which the bird dwells. Generally the nest is composed of moss and various fibres, the finest being always worked into the centre so as to form a warm bed for the nest and young. It is always placed under cover, sometimes being protected by a bunch of reeds drawn together, sometimes built in the midst of some thick bush, and sometimes overshadowed by a tuft of rank aquatic herbage. The eggs are from four to six in number, and their colour is a very light yellow-brown, dappled with a darker hue.

The general colour of this bird is brown of various shades above, pure white on the chin and throat, and buff upon the breast and abdomen. Its total length is rather under five inches.

THE FAN-TAIL WARBLER, which has been referred by Mr. Gould to the genus Salicaria, is a most interesting little bird, and deserving of our best admiration, not only for the elegance of its form and delicate beauty of its colouring, but for the wonderful skill which it displays in the formation of its nest.

The cradle in which is laid the nest of the Fan-tail Warbler is most ingeniously constructed from the living reeds among which the bird loves to make its residence. As it is so minute a creature, it is unable to make use of the thick and sturdy stems, but employs the flat leaf-blades and the smaller grasses in its architectural designs. Each leaf is pierced by the bill, drawn closely to another blade, and secured to it by means of a cottony thread which is passed through the perforation, and secured at each stitch by a knot so elaborately tied, that, in the words of Mr. Gould, "it appears the work of reason." The nest itself is composed of various soft and downy fibres, which are collected from different plants. In this wonderfully constructed nest the Fan-tail Warbler lays four or five eggs of a "bluish-flesh" colour.

The popular name of Fan-tail has been given to this bird on account of the peculiar shape of the expanded tail, which is exhibited by the bird whenever it sets itself in motion. It is a lively little bird, popping in and out of the foliage in a very wren-like manner, spreading and flirting its beautifully loquacious little tail while it darts from spot to spot, as the Spanish ladies flirt their love-speaking fans. When observed, it takes to its wings, and will fly to a considerable distance before alighting. It is a native of the Mediterranean shores, and is found along the northern and eastern parts of Europe, and the adjoining portions of Africa and Asia. At Gibraltar it is found in plenty, enlivening the bushes with its quick and active movements, and its shrill merry notes.

The colour of this pretty little bird is a warm chestnut-brown, each feather being marked with a dark strip running down its centre. The under surface is white with a brown wash, and the tail is brownish-black, each feather being graduated so as to give it the appearance which is presented in the engraving. Each tail-feather is tipped with white, presenting an agreeable contrast

FAN-TAIL WARBLER.—*Salicaria Cisticola.*

to the darker hues of the basal portions. The total length of this bird is about five inches.

A SMALL but very interesting group of birds now claims our attention. These are the *Erythacinæ*, or Redbreast kind, including the Redbreast, the Wheatear, the Chats, the Redstart, and other similar birds.

2.   U

The WHEATEAR, or FALLOW CHAT, is a well-known visitant of the British Isles, and on account of the delicate flavour of its flesh when fat, is sadly persecuted throughout the whole time of its sojourn.

Being in great favour for the table, where it is popularly known as the English ortolan, and consequently fetching a good price in the market, it is caught in great numbers, and sold to the game-dealers of London. The trap by which it is captured is a remarkably simple affair, consisting merely of an oblong piece of turf cut from the soil, and arranged crosswise over the cavity from which it was taken. A horsehair noose is supported under the turf by means of a stick, and the trap is complete, needing no bait or supervision. It is the nature of the Wheatear to run under shelter at the least alarm ; a passing cloud sufficing to drive it under a stone or into a hole in a bank. Seeing, therefore, the sheltering turf, the Wheatear runs beneath it, and is caught in the noose. These simple traps are much used by the shepherds, who can make and attend to four or five hundred in a day, and have been known to catch upwards of a thousand Wheatears within twenty-four hours.

In the northern parts of England, the Wheatear is equally persecuted, but from super-stitious motives ; the ignorant countrymen imagining that its presence foretells the death of the spectator. In order, therefore, to avert so sad an omen, they kill the bird and destroy its eggs on every opportunity.

The chief reason for this absurd practice is, that the Wheatear is in the habit of frequenting any locality where it can find shelter for its eggs and young, and, therefore, may often be found amid old ruins, in burial-grounds, or cairns. " Though it is a very handsome bird," says Mudie, " and in the early season sings sweetly, its haunts have gotten it a bad name. Its common clear note is not unlike the sound made in breaking stones with a hammer ; and as it utters that note from the top of the heap which haply covers the bones of one who perished by the storms or by his own hand ; or from the mound, beneath which there lie

WHEATEAR.—*Saxicola œnanthë.*

the slain of a battle-field, magnified through the mist of years ; or from the rude wall that fences in many generations, it is no very unnatural stretch to the pondering fancy, which dwells in these parts, to associate the Wheatear with all the superstitions that unphilosophically, but not irreverently, belong to the place of graves.

It comes around, too, to meet the traveller, and now running, now flying, seems to pilot him to a place beside the cairn, as if his own bones were soon to be gathered there ; and in that, its note of solemn warning, it is more than usually energetic ; it is seen in the fog too, and, from the contrast of its colours, it is particularly conspicuous even in that. In a highland glen, during a highland mist (which wets but warms you), you hear the *Clacheran* before you see it ; you meet the *Clacheran* before you see the cairn ; so you are at perfect liberty to believe that it is busy breaking the stones that are to cover you—if you choose.

. . . . Beneath that heap of stones there is a little nest, formed of moss and grass, and completely lined with hair, feathers, or wool, with five or six eggs of a delicate bluish-white ; and there is a mate, of whom and her promised brood the bird is as fond, and for them he has as much forsaken the society of other birds, as the most exemplary of the human race could possibly do. He watches early and late, and endeavours to divert any one that passes from the retreat of his charge. He renders it unnecessary for the dam to leave the eggs when they might be injured by the cold and damp, for he fears her moving and roving while the air is cold ; but in order that she may not suffer by the experiment, he takes her place a short while during the warmth of the day, while she exercises her feet and wings a little, and finds a snail, perhaps, from the all-supplying earth.

Yet, for these assiduities to his brood, the Wheatear has been made in the northern parts of Britain, and in places still farther to the north, the victim of superstition. Old and young continue to kill and persecute these birds, and to destroy their eggs, considering the service as one of more than ordinary merit."

As a general rule, the nest of the Wheatear is hidden in the most perfect manner, the bird ordinarily choosing to place its domicile within the recesses of large stone heaps, in deep rocky crannies, and in similar localities ; so that, even if it should be discovered, the work of obtaining it is very severe. In some parts of the cliff-bound sea-coast, the Wheatear's nest is so deeply buried in the rocky crevices, that the only mode of obtaining the eggs is to hook out the nest by means of a bent wire at the end of a long stick. In my early nest-hunting expeditions I used to obtain such deeply hidden eggs by putting a little bird-lime on the end of a fishing-rod, twisting it carefully upon each egg severally, and drawing it out of the hole before it could be disengaged from the tenacious substance by which it was held.

Mr. Yarrell mentions that the Wheatear is not always so cautious, but that it will often make its nest in the interior of rabbit burrows, at no great distance from the opening. It is a curious circumstance that the Wheatear has a habit of collecting little stalks of the common brake, and permitting them to project from the hole in which the nest is built, so that it leaves a sign by which the presence of the nest may be discovered. The eggs of the Wheatear are rather rounded in form, and their colour is of the palest and most delicate blue, in which a few dark spots may be discovered by a close investigation. Their number is from four to six.

The colouring of the Wheatear is bold and pleasing. The upper part of the body is light silver-grey, and the quill-feathers of the wings, together with their coverts, are deep black. The middle tail-feathers and the tips of the various rectrices are of the same hue, and a black streak passes from the edge of the beak to the ear, enveloping the eye, and spreading widely upon the ear-coverts. The breast is buff, with a decided orange tinge, and the abdomen is beautifully white. The female is not quite so handsome ; the wings, tail, and ear-coverts being dark brown, and the lighter portions of the body tinged with brown. The total length of the bird is about six inches and a half.

The STONECHAT is one of the birds that remain in England throughout the year, being seen during the winter months among the furze-covered commons which are now rapidly becoming extinct.

The name of Chat is earned by the bird in consequence of its extreme volubility, for it is one of the noisiest birds in existence. Its song is low and sweet, and may be heard to great advantage, as the bird is not at all shy, and, trusting to its powers of concealment, sings merrily until the spectator has approached within a short distance, and then, dropping among the furze, glides quickly through the prickly maze, and rises at some distance, ready to renew its little song. It is a lively bird, ever on the move, flitting from place to place with restless activity, and ever and anon uttering its sweet strains. Even in the winter months the Stonechat will make itself audible as it flutters about the furze-grown spots in which it loves to live. It is in these localities that it finds its supply of winter food, for the thick furze-bushes afford shelter to various worms and insects, and the little Chat is able to procure a plentiful meal by digging in the damp ground.

It is rather difficult to force a Stonechat to leave its shelter, and a shower of missiles generally has the effect of making it keep closer within its concealment. This little bird has the faculty of making a patch of furze very lively, for it pops in and out in a quick

WHINCHAT.—*Pratincola rubétra.*   STONECHAT.—*Pratincola rubicola.*

cheerful fashion, twitters its pretty song, dives among the rich golden-crowned bushes, and reappears with a toss of the head, and a flirt of the wing, as if exulting in the exuberance of its happiness. In the winter, the same bushes afford it concealment and shelter, for the heavy masses of snow only rest upon the furze, and leave abundant open space beneath, in which the little bird has perfect freedom of movement, and under which it is sure to find worms and insects buried in the soil. Snails, slugs, and other similar creatures always retire for the winter into sheltered spots, and they form many a meal for the Stonechat. Plantations, especially those that are made of young pines or firs, are favourite haunts of the Stonechat; but as the branches are all at some distance from the ground, the bird seldom, if ever, attempts to build its nest under their shade.

The Stonechat resembles the flycatcher in some of its habits, especially in its custom of feeding on an elevated twig, the top of a post, or the highest pebble of a stone-heap, and catching the flies as they pass by its perch. Unlike the flycatcher, however, it does not make choice of one perch, and return to it day after day; but after catching six or seven flies upon one spot, flutters to another, and from that to a third, ever changing its position from time to time.

The nest of the Stonechat is made of mosses, grass of different kinds, and is lined with fine fibres, hairs, and feathers. The number of the eggs is from four to six, and their colour is very pale blue, diversified with numerous minute spots of reddish-brown upon the large end of the shell. The colours of the Stonechat are rather pretty. The head, the neck, the chin, throat, back, and tail, are deep sooty-black, contrasting boldly with the pure white of the tertial wing-coverts, the upper tail-coverts, and the sides of the neck. The remaining wing-coverts are deep brown, and the quill-feathers of the wings are also brown. The breast is chestnut, and the abdomen yellowish-white. The total length of the bird is rather more than five inches.

THE bird which occupies the left-hand of the illustration is called the WHINCHAT, on account of its fondness for the furze or whin. The stonechat has, however, quite as much right to the title, as it frequents the furze as constantly as the Whinchat.

This species may be easily distinguished from the preceding, by the long and bold white streak which passes across the sides of the head, and the absence of white upon the wing-coverts. It is a migratory bird, although it has on one or two occasions been seen in England during the winter. In many of its habits it resembles the stonechat, and might readily be mistaken for that bird by any one who was not well acquainted with the two species. It sings rather constantly, uttering its sweet strains while on the wing, or while perched on some elevated bough. Mr. Yarrell mentions that it is fond of imitating the notes of other birds, and that a caged Whinchat has been heard to mimic the whitethroat, the redstart, willow warbler, missel thrush, and nightingale. The same bird would frequently sing at night.

Like the wheatear, the Whinchat becomes extremely fat in the autumn, and as it is prized as a delicacy for the table, is rather persecuted by the game-dealers and their emissaries. The food of this bird is the same as that of the stonechat. The Whinchat arrives in this country about the middle or towards the end of April, according to the locality and the weather. It builds its nest soon after its arrival, and hatches its young about the end of May or the beginning of June. The nest is placed on the ground, is made after the fashion of the stonechat's habitation, and contains from four to six bluish-green eggs, slightly speckled with reddish-brown. Two broods are hatched in the course of the year.

The colouring of the Whinchat is as follows : The top of the head, the neck, and the back are mottled brown, each feather being lighter at the circumference than in the centre. An irregular broad brown streak extends from the angle of the mouth to the back of the neck, and above the eye is a long and rather wide streak of white. Another white stripe passes immediately below the dark-brown streak, and extends from the chin almost to the shoulder. The tail is white upon the base, and brown at the tip, each feather being edged with a lighter shade of the same hue. The chin is white, the throat and chest are pale fawn, and the abdomen is buff. The length of the bird is not quite five inches.

THE specific title of phœnicura, which is given to the REDSTART, signifies ruddy-tail, and is attributed to the bird in consequence of the light ruddy-chestnut feathers of the tail and upper tail-coverts.

It is a handsomely coloured and elegantly shaped bird, and is a great ornament to our fields and hedgerows. The name of Redstart is a very appropriate one, and has been given to the bird in allusion to the peculiar character of its flight. While walking quietly along the hedgerows, the observer may often see a bird flash suddenly out of the leafage, flirt its tail in the air, displaying strongly a bright gleam of ruddy hue, and after a sharp dash of a few yards, turn into the hedge again with as much suddenness as it had displayed in its exit. These manœuvres it will repeat frequently, always keeping well in front, and at last it will quietly slip through the hedge, double back on the opposite side, and return to the spot from whence it had started.

No one need fancy, from seeing the bird in the hedge, that its nest is in close proximity, for the Redstart seldom builds in such localities, only haunting them for the sake of obtaining food for its young. The nest is almost invariably built in the hole of an old wall, in a crevice of rock, a heap of large stones, in a hollow tree, or in very thick ivy. I have known this bird to make its nest in quite a small hole in a wall of Merton College, Oxford ; the nest looking out upon a passage, and being within five feet of the ground. The eggs are generally five in number, although they vary from four to seven, and are of a beautiful blue, with a slight tinge of green. They are not unlike those of the common hedge sparrow, but are shorter and of a different contour.

The Redstart has a very sweet song, which, although not very powerful, is soft and melodious, bearing some resemblance to that of the nightingale. The bird has a habit of sitting on the top of a wall or some elevated spot, and there pouring forth his song, looking about in every direction, as if inviting a challenge, and spreading and closing his tail at intervals. Presently, without ceasing the song, he will dart off to another spot, in one of the short uncertain flights which characterise the species, and settling upon some fresh perch, sing with new vigour. It often happens that in the breeding season the Redstart continues to sing far into the night, and recommences at the earliest approach of dawn.

REDSTART.—*Ruticilla phœnicùra.*

The food of the Redstart is mostly of an insect nature, and is obtained in various ways. Sometimes the bird dashes from its perch upon a passing insect, after the manner of the flycatcher; sometimes it chases beetles, and other creeping insects upon the leaves and branches of the hedges; sometimes it hunts for worms, grubs, and snails from the ground; and it often picks maggots out of fungi, decaying wood, mosses, and lichens. Soft ripe fruit is also eaten by the Redstart, which, however, ought to be allowed its free range of the garden in recompense for the great service which it has performed in the earlier portion of the year, by devouring the myriad insects that feed upon the blossoms of fruit-trees. The softer berries form part of the Redstart's diet, but the bird does not seem to care about the hard seeds.

The colouring of the Redstart is as follows: The top of the head, the neck, and the back are bluish grey, contrasting finely with the jetty black of the chin, the throat, the face, and the sides of the neck. The wings are rich brown, slightly streaked with a lighter shade of the same hue, and the upper tail-coverts and all the tail-feathers are bright ruddy chestnut, with the exception of the two central tail-feathers, which are striped with the same hue as the wing-feathers. The breast and all the lower surface is very pale chestnut, and the forehead is white. The length of the bird is more than five inches.

ANOTHER species of the same genus is occasionally, though very rarely, seen in England. This is the Black Redstart (*Ruticilla Tithys*), and is readily distinguished from the common species by the sooty black hue of its breast and abdomen. This bird resembles the common Redstart in many of its habits, but is seldom seen on open ground. At a meeting of the British Association a curious anecdote was related of this bird, which well exemplified the force of parental affection.

A railway carriage had been left for some weeks out of use in the station at Giessen, Hesse Darmstadt, in the month of May, 1852, and when the superintendent came to examine the carriage, he found that a Black Redstart had built her nest upon the collision spring; he very humanely retained the carriage in its shed until its use was imperatively demanded, and at last attached it to the train, which ran to Frankfort-on-the-Maine, a distance of nearly forty miles. It remained at Frankfort for thirty-six hours, and was then brought back to Giessen, and after one or two short journeys, came back again to rest at Giessen, after a period of four days. The young birds were by this time partly fledged, and finding that the parent bird had not deserted her offspring, the superintendent carefully removed the nest to a place of safety, whither the parent soon followed. The young were, in process of time, full fledged, and left the nest to shift for themselves. It is evident that one at least of the parent birds must have accompanied the nest in all its journeys, for, putting aside the difficulty which must have been experienced by the parents in watching for every carriage that arrived at Giessen, the nestlings would have perished from hunger during their stay at Frankfort, for every one who has reared young birds is perfectly aware that they need food every two hours. Moreover, the guard of the train repeatedly saw a red-tailed bird flying about that part of the carriage on which the nest was placed.

THERE are few birds which are more familiar to us than the REDBREAST or ROBIN, a bird which is interwoven among our earliest recollections, through the medium of the Children in the Wood, and the mournful ballad of the Death and Burial of Cock Robin.

REDBREAST.—*Erythacus rubecula.*

Although the Redbreast remains in England throughout the winter, it is very susceptible to cold, and one of the first birds to seek for shelter, its appearance among the outhouses being always an indication of coming inclemency. In cold weather, the Redbreast seldom perches upon twigs and branches, but crouches in holes, or sits upon the ground. The bird seems strongly attached to man and his home, and will follow the ploughman over the fields, picking up the worms which he turns up with the ploughshare, or enter his house and partake of his evening meal. Both bold and shy, the Redbreast is a most engaging bird, and seldom fails of receiving the affection of those to whom he attaches himself. One of these birds was exceedingly familiar with all our family, his acquaintance having commenced through the medium of some crumbs from our hands, and would always come to us whenever we called his name, "Bobby." Sometimes he would accompany us on our way to church through the lanes, and I have even seen him keeping pace with us along the one-sided street of Oxford, that is appropriately named Long Wall.

Bread and butter is a very favourite dainty with the Robin, who has often been known to come uninvited, and to peck from the table. "Butter," according to Mr. Thompson, "is so great a dainty to these birds, that in a friend's house, frequented during the winter by one or two of them, the servant was obliged to be very careful in keeping what was in her charge covered, to save it from destruction; if unprotected, it was certain to be eaten. I have known them to visit labourers at breakfast-time to eat butter from their hands, and enter a lantern to feast on the candle. One, as I have been assured, is in the constant habit of entering a house in a tanyard in Belfast, by the window, that it might feed upon tallow, when the men were using this substance in the preparation of hides. But even further than this, I have seen the Redbreast exhibit its partiality for scraps of fat, &c.

Being present one day in December, 1837, when a golden eagle was fed, a Robin, to my surprise, took the eagle's place on the perch the moment that he descended to the ground to eat some food given him, and when there, picked off some little fragments of fat or scraps of flesh; this done, it quite unconcernedly alighted on the chain by which the 'rapacious' bird was fastened.

I at the same time learned that this Robin regularly visited the eagle's abode at feeding-time, though as yet there was no severity of weather. Although the Robin escaped the golden eagle unscathed, as much cannot be said for one which occasionally entered the kitchen at the Falls, and sang there; having one day alighted on a cage in which a toucan was kept, this bird with it huge bill seized and devoured it." Another Robin, mentioned by the same author, was in the habit of attending on a carpenter, stealing the shavings as materials for his nest, and making very free with his grease-pot, pecking from it while in his hand.

The Robin is also remarkably fond of bread and butter on which honey or sugar has been spread, and will eat of this dainty until it is hardly able to fly. One of these birds who had been treated to such a repast, was so pleased with it that he returned, bringing with him three companions, who gorged themselves to such a degree, that they were taken up by hand, and put away for the night into a comfortable recess. After a while, between twenty and thirty Robins came to the house in hopes of obtaining the sweet food. Perhaps they may be instinctively led to sugar and fatty substances, as a means of preserving themselves against the effects of cold. Cream is in great favour with the birds during the winter months, and they have been seen to enter an outhouse which was employed for washing purposes, and to eat the soap.

The Redbreast is a most combative bird, fighting its own species with singular energy, and often killing its opponent. One of these birds killed upwards of twenty of its own kind, merely because they came into a greenhouse which he chose to arrogate to himself. It is very jealous, too, of its human friends, and not only prohibits other individuals from sharing in the friendship, but will often drive away its own young if they approach too closely.

The nest of this bird is generally placed near the ground in a thick leafy bush, or in a bank, and is composed of dry leaves, moss, grass, hair, and feathers. I have seen the nest very well concealed among the thick ivy that had wreathed round a tree-trunk, and placed about eight feet from the ground. The bird seldom flies directly to its nest, or leaves it directly, but alights at a little distance, and creeps through the leaves or branches until it enters its home. When, however, the Redbreast chooses to become familiar with man, it takes advantage of his friendship, and deposits its nest and eggs under his care. The localities which have been chosen for the Robin's home are diverse beyond description, one of the most curious being the centre of a large cabbage growing in a garden.

The bird has been known to make its nest in a workman's tool-basket hanging against the wall, in a fold of a window-curtain, upon a shelf in a greenhouse, in the side of a saw-pit, in a knot-hole of timber used in a ship which was being built, the birds being not in the least discomposed by the constant hammering of the trenails. Mr. Thompson gives the following quaint account of a Redbreast and its nest.

"At Fort William, the seat of a relation, the following circumstance occurred. In a pantry, the window of which was left open during the day, one of these birds constructed its nest early in the summer. The place selected was the corner of a moderately high shelf, among pickle-bottles, which, being four-sided, gave the nest the singular appearance of a perfect square. It was made of green moss, and lined with a little black hair; on the one side which was exposed to view, and that only, were dead beech-leaves. When any article near the nest was sought for by the housekeeper, the bird, instead of flying out of the window, as might have been expected, alighted on the floor, and waited there patiently until the cause of disturbance was over, when it immediately returned to its nest. Five eggs were laid, which, after having been incubated without success for the long period of about five weeks, were forsaken.

The room above this pantry was occupied as a bird-stuffing apartment; after the

Redbreast had deserted the lower story, a bird of this species—doubtless the same individual—visited it daily, and was as often expelled. My friend, finding its expulsion of no avail, for it continued to return, had recourse to a novel and rather comical expedient. Having a short time before received a collection of stuffed Asiatic quadrupeds, he selected the most fierce-looking carnivora, and placed them at the open window, which they nearly filled up, hoping that their formidable aspect might deter the bird from future ingress. It was not, however, to be so frightened 'from its propriety,' but made its *entrée* as usual. The walls of the room, the tables in it, and nearly the entire floor, were occupied by these stuffed quadrupeds.

The perseverance of the Robin was at length rewarded by a free permission to have its own way, when, as if in defiance of the *ruse* that was practised against it, the place chosen for the nest was the head of a shark which hung on the wall (the mouth being gagged may have prevented it being the site); while the tail, &c. of an alligator stuffed, served to screen it from observation. During the operation of forming this nest, the Redbreast did not in the least regard the presence of my friend; but both man and bird worked away within a few feet of each other. On the 1st of June I saw it seated on the eggs, which were five in number; they were all productive, and the whole brood in due time escaped in safety."

The eggs of the Redbreast are generally five in number, as is the case with most of the song birds, and their colour is greyish white, covered with variously sized spots of pale rusty red. The song of this bird is very sweet and pleasing; and it is a pretty sight to observe two or more Redbreasts perched on different trees, and answering each other with their musical cries. Whenever the Redbreast perches on the top of a tree or other elevated spot, and begins to sing merrily, it is an unfailing indication that the weather of the coming day promises to be fair. The bird sings throughout the greater part of the year, beginning early in spring, and continuing it very late into the autumn. Even in the winter months, a bright sunny day is apt to excite the Robin to perch upon a twig, and pour forth a sweet though broken melody.

While hopping and feeding about the ground, it is wonderful to see what large worms and insects the little bird will devour. Should the worm be too large for him to swallow entire, as indeed is mostly the case, he tosses it about with his beak, bangs it against the ground, flings it over his head, jumps on it, and when he has thus mashed it into a pulp, pulls it to bits, and devours it piecemeal.

The colour of the male Robin is bright olive-brown on the back, orange-red on the throat, chin, breast, forehead, and round the eye. A stripe of blue-grey runs round the red, and the abdomen and lower part of the breast are white. The bill and eyes are black. The female is coloured after the same manner, but the tints are not so vivid as in her mate. The total length of the bird is nearly six inches, and its weight about half an ounce.

The BLUE-THROATED WARBLER is very common in the southern parts of Europe, but is extremely rare in the British Isles, only two or three specimens having been obtained in England.

It is a sweet songster, the notes having some resemblance to those of the whinchat, but being more powerful. It prefers to haunt low-lying, marshy grounds, and places its nest among tufts of the rank herbage that generally grows in such localities. The nest is most carefully hidden, and cannot readily be discovered. The materials of which it is composed are dried grass and mosses, and it is lined with grass of a finer character. The eggs are greenish-blue, something like those of the redstart. The bird has a curious habit of rising into the air while singing, spreading its tail widely, and sailing with quivering wings and spread tail to a spot at some distance from that at which it rose. It begins its song early in the morning, and does not cease until late in the evening, being in this respect similar to the redbreast.

The colour of this bird is rather varied, and is briefly as follows: The upper part of the body is rich brown, a colour which extends to the two central tail-feathers, all the other rectrices being bright chestnut at the basal half, and black at the extremity. The

BLUE-THROATED WARBLER.—*Cyanécula Suécica.*

chin, throat, and upper part of the breast are brilliant blue, excepting a spot in the centre, which is white in young birds, but warms into red as they increase in age. A well-marked black bar runs below the blue, followed by a narrow streak of white, and a rather broad band of the same ruddy chestnut as that of the tail. The abdomen is greyish-white. The total length of the bird is about six inches.

THE birds that belong to the genus Copsychus are spread over several portions of India and Africa, where they are rather plentiful in certain favoured localities.

The DAYAL is an inhabitant of India and Ceylon, and in its wild state is a solitary bird, haunting the lower trees and jungle, and often paying visits to gardens and cultivated grounds. It is rather shy, and prefers the thickest foliage for its perch, never caring to rise to any great elevation if observed, but merely flying from tree to tree at a short distance from the ground. Its food consists of insects, which it generally takes upon the ground, jerking its tail upwards in a satisfied kind of manner, and then returning to its post among the bushes. As it regains its perch, it has a curious habit of depressing its tail, as if to counteract the effect of its former elevation. The song of the Dayal is remarkably good, and it possesses the power of mimicking other birds.

As it is readily tamed, it is often caught and caged, and when domesticated is employed in various *rôles*, the most common being that of a combatant. It is a most brave and combative little bird, and will fight to the death with as much courage as any gamecock. Even in its wild state it is constantly engaged in fighting, the male birds challenging each other just as is the case with the domestic fowl, and joining in combat as soon as they can come to close quarters. The native bird-catchers take advantage of this propensity, and employ a tame male for the purpose of decoying the wild birds into captivity; the whole process being singularly like that which is adopted for the capture of elephants in the same country.

The tame bird, on hearing the accustomed signal from his master, pours forth a defiant challenge, which is immediately answered by the nearest male. The decoy bird is then let loose, and the two immediately engage in fight, so fierce that both the combatants are

DAYAL.—*Copsychus sauldris.*

seized by the bird-catcher. It is a very remarkable fact that the tame bird seems to take a pride in aiding its master, and will hold its struggling antagonist by beak and claws in order to prevent it from making its escape. In Ceylon this bird goes by the name of the Magpie Robin.

THE pretty little BLUE-BIRD of America is deservedly a great favourite in the country which it inhabits, not only for its delicate blue back, red bosom, and sweet song, but from the engaging familiarity of its character.

In many respects the Blue-bird takes the place of the redbreast in the affections of bird-loving persons, and fearlessly associates with mankind, even though it be not driven to such companionship by cold or hunger. It is the harbinger of spring, and makes its appearance as soon as the snow begins to melt away from the surface of the earth, and the soil to loose itself from the icy bonds in which it had been held. Sometimes a few days of sharp frost or heavy snow will drive the Blue-bird to its hiding-place, but it soon emerges when the inclemency of the weather is past, and cheers the face of nature with its light-coloured feathers and sweet rich song. Many persons are in the habit of arranging a box with a hole in the side as a nest-box for the Blue-bird, and the grateful little creature never fails to take advantage of the domicile thus offered to it, and to pour forth its thanks in frequent music.

Although, as a rule, the Blue-bird is not seen except in the spring and summer months, it is evident that some specimens must remain throughout the winter, as even in the depth of the cold season, a few days of sunshine and warmth are sure to witness the presence of two or three Blue-birds that have been tempted by the genial warmth to leave for a while the snowy home in which they have been resting. The habits of this pretty bird are very interesting, and not the least so is the extreme care which it takes of its nest and young, sitting near them and singing its best, and occasionally flying off and returning with a caterpillar or other insect for their benefit.

The Blue-bird builds its nest in the hollows of decaying trees and other similar situations, where the eggs and nest are well sheltered from the rain and cold. The eggs are generally from four to six in number, and their colour is a pale blue. Two broods are generally produced in a single season, and it is not uncommon for the bird to rear a third brood later in the year, should the weather be propitious. The food of this bird consists of various insects, chiefly those of the coleopterous order, spiders, small worms, and in the

BLUE-BIRD.—*Siálía síalís.*

autumn of soft fruits and seeds. The bright, cheerful song of the Blue-bird is heard throughout the greater part of the year, commencing at the end of February or the beginning of March, and not ceasing until the end of October. The spring, however, is the season which is most enlivened by the song.

This species is widely and plentifully spread throughout the greater part of North America, and during the cold weather moves southward towards Brazil, Mexico, Guinea and the neighbouring parts, beginning its migration about November. The total length of the Blue-bird is rather more than seven inches, and its colouring is as follows : The head, back of the neck, and the whole upper surface is of a rich azure with purple reflections, excepting the shafts of the quill-feathers of the wing and tail, which are jetty black, and produce a very pleasing effect when contrasted with the blue. The quill-feathers of the wing are also black at their extremities. The throat, breast, and sides are rich ruddy chestnut, and the abdomen is white. The female is similar to her mate in colouring, but the tints are not so bright.

Of the pretty though sober-plumaged Accentors, we have one or two British examples, that which is best known being the HEDGE ACCENTOR, or HEDGE SPARROW, as it is often, though wrongly, called, as it by no means belongs to the same group of birds.

The Hedge Accentor is very common through the whole of England, and may be heard in the gardens, copses, and hedge-rows, chanting its pleasing and plaintive melody without displaying much fear of its auditors. It seems, indeed, to be actually attracted to man, and, in spite of the terrible havoc which is made year after year by young bird-nesters among its homes, it always draws near to human habitations as soon as the cold days of autumn commence, and may be seen flitting about the barns and outhouses in a perfectly unconcerned manner.

It is especially adapted for living among the hedges, as it possesses a singular facility in threading its way through the twigs, stems, and branches. It seems equally at home in dried brushwood, and may often be seen traversing the interior of a woodpile with perfect ease. The nest is one of the earliest to be built, and as it is frequently completed and the

HEDGE SPARROW, OR ACCENTOR.—*Accentor moduldrius.*

eggs laid before the genial warmth of spring has induced the green leaves to burst their inclosures, it is easily seen, and is the first victim of the neophyte bird-nester, who pounces upon its soft mossy walls and delicate blue eggs with exulting eagerness. The more experienced nester, however, will never touch so easy a prey, caring nothing for eggs which the veriest novice can discover.

The nest is generally placed at a very low elevation, seldom more than two or three feet from the ground, and it is rather large in proportion to the size of the bird. The materials of which the structure is made are various mosses, wool, and hair, and the eggs are usually five in number, of a bright bluish green colour. Sometimes, but very rarely, six eggs are found in a single nest. Bird-nesting boys are not the only foes with whom the hedge sparrow has to contend, for the cuckoo profits by the exposed position of the nest, and lays her eggs in the home of the hedge sparrow more often than in that of any other bird. There are generally two broods of young in the year, and when the nest is robbed, the mother bird often lays three sets of five eggs in the course of the season, of which she may think herself fortunate if she succeeds in rearing one.

The song of the Hedge Accentor is sweet, but not varied nor powerful, and has a peculiar plaintive air about it. The bird is a persevering songster, continuing to sing throughout a large portion of the year, and only ceasing during the time of the ordinary moult. Like many other warbling birds, it possesses considerable powers of imitation, and can mock with some success the greater number of British song-birds.

This bird is nearly as bold as the sparrow, and will sometimes take up its residence in cities, where it soon gains the precociously impertinent airs that characterise all town birds, speedily loses the bright rich brown and grey of its plumage, and assumes as dingy a garb as that of the regular city sparrow.

The colour of the Hedge Accentor is bluish grey, covered with small brown streaks upon the head, and the back and sides of the neck. The back and wings are brown streaked with a deeper tint of the same hue, and the quill-feathers of the wings and tail are of a rather darker brown, and not quite so glossy. The chin, the throat, and upper part of the breast are grey, and the lower part of the breast and the abdomen are white with a wash of pale buff. The legs and toes are brown with a decided orange tinge, and the beak is dark brown. The total length of the bird is nearly six inches.

The ALPINE ACCENTOR (*Accentor Alpinus*) is another British representative of this group.

Several specimens of this bird have been killed in England, but it is an extremely rare visitant to this country, and is hardly entitled to take rank as a true British bird. The countries where it is usually found are Italy, France, Germany, and several other parts of Europe. It is a mountain-loving bird, seldom descending to the level of the plains except during the stormy months of winter. It can readily be distinguished from the ordinary Accentor by the throat, which is white spotted with black, and by the chestnut-black and white streaks upon the wing-coverts. The Alpine Accentor is larger than its British relative, being six inches and a half in total length, and its blue-green eggs are larger than those of that bird.

THE group of birds which are distinguished by the name of Parinæ, or TITMICE, are easily recognisable, having all a kind of family resemblance which guards the observer from mistaking them for any other bird. They are all remarkable for their strong, stout little beaks, the boldly defined colour of the plumage, and the quick irregularity of their movements. They are all insect-eaters, and are remarkably fond of the fat of meat, by means of which, used as a bait, they can often be caught. Their feet and claws, though slight and apparently weak, are really extremely strong, enabling the bird to traverse the boughs with great rapidity, and to cling suspended from the branches.

THE first example of these birds is the GREAT TITMOUSE, an inhabitant of England and many parts of Europe.

It does not migrate, finding a sufficiency of winter food in its native land. During the summer it generally haunts the forests, gardens, or shrubberies, and may be seen hopping and running about the branches of the trees in a most adroit manner, searching for insects, and occasionally stocking them out of their hiding-places by sharp blows of the bill. The beak of the Great Titmouse is, although so small, a very formidable one, for the creature has often been known to set upon the smaller birds, and to kill them by repeated blows on the head, afterwards pulling the skull to pieces, and picking out the brains.

During the winter the Great Titmouse draws near to human habitations, and by foraging among the barns and outhouses, seldom fails in discovering an ample supply of food. Mr. White has recorded a curious instance of the ingenuity displayed by this species while searching for food. " In deep snows I have seen this bird, while it hung with its back downwards (to my no small delight and admiration), draw straws lengthwise from out the eaves of thatched houses, in order to pull out the flies that were concealed between them ; and that in such numbers that they quite defaced the thatch, and gave it a ragged appearance." In very severe winters, the birds will even carry away the barley and oat straws from the ricks.

Mudie, in his " British Birds," writes as follows on the same subject : " When the house flies become languid in the autumn, the Tits capture them in vast numbers, and when insects fail, they make prizes of the autumnal spiders. In the dead season, when insect life is chiefly in the egg, though they hunt for the eggs with great diligence, yet they do not find in them a sufficient supply of food. At that time they pick up nuts, acorns, and the capsular fruit of other trees, hold them in their claws, and hammer away with their bill till the hardest shell or toughest capsule is opened. They also eat the seeds of grasses, especially those that are of an oily nature, and of such size that they can hold them with their foot and pick them open, for they do not grind or bruise with the edge of the bill. They feed greedily upon carrion, and when they come upon other birds in a benumbed and exhausted state, they despatch and eat them, first breaking and emptying the skull.

When the snow lies heavy on the ground, they approach houses, and hunt about for any offal that may come in their way. They sometimes draw straws from cornstacks, but they do that much more rarely than some of the birds which feed more exclusively upon vegetable matters. While engaged in these labours, they continue repeating their grating cry ; but as soon as the weather begins to relent, they resort to the trees, preparatory to the labours of the summer, which, in their case, is no idle time."

The voice of the Great Titmouse is exceedingly variable, at one time softening into a kind of song, and at another becoming harsh and grating, resembling, according to some persons, the sound which is given forth by a saw when being sharpened. It is rather a flexible voice, and there is always a chattering element in it.

The nest of the Great Titmouse is always made in some convenient hollow, generally that of a tree, but often in the holes of old walls, and in the cavities that are formed by thick gnarled roots in the sides of a bank. Hollow trees, however, are the favourite nesting-places of this bird, which is able to shape the hollow to its liking, by chiselling away the decaying wood with its sharp, strong beak. The materials of which the nest is made vary according to the locality. Should the hollow be a deep and warm one, the bird takes very little trouble about the nest, merely bringing a few feathers and mosses as a soft bed on which to place the eggs. If, however, the locality be more exposed, the Titmouse builds a regular nest of moss, hair, and feathers, in which to lay its eggs. There are generally from eight to twelve eggs in each nest, and their colour is whitish grey, covered with mottlings of a rusty red, which are thickly gathered towards the larger end.

The colouring of this species is very bold, and is briefly as follows: The top of the head and throat, as far as the middle of the neck, together with a rather broad streak down the centre of the chest and abdomen, are rich purple black, relieved by a spot of pure white on the nape of the neck, and a large flask-shaped patch under each eye. The back and shoulders are ashy-green, the greater wing-coverts are blue-black, each feather being tipped with white, so as to form a bar across the wings. The quill-feathers are dark green-grey, the primaries being edged with greyish-white. The tail-feathers are the same green-grey, except that the extreme feathers are white on their outer ends. The under parts are light sulphurous yellow, and the under tail-coverts are white. The total length of the bird is not quite six inches.

GREAT TITMOUSE.—*Parus major.*

THE little BLUE TITMOUSE is one of the most familiar birds of England, as it is widely spread throughout the land, and is of so bold a nature that it exhibits itself fearlessly to any observer.

In many of its habits it resembles the last-mentioned species, but it nevertheless possesses a very marked character, and has peculiarities which are all its own. As it trips glancingly over the branches, it hardly looks like a bird, for its quick limbs, and strong claws carry it over the twigs with such rapidity that it resembles a blue mouse rather than one of the feathered tribe. Being almost exclusively an insect-eating bird, and a most voracious little creature, it renders invaluable service to the agriculturist and the gardener by discovering and destroying the insects which crowd upon the trees and plants in the early days of spring, and which, if not removed, would effectually injure a very large proportion of the fruit and produce. In the course of a single day a pair of blue Titmice were seen to visit their nest four hundred and seventy-five times, never bringing less than one large caterpillar, and generally two or three small ones.   These birds, therefore, destroyed, on the average, upwards of five hundred caterpillars daily, being a minimum of fifteen thousand during the few weeks employed in rearing their young.

While searching for insects, the Blue Titmouse often bites away the buds of fruit-trees, together with pears and apples, but in almost every case it seeks to devour, not the fruit, but a maggot which lies concealed within it, and which, if not destroyed, would not only injure the particular fruit, but would also destroy many others by means of its future progeny. The food of this bird is of a most multifarious character, for the Blue Titmouse has been known to eat eggs, other birds which it kills when young or disabled, meat of various kinds, for which it always haunts the knackers' yards and country slaughter-houses, peas, oats, and the various kinds of food which are to be found in farmyards. So fond is it of fat meat, that a piece of beef suet is an unfailing bait which always succeeds in attracting the Titmouse into the jaws of the trap.   It has even been known to peck holes in hens' eggs, for the purpose of eating the contents; but on account of the large size of the eggs, it was not able to attain its purpose. I have even seen the Titmice unite against a tame hawk which I kept, assault him simultaneously, and carry off the piece of meat which had just been given to him.

It is a very pugnacious little bird, and is always ready for a combat with any one of its own kind. But in the breeding season its combative character is developed to the fullest extent, and the tiny blue creature will boldly attack a man if he should happen to approach too near the nest. Should the position of the nest be discovered, and the hand inserted in order to feel for the eggs, the mother-bird utters a sharp, angry hiss, and bites so sharply at the intruding fingers that they are generally hastily withdrawn, under the impression that a viper has been the hidden aggressor. Small as is the bird, her beak is so sharp and strong that it can cause considerable pain, and has earned for the bird the provincial name of Billy Biter. I once got the tips of my fingers sadly maltreated by a female Titmouse, while I was trying to feel the position of the eggs.

BLUE TITMOUSE.—*Parus cœruleus.*

The nest of this species may be found in the most extraordinary localities, such as hollow trees, holes in old walls, the interior of disused spouts, sides of gravel-pits, the hat of a scarecrow, the inside of a porcelain jar, or the cylinder of a pump. One bird had actually chosen a bee-hive as its residence, and had succeeded in building its nest and rearing its young while surrounded by the bees going to and returning from their work. Another Titmouse contrived to get into a weathercock on the summit of a spire, and there made its nest in security. The eggs are small and rather numerous, being generally about eight or ten, but sometimes exceeding the latter number.

The bird is readily tamed, as may be seen by the following anecdote related by Mr. Thompson.

" One of the ' Falls,' when let out of the cage in summer, roosted upon the top ; but in winter, although in a warm room, selected the hottest place in which it could remain safely for the night, namely, under the fender, a locality which afforded it at the same time sufficient space and shelter. This bird, from its familiarity and vivacity, was most amusing. The cage was covered with close netting, which it several times cut through, thereby effecting its escape into the room. It then flew to the children, and having taken hold of a piece of bread or cake in the hand of the youngest, would not forego the object of its attack, although shaken with the greatest force the child could exert ; indeed, the latter was so persecuted on one occasion for a piece of apple, that she ran crying out of the apartment. It was particularly fond of sugar.

Confined in the same cage with this bird were some other species, and among them a redbreast, which it sometimes annoyed so much as to bring upon its head severe chastisement. A favourite trick was to pull the feathers out of its fellow-prisoners. A young willow wren was sadly tormented in this way. A similar attempt was made on a song-thrush introduced into its domicile, but it was successfully repelled. This mischievous Tit escaped out of doors several times, but returned without being sought for."

The two little birds which are represented in the accompanying illustration are among the most striking examples of this pretty group, the one for its bold and conspicuous crest, and the other for the curious colouring of the head and neck.

The YELLOW-CHEEKED TITMOUSE inhabits several parts of Asia, and is mostly found among the north-western Himalayas, where it is rather abundant. In its habits it resembles the ordinary Titmouse of Europe. The nest of this species is constructed of moss, hair, and fibres, and is lined softly with feathers. The position in which it is placed is usually a cavity at the bottom of some hollow stump, generally a decaying oak, and it contains four or five eggs of a delicate white blotched with brownish spots. The colouring of this bird is rather peculiar and decidedly bold. The top of the head, the crest, a streak below the eye, and a broad band reaching from the chin to the extremity of the abdomen, are deep jetty black. The cheeks are light yellow, as is the whole of the under surface of the body, with the exception of the flanks, which take a greener

RUFOUS-BELLIED TITMOUSE.—*Parus rubidiventris.*
YELLOW-CHEEKED TITMOUSE.—*Parus xanthógenys.*

hue. The wings are grey, mottled with black and white, and the tail is black with a slight edging of olive-green.

The RUFOUS-BELLIED TITMOUSE inhabits Southern India and Nepâl, and cannot be considered as a rare bird. In this pretty creature the head, the crest, and the throat are jet black, contrasting boldly with the pure white of the ear-coverts and the back of the neck. The back, wings, and tail are ashen grey, washed with a perceptible tinge of blue, and the abdomen is reddish grey, as are the edges of the primary and secondary quill-feathers of the wing.

2. x

The LONG-TAILED TITMOUSE is familiarly known throughout England, and is designated under different titles, according to the locality in which it resides, some of its popular names being derived from its shape, and others from its crest. In some parts of the country it is called "Long Tom," while in others it goes by the name of "Bottle-crested Tit," or "Poke-Pudding," the latter word being a provincial rendering of the useful culinary apparatus termed a pudding-bag.

This pretty little bird is a notable frequenter of trees, hedge-rows, and orchards, and is remarkable for its sociable habits, being generally seen in little troops of six or eight in number. It appears that the young birds always remain with their parents throughout the whole of the first year, so that when the brood happens to be a large one, as many as sixteen Long-tailed Titmice may be seen hopping and skipping about together. The troop is always under the command of one bird, probably one of the parents, who takes the lead, and is copied by the others, so that they seem to be playing at a constant game of "Follow my leader." The leader has a peculiar chirruping cry, which is easily recognisable, and which is always uttered as it flits from one branch to another. I have often seen these birds threading their way among the pea-sticks at the earliest hour of daylight, flirting and coquetting with their long tails, hopping and chirping away most merrily, until the world had fairly begun its morning labours, when they left the gardens and betook themselves to the more retired localities of the fields and lofty hedgerow trees.

During the day the Long-tailed Titmice are always on the move, flitting restlessly from spot to spot, and bidding total defiance to fatigue. At night the whole troop perches on the same spot, and the birds gather themselves into a compact mass, like that which is formed by the wrens under similar circumstances. They seem to be careful of their comfort, for each bird strives to get nearest to the middle, and on a cold evening they fight vigorously until their positions are settled. When sleeping, they form a shapeless mass of soft puffy feathers, in which hardly a tail or a wing can be distinguished.

The wings of this species are rather short, but are more powerful than might be imagined, and the flight of the bird exactly resembles that of a boy's paper-headed dart when thrown into the air.

As far as is known, the Long-tailed Titmouse feeds exclusively on insects, and on account of its microscopical eyes is able to see and to catch the very minutest. The service which is rendered to agriculture by even a single nest of these birds is almost invaluable, for at all seasons of the year they continue to obtain their food, catching the perfect insect in the summer months, and feeding on the eggs, the hidden larvæ, and chrysalides in the winter. They especially are useful in clearing off the larvæ or "grubs" of the various saw-flies, especially that black lanky species which infests the gooseberry bushes, and destroys so many of the blossoms and young berries. Sometimes it has been observed to open the green pods of the broom, and to eat the seeds, carrying away some of the younger and more tender seeds to its nest. Perhaps it may eat these substances medicinally, as its ordinary food is of an animal character. The bird generally prefers the seed of the white flowering broom as an article of diet, but has been noticed in the act of eating those of the yellow variety in default of the more approved white broom.

The voice of this species is a kind of soft and lively twittering chirp, varied according to circumstances. Sometimes it will either pick up from some other bird, or invent a new note, and will be so pleased with its new acquisition that it skips about the branches for an hour or two exulting in its own cleverness, and continually uttering the cry of which it is so proud. But after a while its attention is taken by a peculiarly fat insect, or by one of the little quarrels which, on account of their excitable nature, are so frequent among these birds, and it straightway forgets its lately acquired accomplishment.

The nest of this species is undoubtedly the most wonderful example of bird architecture that is to be found in the British Islands, and is not exceeded in beauty by the home of any bird whatever. In form it somewhat resembles an egg, and it is built of moss, hair, a very little wool, the cocoon webs of spiders, and the silken hammocks of certain caterpillars, all woven into each other in the most admirable manner. The exterior of the nest is spangled with silvery lichens, which generally correspond in colour with the bark of the

LONG-TAILED TITMOUSE.—*Parus caudátus.*

tree on which it is placed, and serve to render it as little conspicuous as possible. The interior of the nest is wonderfully soft and warm, being literally crammed with downy feathers to such an extent that the eggs are deeply buried in the feathery bed, and cannot be counted until the whole lining of the nest is removed. The nest is generally placed rather near the ground, and is so well concealed that it is not easily seen except by experienced eyes.

Its proximity to the ground, together with the great number and minute dimensions of its inmates, may probably be some of the causes which necessitate the thick walls of non-conducting substances of which the nest is made ; thus carrying out the principle laid down on page 256.

The Long-tailed Titmouse is by no means constant in its love for any particular species of tree on which to make its nest, for it builds with equal readiness upon the apple, the blackthorn, the pollard elm, and the sloe, always, however, choosing some thickly branched and heavy foliaged tree for the purpose. The handsomest nest that I ever saw, and one which was for a long time in my possession, was built upon the sprays of a thick furze-bush, and another that I removed from the tree was placed upon a spreading broom-tree in a garden.

The entrance to the nest is usually by a single hole, situated near the roof, but in some instances a second hole has been made near the bottom of the nest on the opposite side. This singular variation has been the subject of much comment, but I believe that it may be caused by the particular locality in which the bird happens to build. Should the nest be in a bleak, exposed situation, there is only the single inlet ; but if it should be placed in a sheltered spot, a second entrance is employed. Moreover, in many instances where notice has been taken of the subject, the aperture has been made towards the south.

Ventilation seems to be one object of this double entrance, as it is never, so far as my own personal experience goes, made when the nest is exposed to the north or east wind. It must be remembered that the exhalations of birds are peculiarly plentiful and unpleasant, and that a closed nest, when made of such warm and thick substances, would be extremely unhealthy without some such precaution.

X 2

The number of eggs which this little bird lays is really surprising. Very seldom does it content itself with eight, and double that number has been frequently counted in a single nest. In consequence, the young birds are packed like so many herrings in a barrel, and the ingenuity which must be exerted by the parent birds in giving each little one its food in proper rotation must be very great indeed. When they have attained to nearly their full growth, and before they venture out of their home, the nest presents a very curious aspect externally, as its sides heave with the movements of the young birds, and it seems as if it were actually breathing. More especially is this the case when the young require fresh air, for they instinctively stretch their legs and necks as far as they can, and by so doing cause the whole nest to expand, and thus to take in a fresh supply of air. When they have produced the desired effect, they subside to their former position, and the nest contracts by the elasticity of its walls ; this curious movement will often take place eight or ten times in an hour.

The colouring of this species is as follows : The upper part of the head, the cheeks, the throat, and the whole of the under surface are greyish white, warming into a rosy hue upon the sides, flanks, and under tail-coverts. A broad stripe of deep black passes over the eye and ear-coverts, and joins a large triangular patch of the same jetty hue, which extends from the shoulders as far as the upper tail-coverts. The shoulders, the scapularies, and the lower part of the back are washed with a decided tinge of a ruddy hue. The wings are mostly black, with the exception of the tertiary quill-feathers, which are edged with white. The long central feathers of the tail are black, and the remainder are black on the inner webs and white on the outer. They are regularly graduated in length, each pair being about half an inch shorter than the preceding pair. Both sexes are similar in their colouring. The total length of the bird is about five inches and a half.

CRESTED TITMOUSE.—*Parus cristátus.*

In personal appearance the CRESTED TITMOUSE is the most conspicuous of the British species, on account of the peculiarity from which it derives its name.

It is a very rare bird in England, but when it does happen to make its appearance, is generally seen in little troops. On several parts of the Continent it is plentifully found, especially frequenting Denmark, Sweden, Russia, Switzerland, and Germany. When it comes over to the British shores, it seems to prefer the pine forests of Scotland to any other locality. It generally builds its nest in the hole of some decaying tree, the oak appearing to be the most favoured. In one singular instance noticed by Sir W. Jardine, the nest was almost wholly lined with the cast exuviæ of snakes. The eggs are about eight or ten in number, and their colour is generally white, spotted with a few light red specks.

The colouring of this bird is mostly black and white, disposed in a manner which can readily be comprehended by examining the accompanying illustration. The feathers of the crest are black at the base and edged with a rather broad band of white ; the back and wings are soft brown the under surface of the chest and abdomen is very pale fawn, and the under surface of the wings and tail is a delicate pearly grey. It is a small bird the total length being only four inches and a half.

ANOTHER British Titmouse is the COLE TITMOUSE (*Parus ater*), so called on account of the dark colouring of its plumage.

It is a tolerably common bird throughout England as well as in Ireland and Scotland. In its habits it is not unlike the Long-tailed Titmouse, being ever restlessly in motion, and constantly running up and down the branches of trees and bushes in search of its insect prey. It is not quite so fearless of man as some of the allied species, and is found in small woods, hedgerows, and copses, rather than in gardens and orchards, so that it frequently escapes the notice of a casual observer. The nest of this species is usually placed above the ground, and is built in some sequestered and sheltered situation, such as the hole of a tree or a wall, the hollow of gnarled or projecting roots, or in the midst of some very thick and shrubby bush. It is composed of moss and wool, and lined with hair. The eggs are generally about seven or eight in number, and are of a pure white, mottled with pale reddish spots.

The voice of the Cole Titmouse is rather peculiar, and is well described by Mudie in his "British Birds:"—"The song of the Cole Tit is not indeed one of many notes, or of mellifluous inflexions; it is little else than the same note repeated four or five times, but with so much variety of pitch and tune as to form a sort of cadence which would make a good variety anywhere, as it is shrill and clear, and one which is particularly welcome and cheering in those mountain woods which the summer warblers but rarely visit. The bird sings in the noontide heat, when most birds, and especially those on the open wastes, with which the haunts of this species are usually interspersed, are silent."

The Cole Titmouse is coloured as follows: The head, chin, throat, ear-coverts, and parts of the sides of the neck are deep black, and the cheeks, sides of the neck, and a patch upon the nape are white. The back is bluish-grey, and the wings are brownish-grey with a little green on some of the feathers, and two narrow bars of white across the tips of the coverts. The breast is greyish white, and the abdomen is pale fawn washed with a slight tinge of green. The total length of the Cole Titmouse measures about four inches and a half.

BEARDED TITMOUSE.—*Paroïdes biármicus.*

The MARSH TITMOUSE is another British species, and may be distinguished from the preceding species, to which it bears a considerable resemblance, by the absence of the black patch upon the throat and the white spot on the nape of the neck. It derives its popular name from its marsh-loving habits, as it is generally to be found near the water meadows and the low-lying banks, hopping about the osiers and willows, or seeking its food in the swampy grounds.

The BEARDED TITMOUSE is nearly as conspicuous a bird as the crested species, and is readily distinguished from any other of its kind by the tuft of soft black feathers which depends from the side of the face.

It is a rare bird in England, and has only lately been admitted into the list of indigenous British birds. It is found in various parts of the Continent, and is extremely plentiful in Holland, where it is captured in great numbers and sent to England as a cage bird. As it always inhabits the marshy spots, and chooses the thickest reed-beds for its home, it is but seldom seen even in places where it has taken up its residence. Moreover, it seldom shows itself at any distance from its reedy home, and does not care even to rise to any great elevation above its covert, preferring just to flit above the waving reed-tops, as it passes from one spot to another in search of the various insects that frequent the blossoms and leaves of aquatic plants. While engaged in this search the birds are very nimble, and frequently hang with their heads downwards, clinging firmly with their claws, and caring nothing for the waving of the slender perch on which they have placed themselves.

The voice of this species is very remarkable, being low, melodious, clear, and metallic, resembling in its sound the distant clang of fairy cymbals. While undisturbed, the bird constantly utters its pretty note, and occasionally shows itself above the reed-tops ;· but if alarmed, it instantly drops to the ground, and by threading its way through the stalks, soon places itself out of danger. Even a human foot would have but little chance of following a Bearded Titmouse through the recesses of its favoured home, for the ground is always so wet and soft that it will not support the weight of any but the lightest of creatures.

The food of this species differs much according to the season of year. In the summer months the bird feeds mostly upon insects, spiders, and similar creatures. It also eats small kinds of molluscs, and, according to Mr. Dykes, is very fond of the delicate *Succinea amphibia*, which it swallows entire, trusting to the action of the gizzard to break up the shells and triturate the food. As many as twenty of these shells were found packed closely together in the crop of a single bird. During winter, when the insects have either died or taken to their winter quarters, the land molluscs retired to their caves, and the aquatic shells sunk deeply below the water, the Bearded Titmouse is thrown upon vegetable substances for its livelihood, and feeds chiefly upon the seeds of the reed.

The nest is a tolerably well-made edifice, cup-shaped, soft and warm in the interior, and covered externally with dried sedge-leaves and grasses. It is always placed near the ground in the midst of a thick tuft of rank grass or dead reeds. The eggs are generally five in number, and are of a pinky-white, covered with specks and streaks of brown irregularly dispersed over the surface. The nest is generally begun in the month of April. It is probable that the young birds remain with their parents for a considerable period of time, as the Bearded Titmice are always to be seen in little troops as they flit over the tops of the reeds.

The plumage of this species is pretty and peculiar. The head, the neck, and the ear-coverts are a delicate grey, contrasting beautifully with the " beard," or pointed tuft of elongated soft feathers of a jetty black, that starts from the space between the opening of the beak and the corners of the eye, and hangs below the neck. The general colour of the upper surface is pale fawn, variegated with a little black upon the wing-coverts. In the young bird there is also a large patch of black upon the back. The tail is prettily marked with black, white, grey, and fawn, and the under surface is whitish-grey on the throat and breast, and yellowish-white on the abdomen, deepening into a more ruddy hue on the flanks. The under tail-coverts are jetty black. The total length of this species is about six inches.

THE *Mniotiltinæ*, or Bush-Creepers, are well represented by the common BUSH-CREEPER of India.

It is a sociable little bird, being generally seen in small troops, and often associating with birds of different species. Although not very shy, it yet loves retired localities, such as woods and thickets, and may there be seen flitting merrily among the foliage and underwood,

COMMON BUSH-CREEPER.—*Zósterops palpebrósus.*

and perpetually engaged in a search after insects. In some of its movements it resembles the honey-eaters, for it often pushes its head completely into the corollas of flowers while endeavouring to capture the minute insects that lurk at the bottom of the cup, and emerges with its forehead covered with yellow pollen. The voice is a low twittering note, constantly uttered while the bird is in motion, but there is no real song.

The nest of the Bush-Creeper is rather variable in its position and structure, sometimes being suspended from the branches, and at others placed in the centre of some thick bush. Generally it is suspended between two twigs, to which it is woven by means of various animal and vegetable fibres, mostly obtained from the cocoon of caterpillars and the fibrous bark of trees. The shape of the nest is cup-like, but the whole structure is so delicately balanced that even in a fierce storm the eggs are not flung out of their places.

There are many species of British Creepers, some of which inhabit Asia, others are found in Africa, and some in Australia. The word Mniotiltinæ is of Greek origin, and signifies "moss-pluckers," while the term Zósterops signifies "girdle-faced," and has been given to this bird in consequence of a well-defined circlet of light coloured feathers which surrounds the eye.

WAGTAILS.

WE now arrive at a small group of birds, which is sufficiently familiar to every observer of nature through the different representatives which inhabit this country. The WAGTAILS, so called from their well-known habit of jerking their tails while running on the ground or on settling immediately after a flight, are found in both hemispheres, and are all well known by the habit from which they derive their popular title. No less than nine species of this group occur in Britain, some of which are nearly as well known as the common sparrow, while others are less familiar to the casual observer.

The PIED WAGTAIL is the most common of all the British examples of this genus, and may be seen at the proper season of the year near almost every pond or brook, or even in the open road, tripping daintily over the ground, pecking away at the insects, and wagging its tail with hearty good will.

GROUP OF BRITISH WAGTAILS.

Mr. Yarrell mentions that this bird is an accomplished fisher, and excels in snapping up the smaller minnows and fry as they come to the surface of the water. It also haunts the fields where sheep, horses, or horned cattle are kept, and hovers confidingly close to their hoofs, pecking away briskly at the little insects which are disturbed by their tread. It also delights in newly mown lawns, and runs over the smooth surface with great agility,

PIED WAGTAIL.—*Motacilla Yarrellii.*

peering between every grass-blade in search of the insects which may be lying concealed in their green shelter. The flight of the Pied Wagtail is short and jerking, the bird rising and falling in a very peculiar manner with every shake of the wings.

Several Wagtails which used to frequent our garden were fond of meat, and, together with the blue titmice, would often assault, or rather pretend to assault the tame sparrow-hawk, and cruelly steal his dinner before his eyes. Indeed, the Wagtails seemed to be quite the rulers over that unfortunate hawk, and led him a sad life.

The Pied Wagtail remains in England throughout the year, but generally retires to the southern counties during the winter, as it would otherwise be unable to obtain its food. Sometimes, however, where the springs are so copious that the water never entirely freezes, the Wagtail may be seen haunting its accustomed spot, and drawing a subsistence from the unfrozen waters. The more northern coasts are a favourite resort of the Wagtails, which run briskly along the edge of the advancing or receding tide, picking up any stray provender that may come within their reach.

The song of the Pied Wagtail is soft, low, and sweet, and is generally uttered in the early morning from the elevation of some lofty spot, such as the summit of a pointed rock, the roof of an outhouse, or the top of a paling. The bird is bold and familiar, coming quite close to human beings without displaying any fear, and even following the ploughman for the purpose of picking up the grubs and insects that are turned out of the soil by the share.

The nest of the Wagtail is generally placed at no great distance from water, and is always built in some retired situation. Holes in walls, the hollows of aged trees, or niches in old gravel-pits are favourite localities with this bird. Heaps of large stones are also in great favour with the Wagtail, and I have generally found that wherever a pile of rough stones has remained for some time in the vicinity of water, a Wagtail's nest is almost invariably somewhere within it. I have also found the nest in heaps of dry brushwood piled up for the purpose of being cut into faggots. In every case the nest is placed at a considerable depth, and no small amount of care and ingenuity is needed to extract the eggs without damaging them. The eggs are generally four or five in number, and their colour is grey-white speckled with a great number of very small brown spots.

The colouring of the Pied Wagtail is almost entirely black and white, very boldly disposed and distributed as follows: The top of the head, the nape of the neck, part of the shoulders, the chin, neck, and throat, are jetty black, contrasting boldly with the pure

WHITE WAGTAIL.—*Motacilla alba.*

snowy white of the sides of the face and the white patch on the sides of the neck. The upper tail-coverts, and the coverts of the wings are also black. The quill-feathers of the wings are black, edged on the outer web with a lighter hue. The two exterior feathers of the tail are pure white, edged on the inner web with white, and the remainder jetty black. The under parts of the body are greyish white, taking a blue tint upon the flanks. The entire length of the bird is between seven and eight inches.

This is the summer plumage of the male bird. In the winter the chin and throat exchange their jetty hue for a pure white, leaving only a collar of black round the throat. The female much resembles her mate in the general colouring of her plumage, but is about half an inch shorter.

As the White Wagtail has often been confounded with the preceding species, I have thought that a figure of each species would be advisable in the present work. According to Mr. Gould, who first determined the characteristics between the two species, they may be distinguished from each other by the following marks of difference.

" The pied Wagtail of England, *Motacilla Yarrellii,* is somewhat more robust in form, and in its full summer dress has the whole of the head, chest, and neck of a full deep jet black ; while in the White Wagtail, *Motacilla alba,* at the same period, the throat and head alone are of this colour ; the back and the rest of the upper surface being of a light ash grey. In winter the two species more nearly assimilate in their colouring, and this circumstance has doubtless been the cause of their being hitherto considered as identical; the black back of *M. Yarrellii* being grey at this season, although never so light as *M. alba.* An additional evidence of their being distinct (and which has doubtless contributed to the confusion) is, that the female of our pied Wagtail never has the beak black as in the male ; this part, even in summer, being dark grey, in which respect it closely resembles the other species." Another distinction may be found in the shape of the beak, which is broader in the White than in the pied species.

The White Wagtail is very common in France, and the southern parts of Europe; but although it may be found plentifully on the shores of Calais, the narrow arm of sea appears to be a boundary which it seldom passes.

The Grey Wagtail is a remarkably pretty and elegant example of this group of birds ; its plumage being delicately marked with various soft colouring, its shape slender and graceful, and its movements light and airy.

YELLOW WAGTAIL.—*Motacilla sulphúrea.*   GREY WAGTAIL.—*Motacilla campestris.*

This species is not quite so common as the Pied Wagtail, and seems to migrate backwards and forwards in England according to the temperature. In the northern counties it is a summer visitant, but is more permanently stationed in the southern parts of the island, and mostly breeds in warm, well-watered localities. About Oxford it is far from uncommon; but although the bird itself might often be seen haunting the river-banks and brooklets which abound in that neighbourhood, I never remember finding its nest. It is a special lover of water, and seldom seems to fly to any great distance from the brook or river in which it finds its food.

Like the Pied Wagtail, it feeds largely on aquatic insects and larvæ, and is also known to eat small water molluscs, not troubling itself to separate the soft body from the hard and sharp-edged shell.

Of the nesting of this species Mr. Thompson speaks as follows: "The situations generally selected for the nest are holes in walls, the preference being given to those of bridges, about mill-wheels, or otherwise contiguous to water. In the romantic glens of the Belfast mountains they also build, and for this purpose a pair generally resort to a fissure in the rock, beside a picturesque cascade of 'the Falls,' just such a place as would be chosen by the water-ousel. On the 18th of March, a pair of Grey Wagtails, 'with black patch on throat,' have been noted, apparently contemplating nidification, at Wolf Hill, by minutely examining their former breeding haunts; and on the 12th of May the young of the first brood were seen on wing, though still requiring their parents' aid to feed them. Occasionally there is a second brood. The nest is generally formed of grasses or other delicate plants, and lined with horsehair. It is singular that they generally manage to pick up enough of this last material for lining. Four nests at Wolf Hill in one season were all lined with it : the eggs were usually four in number, and during incubation, the beautiful and minute bird would admit of a close approach. Throughout the winter the Grey Wagtails generally keep in pairs ; in autumn only have I seen a whole family, and never more together. They may then be seen roosting in company at the base of trees or underwood overhanging the water."

The general colour of this species is grey, a tint which extends over the crown of the head, ear-coverts, neck, and back. Upon the sides of the head are two buff-coloured stripes, one above and the other below the ear-coverts. The quill-feathers of the wings, together with their coverts, are black, sprinkled with white and very pale buff. The six central feathers of the tail are black, and the two utmost black, and the others white with a black line running down the centre web. The throat and chin are black, and the whole of the under surface, together with the upper tail-coverts, are bright yellow. The total length of this species is very nearly eight inches. In the winter months the black of the chin changes to white washed with buff, and the yellow feathers fade into a yellowish-grey.

The YELLOW WAGTAIL, or RAY'S WAGTAIL, as it is sometimes termed, is very common in this country, and is very partial to pasture land, where it revels among the insects that are roused by the tread of cattle.

It is not so partial to water as the pied species, and may often be met with upon the driest lands, far from any stream, busily employed in catching the beetles, flies, and other sun-loving insects. Even upon roads it may frequently be observed, tripping about with great celerity, and ever and anon picking up an insect, and celebrating its success by a triumphant wag of the tail. The name of Yellow Wagtail has been given to it on account of the light yellow hue which tinges the head and the entire under surface of the body. As, however, the preceding species also possesses a considerable amount of yellow in its colouring, the name of Ray's Wagtail has been given to this bird in honour of the illustrious naturalist. It is a gregarious bird, being generally seen in little flocks or troops.

The colouring is as follows : The top of the head, back of the neck, and the whole of the back are olive, brighter upon the head and darker upon the back. The quill-feathers of the wings are dark brown, tipped with yellowish-white, with the exception of the two exterior feathers of the tail, which are white, with a line of black running down the inner web ; all the tail is brownish-black, like the wings. The chin, throat, the whole of the under surface of the body, together with a well-defined stripe over the eye and ear, are bright yellow. In length the bird does not quite reach seven inches.

AUSTRALIA is the habitat of the prettily-marked bird which is known to zoologists by the very long name of WHITE-FACED EPHTHIANURA.

It is tolerably common in several parts of that strange country, and is found in little flocks, as is the case with the Wagtails of England. Of this bird and its habits Mr. Gould speaks as follows. " As the structure of its toes and lengthened tertiaries would lead us to expect, its natural province is the ground, to which it habitually resorts, and decidedly evinces a preference to spots of a sterile and barren character. The male, like many of the saxicoline birds, frequently perches either on the summit of a stone, or on the extremity of a dead and leafless branch. It is rather shy in disposition, and when disturbed flies off with considerable rapidity to the distance of two or three hundred yards before it alights again. I observed it in small companies on the plains near Adelaide, over the hard clayey surface of which it tripped with amazing quickness, with a motion that can neither be described as a hop or a run, but something between the two, with a bobbing action of the tail."

Only the male bird is gifted with the bright contrast of the white throat and banded chest, the female being quite a sombre-plumaged bird. It is always a sprightly and active bird, and is quick of wing as well as of foot.

The PIPITS, or TITLARKS as they are sometimes called, form a well-marked group, which possesses the long hind toe of the hawk, together with very similar plumage, and also bears the long tail which is found in the Wagtails. Several species of Pipit inhabit England, two examples of which will be figured.

The first is the common MEADOW PIPIT, or MEADOW TIT-LING, a bird which may be seen throughout the year upon moors, waste lands, and marshy ground, changing its locality according to the season of year. It is a pretty though rather sombre little bird, and is quick and active in its movements, often jerking its long tail in a fashion that reminds the observer of the Wagtail's habits. It moves with considerable celerity, tripping over the rough and rocky ground which it frequents, and picking up insects with the stroke of its unerring beak. Its food, however, is of a mixed description, as in the crops of several individuals were found seeds, insect and water-shells, some of the latter being entire.

WHITE-FACED EPHTHIANURA.— *Ephthianura albifrons.*

The song of this bird is hardly deserving of the name, being rather a feeble and plaintive "cheeping" than a true song. While uttering its notes, the Meadow Pipit is generally on the wing, but does not begin to sing until it has attained its full elevation, reserving its voice for the gradual descent. The song is begun quite early in the season, but as the bird is so partial to waste lands, it is not heard so commonly as that of rarer birds. It is gregarious in its habits, assembling in little flocks, which generally come to the cultivated grounds about September or October, and roost amicably together on the ground at night.

The nest of this species is placed on the ground, and generally hidden in a large grass-tuft. It appears, from some observations made by Mr. Thompson, that the

MEADOW PIPIT.—*Anthus pratensis.*

bird is in the habit of carrying dead grasses and laying them over her nest whenever she leaves her eggs or young. The object of this precaution is not, however, very evident, as the grass is usually of a different hue from the surrounding foliage, and apparently serves rather as a guide to the nest than a concealment. The eggs are from four to six in number, of a dark brown colour, speckled freely with reddish brown. The cuckoo is said to favour the Meadow Pipit with her society rather more frequently than is agreeable to the bird, and to give it the labour of rearing her voracious young.

The general colour of this Pipit is dark olive-brown, with a wash of green upon the upper parts ; the wings are very dark brown, sprinkled with white, and the tail is also brown, with the exception of a white streak on each exterior feather, and a few white spots towards the extremity. The under surface is brownish-white, and upon the breast of the male there is a pale rosy tinge. Upon the breast there are a number of dark brown spots. The colours of the plumage undergo a decided change in the autumn, and are more showy than those of the summer ; the olive-green on the back becoming more conspicuous and the under surface tinged with yellow.

This bird goes by different names in different parts of England. In many places it is termed the Moss-cheeper, in allusion to its peculiar plaintive note. In other parts it is known by the title of Ling-bird, on account of its habit of haunting the waste moorlands. In Ireland the bird is called the Wekeen, a name which evidently alludes to its note. It has been found in all the British Islands, and in many parts of Europe, extending as far northward as Sweden and Norway in the summer months, and having even been seen in Iceland. Specimens have been taken in Egypt and several parts of Africa, and also in the west of India. It has also been included in the list of Japanese birds, so that it possesses a range of locality which is seldom enjoyed by any single species. Although the bird is so small and delicate, being only six inches in length, it is a strong and daring flyer, a specimen having been taken on board a ship at a distance of nine hundred miles from the nearest shore.

TREE PIPIT.—*Anthus arbóreus.*

The TREE PIPIT derives its name from its habit of perching upon trees, wherein it presents a decided contrast to the meadow Pipit, which chiefly frequents waste lands and marshes.

It is only a summer visitant of this country, arriving towards the end of April and leaving our shores in September, after rearing its brood. Although it can perch on branches, and does so very frequently, it has not a very strong hold of the bough, and is not nearly so agile in hopping or tripping about the branches as is the case with the generality of perching birds. While on the tree it generally settles on the end of some bough, and is not seen to traverse the branches after the fashion of the tree-frequenting birds. Although it is called the Tree Pipit, it seems more at its ease on the ground than among the branches, and runs and trips over the roughest soils with an easy grace that contrasts strongly with its evident insecurity upon the boughs.

The song of this bird is sweeter and more powerful than that of the preceding species, and is generally given in a very curious manner. Taking advantage of some convenient tree, it hops from branch to branch, chirping merrily with each hop, and after reaching the summit of the tree, perches for a few moments and then launches itself into the air, for the purpose of continuing its ascent. Having accomplished this feat, the bird bursts into a triumphant strain of music, and, fluttering downwards as it sings, alights upon the same tree from which it had started, and by successive leaps again reaches the ground.

GIANT BREVE.—*Pitta gigas.*

The nest of the Tree Pipit is almost invariably placed on the ground under the shelter of a tuft of grass, although there are instances where the bird has been known to build in a very low bush. The materials of which the nest is made are moss, roots, and fine grasses, and the lining is mostly of hair. The eggs are five in number, and their colour is a whitish ground covered with reddish-brown spots. There is considerable variation in the colour of the eggs, the spots being larger and more numerous in some examples, and their colour generally possessing different shades of purple intermixed with the brown.

The Tree Pipit may be known from the meadow Pipit by its greater size, its flatter head, larger bill, and shorter hind claws, the last being a very notable distinction. In its general colouring it resembles the meadow Pipit. Besides these two species, others are known to be among the British birds, as the Rock Pipit (*Anthus aquaticus*), and the Richard's Pipit (*Anthus Ricardi*).

THE very large family of the Thrushes now engages our attention. Many of these birds are renowned for their song, and some of them are remarkable for their imitative powers. In general shape there is some resemblance to the crows and the starlings and blackbirds of England, bearing a very great external resemblance to the common starling. This family is divided into five sub-families, all of which will be mentioned in the following pages, and many examples figured. Our own country possesses many representatives of this group of birds.

The ANT-THRUSHES, so called from their ant-eating propensities, form a small but remarkable group of birds, differing greatly in colour and dimensions, but bearing considerable resemblance to each other in their general form. Some species are almost as sombrely clad in black, brown, and white, as the common thrush of England, while the plumage of others glows with a crystalline lustre of animated prismatic hues, as in the black-headed Pitta (*P. melanocephala*), or is gorgeous with the brightest scarlet, blue, and purple, as in the crimson-headed Brachyure (*P. granatica*). All the species, however, bear, in external form, a considerable resemblance to each other, being thick-set, big-bodied, large-headed, long-legged, short-tailed, and strong-billed.

SHORT-TAILED ANT THRUSH.—*Pitta Bengalensis.*

These birds may be separated into two divisions, the Breves and the Ant-thrushes ; the former being found in India, the Indian Islands, and Australia, while the latter inhabit America as well as the Old World.

In whatever part of the world they may be situated, they are most useful birds, as without their assistance the ants which swarm in those lands would increase to a most baneful extent. In allusion to this subject Mr. Swainson makes the following pertinent remarks : "Of all the tribes of insects which swarm in the tropics, the ants are the most numerous ; they are the universal desolators, and in the dry and overgrown parts of the interior, the traveller can scarcely proceed five paces without treading upon their nests. To keep these myriads within due limits, a wise Providence has ordered into existence the Ant-thrushes, and given to them this particular food. Both are proportionate in their geographical range, as far beyond the tropical latitudes the ants suddenly decrease, and their enemies, the Myiotheriæ (*i. e.* the ant-eating creatures) totally disappear."

It must, however, be remembered, that the ants themselves are of the very greatest service in removing and devouring all dead animal substances, and that the great object of the Ant-Thrush is not to extirpate, but to keep within due bounds the insects which might otherwise become absolutely harmful to the bird which is so greatly benefited by their presence in moderate numbers.

The great ANT-THRUSH, which is also called the GIANT PITTA, or the GIANT BREVE, in allusion to its large dimensions, is a native of Surinam, and on account of its bright plumage, its quaint and peculiar shape, its very large head, very long legs, and peculiarly short wren-like tail, which looks exactly as if it had been neatly cropped, is one of the most singular birds of that prolific locality. In size it equals an English rook, but hardly looks so large as that well-known bird, on account of the short tail, which is entirely covered by the wings when they are closed. The general colour of this brilliant bird is a light cobalt blue, which extends over the whole of the back and tail, but is not quite so lustrous upon the wings. The quill-feathers of the wings are black, tipped with sky-blue, and the head, the surface of the neck, together with a stripe that runs partly round the

DIPPER.—*Hydróbates cinclus.*

neck, are also black, a darkish line is drawn through the eyes, the throat is greyish white, and the abdomen and lower surface of the body are brownish grey. In total length the bird measures about nine inches.

The SHORT-TAILED PITTA, so called from the extreme shortness of that member, is a native of India, being most plentifully found in Bengal.

It is a quick, lively bird upon the ground, rarely taking to flight except when absolutely forced so to do, but moving with incredible rapidity over the earth. In its general habits it differs nothing from the remainder of its kind. The plumage of this bird is remarkably pretty, and notable on account of the curious markings of the head and neck, and the beautifully vivid colouring of the wings.

The ground colour of the bird is a soft mouse-brown, which is boldly marked with three jet-black bands, one passing from the forehead over the top of the head to the nape of the neck, and the other two passing from the gape through the eye, and joining the first band in a kind of half-collar between the neck and shoulders. The central stripe suddenly widens upon the crown of the head. Upon each shoulder and upon the basal portion of the tail are a number of glittering verditer-green feathers, that gleam out in the light, and render it a most conspicuous bird. The quill-feathers of the wings are dead black, except a white spot or bar upon their coverts, forming an irregular band when the wing is spread. The throat is a light brown, and the abdomen a pale mouse-colour.

THE Ant Thrushes find an English representative in the well-known DIPPER, or WATER-OUSEL, of our river-banks.

Devoid of brilliant plumage or graceful shape, it is yet one of the most interesting of British birds when watched in its favourite haunts. It always frequents rapid streams and channels, and being a very shy and retiring bird, invariably prefers those spots where the banks overhang the water, and are clothed with thick brushwood. Should the bed of the stream be broken up with rocks or large stones, and the fall be sufficiently sharp to wear away an occasional pool, the Dipper is all the better pleased with its home, and in such a locality may generally be found by a patient observer.

All the movements of this little bird are quick, jerking, and wren-like, a similitude which is enhanced by its habit of continually flirting its apology for a tail. Caring nothing for the frost of winter, so long as the water remains free from ice, the Dipper may be seen throughout the winter months, flitting from stone to stone with the most animated gestures, occasionally stopping to pick up some morsel of food, and ever and anon taking to the water, where it sometimes dives entirely out of sight, and at others merely walks into the shallows, and there flaps about with great rapidity. An interesting account of the proceedings of some Dippers appeared in the formerly celebrated "Annals of Sporting:"—"About four years ago, when on a shooting excursion to the Highlands, I embraced the opportunity (as everybody else who has it ought to do) of visiting the deservedly celebrated falls of the Clyde, and here it was, while viewing the fall of Bonnington, that, happening to cast my eye down below, a little beyond the foot of the cascade where the bed of the river is broken with stones and fragments of rocks, I espied, standing near each other on a large stone, no less than five water-ousels. Thus favourably stationed as I was for a view, myself unseen, I had a fair opportunity for overlooking their manœuvres. I observed, accordingly, that they flirted up their tails, and flew from one stone to another, till at length they mustered again upon the identical one on which I had first espied them. They next entered into the water and disappeared, but they did not all do this at the same time, neither did they do it in the same manner. Three of them plunged over head instantaneously, but the remaining two walked gradually into the water, and, having displayed their wings, spread them upon the surface, and by this means appeared entirely to support themselves. In this position they continued for some time, at one moment quickly spinning themselves, as it were, two or three times round, at another desisting and remaining perfectly motionless on the surface; at length they almost insensibly sank.

What became of them then it is not in my power to state, the water not being sufficiently transparent to enable me to discover the bottom of the river, particularly as I was elevated so much above it. Neither can I say that I perceived any one of them emerge again, although I kept glancing my eye in every direction, in order, if possible, to catch them in the act of re-appearing; the plumage of the bird, indeed, being so much in harmony with the surrounding masses of stone, rendered it not very easily distinguishable. I did, however, afterwards observe two of these birds upon a stone on the opposite side of the stream, and possibly the other three might also have emerged and have escaped my notice."

While employed at the bottom of the stream, the bird keeps itself below the surface by beating rapidly upwards with its wings, just as a human diver beats the water with his hands and feet, while seeking for some object under the water. To an observer at the surface, the bird appears to tumble and scramble about at random in a very comical manner, but in truth the little creature is perfectly capable of directing its course, and of picking up any article of food that may meet its eye. Mr. St. John says of the Dipper, that it walks and runs about on the ground at the bottom of the water, scratching with its feet among the small stones, and pecking at all the insects and animalculæ which it can dislodge. Sometimes the bird has been observed moving about in the water with its head only above the surface.

The food of the Dipper seems to be exclusively of an animal character, and, in the various specimens which have been examined, consists of insects in their different stages, small crustaceæ, and the spawn and fry of various fishes. Its fish-eating propensities have been questioned by some writers, but the matter has been entirely set at rest by the discovery of fish-bones and half-digested fish in the stomachs of Dippers that had been shot. Generally, however, the food consists of water-beetles, particularly of the genus known by the name of Hydrophilus, a flat, oval-shaped insect, with hard wing-cases and oar-like hind-legs. The bird has also been known to pick up the caddis worms, taking them on shore, pulling and knocking to pieces the tough case in which the fat white grub is enveloped, and swallowing the contents with great satisfaction. It does not always dive for the purpose of obtaining its food, but frequently perches upon the water's edge, and pecks its prey out of the muddy soil.

Its flight is remarkably rapid, though generally low, and when the bird is disturbed it flies quickly along the course of the stream, always keeping itself below the shelter of the banks. The general character of the flight is not unlike that of the kingfisher, and in some parts of England the country people firmly believe that the Dipper is the female kingfisher, the blue and red bird being the male. From the drift of a letter addressed to the *Field* newspaper, the Dipper seems to be rather a quarrelsome bird. A gentleman was walking along the bank of a little stream in Pembrokeshire, when he saw a Dipper, shooting along in its usual arrowy flight, divert itself from its course, and dashing itself against a redbreast that was sitting quietly upon a twig overhanging the stream, knock it fairly into the water. The savage little bird was not content with this assault, but continued to attack the poor redbreast as it lay fluttering on the waves, endeavouring to force it below the surface. It twice drove its victim under water, and would have killed it had it not been scared away by the shouts and gestures of the witness. The robin at length succeeded in scrambling to the bank, and got away in safety.

The song of the Dipper is a lively and cheerful performance, and is uttered most frequently in the bright frosty mornings. Sometimes it will stand upon a stone when singing, and accompany its song with the oddest imaginable gestures, hopping and skipping about, twisting its head in all directions, and acting as if it were performing for the amusement of the spectator.

The nest of the Dipper is not unlike that of the wren, being chiefly composed of mosses built into a dome-like shape with a single aperture in the side. The nest is generally placed near the water, and always under some sort of cover, usually a hole in the bank. Sometimes it has been situated in such a position that the water of a little rivulet actually overshot the entrance, and the bird was forced to pass under the falling water in order to enter its nest. Another nest, mentioned by Mr. Thompson, was built in a dark shed erected over a large mill-wheel, nearly forty feet in diameter, and the parent birds were accustomed to shoot through the wheel in passing to and from their nest. They would even perch on the arms of the mill-wheel itself.

The nest is not, however, always so close to the water, for I found one near Swindon, in the side of an old disused pit, at some little distance from the great Swindon reservoir. It was discovered more by accident than by intention, the touch having given the first intimation of its presence. The moss always remains in a green state, as it is placed in a damp locality, so that it can with great difficulty be distinguished from the vegetation of the spot whereon it is situated. The size of the nest varies extremely ; sometimes being enormously large and thick-walled, its whole bulk equalling a man's head, while, on the other hand, it is sometimes small, and scarcely domed at all ; this latter formation being always due to the good shelter afforded by the spot in which it is placed.

The eggs are pure white, and rather long in proportion to their breadth. Their full number is five, and the young remain with their parents for a considerable period, forming little companies of five or six of these curious birds.

The general colour of this bird is brown on the upper surface of the body, the throat and upper parts of the chest are white, and the abdomen is rusty red. The young birds possess a rather variegated plumage of black, brown, ash colour, and white. The total length of the adult bird is about seven inches.

The MOCKING-BIRD of America is universally allowed to be the most wonderful of all songsters, as it not only possesses a very fine and melodious voice, but is also endowed with the capacity for imitating the notes of any other bird, and, indeed, of immediately reproducing with the most astonishing exactness any sound which it may hear.

It is a native of America, and, according to Mr. Webber, there are two varieties, if not two species ; the one an inhabitant of Kentucky, and the other being found in the more southern districts. All persons who come within the sound of a Mocking-bird's voice are fascinated with the thrilling strains that are poured without effort from the melodious throat, and every professed ornithologist who has heard this wonderful bird has exhausted the powers of his language in endeavouring to describe the varied and entrancing melody of the Mocking-bird. Within the compass of one single throat the whole feathered race

Y 2

seems to be comprised, for the Mocking-bird can with equal ease imitate, or rather reproduce, the sweet and gentle twittering of the blue-bird, the rich full song of the thrush, or the harsh ear-piercing scream of the eagle. At night especially, when labour has ceased, "silence has attuned her ear," says Webber, "and earth hears her merry voices singing in her sleep.

Yes, they are all here! Hear then each warble, chirp, and thrill! How they crowd upon each other! You can hear the flutter of soft wings as they come hurrying forth! Hark, that rich clear whistle! 'Bob White, is it you?' Then the sudden scream! is it a hawk? Hey! what a gush, what a rolling limpid gush! Ah, my dainty redbreast, at thy matins early! Mew! what, Pussy! No, the cat-bird; hear its low liquid love-notes linger round the roses by the garden-walk! Hillo! listen to the little wren! he must nearly explode in the climax of that little agony of trills which it is rising on its very tip-toes to reach! What now? Quack, quack! Phut, phut, phut! cock-doodle-doo! What, all the barn-yard! Squeak, squeak, squeak! pigs and all. Hark, that melancholy plaint, Whip-poor-Will, how sadly it comes from out the shadowy distance! What a contrast! the red-bird's lively whistle, shrilly mounting high, higher, highest! Hark, the orchard oriole's gay, delicious, roaring, run-mad, ranting-riot of sweet sounds! Hear that! it is the rain-crow, croaking for a storm! Hey day! Jay, jay, jay! it is the imperial dandy blue-jay. Hear, he has a strange, round, mellow whistle too! There goes the little yellow-throated warbler, the woodpecker's sudden call, the king-bird's woeful clatter, the dove's low plaintive coo, the owl's screeching cry and snapping beak, the tomtit's tiny note, the kingfisher's rattle, the crow, the scream, the cry of love, or hate, or joy, all come rapidly, and in unexpected contrasts, yet with such clear precision, that each bird is fully expressed to my mind in its own individuality."

Yet all these varied notes are uttered by the one single Mocking-bird, as it sits on a lofty spray or flings itself into the air, rising and falling with the cadence of its song, and acting as if absolutely intoxicated with sweet sounds.

Let it but approach the habitation of man, and it straightway adds a new series of sounds to its already vast store, laying up in its most retentive memory the various noises that are produced by man and his surroundings, and introducing among its other imitations the barking of dogs, the harsh "setting" of saws, the whirring buzz of the millstone, the everlasting clack of the hoppers, the dull heavy blow of the mallet, and the cracking of splitting timbers, the fragments of songs whistled by the labourers, the creaking of ungreased wheels, the neighing of horses, the plaintive baa of the sheep, and the deep lowing of the oxen, together with all the innumerable and accidental sounds which are necessarily produced through human means. Unfortunately, the bird is rather apt to spoil his own wonderful song by a sudden introduction of one of these inharmonious sounds, so that the listener, whose ear is being delighted with a succession of the softest and richest-toned vocalists, will suddenly be electrified with the loud shriek of the angry hawk or the grating whirr of the grindstone.

It is impossible to do justice to this most wonderful bird without quoting largely from those writers who speak from personal experience, and I therefore take the following passage from Wilson.

"In measure and accent he faithfully follows his originals; in force and sweetness of expression he greatly improves upon them. In his native groves, mounted on the top of a tall bush or half-grown tree, in the dawn of dewy morning, while the woods are already vocal with a multitude of warblers, his admirable song rises pre-eminent. Over every other competitor the ear can listen to his music alone, to which that of all birds seems a mere accompaniment.

Neither is this strain altogether imitative. His own native notes, which are easily distinguishable by such as are well acquainted with those of our various song-birds, are full and bold, and varied seemingly beyond all limit. They consist of short expressions of two, or three, or at the most four or six syllables, generally interspersed with intonations, and all of them uttered with great emphasis and rapidity, and continued with unlimited ardour for half an hour or an hour at a time. His expanded wings and tail glistening with white, and the buoyant gaiety of his action arresting the eye, as his cry most

MOCKING-BIRD.—*Mimus polyglottus.*

irresistibly does the ear, he sweeps round with enthusiastic ecstasy, as he mounts or descends as his song swells or dies away ; and as my friend Mr. Bartram has beautifully expressed it, 'He bounds aloft with the celerity of an arrow, as if to recover or recall his very soul, expired in the last elevated strain.

While thus exerting himself, a bystander destitute of sight would suppose that the whole feathered tribe had assembled together on a trial of skill, each striving to produce his utmost effect, so perfect are his imitations. He many times deceives the sportsman, and sends him in search of birds that are perhaps not within miles of him, but whose notes he exactly imitates. Even birds themselves are imposed upon by this admirable mimic, and are decoyed by the fancied calls of their mates, or dive with precipitation into the depths of thickets at the scream of what they suppose to be the sparrow hawk."

It is a very remarkable circumstance that one single bird always assumes the mastery in each district, and that whenever he begins to sing, the others cease from their perform-ances, and retire to a distance from the spot where the master bird has taken his stand, so that their voices are only heard as if in distant echoes to his nobler strains. The bird can easily be tamed, and when it turns out to be a good songster, is a most valuable bird, twenty-five pounds having been offered and refused for a good specimen. I knew of one case where a young Mocking-bird was brought over to England, and lived in the family for nearly two years, displaying its imitative talents in a very wonderful manner. It thrived well, and died from the effect of an accident, its legs having been crushed in a doorway.

The male bird can be distinguished from the female by the breadth and pure tint of the white band on the wings. In the adult bird, the white colour ought to spread over all the primary feathers, extending away below the white coverts ; the dark colour of the back is also of a more blackish hue.

The nest of this bird is usually placed in some thick bush, and is in general very carefully concealed. Sometimes, however, when the bird builds in localities where it knows that it will be protected from human interference, it is quite indifferent about the conceal-ment of its home, and trusts to its own prowess for the defence of its mate and young. When engaged in the business of incubation, the Mocking-bird suffers no foe to approach within the charmed circle of its home duties, and jealously attacks hawk, cat, or snake, in defence of its family. The fiercest war is, however, waged against the black snake, a reptile which makes many a meal on the eggs and young of various birds, and is in no wise disposed to spare those of the Mocking-bird. Against this terrible foe both parents aim their fiercest blows, and it often happens that the snake which has writhed its way to the Mocking-bird's nest in hope of devouring the callow young, pays with its life for its temerity, and falls dead to the ground, while the victor bird pours forth a song of triumphant congratulation. The nest is always placed at a short distance from the ground, being seldom seen at an elevation of more than eight feet.

The materials of which the nest is composed are generally dried weeds and very slender twigs as a foundation; straw, hay, wool, dried leaves, and moss, as the main wall; and fine vegetable fibres as the lining. The eggs are four or five in number, and there are often two broods in the course of the year. The colour of the eggs is greenish-blue, spotted with amber-brown.

To bring up a young Mocking-bird is rather a difficult task, as it must be taken from the nest at a very early period of its life, and therefore requires the most unremitting attention. In the work of Mr. Webber, so often mentioned, is a very interesting account of the successful rearing of four very young Mocking-birds. They had been cruelly taken out of their nest by some mischievous and hard-hearted person before they had opened their eyes on life, and left to die on a small piece of carpeting. Mr. Webber and his sister happening to look at the nest, which they had long watched, discovered the poor little things lying cold and apparently lifeless on the carpet. He, however, found that the tiny hearts were still beating, and after inducing his sister to place them in her bosom, rode homewards at full speed. The rest of the story shall be told in his own words.

"We were at home, and we passed hurriedly into the garden. I called a little brother to join us; in a moment we were all three standing beneath the eaves of the summer-house. There was a small hole in the cornice of the eaves, and I knew that in this a pair of blue birds had nested, and supposed that they must be just about hatched now. My sister stood watching my proceedings with great anxiety, for they were entirely mysterious to her. She saw me take my little brother aside and whisper my directions to him; then the little fellow prepared to climb up the columns of the summer-house, and with my assistance reached the cornice. His little hand was inserted into the hole, and with the greatest care not to touch either the sides of the hole or the nest within, he daintily plucks out the young ones, one by one, and hands them down to me. They are the same age with the Mocking-birds, but smaller.

'Now, Sis, give me those little ones; and hurry, dear, for I am afraid the old ones, who have gone out for food, will come back.'

She is so flurried she does not realize what I am about to do, but hastily places the young birds, now warm and fully alive, in my hand. They are reached to my brother. 'Drop them in quick, quick! and come down. Jump! I'll catch you.'

Down he comes, and then after my whispering something more to him, he snatched the young blue-birds from my hand, and ran off among the shrubbery. At this moment we heard the sweet, clear warble of the blue-birds, and I drew my sister a short distance away, where, from behind a tall rose-bush, we could watch the proceedings of the old birds.

'What does all this mean, brother? what do you expect?' she asked, in a low, puzzled voice, for she did not know that the young blue-birds had been taken out—so dexterously had we managed, and only understood that her charge had been transferred to the nest.

'Brother, you surely can't expect that little blue-bird to take care of eight young ones —your fairy will have to help, sure enough!'

'Hush! hush!' said I, all eagerness, for, with an insect in its mouth, one of the old birds, twitting merrily, had alighted near the hole, and without hesitation glided in, and in a moment or two came forth again, without seeming to have observed that there was anything wrong. My heart beat more freely, for I saw that the insect had been left behind, clearly, in the throat of one of the intruders—for the bird plumed himself gaily outside, as if happy in having performed a pleasant duty. But this was the male bird, and it was the arriving of the female that I knew was most to be dreaded—for if the sharp instinct of the mother did not detect the fraud, I felt that it would succeed.

In my elation at my success so far, I had explained my object to my sister, who, as she did not understand about the making away with the young blue-birds, was now infinitely delighted at the probable success of the scheme, and I could scarcely keep within bounds her dancing impatience to see what the mother would do, hear what the mother would say! Here she comes! and in a business-like and straightforward way glided directly into the hole. We held our breaths, and stood on tiptoe. Out she darts with a low cry—still holding the insect in her mouth. Our hearts sunk—she has discovered all, and refuses to adopt the strangers! She flew to her mate, and seemed to communicate some sad intelligence to him. He was busily engaged in trimming his feathers, and merely straightened himself up for a moment, and then, with an air of the coolest indifference, proceeded with his occupation. The poor female seemed to be sadly distressed and puzzled; she flew around the nest, uttering a low, mournful cry—then returned to her philosophical mate for sympathy, which he seemed to be too busy with his feathers to spare just now. Then she would dart into the hole, stay a moment, and out again with the insect still in her mouth. Then she would circle round and round on the wing, as if searching for the cause of the disturbance, the nature of which she evidently did not clearly understand. So she continued to act, until the male, having arranged his feathers to his liking, flew off, with a pleasant call to her, in search of more food. This seemed to decide her uncertainty, for darting now into the nest, she immediately fed the worm to one of those lusty young fellows that had grown so wonderfully since she last went out, and then came forth chirping, and apparently reconciled, and followed her mate.

'There! it succeeds! it succeeds! They are safe now; these birds are more industrious than the Mocking-birds, and will feed them better! good! good!'

'Your fairy spell has succeeded, brother, sure enough!' and she clapped her hands and danced for joy; and I am not sure that I did not join her most obstreperously, for I never was more delighted in my life at the success of any little scheme.

I knew the birds were safe if the female ever fed them once. So it proved; for never did I see little fellows grow with greater lustihood than they. Daily we watched them; and in ten days or two weeks were greatly amused to see the industrious old birds perseveringly labouring to fill gaping throats that were nearly large enough to swallow them bodily whole. I now narrowed the hole with wire, so that the blue-birds could get in and the Mocking-birds could not get out, for they were quite double the size of their foster-parents.

When they were full fledged we took them to the house, and placed them in an aviary I had prepared for them, in a recess which contained a large window and looked out upon the gardens. In two days I found, to my great astonishment, the old blue-birds endeavouring to feed them through the wires. They had found them out, the faithful creatures, and not content with having already spent double the amount of labour upon them that they would have bestowed upon their own offspring, they followed them up with their unwearying solicitude.

I was greatly shocked at first to observe the cool indifference with which the young aristocrats of song surveyed their humble foster-parents. After a while it came—in spite of the shameful ingratitude it exhibited—to be a constant source of merriment with us to watch the lordly and impudent nonchalance with which they would turn their heads to one side, and look down at the poor blue-birds, fluttering against the bars with tender cries to attract their notice with an expression which seemed as plainly as could be to say, 'Who are you, pray? get away, you common fellows!'

A fine pair of old Mocking-birds found them, too; but when they came, our gentry behaved very differently, and seemed crazy to get out. They became very tame, and I finally fulfilled my vow of turning them loose, and for a long time they were so tame that they would take food from our hands anywhere. They lived on the place, and we felt ourselves for years afterwards plentifully, aye, bounteously rewarded for our anxiety on account of the little outcasts, by the glorious songs they sang for us the summer nights to dream by. Thus it was my fair sister helped me out of the scrape with my young Mocking-birds!"

The colour of the Mocking-bird is a dull brown, with a decided ashen tinge. The quill-feathers of the wings are white towards their base, and brown-black towards their extremities, the two central feathers of the tail are dark brownish-black, the two externals are white, and the remainder are white on their inner webs. The chin, throat, and whole of the under part of the body are very pale brown, inclining to grey. As has already been mentioned, the pure white of the wings and the blacker hue of the body afford sufficient indications of the male bird, while the tail is nearly equally white in both sexes. The length of the adult Mocking-bird is about nine inches.

The genus in which the true THRUSHES are placed is one of the largest yet established, containing nearly one hundred and twenty accredited species, which are found in almost all quarters of the globe. In England they are well represented by several familiar birds, together with one or two which, although they are not unfrequently found within our shores, are but little known to the general public.

The first example of this group is the MISSEL THRUSH, one of the largest and handsomest of the species.

It is one of our resident birds, and on account of its great size, its combative nature, its brightly feathered breast ,its rich voice, and gregarious habits, is one of the best known of the British birds. About the beginning of April the Missel Thrush sets about its nest, and in general builds a large, weighty edifice, that can be seen through the leafless bushes from a great distance. Sometimes, however, the nest is concealed with the greatest care, and I cannot but think that in the latter case it is the work of some old bird, who has learnt caution through bitter experience.

The materials of which the nest is composed are the most heterogeneous that can be imagined. Every substance that can be woven into a nest is pressed into the service. Moss, hay, straw, dead leaves, and grasses, are among the ruling substances that are employed for the purpose, and the bird often adds manufactured products, such as scraps of rag, paper or shavings. I once found one of these nests that was ingeniously placed in the crown of an old hat that had evidently been flung into the tree by some traveller. At first, it hardly looked like a nest, but there were a few bits of grass lying over the brim that had a very suspicious aspect, and on climbing the tree, the old hat was proved to have been made the basis of a warm nest, with the proper complement of eggs. As the nest is so conspicuous, and built so early in the season, the eggs of the Missel Thrush generally form, together with those of the hedge accentor, the first-fruits of a nesting expedition.

The nest, although so roughly made on the exterior, is the result of very careful workmanship. The outside walls are made of moss and hay, but there is a fine lining of mud, which, when dry, affords a very perfect resting-place for the eggs. The mud wall is again lined with soft grasses, so that the eggs and young have a warm bed whereon to repose.

At all times a tolerably quarrelsome bird, the Missel Thrush becomes doubly combative at the breeding season, boldly attacking and driving away birds of greater size and strength by the mere force of indomitable courage. Mr. Thompson has given so excellent an account of several of such combats, that I can but quote his words.

" May it not be in some degree to counterbalance the danger to which its nest is subjected from the exposed site (selected according to the dictates of nature) that this bird is endued with the extraordinary courage and perseverance manifested in its defence ? Often have I seen a pair of these birds driving off magpies and occasionally fighting against four of them. The pair to which the first-mentioned nest belonged, attacked a kestrel which appeared in their neighbourhood when the young birds were out, although probably without any felonious intent upon them. One of these thrushes struck the hawk several times and made as many more attempts to do so, but in vain, as the latter, by suddenly rising in the air, escaped the coming blow. This pair of birds followed the kestrel for a great way, until they were lost to our sight in the distance.

MISSEL THRUSH.—*Turdus viscivorus.*

In the wood at Cultra I was once, at the end of April, witness to a single Missel Thrush boldly attacking a kestrel, which fled before it. The courage of the thrush was further evinced by its flying to the summit of the highest pine in the plantation, from which commanding site it for a long time proudly looked defiance against all comers. But by superior numbers Missel Thrushes are, like their betters, sometimes overpowered. This happened at the Falls on one occasion, when a pair of grey crows (*corvus cornix*) joined, or it may be followed in the wake of a pair of magpies, in their assault on a nest, and the thrushes were completely routed.

A pair of these birds, which bred at the residence of a gentleman of my acquaintance near Belfast in the summer of 1837, flew angrily towards himself whenever he walked in the direction of their nest. But the Missel Thrush can exhibit boldness without its nest being attacked. At the end of June, 1848, a friend brought from Scotland to his residence near Belfast, four young peregrine falcons. The first day that these birds, then full grown, were placed out of doors upon their blocks, contiguously, four in a row, they were assailed by a Missel Thrush, which, for several hours, continued dashing down at them, and all but, if not actually, striking them occasionally. No reason, such as having a nest in the vicinity, &c., could be assigned for the thrush's inhospitable welcome to the Scotch falcons."

Towards the end of the summer the Missel Thrushes assemble in flocks of considerable size, and in the autumn often do great harm to gardens and plantations, by devouring the fruit. They are particularly fond of raspberries and cherries, and have been known entirely to ruin the crop of these fruits. They are also fond of the berries of the mountain ash and the arbutus, and are so partial to the viscid berries of the mistletoe plant that they have been called by its name.   Insects of various kinds, caterpillars, and spiders also, form part of the Missel Thrush's diet, and a partly digested lizard has been found in the interior of one of these birds.

The song of the Missel Thrush is rich, loud, clear, and ringing, and is often uttered during the stormiest period of the year, the bird seeming to prefer the roughest and most inclement weather for the exercise of its voice.

Few birds have been known under such a variety of names as this species, its title seeming to vary in different parts of Britain according to the locality.   For example, in allusion to its habit of singing during stormy weather, it is known in many places by the name of Storm-cock; in some counties it goes under the name of Holm Thrush, in others it is confounded with the shrikes and called the Butcher-bird, while in others it is actually termed the Jay.   This curious misnomer holds good in several parts of Ireland as well as in England, and I was once rather victimized by its adoption.   In the year 1849-50 I was engaged in collecting the eggs of the Wiltshire birds, and hearing from a rustic that the "Jay-pie" built in the neighbourhood, I offered him a small sum for every Jay-pie's egg brought to me unbroken.   In a few days the lad came with two hats filled with the eggs of the Missel Thrush, expecting—and receiving—the stipulated sum for each egg.   After that experience, I always made the narrator describe the bird before I commissioned him to procure its eggs.

The colouring of this bird is briefly as follows: The upper parts of the body are a warm reddish brown, excepting the wings, where the brown is of a more sober hue.   The upper surface of the tail is also brown, excepting a patch or two of greyish white upon the outer webs of several of the tail-feathers.   The under surface of the body is yellowish white, covered thickly with jetty black spots, triangular on the neck and throat, and round on the chest and abdomen.   In total length the bird measures very nearly a foot.

ANOTHER large example of the British Thrushes is found in the FIELDFARE.

This bird is one of the migratory species, making only a winter visit to this country, and often meeting a very inhospitable reception from the gun of the winter sportsboy. Very seldom is it seen in this country till November, and is often absent until the cold month of December, when it makes its appearance in great flocks, searching eagerly for food over the fields. At this period of the year they are very wild, and can with difficulty be approached within gunshot, as I have often experienced in my younger days.   I well remember "stalking" a little troop of these birds for several hours, being induced to do so by their extreme shyness, and at last securing one of them by pushing the gun through a drain-hole in an old stone wall, getting rather an uncertain aim through the dried grass stems, and sending the shot within an inch or two of the ground.   When the snow lies heavily upon the fields, this bird betakes itself to the hedgerows and outskirts of woods and copses, and there feeds on the various berries that have survived the autumn.   During this inclement season, the Fieldfare may be approached and shot without much difficulty. Their shyness, however, depends greatly on the amount of persecution which they have sustained.

Although they collect in large flocks, the different individuals always keep themselves rather aloof from their fellows, but as night approaches they close together, and nestle in companies among the hedges or brushwood.   They generally remain in this country until May or June, seldom, however, prolonging their stay to the latter period. In this land they have not been observed to build, but in the northern parts of Europe, such as Norway and Sweden, their nesting is really extraordinary.   A very excellent account of the nidification of these birds is given by Mr. Hewitson.   His attention was aroused by the loud shrieking cries of several birds, "which we at first supposed must be shrikes, but which afterwards proved to be Fieldfares, anxiously watching over their newly-established dwellings.

We were soon delighted by the discovery of several of their nests, and were surprised to find them (so contrary to the habits of other species of *Turdus* with which we are acquainted) herding in society.   Their nests were at various heights from the ground, from four feet to thirty or forty feet or upwards, mixed with old ones of the preceding

FIELDFARE —*Turdus piláris.*

year. They were, for the most part, placed against the trunk of the spruce fir ; some were, however, at a considerable distance from it, upon the upper surface, and towards the smaller end of the thicker branches. They resembled most nearly those of the ring ouzel. The outside is composed of sticks and coarse grasses, and weeds gathered wet, matted together with a small quantity of clay, and lined with a thick bed of fine long grass. None of them yet contained more than three eggs, although we afterwards found that five was more commonly the number than four, and even six was very frequent. They are very similar to those of the blackbird, and even more so to the ring ouzel.

The Fieldfare is the most abundant bird in Norway, and is generally diffused over that part which we visited ; building, as already noticed, in societies ; two hundred nests or more being frequently seen within a very small space."

In their general aspect, the nests are not unlike those of the blackbird, and the eggs are of a light blue ground colour, covered with dark reddish-brown mottlings. Although the bird is essentially a winter visitant to this country, there are seasons which are too cold and stormy even for this hardy bird. In the year 1798, there was a terrible and lengthened storm of sleet, wind, and snow, which killed thousands of the Fieldfares,

RING OUZEL.—*Turdus torquatus.*

and even dashed them into the sea, where they were drowned, and their bodies thrown upon the Devonshire coast for many days afterwards.

In its colour the Fieldfare bears a decided resemblance to the generality of the Thrushes. The upper parts of the body as far as the shoulders are ashen grey, dotted with dark brown spots upon the head; the back and wings are rich brown, and the tail is dark blackish-brown. The chin and throat are a peculiar golden hue, not unlike amber, and covered with numerous black streaks; the breast is reddish brown, also spotted with black, and the abdomen and under parts white, spotted on the flanks and under tail-coverts with brown of various shades. The Fieldfare is not quite so large a bird as the Missel Thrush, being about ten inches in total length.

The RING OUZEL is also only a visitant of England, but its times of arrival and departure are precisely contrary to those of the bird just described.

This species seldom arrives in England until the month of April, and as it generally confines itself to certain districts, is not very common. It is, however, sparingly found in most parts of England, Ireland, and Scotland, though it seems to prefer the more western and southern districts. The name of Ring Ouzel has been universally given to this bird on account of the broad white band that partially surrounds the lower portions of the throat, and is very conspicuous in its contrast with the deep black brown of the rest of the plumage. With the exception of this white band, the general plumage of the Ring Ouzel is very like that of the male blackbird, which it also resembles in size and general form.

It is a shy and wary bird, shunning cultivated grounds and the vicinity of human habitations, and withdrawing itself into the wildest and most hilly districts. It is a quick-flying, lively and active bird, and is said to afford fine sport to the falconer, owing to its singular adroitness and ingenuity in escaping the stroke of the hawk. It will quietly suffer the bird of prey to approach quite closely, screaming a defiance to the enemy, and flitting quietly along a stone wall or rocky ground. Suddenly the hawk makes its swoop, and the Ring Ouzel disappears, having whisked into some hole in the stone, squeezed itself into a convenient crevice, or slipped over the other side of the wall just as the hawk shot past the spot on which it had been sitting.

The song of this bird is loud, clear, and sonorous, but contains a very few notes. The Ring Ouzel can also, when alarmed, utter a loud and hoarse screech, which seems to give warning of danger to every bird within hearing.

The nest of this species is large, and is composed of coarse grasses externally, lined with a thin shell of clay, which is again lined with soft and warm grass. The eggs are of a brightish blue covered with many spots and little dashes of dark reddish brown; their full complement is five. The nest is always placed near the ground in some sheltered situation, a tuft of rank grass, a thick bunch of heather, or the base of a luxuriant bush, being among the most common localities. After the breeding season, the Ring Ouzels assemble towards the southern parts of England, collecting together in flocks preparatory to their departure. During this intermediate period they visit the gardens and orchards, and often commit sad havoc among the fruit.

The general colour of the adult male bird is very dark blackish-brown, slightly varied by the blackish-grey edges of the feathers, and the broad grey outer webs of the wing-feathers. Across the upper part of the chest runs a broad, crescent-shaped mark of the purest white, the points being directed upwards. In the young male this collar is not so broad, and of a decidedly reddish hue, and the whole of the plumage is of a lighter brown. Sometimes the white collar is entirely absent, and in some cases white and pied varieties have been known. The total length of the adult bird is about eleven inches.

ANOTHER well-known example of the British Thrushes is found in the common REDWING, a bird which is plentiful throughout the greater portion of the British Isles.

It is one of the finest songsters even among its own melodious group, rivalling the nightingale in the full sweet tones of its flexible voice. Sometimes the bird sings alone, seated on a favourite perch, but it oftener prefers lifting up its voice in concert with its companions, and fills the air with its harmonious sounds. It has, however, several kinds of voice, sometimes pouring forth its full rich strains, and at other times singing quietly to itself in an under tone that can only be heard at a very short distance. This, however, is only the peculiar sound which is termed "recording" by bird-fanciers, and must not be mistaken for the real song, which, according to Mr. Hewitson, who had every opportunity of hearing this bird, is a loud, wild, and delicious melody. The Redwing partakes so far of the character of the nightingale as to sing after sunset.

The Redwing is less of a fruit-eater than the generality of its kind, feeding principally upon worms, slugs, and insects. In a protracted and severe winter, therefore, when the ground is frozen so hard that the bird's beak cannot penetrate its stony surface, the Redwing is forced to rely for its subsistence on the hibernating molluscs, and the larvæ and pupæ of different insects which may be found in sheltered spots. But when these resources have been exhausted, the poor bird is in a sad plight, and has been known to die of sheer starvation.

During the summer months, the Redwing goes northward, visiting Norway, Sweden, and even Iceland. In these countries it generally builds its nest, which is similar to that of the common blackbird, and is placed in the centre of some thick bush. Occasionally, but very rarely, the Redwing has been known to build in this country, and Mr. Yarrell records two such instances in England, and another in Scotland. The eggs are from four to six in number, and of a blue colour, spotted with black.

The Redwing is, like many of its kind, a sociable bird, gathering together in large flocks, and roosting sociably in company on the thickly matted branches and twigs of

REDWING.—*Turdus iliacus.*

hedgerows and well-wooded plantations. During the winter the birds scatter themselves rather widely; but immediately after their arrival, and before their departure, they gather themselves into societies, and are then sadly persecuted by the fowler.

The general colour of the Redwing is a warm, rich cinnamon-brown upon the upper parts of the body, the wings are rather darker, except the external webs of the quill-feathers, which have a greyish tinge, and over the eye runs a well-defined streak of very pale ashen-brown. The chin, throat, and whole under surface of the body are greyish-white, deepening into a brownish tinge on the sides of the neck, the breast, and the flanks, and profusely studded with longitudinal dashes of the same brown as that of the back. When the wings are closed, the bird very much resembles the common Thrush, but when it spreads its wings for flight, it discloses a large patch of orange-red feathers upon the sides of the body, from which it has derived its name of Redwing. In total length the Redwing nearly reaches nine inches.

The CAT-BIRD, so called from the resemblance which some of its notes bear to the mew and purr of a cat, is a native of America, and one of the most familiar of the birds of that country. As may be seen by the generic title which it bears, it is one of the true Thrushes.

In its character it is one of the most affectionate of birds, as is shown by Wilson in the following passage.

" In passing through the woods in summer, I have sometimes amused myself with imitating the violent chirping or squeaking of young birds, in order to observe what different species were around me; for such sounds at such a season in the woods are no less alarming to the feathered tenants of the bushes, than the cry of fire or murder in the streets is to the inhabitants of a large city.

On such occasions of alarm and consternation, the Cat-Bird is the first to make his appearance, not singly, but sometimes half a dozen at a time, flying from different quarters to the spot. At this time those who are disposed to play with his feelings may almost throw him into fits, his emotion and agitation are so great at the distressful cries of what he supposes to be his suffering young.

CAT-BIRD.—*Turdus félivox.*

Other birds are variously affected, but none show symptoms of such extreme suffering. He hurries backwards and forwards, with hanging wings and open mouth, calling out louder and faster, and actually screaming with distress, till he appears hoarse with his exertions. He attempts no offensive means ; but he bewails, he implores, in the most pathetic terms with which nature has supplied him, and with an agony of feeling which is truly affecting. Every feathered neighbour within hearing hastens to the spot to learn the cause of the alarm, peeping about with looks of consternation and sympathy. But their own powerful parental duties and domestic concerns soon oblige each to withdraw. At any other season the most perfect imitations have no effect whatever on him.

It is a most courageous little creature, and in defence of its young is as bold as the mocking-bird. Snakes especially are the aversion of the Cat-bird, which will generally contrive to drive away any snake that may approach the beloved spot. The voice of this bird is mellow and rich, and according to Audubon is "a compound of many of the gentle trills and sweet undulations of our various woodland choristers, delivered with apparent caution and with all the attention and softness necessary to enable the performer to please the ear of its mate. Each cadence passes on without faltering, and if you are acquainted with the songs of the birds he so sweetly imitates, you are sure to recognise the manner of the different species."

It is a most lively and withal petulant bird in a wild state, performing the most grotesque manœuvres, and being so filled with curiosity that it follows any strange being through the woods as if irresistibly attracted by some magnetic charm. In its disposition the Cat-bird appears to be one of the most sensitively affectionate birds on the face of the earth, as will appear from the following interesting account of a pet Cat-bird, called General Bem. The narrator is Mrs. Webber.

"Well, General Bem went home with us at once, and was immediately given his liberty, which he made use of by peering into every closet, examining and dragging every thing from its proper place, which he could manage, pecking and squalling, dashing hither and thither, until at night he quietly went into his cage as if he was nearly or quite positive that he must commence a new career on the morrow ; it was evident that he had to begin the world over again, yet, as he was not superannuated, and was, withal, ambitious, his case was still not a desperate one, although we had assured him most positively that we would not fall in love with him—we had only invited him there to help us pass the time.

Bem looked wise at the assertion, but said nothing. The next morning we gave him water for a bath, which he immediately used, and then sprang upon my head, very much to my surprise ; then he darted to the window, then back to my head, screaming all the time most vociferously, until finally I went to the window, for peace' sake, and stood in the sunshine, while Bem composedly dressed his feathers, standing on my head first on one foot, then on the other, evidently using my scalp as a sort of foot-stone, and my head as a movable pedestal for his impudent generalship to perch on when he felt disposed to be comfortably elevated ; and had clearly come to the conclusion—as I was so fond of transporting him from his native land, that I should serve as a convenient craft to bear him where his moods commanded.

In a word, he had determined to turn tyrant ; if I had had the deliberate purpose of using him as a mere toy, he had at least the coolness to make me available, and from that time I became the victim of the most unequalled tyranny. Did I neglect his morning bath beyond the instant, my ears were assailed with screams and cries, till I was forced to my duty ; I must bear him into the sunshine, or my hair was pulled ; I must bring him his breakfast, or he pecked my cheek and lips ; in fine, I was compelled to become his constant attendant, while in the meantime he most diligently assailed my heart by endearing confidences. He would sit upon my arm and sleep, he would get into my workbox, and while I watched that he did not pilfer a little, he would quietly seat himself upon its edge, and in a low sweet voice lull my suspicions by such tender melodies, that finally I could no longer say, 'I will not love you, Bem !' but gave him the satisfactory assurance that he was not quite so much of a tease as I had tried to think him ; and he now received my daily offering of small spiders and worms with gestures of evident pleasure.

These were always presented to him enveloped in white paper, which he carefully opened and secured his prey, before it could escape, even though it was sometimes a difficult task to keep his vigilant eye upon so many—apparently escapading—when I was called to the field, and appointing me a station, I was expected to give the alarm when one attempted to get away on my side, which he immediately killed and dropped, and then darted after those on the outskirts of the field of action.

At last, one day, Mr. Webber brought for my sister a Wood-thrush, which was very wild and savage, and was, besides, extremely ugly, but had the reputation of being a good singer, which made us forgive his sullen temper, and hope to win him back to more gentle ways, when he should see that we would be his friends, and that he should be almost free ; besides, General Bem was much inclined to make his acquaintance, and took the first occasion to pay him a visit in his cage-house. This the stranger did not fancy, and drove him out. Bem resented this, by turning on the threshold and pouring forth a torrent of screams and mewings, which came near distracting the poor Thrush, who darted at him and chased him to the bed, under which Bem darted, and was secure for the present.

But from that time there were no more overtures of friendship, they were sworn enemies ; the Thrush from detestation of the impudent fellow who invaded his residence, and finally appropriated it, to the entire desertion of his own, which, by-the-bye, was much larger, and with which the Thrush eventually consoled himself, and Bem continued to occupy, because it amused him to pester the ill-natured fellow, which he had set down the Thrush to be. Many were the quaint scenes which now daily occurred.

If Bem desired to take a bathe, the Thrush would endeavour to push him out ; but Bem was not to be ousted in that style if he could prevent it, and commonly sent the poor Thrush away in consternation, his musical ear stunned by such direful din as threatened to rend his delicate heart as well as tympanum. Never shall I forget one droll scene, One day Bem found on the floor a white grape, which he seemed to be disposed comfortably to discuss, after having rolled it out into the broad sunshine. Just at this moment the

Thrush stepped up in a cool and dignified manner, and carried the grape off; dropped it in the shade, and deliberately drawing up one foot among his feathers, seemed to say, 'I claim the grape as my own; I stand on the defensive; come and get it if you dare!' so closed the 'off' eye and looked as if the matter was settled to his entire satisfaction.

Bem had been in the very act of pecking the grape when it was so unceremoniously withdrawn; he drew himself up on tiptoe fairly with astonishment, his eye seemed to grow larger and rounder, the feathers on his head stood alternately erect and clung close to the scalp; he stood a moment or two, and then with a loud 'mew' darted forward to recapture the stolen fruit, but the Thrush coolly and silently met him with open mouth and body thrown forward, yet still covering the grape. Bem's wit returned to him—he quietly turned off, as if it was a small matter anyhow.

We were astonished. Was Bem a coward after all? would he permit this bird, even if he was larger, to impose upon him in this fashion, and he able to whip mocking-birds at that? We shook our heads; if Bem does that, we shall withdraw his laurels. But see! he comes cautiously about the Thrush—what does he mean? ah, we perceive; Bem has sagaciously only changed his tactics, we will watch him; he thinks the Thrush will want some dinner pretty soon, and then, as Bem disdains to be called quarrelsome, he will quietly appropriate his treasure.

Four hours things retained this position, the Thrush never moving more than six inches from his post, though evidently becoming hungry and weary, while Bem silently wandered about the room, feasting in the most provokingly cool way in both cages, and continually making inadvertent incursions in the neighbourhood of his enemy, as if for the purpose of throwing him off his guard. At last, Bem was on the other side of the room. The Thrush had been eyeing a dainty morsel which Bem had dropped about two feet from him. He looked, Bem was too much engaged to notice him, he could easily venture—he would—he did. Bem, whose keen eye had seen all, darted like lightning and before the Thrush could turn about and seize again the contested treasure, Bem had alighted on the centre of the bed—the only place in the room where the Thrush would not follow him—and there quietly tore the grape to pieces and left it.

But, alas! we had to send our brave sagacious Bem home again. We were to make a long journey to the South, and he must stay behind. Ah, the poor fellow knew as well as we, that we were bidding him adieu. He pecked our fingers in great distress, and bit our lips till the blood came, in the energy of his farewell—while he uttered such sad low cries as made us mourn for many a day in the remembrance.

During our absence we wrote frequently inquiring of Bem, and many an injunction to him, to live and die, if need be, the same brave general we had known him. We never expected to see him again; but, after a year of wandering, we did return to our old home. At once we went to see the general, little dreaming that we should be remembered. What was our surprise then, when we called 'Bem! Bem! General Bem!' to see our dear friend and pet dart down to us from his hiding-place, and most evidently recognise us—his eye sparkling, his scalp-feathers raised, his wings drooping, and that same low cry which had haunted us so long, greeting us again. Our happiness was real; and when we offered him the white paper, he instantly darted upon it, and tore it asunder to get the well-remembered treasure he had always found within.

Again Bem went home with us—this time to fill our hearts with affection by his quaint impish ways and gentle waywardness. Now, he became a privileged character; my paint box was his especial admiration—he treated it with great veneration, having discovered that birds grew out of the little square pebbles, as he doubtless considered them, until one day he perceived I objected to his lifting from its case a black-looking, ill-shaped piece of paint, that I was even decidedly opposed to his meddling with it; from that moment that particular piece became a treasure—its value so great to him, that, hide where I might, it had ever an invisible glitter, which to his eyes was brighter than any gem; he would find and hide it from me, and thus I had at least once every day to search the room over for this indispensable colour.

No matter that I threatened him, he coolly dressed his feathers and commenced so dreamy a song as to soothe my rage at once. He became my constant companion; he bathed with me in the morning, he took his dinner with me from my plate, and perched at night close to my head. He sat on my shoulder or head when I worked, and seemed to express his opinion in regard to my progress in bird-making with quite a connoisseuring air. He grew to be profoundly jealous of all other birds ; and if I talked to a fine mocking-bird, whose cage hung in my room, he would become so enraged, and finally depressed, that I became alarmed—I feared he would die.

One day I had given this bird some water ; my hand was in the cage, the mocking-bird was pecking at my fingers, when with a loud and vicious scream General Bem dashed from the floor up into the cage, and commenced a violent assault on the inmate. The struggle was but for a moment ; he dashed out and I shut the cage-door, while Bem, mounted on the bed-post, sent forth such yells of fury as I never heard from bird's lungs before. I could not pacify him for a long time—several hours ; he hid in the shade of the furniture, and would not be induced to come out. The next day the mocker was flying about the room, Bem assailed him, and the fight became so desperate that I was obliged to send the mocking-bird away, while my poor Bem was seized with convulsions, and I thought him dead after a few moments. But his time had not yet come ; he lived to pass through many such scenes of painful suffering."

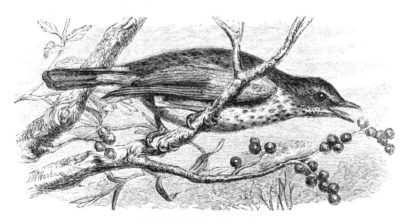

HERMIT THRUSH.—*Turdus solitarius.*

The HERMIT THRUSH is also a native of America, and is generally found in the countries adjoining to the Mississippi during the winter, making a partial migration to Kentucky, Indiana, and Tennessee during the summer.

The name of Hermit is given to this bird on account of its eremitical and retiring habits, for this Thrush withdraws itself from the open country and systematically hides itself in the darkest and most secluded cane-brakes. Even when it ventures into the more open lands in search of food, it does not make itself conspicuous, but keeps ever near the ground, flitting with swift and steady wing to and from the lonely brake where its nest is placed. This species is not known to possess any true song, merely uttering a very low and plaintive cry.

The nest of the Hermit Thrush is always placed in the thickest shelter, and is composed of dried leaves and grasses without any mud wall, and lined with grasses of a fine character. The eggs are about five in number, and their colour a light blue, variegated with black-brown spots on the larger end. There are usually two broods in the year. The food of this species is almost wholly of a vegetable character. The general colour is buffy-brown on the upper parts, warming into a decided ruddy tinge upon the tail and

BLACKBIRD —*Turdus mérula.*

upper tail-coverts. The under parts are greyish-brown covered on the neck and chest with spots of a darker hue. The total length of the adult bird is about seven inches.

AMONG the best known and best loved of our British songsters, the BLACKBIRD is one of the most conspicuous.

This well-known bird derives its popular name from the uniformly black hue of its plumage, which is only relieved by the bright orange-coloured bill of the male bird. The song of this creature is remarkable for its full mellowness of note, and is ever a welcome sound to the lover of nature, and her vocal and visual harmonies. Often the poor bird suffers for its voice; and being kept within the bars of a cage, is forced to sing its wild native notes "in a strange land." In captivity it is sometimes subjected to training, and has been taught to whistle tunes with great spirit and precision. Generally the bird sings in the daytime, but there are times when it encroaches upon the acknowledged province of the nightingale, and makes the night echoes ring with its rich ringing tones.

It is rather curious that even in its native state the Blackbird is something of a mimic, and will imitate the voices of other birds with remarkable skill, even teaching itself to crow like a cock and to cackle like a hen.

The Blackbird feeds usually on insects, but it also possesses a great love of fruit, and in the autumn ravages the gardens and orchards in a most destructive manner, picking out all the best and ripest fruit, and wisely leaving the still immatured produce to ripen on the branches. Perhaps it may be partly carnivorous, as one of these birds was seen to attack and kill a shrew mouse.

As it is so common a bird, and constantly haunts the hedgerows, it is greatly persecuted by juvenile gunners, whom it contrives to draw away from its nest by flitting in and out of the hedge, always taking care to keep out of shot range, and having a curious habit of slipping through the hedge, and flying quietly back to its nest, almost touching the surface of the ground in its rapid progress. It is not a sociable bird, being seldom seen in company with others of its own species, and not often even together with its mate.

z 2

The nest of this bird is made very early in the spring, and is always carefully placed in the centre of some thick bush, a spreading holly-tree being a very favourite locality. It is a large, rough, but carefully constructed habitation, being made externally of grass, stems, and roots, plastered on the interior with a rather thick lining of coarse mud, which when thoroughly dried forms a kind of rude earthenware cup. A lining of fine grass is placed within the earthen cup, and upon this lining the five eggs are laid. These eggs are of a light greyish-blue ground colour, splashed, spotted, and freckled over their entire surface with brown of various shades and intensity. The colouring of these eggs is extremely variable, even those of a single nest being very different in their appearance; and I once took a Blackbird's nest in which the eggs were so curiously marked that no one could have decided whether they belonged to a blackbird or a thrush. Sometimes the spots are almost wholly absent, and at other times the eggs are so covered with reddish-brown markings that the ground colour is hardly discernible.

The Blackbird is very courageous in defence of its nest, and will attack almost any animal that threatens the security of its home. On one occasion a prowling cat was forced to retreat ignominiously from the united assaults of two Blackbirds near whose domicile she had ventured.

The colour of the adult male bird is a uniform deep brown-black over the whole of the plumage, with the exception of the under surface of the wings, which have a decided wash of shining grey. The female is darkish brown upon the upper surface, mouse-brown upon the abdomen and sides, and yellowish brown upon the throat. In total length the bird measures about ten inches.

THE well-known SONG-THRUSH, or THROSTLE, as it is sometimes called, bears a deservedly high rank among our British birds of song.

It is plentifully found in most parts of England, and favours us with its vocal efforts throughout a considerable portion of the year. The song of the Thrush is peculiarly rich, mellow, and sustained, and is remarkable for the full purity of its intonation and the variety of its notes. The Thrush begins to sing as soon as incubation commences, and continues its song from the beginning of the spring until the middle of autumn. In many cases the bird sings to a very late period of the year, and has been heard in the months of November and December. On account of its beautiful voice, it is in great request among bird-fanciers, and is sold in large numbers as a cage songster.

The Thrush is tolerably familiar with mankind, and haunts the neighbourhood of human habitations for the sake of the food which it finds in such localities. It is, nevertheless, rather shy towards human beings, and does not willingly permit itself to be approached. There are, however, exceptions to this general rule, and on occasions an individual bird will overcome its instinctive dread of the human race, and attach itself to some favoured person. In one such instance with which I am acquainted, the bird took a fancy to a man, persisted in following him about, and used to sit on his shoulder and sing with the greatest enthusiasm. By degrees, the bird became accustomed to other human beings, and would accompany its protector into the outhouses. It had a strange predilection for steam, and was fond of perching on the edge of washing-tubs, and would there sit and sing, though so enveloped in the thick vapour as to be hardly visible.

The food of the Thrush is mostly of an animal character, and consists largely of worms, snails, slugs, and similar creatures. In eating snails it is very dexterous, taking them in its bill, battering them against a stone until the shells are entirely crushed, and then swallowing the inclosed mollusc. When a Thrush has found a stone that suits his purpose peculiarly well, he brings all his snails to the spot, and leaves quite a large heap of empty snail-shells under the stone. One of the best examples that I have ever seen, was a large squared boulder-stone, forming part of a rustic stile in Wiltshire. There was a large pile of shells immediately under the stone, and the ground was strewed for some distance with the crushed fragments that had evidently been trodden upon and carried away by the feet of passengers.

The Thrush does not, however, confine itself wholly to this kind of diet, but in the autumn months feeds largely on berries and different fruits, being very fond of cherries,

SONG-THRUSH.—*Turdus musicus.*

and often working great havoc in an orchard or fruit-garden. But in spite of its occasional inroads upon the gardens, it deserves the gratitude of the agriculturist on account of its services in destroying the snails and other garden pests, and may well be allowed to take its autumnal toll of a few of the fruits of which it has been such an efficient preserver. In no case, indeed, does it become us to be over chary of admitting our fellow-creatures to a share of the good things, which are in reality no more the property of the man than of the bird ; remembering that although to man has been given the dominion over every inhabitant of earth, yet the beasts, the birds, and the creeping things have also received their gift of every green herb from the same Divine hand which entrusted man with an authority higher in degree, but not more authentic in origin.

The nest of the Thrush is rather large, and shaped like a basin. The shell of the nest is composed of roots and mosses, inside which is worked a rather thin but wonderfully compact layer of cow-dung and decayed wood, so strongly kneaded that when dry it will hold water almost as well as an earthenware vessel. Sometimes the bird employs rather strange materials for its nest, and I know of an instance where a Thrush carried off a lace cap that was hanging on a clothes-line, and worked it into the sides of its nest. There are usually five eggs, of a beautiful blue spotted with black. The spots are small, round, and well marked, and are extremely variable in size and number ; they are always gathered towards the larger end of the egg.

The fecundity of the Thrush is very great, a single pair having been seen to make five nests in the course of a single season (one of which was destroyed), and to rear seventeen young. The female bird was so tame, that she would permit herself to be fed while sitting on her eggs. I have seen a similar example of confidence in this bird, and have stroked the head of a female Thrush while engaged in the duty of incubation. The birds are very quick about their domestic affairs, as may be seen from the fact that a nest was begun on April 26th, the young hatched on May 19th, and all flown on May 28th. This nest, however, was placed in a very warm situation, which may have had some effect on the rapidity of the process.

Like many other good songsters, the Thrush possesses the power of imitating the notes and even the gestures of other birds, as may be seen by the following letter from a correspondent of the *Field* newspaper: "Some twelve years ago, our gardener caught a young Thrush well able to fly, and caged it. Happening to be in his cottage one day, I heard what I thought was a robin singing, and on looking in the direction of the sound, found that it proceeded from the Thrush. Not only had the bird caught the robin's song, but also its attitudes, the head and tail drooping. On inquiring, I was told that he had other accomplishments, and mocked the sounds of the poultry. These, however, I did not hear, but the robin's song was perfect."

The Thrush can even imitate with great accuracy the mellow sounds of the flute, and has been taught to whistle tunes played on that instrument by its instructor. I have been told of one of these birds by a lady who had it in her possession. The Thrush had been captured when very young, and before it had attained its ordinary juvenile plumage. The person who caught it, placed it in a cage for safety, and the bird soon became tame and loving. Finding that as it approached maturity, it attempted to imitate the notes of a flute which he was in the habit of playing, he determined to teach his bird some melody, and succeeded in making it whistle the "Blue Bells of Scotland" with perfect accuracy.

The colouring of the Thrush, though simple, is very pleasing to the eye. The back and upper surface is brown of slightly different shades, the chin is quite white, and the abdomen and under tail-coverts are greyish-white. The throat, the breast, and flanks, together with the sides of the neck, are yellow, thickly spotted with dark brown. The total length of the bird is about nine inches.

CHESTNUT-CAPPED TIMALIA.—*Timália pileáta.*

ANOTHER sub-family of the Thrushes is named after the genus which is accepted as its type, and is called by the title of Timalínæ. On account of their chattering propensities they are more popularly termed Babblers. Several examples of this group will be given in the following pages, the first of which is the CHESTNUT-CAPPED TIMALIA, a bird which derives its name from the peculiar colouring of the head.

This species is an inhabitant of Java, and is rather common in that country. It is a sufficiently familiar bird, approaching human habitations without much diffidence, and building in close proximity to the barn or the plantation. It is a pleasing songster, possessing a sweet and musical voice, though its song has but little variety, consisting of only five notes. A sixth note is sometimes added, but evidently forms no part of the real song. This melody is repeated at very short intervals, with a peculiarly slow and well-modulated intonation.

SPOTTED GROUND THRUSH, OR GROUND DOVE.—*Cinclosóma punctatum.*

In its habits this bird reminds the observer of the common English Thrush, and in its mode of feeding it also bears a great resemblance to that sweet songster. Its food consists chiefly of insects, which it captures principally on the ground, but it will vary its diet with snails, slugs, and other similar creatures, and will also feed upon berries and fruits. The bill is strong and thick, in order to enable the bird to capture and crush its food, and to disinter it from beneath the surface of the earth. Its feet are employed for the same purpose, and are consequently better developed than in the true Thrushes. The general form of the bird is rather thick and short in proportion to its size.

The colour of the Chestnut-capped Timalia is olive-brown on the upper portions of the body, with the exception of the head, which is coloured on the crown with deep chestnut. The under parts are of a lighter hue than the back. A white band passes over the eye. The throat and cheeks are pure white, and the breast is also white, but is marked by a series of jetty black stripes.

AUSTRALIA possesses a curious and valued specimen of this group, which is popularly called the SPOTTED GROUND THRUSH, or GROUND DOVE.

This bird is found throughout the greater part of Australia and Van Diemen's Land, and on account of the delicacy of its flesh is greatly prized by both natives and colonists. Being always attracted by certain localities, it may be easily found by every one who is acquainted with its habits. Unlike the generality of birds, it cares little for trees or bushes, and seldom is known to perch upon the branches, preferring the tops of low stone-covered hills, or rude and rocky gullies, having a decided predilection for those which are clothed with grass and scrubby brushwood. The spaces between fallen trees are also a favourite haunt of this bird.

The Spotted Ground Thrush is no great flyer, taking to wing with much reluctance, and seldom voluntarily raising itself in the air except to fly from one side of a gully to another. When it does take to flight, especially if alarmed, it rises with a loud fluttering noise, and proceeds through the air in an irregular and dipping manner. To compensate, however, for its imperfect power of wing, its legs are well developed, and render it an exceedingly fast runner, so that it is able to conceal itself with great rapidity as soon as it finds cause of alarm.

The nest of this species is a very loose and negligent kind of structure, made of leaves, the inner bark of trees, and various vegetable substances, laid carelessly together in some casual depression in the ground. The eggs are rather large in proportion to the size of the bird, and their colour is greyish white, covered with large olive-brown mottlings. According to Gould, their number is two, but the author of "Bush Wanderings in Australia" states that this bird lays three eggs. The young are able to run almost as soon as they leave the egg, and in two days their bodies are covered with a soft black down like that of the young water-hen. The flesh of the Ground Dove is remarkably good, and when the bird is fat it meets with a ready sale, and is generally disposed of to the game-dealers together with the painted quail, as it arrives and leaves at the same time with that bird.

The voice of the Ground Dove is not very sweet, its cry or song consisting of a low piping whistle.

The colour of this bird is somewhat similar to that of the Thrush or fieldfare, and is briefly as follows. The back and upper portions are brown, covered with black dashes. Over each eye is a white streak, and there is a white patch on the cheeks. The chest is grey, and the abdomen white, warming into reddish buff upon the flanks, and each feather being marked with a black centre. Between the abdomen and the chest is a black band. The total length of this bird is about ten inches, and its general proportions resemble those of the common fieldfare of England.

BLACK-FACED THRUSH.—*Garrulax Chinensis.*

SEVERAL species of Babblers possess a sweet song, others are admirable mimics, while others are remarkable for the strange oddity of their cry.

One of the best songsters in this group is the BLACK-FACED THRUSH, a native of the mountainous regions of India and China. This bird is very gregarious in its habits, assembling in large flocks and preferring the thickest jungles and deepest ravines to the open country. These flocks, although they are so deeply hidden, are easily discoverable by means of the extraordinary sounds which they emit, and which are said to resemble a chorus of wild laughter. The food of this bird, when at liberty, consists chiefly of fruit and insects; but when tamed, the Black-faced Thrush is rather carnivorous in its character, as will presently be seen. The nest of this bird is rather rudely constructed of little sticks and grasses, worked into some convenient hole in the side of a gully, and generally contains four eggs.

This species is easily tamed, and, as will be seen from the following notes made by Mr. Frith of a Black-faced Thrush that had been for some time in his possession, is a very eccentric and amusing creature.

" The bird was exceedingly tame and familiar, and delighted, like a cockatoo, in being caressed and tickled by the hand, when it would spread out its wings and assume very singular attitudes. It was naturally a fine songster, and a most universal imitator. Whenever chopped meat or other food was put into its cage, it always evinced a propensity to deposit the bits one by one between the wires ; and when a bee or wasp was offered, this bird would seize it instantly, and invariably turn its tail round and make the insect sting this several times successively before eating it. A large beetle it would place before it on the ground, and pierce it with a violent downward stroke of the bill : a small snake, about a foot long, it treated in like manner, transfixing the centre of the head ; it afterwards devoured about half the snake, holding it by one foot while it picked it with the bill, as was its common mode of feeding."

The LAUGHING CROW of India (*Garruláx leucôlophus*) is another species of the same genus, and is remarkable for the singular resemblance which its cry bears to the laughter of human beings. Its name of "leucolophus," or white-crested, has been given to it on account of the white feathers which are found on the crown of the head.

The GOLDEN ORIOLE is an extremely rare visitant of this country, having been but seldom observed within our coasts, but is far from uncommon in many parts of the Continent, especially the more southern portions of Europe, such as the shores of the Mediterranean and Southern Italy.

The Golden Oriole derives it name from the bright golden yellow with which the feathers of the adult male bird are largely tinged ; but as the full glory of its plumage is not displayed until the bird has entered its third year, it is possible that many specimens may have visited this country and again departed without having attracted particular attention. Mr. Yarrell, in his well-known History of the British Birds, mentions several instances where the Golden Oriole has been seen and even bred within the British Isles ; and there is a note from a correspondent of the *Field* newspaper, which records the capture of a pair of Golden Orioles, four young ones, and their nest near Ipswich. Whenever this bird does make its appearance in this country it always comes in the summer months, generally between April and September.

In Italy, this bird is quite common, and by the peasantry is supposed to announce the ripening of the fig, its peculiar cry being translated into a choice Italian sentence, signifying that the fruits have attained maturity. It is rather gregarious in its habits, generally associating in little flocks, and frequenting lofty trees and orchards, where it can obtain abundance of food.

It is an exceedingly shy and timorous bird, keeping carefully from man and his home, and only venturing into cultivated grounds for the sake of obtaining food. Even in such cases it is extremely cautious in its behaviour, and as it always takes the trouble to set sentries on guard, it cannot be approached without the greatest patience and wariness on the part of the sportsman or observer. Being generally found in the loneliest spots, and especially preferring the outskirts of forests, whence it can at once dive into the thick foliage and escape from danger, it often baffles the skill even of the practised fowler, who is forced to trust to the careful imitation of its note for his hope of getting within shot of this cunning bird. Moreover, the imitation must be exceedingly exact, for the ear of the Golden Oriole is wonderfully true and delicate, and if the bird detects the least error in the intonation, it takes instant alarm, and seeks for refuge in the deepest recesses of the forest. According to M. Bechstein, the Golden Oriole is so fearful of exposing itself, that it never perches upon a naked branch, always preferring those boughs which are most thickly covered with foliage, and which will consequently afford it the best shelter.

The food of the Golden Oriole consists chiefly of insects ; and as the bird is rather a voracious one, it is very serviceable in clearing away the caterpillars and other fruit-

GOLDEN ORIOLE.—*Oriŏlus gálbula.*

devouring creatures which are specially rife in the spring, and destroy so much fruit in its earliest stages. As is often the case with the insect-eating birds, the Golden Oriole has a great taste for fruit when it is quite ripe, and in the autumn is very fond of the best and mellowest fruits, having an especial predilection for cherries, figs, and grapes. Perhaps it may be able to detect the larva of some insect within the fruit, and to do good service by destroying it before it has come to maturity.

The nest of this bird is a very elegantly formed and well-constructed edifice, of a shallow cup-like shape, and usually placed in a horizontal fork of a convenient branch. The materials of which it is made are mostly delicate grass-stems interwoven with wool so firmly that the whole structure is strong and warm. The eggs are generally four or five in number, and their colour is purplish white, sparely marked with blotches of a deep red and ashen grey. It is believed that there is but one brood in the year, so that the species does not multiply very rapidly. Sometimes the bird is said to build a deep and purse-like nest, which is suspended from the forked branch instead of being placed upon it.

This species has a very peculiar note, loud, flute-like, and of a singularly articulate character, as may be supposed,from the fact already mentioned, that the Italian peasantry believe it to speak their language. Bechstein considers the note to resemble the word "puhlo," and many writers think that the different names of Oriole, Turiole, Loriot, Pirol, and Bülow are given to the creature in imitation of its cry.

The colour of the adult male is bright yellow over the whole of the head, neck, and body, with the exception of the wings, the two central tail-feathers, and the basal portions of the remaining feathers, which are jetty black, the two colours contrasting finely with each other. Across the eye runs a dark stripe, and the eyes themselves are bright pinky red. In the young bird the yellow is of a dusky greenish hue, and the black feathers are of a dingy brown, and according to Mr. Yarrell the young males after their first moult resemble the old females. In the second year the yellow of the back is more decided, and the wings and tail are of a deeper black, and in many of the remaining feathers the colours are less brilliant than in the bird of full plumage. It is rather curious that as the bird breeds in its second year, it is hardly possible to distinguish the sexes, both wearing

YELLOW-BREASTED CHAT.—*Ictéria víridis.*

the same greenish-yellow and brownish-black apparel. The total length of the Golden Oriole is not quite ten inches.

There are many other Orioles known to ornithologists which cannot be described in these pages for lack of space, and it must be sufficient to record the Mango Bird of India (*Oriolus Kundoo*), remarkable for its peculiarly melancholy cry, and the Black-Headed Oriole of Bengal (*Oriolus melanocéphalus*), notable for its lengthened monotonic flute-like note. None of the true Orioles are found in America; and the reader must be careful not to confound those birds, which are nearly allied to the starlings, with the Orioles of the eastern hemisphere, the only resemblance between them being a similarity of colouring.

Before quitting this interesting family of birds, we must give a passing notice to the Bulbuls, so well known by the repeated references to them in Oriental writings. Some of the species are possessed of remarkably sweet voices, and are popularly called nightingales. They are easily tamed, becoming very fond of a kind owner, and can be taught to perform many interesting tricks. One species is kept for the purpose of fighting, and is trained for this object as carefully as gamecocks were formerly trained for the cruel amusement of their owners. In a wild state they are generally found in the woods and jungles, and are in the habit of visiting gardens for the sake of preying upon the ripened fruits and insects. They are all exotic birds, and are only found in the eastern hemisphere.

The interesting family of the Flycatchers is composed of a large number of species, extremely variable in size, form, and colour. The average dimensions of these birds are about equal to those of a large sparrow, and many are smaller than that bird, although two or three species nearly equal the thrush in size. Their shape is always neat and elegant, and their plumage sits closely on the body in order to permit the short but rapid evolutions which they make in pursuit of their active prey. One or two, such as the Paradise and Fork-tailed Flycatchers, are remarkable for the mode in which the tail is elongated into a graceful and elegant train, and in other species the tail is broad and fan-like. In colour the Flycatchers are mostly of sober but pleasing tints, but there are several notable exceptions to the rule, such as the Crested Flycatcher (*Pyrocéphalus coronátus*), remarkable for the crown of fiery scarlet feathers which decorates the top of the head, the Blue Niltava (*Niltáva sundara*), which has its broad back and tail of a brilliant azure, and the *Selóphagus picta*, whose abdomen is of a bright scarlet.

The bill of the Flycatcher is of various lengths, but is almost invariably rather hard and flattened at the base, slightly curved at the point, and compressed towards the tip. At the corners of the mouth are generally several long bristles like those of the nightjar and probably placed there for the same purpose, *i. e.* to aid the bird in the capture of its insect prey. The wings are long and firmly made, and the feet are slender and feeble in comparison with the dimensions of the body.

ONE of the sub-families into which the Flycatchers are divided is known by the name of *Vireonínœ*, or GREENLETS, on account of the constant presence of green in some part of their plumage. They are all little birds, and are confined to the New World, inhabiting America, Brazil, Guiana, and the West Indian Islands. They are mostly insect-feeders, though they will vary their diet with fruits, berries, and other vegetable food. Many species of Greenlets are known to ornithologists, and some of them are remarkable for their eccentric habits and their curious mode of nesting.

The YELLOW-BREASTED CHAT of America is well worthy of notice, as it possesses a very prettily coloured plumage and elegant form, and at the same time is one of the most eccentrically behaved of the feathered creation, even surpassing in the whimsical oddity of its manners the mirth-provoking evolutions of the Demoiselle crane.

It is a partially migratory bird, having rather an extensive range in its native country, and passing from north to south according to the season of the year, and the warmth or inclemency of the weather. According to Wilson, it arrives in Pennsylvania about the first week in May, and departs for the south in the month of August. As is usually if not invariably the case with birds, its migrations are restricted to a narrow line, which runs almost due north and south, and the male birds always make their appearance before their mates. Of the habits of this bird, Wilson gives the following interesting description.

" When he has once taken up his residence in a favourite situation, which is almost always in close thickets of hazel, brambles, vines, and thick underwood, he becomes very jealous of his possessions, and seems offended at the least intrusion ; scolding every passenger as soon as they come within view in a great variety of odd and uncouth monosyllables, which it is difficult to describe, but which may be readily imitated so as to deceive the bird himself, and draw him after you for half a quarter of a mile at a time, as I have sometimes amused myself in doing, and frequently without once seeing him. On these occasions his responses are constant and rapid, strongly expressive of anger and anxiety, and while the bird itself remains unseen, the voice shifts from place to place among the bushes as if it proceeded from a spirit.

First is heard a repetition of short notes resembling the whistling of the wings of a duck or a teal, beginning loud and rapid, and falling lower and slower, till they end in detached notes ; then a succession of others, something like the barking of young puppies, is followed by a variety of hollow, guttural sounds, each eight or ten times repeated, more like those proceeding from the throat of a quadruped than that of a bird, which are succeeded by others not unlike the mewing of a cat, but considerably hoarser.

All these are uttered with great vehemence, in such different keys and with such peculiar modulations of voice as sometimes to seem at a considerable distance, and instantly as if just beside you ; now on this hand, now on that ; so that from these manœuvres of ventriloquism you are utterly at a loss to ascertain from what particular spot or quarter they proceed. If the weather be mild and serene, with clear moonlight, he continues gabbling in the same strange dialect, with very little intermission, during the whole night, as if disputing with his own echoes, but probably with a desire of inviting the passing females to his retreat ; for when the season is further advanced, they are seldom heard during the night."

CUNNINGHAM'S BUSH SHRIKE.—*Gubernétes Yetapa.*

It is a very retiring bird, keeping itself completely out of view, but if once detected, flinging itself into a state of ludicrous alarm at the sight of a human being. It generally restricts itself to the brushwood, and flits quietly among the densest shade; but if it should be discovered, it immediately dashes upwards to a height of some forty or fifty feet, drops as suddenly as it had mounted, then rises again, letting its legs dangle at full length, and uttering a succession of terrified squeaks and yells. So quick are its movements, and so wary are its habits, that a single gunner can seldom succeed in shooting one of these little birds, and the aid of a second sportsman is required before the crafty and active little creature falls to the shot.

The food of the Yellow-breasted Chat consists principally of insects, and it has a special predilection for the larger beetles, which it eats of such great dimensions, that the spectator instinctively wonders how so small a bird can eat so large an insect. It will also feed upon berries and many kinds of fruit. The colour of this bird is dark olive-green upon the upper portions of the body; the breast and the under surface of the wings are light yellow, and the abdomen is nearly white.

ANOTHER group of the Flycatchers is denominated the Alectrurinæ or Cock-tailed birds, a name which has been given to them on account of their habit of raising their long and curiously formed tails in a manner similar to that of the domestic fowl. These birds are only found in South America, and are all of small dimensions, the average length being about six inches. There are many species of this group, and they differ considerably in their habits and in the localities which they frequent. Some are fond of forest lands, perching upon lofty branches, and fluttering from their post in chase of passing insects, while others shun the wooded districts and are only found upon the low-lying lands where water is plentiful, and where they find their insect food upon the leaves and stems of aquatic plants. In all the species the bill is flattened towards the base and rather convex at the point.

One of the most interesting of the Alectrurine birds is the CUNNINGHAM'S BUSH SHRIKE, which from the strong and slightly hooked beak was formerly supposed to belong to the Shrike family. It is a native of South America, and haunts the thickly wooded districts, foraging in many directions in search of its prey, which generally consists of the larger insects. It is possessed of strong and firmly vaned wings, and is able to fly with remarkable rapidity. The tail of this species is extremely elongated and deeply forked, the two exterior feathers being the longest, and the others decreasing rapidly in length. Even when the bird is stationary, this long tail renders it very conspicuous, but when it is living and in motion, it renders the tail a very ornamental appendage, by raising it so as to droop like the feathers of a cock's tail, and permitting the long plumes to wave gracefully in the breeze.

The general colour of this bird is a uniform ashen grey over the upper surface, covered with numerous longitudinal streaks of brown. The throat is white, with the exception of a rather broad semilunar band of deep purple-brown, which marks the division between the chest and the throat. The wings and tail are blackish-brown, and the quill-feathers of the wing are marked with a series of longitudinal ruddy bands.

A SECOND group of the Flycatchers is distinguished by the name of Tyranninæ, a title that has been applied to them on account of their exceedingly combative habits during the season of incubation, and the tyrannous sway which they exercise over birds of far greater size, powers, and armature. They are all inhabitants of America, and for the greater part are found in the more tropical regions of that land. They have a very shrike-like bill, and many of the shrike habits, preying not only upon insects, but pouncing upon young birds, animals, and reptiles, and even adding fish to their scale of diet. The beak of these birds is very large, wide at the base, and narrowing gradually to the tip, where it is boldly compressed and rather strongly hooked. The angle of the mouth is furnished abundantly with long bristles, and the small nostrils are almost entirely hidden by the feathers of the forehead.

Several species are included in this group, among which the two birds which will be described in the following pages are the most remarkable.

The first of the Tyranninæ is the well-known KING BIRD, or TYRANT FLYCATCHER, celebrated by Audubon, Wilson, and many other writers on the ornithology of America.

This very interesting bird is one of the migratory species, arriving in the United States about the month of April, and remaining until the end of the autumn, when its young are fully fledged, and able to shift for themselves. The name of King Bird has been given to this species not only on account of the regal sway which it wields over most of the feathered race, but also on account of the flame-coloured crest which appears whenever the bird raises the feathers of the head.

The habits of this Flycatcher are very remarkable, and have been so ably narrated by Wilson, that they must be given in his own graphic language.

"The trivial name King, as well as Tyrant, has been bestowed on this bird for its extraordinary behaviour, and the authority it assumes over all others during the time of breeding. At that season, his extreme affection for his mate, and for his nest and young, makes him suspicious of any bird that happens to pass near his residence, so that he attacks, without discrimination, any intruder. In the months of May and June, and part of July, his life is one continued scene of broils and battles, in which, however, he generally comes off conqueror. Hawks and crows, the bald eagle and the great black eagle, all equally dread a *rencontre* with this dauntless little champion, who, as soon as he perceives one of these last approaching, launches into the air to meet him, mounts to a considerable height above him, and darts down upon his back, sometimes fixing there, to the great annoyance of his sovereign, who, if no convenient retreat or resting-place be near, endeavours by various evolutions to rid himself of his merciless adversary.

But the King Bird is not so easily dismounted. He teases the eagle incessantly, sweeps upon him from right and left, remounts, that he may descend on his back with the greater violence; all the while keeping up a shrill and rapid twittering; and continuing the attack sometimes for more than a mile, till he is relieved by some other of his tribe equally eager for the combat.

There is one bird, however, which, by its superior rapidity of flight, is sometimes more than a match for him; and I have several times witnessed his precipitous retreat

KING BIRD.—*Tyrannus intrepidus.*

before this active antagonist. This is the purple martin, one whose food and disposition is pretty similar to his own, but who has greatly the advantage of him on the wing, in eluding all his attacks, and teasing him as he pleases. I have also seen the red-hooded woodpecker, while clinging on a rail of the fence, amuse himself with the violence of the King Bird, and play *bo-peep* with him round the rail, while the latter, highly irritated, made every attempt, as he swept from side to side, to strike him, but in vain. All this turbulence, however, vanishes as soon as his young are able to shift for themselves, and he is then as mild and peaceable as any other bird."

Audubon relates an account of a battle between a martin and King Bird, wherein the former proved victorious. The martin had long held sole possession of a farmyard, and when a King Bird came to build its nest within the same locality, it assaulted the intruder with the utmost fury. The act of building on the forbidden ground aroused the anger of the martin to such an extent, that whenever the male King Bird passed with materials, the martin attacked, and by force of superior agility dashed its foe to the ground. At last the poor King Bird died, being worn out with continual struggles, and its mate was forced to leave that spot.

The flesh of the King Bird is held in some estimation in one or two of the States, and the bird is shot in order to supply the table.

The narrator further proceeds to observe, that the King Bird is in great disfavour with the farmers, who are in the habit of shooting it whenever they can find an opportunity, on account of its fondness for bees. It cannot be denied that the suspicions of the bee-owner are not without foundation, for the King Bird will perch upon a rail or fence near the hives, and from that elevated post pounce upon the bees as they leave or return to their homes. Many persons, however, think that it does not devour the working bees, but merely singles out the drones, thus sparing the workers the trouble of killing those idle members of the community at the end of the season. This supposition derives some force from the well-known fact, that the King Bird is very fastidious in its taste, and that it will watch the flight of many insects in succession before it can select one to its taste. Even if it should destroy a few hundred bees annually, it repays the loss a thousand-fold by the enormous destruction which it works among the caterpillars and other noxious insects during the earlier parts of the year; and, according to Wilson, every King Bird shot is a clear loss to the farmer.

The food of the King Bird, although mostly of an insect character, and perhaps wholly so in the spring and summer, is sometimes mixed with vegetable substances, and in the autumn the bird delights in berries and ripe fruits, the blackberry being one especial favourite. It often hovers over streams and rivers, chasing insects like the swallow, and occasionally dashing into the water for a bath, and then sitting to plume its feathers on some convenient branch overhanging the water.

The flight of the King Bird varies according to circumstances. When it is migrating it flaps its wings rapidly six or seven times in succession, and then sails onwards for a considerable distance, repeating this process continually as it proceeds on its long voyage. During the flight it is perfectly silent, and associates in bands of twenty or thirty in number. But in the season of love the bird dashes some thirty yards aloft, and there hangs with quivering wings and ruffled plumes, uttering the while a continual low shriek.

The nest of the bird, which is so valiantly defended by the parent, is generally begun in the beginning of May, and is placed among the branches of a tree. The substances of which it is composed are slender twigs, wood, vegetable fibres, fine grasses, and horsehair. There is another species of tyrant, the CRESTED TYRANT (*Tyrannus cristátus*), which employs many similar materials for its nest, hay, feathers, hogs' bristles, dogs' hair, and the cast exuviæ of snakes. This last substance seems to be absolutely essential to the birds' comfort, for Wilson says that of all the numerous nests which he discovered, he never found one without some of this curious material. The eggs of the King Bird are generally five in number, and there are mostly two broods in the year.

With the exception of the few bright feathers of the crest, the plumage of the King Bird is of a rather sombre character. The head is black, but when the bird raises the crest feathers, their bases are seen to be of a bright orange or flame colour. This appearance is never seen unless the bird is excited. The tail is also black, but is tipped with white. The general colour of the upper parts of the body is ashen grey, and the quill-feathers and coverts of the wing are marked with dull white. The under parts of the body are white with the exception of a large grey patch on the breast. The total length of the bird is about eight inches. In the southern States of America the King Bird is called the Field Martin.

Our second example of the Tyrant Birds is the curious FORK-TAILED FLYCATCHER.

This remarkable species is an inhabitant of tropical America, and is rather frequently found in Guiana, where it is popularly, but erroneously, called the Widow Bird, that appellation belonging by right to one of the finches. Sometimes the bird is quite solitary, but at other times it assembles in little flocks on the branches, and from thence darts on the passing insects. It is also fond of frequenting the low flooded lands, and of perching upon the tufts of rank herbage that appear above the water, opening and shutting its long tail like a pair of shears. Its food is mostly of an insect character, but it will feed upon various fruits and berries.

It is quick and agile of wing, and by means of its long and firmly set tail is enabled to make many sharp turns in the air, an accomplishment which is needful for the purpose of overtaking the large winged insects on which it loves to feed ; and while engaged in these aerial manœuvres it constantly spreads or closes its tail. Except immediately after moulting, the long scissor-like feathers of the tail are seldom in a perfect state, as the bird is very vivacious in its movements, and in its quick glancing flight among the branches is apt to fray the beautiful plumes against the boughs, and often rubs the webs entirely away, leaving the long shafts protruding, clothed only with little ragged fragments of web. The Fork-tailed Flycatcher is quite as brave a bird as the preceding species, and is frequently seen to attack and defeat birds that are far superior in size and bodily strength, but inferior in dashing courage.

The colouring of this bird is briefly as follows : The top of the head is velvety black when the bird is at rest ; but when it becomes excited, it raises the feathers of its head into a kind of crest, and displays a bright orange spot, caused by the orange hue which tinges the basal parts of each feather. The neck, back, and upper parts of the body are dark grey, deepening gradually towards the tail, which is jetty black with the exception of the white outer web of the exterior quill-feathers. The under portions of the bird are white. The total length of this bird is about fourteen inches, of which the tail occupies ten, so that the dimensions of the bird itself are really small.

WE now arrive at the typical Flycatchers, named, in allusion to their insect-eating habits, the Muscicapine birds.

This group includes many curious and interesting species, one of the most remarkable being the WHITE-SHAFTED FAN-TAIL of Australia. Beyond elegance of form and pleasing arrangement of rather sombre colouring, this bird possesses no great external attractions ; but for the singular form of its nest, and the eccentricity of some of its habits, it is well worthy a short memoir. It is a native of the southern and western portions of Australia as well as of Van Diemen's Land, and seems to be a permanent resident, merely shifting its quarters to different portions of the same country according to the season of year. It is by no means a gregarious bird, being seldom seen associated with any other companions except its mate.

FORK-TAILED FLYCATCHER.—*Melvulus tyrannus.*

2.   A A

In its habits it is brisk, cheerful, and lively, mounting high into the air with a few rapid strokes of the wings, and then descending upon some convenient bank in a headlong, reckless style, after turning completely over in the air after the fashion of the tumbler pigeons. While descending it spreads its wings and tail widely, the latter organ being so broad as to resemble a feather fan. It is daring and confiding in its nature, permitting the close approach of human beings, haunting the neighbourhood of human habitations, and even boldly entering houses in chase of flies and other insects. Its song is not powerful or varied, but is full and pleasing, consisting of a soft and sweet twittering sound.

During the breeding season it becomes suddenly shy, wary, and restless, and should it perceive an enemy in too close proximity to its nest, puts in practice a series of rather transparent wiles in order to induce the intruder on its domestic joys to leave the vicinity. For this purpose it feigns lameness, and flutters before the supposed foe in a manner that is intended to induce a belief in its easy capture, and to lure him from the cherished spot where all its loves and hopes are concentrated.

The nest is of a most remarkable shape, as may be seen in the illustration, being notable for a long and apparently useless tail that hangs far below the branch to which it is attached, and which, owing to its narrow dimensions and slight weight, can be of no service in preserving the balance of the structure. I would offer a suggestion that this singular form may have reference to the electrical conditions of the atmosphere, and serve as a conductor whereby the superabundant electricity is carried off from the eggs or young birds, which are placed in an open and undefended nest, and conveyed harmlessly to the ground. The materials of which the nest is made are the inner bark of the gum-tree, mixed with moss and the down of the tree fern, and woven together with spiders' webs.

WHITE-SHAFTED FAN-TAIL.—*Rhipidúra albiscapa.*

The position of the nest is invariably at a low elevation, and it is found either hanging from a branch near the water at no great height from its surface, or suspended from some low branch in a forest. The eggs are two in number, and their colour is greyish-white, covered with olive-brown blotches. There are generally two broods in the year, and a third brood is sometimes known to be successfully reared.

In its colouring the White-shafted Fantail is a dusky olive-black above, and there is a white dash above the eye, and another curved white streak below the eye. The throat, the ends of the wing-coverts, the edges of the secondaries, together with the outer webs, the tips and the shafts of the tail-feathers, are pure white, with the exception of the two central quill-feathers of the tail, which retain their dusky hue. The total length of this bird is five or six inches. The name Rhipidura is of Greek origin, signifying "fan-tail," and is therefore applied to this and other species of the same genus, whose tails are capable of being spread in a fan-like fashion ; the name albiscapa refers to the white shafts of the tail-feathers, and is therefore only applied to this particular species.

THE most elegant and striking of all the Flycatchers is undoubtedly the bird which is figured in the accompanying illustration.

The PARADISE FLYCATCHER is an Asiatic bird, being found spread over the greater portion of India, where it is far from uncommon. It is generally found in thick clusters of tall bamboos, and is in the habit of frequenting gardens, shrubberies, and plantations in search of its prey. Its mode of feeding is rather variable. Generally it perches upon some lofty branch, and when it sees an insect passing within easy reach, makes a sudden swoop upon it, catches its prey with a hard snap of the beak, which can be heard at some distance, and returns to its post in readiness for another swoop. Sometimes, however, it searches upon the branches for the various insects that are found crawling on the bark or hidden beneath its irregularities, and picks them off with great certainty of aim. According to Colonel Sykes, it has even been known to alight on the ground and to seek its food upon the soil.

It is a most restless bird, ever on the move, flitting from branch to branch, or darting after its winged prey with ceaseless activity. Like many predaceous creatures, it is rather solitary in its habits, being generally seen singly or in pairs, or at all events in no greater numbers than may be accounted for by the presence of the two parents and their young.

There are several species closely allied to each other, which are found both in India and Africa ; and even the present species was once supposed to be separated into three, the adult male, the female, and

A A 2

PARADISE FLYCATCHER.—*Tchitrea Paradisi.*

the young being so different in form and colour, that each was set down as a distinct species. It is now known that the long-tailed birds, of whatever colour they may be, are the adults of either sex, while the comparatively short-tailed bird is the young male or female. When these distinctions are once known, it is very easy to discriminate between the birds, the white long-tailed bird being always the adult male, the reddish buff long-tailed bird the adult female, and the short-tailed bird the young male or female, as the case may be.

The colouring of this species is remarkably bold and pleasing, and may be briefly described as follows.

The head and crest of the male are bright steely green, and the whole of the upper surface is pure white, curiously streaked with a narrow black line down the centre of each feather. The primary quill-feathers of the wings are jetty black with a narrow edge of white, and the secondaries are also black, edged with white on both webs. The beautiful tail is more than double the length of the body, as it measures thirteen or fourteen inches in length, while the bird itself is only some six inches long. The colour of the tail-feathers is pure white, with black shafts, except the two central feathers, where the black colour of the shaft only extends half their length. The whole of the under surface is white. The adult female has the head and neck feathers steely green as in the male, but not of quite so brilliant a hue. The back and tail are ruddy chestnut, the throat, breast, and nape of the neck are dark grey, and the abdomen and remainder of the under parts are white. The young bird is coloured like the female, but the white of the abdomen is tinged with buff.

On account of the peculiar shape of this species, it is sometimes called the Rocket Bird.

ENGLAND possesses some examples of the Flycatchers, the two birds represented in the illustration being familiar to every one who has noticed the manners and customs of native birds.

The SPOTTED FLYCATCHER is by far the more common of the two species, and has received several local names in allusion to its habits, the titles WALL BIRD and BEAM BIRD being those by which it is most frequently designated. It is one of the migrating birds, arriving in this country at a rather late season, being seldom seen before the middle or even towards the end of May. The reason for this late arrival is probably that, if the bird were to make an earlier appearance, the flying insects on which it feeds would not be hatched in sufficient numbers to insure a proper supply of food for itself and young. It is a common bird throughout the whole of England and Ireland, and is also seen, but not so frequently, in Scotland. It has a rather wide range of locality, having been observed in different parts of Europe, and extending its flight even to Southern Africa.

This bird is fond of haunting parks, gardens, meadows, and shrubberies, always choosing those spots where flies are most common, and attaching itself to the same perch for many days in succession. When the Flycatcher inhabits any place where it has been accustomed to live undisturbed, it is a remarkably trustful bird, and permits the near approach of man, even availing itself of his assistance.

I well remember a curious instance of this exceeding tameness on the part of one of these birds. In the grounds of a large estate in Oxfordshire, I was sitting in a gig, waiting for a friend, and as the sun was shining very powerfully, I moved the vehicle under the shade of a tree. On one of the lower branches a Flycatcher was sitting, watching the flies, and occasionally fluttering in chase of an insect, and then returning to its post in true Flycatcher fashion. After watching the bird for some little time, I struck with the whip at a clover blossom, thereby starting a number of flies, which rose into the air. To my surprise, the bird instantly left the branch, darted among the flies, captured one of them, and returned to its perch. I again drove some flies into the air with the whip, and again the bird came and snapped them up within stroke of the lash. This proceeding was continued until my friend rejoined me and we drove away, leaving the bird in sole possession.

The Spotted Flycatcher builds a very neatly made nest, and is in the habit of fixing its home in the most curious and unsuspected localities. The hinge of a door has on more than one occasion been selected for the purpose, and in one instance the nest retained its position although the door was repeatedly opened and closed, until a more severe shock than ordinary shook the eggs out of the nest and broke them. It is fond of selecting some human habitation for the locality in which it builds its nest, and its titles of Beam Bird and Wall Bird have been given to it because it is in the habit of making its home on beams or the holes of walls. The branches of a pear apricot, vine, or honeysuckle are

favourite resorts of the Spotted Flycatcher, when the tree has been trained against a wall. The bird seems to be in the habit of returning to the same spot year after year; and as in one case the same locality was occupied for a series of twenty consecutive years, it is most probable that the young may have succeeded to the domains of their parents.

The nest is generally round and cup-shaped, and is made of fine grasses, moss, roots, hair, and feathers, the harder materials forming the walls of the nest, and the softer being employed as lining.

I once watched one of these birds in the act of building her nest, and was greatly interested by the manner in which the business was conducted. First she arranged a rather large bundle of fine dry grass in the thick fork of some branches, and having pecked it about for some little time as if to shake it up regularly, she sat in the middle of it, and by a rapid movement of her wings spun round and round like a top, so as to produce a shallow, cup-like hollow. She then fetched some more grasses, and after arranging them partly around the edge and partly on the bottom, repeated the spinning process. A few hairs and some moss were then stuck about the nest, and woven in very neatly, the hairs and some slender vegetable fibres being the threads, so to speak, with which the moss was fastened to the nest.

PIED FLYCATCHER.—*Muscicapa atricapilla.*
SPOTTED FLYCATCHER.—*Muscicapa grisola.*

In working out the long hairs and grasses, she generally moved backwards, laying them with her bill, and continually walking round the nest, a circumstance which has also been noted by Mr. Yarrell. I cannot say, however, whether, as is related by that writer, the male brings all the materials, nor can I give any further personal description of the architectural powers of the bird, as when the nest had reached the stage which has been described, I was forced to return home, and on my next visit the nest was finished and the mother bird sitting in it. I was close to the bird during her labours, being sheltered from observation by a thick bush and the trunk of an ivy-covered tree, and could even see the colour of the bright glancing eyes, and note the self-satisfied ruffle of her feathers whenever she had made a stroke to her satisfaction.

The eggs of the Spotted Flycatcher are four or five in number, and their colour is a very pale bluish white, spotted with ruddy speckles. As the nest is made at so late a period of the year, being but just begun when some birds have hatched their first brood, there is not often more than a single family in the course of the season. Sometimes, however, it has been known to hatch and rear a second brood in safety. The young are seldom hatched until the tenth or twelfth of June, and they seem to follow their parents longer than is the case with most birds.

The food of the Spotted Flycatcher is almost, if not exclusively, composed of insects, mostly flies and other winged members of the insect world. It seldom descends to the ground for the purpose of procuring its prey, nor does it seem to pick caterpillars off the leaves and branches, but, standing on some chosen perch, it darts at the passing insects, and returns to the same spot. Fruit seems to form no part of the Flycatcher's food, although it has often been observed on the fruit-trees, having in all probability been attracted to the tree by the many winged insects which feed on fruit.

The general colour of the Spotted Flycatcher is a delicate brown on the upper parts of the body, the quill-feathers of the wings and tail being, as is usually the case, of a blacker hue than the feathers of the back. There are a few dark spots on the top of the head, and the tertial feathers of the wings are edged with light brown. The breast is white, with a patch of very light dull brown across its upper portion, and both the chin and breast are marked with dark brown longitudinal streaks. Upon the sides and flanks, the dull white deepens into a yellowish brown. The total length of this bird is about five inches and a half. When young, the plumage is largely spotted with buff and brown of different tints. This species has no song, but only a few low twittering notes.

THE other species of British Flycatchers is much more rare than the bird just described, and may easily be distinguished from it by the peculiarity of plumage, from which it derives its popular title. The PIED FLYCATCHER has been observed in most parts of England, but seems to be of very rare occurrence, except in the counties of Cumberland and Westmoreland, where it is found in the vicinity of the lakes. There are many other localities where it has also been seen, but to enumerate them would be a needless task. It is known to be a frequenter of many parts of Europe, even visiting Norway and Sweden in the summer months, while on the coasts of the Mediterranean it is very plentiful. It is a migratory bird, generally arriving in England about the middle of April, and leaving us for a warmer climate in September.

The habits of this bird, its mode of flight, and fly-devouring propensities, closely resemble those of the preceding species, from which, however, it differs greatly in the locality of its nest, and the number of the eggs. Instead of placing a simply constructed nest upon a branch or other convenient spot, the Pied Flycatcher always chooses a hollow in some decaying tree for its home, and there deposits no less than seven or eight eggs.

In one instance, noticed by Mr. Yarrell, the eggs were disposed in a very curious manner. "In the season of 1830, a pair had a nest in the identical hole where this species had bred for four successive years. On the 16th of May, this nest contained eight eggs, arranged in the following manner: one lay at the bottom, and the remainder were all severally placed perpendicularly round the sides of the nest with the smaller end resting upon it, the effect of which was exceedingly beautiful." The author of this well-observed note further remarks that the eggs from different nests are found to vary greatly in size. The nest itself is made of dried leaves, moss, and hair, and is rather loosely built, and the young make their appearance about the middle of June.

The colouring of this bird is as follows: In the adult male, the top of the head, back of the neck, back and wings are dark blackish brown, with the exception of a white patch upon the forehead, and a broad stripe of white on the tertiary and greater wing-coverts. The tail is black except some bold white marks on some of the outer feathers, and the whole of the under surface is pure white. The female is of a delicate brown on the upper parts of the body, and those portions which in the male are pure white, are in the female of a dull whitish grey. In dimensions the bird is not equal to the spotted Flycatcher, barely exceeding five inches in total length.

THE singular and beautiful bird which is known by the name of KING TODY, or ROYAL GREAT CREST, is a native of Brazil, and may challenge competition with many of the flycatchers for elegance of form and beauty of colouring.

It is a very rare bird, being seldom brought to England, and to all appearance but little known in its native land. This species is chiefly remarkable for its splendid crest, which is capable of being lowered upon the neck, or raised almost perpendicularly, in which latter position it assumes a spreading and rounded form, like an open fan. The feathers of the crest are long and slender, and spoon-shaped at their extremities. Each feather is bright chestnut-red for the greater part of its length, a narrow stripe of rich orange

KING TODY. *Muscivora régia.*

succeeds, and the tip is velvet black, encircled by a band of steel-blue. As may be supposed, the effect of its spread crest is remarkably fine and striking. The upper parts of the body are dark chestnut-brown, rather deeper on the quill-feathers of the wings. The throat, chest, and abdomen are pale fawn, warming towards chestnut on the central line. The total length of this bird is six inches and a half.

THE family of the AMPELIDÆ, or Chatterers, is one of considerable size, and includes some very beautiful and interesting birds. In all these species the beak is rather broad and short, curved on the upper mandible, and well notched at the tip. The claws are sharp and hooked, and are grooved underneath. The Chatterers are found in all the warm portions of the world, and even our own country is sometimes favoured by a visit from the typical species, the Waxen Chatterer. They are divided into several groups or sub-families, the first of which is the Pachycephalinæ or Thick-heads, so called from the heavy make and great comparative size of their heads.

Of this group we find an excellent example in the well-known DIAMOND BIRD of Australia.

This pretty little creature inhabits Van Diemen's Land and the whole of the southern portions of Australia, and is generally found upon trees and bushes, skipping about the branches with the greatest velocity, and peering into every crevice after the insects on which it feeds. It possesses great activity of limb and strength of claw, and is able to traverse the boughs while hanging suspended beneath them by its feet. It is not restricted to any particular tree, but may be seen on the hard scrubby bushes as often as upon the lofty trees. The voice of the Diamond Bird is rather harsh and piping, and consists of two notes constantly repeated, from which circumstance the natives call it by the name of " We-deep, we-deep."

The most remarkable peculiarity in this bird is its nest and the position which it chooses for its home. Instead of placing its nest among the branches, or even in the hollow of one of the innumerable decaying trees that abound in its native country, the Diamond Bird makes a deep burrow in the face of some bank, usually on the margin of a stream, and builds its nest at the extremity of the hole. The tunnel slopes slightly upwards, and is about two or three feet in length, the nest being placed in a chamber at its extremity. Contrary to the usual custom of burrowing birds, the Diamond Bird builds a most neat and elaborately constructed nest in its burrow the marvel being increased by the evident difficulty of working in the dark. The structure is almost globular in form, and is entered by means of a hole left in the side. The materials of which it is composed are principally strips of the inner gum-tree bark, and it is lined with finer portions of the same substance.

DIAMOND BIRD.—*Pardalótus punctátus.*

The Diamond Bird is a pretty little creature, and decorated with most vivacious colouring. The crown of the head, the wings and the tail are black, speckled with pure white, each feather having a snowy white spot at its extremity. A white streak begins at the nostrils, crosses the face, and passes over each eye. The back is curiously diversified with several harmonising tints, each feather being grey at its base, and having at its extremity a triangular spot of fawn edged with black. The upper tail-coverts are ruddy brown, becoming redder towards the tail; the chin, throat, and chest are bright golden orange, and the abdomen is tawny. The female has a browner head, and no golden orange streak on the breast. The bird is about as large as a wren.

The Manakins, or Piprinæ, form a moderately large group of birds, many of which are of very beautiful and curious plumage. With very few exceptions they are inhabitants of America, and are found only in the hottest portions of the tropical regions of that vast country. They feed indiscriminately on animal and vegetable substances, are very active in their movements, and frequent the hottest and moistest forests, where vegetation grows most luxuriantly, as in such situations they find the greatest abundance of food.

One very beautiful species is the GOLDEN-WINGED MANAKIN of America. This bird is always to be found on the skirts of forests, where it chooses the hot and marshy grounds that are often formed in such localities, and there plies its busy search for food, unharmed by the noxious and miasmatic exhalations of the decaying vegetation that are continually steaming upwards, whilst the burning rays of the tropical sun convert the moisture into vapour, and cover the earth with a heavy, warm, and poisonous mist. The bird is remarkably vivacious in its movements, and may often be seen on the ground, peering and feeding in every direction, or perched in large flocks on the top of some lofty tree.

The Golden-winged Manakin is a very pretty bird, its plumage being brightly mottled with black, yellow, and orange, which tints are arranged in a manner both bold and soft. The wings are remarkable for the bright yellow feathers from which the bird derives its popular name, and the crown of the head is decorated with a beautiful series of gradually deepening plumes, of a golden yellow at the base of the bill and on the forehead, and warming into a rich ruddy orange towards the back of the neck, something like the crown of the fire-crested Regulus.

GOLDEN-WINGED MANAKIN.—*Pipra chrysoptera.*

THE largest and the most showy of all the Manakins is the COCK OF THE ROCK, so termed on account of a slight external resemblance which it bears to the gallinaceous form.

It is a native of Southern America and Guinea, and, as it is a solitary and extremely retiring bird, is but seldom seen except by those who go in special search of it. This bird is remarkable, not only for the bright orange-coloured plumage with which its whole body is covered, but for its beautiful crest, which extends over the head like the plume of an ancient helmet. It generally frequents the banks of rocky streams and deep sombre ravines, where it traverses the ground with much rapidity, by means of its powerful and well-developed legs. As it is a solitary and very wary bird, it is seldom shot by white men, the greater number of existing specimens having been procured by means of the poisoned arrow thrown through the deadly sumpitan, or blow-pipe, of the Macoushi Indians. As the skin commands a high price in the market, the Indians kill great numbers of the birds, and are gradually thinning their ranks.

During the daytime the Cock of the Rock retires into its dark hiding-place among the rocks, and only comes out to feed before sunrise and just after sunset. Not only is it never found in company with other birds, but it does not even seem to associate with those of its own kind. The nest of this species is of a very slight description, and is composed of little sticks, splinters of wood, and dry grasses, laid loosely in a hole of some rock, and containing two white eggs.

The colour of the Cock of the Rock is remarkably beautiful, and consists of a rich orange tint, which dyes the whole of the plumage with the exception of the quill-feathers of the wings, which are of a sooty-black hue, and those of the tail, which are brown, tipped with yellow. The feathers of the head stand erect in a double row, with their extremities uniting in a line corresponding with the central line of the head, and consequently form a peculiar fan-like crest, which overhangs the forehead and extends quite to the back of

COCK OF THE ROCK —*Rupicola aurántia.*

the head. The tips of the crest-feathers are tinged with brown and yellow. Upon the wing-coverts and the upper tail-coverts, the feathers are modified into flowing plumes, which droop in a very graceful manner over the firmer feathers of the tail and sides. In size the Cock of the Rock about equals a common pigeon.

The female bird is not nearly so beautiful as her mate, being of a yellowish-brown colour, and having only a small and inconspicuous crest.

ANOTHER species of Manakin which belongs to the same genus is the PERUVIAN COCK OF THE ROCK (*Rupícola Peruviána*), a bird which is possessed of considerable beauty, though it is not quite so splendid as the previous species. Like that bird, its plumage is of a bright orange colour, but its crest wants the curious fan-like form which is so conspicuous in the Cock of the Rock, and the quill and tail-feathers are jetty black, and the wing-coverts are ashen grey. Moreover, the feathers of the wing-coverts and upper tail-coverts are not so loose and flowing, and its tail is longer in proportion.

THERE is one species of Manakin which does not, so far as is known, inhabit America, but is found in Singapore and the interior of Sumatra. This is the GREEN CALYPTOMENA (*Calyptómena víridis*), a very beautiful, though not very large bird. Like the Cock of the Rock, it is extremely shy and solitary in its habits, but instead of retiring into the deep recesses of rocky ground, it shrouds itself among the heavy verdure of the forest trees, where its bright green feathers harmonize so well with the foliage, that it is hardly percep-tible even to a practised eye. The food of this bird seems to be entirely of a vegetable character.

This bird possesses a fine and well-marked crest, which curves so boldly that it nearly hides the short, wide, and hooked beak under its feathers. According to Sir S. Raffles, the colouring of this species is as follows : "The general colour of this bird is a brilliant emerald-green. . . . A little above and before the eyes, the feathers are of a deep velvet-black at their base, and only tipped with green, but crossed in the coverts by three velvet-black bands. The primary feathers, as well as the whole under side of the wings, are dusky, approaching to black, with the exception of the outer margins of some, which are edged with green. The tail is short, rounded, and composed of ten feathers, which are green above and bluish-black below. The whole of the under parts are green ; this colour is brightest on the sides of the neck and round the eyes."

The total length of this species is about six inches, and the bird resembles a thrush in the general contour of its body.

BOHEMIAN WAXWING, OR WAXEN CHATTERER.—*Ampelis gárrula.*

A SMALL but interesting group of birds has been designated by the name of Ampelinæ, or Chatterers, in allusion to the loquacity for which some of the species are remarkable. They all have a wide mouth, opening nearly as far as the eyes, but without the bristly appendages which so often accompany a large extent of gape. Several of the species are celebrated for the singular hairy appendage to the secondary and tertiary quill-feathers of the wings, which closely resemble spots of red sealing-wax, and have given rise to the title of Waxen, which has been almost invariably applied to these birds.

ONE well-known species, the WAXEN CHATTERER, is a tolerably frequent visitor of England, though it cannot be reckoned among the common British birds. It is also known by the name of the BOHEMIAN CHATTERER, the latter name being singularly inappropriate, as the bird is quite as rare in Bohemia as in England.

It is a very gregarious bird, assembling in very large flocks, and congregating so closely together, that great numbers have been killed at a single discharge of a gun. A corre-spondent of the *Field* newspaper, dating from Christiana, in Norway, gives the following interesting particulars of this curious bird : "For the last month there have been, and indeed still are, immense flocks of Waxwing Chatterers quite close to the house. They are not at all shy, allowing a person to approach easily within shot. They come into all the gardens round by thousands, in quest of the berries of a tree, which I believe is the mountain ash, having been driven south, as I suppose, either on account of the cold or in search of food. Some of the flocks contained several thousands, but are now much diminished in numbers, on account of some having gone southwards, and others been killed. They make a great noise when sitting together, which they do in great numbers, making a tree look quite black with them. On one occasion I killed twenty at one shot, at another eighteen, and at another seventeen. One of these birds I shot had the wax at the tip of the tail, as well as on the wings." This curious divergence from the usual formation has been noticed in the cedar bird (an American species of the same genus), by Wilson, as will be mentioned in the account of that bird. Perhaps the waxen appendage of the tail may rather be termed a full development of the original idea, than a divergence from the usual form.

The long, flat, scarlet appendages to the wings, and, as we have seen, to the tail also, are usually confined to the secondaries and tertiaries, at whose extremities they dangle as if they had been formed separately, and fastened to the feathers as an afterthought. Indeed, they so precisely resemble red sealing-wax, that any one on seeing the bird for the first time would probably suppose that a trick had been played upon him by some one who desired to tax his credulity to a very great extent. The full number of these appendages is eight, four on the secondaries and the same number on the tertiaries, but they vary according to the age of the bird, the secondaries keeping their full complement, and the tertiaries having from one to four, according to age and development. None of the wax-like appendages are developed until the second year.

Although the migratory habits of this bird are well known, and many of the localities which it frequents have been recorded by various writers, no one seems to have any certain information as to its true home, or the country wherein it breeds, although it is so numerous a species in its own locality that its hiding-places could hardly have escaped notice had they occurred within the ordinary limits of scientific observation.

Some authors place its residence in Central Asia, upon the elevated table-land of that region, others think that it builds in Tartary, others place its home in the eastern parts of Northern Europe, others in the Arctic regions, while Dr. Richardson believes that it may be traced to America: "The mountainous nature of the country skirting the Northern Pacific Ocean being congenial to the habits of this species, it is probably more generally diffused in New Caledonia and the Russian-American territories, than to the westward of the Rocky Mountain chain. It appears in flocks at Great Bear Lake about the twenty-fifth of May, when the spring thaw has exposed the berries of the alpine arbutus, marsh vaccinium, &c. that have been frozen and covered during winter. It stays only for a few days, and none of the Indians of that quarter, with whom I conversed, had seen its nest; but I have reason to believe that it retires in the breeding season to the rugged and secluded mountain-limestone district in the sixty-seventh and sixty-eighth parallels, where it feeds on the fruit of the common juniper which abounds in those places."

To this country it only comes in the winter months, although there has been an example of its appearance as early as August.

In its plumage the Bohemian Waxwing is a very pretty and striking bird, being as notable for the silken softness of its feathers, as for its pleasingly blended colours and the remarkable appendage from which it derives its popular name. The colouring of the bird is very varied, but may briefly be described as follows: The top of the head and crest are a light soft brown, warming into ruddy chestnut on the forehead. A well-defined band of black passes over the upper base of the beak, and runs round the back of the head, enveloping the eyes on each side, and there is a patch of the same jetty hue on the chin. The general colour of the bird is grey-brown, the primary and secondary feathers of the wings and tail are black, tipped with yellow, the primary wing-coverts are tipped with white, and the tertiaries are purplish brown, also tipped with white. The under surface of the bird is sober grey, and the under tail-coverts are rich ruddy brown. The length of the Waxen Chatterer is about eight inches.

The flesh of this bird is held in great estimation in the countries where it appears in greatest numbers, and in Norway it is regularly killed and exposed for sale at the average price of one penny.

A CLOSELY allied species is found in America, where it has been taken for a variety of the preceding species, but is clearly distinct from that bird. On account of its fondness for cedar berries, it goes by the popular name of the CEDAR BIRD, or CHATTERER, the latter name being not at all appropriate to this species, as it is one of the most silent of birds, not even raising its voice in the season of love.

This bird is found in different parts of America, migrating to and fro according to the season of year. Wilson tells us that in the months of July and August it associates together in great flocks, and retires to the hilly parts of the Blue Mountains for the purpose

CEDAR BIRD.—'Ampelis cedrorum.

of feeding on the whortleberries which grow in those localities so plentifully that the mountains are covered with them for miles. In October they descend to the lower parts of the country, and there feed on various berries, especially those of the red cedar, which they devour so greedily that no less than fifteen cedar berries have been found in the throat of a single bird. They also eat the fruit of the persimmon, cherries, and many other fruits, and aid greatly in the vegetation of the country by transporting to different localities the seeds of the plants on which they subsist.

Unlike the Waxen Chatterer, the Cedar Bird carries with it no mystery respecting its dwelling-place, but openly builds in the month of June upon various trees, sometimes choosing the cedar, and at other times fixing on different orchard trees.

Wilson makes the following remarks upon the nest and general habits of the bird during the breeding season. " The nest is large for the size of the bird, fixed in the forked or horizontal branch of an apple tree, ten or twelve feet from the ground ; outwardly and at bottom is laid a mass of coarse, dry stalks of grass, and the inside is lined wholly with very fine stalks of the same material. The eggs are three or four, of a dingy bluish white, thick at the great end, tapering suddenly, and becoming very narrow at the other ; marked with small roundish spots of black of various sizes and shades, and the great end is of a pale dull purple tinge, marked likewise with various shades of purple and black. About the last week in June the young are hatched, and are at first fed on insects and their larvæ, but as they advance in growth, on berries of various kinds. These facts I have myself been an eye-witness to.

The female, if disturbed, darts from the nest in alarm to a considerable distance ; no notes of wailing or lamentation are heard from either parent, nor are they ever seen, notwithstanding you are in the tree examining the nest and young. These nests are less frequently found than many others, owing not only to the comparatively few numbers of the bird, but to the remarkable muteness of the species. The season of love, which makes almost every other small bird musical, has no such effect on them, for they continue at that interesting period as silent as before."

Like the waxen Chatterer, the Cedar Bird is held in great estimation as an article of food; and as in the autumn and end of summer it becomes very fat in consequence of the enormous amount of berries and other food which it consumes, it is in great requisition in the markets, being sold in large numbers and for a very small price. Even as early as May the Cedar Bird begins its depredations on the cherries, always choosing the best and ripest fruit, and continues its robberies, undisturbed by scarecrows or any other means except the loaded gun. The Cedar Bird does not limit itself to fruits and berries, but also feeds largely on insects, chasing and devouring flies and other winged insects in a manner very similar to that of the flycatchers, but not exhibiting the airy liveliness and quick vivacity of those birds.

The general colour of the Cedar Bird is yellowish brown, the upper parts of the body being fawn-coloured, rather darker on the head, which is surmounted with a long and pointed crest, which can be raised almost perpendicularly from the head. The chin is black, the breast and abdomen yellow, and the under tail-coverts white. The wings are deep slaty blue, and the upper tail-coverts are slate-blue, deepening into black, which also extends over the greater part of the tail. The extremities of the tail-feathers are rich yellow. A rather broad line of black crosses the forehead, and passes round the head, enveloping the eyes in its course. The secondary feathers of the wings are adorned with wax-like appendages resembling those of the Bohemian chatterer, and their number is variable, sometimes being only four or five, and sometimes as many as nine. Wilson supposes that their object is to guard the tips of the feathers from being worn away, but this conjecture does not seem to carry much weight with it. The appendages are nothing more than horny expansions of the shafts. As some female birds are without these wax-like ornaments, it was once supposed that they only belonged to the male bird ; but it is now ascertained that they are found in both sexes alike. On several occasions Wilson found one of the tail-feathers decorated with a waxen tip similar to those of the wing. The colour of the female is similar to that of the male, but the tints are not so brilliant. This bird is much smaller than the European species, being only six inches and a half in length, and very slenderly built.

As the numbers of acknowledged species among birds amount to several thousands, it is evident that in a comprehensive work of this character it will be impossible to mention the whole of the feathered tribe, and that only those birds can be described which act as representatives of the several groups into which the division has been separated. Passing over, therefore, many remarkable species, we arrive at one which is perhaps as extraordinary a bird as any that has hitherto been figured. This is the celebrated BELL BIRD, or CAMPANERO of America, so called on account of the singular resemblance which its note bears to the slow, solemn tolling of a church bell.

The Bell Bird is about the size of an ordinary pigeon, and its plumage is quite white. From a pigeon it can, however, be readily distinguished, even at some distance, by the curious horn-like structure which grows from its forehead, and rises to a height of some three inches when disturbed. This "horn" is jetty black in colour, sprinkled very sparingly with little tufts of snowy-white down, and as it has a communication with the palate, has probably something to do with the bell-like sound of the voice. The song or cry of the Campanero has been admirably described by Waterton, in his well-known "Wanderings in South America."

"His note is loud and clear, like the sound of a bell, and may be heard at the distance of three miles. In the midst of these extensive wilds, generally on the dried top of an aged mora, almost out of your reach, you see the Campanero. No sound or song from

BELL BIRD.—*Aropunga alba.*

any of the winged inhabitants of the forest, not even the clearly pronounced 'Whip-poor-Will!' from the goatsucker, causes such astonishment as the toll of the Campanero.

With many of the feathered race, he pays the common tribute of a morning and evening song; and even when the meridian sun has shut in silence the mouths of almost the whole of animated nature, the Campanero still cheers the forest. You hear his toll, and then a pause for a minute, then another toll, and then a pause again, and then a toll, and again a pause. Then he is silent for six or eight minutes, and then another toll, and so on. Actæon would stop in mid chase, Maria would defer her evening song, and Orpheus himself would drop his lute to listen to him, so sweet, so novel and romantic, is the toll of the pretty snow-white Campanero."

The "horn" of the Bell Bird is only erect while the creature is excited and during the resonant cry, and when the bird is at rest it hangs loosely on the side of the face. It is supposed that the Bell Bird builds in Guiana, but its nest and locality of breeding are at present unknown.

To the Chatterers succeed the Campephaginæ, or Caterpillar-eaters, which are nearly all found in the various countries of the Old World. As their name imports, they live chiefly upon caterpillars and other insects, preferring those that are still in the larval state, and assiduously examining each leaf and branch in search of their prey. They also eat ants, beetles, and other ground-living insects, and are quite as active in chasing them upon the earth as in their haunts among the branches. They also eat fruit and berries in the autumn.

ONE of the most remarkable birds of this group is the GREAT PERICROCÓTUS, the largest of its genus.

This bird is a native of India, where it is found spread over the greater portion of that country, and on account of its splendid plumage it attracts great notice even from unscientific and casual observers. It seems to be solitary in its habits, being generally found alone or in very small societies, in all probability consisting merely of the parents and their young. It is almost exclusively an insect-feeder, eating caterpillars, flies, ants, and various kinds of the insect tribe, preferring, however, the beetles, of which it devours very great numbers. It is a suspicious and timorous bird, carefully avoiding the presence of human beings, and thus ranking as a very scarce bird, although it probably exists in considerable numbers, in its own peculiar localities.

GREAT PERICROCOTUS.—*Pericrocótus speciósus.*

As it is so beautiful a species, it has several times been captured and caged, but it seems to defy the powers of the tamer, pines away under confinement, and soon dies.

The sexes of the Great Pericrocotus are so different in their external appearance, that they might easily be mistaken for two distinct species. The adult male is a truly beautiful bird, and is thus coloured. The ground colour of the bird is the deepest imaginable steely blue, so deep, indeed, as to appear black except in certain lights. The head, neck, back, wings, the two central tail-feathers, and the base of the remaining tail-feathers, are rich glowing scarlet. The bill and legs are black, and the eyes dark brown. The female, although a very pretty bird, cannot lay claim to the gorgeous colouring which decorates her mate. In her, the parts which in the male are scarlet, are bright golden yellow, and the back of the head and the scapularies are grey. The greater coverts are olive-brown.

THIS genus contains many species, several of which are remarkable for the rich beauty of their plumage. They are gregarious, assembling in little flocks, and as they are extremely loquacious, they make a considerable noise as they sit chattering and whistling in groups upon the topmost branch of some lofty tree. In their habits they are similar to the last-mentioned bird, being insect-feeders, and preferring the beetles, or more rigidly speaking, the coleopterous insects, to any other food. Their nest is generally placed in the branches of some tall tree, at a considerable elevation from the ground, is small in size, is composed of grasses and lichens, and generally contains about two small streaked eggs.

NEXT in order comes a group of birds, called, from the peculiar form of their tails, the Dicrurinæ, or Double-tailed Birds, and also known by the title of DRONGO SHRIKES. These birds are so very like the shrikes, or butcher birds, that they have often been confounded with them; and, as may be seen from the popular title of the group, have been ranked with these birds in some systems. They are not, as a rule, of large dimensions, their average size being that of a common blackbird, and many of them are remarkable for beauty of plumage and grace of form.

WOOD SWALLOW.—*Artamus sordidus.*

The WOOD SWALLOWS are spread over a large portion of the globe; some species being found in India and the islands of the Indian seas, and others being inhabitants of Australia. Owing to their shrike-like form, and their swift flight, they have been termed Swift Shrikes by some naturalists. Several species of this genus are found in Australia, and that which is most frequently noticed is the common WOOD SWALLOW, or SORDID THRUSH. This species is common in many parts of Australia, and is migratory in its habits, arriving in and leaving Van Diemen's Land at regular intervals, and making a partial migration on the Australian continent. Some individuals, however, remain in the same country throughout the year, as they find abundance of food without the absolute need of repairing to another climate. The habits of the Wood Swallow are very curious and interesting, and are well described by Mr. Gould, in his well-known work on the Birds of Australia :—

"This Wood Swallow, besides being the commonest species of the genus, must, I think, be considered a general favourite with the Australians, not only from its singular and pleasing actions, but by its often taking up its abode and incubating near the houses,

2. B B

particularly such as are surrounded by paddocks and open pasture-lands skirted by large trees. It was in such situations as these, in Van Diemen's Land, at the commencement of spring, that I first had the opportunity of observing this species; it is there very numerous on all the cleared estates on the south side of the Derwent, about eight or ten being seen on a single tree, and half as many crowding one against another on the same dead branch, but never in such numbers as to deserve the appellation of flocks. Each bird appeared to act independently of the other; each, as the desire for food prompted it, sallying forth from the branch to capture a passing insect, or to soar around the tree, and return again to the same spot. On alighting, it repeatedly throws up and closes one wing at a time, and spreads its tail obliquely prior to settling.

At other times a few were seen perched on the fence surrounding the paddock, on which they frequently descended, like starlings, in search of coleoptera and other insects.

It is not, however, in this state of comparative quiescence that this graceful bird is seen to the best advantage; neither is it that kind of existence for which its form is especially adapted; for, although its structure is more equally suited for terrestrial, arboreal, and aërial habits than of any other species I have examined, the form of the wing at once points out the air as its peculiar province. Here it is that, when engaged in pursuit of the insects which the serene and warm weather has enticed from their lurking-places among the foliage to sport in higher regions, this beautiful species in its aërial flights displays its greatest beauty, while soaring above in a variety of easy positions, with its white-tipped tail widely spread.

It was very numerous in the town of Perth until about the middle of April, when I missed it suddenly, nor did I observe it again until near the end of May, when I saw it in countless numbers flying, in company with the common swallows and martins, over a lake about ten miles north of the town; so numerous, in fact, were they, that they darkened the water as they flew over it. Its voice greatly resembles that of the common swallow in character, but is much louder."

This Wood Swallow is remarkable for a habit which is perhaps unique among birds, and hitherto has only been observed in certain insects. A large flock of these birds will settle upon the branches of a tree, and gather together in a large cluster, precisely like bees when they swarm. Four or five birds suspend themselves to the under side of the bough, others come and cling to them, and in a short time the whole flock is hanging to the bough like a large swarm of bees. Mr. Gilbert, who first noticed this curious habit, states that he has seen the swarms as large as an ordinary bushel measure.

The nest of the Wood Swallow is cup-shaped and rather shallow, and is made of very slender twigs bound and lined with delicate fibrous roots. The locality in which the nest is placed is extremely variable, the bird seeming to be wonderfully capricious in its choice of a fit spot whereon to fix its residence. Sometimes it is placed in a low forked branch, at another time it will be buried in thick massy foliage, while it is sometimes found fixed against the trunk of a tree, resting on some protuberance of the bark, or looped within some conical cavity. The eggs are about four in number, and are greyish white, speckled and mottled very variably with grey and white.

The colour of this species is very simple, the general tint being black, the abdomen white, and the tail-feathers, excepting the two central, which retain their jetty hue throughout, tipped with the same colour.

THE largest of the Australian species, the CINEREOUS WOOD SWALLOW ('Artamus cinereus), is found both at Timor and the eastern and western coasts of Australia, thus having a very large range. This bird, although not at all uncommon, seems to be rather local, preferring certain spots for its residence, and keeping itself within some peculiar boundary of its own choosing.

It inhabits the banks of the Swan River and parts of the interior, and varies in its habits with the locality in which it happens to reside. Wherever the grass-tree (Xanthorhœa) grows, there may be found the Cinereous Wood Swallow, feeding with the greatest avidity upon the seeds, and absolutely crowding each other upon the upright

GREAT DICRURUS.—*Dicrurus grandis.*

seed-stalks while engaged in digging out their food. It does not, however, depend upon the grass-tree for its subsistence, as it feeds largely upon insects, chasing them in the air with nearly as much activity as the ordinary swallow, or pursuing the quick-limbed beetles on the ground, digging out the hidden larvæ from beneath bark, or under the soil, and picking them from the leaves on which they feed. The nest of this species is deeper than that of the common wood swallow, and the mottlings of the eggs have more of a ruddy hue. The position of the nest is generally in a thick low bush, or among the foliage of the grass-tree.

THE beautiful bird which is represented in the illustration is the GREAT DICRURUS of the East Indies.

In its general outline this beautiful bird bears some analogy to the Leona nightjar which has been figured on page 126 of the present work, having two long feathery appendages, naked throughout the greater portion of their length and webbed only at their extremity. There is, however, this great difference, that in the Leona nightjar they proceed from the wings, whereas in the Great Dicrurus they are merely prolongations of the external tail-feathers. The colour of this bird is deep blue-black, like that of the raven, and its weird-like aspect is further strengthened by a large and well-developed crest that starts from the top of the head and bends backward over the neck. A few of its feathers project slightly forwards so as to come beyond the base of the beak.

ANOTHER species of Dicrurus is well known in India by the name of KING CROW (*Dicrúrus macrocercus*), a title which it has earned by its boldness in attacking the crows, and maintaining royal dominion over them. In their habits the birds of this genus differ but very slightly from each other, and in all essentials they agree. They are insect-feeders, preferring grasshoppers to any other prey, and often pouncing upon the backs of cattle for the sake of capturing the flies that are so fond of attacking the poor beasts in the warm weather. They will even take their posts with perfect composure on the back of a cow or goat, and consider the animal as their especial property for the time being. As they are swift of flight, they constantly dart from their perches and capture insects on the wing.

The nest of the King Crow is placed in the trees which grow in the thickest jungle, and is made of slender twigs interwoven with grasses, moss, and lichens. Some of the Dicrurine birds build a very beautiful and elaborately constructed nest, while others are content with a negligent and slovenly residence. The eggs are generally three or four in number.

WE now arrive at the family of LANIDÆ or SHRIKES, or BUTCHER BIRDS, whose character is given in the names by which they are distinguished. The scientific term Lanidæ is of Latin origin, and is derived from a word which signifies lacerating or tearing, in allusion to the habits of the bird. These birds are found in all parts of the globe, and in all countries are celebrated for their sanguinary and savage character. They are quite as rapacious as any of the hawk tribe, and in proportion to their size are much more destructive and bloodthirsty. They feed upon small and disabled mammalia, and birds of various kinds, especially preferring them while young and still unfledged, and upon several kinds of reptiles, and also find great part of their subsistence among the members of the insect world.

In order to fit them for these rapacious pursuits, the bill is strong, rather elongated, sharp-edged, curved at the tip, and armed on each side with a well-marked tooth. The wings are powerful, the plumage closely set, and the claws strong, curved and sharp. The Shrikes are separated for convenience of reference into two groups or sub-families, namely the true Shrikes, or LANÍNÆ, and the Bush Shrikes, or THAMNOPHILÍNÆ.

OF the true Shrikes we find an excellent example in the well-known GREAT GREY SHRIKE, a bird which is very common in many parts of Europe, especially in the more southern and warmer regions, but is generally scarce in England, visiting us, whenever it does make its appearance, in the winter season.

This bird eats mice, shrews, small or young birds, frogs, lizards, beetles, grasshoppers, and many other creatures. It generally, if not always, destroys its prey by a severe bite across the head, crushing in the skull, and usually commences its meal with the head. This, together with the other Shrikes, has a curious habit of hanging its food upon some convenient spot, such as a forked branch, a thorn or sharp broken end of a bough, and will frequently leave its prey thus suspended for a considerable period. Even insects are served in this manner, being impaled upon thorns and left hanging in the branches. The object of this curious custom is extremely dubious. It cannot be merely for the purpose of holding the prey securely while it feeds, for the Shrike will frequently commence eating a bird immediately after its capture, holding the prey tightly with its claws after the manner of the hawks, and tearing it to pieces with its powerfully hooked bill. Nor can it be with the object of making it tender by hanging in the air, as the bird often devours the prey at once. Moreover, insects would not become more tender by exposure, but would rapidly dry up in the sunshine and become hard and useless for food.

Even when tame it continues this habit, and has been known to make constant use of a spike driven into a wall for that purpose by its owner, always carrying its food, whether it consisted of meat or small birds, and impaling it upon the accustomed spike. A caged bird, mentioned in Mr. Yarrell's work, was in the habit of employing the spaces between the wires for the same purpose, always hanging the remnants of its meal between the wires, and pushing the prey through the bars while eating.

Its name of Excúbitor, or Watchman, has been given to it from the services which it enders to the fowler.

Fierce and powerful as it is, it holds the falcon in the greatest terror, and is gifted

GROUP OF SHRIKES.

with so true an eye for its enemy that it can perceive a falcon when at an immense distance. Taking advantage of this peculiarity, the fowlers who set their nests for falcons always take with them a Grey Shrike, and after setting their nets, fasten the string to which the bird is tied to a peg near the nets. A little turf hut is built as a place of refuge for the Shrike, and a small mound or hillock raised, on which it perches. The

GREAT GREY SHRIKE.—*Lánius excúbitor.*

fowler then retires to his own little hut, places the strings which draw the net within reach of his bird, and watches the Shrike out of a small window which commands the mound where it is perched. Feeling secure that the Shrike will not suffer a hawk to come within sight without giving notice, the fowler takes out his netting or other sedentary work, and continues his labour.

Hundreds of birds may pass over the net without the Shrike giving the least alarm, but as soon as it can see a falcon, it flutters about, gets uneasy, and at last begins to kick and squall with terror. Roused by the sounds, the fowler jerks some strings communicating with perches on which living pigeons are perched, and the flutter thus occasioned attract the falcon's attention and induces him to stoop for a prey that appears so easy. As the foe approaches nearer, the Shrike's terror increases, and as the falcon swoops at the pigeons, the Shrike screams with fear and runs for shelter under the tiny hut. This movement is a signal for the fowler, who draws the strings of his net and incloses the falcon as he makes his dart on the pigeons.

The voice of the Shrike, although sufficiently harsh on occasions, is capable of great modulation, so that the creature can imitate the cries of many birds, and even copy witn some success the sweet notes of the songsters.

The nest of the Grey Shrike **is** situated in forests, and is placed in the lofty branches of some tall tree. The substances of which it is made are fine grass, roots, mosses, down, and wool. The eggs are from four to six in number, and are bluish-grey, spotted at the large end with deep grey and brown. The colour of this species is pearl-grey on the upper part of the body; the chin, breast, and abdomen are white; the quill-feathers of the tail black variegated and tipped with white; and a black band crosses the forehead,

RED-BACKED SHRIKE.—*Enneóctonus collúrio.*

runs under the eyes and then expands into a black patch on the ear-coverts. The total length of the bird is about ten inches.

The RED-BACKED SHRIKE is a summer visitant to this country, and is very much more common than the last-mentioned species. Its winter quarters seem to be situated in Africa, and it reaches us at the end of April or the begining of May, passing through Italy on its passage.

During the time of its residence it may often be seen flitting about the tops of hedges and small trees, evidently in search of its prey, and even at a considerable distance may be recognised by its habit of wagging its tail up and down whenever it settles, in a manner very similar to that of the wagtails. Usually it is seen in pairs; but when the eggs are laid, the male bird is generally engaged in procuring food while the mother bird stays at home and attends to her domestic affairs.

The food of the Red-backed Shrike chiefly consists of the larger insects, such as grasshoppers, beetles, and chaffers, and it is in the habit of impaling them on the thorns near its nest, probably to save the mother bird the trouble of going to look for her own meals. I have elsewhere mentioned that a not very common species of cocktail beetle, the *Staphylinus erythropterus* is a very favourite prey of this bird; and when I was making up my collection of Wiltshire insects, I used to derive considerable assistance from the labours of the Red-backed Shrike. These impaled insects are stuck about the bush in such numbers, and in so very open a fashion, that they form a ready guide to the position of the Shrike's nest. Moreover, the parent birds are so solicitous about their home, that as soon as they see a human being approaching their nest, they set up such a shrieking and fluttering that they intimate the position of their nest to the least experienced observer.

Although the chief food of this bird consists of insects, it occasionally takes to larger game, and has been known to destroy other birds, generally while in their nestling state. It has sometimes been caught in fowlers' nets while striking at their decoy birds, and has been detected in dragging young and weakly pheasants through the bars of the cage in which they had been confined.

I can also add the testimony of personal observation to the bird-destroying capabilities of this Shrike. A few months ago a lady kindly presented to me a box containing several nestling birds, each pierced by a thorn, which she said had been killed and stuck there by the Red-backed Shrike. Thinking that there might possibly have been some mistake about the slayer, I asked if it could be procured, and in a few more days another box was sent, containing a fine Red-backed Shrike and another impaled victim. Most of the dead birds were headless, and in every case the thorn, instead of transfixing the body, had been thrust between the skin and the muscles, but in so firm a manner that to draw it out again required considerable force. The victims were very small, and too much dilapidated for me to ascertain their species.

In many parts of England, and indeed in most countries where it dwells, the Shrike is termed "Nine-killer," from a notion that it always kills and impales nine creatures before it begins its meal. The generic name enneóctonus bears the same signification, and has been applied to the bird in allusion to this idea. Mr. Blyth says that wherever food is very abundant, the Red-backed Shrike only eats the soft abdomen of the impaled insect, leaving the wings, limbs, and hard parts on the thorns. I have never observed this practice, although I have seen very many Shrikes, their nests, eggs, and young. Still, however, it may be the case with individual birds.

The nest of this Shrike is situated in hedges or bushes, generally from five to ten feet from the ground, the average elevation being about seven feet. It is large, rather clumsy, and very easily seen through the foliage, being made of thick grass-stems, moss, and roots on the exterior, and very fine grasses and hair. In some places the nests are quite common, and I have found three in a hedge surrounding a single field of no very great extent. The eggs are generally five in number, and are rather variable in colouring, their ground colour being always white, tinged in some cases with blue, in others with green, and in a few specimens with rusty red. The spots with which they are marked are quite as variable, sometimes being numerous, dark, and gathered into a ring at the large end of the egg, and sometimes being only grey and light brown scattered irregularly. In all cases, however, they are gathered upon the large end of the egg.

In the adult male, the head, neck, and upper parts of the shoulders are pearly-grey, with a black stripe across the base of the beak and running through the eye. The back and wing-coverts are ruddy chestnut, fading into reddish-grey upon the upper tail-coverts. The quill-feathers of the wings are black, edged with red upon their outer webs, and the quill-feathers of the tail are white at the basal half, and the remainder of each feather is black tipped with a very narrow line of white. The chin and under tail-coverts are white, and the rest of the under surface is pale rusty red. The strongly notched and hooked beak is deep shining black. The female bird may at once be known by the absence of the black streak across the eye, which in her case is replaced by a light coloured stripe over the eye. The head and all the upper parts of the body are reddish-brown, and the red edges of the wing-feathers are narrower than in the male. The under side of the body is wholly greyish white, covered with very numerous transverse lines of a darker hue. The young male is similarly coloured, but is distinguished by the back being also covered with transverse bars of dark grey. The length of the adult bird is between seven and eight inches.

ANOTHER species of the same genus, the WOODCHAT SHRIKE, has occasionally but very rarely been found in England. It is about the same size as the red-backed Shrike, and possesses many of the same habits, but may readily be distinguished from that bird by the difference of colouring.

WOODCHAT SHRIKE.—*Enneóctonus rufus.*

In many districts of the Continent the Woodchat Shrike is a common bird, especially preferring the warmer and more southern districts. In many parts of Africa it is extremely plentiful, being particularly abundant in Northern Africa. It is also seen at the Cape of Good Hope. On account of their habit of hanging and impaling, the Shrikes are known at the Cape by the popular name of Magistrate Birds. The nest of the Woodchat Shrike is made rather more neatly than that of the red-backed species, and is always placed on the branch of a tree, the oak being preferred for this purpose. The materials of which it is made are pine-twigs, moss, and wool, and it is lined with wool and slender grasses. The eggs are smaller than those of the last-mentioned bird and are quite as variable in their markings, the general colour being very pale bluish white speckled with rusty brown.

The colouring of the Woodchat Shrike is as follows: The top of the head and back of the neck are rich chestnut-red, a white streak runs across the base of the upper mandible, and a broad black band crosses the forehead and reaches as far as the ear-coverts, enveloping the eye in its progress. The back-wings and wing-coverts are black, relieved by the white feathers of the scapularies and upper tail-coverts. The primary feathers of the wings are also white at their base, and the secondaries are tipped with white. The two central tail-feathers are black, the two exterior feathers white, and the remainder are partly of one colour and partly of the other. The whole of the under surface is white. In the female, the head and neck are dusky red, the back is brown black, the wing-coverts are marked with rusty red, and the breast is greyish white.

VIGORS' BUSH SHRIKE.—*Thamnóphilus Vigorsii.*

The second sub-family of the Butcher Birds, namely the THAMNOPHILINÆ, or Bush Shrikes, are well represented by the beautiful VIGORS' BUSH SHRIKE.

This bird is a native of Southern America, and is generally found in forests and thick brushwood, where it passes its time in a constant search after the small mammalia, birds, reptiles, and insects, on which it feeds. It is a large and rather powerful bird, and as it possesses a strong and sharply hooked beak, is a very formidable foe to any creature which it may attack. Its claws are also powerful, curved and very sharp, so that the bird is aided by its feet as well as by its beak in the demolition of its prey. In order to enable the bird to prey among the rank herbage and thick massy foliage of the localities in which it dwells, its legs are long in proportion to the size of its body, and the grasp of its feet very strong, so that it is able to perch upon a bough or on the ground, and raise its head to some height while surveying the locality with its piercing glance. The wings are rather short and rounded, as long and sharply-pointed wings would be of little use in threading the network of leaves and branches among which it takes up its residence.

The tail of the Vigors' Bush Shrike is long in proportion to the size of the body, and extends far beyond the closed wings. The general colour of the male bird is sooty black upon the head and the whole of the upper surface, diversified with numerous transverse bars of rich red chestnut. The under parts of the body are pale greyish brown. The head is decorated with a crest of erectile feathers, ruddy throughout the greater part of their length, and marked with black at their tips. The female is distinguished from the male by the blacker crest, the paler tint of the transverse bars, and the uniform ashen-grey of the under parts. The total length of this species is about thirteen inches.

There are many species of Bush Shrikes, the greater number being inhabitants of the eastern hemisphere. As their name imports, they all live among the thickest brushwood and in forests, and their food consists chiefly of insects. They are rather silent birds, their note being merely a single syllable constantly repeated, and only uttered during the breeding season.

WE now arrive at a very large and important group, called from the shape of their beaks the CONIROSTRES, or CONE-BILLED BIRDS. In these birds the bill varies in length and development, in some being exceedingly short, while in others it is much elongated ; in some being straight and simple, while in others it is curiously curved and furnished with singular appendages ; in some being toothless, while in others there is a small but perceptible tooth near the tip. In all, however, the bill is more or less conical in form, being very thick and rounded at the base, and diminishing to a point at the extremity. There are no less than eight recognised families of this large group, containing some of the most important and most remarkable members of the feathered race.

THE first family is that which is well known under the title of CORVIDÆ or Crows, containing the crows, rooks, magpies, starlings, and other familiar birds, together with the equally celebrated but less.known paradise birds, bower birds, troopials, and orioles. The beak of all these birds is long, powerful, and somewhat compressed,—i. e. flattened at the sides,—curved more or less on the ridge of the upper mandible, and with a notch at the extremity. This family is divided into several smaller groups or sub-families, the first of which is the PHONYGAMINÆ, or PIPING CROWS. These birds are inhabitants of Australia, New Holland, New Guinea, and several adjacent islands, and may be distinguished by the long, narrow, and naked nostrils.

The PIPING CROW SHRIKE, sometimes called the Magpie by the colonists, on account of its magpie-like white and black plumage, is a native of New South Wales, and towards the interior is very plentiful.

This bird is found in almost every part of the country, preferring, however, the open localities to the wooded districts, especially if they are cleared by artificial means. For the Piping Crow Shrike is a wonderfully trustful bird, attaching itself instinctively to mankind, and haunting the vicinity of barns and farmyards. On the very slightest encouragement the bird will take possession of a barn, garden, or plantation, and with the exception of a favoured few, will not suffer any of his friends to intrude upon his property. The owner of the garden is well repaid for his hospitality by the rich and varied song which the bird pours forth in the early morning and towards evening, as if in gratitude for the protection which has been afforded it.

The notes of this bird are peculiarly rich and mellow, and in speaking of them the author of " Bush Wanderings in Australia " remarks, " No bush-bird, to my fancy, had a clearer or richer note than the Magpie : one of the earliest birds of morning, it was also one of the latest at night, and the deep flute-like evening song of the Magpie was heard in the forest long after all the other birds of day had retired to roost. The Magpie is a very common bird throughout the land during the whole year, often in small companies, and in the autumn the old and young birds congregate in flocks. The young Magpies are excellent eating." The name " tibicen " signifies a flute player.

As it is a very hardy bird and bears captivity well, Mr. Gould thinks that it would be easily adapted to the English climate, and in that case would be a very valuable addition to our list of cage-birds. In its native country it has little of the migrating spirit, generally remaining stationary throughout the year in the spot which it has adopted for its home. The food of the Piping Crow consists mostly of insects, the large grass-hoppers being especial dainties. The bird is an excellent hunter, pursuing its active prey over the ground with considerable agility, and pouncing upon it at last with remarkable accuracy of aim. In captivity it will eat almost any description of animal food, and also feeds upon different fruits and berries.

PIPING CROW.—*Gymnorhina tibicen.*

The nest of the Piping Crow is a large and not very neatly constructed edifice, made principally of sticks, leaves, and small grasses. It is loosely placed among the branches of a lofty tree at a considerable elevation above the ground, and contains from two to four eggs. There are generally two broods in the year.

The colouring of this bird is remarkable for its boldness and simplicity, consisting only of two opposite tints, disposed in large and contrasting masses. The greater part of the body and wings is rich jetty black, as deep as that of the raven, and the whole back of the neck, the wing-coverts, the upper and under tail-coverts, and the basal portions of the tail-feathers, are pure snowy white, so that the colonists are quite justified in the use of their popular title. In dimensions it about equals our common magpie, but does not appear to be so large a bird on account of its comparatively short tail. The bill is blue-black, and the eyes are deep ruddy hazel.

The PIED CROW SHRIKE is an inhabitant of New South Wales, and is very widely spread throughout that country.

It is by no means a local bird, finding subsistence in almost every district, and being equally found in the bushes of the coast, the mountains, and the forests. Its food is chiefly of a vegetable character, consisting of berries, fruits, and seeds, and the bird is in consequence of a more arboreal character than the preceding species, which finds its greater part of its nourishment on the ground. It is a stationary bird, only moving from one district to another according to the season of the year, and is generally seen in little parties of five or six in number, which are supposed to be the parents and their young family.

The flight of this bird is neither strong nor sustained, and it seldom takes to wing without being forced to do so. Even when it has been obliged to entrust itself to the air, it rarely flies farther than from one clump of trees to another, or across one of the deep gullies that are so common in its native land. As a general rule, it contents itself with merely flitting from one tree to another, and avoids any open space with great solicitude. While flying, the beautiful black and white markings of its plumage are very conspicuous. It is a most noisy and loquacious bird, possessing a loud and curiously ringing voice, and being so fond of exercising its vocal powers that it is generally heard long before it is seen. Like the piping crow, it is killed for the purposes of the table, and is held in some estimation as an article of food.

The nest of the Pied Crow Shrike is very large in proportion to the size of the bird, round in form, and cup-shaped in the interior. It is almost wholly constructed of very little sticks, and is lined with dried grasses. Unlike the nest of the preceding species, it is placed in some low branch of a tree. The number of eggs is three or four. The colour of this species is a rich deep blue-black, with the exception of the basal halves of the primary quill-feathers of the wings and tail, and the tips of the tail-feathers, which are snowy white. The bill is black, and the eye bright topaz yellow.

INTERMEDIATE between the piping crow shrikes and the true crows, comes a group of birds well known by the popular title of JAYS, or the scientific name of Garrulinæ or talkative birds, so called from their exceeding loquacity. The birds of this group have bills with a little notch near the extremity, but they may be distinguished from their relatives by the fact that the nostrils are covered by the feathers of the forehead. Their tails are generally rather long in proportion to the size of the bird, and the wings are short and rounded. In some instances the colouring of these birds is very fine, and it is curious that blue seems to hold predominance throughout the group.

PIED CROW SHRIKE.—*Strépera graculina.*

THE best known of this group is our common English JAY, one of the handsomest of our resident birds.

Although distributed with tolerable regularity over the greater part of England, it is nowhere plentiful, seeming instinctively to seek some home far removed from those of its own species.   The localities which it best loves are thick woods and plantations, particularly those where heavily foliaged trees are found.   Sometimes, however, it is not so careful, and I have seen it near Oxford, flitting about the topmost branches of the trees in the early morning, and pecking at the beech mast with perfect unconcern, even though within a few hundred yards of houses.   In general, however, the Jay is seldom seen, as it is much afraid of human beings, and conceals itself in the thickest covert on the slightest alarm.

The ordinary note of the Jay is a rather soft cry, but the bird is a most adroit imitator of various sounds, particularly those of a harsh character.   It has one especial harsh scream, which is its note of alarm, and serves to set on the alert not only its own kind, but every other bird that happens to be within hearing. The sportsman is often baffled in his endeavours to get a shot at his game by the mingled curiosity and timidity of the Jay, which cannot hear a strange rustling or see an unaccustomed object without sneaking silently up to inspect it, and is so terribly frightened at the sight of a man, a dog, and a gun, that it dashes off in alarm, uttering its loud "squawk," which indicates to every bird and beast that danger is abroad.   In captivity the Jay soon learns to talk, and even when caged displays its imitative powers with considerable success, mocking the bleating of sheep, the cackling of poultry, the grunting of pigs, and even the neighing of horses with wonderful truth.

The Jay, like all the crow tribe, will eat animal or vegetable substances with equal zest, and will plunder the hoards of small quadrupeds or swallow the owner with perfect impartiality.   Young birds are a favourite food of the Jay, which is wonderfully clever at discovering nests and devouring the fledglings.   Occasionally it even feeds upon birds, and has been seen to catch a full-grown thrush.   Eggs also are great dainties with this bird, particularly those of pheasants and partridges, so that it is ranked among the "vermin" by all gamekeepers or owners of preserves.   So fond is it of eggs, that it can almost invariably be enticed into a trap by means of an egg or two placed as bait, and it is a curious fact that the Jay does not seem to be aware of the right season

JAY.—*Gárrulus glandárius.*

for eggs, and suspects no guile even when it finds a nest full of fine eggs in the depth of winter.

It also eats caterpillars, moths, beetles, and various similar insects, preferring the soft, fat, and full-bodied species to those of a more slender shape. Fruits and berries form a considerable portion of the autumnal food of this bird, and it occasionally makes great havoc in the cherry orchards, slipping in quietly at the early dawn, accompanied by its mate and young family, and stripping the branches of the bark and finest fruit. The kitchen garden also suffers severely from the attacks of the Jay, which has a great liking for young peas and beans. It also eats chestnuts, nuts, and acorns, being so fond of the last-mentioned fruit as to have received the title of "glandarius," meaning a lover of acorns. Sometimes it becomes more refined in its taste, and eats the flowers of several cruciferous plants, which, according to Mudie, it plucks slowly and carefully, petal by petal.

The nest of the Jay is a flattish kind of edifice, constructed of sticks, grass, and roots, the sticks acting as the foundations, and a rude superstructure of the softer substances being placed upon them. It is always situated at a considerable elevation from the ground. There are generally four or five eggs, and the bird mostly brings up two broods in the year. During the earlier portion of their existence the young birds accompany their parents, and as they wander in concert, often do great damage among the gardens and orchards which they visit.

One mode of taking the Jay has already been mentioned. Fowlers, however, employ several methods for the capture of this pretty bird, and find that they can catch Jays better by working on their curiosity than on their appetite. None of the crow tribe seem to be able to pass an owl without dashing at it; and the birdcatchers take advantage of this propensity by laying their snares in the branches of a thick bush, and fastening a common barn owl in such a manner, that when the Jay makes its attack, it is arrested and secured by the snare. Should an owl not be attainable, a white ferret will answer the purpose equally well, the Jay having a great objection to all the weasel tribe, and invariably attacking ferret, polecat, stoat, or weasel with the greatest virulence and perseverance.

In size, the Jay equals a rather large pigeon; and the colouring of its plumage is very attractive. The general tint of the upper part of the body is light reddish brown, with a perceptible purple tinge, varying in intensity in different specimens. The primary wing-coverts are bright azure, banded with jetty black, and form a most conspicuous ornament on the sides, as the bird sits with closed wings. The head is decorated with a crest, which can be raised or lowered at pleasure, and the feathers of which it is composed are whitish grey, spotted with black. There is a black streak on each side of the chin, and the quill-feathers of the wings and tail are also black. The eye is a bright blue-grey, which, when the bird is excited, can gleam with fiery rage, and together with the rapidly moved crest and harsh screams gives an angry Jay a very savage aspect.

In many points, the BLUE JAY of America closely resembles its English relative, but as it possesses a decided individuality of its own, it is well worthy of a short memoir.

The Blue Jay seems to be peculiar to Northern America, and may be found among the woods, where it is very plentiful, but never seems to associate in great numbers, the largest flocks amounting merely to some thirty or forty members, and these only being seen during a small portion of the year. Like the European Jay, it is both inquisitive and suspicious, and never fails to give the alarm as soon as it sees a sportsman among the trees. Many a deer has been lost to the anxious hunter through the warning cry of the Jay, for the deer understand bird language quite well enough to know what is meant when a Jay sets up its loud dissonant scream, and many a Jay falls a victim to the bullet that had been intended for the heart of the escaped deer. Indeed, some hunters have taken so rooted a dislike to this bird, that they always shoot it whenever they see it.

The voice of the Blue Jay is remarkably flexible, being attuned either to soft and musical notes, to the harshest screamings of the hawk tribe, or the most ear-rending shrieks, resembling nothing so much as the piercing creaks of an ungreased wheel. It is well adapted for imitation, and there is hardly a bird of the forest whose voice is not mocked by the Jay with a fidelity that even deceives the species whose notes are thus wonderfully reproduced. Being a bird of some humour, it is greatly delighted by mimicking the scream of a hawk, and the terrified cry of a little bird in distress, thereby setting all the small birds in a turmoil, under the impression that one of their number has just been carried off by a hawk.

The Blue Jay attacks owls whenever he meets with them, and never can see a hawk without giving the alarm, and rushing to the attack, backed up by other Jays, who never fail to offer their assistance to their comrade. Often they will assemble in some numbers, and buffet the unfortunate hawk with such relentless perseverance that they fairly drive him out of the neighbourhood; but sometimes the tables are reversed, and the hawk, turning suddenly on his persecutors, snaps up the foremost and boldest, and silently sails away into the thickest covert, bearing his screaming prey in his talons.

As the Blue Jay is very fond of fruit and seeds, it often does great harm to the agriculturist, robbing his fruit-trees in a very complete and systematic manner, and doing no small amount of harm to the crops. Yet the bird is not without its use, for in replenishing its winter stores, which consist of nuts, mast, chestnuts, and similar provisions, the Jay drops many of them in its passage, and thus unconsciously succeeds in planting many a useful tree. One careful observer of this bird and its habits says that in a few years' time the Jays alone would replant all the cleared lands.

The diet of the Jay is, however, by no means restricted to vegetable substances, as the bird lives more upon animal than on vegetable food. In the spring and early summer, young birds form a large portion of its sustenance, and it robs many a nest of its eggs, or even when pressed by hunger makes an attack on the parent bird.

In captivity, the Blue Jay is equally mischievous with its European relative, being attracted by anything that glitters or that he thinks is valued by its owner, and hiding it in some of his especial treasure-houses. He will also learn to talk, and becomes very proud of his accomplishment, displaying his newly-acquired talents to every one who will listen, and being extremely loquacious when excited by the presence of several persons at the same time. If kindly treated, the Blue Jay becomes very affectionate to its owner, and can even be taught to live in loving communion with creatures whom it would in a wild state immediately devour. One of these birds, kept for some time by Wilson, was on terms of intimate friendship with one of the Baltimore orioles, and would permit her to take all kinds of liberties, such as pulling its whiskers, jumping into the water and splashing it whenever it desired to drink.

The nest of the Blue Jay is large, and rather clumsily made, and is placed in a lofty branch of some tall tree, the cedar being in principal request for this office. It is lined with fine fibrous roots, and contains four or five eggs of a dull olive, spotted with brown. The male bird is very cautious in his approaches to the nest, always gliding secretly and silently to the spot where his mate and young have made their home, carrying with him the results of his foraging expedition.

Much more might be said of this bird, but its character has been so well described by Webber in a few graphic passages, that I should do it injustice, were not his account to be presented in his own words.

"See him of a fine spring morning in love-making time! See him rise up and down upon the mossy limb, his gay crest bent in quick and frequent salutation, while a rich, round, thrilling love-note rolls liquidly from off his honeyed tongue. Then see him spring in air with his wide wings, azure and white, and dark-barred, graceful tail, spread to the admiring gaze of her he woos, float round and round her fairer form, then to return again in rapturous fervour to her side, to overwhelm his glowing charms with yet more subduing graces.

But the fun of it all is, to see our euphuist practising these seductive arts by himself. You will often catch him alone, thus making love to his own beauty with an ardour fully equal to that of the scene we have just described; indeed, I am not sure that it does not surpass it; for, like other dandies, he is most in love with his own beauty. It is the

AMERICAN BLUE JAY.—*Cyanócorax cristátus.*

richest and most fantastic scene I know of among the comicalities of the natural world, to catch him in one of these practising humours; he does court to his own charms with such a gay and earnest enthusiasm; he apes all the gestures and lovelorn notes of his seemingly volcanic amours, and turning his head back, gazes on his own fine coat with such fantastic earnest, that one can hardly resist roaring with laughter.

So jealous is he of his sole prerogative of supervision over the interest and welfare of his neighbours, that he is for ever on the look-out for all interloping stragglers. Every racoon that shows his inquisitive nose is assailed with vehement clamours and angry snappings of beaks, which compel him, in terror for his eyes, to return to his home. Our friend Jay is said to attribute the nocturnal habits of racoons, wild cats, opossums, owls, &c., to their apprehension of his valorous vigilance by daylight. Be the facts of the case what they may, no one of these gentry, nor mole, nor mink, nor weasel, can make its appearance without being beset by the obstreperous screams of this audacious knave. Nor does he confine his operations to the defence of his foraging-ground from these depredators, from whom he has little to fear of personal danger, on account of his superior activity. But he even sometimes does assail the lightning-winged and lordly hawk;

2.   C C

these scenes are very characteristic and very amusing, and I have frequently witnessed them.

The pine-log cutters of the north know him well, and bestow on him many a blessing from the wrong side of the mouth. The deep snow is raked away, and the camp is pitched beneath the gloomy shelter of the heaving pines. Scarcely has the odour of the first roast streamed through the air and freighted every biting wind, when, with hungry cries from every side, the Jays come gathering in. They swarm about the camp in hundreds, and such is their audacity when hard pinched with hunger, that they are frequently seen to dash at the meat roasting before the fire, and, hot as it is, bear pieces off till they can cool it in the snow. They are regarded with singular aversion by these lonely men; for, take what precaution they may, they are often robbed to such serious extent by their persevering depredators as to be reduced to suffering. They dare not leave any article that can be carried off within their reach; when they kill game, and leave it hung up until the hunt is over, the Jays assemble in thousands, and frequently tear it in pieces before their return.

The Blue Jay has many of the traits of the magpie, and like him possesses an inveterate propensity for hiding everything he can lay hold of in the shape of food. The magpie hides things that are of no value, but a Jay is in every respect a utilitarian, and when, after feeding to repletion, he is seen to busy himself for hours in sticking an acorn here, or a beech-nut there, in a dust-hole, or wedging snails between the splinters of some lightning-shivered trunk, or making deposits beneath the sides of decaying logs, naturalists wonder what he is doing it for. But our Euphuist knows well enough, and you may rest assured, if you see him along that way next winter, as you will be apt to do if you watch, you will find that he has not forgotten the place of one single deposit, and that with a shrewder economy than the ant or the squirrel, instead of heaping up his winter corn in one granary, where a single accident may deprive him of it all, he has scattered them here and there in a thousand different spots, the record of which is kept in his own memory. So it cannot be denied, whatever may be said of his thieving and other dubious propensities, that the Blue Jay is a decidedly sagacious personage.

So universal is the Blue Jay's reputation for mischievous and impish tricks of any kind, that the negroes of the South regard them with a strange mixture of superstition and deadly hate. The belief among them is, that it is the special agent of the devil here on earth—carries tales to him of all kinds of slanderous gossip, particularly about negroes —and more especially, that they supply him with fuel to burn them with. Their animosity is entirely genuine and implacable.

When a boy, I caught many of them in traps during the snows, and the negro boys, who generally accompanied me on my rounds to the traps, always begged eagerly for the Jay birds we captured to be surrendered to them, and the next instant their necks were wrung, amid shouts of laughter.

Alas for the fate of our feathered Euphuist—yet he was a 'fellow of infinite wit.'"

The colouring of the Blue Jay is as follows: The upper portions of the body are a light bluish purple, and the head is adorned with a moveable crest of bright blue or purplish feathers. On each side of the head runs a narrow black line, rising higher than the eye, but not passing it, and a collar of the same jetty hue is drawn from the back of the head down each side of the neck to the upper part of the breast. The chin, cheeks, and throat are bluish white, and the abdomen is pure white. The greater wing-coverts are rich azure, the secondary coverts are purple-blue, and nearly all are richly barred with semilunar black streaks and tipped with white. The two middle feathers of the tail are light blue, deepening into purple at the tip, and the remaining feathers are also light blue, barred with black and tipped with white. The eye is hazel. The length of this species is about eleven inches.

ASIA presents a most beautiful and interesting example of this group of birds in the HUNTING CISSA, or HUNTING CROW of India.

This lovely bird is a native of Nepâl, and is spread throughout the south-eastern part of the Himalayas, and in its own favoured locality is far from scarce. Owing, however

HUNTING CISSA.—*Cissa Sinensis.*

to certain peculiarities in the colouring, hereafter to be described, a specimen is very seldom obtained in first-rate condition, and never takes its place in our museums glowing in all the resplendent tints with which it is so liberally gifted. It is a very brisk and lively bird, and, like many others of the same group, is much given to imitating other birds, performing its mimicry with wonderful truth, and copying not only their voices, but even their peculiar gestures.

It is much more carnivorous in its tastes than would be imagined from an inspection of its form and plumage, and it possesses many of the habits of the shrikes, not only killing and eating the smaller birds, but hanging its food upon branches in true shrike fashion. It is an excellent hunter, and as it can be easily tamed and taught to hunt after small birds for the amusement of its owner, it has earned the name of Hunting Crow. In its native country it is very commonly kept in captivity, and even in England has lived for a considerable time in a cage in the gardens of the Zoological Society. The voice of the Hunting Cissa is loud and screeching, but possesses withal a certain joviality of utterance that renders it far from unpleasing.

The colour of this bird is singularly beautiful, and may challenge comparison with that of any other bird of either hemisphere. The general hue is a pale but bright grassy green, very vivid upon the upper parts, and taking a yellowish tint below; there is also a dash of yellow across the forehead and the sides of the crest. A broad black band crosses the forehead, and, enveloping the eye in its progress, passes round the back of the neck. The quill-feathers of the wing are mostly bright chestnut-red, and the tips of the inner quill-feathers are grey, diversified with a bold semilunar black band near their extremities. The central feathers of the tail are green, taking a greyer tinge at their extremities, and all the other tail-feathers are bright green for the first two-thirds of their length, are then crossed with a bold broad black band, and the tips are greyish white. The legs, bill, and feet are bright scarlet. The size of the Hunting Cissa is about equal to that of a common magpie.

These beautiful colours are unfortunately never seen except for a very short time after moulting, as they rapidly fade by exposure to light, even during the life of the bird, and after its death become comparatively dingy. The delicate and brilliant grass-green of the upper surface soon takes a more sober hue, and before many days have elapsed, the general colour of the bird is simply grey with a greenish wash, in place of the rich resplendent tints which it had so lately boasted.

BENTEOT.—*Crypsirhina varia.*

BETWEEN the true Crows and the Jays, another small sub-family has been placed by the authors whose arrangement we follow, and is known by the title of Calleatinæ, or Tree Crows. In these birds there is no tooth in the upper mandible, and the bill is comparatively short, curved and rather rounded above. They are only to be found in the warmer parts of the eastern hemisphere, and many of them are quite as carnivorous as any of the preceding Corvidæ, some feeding chiefly upon insects of various kinds, and others varying their diet with small birds and quadrupeds.

The BENTEOT, one of these birds, is a native of Java, where it is not very scarce, but is seldom seen except by those who go to search for it, as it is extremely timid, and is never known to approach within a considerable distance of human habitations, as is the case with the generality of the Crow tribe. Sometimes it may be seen cautiously making its way towards some newly cleared ground, in the hope of making a meal on the worms, grubs, and other earth-living creatures that are generally to be found in freshly-turned soil, and also for the sake of feeding upon the fruits of the trees that skirt the field. Should, however, the land be near a house, the Benteot holds aloof, and declines to put itself into danger.

Part of this excessive timidity may, perhaps, be owing to the fact that it is by no means a strong or rapid flyer, its wings being short and rounded, and its flight in consequence weak and not capable of long duration. It usually flies by day, and, according to Mr. Horsfield, "may be seen about noon, sailing heavily through the air in a right line towards the trees surrounding the openings in the forest." The strong bill and powerful claws show that the bird is well adapted for the capture of insects and disinterring them from their subterranean hiding-places, as well as for eating the various hard-shelled fruits on which it partly subsists. In colour the Benteot appears at a little distance to be nearly black, but on a close approach its plumage is seen to be a very dark and rather dull green, "shot" plentifully with a deeper hue of bronze.

ANOTHER and more beautiful member of this group is an Asiatic bird, very common in the naturalist's shop and in glass cases, and known by the popular and very appropriate name of the "WANDERING PIE."

This bird is a native of the Himalayas, and is found in some numbers spread over a large part of India. It is called the Wandering Pie on account of its habit of wandering over a very large extent of country, travelling from place to place and finding its food as it best may, after the fashion of a mendicant friar. This custom is quite opposed to the general habits of the Pies, who are remarkable for their attachment to definite localities, and can generally be found wherever the observer has discovered the particular spot which they have selected for their home. Mr. Gould suggests that its wandering habit may be occasioned by the necessity for obtaining subsistence, the Wandering Pie feeding more exclusively on fruits and other vegetable nutriment than is generally the case with the Crow tribe, and being therefore forced to range over a large extent of land in search of its food. Indeed, the short legs and very long tail of this species would quite unfit it for seeking its living on the ground, and clearly point out its arboreal habits.

The shape of this species is very remarkable, on account of the greatly elongated and elegantly shaped tail, which is coloured in a manner equally bold with its form. The general colour of this bird is blackish grey upon the upper parts, warming into cinnamon upon the back. The quill-feathers of the wings are jetty black, the wings themselves grey, and the tail-feathers grey, with a large bold bar of black at their extremities. The under surface of the bird is light greyish fawn. The two central feathers of the tail are extremely long, and the others are graduated in a manner which is well exemplified in the accompanying illustration. Although it appears to be a rather large bird, the aspect is a deceptive one, on account of the long tail, which is ten inches in length, the remainder of the head and body being only six inches long.

WE now arrive at the true Crows, which, like the preceding group, have

WANDERING PIE.—*Temnúrus vagabundus.*

no tooth in the upper mandible, but may be distinguished from them by the greater comparative length of the wings.

THE first of these birds on our list is the celebrated RAVEN, our finest representative of the family.

This truly handsome bird is spread over almost all portions of the habitable globe, finding a livelihood wherever there are wide expanses of uncultivated ground, and only being driven from its home by the advance of cultivation and the consequent inhabitance of the soil by human beings. It is a solitary bird, living in the wildest district that it can find, and especially preferring those that are intersected with hills. In such localities the Raven reigns supreme, hardly the eagle himself daring to contest the supremacy with so powerful, crafty, and strong-beaked a bird.

The food of the Raven is almost entirely of an animal nature, and there are few living things which the Raven will not eat whenever it finds an opportunity of so doing. Worms, grubs, caterpillars, and insects of all kinds are swallowed by hundreds, but the diet in which the Raven most delights is dead carrion. In consequence of this taste, the Raven may be found rather plentifully on the Scottish sheep-feeding grounds, where the flocks are of such immense size that the bird is sure to find a sufficiency of food among the daily dead ; for its wings are large and powerful, and its daily range of flight is so great, that many thousands of sheep pass daily under its ken, and it is tolerably sure in the course of the day to find at least one dead sheep or lamb. Sometimes the Raven accelerates matters, for if it should find an unfortunate sheep lying in a ditch, a misfortune to which these animals are especially prone, it is sure to cause the speedy death of the poor creature by repeated attacks upon its eyes. Weakly or ailing sheep are also favourite subjects with the Raven, who soon puts an end to their sufferings by the strokes of his long and powerful beak. Even the larger cattle are not free from the assaults of this voracious bird, which performs in every case the office of a vulture.

So strongly is the desire for attacking wounded or dying animals implanted in the breast of the Raven, that, according to Mudie, the best method of attracting one of these birds within gunshot is to lie on the back on some exposed part of a hill, with the gun concealed and close at hand. It is needful to remain perfectly quiet, because if there is the slightest sign of life the Raven will not approach, for, as Mudie rather quaintly observes, "he is shy of man and of all large animals in nature ; because, though glad to find others carrion, or to make carrion of them if he can do it with impunity, he takes good care that none shall make carrion of him." It is equally needful to watch carefully and not to be overcome by sleep, as the first indication of the Raven's approach would to a certainty be the loss of an eye.

"But if you lie on your back," says Mudie, "he will come you know not whence, and hovering round you on slow wing, examine you from all points. If you do not stir, he will drop down at a little distance, and begin to hop in an *échellon* fashion, bringing his shoulders forward alternately, after a few hops on each line of the zigzag. Sometimes he will utter his ' cruck-cruck,' and pause to see if that makes you stir, and if it does not, he will accelerate his advance."

Sheep and cattle do not, however, form the whole of a Raven's diet, for besides the insects which have already been mentioned, this bird eats mice, rabbits, birds of various kinds, including young partridges and pheasants, and will invade the farmyard when pressed by hunger, and carry off the young poultry. Even the hedgehog falls a victim to the Raven, who cares nothing for his spiked armour, but drives his sharp bill through the poor beast, tears away the prickly skin, and devours the carcase at his leisure. In Northern America, and indeed in many other countries, the Raven is a regular attendant on the hunters, and follows them for the purpose of feeding upon the offal of the creatures which they kill.

The tongue of the Raven is rather curiously formed, being broad, flat, covered with a horny kind of shield, and deeply cleft at the extremity. At the root are four rather large projections or spines, the points being directed backwards. The use of these spines is not known, though Mr. Buckland suggests that they may be for the purpose of

RAVEN.—*Corvus Corax.*

preventing the food from being thrown back into the mouth. I do not, however, think that this suggestion is sufficient, as there is no reason why the Raven should regurgitate its food more than other birds which feed on similar substances. If the bird were in the habit of eating living prey, such as lizards and other reptiles which retain life for a considerable period and after considerable injuries, this idea might be a good one, but as the Raven always kills its prey before eating it, the theory will not hold its ground.

The cunning of the Raven is proverbial, and anecdotes of its extraordinary intellectual powers abound in various works. From the great mass of these stories I can only select one or two which are not generally known.

One of these birds struck up a great friendship for a terrier dog belonging to the landlord of an inn, and carried his friendship so far as to accompany his ally in little hunting expeditions. In these affairs the two comrades used to kill an astonishing number of hares, rabbits, and other game, each taking his own share of the work. As soon as they came to a covert, the Raven would station himself outside, while the dog would enter the covert and drive out the hares from their concealment, taking care to send them in the direction of the watchful bird. On his part the Raven always posted himself close to one of the outlets, and as soon as any living creature passed within reach, he would pounce upon it, and either destroy it at once or wait until the dog came to his assistance, when by their united efforts the prey was soon killed. Rat-hunting was a favourite sport of these strange allies, and it was said by those who witnessed their proceedings, that the Raven was even more useful than a ferret would have been.

Another and a very amusing anecdote of the Raven and its cunning is related by Captain McClure, the well-known Arctic voyager. Speaking of the behaviour of various birds and beasts during the winter, he remarks that the Raven is the hardiest of the feathered tribe, and even in the depths of winter, when wine freezes within a yard of the fire, the Raven may be seen winging his way through the icy atmosphere and uttering his strange rough, croaking cry, as unconcernedly as if the weather were soft and warm as an English spring. "Two Ravens," he observes, "once established themselves as friends of the family in Mercer Bay, living mainly by what little scraps the men might have thrown away after meal times.

The ship's dog, however, looked upon them as his especial perquisites, and exhibited considerable energy in maintaining his rights against the Ravens, who nevertheless out-witted him in a way which amused every one. Observing that he appeared quite willing to make a mouthful of their own sable persons, they used to throw themselves intentionally in his way just as the mess-tins were being cleared out on the dust heap outside the ship. The dog would immediately run at them, and they would just fly a few yards; the dog then made another run, and again they would appear to escape him but by an inch, and so on, until they had tempted and provoked him to the shore a con-siderable distance off. Then the Ravens would make a direct flight for the ship, and had generally done good execution before the mortified-looking dog detected the imposition that had been practised upon him, and rushed back again."

Not long ago, I saw a Raven in one of the great London breweries, holding a large sausage in his beak, and flapping about the yard just in front of one of the draymen, to whom the stolen dainty had evidently belonged. The bird would not trouble itself to make its escape, but in the most provoking manner hopped along just a yard or so before its pursuer, and from all appearance as likely to carry on the same game for an hour or two; for while I was sitting, the relative positions of the parties did not alter in the least. If the man stopped, the bird stopped too, and began to make such evident preparations for swallowing the sausage that the drayman rushed at it again, and again the bird would just flap a yard or two in advance.

In captivity the Raven is a most amusing, although a terribly mischievous creature, and displays a talent for the invention of mischief which can only be equalled by its rapidity of execution and audacity of demeanour. Except when placed in an inclosed yard where there is nothing that is capable of damage, a single Raven will get through more mischief in one hour than a posse of boys in twelve, and as he always seems to imagine himself engaged in the performance of some extremely exemplary duty, and works his wicked will as methodically as if he had been regularly trained to the task and very well paid for it, he excites no small amount of rage on the part of the aggrieved person. I have personally known several tame Ravens, but as I have already recorded their performances elsewhere, I shall not here repeat the story of their ill deeds.

The Raven is an excellent linguist, acquiring the art of conversation with wonderful rapidity, and retaining with a singularly powerful memory many sounds which it has once learned. Whole sentences are acquired by this strange bird, and repeated with great accuracy of intonation, the voice being a good imitation of human speech, but always sounding as if spoken from behind a thick woollen wrapper. So remarkable is the cunning of this bird, and so weird-like its aspect, that the ancient Scandinavians had good cause for the trembling respect which they paid to the sullen " Bird of Odin." Their idea of the Raven was, that it was accustomed to watch for Odin's return every evening, and, perched upon his shoulder, to relate all the incidents that had taken place on earth within its ken.

As the bird is so crafty, its capture would seem to be a very difficult business, and the number of tame Ravens now existing in England seems to be almost remarkable. The fact is, that while still unfledged the young Ravens have a strange habit of falling out of their nests, and flapping their wings heavily to the ground. Next morning they are found by the shepherds, sitting croaking on the ground beneath their former homes, and are then captured and taken away with comparative ease. Even in this case, however, to secure one of the young Ravens is no slight task, for, on seeing that escape is impossible, it turns boldly to bay, and makes such fierce attacks with its powerful beak that it must be enveloped in a cloth or a plaid before it can safely be held. It is remarkable that when a Raven makes its assault it does not merely peck with its beak, but flings its whole weight upon the blow.

The Raven is also celebrated for its longevity, many instances being known where it has attained the age of seventy or eighty years, without losing one jot of its activity, or the fading of one spark from its eyes. What may be the duration of a Raven's life in its wild state is quite unknown.

The colour of the Raven is a uniform blue-black, with green reflections in certain lights. The female is always larger than her mate.

CROW. – *Corvus Coróne.*

THE common CARRION CROW, so plentiful in this country, much resembles in habits and appearance the bird which has just been described, and may almost be reckoned as a miniature raven.

In many of its customs the Crow is very raven-like, especially in its love for carrion, and its propensity for attacking the eyes of any dead or dying animal. Like the raven, it has been known to attack game of various kinds, although its inferior size forces it to call to its assistance the aid of one or more of its fellows before it can successfully cope with the larger creatures. Rabbits and hares are frequently the prey of this bird, which pounces on them as they steal abroad to feed, and while they are young is able to kill and carry them off without difficulty. The Crow also eats reptiles of various sorts, frogs and lizards being common dainties, and is a confirmed plunderer of other birds' nests, even carrying away the eggs of game and poultry by the simple device of driving the beak through them and flying away with them thus impaled. Even the large egg of the duck has thus been stolen by the Crow. Sometimes it goes to feed on the seashore, and there finds plenty of food among the crabs, shrimps, and shells that are found near low-water mark, and ingeniously cracks the harder shelled creatures by flying with them to a great height and letting them fall upon a convenient rock.

The Crow, unlike the rook, is not a gregarious bird, being generally seen either single or in pairs, or at the most only in little bands of four or five, consisting of the parents and their children. In the autumn evenings, however, they assemble in bands of ten or twelve before going to roost, and make a wonderful chattering, as if comparing notes of the events which have occurred during the day, and communicating to each other their latest experiences, for the benefit of the rising generation.

The nest of the Crow is invariably placed in some tree remote from the habitations of other birds, and is a structure of considerable dimensions, and very conspicuous at a distance. It is always fixed upon one of the topmost branches, so that to obtain the eggs safely requires a steady head, a practised foot, and a ready hand, the uncultivated germs of the professional acrobat.

Generally the nest is rather loosely constructed, and more saucer than cup shaped; but I remember an instance where it was very firmly made and quite deep. In a little copse in

Derbyshire, that was planted along one side of a valley, an oak-tree had sprung up about half-way down the declivity, and, as is the custom with trees in such situations, had grown inclining towards the somewhat abrupt angle formed by the shape of the ground on which it stood. As there had been formerly many other trees around it, it had been drawn up like a maypole, being long, slender, and swinging about with every breeze. The tree was not more than forty feet high; but as it was bent in the middle and bowed over the valley, its summit was nearly a hundred feet from the ground below.

It was with the greatest difficulty that I reached the nest, for the tree yielded like a carter's whip with my weight, although I could not approach nearer than arm's length to the nest, and after three attempts I was finally baffled in my endeavour to obtain the eggs. Although the top of the tree was then nearly level with the horizon, and swinging about most alarmingly in the wind that rushed through the valley, not one egg was thrown out of its place, and the nest was so much deeper than ordinary, that I could not succeed in withdrawing the eggs from their cradle. It seems an easy matter to take eggs out of a nest; but if the reader will bear in mind that when the slender tree stem to which one is clinging bends nearly double with one's weight, that the elasticity of the wood dances one up and down through an arc of four or five feet, and that a strong wind is at the same time acting on the foliage of the tree and swaying it from side to side, and that there is a clear fall of some hundred feet below, he will comprehend that it is not so simple a matter to spare a hand long enough to take an egg from a rather distant spot, and to do so in so delicate a manner that the egg remains unbroken.

The materials of which the Crow's nest is made are very various, but always consist of a foundation of sticks, upon which the softer substances are laid. The interior of the nest is made of grasses, fibrous roots, the hair of cows and horses, which the Crow mostly obtains from trees and posts where the cattle are in the habit of rubbing themselves, mosses, and wool. The eggs are extremely variable, or rather individual, in their markings and even in their size, those in my own collection being so different from each other that an inexperienced person would set them down as belonging to different species. The Crow very seldom uses the same nest for a second breeding season, although it often repairs to the same locality year after year. Once or twice it has been known to lay its eggs on the same foundation as it had employed during the previous season, but in general it pulls the former nest to pieces, and constructs a fresh one on its site.

This bird is remarkable for its attachment to its mate and young, far surpassing the fawn and turtle dove in matrimonial courtesy.

The Somali Arabs bear a deadly hatred towards the Crow, and kill it whenever they meet with it. The origin of their detestation is as follows : During the flight of Mohammed from his enemies, he hid himself in a cave, where he was perceived by the Crow, at that time a light-plumaged bird, who, when it saw the pursuers approach the spot, sat over Mohammed's hiding-place, and screamed, "Ghar! ghar!" i. e. "Cave! cave!" so as to indicate the place of concealment. His enemies, however, did not understand the bird, and passed on, and Mohammed, when he came out of the cave, clothed the Crow in perpetual black, and ordered it to cry "ghar" as long as Crows should live. When they have killed a Crow, the Arabs remove the gall, employing it for the manufacture of collyrium, or dye for the eyelids.

The colour of the Crow is a uniform blue-black, like that of the raven, but varieties are known in which the feathers have been pied or even cream-white.

THE most familiar of all the British Corvidæ is the common ROOK, a bird which has attached itself to the habitations of mankind, and in course of time has partially domesticated itself in his dominions.

The Rook may claim the doubtful honour of having originated two of the most pertinacious and persistent ornithological controversies on record. The subject of the first is its conduct towards man—whether it is to be looked upon as a feathered benefactor or must be ranked among the "vermin." This dispute has now been carried on for many years, and finds as many and as eager advocates on both sides of the question as on the day on which it was started. The second controversy is quite as fierce as the

ROOK.—*Corvus frugilegus.*

former, and has lately revived with tenfold vigour, the subject being the cause and effect of the naked white skin which is found at the base of the Rook's beak. Before proceeding further, we will just say a word or two on these interesting discussion.

Firstly, as to the relation in which the Rook stands to mankind with regard to its conduct. It is thought by many persons of practical experience that the Rook is one of the greatest enemies to the farmer, eating up his grain as soon as planted, pecking up his potatoes and devouring all the "sets," boring holes in his turnips, and altogether doing exceeding mischief in the fields. The farmer, therefore, detests the "blackening train" of Rooks with a very heartfelt hatred, and endeavours by all kinds of contrivances, such as scarecrows, boys with noisy clappers, and loud voices, or even the gibbeted dead bodies of slaughtered Rooks, to keep them off his grounds. Whenever he can find a chance he shoots them, but the bird is so cautious that very few Rooks fall victims to the agricultural gun. The gamekeepers also hate the Rook as a persecutor of their charge, and in truth the Rooks have been actually seen engaged in the destruction of young partridges, and one of them was shot with the prey still in its beak.

Moreover, the Rook has been seen to attack a hen pheasant while sitting on her eggs, to pull the feathers out of the mother bird, and to destroy the eggs, having evidently been attracted to the spot by the large bunch of hay-grass amid which the nest had been placed, and which had been left standing by the mowers in order to afford a shelter to the poor bird.

So much for the one side of the question; we will now proceed to view the Rook from a more favourable point of view.

The advocates of this bird (of whom I confess myself to be one) do not deny that the Rook is on occasions somewhat of a brigand, and that it has small scruples when pressed by hunger in eating eggs or the young of other birds. Also they fully admit that it pulls up a great number of green corn blades almost as soon as they show their emerald tops above the dark soil, that it digs up the potatoes, and throws the fragments about the ground, eating no small number of them, and that it often bores a turnip so full of holes that it pines away and dies. But although granting thus much, they yet think the Rook a most beneficial bird to the agriculturist.

For its depredations on game they attempt no excuse, but only offer an apology on the ground that the affair is very rare, and that condonation may be granted to the bird in consideration of the great services rendered in other parts of the year. They aver that its object in pulling up the young corn sprouts is not so much to eat the corn as to devour that pest of the farmer, the terrible wireworm, which lurks at the root of the corn, and infallibly destroys every plant which it has once attacked. That such has been the case may often be seen by the yellow and unhealthy aspect of the destroyed blades which are left scattered on the ground after the extraction of the wireworm. Potatoes again are attacked by numerous insect foes, and it is to eat these that the Rook unearths the "sets." It is true that bits of potato have been found in the Rook's crop, but in all probability they have been casually eaten together with the insects that are lurking within. The same remark may be made of the turnips.

Besides performing these services, the Rook saves acres of grass annually from being destroyed by the grub of the common cockchaffer beetle. The grub or larva of this insect is one of the most destructive foes to grass lands, feeding upon the roots and shearing them very nearly level with the surface of the ground by means of its scissor-like jaws. So destructive are these insects, and so complete are their ravages, that a person has been able to take in his hands the turf under which they had been living and to roll it up as if it had been cut with a spade. In one place, the grubs were so numerous that they were counted by the bushel. When it is remembered that this creature lives for three years underground, is furnished with a huge stomach, a wonderful capability of digestion, and a formidable cutting apparatus for obtaining its food, the services of the Rook in destroying it may be better imagined. Moreover, the beetle is just as destructive as the grub, settling upon trees and fairly stripping them of their leaves. I have dissected many of these grubs, and always found their stomachs distended to the utmost with a mixture of black earth and vegetable matter.

Again, when the ploughman is turning up the soil, how common, or rather how invariable, a sight it is to see the Rooks settling around him, alighting in the furrow which he makes, and seizing the grubs and worms as they are turned up by the share. Not a single worm, grub, or other insect escapes the keen eye and ready bill of this useful bird. Some idea of the extensive character of its operations may be formed from the following remarks by Mr. Simeon, in his interesting work entitled " Stray Notes on Fishing and Natural History."

" I was walking one day with a gentleman on his home farm, when we observed the grass on about an acre of meadow land to be so completely rooted up and scarified that he took it for granted it had been done under the bailiff's direction to clear it from moss, and on arriving at the farm, inquired whether such was not the case. The answer was, however, ' Oh no, sir, we have not been at work there at all ; it's the Rooks done all that.' The mistake was a very natural one, for though I have often seen places where grass has been pulled up by Rooks, yet I never saw such clean and wholesale work done by them as on this occasion. It could not apparently have been executed more systematically or perfectly by the most elaborate ' scarifier' that Croskill or Ransome could turn out.

On examining the spot afterwards, I found that the object of the Rooks' researches had doubtless been a small white grub, numbers of which still remained in the ground a short distance below the surface. In the following spring I noticed that the part of the field where this had taken place was densely covered with cowslips, much more so than the rest of it. Possibly the roots of these plants may have been the proper food for the grubs, and therefore selected by the parent insect as receptacles for her eggs."

The Rook also feeds upon berries and various fruits, being especially fond of oak-nuts, and having a curious habit of burying them in the earth before eating them, by which means, no doubt, many a noble oak-tree is planted. It also eats walnuts, and is fond of driving its bill through them and so taking them from the tree. The cones of the Scotch fir are also favourite food with the Rook, which seizes them in its beak, and tries to pull them from the bough by main force ; but if it should fail in this attempt, it drags the branch forcibly upwards, and then suddenly releases it, so as to jerk the cones from their stems by the recoil.

The practice of terrifying Rooks by means of scarecrows has already been mentioned, together with its usual failure. Even the bodies of slaughtered Rooks suspended from sticks have but little effect on these audacious birds, who may be seen very unconcernedly searching below the carcases for the beetles and other carrion-eating insects that are always found in such localities. The surest way to frighten the Rooks by means of dead comrades is not to hang them up in a position which every Rook knows is not likely to be assumed by any of its friends, and therefore conveys no intimation of alarm to its logical mind, but to lay them flat upon the earth with outstretched neck and spread wings as if they had fallen dead from something evil in the locality. Another useful method is to post a number of sticks in double rows and connect them with each other by strings tied in zigzag fashion, when it will be found that the Rooks are so suspicious of a trap, that they will not venture to enter any of the angles so formed.

The second subject of controversy is the presence of a bare white skin upon the forehead of the adult Rook and the base of its neck, those portions being clothed with feathers during the bird's youth.

The general opinion was, that the bird, by constantly delving in the soil, wore off all the feathers, only leaving the white skin behind. This solution of the problem was current for a long time, until some observer remarked that the base of the bill showed no particular marks of hard wear; that the bald space extended behind the line of the eyes, so that the bird could not possibly plunge its beak to so great a depth; that the white skin was evidently an intentional arrangement, and was too well defined at the edges to have been produced by the operation of digging, and must in that case always vary with the soil and the kind of food; moreover, there are many other birds which have bald spaces on their persons, such as the vultures and the turkey, and that in their case no theory of friction is required by which the phenomenon can be accounted for.

Matters having proceeded thus far, dissection was next employed, and it was observed that although feather bulbs could be found within the white skin, they were shrivelled and useless for the production of feathers. Experiments were then tried, wherein sundry young Rooks were kept caged, and denied access to any earth or mouldy substances; and in every case except one (and probably in that case also when the bird had attained maturity) the feathers with which the base of the back were covered fell off in the course of moulting, and were never replaced by fresh plumage. Every ornithologist knows well that many birds when young are distinguished by feathery or hairy tufts, as in the case of the Leatherhead, figured in page 219 of this work, which, when young, is decorated with a tuft of plumy hair upon its head; but after the moult, loses its cranial ornament. Mr. Simeon pertinently remarks, in allusion to this controversy, that a similar phenomenon may be seen in the human race, the forehead of a baby being often covered with fine downy hairs, which fall off as the child grows; and that in the elephants of Ceylon, the young is often clothed with a thick woolly fur over its head and fore-parts when born, but loses its covering as it approaches maturity. Altogether it seems that those who advocate the naturally bare forehead and beak have the best of the argument.

The habits of the Rook are very interesting, and easily watched. Its extreme caution is very remarkable, when combined with its attachment to human homes. A colony of a thousand birds may form a rookery in a park, placing themselves under the protection of its owner; and yet, if they see a man with a gun, or even with a suspicious-looking stick, they fly off their nests with astounding clamour, and will not return until the cause of their alarm is dissipated. During the "Rook-shooting" time, all the strong-winged birds leave their nests at the first report of the gun, and, rising to an enormous elevation, sail about like so many black midges over their deserted homes, and pour out their complaints in loud and doleful cries, which are plainly audible even from the great height at which they are soaring. The voice of the Rook is too well known to need description, and the bird is rather capricious in the utterance of its hoarse cry, sometimes keeping a prolonged silence, and at other times cawing about incessantly.

In captivity the Rook retains many of its wild customs, and in one instance was in the habit of going round the hens' nests and eating the eggs as soon as they were laid. The Rook is not often kept as a domestic pet, as it is with difficulty reared when young. Before rain, the Rook has a curious custom of ascending to a considerable height, and then shooting obliquely through the air, in a manner somewhat similar to a hawk making its stoop. During the daytime, the Rooks are widely dispersed throughout the fields, occasionally visiting their homes and then returning to their feeding-places; but, as the evening approaches, they cease feeding with one accord, and seek their nests, flying in long trains to the spot where they have made their residence.

The nest of the Rook is large, and rather clumsily built; consisting chiefly of sticks, upon which are laid sundry softer materials as a resting-place for the eggs. The Rook is a very gregarious bird, building in numbers on the boughs of contiguous trees, and having a kind of social compact that often arises into the dignity of law. For example, the elder Rooks will not permit the younger members of the community to build their nests upon an isolated tree at a distance from the general assemblage; and if they attempt to infringe this regulation, always attack the offending nest in a body, and tear it to pieces. They are even clever enough to notice the marks that are made on the trunks of trees that are to be felled, and will neither build on those doomed branches nor permit their young friends to do so. They also have a kind of criminal code, for they have been seen to hold a sort of trial, ending in the condemnation and execution of the culprit; and they unanimously punish those lazy Rooks which, instead of going out to fetch sticks for their nests, stay quietly at home and rob those of other Rooks.

The number of birds that are to be found in such rookeries is enormously great, several thousands having been counted in a single assemblage. In such cases they do great damage to the upper branches of the trees, and in some instances have been known to kill the tree, by the continual destruction of the growing boughs.

The colour of the Rook is a glossy, deep blue-black; the blue being more conspicuous on the wing-coverts and the sides of the head and neck. The bird may be easily recognised, even at a distance, by the conspicuous greyish-white skin, which serves to distinguish it from the crow. The length of an adult Rook is about eighteen or nineteen inches.

THE smallest of the British Corvidæ is the well-known JACKDAW, a bird of infinite wit and humour, and one that has an extraordinary attachment for man and his habitations.

Although of similar form, and black of plumage, the Jackdaw may easily be distinguished from either the rook or the crow by the grey patch upon the crown of the head and back of the neck, which is very conspicuous, and can be seen at a considerable distance. The voice, too, is entirely different from the caw of the rook or the hoarse cry of the crow, and as the bird is very loquacious, it soon announces itself by the tone of its voice. It generally takes up its home near houses, and is fond of nesting in old buildings, especially preferring the steeples and towers of churches and similar edifices, where its nest and young are safe from the depredations of stoats, weasels, and other destroyers. Indeed, there are few places where Jackdaws will not build, provided that they are tolerably steep and high; and there are many curious circumstances in connexion with its nesting, which will presently be mentioned.

In its wild state the Jackdaw has many of the rook habits, and therefore needs no particular description. Mudie, however, mentions a curious circumstance, which seems to point out a closer relationship between rooks and Jackdaws than could be supposed. "In the latter part of the season, when the rooks from one of the most extensive rookeries in Britain (in the woods of Panmure) made daily excursions of about six miles to the warm grounds by the seaside, and in their flight passed over a deep ravine in the rocky sides— or rather side, for they only inhabited the sunny one—on which there were many Jackdaws, I have observed that when the cawing of the rooks in their morning flight was heard at the ravine, the Jackdaws, who had previously been still and quiet, instantly raised their shriller notes and flew up to join the rooks, both parties clamouring loudly as if welcoming

JACKDAW.—*Corvus monedula.*

each other, and that on the return, the time of which was no bad augury of the weather of the succeeding day, the Daws accompanied the rooks a little past the ravine ; then both cawed their farewell and departed.

What is more singular, I have seen, too frequently for its being merely accidental, a Daw return for a short time to the rooks, a rook to the Daws, or one from each race meet between and be noisy together for a space after the bands had separated. With the reason I do not interfere, not being in the secrets of either party, but the fact is as certain as it is curious."

In captivity, to which it accommodates itself with most philosophical composure, the Jackdaw is a very amusing bird, and soon learns many curious tricks. I have already recorded many anecdotes of some tame Jackdaws in "My Feathered Friends," published by Messrs. Routledge, to which the reader is referred, as well as for a more detailed history of the rook, magpie, and many others of the same tribe. I will therefore refrain from repeating them, and only give one or two anecdotes of a Jackdaw that belonged to one of my friends, and which was to the full as remarkable a bird as any that I have met with.

He was imitative in the extreme, and more than once had put the house in danger by his passion for lighting lucifer matches, of which amusement he was as fond as any child. On one occasion he lighted the kitchen fire in the course of the night. The cook had laid the fire over-night, intending to apply the match early in the morning. The Jackdaw contrived to get hold of the lucifer box, and had evidently rubbed the match upon the bars and so set fire to the combustibles, as the cook found the fire nearly burnt out, the Jackdaw in the kitchen, and some eighteen or nineteen exploded matches lying in the fender.

The first time that this Jackdaw lighted a match he was so frightened at the sharp crackling report that he ran away as fast as he could go, coughing and sneezing after his fashion from the fumes of the sulphur, he having held the match close to the phosphoric end. He never seemed to distinguish the ignitible end of the match, and would rub away with great perseverance on the blank end, without discovering the cause of his failure. By degrees he contrived to singe all the feathers from his forehead and nostrils, and once burned his foot rather severely.

He was greatly afraid of thunder, and had a singular power of predicting a coming storm. In such a case he would retire to some favourite hiding-place, generally a dark hole in a wall, or a cavity in an old yew which exactly contained him, and would there tuck himself into a very compact form so as to suit the dimensions of his hiding-place, his body being tightly squeezed into the cavity and his tail projecting along the side. In this odd position he would remain until the storm had passed over, but if he were called by any one whom he knew, his confidence would return, and he would come out of his hole very joyously in spite of the thunder, crying out, "Jack's a brave bird!" as if he entirely understood the meaning of the sentence. He may possibly have had some idea of the sense of words, for he hated being called a coward, and would resent the term with all the indignation at his command.

There are, however, few birds which are possessed of the ingenuity which characterised this Jackdaw, for it may be noticed that every bird has its own individuality strongly marked, even though the same type of intellectual power may characterise it in common with all others of the same tribe.

Another Jackdaw, belonging to one of my friends, was a most inveterate poacher, having taken to himself an associate or accomplice in the person of the cat belonging to the house. This oddly matched couple used to make their egress and ingress through a hole in the bottom of a very thick quickset hedge, and as soon as they emerged into the open fields, would immediately hunt for game. Their mode of catching and killing game was not clearly ascertained, but its successful results were evident from the frequency with which they used to bring home dead hares, often as large as the cat, but generally small. On one occasion a singular fluttering of wings and scratching of claws was heard in the hedge, and when the owner of the two animals went to ascertain its cause, he found that they had brought home a hare so large that they could not drag it through the hole in the hedge, and were quite frantic in their eagerness to attain their object, the cat pulling from within, and the Jackdaw pushing from without.

In the "Annals of Sporting" is recorded a curious anecdote of the attachment displayed by the Jackdaw to its owner. The relator of the anecdote, after making a few casual remarks, proceeds as follows: "I pulled up for the first time to bait at the King's Head, Egham, and soon after my arrival a young man rode into the inn-yard from the opposite direction, and dismounted at the door of the taproom belonging to the hotel. Almost immediately following this common event, a Jackdaw alighted on a shed adjoining, which, however, as those birds are frequently kept at such places, did not attract any particular attention, till the ostler called out, 'Ah! here you are then again, true to the old house and young master.' I immediately asked whom he meant. 'Why Jack, Sir, yonder!' pointing to the Daw. 'And what of him?' I went on to inquire. 'Oh! Sir, he is a most 'cute and cunning fellow, and follows his master wherever he goes, either on horseback or on foot.' This awakened my interest, and I received these further particulars of this extraordinary bird.

He belonged to the son of the ostler of the 'Bush' at Staines, and was constantly fed and taken care of by him, until he became quite his familiar friend; so much so, indeed, that the circumstance created wonder in the vicinity of its home. So convinced was the ostler of the faith and devotion of his feathered acquaintance, that on one particular occasion, as he was setting off from Staines to Hounslow on horseback, he made a wager —a large one for him—of two bowls of punch with a person who doubted that the bird would obey the call of his master and follow his route. He then mounted, and exclaiming, 'Come, Jack, I'm going!' put his horse in motion. In a very short time the bird's wings were extended, and he attended the progress and return of his feeder, leaving not the shadow of a plea for the non-payment of the bet which the sceptic had so unwittingly ventured.

This, and some other circumstances which my informant mentioned, induced me to watch more narrowly the motions of the bird, and I observed him constantly hopping from place to place, and every now and then pitching upon the sill of the window that lighted

GREAT-BILLED CROW.—*Corvus crassirostris.*

the taproom, in order to ascertain if his travelling companion was still within. On one occasion, indeed, he pressed quite anxiously into the room, and observing him he sought not inclined immediately to move, he took a flight in a circular direction for nearly half a mile, returning again to his former station. Soon after this the man prepared his horse, Jack mounted upon the sign-post, and as soon as the former had ridden about a hundred yards on his road to Staines, he fluttered his dark pinions and followed the well-remembered track of the ostler-boy of Staines."

The grey patch on the head and neck is not seen until the bird attains maturity, the feathers being of the same black hue as on the remainder of the body until the first moult, when the juvenile plumage is shed and the adult garments assumed.

The nest of the Jackdaw is a very rude structure of sticks, lined, or rather covered, with hay, wool, feathers, and all kinds of miscellaneous substances of a warm kind for the eggs and young. It is placed in various localities, generally in buildings or rocks, but has often been found in hollow trees and even in the holes of rabbit-warrens, the last-mentioned locality being a very remarkable one, as the young birds must be in constant danger of marauding stoats and weasels. In one instance a quantity of broken glass was employed

2.   D D

in the foundation of the nest. The Jackdaw is not choice in the selection of feathered neighbours, for I have found in the same tower the nests of pigeons, Jackdaws, and starlings, in amicable proximity to each other. The eggs are smaller and much paler than those of the rook or crow, but have a similar general aspect. Their number is about five.

The general colour of the Jackdaw is black, with the exception of the back of the head and the nape of the neck, which are grey. A decided tinge of glossy blue is perceptible on the wings. The total length of the adult Jackdaw is about fourteen inches. The female is distinguishable from her mate by the darker colour of the grey hood. In both sexes the wings are short, and when closed do not reach within an inch of the tip of the tail. As is the case with many others of the same tribe, there are instances of pale and pied Jackdaws, the wings and tail being generally darker than the remainder of the body. In the British Museum there is a good specimen of an albino jay, the body being creamy white, while the wings retain the barred blue and black so characteristic of the species, but extremely pale.

Sometimes the Jackdaw will take possession of the deserted nest of a rook or crow, and laying a substratum of hay and wool upon the original fabric, deposit its eggs and rear its young upon this easily gotten property. Occasionally, but very rarely, the Jackdaw has been seen to build a regular nest in the branches of trees, rocks, or rabbit-warrens, —ruins and church-towers being wanting.

THE remarkable bird which has very appropriately been called the GREAT-BILLED CROW is, undoubtedly, the most singular example of the whole tribe.

In its dimensions it is much larger than an ordinary crow, and rather smaller than a raven, for which bird it might be taken but for the extraordinary beak. The bill of this species is so large as to remind the observer of a toucan or a puffin, and the bite of such a powerful weapon must be most formidable. It is very deep, thick and rounded, becoming wider at the top and deeply ridged, curving suddenly to a point, and very sharp at its extremity. In colour it is jetty black, except the extreme tip, which is white.

The colour of this bird is deep shining black on the upper parts, like that of the raven, having a slight purplish gloss upon the sides. Upon the back of the neck there is a pure white oval mark, and upon the shoulders there is another white patch of a crescentic shape ; the two being connected with a narrow line of white down the back of the neck, so that the whole shape of the mark resembles an orange in a wineglass. The Great-billed Crow is a native of Abyssinia.

The ROYSTON CROW, or HOODED CROW, or GREY CROW, is a very conspicuous bird, on account of the curiously pied plumage with which it is invested.

This bird is not very common in England, but is plentifully found both in Ireland, Scotland, and the Scottish isles, having been seen in large flocks of several hundred in number on the east coast of Jura. Generally it is not very gregarious, the male and female only being found in company ; but it sometimes chooses to associate in little flocks of fifteen or sixteen in number. It seems to prefer the sea-coast to any inland locality, as it there finds a great variety of food, and is not much exposed to danger. I have often seen these birds in the Bay of Dublin, perching upon the rocks at low water, and searching for food among the dank seaweed, and in the rock-puddles that are left by the retreating tide. They seemed always extremely bold, and would permit a very close approach without exhibiting any alarm.

The food of the Hooded Crow is almost wholly of an animal nature, and consists of small quadrupeds, carrion of every kind, worms, insects, marine animals, and the miscellaneous mass of animal substances which are cast up by the tide and left upon shore. Limpets form a considerable portion of its food, and are detached from the rock, to which they cling so tightly, by an adroit peck and wrench of the bill. Sometimes it is said that the poor Crow is not quick enough in its movements, and is held so firmly to the rock by the alarmed limpet, that it is retained in that unpleasant position until the returning tide overwhelms both mollusc and bird.

Mussels and cockles are also favourite dainties with the Hooded Crow, which, however, is unable to open their tightly closed shells with its beak, and has recourse

ROYSTON OR HOODED CROW.—*Corvus cornix.*

to the expedient of carrying its prey to a great height in the air, and smashing the shell by letting it fall upon a rock. Sometimes its ingenuity has been very ill rewarded by the loss of its dinner, for no sooner has the shell struck upon the rock than it is seized and carried off by another Hooded Crow which has concealed itself near the spot. As the mussels are often very firmly bound to the rock, this ingenious bird employs another mode of breaking their shells. He takes a tolerably large stone in 'his beak, rises perpendicularly above the mussel bed, drops the stone at random upon the black mass of molluscs below, and descends to feed on the bodies of those which have been crushed by the missile.

Not only does the Hooded Crow feed upon such harmless diet as has already been mentioned, but it makes great havoc among small and young birds, and has often been known to hover about the shore sportsman, and carry off the dead and wounded birds as they fall to the gun. It also haunts the farmyard when it finds a deficiency of food in the open country, and darting among the poultry, kills and carries off young chickens, or breaks and drains the eggs on which the hens are sitting. Sickly and very young lambs are also persecuted by this voracious bird, who goes its rounds among the flocks as regularly as a sentinel, and if its watchful eye should discover a lamb or sheep lying on its back in a ditch, is sure to hasten its death by punching out its eyes with its long and powerful bill. For these reasons, the Hooded Crow is entirely detested by the country people, many of whom are imbued with sundry superstitions concerning its origin and object.

Like many of the same tribe, it is a most annoying neighbour to the larger birds, especially those of a rapacious nature, and never can allow a hawk, heron, or owl to pass within ken without mobbing it in a very persevering manner.

The Hooded Crow never breeds in society, but always builds its nest at some distance from the homes of any other of the same species, so that, although a forest or a range of cliffs may be inhabited by these birds, the nests are scattered very sparingly over the whole extent. The structure of the nest is somewhat similar to that of the crows and rooks, being a mass of sticks and heather stalks as a foundation, upon which is placed a layer of wool, hair, and other soft substances. Sometimes the bird builds a better and more compact nest with the bark of trees; and in all cases this species breeds very early in the season.

It is said that the Hooded Crow will sometimes breed with the common species, and the following curious observations are recorded in the "Field Naturalist," and quoted by Mr. Yarrell in his history of the bird.

"For four successive years I have had opportunities of witnessing the pairing of the carrion Crow and the Hooded Crow upon some large beech-trees which surrounded my house in Forfarshire. They never re-occupied the old nest, nor did they always build their nest on the same tree; nor was I positively certain that they were the same individuals who returned every year to these trees, though it is probable that they were, for they were never molested. Knowing the predatory propensities of the carrion Crow on hens' eggs, young chickens, and even turkey poults, I would have shot them had they been a pair of carrion Crows; but I was anxious to watch the result of what appeared to me at the time a remarkable union.

PHILIPPINE CROW.—*Corvus Sinensis.*

Judging from the manners of the two birds, the almost evident incubations and carefulness exhibited, I should say that the Hooded Crow is the female. though the carrion Crow did frequently sit upon the eggs. After the young of the first year took wing, I perceived that the one was a carrion and the other a Hooded Crow, and this distinctive character was maintained in the young which were hatched every year, so long as I remained in that part of the country. I shot the first young pair, and ascertained that the hooded one was the female, and the carrion was the male, which confirmed me in my conjecture of the sexes of the parents. Ever after, old and young were unmolested by me; but notwithstanding the increase of number every year after the first one, only one pair came annually to build in these beech-trees."

This species has often been tamed, and displays much affection for its owner. One of these birds, which had been wounded and captured, was placed in a walled garden together with the poultry, with whom it soon made friends. In process of time it recovered from its wound, took flight and disappeared. But after an absence of some months it returned to its old quarters, and voluntarily took its place again with the poultry

FISH CROW.—*Corvus ossifragus*

in the well-remembered spot, and was quite as familiar with the owner of the house as any of the hens.

The Hooded Crow is boldly and conspicuously pied with grey and black, distributed as follows : The head, back of the neck, and throat, together with the wings and tail, are glossy bluish black, while the remainder of the body is a very peculiar grey with a slight blackish wash. The length of the bird is about nineteen or twenty inches. It goes by many names in different parts of the country, among which Dun Crow, Hoody, and Hoddy are the most common.

The PHILIPPINE CROW derives its name from the locality in which it is found, its place of residence being the Philippine Islands.

It is a striking and handsome bird on account of the elegant crest which decorates its head and the general hue of its plumage. It is not a large bird, measuring only eleven inches in total length. The colour of the upper parts of the body is pale green dashed with yellow here and there, according to the direction of the light, and a similar tint, but with more yellow, under the throat. A black band runs round the head enveloping the eye in its progress, and is partially covered by the loose flowing feathers of the crest. The dense wing-coverts are brown, the quill-feathers are deep olive-green on their exterior sides, and the secondaries are tipped with white with a slight dash of green. The bill and legs are of a reddish hue.

The FISH CROW of America is about the size of a common jackdaw, its length being generally about sixteen inches. Our chief information of this bird and its habits is derived from Wilson ; and as his account cannot be condensed without great loss of its original vigour and freshness, it is here given at length.

" I first met with this species on the sea-coast of Georgia, and observed that they regularly retired to the interior as evening approached, and came down to the shores of the river Savannah by the first appearance of day. Their voice first attracted my notice, being very different from that of the common Crow, more hoarse and guttural, uttered as if something stuck in their throat, and varied into several undulations as they flew along. Their manner of flying was also unlike the others, as they frequently sailed about without flapping their wings, something in the manner of the raven; and I soon perceived that their food and their mode of procuring it were also different, their favourite haunts being about the banks of the river, along which they usually sailed, dexterously snatching up with their claws dead fish or other garbage that floated on the surface. At the country seat of Stephen Elliot, Esq. near the Ogechee river, I took notice of these Crows frequently perching on the backs of the cattle, like the magpie and jackdaw of Britain, but never mingling with the common Crows, and differing from them in this particular, that the latter generally retire to the shore, the reeds, and marshes, to roost, while the Fish Crow always a little before sunset seeks the interior high woods to repose in.

On my journey through the Mississippi territory last year, I resided for some time at the seat of my hospitable friend Dr. Samuel Brown, a few miles from Fort Adams on the Mississippi. In my various excursions there, among the lofty fragrance-breathing magnolia woods and magnificent scenery that adorn the luxuriant face of nature in these southern regions, this species of Crow frequently make its appearance, distinguished by the same voice and habits it had in Georgia.

There is in many of the ponds there, a singular kind of lizard, that swims about with its head above the surface, making a loud sound not unlike the harsh jarring of a door. These, the Crow now before us would frequently seize with his claws as he flew along the surface, and retire to the summit of a dead tree to enjoy his repast. Here I also observed him a pretty constant attendant at the pens where the cows were usually milked, and much less shy, less suspicious, and more solitary than the common Crow. In the county of Cape May, New Jersey, I again met with these Crows, particularly along Egg-Harbour river, and latterly on the Schuylkill and Delaware near Philadelphia, during the season of shad and herring fishing, viz. from the middle of March until the beginning of June. A small party of these Crows during this period regularly passed Mr. Bertram's gardens to the high woods to roost every evening a little before sunset, and as regularly returned at a little before sunrise every morning, directing their course towards the river. The fishermen along these rivers also inform me that they have particularly remarked this Crow by his croaking voice and his fondness for fish; almost always hovering about their fishing-places to glean up the refuse.

Of their manner of breeding I can only say that they separate into pairs and build in tall trees near the sea or river shore; one of their nests having been built this season in a piece of tall woods near Mr. Beasley's, at Great Egg Harbour. From the circumstance of six or seven being usually seen here together, in the month of July, it is probable that they have at least four or five young at a time."

The colour of the Fish Crow is deep steel-blue, appearing black in certain lights, and glazed in many places with rich purple. When closed, the tips of the wings do not reach within two inches of the end of the tail.

THE very curious bird which is known by the appropriate name of the BALD CROW is so different in aspect from the remainder of the genus, that it has been separated from them by common consent.

It hardly looks like a Crow, but reminds the observer of a cross between the hooded Crow, the darter, and the leather-head. Although small and slender in make, it is longer than many birds of much greater proportions, on account of the extremely long neck. The legs, too, are much more elongated than in any of the true Crows. The head is entirely denuded of feathers, or even hairs, and is covered with a blackish brown skin, drawn closely over the skull and bones of the head. On the back of the neck and head, the place of feathers is supplied by a scanty covering of white down. The back is black-brown, as is also the tail, and the wings are of a remarkably pure and beautiful mouse-brown. The neck, throat, and under surface are yellowish white. The total length of this bird is about fifteen inches.

THE small but handsome and striking bird which is popularly called the NUTCRACKER CROW is extremely scarce in England, having but seldom been discovered upon the British Islands.

As it is so conspicuous a bird, it would not escape the notice of even the most careless observer, and we may be sure that it has very seldom, if ever, visited England without its arrival being duly noted. It is tolerably common in several parts of Europe, and has been seen in Switzerland in large flocks, feeding upon the seeds of the pine-trees after the fashion that has gained for the bird its name of Nutcracker. This species feeds mostly upon seeds, especially those of the pine, the beech, and various nuts, and it breaks the hard shells by fixing the nut or pine-cone in a convenient crevice, and hammering with its beak until it has exposed the kernel. Indeed, while engaged in this pursuit, its movements are almost precisely those of the common nuthatch. It is a rather shy and suspicious bird, keeping closely to the tops of trees, and mostly being beyond the range of an ordinary shot-gun.

It does not, however, feed wholly on seeds, but varies its diet with insect food, in pursuit of which it ranges for a considerable distance over the country, seeking the insects either on the ground or on the trees—generally the latter.

By means of the powerful bill and neck muscles, the Nutcracker is able to dig out the large-bodied grubs which are found deeply buried in the wood of various trees, and which it discovers through its quick sense of sight and hearing.

The Nutcracker is common in Southern Europe, and is also a visitant of the more northern regions of that continent, being frequently seen in Norway, and even migrating so far north as the great pine-forests of Russia, Siberia, and Kamtschatka.

BALD CROW.—*Picathartes gymnocéphalus.*

Sometimes the Nutcracker becomes carnivorous in its taste, after the manner of the corvidæ in general, and robs sitting birds of their eggs, or even seizes and eats their callow young. It is a very active bird, traversing the branches with great rapidity, and being able to climb the perpendicular trunk of a tree almost as well as the creepers. That it frequently puts this accomplishment in practice is evident from the fact that the tips of the tail-feathers are often found to be worn away, evidently by the pressure which they have exerted against the rough bark of the trees.

NUTCRACKER.—*Nucifraga caryocatactes.*

The nest of this species is made at the extremity of a long tunnel cut in the wood of some decaying tree, and either originally dug by the bird, with the express object of making a resting-place for its eggs, or altered and adapted from an already existing hole. In this respect, as well as in other habits, the Nutcracker bears some resemblance to the woodpeckers. The eggs are said to be from five to six in number, and greyish yellow in colour, spotted with a darker hue. The bird seems to require a large supply of nourishment; and although it cannot be ranked among the true migrators, it can range over a large extent of country in search of food, being instinctively able to discover the localities where its wants can be best supplied.

The colouring of this bird is peculiar, and rather complicated. The crown of the head is deep brown, and the space between the beak and the eye is greyish white. The back, the sides of the head, the scapularies, the wing-coverts, and the whole of the under surface of the body are of a warm brown, covered thickly with elongated white spots, caused by the white tips with which each feather is furnished. The throat and chin take a darker hue, but are still marked with the conspicuous white spots. The wings and upper tail-coverts are dark black-brown, the black being less marked at the extremities of the feathers. The tail is mostly blackish brown, diversified with white, the two central feathers being totally dark, and the remainder taking more white as they approach the two external feathers. The under surface of the tail is light grey-brown. The total length of the Nutcracker is about thirteen or fourteen inches.

AMERICA possesses a very pretty example of this genus in the CLARK'S NUTCRACKER. This bird is notable for the diversified beauty of its plumage, and for the extremely formidable claws with which it is armed; the latter peculiarity leading to the idea that the bird preys on various living animals, after the manner of the fish crow. It frequents the rivers and sea-shore in considerable numbers, assembling in flocks, like the rook of Europe, and pouncing continually upon various substances which it immediately swallows. It is a very noisy as well as gregarious bird, chattering continually while feeding.

The wings, the two central tail-feathers, and several of the remaining feathers of the tail are deep glossy blue-black, and the secondaries are also black, but are marked with

CLARK'S NUTCRACKER.—*Nucifraga Columbiána.*

a large patch of white. The head, neck, and greater part of the body is light fawn, changing to a pearly grey upon the breast and abdomen. The total length of this bird is about thirteen inches.

Who does not know the Magpie, the pert, the gay, the mischievous? What denizen of the country is not familiar with his many exploits in the way of barefaced and audacious theft, his dipping flight, and his ingenuity in baffling the devices of the fowler and the gunner? What inhabitant of the town has not seen him cooped in his wicker dwelling, dull and begrimed with the daily smoke, but yet pert as ever; talkative, and a wonderful admirer of his dingy plumage and ragged tail?

The Magpie is found in very many parts of the world, and is plentiful throughout England, always keeping to well-wooded districts, as if distrusting its power of flight in the open country; for the larger hawks are prone to fly at the Magpie, which has but little chance of escape upon the plain, but can always evade his foe among hedgerows and plantations, by slipping among the branches and dodging through the foliage. Even a trained falcon fails to catch a Magpie when it has once reached such an asylum, and the falconer is forced to drive it from its refuge before the hawk can secure its prey. In some parts of England, Magpie hawking is a favourite amusement, for the Magpie is to the full as cunning as a fox, and in spite of all the array of beaters, hounds, and horsemen, not unfrequently baffles its pursuers, and makes its escape in safety.

The food of the Magpie is as multifarious as that of the crow or raven, and consists of various animal and vegetable substances. It is a determined robber of other birds' nests, dragging the unfledged young out of their homes, or driving its bill through their eggs and thus carrying them away. Even hens' nests are not spared by this bold and voracious bird, who, however, sometimes falls a victim to its marauding propensities. The aggrieved poultry-owner, after removing the eggs from all the hens' nests, empties one of the eggs, and fills it again with bird-lime. This prepared egg is then placed in the nest as a bait for the Magpie, who soon returns to the scene of its former robberies, drives its beak into the egg, and makes off with its booty. Its triumph is, however, very short-lived, for the bird soon finds itself unable to get rid of the stolen egg in the usual manner, and at last batters it against a stone or branch for the purpose of breaking the shell. The natural consequence is, that the birdlime immediately clings to the beak, and the broken fragments of shell, which fly in every direction, cover the wings and plumage as the bird tries to shake itself loose from its impediments, and the Magpie falls to the ground in a hopelessly crippled state, and becomes an easy victim to the author of the snare.

The Magpie also attacks full-grown birds, mice, reptiles of various kinds, and has been observed in the act of killing a common grass snake. Beetles it eats in very large quantities, and also feeds upon worms, snails, and various similar creatures, so that the harm which it does to the game and poultry is probably more than compensated by its good offices in ridding the gardens and cultivated grounds of their varied foes. It also eats fruits, and has been seen to feast eagerly on the light succulent berries of the mountain ash.

Like the crow, the Magpie is a determined persecutor of various birds and beasts of prey, scarcely allowing a hawk to pass within ken, or a weasel or stoat to glide along the bottom of a hedge without screeching forth an alarm and a summons to its allies, and dashing at once to the attack. Mr. Metcalf relates, that while in Norway, he saw his dog pursued and mobbed by at least forty of these birds. The same writer also remarks, that he captured a Magpie by means of a piece of meat on a hook. The bird took the bait as eagerly as any perch would have done, and, to its profound astonishment, was immediately hooked. Mr. Metcalf amused himself for a little while in "playing" the bird as if it had been a fish, with this difference, that the Magpie was trying to escape by flight, and poured forth a succession of most dismal yells, which sent off all its formerly valiant companions screaming with terror at the unexpected sight.

The nest of the Magpie is a rather complicated edifice, domed, with an entrance at the side, and mostly formed on the exterior of three branches, so as to afford an effectual protection against any foe who endeavours to force admittance into so strong a fortress. Generally the nest is placed at the very summit of some lofty tree, the bird usually preferring those trees which run for many feet without a branch. The tops of tall pines are favourite localities for the Magpie's nest, as the trunk of these trees is bare of branches except at the summit, and the dark green foliage of the spreading branches is so thick that it affords an effectual shelter to the large and conspicuous edifice which rests upon the boughs. Sometimes, however, when the Magpie has been protected, and accustomed itself to the vicinity of human habitations, it has fixed its nest in a low bush near the ground, as if trusting to the kindly feelings of its human neighbours.

Although displaying great attachment to its mate, and the most dauntless courage in defending its nest and young, its affections seems to be rather transient in their character, and quite unable to withstand the test of absence. For example, if one Magpie of a pair be shot, the survivor never fails to find another mate within the space of two or three days. Sometimes the period of widowhood exists only for some twenty-four hours, and there have been instances where a Magpie has found another mate within a few hours after the decease of its former spouse.

When tame, it is a most amusing bird, teaching itself all kinds of odd tricks, and learning to talk with an accuracy and volubility little inferior to that of the parrot. It is, however, a most incorrigibly mischievous bird, and unless subjected to the most careful supervision is capable of doing a very great amount of damage in a wonderfully short space of time. I have witnessed a multitude of these exploits, but as I have already related many of them in my "Feathered Friends," the reader is referred to the pages of that little work for a tolerably long series of new and original Magpie anecdotes.

Mr. Thompson tells an amusing story of a tame Magpie which struck up a friend-ship with a peculiarly long-wooled sheep. The bird was accustomed to sit on the back of its friend, couching luxuriantly upon the long thick fleece, and making short excursions among the sheep for the purpose of pecking their legs, and making them run about. He also employed the fleece of his friend as a treasury of stolen goods, being accustomed to hide his pilferings among the thick wool and mount guard over them. The same writer has published the following interesting account of a tame Magpie belonging to

MAGPIE.—*Pica caudáta.*

Dr. Stevelly, of Belfast, who communicated to him the story, from which the following is an extract.

"He was particularly fond of any shining article, such as spoons and trinkets ; these he frequently stole, and we came upon his treasure-house in a remarkable way. There was an old gentleman, a great friend of my father's, who resided with us almost continually. He was of a peculiarly studious disposition, but from a deformity in his person used generally to read standing, with his arms and breast resting on the back of a chair, and the book placed on a table before him. After having read for a while, it was his habit to take off his spectacles, lay them beside him, blow his nose, take a pinch of snuff, and after a few moments pondering what he had been reading, resume the spectacles and proceed.

One very warm day I lay reading at the end of a room in which there was an open glass door leading to the greenhouse ; in this room the old gentleman was most intently pursuing his studies at a little distance from me. My attention was soon arrested by seeing the Magpie perched upon the chair near him, eyeing him most intently, and with a very arch expression, and at length, in an instant, he had with a most active hop reached the table, secured the red leathern spectacle-case, and was out of the glass door with the most noiseless wing, and with a very graceful motion.

I remained quiet, resolved to see the end of the joke. After a few seconds' absence. 'Jack' was again at his post, eyeing the old gentleman with a most inquisitive and yet business-like glance; it was nearly impossible to resist the ludicrous impression produced by the entire scene. At length off came the spectacles, and out came the pocket-handkerchief and snuff-box; quick as thought Jack had visited the table and was out of the open door with the prize, which I have no doubt had from the beginning been the object of his covetous admiration while they were on the nose of the old gentleman. This time the Magpie did not return, either because he found it more difficult to reach his store-house with the spectacles than with the case, or because, having gained the object of his ambition, he conceived his presence no longer necessary.

At length, the period of rumination having elapsed, the old gentleman was about replacing the spectacles. As soon as his surprise had abated at not finding them with his hand beside him on the table, he removed the chair and groped about on the carpet, then raised the book and examined every part of the table. Not being able to restrain myself any longer, I exploded in laughter, and of course I was instantly suspected of playing off a practical joke, and charged with taking the spectacles, but at length succeeded in convincing him that I had never risen from the sofa on which I reclined. After a good deal of laughing, and two or three other members of the family having been attracted to the room by the hubbub, I was compelled under cross-examination to own that I had witnessed Jack's abstractions.

The question then became serious how the articles were to be recovered, and some person suggested to leave a teaspoon near him and watch him. This was accordingly done, but his motions were so rapid that he eluded us all, seeming at first to pass completely over the house. At length, by placing two or three persons in favourable positions, he was 'marked' in a leaden valley between a double part of the roof, and this having been closely searched, a deposit was discovered, not only of the things which Jack had that day carried off, but also of some articles which had been for some time supposed to be lost, but respecting which a breath of suspicion as to him had never been entertained. This day's successful foray led to his losing his entire store, no doubt in the midst of his triumphant rejoicing."

Although imported into Ireland at a comparatively late period, the Magpie has taken complete possession of that island, and is found in very great numbers, four or five being often seen within a few hundred yards of each other. The beautiful wings of this bird are much sought after for the purpose of being dried, flattened, and mounted on hand-screens; the two wings being mounted with their external quills together, so as to form a screen of a heart-like shape.

Many superstitious ideas have always been current respecting the Magpie, its appearance singly, doubly, or trebly being held as an omen of bad or good luck, and various predictions being made from the direction in which it appeared to the observer. Excepting the illiterate, however, there are few in our land who now give any credence to such tales; but in many other countries these doctrines are held by high and low alike. In the latter part of 1860, an official despatch was presented to the Chamber of Deputies at Dresden, requesting a supply of Magpies for the purpose of manufacturing a powder all-potent against epilepsy. Great stress was laid upon the two points, that the birds must be neither deficient in claws nor feathers, and that they must be shot between the 24th of December and the 18th of January. This extraordinary document was not only presented and read in good faith, but was backed by many noble names.

The plumage of this bird is remarkably handsome both in colour and form. The head, neck, back, and upper tail-coverts are deep black, with a slight green gloss in certain lights; and the same colour is found on the chin, the throat, the upper part of the breast, and the base, tips, and outer edges of the primary quill-feathers. The secondaries are also black, but with a blue gloss, which becomes peculiarly rich on the tertials and wing-coverts. The inner web of the primaries is white for a considerable portion of its length, presenting a bold and conspicuous appearance when the bird spreads its wings. The central feathers of the tail are nearly eleven inches in length, and they decrease gradually in size; those on the exterior being hardly five inches long. Their colour is a wonderfully rich mixture of the deepest blue, purple, and green, the green being towards the base, and the blue and purple towards the extremity. The under surface of the tail-feathers is dull black. The lower parts of the breast, abdomen, and flanks are snowy white. The total length of the adult male bird is about eighteen inches, the female being rather smaller and with a shorter tail.

THE Fruit Crows are placed by some systematic authors among the chatterers, while others, as in the catalogue which we follow, have considered them to be nearly related to the true Crows. They are all natives of Southern America, and are distinguished by their straight flattened beak, with its upper mandible round, and a notch at its extremity. The nostrils are placed in two membranous groves at each side of the bill. Most of the Fruit Crows are of considerable dimensions, some species equalling the Crows of Europe, while others are a little less.

The BARE-NECKED FRUIT CROW inhabits Brazil and Guiana, and is far from uncommon in those countries. It is not a very large bird, hardly equalling the common jackdaw in size, but is worthy of notice on account of the peculiarity from which it derives its popular name. Instead of being covered with the usual plumage, the upper part of the head, the back of the neck, and the throat are clothed with very minute and closely-set feathers of a very deep black, so that the bird looks as if the neck had been denuded of feathers, and covered with a piece of neatly sewn black velvet. On the sides of the neck even this slight clothing is absent, the

BARE-NECKED FRUIT CROW.—*Gymnoderus fœtidus.*

plumage being represented merely by a few scanty feathers of down. The general colour of the feathers is black in the male, and brownish grey in the female, excepting the wing-coverts and the edges of the central quill-feathers, which are slaty grey. The Bare-necked Fruit Crow is not at all an elegant bird in its form, being heavily made and thick-set.

THERE are several other members of this curious group, such as the BALD FRUIT CROW, called also the CAPUCHIN BALD HEAD, on account of the peculiarity which has

BALD FRUIT CROW.—*Gymnocephalus calvus.*

earned for it the popular titles by which it is known. In allusion to the monk-like aspect of the head, the Creoles of Cayenne call it "Oiseau mon Père."

This species is larger than the preceding, being quite equal to the English Crow in size, and being altogether of a thicker and larger make. It is very common in Guinea. The head of the Bald Fruit Crow is very large and heavily made, and the whole front of the bird is totally bare, like that of the leatherhead, already described and figured. Many naturalists think that while the Bald Fruit Crow is still young, its head is clothed with feathers, together with the remainder of the body, and that, like the rook of Europe, it loses the feathers when it attains maturity. There has been considerable argument on the subject, but it seems to have been tolerably well settled that the young bird is feathered and the old bird bare.

This bird, together with the other Fruit Crows, makes the greater part of its meals on berries, fruits, and other vegetable substances; but will often vary its diet by an admixture of insects, snails, and many similar creatures. It is seldom seen upon the ground, finding its food among the branches, and confining itself almost exclusively to their shelter. The generic names Gymnoderus and Gymnocephalus, which have been appropriately given to these birds, are of Greek origin, and signify, the former, "naked-necked," and the latter "naked-headed." The general colour of the Bald-headed Fruit Crow is dark brown, something like the dingy brown of a capuchin's cloak, thus giving to the bird the popular title of Capuchin. This colour is, however, relieved by the darker hue of the quill-feathers of the wings and tail, which are deep black.

UMBRELLA BIRD.—*Cephalópterus ornátus.*

THE group of the Fruit Crows may lay claim to the credit of reckoning among their number one of the most singular of the feathered tribe. The UMBRELLA BIRD, so well depicted in the accompanying illustration, is a truly remarkable creature, and from the extraordinary mode in which its plumage is arranged, never fails of attracting the attention of the most casual spectator.

The bird is a native of the islands of the South American rivers—being seldom if ever seen on the main land—from whence it is not unfrequently brought by collectors, as there is always a ready sale for its skin, either to serve as an ornament in glass cases, or a specimen for a museum. In dimensions the Umbrella Bird equals the common crow of England, and but for the curious plume which adorns its head, and the tuft which hangs from its breast, might be mistaken at a distance for that bird. The general colour of this species is rich shining black, glazed with varying tints of blue and purple like the feathers of the magpie's tail.

Very little is known of the habits of the bird; but a very good description of its appearance when living has been given by Mr. Wallace in the following words: "Its crest is, perhaps, the most fully developed and beautiful of any bird known. It is composed of long slender feathers, rising from a contractile skin on the top of the head. The shafts are white, and the plume glossy blue, hair-like, and curved outward at the tip. When the crest is laid back, the shafts form a compact white mass, sloping up from the top of the head, and surmounted by the dense hairy plumes. Even in this position it is not an inelegant crest, but it is when it is fully spread that its peculiar character is developed. The shafts then radiate on all sides from the top of the head, reaching in front beyond and below the tip of the beak, which is completely hidden from view. The top then forms a perfect, slightly elongated dome, of a beautiful shining blue colour, having a point of divergence rather behind the centre, like that in the human head. The length of this dome from front to back is about five inches, the breadth four to four and a half inches.'

Scarcely less curious than the "umbrella," as this overhanging plume is very appropriately named, is a bunch of elongated feathers that hang from the breast in a tuft, perfectly distinct from the rest of the plumage. The peculiarity in this tuft is, that the feathers of which it is composed do not grow from the neck, but from a cylindrical fleshy growth, about as thick as an ordinary goose-quill and an inch and a half long. The whole of this curious appendage is covered with feathers, so that the breast tuft is wholly distinct from the feathers of the neck and breast. The entire skin of the neck is extremely loose, more so than in any other bird, according to Mr. Wallace. The feathers of this tuft are edged with a beautiful and resplendent blue, and lap over each other like so many scales.

The food of the Umbrella Bird consists chiefly of berries and various fruits, and it always rejects the hard stones of stone fruit. As its cry is extremely loud and deep, the natives call the bird by a name which signifies a pipe.

OF the next little group of Corvidæ, named the Pyrrhocoracinæ, or Scarlet Crows, in allusion to the red bill and legs of some of the species, England possesses a good example in the common CHOUGH. In all these birds the beak is long, slender, slightly curved downwards, and with a small notch at the extremity.

The Chough is essentially a coast bird, loving rocks and stones, and having a great dislike to grass or hedges of every kind. When in search of food it will venture for some little distance inland, and has been observed in the act of following the ploughman after the manner of the rook, busily engaged in picking up the grubs that are unearthed. Sometimes it will feed upon berries and grain, but evidently prefers animal food, pecking its prey out of the crevices among the rocks with great rapidity and certainty of aim, its long and curved beak aiding it in drawing the concealed insects out of their hiding-places. Cornwall is the chief nesting-place of the Chough, but it is also found in many other portions of England and the British Isles; and the celebrated lines in "King Lear" are too familiarly known to need quoting as a proof that the Chough was in Shakespeare's time an inhabitant of the Dover cliffs. It is also found in many other parts of the world, having been observed in Europe, Asia, and several districts of Africa.

The character of the Chough is not unlike that of the Magpie, and is so admirably delineated by Montagu in an account of a tame specimen in his possession, that it must be related in his own words.

" His curiosity is beyond bounds, never failing to examine anything new to him. If the gardener is pruning, he examines the nail-box, carries off the nails, and scatters the shreds about. Should a ladder be left against the wall, he instantly mounts and goes all round the top of the wall; and if hungry, descends at a convenient place and immediately travels to the kitchen-window, where he makes an incessant knocking with his bill till he is fed or let in: if allowed to enter, his first endeavour is to get upstairs, and if not interrupted, goes as high as he can, and gets into any room in the attic storey; but his intention is to get upon the top of the house. He is excessively fond of being caressed, and would stand quietly by the hour to be smoothed, but resents an affront with violence and effect both by bill and claws, and will hold so fast by the latter that he is with difficulty disengaged. Is extremely attached to one lady, upon the back of whose chair he will sit for hours, and is particularly fond of making one in a party at breakfast, or in a summer's evening at the tea-table in the shrubbery.

His natural food is evidently the smallest insects; even the minute species he picks out of the crevices of the walls, and searches for them in summer with great diligence. The common grasshopper is a great dainty, and the fern chaffer is another favourite morsel: these are swallowed whole; but if the great chaffer be given to him, he places it under one foot, pulls it to pieces and eats it by piecemeal. Worms are wholly rejected, but flesh, raw or dressed, and bread he eats greedily, and sometimes barley, with the pheasants and other granivorous birds occasionally turned into the garden, and never refuses hempseed. He seldom attempts to hide the remainder of a meal.

With a very considerable share of attachment, he is naturally pugnacious, and the hand that the moment before had tendered him food and caresses will repent an attempt to take him up. To children he has an utter aversion, and will scarcely suffer them to enter the garden. Even strangers of any age are challenged with impunity; he approaches all with daring impudence, and so completely does the sight of strangers change his affection for the time, that even his favourites and best benefactors cannot touch him with impunity in these moments of evident displeasure."

As is the case with nearly all coast birds, the Chough builds its nest at no great distance from the sea, generally choosing some convenient crevice in a cliff, or an old ruin near the sea-shore. The nest is always placed at a considerable elevation from the ground, and is made of sticks lined with wool, hair, and other soft substances. The eggs are usually five in number, and in colour they are yellower than those of the crow or rook, but are spotted with similar tints. The general colour of the Chough is black, with a rich blue gloss, contrasting well with the vermilion-red of the beak, legs, and toes. The claws are black, and the eyes are curiously coloured with red and blue in concentric circles. The total length of the adult male Chough is about seventeen inches, and the female is about three inches shorter.

CHOUGH.—*Coracia gracula.*

THE supremely glorious members of the feathered tribe which have by common consent been termed BIRDS OF PARADISE are not very numerous in species, but are so different in form and colour, according to the sex and age, that they have been considered far more numerous than is really the case. The plumage of these birds is wonderfully rich and varied, and not even the humming-birds themselves present such an inexhaustible treasury of form and colour as is found among the comparatively few species of the Birds of Paradise. In all, the feathers glow with resplendent radiance, in nearly all there is some strange and altogether unique arrangement of the plumage, and in many the feathers

2.   E E

are modified into plumes, ribbons, and streamers, that produce the most surprising and lovely effects.

Various strange tales were told of these birds by the ancient writers. The Paradise Bird was thought to have no legs or feet, or use for them, but to pass its time floating in the air, and only taking a little occasional rest by suspending itself from the branches of a tree by the feathers of its tail. The obvious difficulty of hatching the eggs was accounted for by saying that they were laid in the hollow formed by the plumage of the male, and that the mother bird sat upon them while resting on the back of her husband, both birds floating luxuriously in the breeze, and feeding on the soft dews of heaven. This fable found its origin in the fact that the natives of the country where the Paradise Bird resides, always cut off its legs before preserving the skin, so that all the specimens which reached Europe were legless. The plan of preservation adopted was simple in the extreme. The interior organs of the bird were removed, the legs cut off, a hot iron thrust into the body, and the bird dried over a fire without any further care.

The food and habits of these birds seem to be very similar, although the plumage is so distinct. I shall, therefore, give a detailed account of one species, and merely present the reader with good figures and short descriptions of the others. These birds had formerly been broken up into many genera, but are now very rightly shown to be members of the same genus.

THE first species on our list is the MANUCODE, or KING BIRD OF PARADISE, so called because it was thought to exercise a regal sway over the other species, and to hold itself aloof from them like a proud and imperious monarch.

It is a very little bird, the body being scarcely larger than that of a common sparrow, and is remarkable for the very eccentric way in which its plumage is arranged, as will be seen from an examination of the engraving. The natives of the country which it inhabits say that it lives in flocks of thirty or forty in number, under the guidance of one bird which is termed the king, and which is known by the eyes at the extremity of the long tail-feathers. They further relate that the whole troop perches together on the branches of a tree, and that if the king can only be shot, the whole of his subjects remain around his dead body and can be slain without difficulty.

Le Vaillant, in treating of this subject, remarks that the King Bird of Paradise very often gets among a flock of another species, and would therefore hold, and be held, rather aloof from them. Moreover, this species is solitary, and is by no means fond of tall trees, preferring to flit among the low bushes in search of the berries and other food on which it subsists. The natives of New Guinea are in the habit of capturing the King Bird of Paradise by means of a kind of bird-lime, which they make from one of their native plants, and which they lay along the branches which the bird is known to frequent. During the greater part of the year this species remains in New Guinea, but during the western monsoon it migrates to the Aroo islands, returning as soon as the rainy season sets in.

Lesson, who had the opportunity of a few days' visit to New Guinea, and who, like an enthusiastic naturalist, certainly made the very best use of his limited time, makes the following remarks upon this bird : " The Manucode presented itself twice in our shooting excursions, and we killed the male and the female. This species would seem to be monogamous, or perhaps it is only separated into pairs at the period of laying. In the woods this bird has no brilliancy ; its fine coloured plumage is not discovered, and the tints of the female are dull. It loves to take its station on the teak-trees, whose ample foliage shelters it, and whose small fruit forms its nourishment. Its irides are brown, and the feet are of a delicate azure. The Papuans call it Saya."

The King Bird of Paradise is as beautiful as it is rare. The whole upper parts of the body are rich chestnut with a wash of purple, and the under portions of the body are pure white. Across the chest is drawn a band of light golden-green, and from the sides and below the shoulders spring a series of feathers, disposed so as to form a plume, their colour being dusky brown tipped with vivid green. From the upper tail-coverts spring

EMERALD BIRD OF PARADISE.—*Paradisea ápoda.* (Male and Female.)　　KING BIRD OF PARADISE.—*Paradisea régia.*

the very long and very slender shafts, which are bare, excepting at their extremities, when they suddenly expand into a light emerald-green web, which is curled as if it had been just coiled into a spiral form and then flattened. The green web only belongs to one side of the shaft.

INCOMPARABLE BIRD OF PARADISE.—*Paradisea gularis.*

ANOTHER species of these wonderful birds is well represented in the accompanying illustration. This is the INCOMPARABLE BIRD OF PARADISE, also an inhabitant of New Guinea.

This bird is remarkable not only fo the glorious iridescent splendour of its robes, but for the extraordinary development of its tail and the velvety crest on its head, which would render it a truly beautiful bird even were the plumage a sober black or brown. Indeed, on first seeing one of these birds, it is difficult to believe that it is not altogether a "made-up" specimen, composed, like the many mermaids now in existence, of portions taken from different species and ingeniously put together. We are accustomed by our knowledge of the peacock to see a bird with a disproportionately long train, but in this case the true tail-feathers are developed both in length and width to such an extent that they hardly seem to have started from the little body to which they belong.

The true position of this species has been much doubted by naturalists, some having considered it to be analogous to the thrushes, and having accordingly placed it near those birds, while others have ranked it among the Paradise Birds, but have made it into a fresh genus. There seems, however, no real cause for removing it from the other Paradise Birds, and thus increasing the number of genera, which is already allowed to be far too large. As it is by no means a common bird, and the natives of New Guinea are not observant naturalists, caring nothing for the birds but the price which is paid for their skins, very little is known of its habits. The tail of this species is fully three times as long as the body, the head is ornamented with a double crest of glittering feathers, and its whole plumage glows with an effulgence of varied hues that almost baffle description. This gorgeous plumage belongs only to the male; the female being smaller, comparatively sober in hue, and devoid of the beautiful crest which adorns the head of her mate.

IT is hardly possible to conceive a more singular arrangement of plumage than is presented in the GOLDEN BIRD OF PARADISE, although in many species

GOLDEN BIRD OF PARADISE.—*Paradisea sexsetácea.*

there is something so remarkable and unexpected that we believe the extreme of uniqueness to have been reached, until we come across another species which equally raises our wonder and admiration.

In the king Bird of Paradise we have already seen two long bare shafts springing from the upper tail-coverts and extending beyond the tail. Such an arrangement is not, however, without a parallel in other members of the feathered race—as the Leona nightjar and the great Dicrurus both possess a similar development of feathers, the wing of the one and the tail of the other being thus decorated. But in the species which we are now examining six long slender shafts start from the head, three on each side, bare for the greater part of their length, and furnished with a little patch of web at their extremities. These curious shafts are movable, as the bird possesses the power of raising them so as to stand out horizontally on each side of the head, or of permitting them to hang loosely down the sides of the neck. The flanks are decorated with massive plumes of a jetty black, that are also capable of being raised or lowered at the pleasure of the bird, and that fall over the wings and tail so as nearly to conceal them.

The general colour of this curious species is deep velvety-black, changing into gray on the top of the head, and into the richest changeable golden-green on the back of the neck. The throat is most gorgeous in the sunshine, being covered with scale-like feathers of glittering green edged with gold. The feathers of the tail are also velvet-like, and some of the shafts are long and filamentous. The total length of this bird is rather under a foot.

The Emerald Bird of Paradise is the species which is most generally known, and is the one of which were related the absurd tales which have already been mentioned. The specific term, *apoda*, signifies "footless," and was given to the bird by Linnæus in allusion to those fables which were then current, but which he did not believe.

This most lovely bird is a native of New Guinea, where it is far from uncommon, and is annually killed in great numbers for the sake of its plumage, which always commands a high price in the market. It is a very retiring bird, concealing itself during the day in the thick foliage of the teak-tree, and only coming from the green shelter at the rising and setting of the sun, for the purpose of obtaining food. Almost the only successful method of shooting the Emerald Paradise Bird is to visit a teak or fig tree before dawn, take up a position under the branches, and there wait patiently until one of the birds comes to settle upon the branches, or leaves the spot which has sheltered it during the night. This bird is rather tenacious of life, and unless killed instantly is sure to make its escape amid the dense brushwood that grows luxuriantly beneath the trees, and if the sportsman ventured to chase a wounded bird amid the bushes, he would, in all probability, lose his way and perish of hunger. Those sportsmen, therefore, who desire to shoot this bird always provide themselves with guns that will carry their charge to a great distance, and employ very large shot for the purpose, as the bird always perches on the summits of the loftiest trees of the neighbourhood, and would not be much damaged by the shot ordinarily used in shooting.

This species is very suspicious, so that the sportsman must maintain a profound silence, or not a bird will show itself or utter its loud full cry, by which the hunter's attention is directed to his victim.

The following valuable account of an Emerald Paradise Bird may be found in "Bennett's Wanderings," and as it is highly descriptive of the habits which characterise the Paradise Birds, it must be given in the words of the narrator :—

"This elegant creature has a light, playful, and graceful manner, with an arch and impudent look ; dances about when a visitor approaches the cage, and seems delighted at being made an object of admiration : its notes are very peculiar, resembling the cawing of the raven, but its tones are by far more varied. During four months of the year, from May to August, it moults. It washes itself regularly twice daily, and after having performed its ablutions, throws its delicate feathers up nearly over the head, the quills of which feathers have a peculiar structure, so as to enable the bird to effect this object. Its food during confinement is boiled rice, mixed up with soft eggs together with plantains, and living insects of the grasshopper tribe ; these insects, when thrown to him, the bird contrives to catch in its beak with great celerity ; it will eat insects in a living state, but will not touch them when dead.

I observed the bird, previously to eating a grasshopper given him in an entire or unmu-tilated state, place the insect upon the perch, keep it firmly fixed with the claws, and, divesting it of the legs, wings, &c., devour it, with the head always placed first ; the servant who attends upon him to clean the cage, give him food, &c. strips off the legs, wings, &c. of the insects when alive, giving them to the bird as fast as he can devour them. It rarely alights upon the ground ; and so proud is the creature of its elegant dress, that it never permits a soil to remain upon it, and it may frequently be seen spreading out its wings and feathers, and regarding its splendid self in every direction, to observe whether the whole of its plumage is in an unsullied condition. It does not suffer from the cold weather during the winter season at Macao, though exposing the elegant bird to the bleak northerly wind is always very particularly avoided. Mr. Beale is very desirous of procuring a living female specimen of this bird; to endeavour, if possible, to breed them in his aviary.

The sounds uttered by this bird are very peculiar; that which appears to be a note of congratulation resembles somewhat the cawing of a raven, but changes to a various scale in musical gradation, as *he! hi! ho! haw!* repeated rapidly and frequently, as lively and playfully he hops round and along his perch, dismounting to the second perch to be admired, and congratulate the stranger who has made a visit to inspect him; he frequently raises his voice, screeching forth notes of such power as to be heard at a long distance and as it would scarcely be supposed so delicate a bird could utter; these notes are *whock! whock! whock! whock!* uttered in a barking tone, the last being uttered in a low note as a conclusion.

A drawing of the bird of the natural size was made by a Chinese artist; this was taken one morning to the original, who paid a compliment to the artist by considering it one of his own species. The bird advanced stedfastly towards the picture, uttering at the same time its cawing, congratulatory notes; it did not appear excited by rage, but pecked gently at the representation, jumping about the perch, knocking its mandibles together with a clattering noise, and cleaning them against the perch as if welcoming the arrival of a companion.

After the trial with the picture, a looking-glass was brought to see what effect it would produce upon the bird, and the result was nearly the same; he regarded the reflection of himself most stedfastly in the mirror, never quitting it during the time it remained before him. When the glass was removed to the lower from the upper perch, he instantly followed, but would not descend upon the floor of the cage when it was placed so low.

It seemed impatient, hopping about without withdrawing its gaze from the mirror, uttering the usual cawing notes, but with evident surprise that the reflected figure, or, as he seemed to regard it, his opponent, imitated so closely all his actions and was as watchful as himself. There was, however, on his part no indication of combativeness by any elevation of the feathers, nor was any irritation displayed at not being able to approach nearer to the supposed new comer from his own native land. His attention was directed to the mirror during the time it remained before him; but when removed, he went quietly and composed himself upon the upper perch, as if nothing had excited him.

One of the best opportunities of seeing this splendid bird in all its beauty of actions, as well as display of plumage, is early in the morning, when he makes his toilet; the beautiful sub-alar plumage is then thrown out and cleaned from any spot that may sully its purity, by being passed gently through the bill; the short chocolate-coloured wings are extended to the utmost, and he keeps them in a steady flapping motion as if in imitation of their use in flight, at the same time raising up the delicate long feathers over the back, which are spread in a chaste and elegant manner, floating like films in the ambient air. In this position the bird would remain for a short time seemingly proud of its heavenly beauty, and in raptures of delight with its most enchanting self; it will then assume various attitudes, so as to regard its plumage in every direction.

I never yet beheld a soil on its feathers. After expanding the wings it would bring them together so as to conceal the head, then bending it gracefully it would inspect the state of its plumage underneath. This action it repeats in quick succession, uttering at the same time its croaking notes; it then pecks and cleans its plumage in every part within reach, and throwing out the elegant and delicate tuft of feathers underneath the wings, seemingly with much care and with not a little pride, they are cleaned in succession, if required, by throwing them abroad, elevating, and passing them in succession through the bill. Then turning its back to the spectator, the actions above mentioned are repeated, but not in so careful a manner; elevating its tail and long shaft-feathers, it raises the delicate plumage of a similar character to the sub-alar, forming a beautiful dorsal crest, and throwing its feathers up with much grace, appears as proud as a lady dressed in her full ball-dress. Having completed his toilet he utters the usual cawing notes, at the same time looking archly at the spectators, as if ready to receive all the admiration that it considers its elegant form and display of plumage demands; it then takes exercise by hopping in a rapid but graceful manner from one end of the upper perch to the other, and descends suddenly upon the second perch close to the bars of the cage, looking out for the grasshoppers which it is accustomed to receive about this time.

Should any person place his finger into the cage, the bird darts at it rapidly : if it is inserted and withdrawn quickly, the slightly curved extremity of the upper mandible causes the intruder to receive a sharp peck ; but if the finger is placed quietly in the cage close to the beak of the animal, he grasps and thrusts it out as if hinting that he dislikes the intrusion.

His prehensile power in the feet is very strong, and still retaining his hold, the bird will turn himself round upon the perch. He delights to be sheltered from the glare of the sun, as that luminary is a great source of annoyance to him if permitted to dart its fervent rays directly upon the cage. The iris frequently expanding and contracting, adds to the arch, wicked look of this animated bird, as he throws the head on one side to glance at visitors, uttering the cawing notes or barking aloud, to the astonishment of the auditors, who regard the bird as being in a very great rage at something or other beyond their conception. Having concluded, he jumps down to the lower perch in search of donations of living grasshoppers, seemingly in the most happy and good-humoured manner.

The bird is not at all ravenous in its habits of feeding ; but it eats rice leisurely, almost grain by grain ; should any of the insects thrown into his cage fall on the floor he will not descend to them, appearing to be fearful that in so doing he should soil his delicate plumage ; he therefore seldom or never descends, except to perform his ablution in the pan of water placed at the bottom of the cage expressly for his use."

Several other specimens which were seen by M. Lesson in Amboyna were generally fed on boiled rice, but were exceedingly fond of cockroaches. They had been kept in the same cage for more than six months, but had thriven well and were extremely lively, always on the move. The bird described by Mr. Bennett had been in captivity nine years.

The general colour of the upper part of the body is rich chocolate brown, the whole of the front being covered with velvet-like feathers of the deepest green, at one moment sinking into black, and at another flashing forth with glittering emerald. The upper part of the throat is bright golden green, and the upper part of the neck a delicate yellow. The most wonderful part of this bird's plumage is the mass of loose floating plumes that rises from the flanks, and extends in a most graceful manner far beyond the tail. Even in the absolute quiet of a stuffed skin under a glass case, these plumes are full of astonishing beauty, their translucent golden-white vanelets producing a most superb effect as they cross and recross each other, forming every imaginable shade of white, gold, and orange, and then deepening towards their extremities into a soft purplish red. There is a magnificently arranged specimen of this bird in the British Museum, placed in a separate case, and worthy of a journey merely for that one object.

But when the bird is living and healthy, no pen can describe the varied and changeful beauties that develop themselves at every moment, for the creature seems to comprehend within its own single form the united beauties of all other members of the feathered tribe. There is delicacy of shape, grace of movement, and elegance of form, which alone would render it a most beautiful bird. But in addition to all these qualities, it possesses plumage which in one part glitters with all the dazzling gem-like hues of the humming-bird's wing, in another is soft, warm, and delicately tinted, and in another is dyed with a rich intensity of colouring that needs a strong light to bring out its depth of power. Yet more, the torrent of graceful and softly tinted plumes that flow with such luxuriant redundance of changeful curves over the body, are in themselves sufficient to place the Emerald Paradise Bird in the first rank of beauty. What, then, must be the effect of these manifold perfections when exhibited under the pure bright air and cloudless sky of its native land, its velvet feathers flashing with emerald rays at every change of position, and its wondrous plumes waved according to the mental emotions which flit through the bird's mind, and swaying gracefully with the slightest breath of air ?

In addition to these beautiful feathers, there are two long horn-like shafts which rise from the upper part of the tail, and are prolonged to nearly two feet in length, taking a

SUPERB BIRD OF PARADISE.—*Paradisea superba.*

bold and elegant curve in their course. Subtracting these elongated shafts, the total length of the adult male is about thirteen inches. The female is comparatively a sober-plumaged bird, having no gem-like feathers or floating plumes, and being coloured of a deep chocolate, a reddish-brown, and pure white.

ALTHOUGH undistinguished by the wonderful floating plumes which form so conspicuous a decoration of the preceding species, and not being equal to it in dimensions, the SUPERB BIRD OF PARADISE is by no means the least curious of this group.

In this bird, the scapulary feathers are greatly developed, being elongated and widened so as to form a very large double plume or crest, which lies along the back and sides when the bird is at rest, but can be raised at will, and then overtops the head on each side in the manner shown in the engraving. As if to balance this shoulder-crest, another curious tuft of feathers hangs from the breast, spreading into a doubly pointed form, the extremities being directed downwards. The general colour of the plumage is the deepest imaginable violet "shot" with green, appearing of a velvety blackness from its very intensity, and only flashing forth in the brighter hues as the light falls upon the edge of

RED BIRD OF PARADISE.—*Paradisea rubra.*

each feather. The breast tuft, however, forms an exception to this rule, being of the most brilliant steely green, glittering with gem-like radiance in the sunbeams. Although it is not a very large bird, measuring only nine inches in total length, it is really not so very inferior in size to the emerald Paradise Bird, as its tail is short and its plumage closely set.

THE last species of these birds which will be mentioned in these pages is the RED BIRD OF PARADISE.

Although not possessed of such dazzling and refulgent plumage as characterises several of its kin, it is yet a most beautiful bird, and both for the soft delicate purity of the tints with which it is adorned, and the harmony of their arrangement, may challenge competition with any of the feathered race. In size it is about equal to a small pigeon. The forehead and chin are clothed with velvet-like feathers of the intensest green, so arranged as to form a kind of double crest on the forehead, and a sharply defined gorget on the throat. The head, back, and shoulders, together with a band round the neck immediately below the green gorget, are rich orange-yellow, golden in the centre and tinged with

SATIN BOWER BIRD.—*Ptilonorhynchus holosericeus.*

carmine on the margins. The wings, chest, and abdomen are a deep warm chocolate-brown, and the tail is somewhat of the same tint, but not quite dark. Over the tail falls a long double tuft of loose plumy feathers of a beautiful carmine, and two long black filamentous appendages also hang from the tail, and extend to a considerable length.

THE large and important family of the STARLINGS now claims our attention. These birds are seldom of great size, the common Starling of England being about an average example of their dimensions. The bill of the Starling tribe is straight until near its extremity, when it suddenly curves downward, and is generally armed with a slight notch. The first sub-family of these birds is that which is known by the name of Ptilono-rhynchinæ, or Glossy Starlings, so called on account of the silken sheen of their plumage.

The best representative of this little group is the celebrated SATIN BOWER BIRD of Australia.

This beautiful and remarkable bird is found in many parts of New South Wales, and although it is by no means uncommon, is so cautious in the concealment of its home, that even the hawk-eyed natives seem never to have discovered its nest. Perhaps they may be actuated by some superstitious reverence for the bird, and have therefore feigned ignorance of its residence, for it is well known that the voracious native, who will eat almost anything which is not poisonous and will yield to his sharp and powerful teeth, has in many portions of the country so great an awe for this bird that he will never kill it.

The chief peculiarity for which this bird is famous is a kind of bower or arbour, which it constructs from twigs in a manner almost unique among the feathered tribes. The form of this bower may be seen in the illustration, and the mode of construction, together with the use to which the bird puts the building, may be learned from Mr. Gould's account.

"On visiting the Cedar Brushes of the Liverpool range, I discovered several of these bowers or playing-places; they are usually placed under the shelter of the branches of some overhanging tree in the most retired part of the forest; they differ considerably in size, some being larger, while others are much smaller. The base consists of an exterior and rather convex platform of sticks, firmly interwoven, on the centre of which the bower itself is built. This, like the platform on which it is placed and with which it is interwoven, is formed of sticks and twigs, but of a more slender and flexible description, the tips of the twigs being so arranged as to curve inwards and nearly meet at the top; in the interior of the bower, the materials are so placed that the forks of the twigs are always presented outwards, by which arrangement not the slightest obstruction is offered to the passage of the birds.

For what purpose these curious bowers are made is not yet, perhaps, fully understood; they are certainly not used as a nest, but as a place of resort for many individuals of both sexes, who when there assembled run through and round the bower in a sportive and playful manner, and that so frequently that it is seldom entirely deserted.

The interest of this curious bower is much enhanced by the manner in which it is decorated, at and near the entrance, with the most gaily coloured articles that can be collected, such as the blue tail-feathers of the Rose Hill and Lory Parrots, bleached bones, the shells of snails, &c. Some of the feathers are stuck in among the twigs, while others, with the bones and shells, are strewed about near the entrance. The propensity of these birds to fly off with any attractive object is so well known, that the blacks always search the runs for any missing article."

So persevering are these birds in carrying off anything that may strike their fancy, that they have been known to steal a stone tomahawk, some blue cotton rags, and an old tobacco-pipe. Two of these bowers are now in the nest room of the British Museum, and at tne Zoological Gardens the Bower Bird may be seen hard at work at its surface, fastening the twigs or adorning the entrances, and ever and anon running through the edifice with a curious loud full cry that always attracts the attention of a passer-by. The Satin Bower Bird bears confinement well, and although it will not breed in captivity, it is very industrious in building bowers for recreation.

The food of this bird seems to consist chiefly of fruit and berries, as the stomachs of several specimens were found to contain nothing but vegetable remains. Those which are caged in Australia are fed upon rice, fruit, moistened bread, and a very little meat at intervals, a diet on which they thrive well. It is rather a gregarious bird, assembling in flocks led by a few adult males in their full plumage, and a great number of young males and females. They are said to migrate from the Murrumbidgee in the summer, and to return in the autumn.

The plumage of the adult male is a very glossy satin-like purple, so deep as to appear black in a faint light, but the young males and the females are almost entirely of an olive-green.

HARDLY less beautiful in form and plumage, and quite as interesting in habits, the SPOTTED BOWER BIRD now comes before our notice.

This species is an inhabitant of the plains in the interior of New South Wales, and is thought by Mr. Gould to be sufficiently distinct from the preceding species to warrant its introduction into a separate genus. Of this species Mr. Gould makes the following valuable remarks:—

"It is as exclusively an inhabitant of the forests in the interior of the country, as the Satin Bower Bird is of the bushes between the mountain ranges of the coast. From the extreme shyness of its disposition, the bird is seldom seen by ordinary travellers, and it must be under very peculiar circumstances that it can be approached so as to observe its

SPOTTED BOWER BIRD.—*Chlamydéra maculáta.*

colours. It has a harsh, grating, scolding note, generally uttered when its haunts are intruded on, by which its presence is detected.

The situation of its runs or bowers varies much. They are considerably larger and more avenue-like than those of the Satin Bower Bird, being in many instances three feet in length. They are outwardly built of twigs and lined with tall grasses; the decorations are very profuse, consisting of bivalve shells, crania of small mammalia, and other bones. Evident indications of contrivance are manifest throughout the whole of the bower and its decorations, particularly in the manner in which the stones are arranged, apparently to keep the grasses with which it is lined firmly in their places. A row of stones diverges from the mouth of the run on each side, so as to form little paths, while the heap of decorative materials is placed at a heap before the entrance; this arrangement is the same at both ends. Some of the larger bowers, which had evidently been resorted to for many years, contained nearly half a bushel of bones and shells."

The colour of this bird is a rich brown covered with buff spots, and upon the back of the neck there is a band of lengthened feathers of a beautiful rose-pink and glistening with a satin-like sheen. For more detailed information of these curious birds, as well as for some admirable coloured engravings of themselves and their bowers, the reader is referred to Gould's "Birds of Australia."

The account of the Glossy Starlings would be incomplete without a passing mention of the JUIDA; a bird rather larger than our English starling, with an elongated tail, and a plumage that is most singularly covered with every imaginable shade of shining copper, purple, violet, and blue, intermixed in such a manner as to defy description, and seeming

MINO BIRD.—*Grácula música.*

as if the hues had been splashed at random upon the feathers, and then rubbed in and polished. There are several species of Juida, some inhabiting Australia, and others being found in India and Africa. They live in flocks, often attacking the gardens and making great havoc among the fruit. They also are in the habit of perching upon the backs of cattle for the sake of catching the various insects which are always to be found upon a cow's back. Their nest is usually made in the clefts of rocks.

THE Graculinæ, or Grakles, form the next group of birds. Formerly a very large number of species were ranked among the members of this group, but the naturalists of the present day have restricted the appellation to comparatively few birds. In all the species the bill is broad at the base, with the ridge of the upper mandible slightly curved, and there is a little notch near the extremity.

The MINO BIRD is very common in many parts of India and the Indian islands, where it is frequently captured and domesticated.

It is a bright and lively little bird, wonderfully intelligent, and even conceiving so great an affection for its master, that it is permitted to fly about at will. Many amusing tricks are often taught to the Mino Bird, and it possesses a talent for talking equal to that enjoyed by the magpie, the raven, the starling, or the parrot. So admirable a conversationalist is the Mino Bird that some writers who have had personal experience of its capabilities, think that it surpasses even the grey parrot in its powers of imitating the human voice. It will repeat many words with extraordinary accuracy, and some specimens have learned phrases and sentences of considerable length.

The colour of the Mino Bird is a deep velvet-like black, with the exception of a white mark on the base of the quill-feathers of the wing. Around the base of the beak and the forehead the feathers are extremely short and have a velvety sensation to the touch. The bill and the feet are yellow, and on the back of the head are two wattles of a bright yellow colour. The food of this bird consists chiefly of berries, fruits, and insects, and in dimensions it is about equal to a common thrush. By the Javanese it is known by the titles of Beö and Mencho, and the Sumatrans call it Teeong.

In one solitary instance the Mino Bird has been known to visit England, having been seen on the Norfolk coast. In all probability the bird had escaped either from an aviary, or from some homeward-bound ship.

The CROWNED GRAKLE is one of the handsomest of the genus to which it belongs, and on account of the peculiar colouring from which it derives its name is a very conspicuous bird.

It inhabits the parts of the jungle where the vegetation is thickest, and interspersed with tall trees, on whose topmost branches the Crowned Grakle loves to settle while engaged in its search after berries, fruits, and the various substances on which it feeds. It is not a very timid bird, and will frequently haunt human habitations, entering the gardens wherever tall trees have been left standing, and whistling cheerily as it flies from one tree or bough to another. When frightened it signifies its alarm by a harsh, rough screech, but its ordinary notes are full and melodious.

The top of the head and part of the nape, together with the chin and a mark on the centre of several of the primary feathers of the wings, are bright " king's " yellow. Round the eye is a large, comma-shaped patch of bare pink skin, the point of the comma being directed towards the ear. The general colour of the body, as well as the short and square-tipped tail, which looks as if it had been snipped off abruptly by a pair of shears, is a very deep green, "shot" with blue in certain lights, and sooty black in others.

CROWNED GRAKLE.—*Grácula coronáta.*

ANOTHER curious group of this large family is known by the name of Buphaginæ, *i.e.* Beef-eaters or Ox-peckers; a title which they have earned by their habits. They may be easily known by their remarkably shaped bill, which is wonderfully adapted for the peculiar duties which it has to perform. One of the most common species of this group is the AFRICAN BEEF-EATER, a bird which is found in great numbers both in Southern and Western Africa.

It generally assembles in flocks, and haunts the spots where cattle are kept, alighting upon their backs and setting vigorously to work in digging from beneath their skins the larvæ of the bot-flies which burrow beneath the hide, and may often be seen on the backs of our English cattle, by means of the little hillock of skin which they raise. To extract

AFRICAN BEEF-EATER.—*Búphaga Africána.*

these deeply buried creatures would seem to be a matter of considerable difficulty, but the Beef-eater manages the matter easily enough, by fixing itself tightly on the animal's back by means of its extremely powerful claws, and working with its strong and oddly shaped beak. Other animals beside oxen are subject to the attacks of these insect foes, and are equally visited by the Beef-eater, who pursues his beneficial avocation without the least opposition on the part of the suffering animal.

The general colour of the African Beef-eater is a dull brown upon the whole of the upper portions of the body, the chin, and the throat. The breast and abdomen are buff-coloured, and the upper and under tail-coverts are nearly of the same hue. The tail is wedge-shaped, and of a greyish brown colour, warming into reddish brown on the inner webs of the exterior feathers. The basal half of the bill is rich orange, and the curiously squared extremity is scarlet. The total length of the bird is between nine and ten inches.

WE now come to the true Starlings, or Sturninæ, as they are scientifically termed. In these birds the bill is almost straight, tapering and elongated, slightly flattened at the top, and with a hardly perceptible notch. Two examples of this group are found in England, the first and rarest of which is the ROSE-COLOURED PASTOR.

Although of rare occurrence in England, these birds are very common in many other countries, and in some parts of India are so numerous that forty or fifty have been killed at a single shot, and they are said by the agriculturists to be hardly less destructive than locusts. Like the common Starling, the Rose-coloured Pastor always flies in flocks, and seems to possess many of the habits which belong to the beef-eaters, perching on the backs of cattle and feeding on the parasitic insects and grubs which are generally found in such situations. On account of this habit of frequenting the cattle-field and the sheep-fold, the bird has received the title of Pastor, or shepherd. It feeds chiefly on insects, but in the autumn months varies its diet with ripe fruits.

The Rose-coloured Pastor possesses a rather flexible voice; its ordinary cry is rather harsh and grating, but the bird is able to modulate its voice so as to imitate the tones of various other members of the feathered tribe. One of these birds, that was domesticated by a person who had slightly wounded it and afterwards tended it until it had recovered, was so good a mimic that an excellent judge of songsters, who had heard its voice without seeing the bird from which it proceeded, thought that he was listening to a concert of two starlings, two goldfinches, and some songster, probably a siskin. This bird was fed upon insects and barley-meal moistened with milk.

It is a remarkably pretty and conspicuous bird; the beautiful crest which decorates the crown and the delicate tints of the plumage rendering it easily distinguishable from any of its kin. The head is orna-mented with a crest of long flow-ing feathers, which are of a jetty black glossed with violet; and the neck, wings, and tail are of the same hue. The chin, throat, front of the neck, thighs, and under tail-coverts are also black, but without the blue gloss. The back, scapularies, breast, sides, and abdomen are of a beautiful rose-pink; the legs and toes are yellowish brown, and the beak yellow with a dash of rose. The total length of this species is between eight and nine inches. The bird does not attain this beautiful plumage until the third year; in the first year there is no crest at all, and the plumage is simply coloured with different shades of brown and white; in the second year the crest is com-paratively small and scanty, the dark parts of the plumage have a brown tinge, and the rosy parts are dull and washed with grey or brown.

THE common STARLING is one of the handsomest of our British birds, the bright mottlings of its plumage, the vivacity of its move-ments, and the elegance of its form, rendering it a truly beauti-ful bird.

It is very common in all parts of the British Isles, as well as in many other countries, and assembles in vast flocks of many thousands in number, enormous accessions being made to their ranks after the breeding season. These vast assemblies are seen to

ROSE-COLOURED PASTOR.—*Pastor róseus.*

best advantage in the fenny districts, where they couch for the night amid the osiers and aquatic plants, and often crush whole acres to the ground by their united weight. In their flight the Starlings are most wonderful birds, each flock, no matter how large its dimensions, seeming to be under the command of one single bird, and to obey his voice with an instantaneous action which appears little short of a miracle. A whole cloud of

Starlings may often be seen flying along at a considerable elevation from the ground, darkening the sky as they pass overhead, when of a sudden the flock becomes momentarily indistinguishable, every bird having simultaneously turned itself on its side so as to present only the edge of its wings to the eye. The whole body will then separate into several divisions, each division wheeling with the most wonderful accuracy, and after again uniting their forces they will execute some singular manœuvre, and then resume their onward progress to the feeding-ground or resting-place. On one occasion, a pheasant, rising from the ground, flew across the path of a flock of Starlings, and was killed on the spot by the shock. This circumstance occurred near Wootton Bassett.

The migration of the Starling has been well worked out by Mr. Thompson. "In that portion of the north of Ireland with which I am myself best acquainted, there is nothing irregular in the migration of Starlings; they do not await any severity of weather, and although they may occasionally change their quarters when within the island, yet of all our birds they present the clearest evidence of migration, as they are annually observed for several weeks to pass into Ireland from the north and wing their way southward.

To myself they have frequently so appeared; but I prefer giving the more full and satisfactory testimony of trustworthy and intelligent 'shore-shooters,' three of whom, being consulted, agree upon the subject. They state that the general autumnal migration of Starlings commences towards the middle or end of September, according to the season, and continues daily for about six or eight weeks. So early as the middle of July, a flock was once observed flying southerly in the autumnal course. When the weather is moderate, flocks consisting of from half a dozen to two hundred individuals are seen every morning, coming from the north-east, passing over a point of land where a river enters Belfast Bay about a mile from the town, and continuing in the same course until lost to view. They are generally seen only for one and a half or two hours—from eight to ten o'clock, A.M.—never appearing before the former hour, and rarely many after the latter, except when the wind is high, and then the flight is protracted until noon; if very stormy, they do not come at all.

When they commence migration unusually late in the season, as was the case in 1838, they make up for lost time by an increase of numbers. Thus, they were first seen in that year on the 23d of October, when they made their appearance at half-past eight o'clock A.M., and continued passing in flocks of from twenty to one and two hundred individuals until two o'clock. At the season of their earliest appearance there is daylight between four and five in the morning; and the fact of their not being seen before eight o'clock, leads to the belief that they have left some distant place at an early hour. On the same morning the flocks all take the same line of flight, but the direction varies when the wind is sufficiently strong to affect their movements.

The number of birds that come in this course is not very great. The average of five or six flocks seen in a morning perhaps consisted of two hundred and fifty individuals, the greatest number ever seen in one day probably amounted to fifteen hundred, and those altogether seen throughout the migratory period may be estimated at about fifteen thousand. Of my three informants, two lived in the district over which the Starlings flew, and consequently had daily opportunities of seeing them in their season. One has indeed done so for the last half-century, and the other was in the habit of going to the place every morning, in the hope that the flocks would pass over within shot, which they often did. In only one instance did any of these persons see Starlings return this way in spring, when on the 13th of March a flock appeared passing north-eastward in the direction whence they came in autumn;—on the 23d of that month a flock, consisting of sixty, was once observed by myself, returning by this course."

The nest of the Starling is a very loose kind of affair, composed of straw, roots, and grasses, thrust carelessly together, and hardly deserving the name of a nest. In many cases the bird is so heedless that it allows bits of straw or grass to hang from the hole in which the nest is placed, just as if it had intentionally furnished the bird-nesting boy with a clue to the position of the nest. Although this bird makes its home in some retired spot, such as the cleft of a rock, a niche in some old ruin, a ledge in a church-tower, or a hole in a decaying tree, there are few nests more easy to discover; for not only does the bird leave indications of its home in the manner already described, but is so very loquacious that it cannot resist the temptation of squalling loudly at intervals, especially when returning to its domicile laden with food for its young, and so betrays the position of its home. The eggs are generally five in number, and of the faintest imaginable blue. I often used to get these eggs out of the deep holes in which they had been placed by means of a little bird-lime at the end of a fishing-rod.

Oftentimes the Starling makes its home in the vicinity of man, and it is a very common thing to find Starlings and pigeons occupying the same cote. In such cases the neighbours seem to be perfectly friendly, and there seem to be no grounds for the assertion that Starlings kill and eat the young pigeons. The Starling is, however, very eccentric in its choice of a locality for its domicile, having been known to breed in the most unexpected places, such as chimneys, empty flower-pots, and deserted rabbit-burrows. The affection of the Starling for its young is very great, as was shown some years ago. A barn was on fire, and a poor Starling had her nest full of young in the thatch. As the flames neared her residence the bird darted about in great consternation, and at last flew to the nest, drew out one of the young birds, and removed it to a place of safety. She then returned for another load, and before her nest had been reached by the fire, she had placed the whole five of her young in security.

I once remember seeing some Starlings in a great state of perplexity. A few very cold days had suddenly come at the beginning of autumn, and a heavy fall of snow descended while some Starlings, who had built in the roof of an adjoining barn, were in the fields collecting food for their young. On their return they were quite unable to discern the

COMMON STARLING.—*Sturnus vulgari.*

entrances to their nests, and flapped about in the soft snow in the most pitiable manner, screaming with alarm and half frantic at hearing the cries of their young, without being able to reach them.

The food of the Starling is very varied, but consists chiefly of insects. These birds have a habit of following cows, sheep, and horses, fluttering about them as they move for the purpose of preying upon the insects which are put to flight by their feet. The Starlings also perch upon the backs of the cattle, and rid them of the parasitic insects

GREAT BOAT-TAIL.—*Quiscalus major.*

that infest them. From the sheep the Starling often takes toll, pulling out a beakful of wool now and then, and carrying it away to its nest. It is a voracious bird, the stomach of one of these birds having been found to contain more than twenty shells, some of no small size and all nearly perfect; a great number of insects, and some grain. Another Starling had eaten fifteen molluscs of different kinds, a number of perfect beetles, and many grubs.

The Starling is easily tamed, and is a most amusing as well as a graceful pet. It is an admirable talker, and can be taught to repeat words and phrases nearly as well as a parrot. Some ignorant and cruel people have a horrid custom of slitting the tongue of the Starling, in order to enable it to speak; a proceeding which is not only entirely ineffectual, but often causes the death of the poor bird.

The colour of this bird is very beautiful, and is briefly as follows : The general tint is an extremely dark purplish green, having an almost metallic glitter in a strong light. The feathers of the shoulders are tipped with buff, and the wing-coverts, together with the quill-feathers of the tail and wings, are edged with pale reddish brown. The beak is a fine yellow. The feathers of the upper part of the breast are elongated and pointed. This is the plumage of the adult male, and is not brought to its perfection until three years have elapsed. The first year's bird, before its autumnal moult, is almost wholly of a brownish grey, and after its moult is partly brown and partly purple and green. In the second year the plumage is more decided in its tints, but is variegated with a great number of light-coloured spots on the under and upper surfaces, and the beak does not attain its beautiful yellow tinge.

THE Quiscalinæ, or Boat-tailed Birds, are so named from the peculiar formation of their tails, which, as may be seen on reference to the illustration, are hollowed in a manner somewhat similar to the interior of a canoe. There are several species of Boat-tails, all being natives of America, and being spread over the greater part of that vast country. One of the best known species is the GREAT BOAT-TAIL, or GREAT CROW BLACKBIRD, as it is sometimes called.

This bird is rather a large one, being between sixteen and seventeen inches in length, and twenty-two inches across the outspread wings. Its general colour is black, glossed with blue, green, and purple, in different lights. It is mostly found in the southern portions of the United States, where it passes under the name of jackdaw, and is seen in vast flocks among the sea islands and marine marshes, busily engaged in finding out the various substances that are left by the retiring tide. It preserves its social disposition even in its nesting, and builds in company among reeds and bushes in the neighbourhood of forests and marshy lands. The eggs are of a whitish colour and generally five in number. It is a migratory bird, leaving America for winter quarters about the latter end of November, and returning in February and March.

A CLOSELY allied species, popularly called the PURPLE GRAKLE (*Quiscalus versicolor*), inhabits the northern states of America, where it lives in almost incredible numbers ; and although it is very beneficial to the agriculturist during the earlier parts of the year by devouring the grubs and caterpillars, it is so terribly destructive when the crops are ripening, that it seems to counterbalance by its voracity the good which it had previously done. Wilson, in speaking of this bird, makes the following remarks :—

"The trees where these birds build are often at a great distance from the farmhouse, and overlook the plantations. From thence they issue in all directions, and with as much confidence, to make their daily depredations among the surrounding fields, as if the whole were intended for their use alone. Their chief attention, however, is directed to the Indian corn in all its progressive stages. As soon as the infant blade of its grain begins to make its appearance above ground, the Grakles hail the welcome signal with screams of peculiar satisfaction, and, without waiting for a formal invitation from the proprietor, descend on the fields and begin to pull up and regale themselves on the seed, scattering the green blades around. While thus eagerly employed, the vengeance of the gun sometimes overtakes them, but these disasters are soon forgotten.

About the middle of August, when the young ears are in their milky state, they are attacked with redoubled vigour by the Grakles and redwings in formidable and combined bodies. They descend like a blackening, sweeping tempest in the corn, drag off the external covering of twelve or fifteen coats of leaves as dexterously as if done by the hand of man, and having laid bare the corn, leave little behind to the farmer but the cobs and shrivelled skins that contained their favourite fare. I have seen fields of corn of many acres, where more than one-half was thus ruined.

A few miles from the bank of the Roanoke, on the 20th of January, I met with one of these prodigious armies of Grakles. They rose from the surrounding fields with a noise like thunder, and descending on the length of road before me, covered it and the fences completely with black, and when they again rose, and after a few evolutions descended on the skirts of the high-timbered woods, at that time destitute of leaves, they produced a most singular and striking effect ; the whole trees for a considerable extent, from the top to the lowest branches, seeming as if hung in mourning ; their notes and screaming the meanwhile resembling the distant sound of a great cataract, but in more musical cadence, swelling and dying away in the air, according to the fluctuation of the breeze.

A singular attachment frequently takes place between this bird and the fish-hawk. The nest of this latter is of very large dimensions, often from three to four feet in breadth, and from four to five feet high ; composed externally of large sticks or faggots, among the interstices of which sometimes three or four pairs of Crow Blackbirds will construct their nests, while the hawk is sitting or hatching above. Here each pursues the duties of incubation and rearing the young, living in the greatest harmony, and protecting each other's property from depredators."

The colour of this species resembles that of the great Boat-tail, the violet gloss being remarkably conspicuous on the head and breast, the green on the back of the neck, and a coppery red upon the back and abdomen. All the plumage is extremely glossy and shining, and varies according to the angle at which the light falls upon each fibre. In dimensions the Purple Grakle is not so large as the Great Boat-tail, being only twelve inches in length, and eighteen across the outspread wings.

THE ICTERINÆ, or Hang-nest birds, now claim our attention. These birds are remarkable for the hammock-like nest which they construct, and the wonderful skill with which they adapt its structure to the exigencies of the climate or locality.

ONE of the most familiar examples of these birds is the ORCHARD ORIOLE, popularly known by the title of Bob o' Link throughout the countries which it inhabits.

ORCHARD ORIOLE.—*Xanthornis várius.*

This bird, in common with other allied species, is so extremely varied in its plumage according to its age and sex, that several species were confounded together in the most perplexing manner, until Wilson succeeded, by dint of patient observation, in unravelling the tangled web which had been woven by other writers.

The nest of the Orchard Oriole is a truly wonderful structure, woven into a bag or purse-like shape from long grasses, almost as if it had been fashioned in a loom, and so firmly constructed that it will withstand no small amount of rough treatment before its texture gives way. In one of these purse-like nests now lying before me, I find that the bird often employs two and sometimes three threads simultaneously, and that several of these double threads pass over the branch to which the nest is hung, and are then carried to the very bottom of the purse, so as to support the structure in the firmest possible manner. The entrance is from above, and near the mouth; the nest is comparatively slight in texture, becoming thicker and more compact near the foot, where the eggs and young are laid. The interior of the nest is generally lined with some soft downy seeds. So admirably does the bird's beak weave this remarkable nest, that an old lady to whom Wilson exhibited one of these structures, remarked that the Orchard Oriole might learn to darn stockings.

The size and form of the nest may vary very greatly according to the climate in which the bird lives, and the kind of tree on which its home is placed. Should the nest be suspended to the firm stiff boughs of the apple or other strong-branched tree, it is comparatively shallow, being hardly three inches in length, and rather wider than it is deep. But if it should be hung to the long and slender twigs of the weeping willow, as is often the case, the nest is lengthened until it is four or five inches in depth, the size of the entrance remaining the same as in the shallower nest. This variation in structure is evidently intended to prevent the eggs or young from being shaken out of their home by the swaying of the boughs in the wind. The same amount of material appears to be used in either case, so that the elongated nest is not so thick as the short one. My own specimen is an example of the elongated structure. Moreover, in the warmer parts of America, the nest is always much slighter than in the colder regions, permitting a free circulation of air through its walls.

The habits of this bird are very curious and interesting, and are well described by Wilson in his well-known work on the Birds of America :—

"The Orchard Oriole, though partly a dependent on the industry of the farmer, is no sneaking pilferer, but an open and truly beneficent friend. To all those countless multitudes of destructive bugs and caterpillars that infest the fruit-trees in spring and summer, preying on the leaves, blossoms, and embryo of the fruit, he is a deadly enemy ; devastating them wherever he can find them, and destroying on an average some hundreds of them every day without offering the slightest injury to the fruit, however much it may stand in his way. I have witnessed instances where the entrance to his nest was more than half closed by a cluster of apples, which he could easily have demolished in half a minute ; but, as if holding the property of his patron sacred, or considering it a natural bulwark to his own, he slid out and in with the greatest gentleness and caution.

I am not sufficiently conversant with entomology to particularize the different species on which he feeds, but I have good reason for believing that they are almost altogether such as commit the greatest depredations on the fruits of the orchard ; and, as he visits us at a time when his services are of the greatest value, and, like a faithful guardian, takes up his station where the enemy is most to be expected, he ought to be held in respectful esteem, and protected by every considerate husbandman. Nor is the gaiety of his song one of his least recommendations. Being an exceedingly active, sprightly, and restless bird, he is on the ground—in the trees—flying and carolling in his hurried manner, in almost one and the same instant. His notes are shrill and lively, but uttered with such rapidity and seeming confusion, that the ear is unable to follow them distinctly. Between these he has a single note, which is agreeable and interesting.

Wherever he is protected, he shows his confidence and gratitude by his numbers and familiarity. In the Botanic Gardens of my worthy and scientific friends, the Messrs. Bartrams, of Kingsess, which present an epitome of everything that is rare, useful and beautiful in the vegetable kingdom of this western continent, and where the murderous gun scarce ever intrudes, the Orchard Oriole revels without restraint through thickets of aromatic flowers and blossoms, and, heedless of the busy gardener that labours below, hangs his nest in perfect security on the branches over his head."

Audubon, also, has taken great interest in this bird, and has devoted a considerable portion of his work to the elucidation of its habits.

"No sooner have they reached that portion of the country in which they intend to remain during the time of raising their young, than the birds exhibit all the liveliness and vivacity belonging to their nature. The male is seen rising in the air from ten to twenty yards in a violent manner, jerking his tail and body, flapping his wings, and singing with remarkable impetuosity, as if under the influence of haste, and anxious to return to the tree from which he has departed. He accordingly descends with the same motions of the body and tail, repeating his pleasing song as he alights.

These gambols and warblings are performed frequently during the day, the intervals being employed in ascending or descending along the branches and twigs of different trees in search of insects or larvæ. In doing this they rise on their legs, seldom without jetting the tail, stretch the neck, seize the prey, and emit a single note, which is sweet and mellow, although in power much inferior to that of the Baltimore. At other times it is seen bending its body downwards in a curved posture, with the head gently inclined upwards, to peep at the outer part of the leaves, so as not to suffer any part to escape its vigilance. It soon alights on the ground when it has espied a crawling insect, and again flies towards the blossoms, in which are many lurking, and devours hundreds of them each day, thus contributing to secure to the farmer the hopes which he has of the productiveness of his orchard."

One of these birds that was kept in a cage by Wilson proved to be a very interesting creature, chanting its wild clear notes at an early age, and accommodating itself to its captivity with perfect ease. It had a curious love for artificial light, fluttering about its cage, and becoming uneasy at the sight of a lighted candle, and not being satisfied when its cage was placed close to the object of attraction. In that case, it would sit close to the side of the cage, dress its plumage, and occasionally break into snatches of song.

The adult male is nearly black upon its head, neck, back, wings, and tail, a brownish tint being perceptible in the wings. The lower part of the breast, the abdomen, tail-coverts, and some of the wing-coverts, are light reddish brown, and the greater wing-coverts are tipped with white. The adult female is yellowish olive above, with a brown tinge on the back, and a brown wash over the wings. The whole of the lower parts are yellow, the primary feathers of the wings are slightly edged with yellowish white, and the same colour is found on the edges of the secondaries and greater coverts, and on the tips of the lesser coverts. The length of the bird is between six and seven inches. The young male is like the female during his first year, but in his second year sundry feathers of black make their appearance in various parts of the body, and in the third year they spread over the upper surface and breast, as has already been mentioned.

The BALTIMORE ORIOLE is an inhabitant of the whole of Northern America, its range extending from Canada to Mexico—even as far south as Brazil.

It is a migratory bird, arriving about the beginning of May, and departing towards the end of August or the beginning of September. The name of Baltimore Oriole has been given to it because its colours of black and orange are those of the arms belonging to Lord Baltimore, to whom Maryland formerly belonged. This species is remarkably familiar and fearless of man, hanging its beautiful pensile nest upon the garden-trees, and even venturing into the streets wherever a green tree flourishes, and chanting its wild mellow notes in close proximity to the sounds and sights of a populous city.

The nest of the Baltimore Oriole is somewhat similar to that of the preceding species, although it is generally of a thicker and tougher substance, and more ingeniously woven. The materials of which this beautiful habitation is made are flax, various kinds of vegetable fibres, wool, and hair, matted together, so as to resemble felt in consistency. A number of long horsehairs are passed completely through the fibres, sewing it firmly together with large and irregular, but strong and judiciously placed stitching. In one of these nests Wilson found that several of the hairs used for this purpose measured two feet in length. The nest is in the form of a long purse, and at the bottom is arranged a heap of soft cows' hair and similar substances, in which the eggs find a warm resting-place. The female bird seems to be the chief architect, receiving a constant supply of materials from her mate, and occasionally rejecting the fibres or hairs which he may bring, and sending him off for another load better to her taste.

Since the advent of civilization, the Baltimore Oriole has availed himself largely of his advantages, and instead of troubling himself with a painful search after individual hairs, wherewith to sew his hammock together, keeps a look-out for any bits of stray thread that may be thrown away by human sempstresses, and makes use of them in the place of the hairs. So sharpsighted is the bird, and so quick are his movements, that during the bleaching season the owners of the thread are forced to keep a constant watch upon their property as it lies upon the grass, or hangs upon the boughs, knowing that the Oriole is ever ready to pounce upon such valuable material, and straightway to weave it into his nest. Pieces of loose string, skeins of silk, or even the bands with which young grafts are tied, are equally sought by this ingenious bird, and often purloined to the discomfiture of the needlewoman or the gardener. The average size of the nest is six or seven inches in depth, and three or four in diameter. Wilson thinks that the bird improves in nest-building by practice, and that the best specimens of architecture are the work of the oldest birds.

The eggs are five in number, and their general colour is whitish pink, dotted at the larger end with purplish spots, and covered at the smaller end with a great number of fine intersecting lines of the same hue. The food of the Baltimore Oriole seems to be almost

BALTIMORE ORIOLE.—*Yphantes Baltimore.*

entirely of an animal nature, and to consist of caterpillars, beetles, and other insects, most of them injurious to the farmer or the gardener.

The colouring of this bird is as follows : The head and throat, together with the upper part of the back and the wings, are deep black, with the exception of an orange bar upon the shoulders. The lower part of the back and the whole of the under surface are bright orange, warming into scarlet on the breast. The edges of the secondaries, the exterior edges of the greater wing-coverts, and part of those of the primaries, are white The tail is rather curiously coloured, and thus described by Wilson: " The tail-feathers under the coverts, orange ; the two middle ones from thence to the tips are black, the next five on each side black near the coverts, and orange towards the extremities, so disposed that when the tail is expanded and the coverts removed, the black appears in the form of a pyramid supported on an arch of orange." The female is dull black upon the upper parts and mottled with brownish yellow, each feather being marked with that tint upon the edges. The lower part of the back and all the under portions of the body are dull orange, and the tail is mostly olive yellow. The wings are dull brown, and marked with yellowish white upon the coverts.

From these colours the bird has derived the names of Golden Robin, and Fire Bird. Its total length is about seven inches.

ONE of the most curious and handsome birds of this group is termed the CRESTED ORIOLE on account of the sharp pointed crest which rises from its head.

CRESTED ORIOLE.—*Cácicus cristátus.*

It is a native of tropical America, and seems to be rather a familiar bird, often leaving the forests where it usually dwells, and making its home near the habitations of man. Whether in the vast woods of its native land, or whether in the cultivated grounds, it is always to be found in the loftiest trees, traversing their branches in search of food, and suspending its nest from the extremity of the slenderest twigs. It is a very active bird both on foot and in the air, one quality being needful for its movements among the boughs while getting berries, and the other for the chase of the various insects with which it varies its diet.

The nest of the Crested Oriole is a very elegant structure, much larger than that of either of the preceding species, being sometimes not less than three feet in length. It is always hung from the very extremity of some delicate twig, so as to escape the marauding hand of the monkey or the dreaded fangs of the snake; and as a great number of these are generally found upon one tree, the combined effect, together with the busy scene of the parent birds continually going and returning from their homes, is remarkably fine. The shape of the nest is cylindrical, swelling into a somewhat spherical form at the bottom; and it is found that both birds take an equal share of work in its construction.

The Crested Oriole is very beautifully as well as curiously coloured. The head, shoulders, breast, and abdomen are warm chocolate-brown, and the wings are dark green, changing gradually into brown at their tips. The central feathers of the tail are dark brown, and the remaining feathers are bright yellow. There is also a green tinge upon the thighs and the middle of the breast. Upon the top of the head there is a long and pointed crest, and the horny portion of the bill is green, and, as may be seen in the illustration, extends above the eye. The legs and feet are black. The Crested Oriole is larger than either of the preceding species, being about the size of a common jackdaw.

In the COW BIRD or COW TROOPIAL of America we have a curious instance of the frequency with which a remarkable habit, supposed to be almost unique, and especially characteristic of some particular species, is found to occur in a totally distinct species inhabiting another continent. That the cuckoo of Europe is no nest-maker, but only usurps the homes of other birds, and forces them to take care of its progeny, is a well-known fact, and it is really remarkable that the Cow Bird, which inhabits the opposite quarter of the globe, and belongs to an entirely different order of birds, should follow the same principle.

Before commencing the description of this bird, I must caution the reader against mistaking the present species for the American cuckoo, which is by many persons called the Cow Bird on account of its cry, which resembles the word "cow, cow," frequently repeated. The American cuckoo is free from the intrusive habits of the Cow Troopial, and not only builds its own nest, but rears and tends its young with great affection.

COW TROOPIAL.—*Mólothrus pécoris.*

The Cow Bird is one of the migrators, arriving in Pennsylvania about the end of March or the beginning of April, and is somewhat gregarious, being found in little parties, generally accompanied by the red-winged starling, which bird will soon be described. Towards the middle or end of October the Cow Birds begin to leave the place of their temporary residence, and again assembling in flocks, together with the red-winged starling, take their departure for their winter quarters in Carolina and Georgia. While remaining in the country, they are generally seen near streams, perched on the trees that skirt rivers and creeks. It is a rather curious fact that during the months of July and August the Cow Troopials suddenly vanish, and are not seen again until September, when they make their appearance in considerable numbers. Whether they take a journey during that time, or whether they retire into the depths of the forest, is not clearly ascertained.

Unlike the generality of birds, the Cow Bird seems to be actuated by no attachment to those of the opposite sex. No pairing has yet been observed, nor does the male bird take possession of a number of females as is the case with many species. Indeed, there would be no need for such an alliance, for the female Cow Bird makes no nest, neither does she trouble herself about rearing her young, but searching out for the nest of some little bird, she deposits her own egg among the number, and then leaves it to its fate. The remarkable feature in the matter is, that the poor bird on whom this intruder has been foisted invariably takes charge of it in preference to its own offspring, and will always rear the young Cow Bird, even though the whole of its own offspring perish.

There seems to be in the Cow Bird an irresistible attractive power, forcing other birds to take charge of it and attend to its wants. This supposition is strengthened by the conduct of a cardinal grosbeak, kept by Wilson, into whose cage was introduced a young Cow Bird just taken out of the nest of a Maryland yellow throat. At first, the grosbeak examined the intruder with some reserve, but as soon as the stranger began to cry for food, the grosbeak took it under its protection, tended it carefully, brought it food, tore large insects to pieces in order to suit the capacity of the young bird's mouth, cleaned its plumage, taught it to feed itself, and exhibited towards it all a mother's care.

Several species of the smaller birds are especially selected by the Cow Troopial for the purpose of acting as foster parents towards its young. Among these the favourites are the Maryland yellow-throat, the red-eyed flycatcher, the white-eyed flycatcher, the blue-grey flycatcher, the blue-bird, the chipping sparrow, and the golden-crowned thrush. Why these birds should be thus distinguished is not easy to determine, for they are of very different sizes, and build very different nests ; some making their home in the fork of a branch, some on the ground, and others building a regular pensile nest suspended to the end of a slender twig. The character of this bird and its curious habits have been admirably worked out by Dr. Potter, who communicates his information to Mr. Wilson in the following words :—

"That the 'fringilla' never builds a nest for itself you may assert without the hazard of a refutation. I once offered a guinea for the nest, and the negroes in the neighbourhood brought me a variety of nests, but they were always traced to some other bird. The time of depositing the eggs is from the middle of April to the last of May, or nearly so, corresponding with the season of laying observed by the small birds on whose property it encroaches. It never deposits but one egg in the same nest, and this is generally after the rightful tenant begins to deposit hers, but never, I believe, after she has commenced the process of incubation. It is impossible to say how many they lay in a season unless they could be hatched when confined in an aviary.

By a minute attention to a number of these birds when they feed in a particular field in the laying season, the deportment of the female when the time of laying draws near becomes particularly interesting. She deserts her associates, assumes a drooping, sickly aspect, and perches upon some eminence where she can reconnoitre the operations of other birds in the process of nidification. If a discovery suitable to her purpose cannot be made from her stand, she becomes more restless, and is seen flitting from tree to tree till a place of deposit can be found.

I once had an opportunity of witnessing a scene of this sort which I cannot forbear to relate. Seeing a female prying into a bunch of bushes in search of a nest, I determined to see the result if practicable, and knowing how easily they are disconcerted by the near approach of man, I mounted my horse and proceeded slowly ; sometimes seeing and sometimes losing sight of her, till I had travelled nearly two miles along the margin of the creek. She entered every thick place, prying with the strictest scrutiny into places where the small birds usually build, and at last darted suddenly into a thick copse of alders and briars, where she remained five or six minutes, when she returned, soaring above the underwood, and returned to the company she had left feeding in the field. Upon entering the covert I found the nest of a yellow-throat with an egg of each. Knowing the precise time of deposit, I noticed the spot and date, with a view of determining a question of importance—the time required to hatch the egg of the Cow Bird, which I supposed to commence from the time of the yellow-throat laying the last egg. A few days after the nest was removed, I know not how, and I was disappointed.

In the progress of the Cow Bird along the creek's side, she entered the thick boughs of a small cedar, and returned several times before she could prevail upon herself to quit the place, and upon examination I found a sparrow sitting on its nest, on which she no doubt would have stolen in the absence of the owner. It is, I believe, certain that the Cow Troopial never makes a forcible entry upon the premises by attacking other birds and ejecting them from their rightful tenements, although they are all, perhaps, inferior in strength except the blue-bird, which, although of a mild as well as affectionate disposition, makes a vigorous resistance. When assaulted, like most other tyrants and thieves, they are cowardly, and accomplish by stealth what they cannot obtain by force.

The deportment of the yellow-throat on this occasion is not to be omitted. She returned while I waited near the spot, and darted into her nest, but returned immediately and perched upon a bough near the place; remained a minute or two, and entered it again, returned, and disappeared. In ten minutes she returned with her mate. They chattered with great agitation for half an hour, seeming to participate in the affront, and then left the place. I believe all the birds thus intruded upon manifest more or less concern at finding the eggs of a stranger in their own nests. Among others, the sparrow is exceedingly punctilious; for she sometimes chirps her complaints for a day or two, and often deserts the premises altogether, even after she has deposited one or more eggs.

The following anecdote will show not only that the Cow Troopial insinuates herself slyly into the nests of other birds, but that even the most pacific of them will resent the insult. A blue-bird had built for three successive seasons in the cavity of a mulberry-tree near my dwelling. One day when the nest was nearly finished I discovered a female Cow Bird perched upon a stake fence near, her eyes apparently fixed upon the spot, while the builder was busy adjusting her nest. The moment she left it the intruder dashed into it, and in five minutes returned and rushed off to her companions with noisy delight, which she expressed by her gestures and notes. The blue-bird soon returned, and entered the nest, but instantaneously fluttered back with much apparent hesitation and perched upon the highest branch of the tree, uttering a rapidly repeated note of complaint and resentment, which soon brought the male, who vociferated his feelings by every demonstration of the most vindictive resentment. They entered the nest together, and returned a second time, uttering their uninterrupted complaint for ten or fifteen minutes. The male then dashed away to the neighbouring trees, as if in quest of the offender, and fell upon a cat-bird, which he chastised severely and then attacked an innocent sparrow that was chirping its ditty in a beech-tree. Notwithstanding the affront was so passionately resented, I found the Cow Bird had laid an egg the next day. Perhaps, a tenant less attached to a favourite spot would have acted more fastidiously by deserting the premises altogether. In this instance also I determined to watch the occurrences that were to follow; but on one of my morning visits I found the common enemy of the eggs and young of all the small birds had despoiled the nest—a coluber was found coiled at the bottom, and the eggs sucked.

Agreeably to my observation, all the young birds destined to cherish the young Cow Bird are of a mild and affectionate disposition, and it is not less remarkable that they are all smaller than the intruder: the blue-bird is the only one nearly as large. This is a good-natured creature, although it makes a vigorous defence when assaulted. The yellow-throat, the sparrow, the goldfinch, the indigo-bird, and the blue-bird, are the only kinds in whose nests I have found the eggs or young of the Cow Troopial, though doubtless there are some others.

What becomes of the eggs or young of the proprietors? This is the most interesting question that appertains to this subject. There must be some especial law of nature which determines that the young of the proprietors are never to be found tenants in common with the young Cow Bird. I shall offer the result of my own experience on this point, and leave it to you and others better versed in the mysteries of nature than I am to draw your own conclusions. Whatever theory may be adopted, the fact must remain the same. Having discovered a sparrow's nest with five eggs, four and one, and the sparrow sitting, I watched the nest daily. The egg of the Cow Bird occupied the centre, and those of the sparrow were pushed a little up the side of the nest. Five days after the discovery, I perceived the shell of the finch's egg broken, and the next the bird was hatched. The sparrow returned while I was near the nest with her mouth full of food, with which she fed the young Cow Bird with every possible mark of affection and discovered the usual concern at my approach. On the succeeding day only two of the sparrow's eggs remained, and the next day there were none. I sought in vain for them on the ground in every direction.

Having found the eggs of the Cow Bird in the nest of a yellow-throat, I repeated my observations. The process of incubation had commenced, and on the seventh day from the discovery, I found a young Cow Bird that had been hatched during my absence of twenty-four hours, all the eggs of the proprietor remaining. I had not an opportunity of visiting the nest for three days, and on my return there was only one egg remaining, and that rotten. The yellow-throat attended the young interloper with the same apparent care and affection as if it had been its own offspring.

The next year after, my first discovery was in a blue-bird's nest, built in a hollow stump. The nest contained six eggs, and the process of incubation was going on. Then a few days after my visit I found a young Cow Bird and three eggs remaining. I took the eggs out: two contained young birds apparently come to their full time, and the other was rotten. I found one of the other eggs on the ground at the foot of the stump, differing in no respect from those in the nest, no signs of life being discernible in either.

Soon after this I found a goldfinch's nest, with an egg of each only, and I attended it carefully till the usual complement of the owner were laid. Being away at the time, I could not ascertain precisely when the process of incubation commenced, but from my reckoning, I think the egg of the Cow Bird must have been hatched in nine or ten days from the commencement of incubation. On my return, I found the young Cow Bird occupying nearly the whole nest, and the foster-mother as attentive to it as she could have been to her own.

I ought to acknowledge here that in none of these instances could I ascertain exactly the time required to hatch the Cow Bird's eggs, and that, of course, none of them are decisive; but is it not strange that the eggs of the intruder should have been so uniformly the first hatched? The idea of the egg being larger, and therefore from its own gravity filling the centre of the nest, is not sufficient to explain the phenomenon; for in this supposition the other eggs must be proportionably elevated at the sides, and therefore have as much or more warmth from the body of the incumbent than the others. This principle would simply apply to the eggs of the blue-bird, for they are nearly of the same size; if there be any difference, it would be in favour of the builder of the nest. How do the eggs get out of the nest? Is it by the size and nestling of the young Cow Bird? This cannot always be the case, because in the instance of the blue-bird's nest in the fallen stump the cavity was a foot deep, the nest at the bottom, and the ascent perpendicular; nevertheless, the eggs were removed, although filled with young ones. Moreover, a young Cow-hen finch is as helpless as any other young bird, and so far from having the power of ejecting others from the nest, or even the eggs, that they are sometimes found on the ground under the nest, especially when the nest happens to be very small.

I will not assert that the eggs of the builder of the nest are never hatched; but I can assert that I have never been able to find one instance to prove the affirmative. If all the eggs of both birds were to be hatched, in some cases the nest would not hold half of them: for instance, those of the sparrow or yellow-breast. I will not assert that the supposititious egg is brought to perfection in less time than those of the bird to which the nest belongs, but from the facts stated I am inclined to adopt such an opinion. How are the eggs removed after the accouchement of the spurious occupant? By the proprietor of the nest unquestionably, for this is consistent with the rest of her economy. After the process of hatching, she is taken away by her attention to the young stranger; the eggs would be only an incumbrance, and therefore instinct prompts her to remove them. I might add, that I have mostly found the eggs of the sparrow, in which were remarked young ones, lying near the nest containing a Cow Bird, and therefore I cannot resist this conclusion.

Would the future parent feed two species of young at the same time? I believe not. I have never seen an instance of any bird feeding the young of another immediately after rearing their own. I should think the sooty-looking stranger would scarcely interest a mother while the cries of her own offspring, always intelligible, were to be heard. Should such a contradiction ever take place, I judge the stranger would be the sufferer, and probably the sparrows removed soon become extinct. Why the *lex naturæ conservatrix* should decide in favour of the supposititious fledgling is not for me to determine."

The Cow Bird derives its name from its habit of haunting the pasture-lands for the purpose of feeding upon the numerous flies and other insects that are always to be found in the vicinity of cattle; it is also known under the titles of Cow Bunting, and Cow-pen Bird.

The colouring of the Cow Bird is pleasing, though not brilliant. The head and neck are of a dark drab, and have a kind of silken gloss; the whole of the upper surface and abdomen are black "shot" with green, and the upper part of the breast is dark violet. When young it is altogether brown, and the darker tints make their appearance by degrees, showing themselves in patches here and there, which enlarge as the bird grows older, and finally overspread the entire body. The length of the bird is about seven inches.

THE RED-WINGED STARLING is one of those birds which may either be looked upon as most beneficial or most hurtful to the coasts in which they live, according to the light in which they are viewed.

From the farmer's point of view, it is one of his worst enemies, as it eats vast amounts of grain, and assembles in such enormous flocks that the fields are black with their presence, and the sun is obscured by the multitude of their wings. The soft immature grain of the Indian corn is a favourite food with the Red-winged Starlings, and according to Wilson, "reinforced by numerous and daily flocks from all parts of the interior, they pour down on the low countries in prodigious multitudes. Hence

RED-WINGED STARLING.—*Agelaius phœniceus*

they are seen like vast clouds, wheeling and driving over the meadows and deserted corn-fields, darkening the air with their numbers. They commence the work of destruction on the corn, the husks of which, though composed of numerous envelopments of closely wrapped leaves, are soon completely torn off; while from all quarters myriads continue to pour down like a tempest, blackening half an acre at a time, and if not disturbed repeat their depredations till little remains but the cob and the shrivelled skins of the grain. From dawn to nearly sunset this open and daring devastation is carried on, under the eye of the proprietor; and a farmer who has very considerable extents of corn would require half a dozen men at least with guns to guard it, and even then all their vigilance and activity could not prevent a ground-tithe of it from becoming the prey of the black-birds."

In consequence of their depredations the Red-winged Starling is persecuted in every possible way. Every man and boy who has a gun takes it and shoots at the "blackbirds," every urchin who can throw a stone hurls it at their blackening flocks, and even the hawks come from far and near to the spot where these birds are assembled, and make great havoc among them. As they are in the habit of resting at night among the reeds that grow in profusion upon the morasses, the farmers destroy great multitudes of them by stealing quietly upon their roosting-places at night and setting fire to the dry reeds. The poor birds being suddenly awakened by the noise and flames, dart wildly about, and those who escape the fire generally fall victims to the guns of the watchful farmer and his men. Thousands of birds are thus killed in a single night, and as their flesh is eatable, though not remarkable for its excellence, the party return on the following morning for the purpose of picking up the game.

Such are the devastations wrought by the Red-winged Starling, and on the first glance they appear so disastrous as to place the bird in the front rank of winged pestilences. But there is another side of the question, which we will now examine.

During the spring months these birds feed almost exclusively upon insects, especially preferring those which are in their larval state, and devour the young leaves of growing crops. These destructive grubs are hunted by the Red-winged Starling with the greatest perseverance, seeing that upon these the existence of themselves and their young entirely depends. Whether a grub be deeply buried in the earth, eating away the root of some doomed plant, whether it be concealed among the thick foliage which it is consuming, or whether it be tunnelling a passage into the living trunk of the tree, the Red-winged Starling detects its presence, and drags it from its hiding-place. From many dissections which he made, Wilson calculated that on the very smallest average each bird devours at least fifty larvæ per diem, and that it probably eats double that number. But, taking the former average as the true one, and multiplying it by the number of Red-winged Starlings which are known to visit the country, he calculates that these birds destroy sixteen thousand millions of noxious insects in the course of each breeding season, even supposing that they do not eat a single insect after the young are able to shift for themselves.

The nest of this bird is made among the rank foliage of marshy and low-lying soils, and is not unfrequently placed upon the bare ground. The materials of which it is made are fine reeds, roots, and grasses, lined with soft herbs. In order to keep the nest in its place among the loose and yielding substances in which it is placed, the bird fastens the twigs or herbage together by intertwining them with the exterior rushes which edge the nest, and sometimes fastens the tops of the grass-tufts together. The eggs are five in number, pale blue in colour, and marked with pale purplish blotches and many lines and shades of black. The male bird is extremely anxious about his home, and whenever he fears danger from an intruder, he enacts a part like that which is so often played by the lapwing of England, and by feigning lameness and uttering pitiful cries as he flutters along, endeavours to entice the enemy from the vicinity of its nest. The young birds are able to fly about the middle of August, and then unite in large flocks.

When captured young it soon accommodates itself to its new course of life, becomes very familiar with its owner, and is fond of uttering its curious song, puffing out its feathers and seeming in great spirits with its own performance.

The colour of the adult male is deep glossy black over the greater part of the body, reddish brown upon the first row of the wing-coverts, and a rich bright scarlet decorating the remaining coverts. In length it measures about nine inches. The female is much smaller than her mate, being only seven inches long, and is coloured in a very different manner. The greater part of the plumage is black, each feather being edged with light brown, white, or bay, so that she presents a curiously mottled aspect. The chin is cream, also with a dash of red; two stripes of the same colour, but dotted with black, extend from the nostrils over the eyes, and from the lower mandible across the head.

There is a stripe of brown-black passing from the eye over the ear-coverts, and the whole of the lower parts are black streaked with creamy white. The young males resemble the females in their colouring, and as they advance in age present feathers of the characteristic black and red in different parts of their plumage. Not until several years have elapsed is the male joyous in his full plumage, and it is seldom that a perfectly black and scarlet bird is found, some of the feathers generally retaining their juvenile brown and bay.

FEW of the American birds are better known than the RICE TROOPIAL, which is familiar over the greater part of that continent.

No American zoologist omits a notice of the Rice Troopial, and there are few writers on country life who do not mention this little bird under one of the many names by which it is known. In some parts of the States it is called the RICE BIRD, in another the RED BIRD, in another the RICE or REED BUNTING, while its more familiar title, by which it is called throughout the greater part of America, is BOB-O-LINK or BOB-LINKUM. It also occasionally visits Jamaica, where it gets very fat, and is in consequence called the BUTTER BIRD. Its title of Rice Troopial is earned by the depredations which it annually makes upon the rice crops, though its food is by no means restricted to that seed, but consists in a very large degree of insects, grubs, and various wild grasses.

Like the preceding species it is a migratory bird, residing during the winter months in the southern parts of America and the West Indian Islands, and passing in vast flocks northwards at the commencement of the spring. Few birds have so extensive a range as the Rice Troopial, for it is equally able to exist in the warm climates of tropical America and the adjacent islands, and in the northerly regions of the shores of the St. Lawrence.

RICE TROOPIAL, OR BOB-O-LINK.—*Dolichonyx oryzivorus.*

According to Wilson their course of migration is as follows : " In the month of April, or very early in May, the Rice Buntings, male and female, arrive within the southern boundaries of the United States, and are seen around the town of Savannah in Georgia about the fourth of May, sometimes in separate parties of males and females, but more generally promiscuously. They remain there but a short time, and about the twelfth of May make their appearance in the lower part of Pennsylvania as they did in Savannah.

2. G G

While here, the males are extremely gay and full of song, frequenting meadows, newly ploughed fields, sides of creeks, rivers, and watery places, feeding on mayflies and cater-pillars, of which they destroy great quantities. In their passage, however, through Virginia at this season, they do great damage to the early wheat and barley while in its milky state. About the 20th of May, they disappear on their way to the north. Nearly at the same time they arrive in the State of New York, spread over the whole New England States, as far as the river St. Lawrence, from Lake Ontario to the sea, in all of which places, north of Pennsylvania, they remain during the summer, building and rearing their young."

As soon as the young are able to fly, the Rice Troopials collect in vast flocks, and settling down upon the reeds and wild oats, feed so largely that they become very fat, and are thought to be equal in flavour to the celebrated ortolan of Europe. Multitudes of these birds are killed for sale, and are exposed in the dealer's shop tied together in long strings.

The song of this bird is very peculiar and varies greatly with the occasion. As the flocks of Rice Birds pass southwards after the breeding season, their cry is nothing but a kind of clinking note frequently repeated ; but the love song addressed to the actual or intended mate is of a very different character, excellent, voluble, and fervent. Filled with happiness, and uttering the joy of his heart in blithe and merry notes, the Rice Bird flings himself into the air, hovers for a while over the object of his love, pours forth a volley of wild and rapid notes whose exulting strains can be interpreted even by a human ear, and, alternately rising and sinking on the wing, thus pays his court with the wildest medley of melodious notes that ever issued from a feathered throat. The rapidity with which the sounds succeed each other is so great that the ear can hardly distinguish them, the high and low notes being uttered apparently at random, making such a singular jumble of musical sounds that they seem to be produced by at least six or seven throats instead of one. Although so unconnected with each other, these short and variable notes harmonise well when several birds are singing simultaneously. The song lasts for a considerable time, one of these birds when kept in a cage having been known to sing continuously from April to the end of June.

Sometimes a party of thirty or forty will begin to sing altogether, as if following the direction of some leader, each taking up the strain in proper succession, and the whole flock ceasing suddenly as if by some preconcerted signal. These curious concerts are performed as often as the little flock perches upon a tree.

The male Rice Troopial is a handsome little bird, and is thus coloured: The upper part of the head, the sides of the neck, the wings, tail, and lower surface of the body, are deep black, the feathers being sometimes edged with yellow if the bird is not an old one. The back is also black, but diversified with streaks of brownish yellow. The scapularies and tail-coverts are white, and the lower part of the back is also white, with a bluish tinge. The female is streaked with brownish yellow upon the back, and the whole of the lower parts are dull yellow. This plumage is also assumed by the male as soon as the breeding season is over, and is not laid aside for his brilliant vestments until the next season of love. Sometimes the change is rather tedious, and then the poor bird seems to be greatly sensible of the disadvantages under which he is labouring, sitting idly in a disconsolate kind of fashion, until his new clothes arrive and his vivacity with them. The total length of the Rice Troopial is between seven and eight inches.

WE now arrive at the large and important families of the FINCHES, in which group is contained very many of the more familiar British birds, which are popularly known by the title of Finch, together with some distinctive prefix, as well as a large number of less known but not less interesting natives of foreign lands. In all these birds the bill is conical, short and stout, sharp at the extremity and without any notch in the upper mandible.

The first group of the Finches is composed of a number of species, which, although for the most part not conspicuous either for size, beauty of form, or brilliancy of colour, are yet among the most remarkable of the feathered tribe. The nests of the Baltimore and orchard oriole are sufficiently curious examples of bird architecture, but those of the Weaver Birds are even more wonderful. Dissimilar in shape, form and material, there is yet a nameless something in the construction of their edifices, which at once points them out as the workmanship of the Weaver Birds. Some of them are huge, heavy, and massive, clustered together in vast multitudes, and bearing down the branches with their weight.

Others are light, delicate, and airy, woven so thinly as to permit the breeze to pass through their net-like interior, and dangling daintily from the extremity of some slender twig. Others, again, are so firmly built of flattened reeds and grass blades, that they can be detached from their branches and subjected to very rough handling without losing their shape, while others are so curiously formed of stiff grass-stalks that their exterior bristles with sharp points like the skin of a hedgehog.

The true Weaver Birds all inhabit the hotter portions of the Old World, the greater number of them being found in Africa, and the remainder in various parts of India.

The SOCIABLE WEAVER BIRD is found in several parts of Africa, and has always attracted the attention of travellers from the very remarkable edifice which it constructs. The large social nests of this bird are so conspicuous as to be notable objects at many miles' distance, and it is found that they are generally built in the branches of the giraffe thorn or "kameel-dorn," one of the acacia tribe, on which the giraffe is fond of feeding, and which is especially valuable in Southern Africa for the hardness of its wood, from which the axletrees of waggons, handles of agricultural tools, and the strongest timbers of houses are made. This tree only grows in the most arid

SOCIABLE WEAVER BIRD.—*Philetærus socius.*

districts, and is therefore very suitable for the purposes of the Sociable Weaver Bird, which has a curious attachment to dry localities far from water.

The Sociable Weaver Bird, which is by some writers termed the Sociable Grosbeak, in choosing a place for its residence, is careful to select a tree which grows in a retired and sheltered situation, secluded as far as may be from the fierce wind storms which are so common in hot countries. When a pair of these birds have determined to make a new habitation, they proceed after the following fashion. They gather a vast amount of dry grasses, the favourite being a long, tough, and wiry species, called "Booschmaneesgrass," and by hanging the long stems over the branches and ingeniously interweaving them, they make a kind of roof, or thatch, which is destined to shelter the habitations of the community.

In the under sides of this thatch they fasten a number of separate nests, each being inhabited by a single pair of birds, and only divided by its walls from the neighbouring habitation. All these nests are placed with their mouths downward, so that when the entire edifice is completed, it reminds the observer very strongly of a common wasp's nest This curious resemblance is often further strengthened by the manner in which these birds will build one row of nests immediately above or below another, so that the nest-groups are arranged in layers precisely similar to those of the wasp or hornet. The number of habitations thus placed under a single roof is often very great. Le Vaillant mentions that in one nest which he examined, there were three hundred and twenty inhabited cells, each of which was in the possession of a distinct pair of birds, and would at the close of the breeding season have quadrupled their numbers.

The Sociable Weaver Bird will not use the same nest in the following season, but builds a new house, which it fastens to the under side of its previous domicile. As, moreover, the numbers of the nests are always greatly increased year by year, the Weaver Birds are forced to enlarge their thatched covering to a proportionate extent, and in course of years they heap so enormous a quantity of grass upon the branches that it fairly gives way with the weight, and they are forced to build another habitation. So large is this thatch-like covering, that Harris was once deluded by the distant view of one of these large nests with the belief that he was approaching a thatched house, and was only undeceived to his very great disappointment on a closer approach.

The object of this remarkable social quality in the bird is very obscure. As in many instances the nests of the Weaver Birds are evidently constructed for the purpose of guarding them from the attacks of snakes and monkeys, the two most terrible foes against which they have to contend, it is not improbable that the Sociable Weaver Birds may find in mutual association a safeguard against their adversaries, who might not choose to face the united attacks of so many bold though diminutive antagonists. The shape and general aspect of the nest varies greatly with their age, those of recent construction being comparatively narrow in diameter, while the older nests are often spread in umbrella fashion over the branches, enveloping them in their substance, and are sometimes only to be recognised as a heap of ruins from which the inhabitants have long fled.

In general the Sociable Weaver Bird prefers to build its nest on the branches of some strong and lofty tree, like the giraffe thorn above mentioned, which also has the advantage of massive and heavy foliage, disposed in masses not unlike the general shape of the Weaver Bird's nest. Sometimes, however, and especially near the banks of the Orange River, the bird is obliged to put up with a more lowly seat, and contents itself with the arborescent aloe. The number of eggs in each nest is usually from three to five, and their colour is bluish white, dotted towards the larger end with small brown spots. The food of this bird seems to consist mostly of insects, as when the nests are pulled to pieces, wings, legs, and other hard portions of various insects are often found in the interior of the cells. It is said that the Sociable Weaver Birds have but one enemy to fear in the persons of the small parrots who also delight in assembling together in society, and will sometimes make forcible entries into the Weaver Bird's nest and disperse the rightful inhabitants.

The colour of the Sociable Weaver Bird is brown, taking a pale buff tint on the under surface of the body, and mottled on the back with the same hue. It is quite a small bird, measuring only five inches in length.

The MAHALI WEAVER BIRD is also an inhabitant of Africa, and has a rather large range of country, being found spread over the land as far south as the tropic of Capricorn, and probably to a still farther extent.

The nest of this bird is quite as remarkable as that of the preceding species. In general shape and size it somewhat resembles the reed-covered bottles which are often to be seen in the windows of wine importers, being shaped somewhat like a flask, or perhaps more like a common skittle, and being composed of a number of very thick grass stems laid longitudinally, and interwoven in a manner that can hardly be understood without an illustration. Contrary to the usual custom of nests, in which the materials are woven very smoothly, the nest of this bird is purposely constructed so as to present the roughest possible exterior, all the grass stems being so arranged that their broken ends protrude for several inches in a manner that reminds the observer of a military " abattis," a defence formed by prostrate trees with the ends of the branches cut off and sharpened. Probably this structure is for the same purpose as the abattis, and is meant to protect the bird from the inroad of its enemies. Several of their curious edifices may be seen in the nest-room of the British Museum. The interior of the nest is sufficiently soft and warm, more so, indeed, than would be supposed from the porcupine-like aspect of the exterior walls.

The Mahali Weaver Bird is very social in its habits, the nests being placed in close proximity to each other, a single tree often containing from thirty to forty of these ingenious domiciles. The birds associate in flocks of different sizes, varying probably according to the season of the year, as is often the case with gregarious birds, whose numbers are greatly augmented after the breeding season by the presence of the yearling young. It feeds upon insects and fruit of different kinds, and a whole flock may often be seen on the ground, feeding away with great zest, and rapidly traversing the soil in search of food. Its disposition is wary, and at the first symptom of danger the whole flock takes to wing and flies to a more distant tree, where the birds sit amicably together until the cause of their alarm has disappeared, and then return in a body to their feeding ground.

The colour of the Mahali Weaver Bird is brown of different depths, variegated with yellowish white. The breast and abdomen are of a creamy yellow, and the chin, throat, and under tail-coverts are white. It is rather larger than the preceding species, measuring between six and seven inches in total length.

TAHA WEAVER BIRD.—*Plóceus Taha*

ANOTHER small species belonging to this group, called the TAHA WEAVER BIRD, does not extend its range so far north as the preceding birds, being seldom if ever seen beyond the twenty-sixth degree of north latitude.

This species is very common, and is found plentifully in many districts north of the line which has just been mentioned. Perhaps it is rather too plentiful in the estimation of the agricultural population of the land which it inhabits, as it is in the habit of associating in large flocks, and descending upon the cultivated lands, where it feeds upon the growing crops and does considerable damage. In many places, these birds are so numerous, that in order to preserve their fruits or grain, the owners of gardens or corn-fields are obliged to keep a constant watch over their crops during the whole day. For the greater part of the year the Taha Weaver Bird frequents trees near the banks of rivers, and may be seen in considerable numbers disporting among the branches. But at the moment of the breeding season the birds retire from the trees and betake themselves to the rank and luxuriant reeds which heavily fringe the edges and shoals of rivers, and which form the supports on which the pensile nests are suspended.

The colour of the adult male is yellowish brown, grey, and black, disposed rather pleasingly over the whole plumage. The female is yellowish brown above, covered with dashes of deep blackish brown. The chest is yellowish white with a slight rusty red tinge, and the abdomen and remainder of the under portion of the body are greyish white. During the winter the male lays aside his handsome vestments, and assumes the more sober tints of the female, so that the two sexes are almost undistinguishable.

RUFOUS-NECKED WEAVER BIRD.—*Hyphantornis textor.*

The RUFOUS-NECKED WEAVER is also an inhabitant of Africa, being found in Senegal, Congo, and other hot portions of that continent.

By many persons this species is known by the name of the Capmore Weaver, a term which is evidently nothing but a corruption of Buffon's name for the same bird, namely "Le Cap-noir," or Blackcap Weaver. It is a brisk and lively bird, and possesses a cheerful though not very melodious song. It has often been brought to Europe, and is able to withstand the effects of confinement with some hardihood, living for several years in a cage. Some of these caged birds carried into captivity the habits of freedom, and as soon as the spring made its welcome appearance, they gathered together every stem of grass or blade of hay, and by interweaving these materials among the wires of their cage, did their utmost to construct a nest. The food of this bird consists mostly of beetles and other hard-shelled insects; and in order to enable it to crush their defensive armour, which is extremely strong in many of the African beetles, its beak is powerful and its edges somewhat curved. Seeds of various kinds also form part of its diet; and the undulating edge of the bill is quite as useful in shelling the seeds as in crushing the insects.

The general colour of this species is orange-yellow, variegated with black upon the upper surface. The head, chin, and part of the throat are black, and a ruddy chestnut band crosses the nape of the neck. Like many other birds, however, it changes the colour of its plumage according to the time of year, and after the breeding season is over, its head assumes a tint somewhat like that of the back. It is by no means a large bird, its total length being a little more than six inches.

ONE of the best known of these curious birds is the RED-BILLED WEAVER BIRD.

This species is common in Southern Africa, and is notable for its habit of attending the herds of buffaloes in a manner somewhat similar to that of the African beef-eater, which has already been described. It does not, however, peck the deep-seated grubs from the hide, as its bill is not sufficiently strong for that purpose, but devotes itself to the easier task of capturing and eating the numerous parasitic insects which always infest those large quadrupeds. The buffaloes are quite sensible of the benefit which is conferred upon them by their feathered allies, and move about quite unconcernedly while serving as pasture-grounds for the Weaver Birds.

RED-BILLED WEAVER BIRD.—*Textor erythrorhynchus.*

Another important service is rendered to the buffalo by this Weaver Bird. It is a watchful and suspicious creature, and at the first intimation of danger it flies abruptly into the air from the buffalo's back. The beast, who as long as the Weaver Bird remained quietly on his back, continued to feed calmly, is roused by the sudden flutter of the wings and raises its head to ascertain the cause of the disturbance. Should it see grounds for apprehension, the alarm is given, and the whole herd dash off to a place of safety, accompanied by their watchful feathered friends.

This species has also been brought to Europe, and accommodates itself so well to the climate that the Parisian bird-dealers are able to breed it like the canary, though not with quite such success. The general colour of this bird is blackish brown, variegated with white on the primaries, and reddish brown below. The chin is black, as is a patch on the ears, and the beak is crimson, with a dash of purple on the sides. Sometimes the plumage varies slightly, and when the bird is in peculiarly fine condition and has arrived at its full maturity, a roseate hue appears on several parts of the body, and gives to it a very pleasing aspect.

AMONG the birds which are grouped together under the title of Weavers, none are more curious than those species which are popularly known by the title of Widow Birds, and more rightly by the name of Whidah Birds.

PARADISE WHIDAH BIRD.—*Vidua paradisea.*

The PARADISE or BROAD-SHAFTED WHIDAH BIRD is the species that is most familiar in cages and menageries, as it is by no means an uncommon bird in its native land, and bears confinement better than most inhabitants of a tropical land. It is an inhabitant of Western Africa, being found throughout the whole district from Senegal to Angola; and as it is of a light and airy disposition, it gives a lively aspect to the trees among which it lives. It is perpetually in motion, flitting from bough to bough with graceful lightness, pecking here and there after a casual insect, and evidently admiring its own beautiful tail with thorough appreciation.

The name Widow Bird is altogether an erroneous title, although it is supposed by many persons to have been given to the bird on account of its dark colour and long train, as well as in consequence of its evidently disconsolate state when the beautiful tail-feathers have fallen off after the breeding season. Certainly a caged Whidah Bird in such a condition exhibits the sincerest grief for his loss, and conducts himself as if labouring under the most poignant sorrow. Instead of boldly skipping among the highest forks, and flirting his long tail for the admiration of every spectator, he sits humbly on the lowest perches, or even on the floor of the cage, backs himself into a corner, and seems thoroughly ashamed of his undress. In point of fact, however, the proper name is Whidah Bird, a title that was originally given to it by the Portuguese, because the first specimens that were brought to Europe came from the kingdom of Whidah on the eastern coast of Africa.

Of late years the Whidah Bird has come into fashion in England as an inhabitant of the aviary, and in France has been common for a very long time. Many of the French dealers have succeeded in breeding these birds. On account of the peculiarly long tail, the Whidah Bird requires a very roomy cage, with perches of considerable height, and so arranged as not to interfere with the movements of the bird. It is very fond of bathing, and like many other tame birds, bursts into a cry of gratitude when supplied with fresh water. Of its habits in a wild state little or nothing is known,

save that its nest is ingeniously woven
from vegetable fibres, said to be wholly
those of cotton down, and is divided into
two compartments, one being for the use
of the female and her eggs or young,
and the other as a seat for the male
whereon he may perch himself to sing
to his family.

Although not very brilliant in hue,
the Paradise Whidah Bird is very beautifully clothed with softly tinted and
gracefully shaped plumage. The general
colour of the adult male bird in his full
dress is very deep brown-black, the
former tint prevailing on the wings and
tail, and the latter on the back. The
head, chin, and throat are also black,
and a line of the same colour starts from
the chin down the centre of the breast.
Round the neck runs a collar of rich
ruddy brown, which edges the black
line down the breast, and softly melts
into the pale buff of the abdomen and
under portions of the body. The tail
of this bird is most singularly formed.
Both webs of the two central feathers
are extremely broad for about three
inches, and then suddenly disappear,
leaving the bare slender shaft to project
for two or three inches. The two next
feathers are equally elongated, and rather
broadly webbed, being nearly three quarters of an inch in width. They are often
more than eleven inches long, and sweep
in a graceful curve from the insertion of
their quills to the extremity of their
points. All the feathers of the tail are
set vertically, so that the profile is more
striking than the full view.

The Broad-shafted Whidah Bird is
about the size of a sparrow, measuring
between five and six inches exclusive of
the elongated tail-feathers. After the
breeding season, the beautiful plumes
fall out, and the whole colouring of the
bird is changed from the deep black
and orange into rusty brown and dull
white.

THERE are many species of these
pretty little creatures, all being remarkable for some peculiarity in their form
or colouring. Space will not permit us
to engrave more than one other species,
the SHAFT-TAILED WHIDAH BIRD.

This exquisite bird is found along the
African coasts, and is in great favour in

SHAFT-TAILED WHIDAH BIRD.—*Vidua regia.*

Europe as a cage bird. Its voice is superior to that of the preceding species, although none of the Whidah Birds are remarkable for the musical power or brilliancy of their song. It is bright and sprightly in all its movements, flitting about its cage with a restless activity and fearless demeanour that endear it to its owner. From the Paradise Whidah Bird it may be distinguished not only by its colouring, but by the curious arrangement of its tail-feathers, which are very short, with the exception of the four central feathers, which are most singularly elongated, each feather presenting to the eye little but the bare shaft for the greater part of its length, and then slightly widening towards the extremities. The sides of the head and around the neck are deep rusty red, and the back of the neck and top of the head are mottled black. The total length of the Shaft-tailed Whidah Bird is from nine to ten inches.

CARDINAL GROSBEAK.—*Cardinális Virginiánus.*

THE Grosbeaks or Hawfinches now claim our attention. They are all remarkable for their very large, broad, and thick beaks, a peculiarity of construction which is intended to serve them in their seed-crushing habits.

The most magnificent example of this group is the CARDINAL or SCARLET GROS-BEAK, an inhabitant of various parts of America, where it is known under the titles of Red Bird, Crested Red Bird, and Virginian Nightingale.

It is rather a large bird, measuring about eight inches in total length, and is coloured in a most gorgeous fashion. The back is dusky red, and the whole of the rest of the plumage is bright vivid scarlet with the exception of a patch of jetty black short feathers that decorate the chin, forehead, and base of the beak. Upon the head there is a high pointed crest, which can be raised or lowered at pleasure. Even the bill is bright scarlet. The female is a smaller bird, and is not nearly so handsome as her mate. The upper parts of the body are brown-olive, and the tail, tip of the crest, and the wings are scarlet. The chin and forehead are ashen grey, and the breast and abdomen are drab with a dash of red. The bill is scarlet like that of the male.

The voice of the Cardinal Grosbeak is naturally fine, though the song is apt to be rather too monotonous, the bird repeating the same phrase twenty or thirty times before proceeding to another. Still its musical powers are sufficiently marked to earn for the bird the title of Virginian Nightingale, and it is a curious fact that the female often sings nearly as well as her mate.

This bird seems to be of a very tender-hearted disposition, and given to the adoption of other birds when young and helpless. Wilson mentions that he placed a young cow bird in the same cage with a Cardinal Grosbeak, which the latter immediately adopted, and reared the poor helpless little creature that had appealed so suddenly to its compassionate feelings. Mr. Webber moreover, in his account of the Birds of America, gives an anecdote of a Scarlet Grosbeak belonging to an old woman in Washington city, which

HAWFINCH OR GROSBEAK.—*Coccothraustes vulgáris.*

used to make a regular business of rearing the young of other birds which were placed under his charge, and thereby earning a considerable sum of money in the course of a season. She had often been offered a high price for her bird, but always refused to sell him, impelled either by hope of gain or by love of the bird: we may hope that the latter feeling predominated.

In its native land the Cardinal Grosbeak is most common in the Southern States, and in some localities is migratory, while in others it remains throughout the year. "In the Northern States," says Wilson, "they are migratory, but in the lower parts of Pennsylvania they reside during the whole year, frequenting the borders of creeks and rivulets, in sheltered hollows covered with holly, laurel, and other evergreens. They love also to reside in the vicinity of fields of Indian corn, a grain that contributes their chief and favourite food. The seeds of apples, cherries, and of many other sorts of fruit are also eaten by them, and they are accused of destroying bees."

Many of these splendid birds are now brought to England as inhabitants of the aviary, and are found to be hardy birds, able to withstand the inclemency of the English climate. Many of them may be seen living in the Crystal Palace and in other buildings where caged birds are kept. It is a remarkable fact, that in confinement the Cardinal Grosbeak is very apt to change its colour, the bright scarlet and vermilion fading to a dull whitish red; probably the effect of insufficient or improper food. When carefully tended, it is a really healthy and long-lived bird, having been known to survive for a space of twenty years in a cage.

The nest of the Cardinal Grosbeak is generally placed in a holly, cedar, laurel, or other thick evergreen, and is made of slender sticks, weeds, strips of bark, and fine grass-stems. The eggs are generally five in number, and their colour is dull grey-white covered with numerous blotches of brownish olive. There are generally two broods in the season.

ENGLAND possesses a good example of this group in the well-known HAWFINCH, or GROSBEAK.

This bird was once thought to be exceedingly scarce, but is now known to be anything but uncommon, although it is rarely seen, owing to its very shy and retiring habits, which lead it to eschew the vicinity of man and to bury itself in the recesses of forests. So extremely wary is the Hawfinch, that to approach within gunshot is a very difficult matter, and can seldom be accomplished without the assistance of a decoy-bird, or by imitating the call-note, which bears some resemblance to that of a robin. It feeds chiefly on the various wild berries, not rejecting even the hard stones of plums and the laurel berries. In the spring, it is apt to make inroads in the early dawn upon the cultivated grounds, and has an especial liking for peas, among which it often works dire havoc.

It is a gregarious bird, associating in flocks varying in number from ten to two hundred, and always being greatest after the breeding season. According to Mr. Doubleday, it is not migratory, but remains in England throughout the year, Epping Forest being one of its chief strongholds, as it abounds in berries of various kinds, is within a reasonable distance of cultivated grounds and affords an excellent retreat. When in the forest, the bird generally perches upon the extreme top of some lofty tree, from whence it keeps so complete a watch that hardly a weasel could steal upon it without being perceived and its presence reported by an alarm note, which is perfectly understood not only by other Grosbeaks, but by all the feathered and some of the furred tribes.

The nest of the Hawfinch is not remarkable either for elegance or peculiarity of form. It is very simply built of slender twigs, bits of dried creepers, grey lichens, roots, and hair, and is so carelessly put together that it can hardly be moved entire. The eggs are from four to six in number, and their colour is very pale olive-green, streaked with grey and spotted with black dots. The birds pair

BLACK AND YELLOW GROSBEAK.—*Coccothraustes melanoxanthus.*

in the middle of April, begin to build their nests about the end of that month, and the young are hatched about the third week in May.

The colour of the adult male Grosbeak is briefly as follows: The head and nape of the neck are fawn colour, deepening towards the shoulders and fading into grey on the other portions of the neck, and the chin and throat are velvety black. The upper part of the body is chestnut-brown, and the wing-coverts are variegated with white, black, and fawn. The primary feathers of the wing are deep blue-black, white on the inner webs. The upper tail-coverts are fawn, and the tail itself is black and white, with the exception of the two central feathers, which are greyish brown, tipped with white. The sides of the neck, the breast, abdomen, and whole of the under parts are brown of a lighter and paler hue than that of the back, and the under tail-coverts are white. The female is similarly coloured, but the hues are much duller than in her mate. The total length of the bird is seven inches.

On examining the wings of this bird, the observer will be struck with the curious shape of the fifth, sixth, seventh, eighth, ninth, and tenth primary feathers, which are not pointed, but are larger at the ends, rounded and hooked in a manner which is well compared by Yarrell to the head of an ancient battle-axe. Perhaps the Jedburgh axe is more like the shape of these curious feathers.

ALTHOUGH not possessed of the glowing scarlet hue which decorates the cardinal Grosbeak, the BLACK AND YELLOW GROSBEAK is quite as remarkable and scarcely a less handsome bird.

Its ordinary habitation is in the northern parts of India, but it is a bird of strong wing, and often wanders as far as Central India in search of food. Like others of the same group, it mostly feeds on berries and various stone-fruits, crushing even the hard-shelled seeds and stones in its thick and powerful beak. Even at a distance, this bird is very conspicuous on account of the bold and dashing manner in which the whole of the plumage is variegated with black, white, and yellow, all these colours being of the purest and brightest quality.

The whole of the upper surface and the breast are deep jetty black, with a slight silken gloss when the bird is in good condition. A few snowy white spots appear on the basal portions of the four central primary feathers of the wing, and several of the primaries, together with the whole of the secondaries, are edged with the same hue, thus presenting a very strong contrast to the jetty feathers of the back. The lower part of the breast and the abdomen are bright golden yellow, so that the bird is coloured only with these three decided hues, without any gradation through intermediate hues, as is generally the case in birds of bright plumage. The female is easily distinguished from her mate, as the upper surface is dusky black, largely mottled with yellow upon the head, neck, and back. The breast and abdomen are greyish yellow, profusely covered with black spots resembling the "tears" in heraldry. The quality of the hue is rather variable, as in some specimens the black is of the deepest, and the yellow of the richest, glossiest gold, whereas in some individuals—probably the young male just entering his perfect plumage, or the old male getting feeble with age—the black has a dirty look, and the yellow is nearly white. In size this bird is about equal to the cardinal Grosbeak.

The Tanagrine birds are well represented by the SCARLET TANAGER of America.

It is a very handsome bird, decorated with lively scarlet and deep black, and is possessed of a tolerable, though not especially musical voice. This is one of the migratory species, arriving in the northern portions of the United States about the end of April, and remaining until the breeding season is over. The nest is made of rather rough materials, such as flax-stalks and dry grass, and is so loosely put together that the light is perceptible through the interstices of the walls. The number of eggs is generally three, and their colour is dullish blue, variegated with brown and purplish spots. While engaged in the business of incubation, both birds are extremely terrified at the presence of any strange object, and if a human being approaches the nest, the male flies to a little distance and keeps cautiously aloof, peering through the boughs at the foe, and constantly fearful of being seen. The female also leaves the nest, but continues to fly restlessly about her home, hovering over the eggs or young in great distress. When, however, the young are hatched, the male parent takes his full share in attending upon them, and cares nothing for being seen.

The attachment of the male bird to his young seems to be very strong, as is shown by the following account, extracted from Wilson.

"Passing through an orchard, and seeing one of these young birds that had but lately left the nest, I carried it with me about half a mile to show it to my friend, Mr. William Bartram, and having procured a cage, hung it up on one of the large pine-trees in the Botanic Garden, within a few feet of the nest of an orchard oriole, which also contained young, hopeful that the charity and kindness of the orioles would induce them

SCARLET TANAGER.—*Pyranga rubra.*

to supply the cravings of the stranger. But charity with them, as with too many of the human race, began and ended at home.

The poor orphan was altogether neglected, notwithstanding it plaintive cries, and as it refused to be fed by me, I was about to return it to the place where I found it, when towards the afternoon a Scarlet Tanager, no doubt its own parent, was seen fluttering round the cage, endeavouring to get in. Finding this impracticable, he flew off and soon returned with food in his bill, and continued to feed it till after sunset, taking up his lodgings on the higher branches of the same tree. In the morning, almost as soon as day broke, he was again seen most actively engaged in the same affectionate manner, and, notwithstanding the insolence of the orioles, continued his benevolent offices the whole day, roosting at night as before.

On the third or fourth day he appeared extremely solicitous for the liberation of his charge, using every expression of distressful anxiety, and every call and invitation that nature had put in his power for him to come out. This was too much for the feelings of my venerable friend; he procured a ladder, and mounting to the spot where the bird was suspended, opened the cage, took out the prisoner, and restored him to liberty and to his parent, who, with notes of great exultation, accompanied his flight to the woods."

The colour of the adult male bird is brilliant scarlet, with the exception of the wings and tail, which are deep black. The tail is forked, and very slightly tipped with white. This plumage is, however, only donned during the breeding season, for in the autumnal moult a number of greenish yellow feathers make their appearance, giving the bird a uniformly dappled or mottled aspect. The female is a comparatively soberly clad bird, being green above and yellow beneath, with wings and tail brownish black, edged with green. The total length of the Scarlet Tanager is between six and seven inches.

All the Tanagers inhabit America, and are mostly confined to the southern portions of that land. They may readily be distinguished from the other Fringillidæ by the notched upper mandible, and by the triangular base and arched ridge of the beak. Most of them are possessed of musical powers; one species, the Organist Tanager (*Euphonia musica*), deriving its popular and scientific title from its rich full tones. The colours of the Tanagers are generally brilliant, scarlet, black, and orange being the ordinary hues with which their plumage is bedecked.

GROUP OF FINCHES.

THE true Finches are known by their rather short and conical beak, their long and pointed wings, and the absence of nostrils in the beak. England possesses many examples of these birds, several of which are celebrated for their beauty of plumage and powers of song.

AMONG the most beautiful of these birds, the GOULDIAN FINCH holds a high place, its plumage being decorated with the softest and most harmonious hues, the feathers glowing with delicately opalescent shades of lilac, green, and golden yellow.

This exquisite little bird is a native of New South Wales, and although not very scarce in the district which it frequents, is yet decidedly local in its habits. It is seldom seen in the open country, preferring to haunt the thicket and edges of forests, where it may be seen hopping easily among the branches, in little bands of from four to seven or eight in number. The voice of this finch is not remarkable for force or beauty, being little more than a querulous kind of twitter, which it utters mournfully when disturbed, at the same time flying to the summit of the nearest tree, and there sitting until the cause of alarm is removed.

The colour of this bird is as follows: The head and throat are deep velvet-black, the back and wings are soft yellowish green, and a stripe of bright verditer green runs from behind the eye down the sides of the neck, until it is merged in the yellow-green of the back. Across the breast runs a broad band of purple, yellow, or lilac, and the whole of the under surface is golden yellow with a kind of waxen gloss. The bill is scarlet at the tip, and white at the base. These tints belong only to the adult bird, the young being soberly clad in grey, buff, and olive.

GOULDIAN FINCH.—*Amadina Gouldiæ.*

ON the opposite illustration are two of the most familiar and prettiest of the British songsters, the lower figure representing the CHAFFINCH, and the upper the GOLDFINCH.

The CHAFFINCH is one of our commonest field birds, being spread over the whole of England in very great numbers, and frequenting hedges, fields, and gardens with equal impartiality. It is a most gay and lively little bird, and whether singly, or assembled in large flocks, it always adds much life to the landscape, and delights the eye of every one who is not a farmer or a gardener, both of which personages wage deadly war against the bright little bird. For the Chaffinch is apt at times to be a sad thief, and has so strong a liking for young and tender vegetables that it pounces upon the green blades of corn, turnips, radishes, and similar plants, as soon as they push their way through the soil, and in a few hours destroys the whole of the seedlings. In one instance, a few Chaffinches settled upon a piece of ground about one hundred and twenty yards square, that had been planted with turnips, and before the day had closed, they had pulled up every young shoot, and eaten a considerable amount of them.

As, however, is the case of the rook, the chief food of the Chaffinch consists of insects which would be most noxious to the agriculturist, and in all probability the harm which they do in eating young plants and buds is more than counterbalanced by the benefit which they confer in destroying myriads of dangerous insects.

CHAFFINCH.—*Fringilla cœlebs.*　　　　GOLDFINCH.—*Fringilla carduélis.*

The specific title of Cœlebs, which is given to the Chaffinch, signifies a bachelor, and refers to the annual separation of the sexes, which takes place in the autumn, the females departing to some other region, and the males congregating in vast multitudes, consoling themselves as they best can by the pleasures of society for the absence of the gentler portion of the community. Very large flocks of these birds appear towards the end of autumn, and seem at first to be wholly composed of females. It is, however, more than probable that they consist of the females together with their young families of both sexes, and that the immature males have not as yet assumed their perfect plumage. The flocks are generally seen about hedgerows and stubble-fields; and if the weather should be very severe, they adjourn to the vicinity of human habitations, haunting the gardens and farm-yards, and often rivalling the sparrows in their boldness of demeanour.

The note of this bird is a merry kind of whistle, and the call-note is very musical and ringing, somewhat resembling the word "pinck," which has therefore been often applied to the bird as its provincial name.

The nest of the Chaffinch is one of the prettiest and neatest among the British nests. It is deeply cup-shaped, and the materials of which it is composed are moss, wool, hair, and lichens, the latter substances being always stuck profusely over the surface, so as to give it a resemblance to the bough on which it has been built. The nest is almost invariably made in the upright fork of a branch, just at its junction with the main stem. or bough from which it sprang, and is so beautifully worked into harmony with the bark of the particular tree on which it is placed, that it escapes the eye of any but a practised observer. Great pains are taken by the female in making her nest, and the structure occupies her about three weeks. The eggs are from four to five in number, and their colour is pale brownish buff, decorated with several largish spots and streaks of very dark brown.

The colour of this pretty bird is as follows: At the base of the beak the feathers are jetty black, and the same hue, but with a slight dash of brown, is found on the wings and the greater wing-coverts. The top of the head and back of the neck are slaty grey, the back is chestnut, and the sides of the head, the chin, throat, and breast are bright ruddy chestnut, fading into a colder tint upon the abdomen. The larger wing-coverts are tipped with white, the lesser coverts are entirely of the same hue, and the tertials are edged with yellowish white. The tail has the two central feathers greyish black, the next three pairs black, and the remaining feathers variegated with black and white. The total length of the bird is six inches. The female is coloured something like the male, but not so brilliantly.

OF all the British Finches, none is so truly handsome as the GOLDFINCH, a bird whose bright yellow orange hues suffer but little even when it is placed in close proximity to the more gaudy Finches of tropical climates. Like the chaffinch, it is spread over the whole of England, and may be seen in great numbers feeding on the white thistledown. There are few prettier sights than to watch a cloud of Goldfinches fluttering along a hedge, chasing the thistledown as it is whirled away by the breeze, and uttering all the while their sweet merry notes.

The birds are not very shy, and by lying quietly in the hedge the observer may watch them as they come flying along, ever and anon perching upon the thistle-tops, dragging out a beakful of down, and biting off the seeds with infinite satisfaction. Sometimes a Goldfinch will make a dart at a thistle or burdock, and without perching snatch several of the seeds from their bed, and then alighting on the stem, will run up it as nimbly as a squirrel, and peck away at the seeds, quite careless as to the attitude it may be forced to adopt. These beautiful little birds are most useful to the farmer, for they not only devour multitudes of insects during the spring months, but in the autumn they turn their attention to the thistle, burdock, groundsel, plantain, and other weeds, and work more effectual destruction than the farmer could hope to attain with all his labourers. Several Goldfinches may often be seen at one time on the stem and top of a single thistle, and two or three are frequently busily engaged on the same plant of groundsel.

Like the preceding species, the Goldfinch keeps to the open ground or hedgerows during the summer and autumn, but during the winter is often forced to seek for food nearer to man, and accordingly vacates the fields and proceeds in flocks to the farmyards, where it makes the most of its opportunities. In Mr. Thompson's History of the Birds of Ireland, there is a communication from Mr. R. Patterson respecting the conduct of this bird during a very severe winter.

"When at Limerick, in August, 1843, I had the pleasure of meeting Randal Burough, Esq. of Cappa Lodge, Kilrush, county of Clare, who communicated to me the following particulars respecting an unusual assemblage of Goldfinches observed at his residence in the winter of 1836.

Mr. B. had two pet Goldfinches, which were allowed not only to fly about the room, but also through the open window. The winter was beginning to be severe, and the food suitable for small birds consequently scarce, when one day the two Goldfinches brought with them a stranger of their own species, who made bold to go into the two cages that were always left open, and regale itself on the hospitality of its new friends, and then took his departure. He returned again and brought others with him, so that in a few days half a dozen of these pretty warblers were enjoying the food bountifully provided for them. The window was now kept up, and the open cages with plenty of seed were placed on a table close to it, instead of on the sill as previously.

The birds soon learned to come into the room without fear. The table was by degrees shifted from the window to the centre of the room, and as the number of the birds had continued gradually to increase, there was soon a flock of not less than twenty visiting the apartment daily, and perfectly undisturbed by the presence of the members of the family. As the inclemency of the weather decreased, the number of birds gradually diminished, until at length, when the severe weather had quite passed away, there remained none. except the original pair."

After perusing the foregoing note, Mr. Burough added :—

" This account is perfectly correct; and I have only to remark that it was the innocent cause of making many idlers, for several strange gentlemen were in the habit of stopping for hours in amazement at the novel scene. My house is situated immediately on the banks of the Shannon, the road only separating it from the shore, and scarcely any timber growing nearer than a mile. The two pet Goldfinches must, in their flight, have gone a considerable distance to make out new acquaintances, for they (Goldfinches) are very scarce indeed close to the sea—as Cappa Lodge is situated."

There are few birds in so much favour as the Goldfinch, for it soon accommodates itself to captivity, is of a most loving disposition, and being extremely intelligent and docile, learns to play all kinds of tricks. I remember one of these birds that used to live in a drawer in the surgery of my father's house, and was a most amusing little creature, hauling up its daily supply of water by means of a little metal bucket, perching on the finger and whistling, and firing a little cannon with perfect steadiness and absence of alarm. Many professional exhibitors are in the habit of teaching their Goldfinches to go through a regular performance, and even to enact a sort of play or pantomime, which is quite intelligible even without the somewhat involved explanations of the proprietor.

The nest of the Goldfinch is very neat and prettily made, sometimes built in a hedge or thick bush, but mostly placed towards the extremity of a thickly foliaged tree-branch, such being a favourite for this purpose. In this position, the nest is so ingeniously concealed from the gaze of every one beneath, by the disposition of the branches and leaves and by the manner in which the exterior of the nest is made to harmonize in tint with the bark, that it can scarcely be discerned even when the observer has climbed the tree and is looking down upon the nest. The bird, too, seldom flies directly in or out of the nest, but alights at a little distance from her tree, and then slips quietly through the leaves until she reaches her eggs or young.

The materials of which the exterior of the nest is made differ according to the tree in which it is placed. In general, fine grasses, wool, hairs, and very slender twigs are employed in constructing the walls, and the interior is softly lined with feathers, down, and hairs. The eggs are generally four or five in number, and delicately marked with small dots and streaks of light purplish brown upon a white ground, having a tinge of blue, something like " sky-blue " milk.

This pretty bird is coloured after a very beautiful and rather complicated fashion. Around the base of the bill is a band of bright crimson, and the top of the head is jetty black, continued down the sides of the face and ending in the point of the shoulders. The sides of the face are white, and a stripe of the same hue comes below the black cap where it joins the nape of the neck. The back and upper surface are greyish brown, and the throat, breast, sides of the body, and thighs are tinged with the same colour. The greater wing-coverts and the first half of the outer edge of the primaries are light yellow, and the remainder of the primaries are black. Each of the tertials has a white spot at the extremity, and the upper tail-coverts are white with a grey tinge. The tail is black, with a spot or two of white at the tip of several of the feathers, and the abdomen is greyish white. The female is coloured in a somewhat similar fashion, but the black takes a browner tinge, the white and yellow are dashed with grey, and the scarlet band round the beak is much narrower and spotted with black. The total length of the bird is about five inches.

The SISKIN, or ABERDEVINE, is one of the British birds which performs an annual migration either partial or complete, a question about which there has been some controversy, and one which may fully receive a solution from the supposition that some birds remain in this country throughout the year, retiring no farther to the north than Scotland, while others pass to Norway and Sweden for the purpose of nidification, and do not return to this land until the autumn.

H H 2

SISKIN, OR ABERDEVINE.—*Fringilla spinus.*

They are lively little birds, assembling in small flocks of eight or ten in number, and haunting the edges of brooks and streams for the purpose of seeking the seeds of the elder and other trees, on which they chiefly feed. Along the banks they are quick and active, fluttering from one bough to another, and clinging in every imaginable attitude, with a strength of limb and briskness of gesture much resembling the movements of the titmice. While thus engaged, they constantly utter their sweet and gentle call-note, which is so soft that bird-dealers are in the habit of pairing the Siskin with the canary, in order to obtain a song-bird whose voice is not so ear-piercing as that of the pure canary.

The nest of this bird is usually placed on the fork of a bough, and is composed of wool, grass, hairs, and similar substances. The eggs are from three to five in number, and the young are hatched in a fortnight, being able to take to wing in a month or five weeks.

The colouring of this bird is remarkable for the very peculiar green with which most of its plumage is tinged, and which is spread over the whole of its back and the upper portions of its body. The centre of each feather of its back is dark olive-green. The top of the head and the chin are black, and a light sulphur-yellow is found behind the ear, on the neck, breast, and edge of the quill-feathers of the wings, upon the greater coverts, and the tail. This yellow varies in quality and intensity, in some parts being sulphurous, and in others of a rich golden hue. The middle of the quill-feathers of the wings and tail are black washed with olive; the abdomen and under tail-coverts are white, deepening into grey on the flanks. The female is similarly coloured, but the hues are not nearly so bright. In total length the Siskin measures rather more than five and a half inches, the female being a little less.

The GREENFINCH is one of our commonest birds, being resident in this country throughout the year, and not even requiring a partial migration.

It is mostly found in hedges, bushes, and copses, and as it is a bold and familiar bird, is in the habit of frequenting the habitations of men, and even building its nest within close proximity to houses or gardens. During the mild weather, the Greenfinch remains

GREENFINCH.—*Fringilla chloris.* LINNET.—*Fringilla cannábina.*

in the open country, but in the severe winter months it crowds to the farmhouses, and boldly disputes with the sparrows the chance grains of food that it may find. When young, the bird is fed almost wholly upon caterpillars and various insects, and not until it has attained its full growth does it try upon the hard seeds the large bill which has obtained for it the title of Green Grosbeak.

The voice of the Greenfinch is very ordinary, being possessed neither of strength nor melody, so that the bird is in very little demand as an inhabitant of the aviary.

The nest of this bird is generally built rather later than is usual with the Finches, and is seldom completed until May has fairly set in. Its substance is not unlike that of the chaffinch, being composed of roots, wool, moss, and feathers. It is not, however, so neatly made, nor so finely woven together, as the nest of that bird. The eggs are from three to five in number, and the colour is bluish white covered at the larger end with spots of brown and grey.

In the adult male bird, the head, neck, and all the upper parts of the body are yellow, with a green wash, and the wings are partly edged with bright yellow. The primary feathers of the wings are grey-black, edged for a considerable portion of their length with brilliant yellow. The greater wing-coverts, together with the tertiaries, are grey; the chin, throat, breast, and under parts of the body are yellow, falling into grey on the flanks. With the exception of the two short middle feathers, which are grey-brown throughout, the tail-feathers are yellow for the first half of their length, and grey-brown for the remainder. The female is of much more sober colours, being greenish brown on the back and under surface, and the yellow of the wings being very dull. The total length of the bird is about six inches, the female being little less than her mate.

THE lower figure of the illustration represents the common LINNET, sometimes called the BROWN LINNET, in contradistinction to the preceding species, or the GREATER REDFINCH, in allusion to the vermilion-tipped feathers of the crown.

Few birds are better known than the Linnet, although the change of plumage to which it is subject in the different seasons of the year has caused the same bird, while in its winter plumage, to be considered as distinct from the same individual in its summer dress. Except during the breeding season, the Linnets associate in flocks, flying from spot to spot, and feeding upon the seeds of various plants, evidently preferring those of the thistle, dandelion, and various cruciferous plants. It is a very lively bird, and is possessed of a sweet and agreeable, though not very powerful song.

The nest of this bird is strangely variable in the positions which it occupies, sometimes being placed at a considerable height upon a tree, and at other times built in some bush quite close to the ground, the latter being the usual locality chosen by the bird. The full number of the eggs is five, and the colour is mostly blue, spotted with dark brown, and a rather faint and undecided purple. The Linnet is not at all successful in concealing her nest, so that the eggs are always found in great numbers on the string of the bucolic nest-seeker.

The summer plumage of the male bird is as follows: On the top of the head, the feathers are greyish brown at their base, but are tipped with bright vermilion, a tint which contrasts well with the ashen grey-brown of the face and back of the neck. The upper parts of the body are warm chestnut, and the wing is black, excepting the narrow exterior webs, which are white. The chin and throat are grey, the breast bright red slightly dappled with brown, and the under portions of the body are grey-brown, taking a yellowish tint on the flanks. The tail is rather forked, and the feathers are black, edged with white.

During the winter the vermilion-red of the head and breast vanishes, and is replaced by simple brown. It is a curious fact that the fine red tint is only to be found in the wild birds. The female is lighter in colour than her mate, and the under surface of the body is brown, slightly dashed with red. The total length of the bird is nearly six inches, the female being about half an inch shorter.

The SNOW BIRD, which is not to be confounded with the Snow Bunting, hereafter to be described, is an inhabitant of America, and has a very large range of country.

According to Wilson's lively description of this bird, " at first they are most generally seen on the borders of woods among the falling and decayed leaves, in loose flocks of thirty or forty together, always taking to the trees when disturbed. As the weather sets in colder, they venture nearer the farmhouses and villages, and on the approach of what is usually called 'falling weather,' assemble in larger flocks, and seem doubly diligent in searching for food. This increased activity is generally a sure prognostic of a storm.

When deep snow covers the ground, they become almost half domesticated. They collect about the barns, stables, and other outhouses, spread over the yard, and even round the steps of the door; not only in the country and villages, but in the heart of our large cities; crowding around the threshold early in the morning, gleaning up the crumbs, and appearing very lively and familiar. They also have recourse at this severe season, when the face of the earth is shut up from them, to the seeds of many kinds of weeds that still rise above the snow, in corners of fields, and low sheltered situations, along the borders of creeks and fences, where they unite with several species of sparrow. They are at this time easily caught with almost any kind of trap, are generally fat, and it is said are excellent eating."

At the very beginning of summer, as soon as the weather begins to be warm, the Snow Bird retires from its winter quarters, and migrates to the higher regions of the

SNOW BIRD.—*Fringilla hyemalis.*

earth, for the purpose of breeding. Even in the business of rearing their young, the Snow Birds are very gregarious, placing the nest upon the ground, or on the grass, in close proximity to each other.

The head, neck, and upper parts of the body, and the wings, are very deep slaty brown, either colour predominating according to the age of the individual and the season of the year. The lower parts of the breast and the abdomen are pure snowy white, and the two exterior tail-feathers are of the same hue, the secondaries being dark slate. The female has but little of the slaty blue, and is almost wholly brown. The total length of this species is about six inches.

THERE are many other British Birds which are worthy of notice, such as the Lesser and Mealy Redpoles, the Mountain Linnet, or Twite, and others; but as our space will not permit us to give a history of all the British birds, we must now pass on to another species, which, although not indigenous to England, has become so far naturalized, that to many eyes it is even more familiar than the sparrow.

THE pretty little CANARY BIRD, so prized as a domestic pet, derives its name from the locality whence it was originally brought.

Rather more than three hundred years ago, a ship was partly laden with little green birds captured in the Canary Islands, and having been wrecked near Elba, the birds made their escape, flew to the island and there settled themselves. Numbers of them were caught by the inhabitants, and on account of their sprightly vivacity and the brilliancy of their voice they soon became great favourites, and rapidly spread over Europe.

The original colour of the Canary is not the bright yellow with which its feathers are generally tinted, but a kind of dappled olive-green, black, and yellow, either colour predominating according to circumstances. By careful management, however, the bird-fanciers are able to procure Canaries of every tint between the three colours, and have instituted a set of rules by which the quality and arrangement of the colouring is reduced to a regular system. Still, the original dappled green is always apt to make its appearance; and even when two light-coloured birds are mated, a green young one is pretty sure to be found in the nest. For my own part, I care little for the artificial varieties produced by the fanciers according to their arbitrary rules, always subject to variation; and to my mind, an intelligent bird and a good songster is not one whit the less attractive because the colours of its plumage are not arranged precisely according to the fanciers' rules.

I have kept Canaries for many years, and could fill pages with anecdotes and histories of them and their habits, but as I have already written rather a long biography of my Canaries in "My Feathered Friends," together with instructions for the management and rearing of these pretty birds, there is no need to repeat the account in the present pages.

Several accounts of talking Canaries have come before the public, but none of them very reliable until Mr. S. L. Sotheby addressed the following communication to Mr. Gray, Vice-President of the Zoological Society.

"Touching that marvellous little specimen of the feathered tribe, a talking Canary, of which I had the pleasure a few days since of telling you, I now send you all the information I can obtain respecting it, from the lady by whom it was brought up and educated at this our homestead.

Its parents had previously and successfully reared many young ones; but three years ago they hatched only *one* out of four eggs, the which they immediately neglected, by commencing the rebuilding of a nest upon the top of it. Upon this discovery, the unfledged and forsaken bird, all but dead, was taken away and placed in flannel by the fire, when after much attention it was restored and then brought up by hand. Thus treated, and away from all other birds, it became familiarised with those only who fed it; consequently its first singing notes were of a character totally different to those usual with the Canary.

Constantly being talked to, the bird, when about three months old, astonished its mistress by repeating the endearing terms used in talking to it, such as '*Kissie, kissie,*' with its significant sounds. This went on, and from time to time the little bird repeated other words; and now, for hours together, except during the moulting season, astonishes us by ringing the changes, according to its own fancy, and as plain as any human voice can articulate them, on the several words—'Dear sweet Titchie' (its name), 'Kiss Minnie, 'Kiss me then, dear Minnie,' 'Sweet pretty little Titchie,' 'Kissie, kissie, kissie,' 'Dear Titchie,' 'Titchie wee, gee, gee, gee, Titchie, Titchie.' The usual singing notes of the bird are more of the character of the nightingale, mingled occasionally with the dog-whistle used about the house. It sometimes whistles, very clearly, the first bar of 'God save the Queen.'

It is hardly necessary to add that the bird is, of course, by nature remarkably tame; so much so that, during its season, it will perch down from its cage on my finger, shouting and talking in the most excited state. Our friend Mr. B. Waterhouse Hawkins, who has heard the bird, tells me that about twenty years ago a Canary that spoke a few words was exhibited in Regent-street, the only other instance, I believe, publicly known."

This very interesting communication was printed in the Proceedings of the Zoological Society, and may be found in No. CCCLXI. 1858.

One word of advice I will just give to possessors of Canaries: It often happens that the birds become dispirited, sit drooping on the ground or their perches, and have every symptom of severe illness. In nine cases out of ten, this is caused by the red-mite, a tiny parasitic creature, almost invisible to the naked eye, which attacks the Canaries, and by the continual irritation and want of rest which it occasions, especially during the night, gives rise to many dangerous complaints. Formerly I lost several birds by this pest, but have now succeeded in its almost complete extirpation.

The red-mites are haters of light, and during the daytime they generally retire from the birds, and conceal themselves in the cracks and crevices of the cage, their minute dimensions enabling them to congregate in immense numbers within a very small space. I am accustomed, therefore, at the brightest moments of noonday, when the mites have entirely retreated into their hiding-places, to remove the birds from the cage, and to apply neat's-foot oil to every part of the cage where a mite could take shelter. This plan has the advantage over all others, that it not only kills the mites, but also destroys their eggs, and so prevents a fresh supply from being raised. I then take each bird separately, and, after rubbing some Persian Insect Powder well into the feathers, I scatter some powder in a piece of calico, wrap the bird in it, and let it lie for a quarter of an hour. A feather

CANARY.—*Carduélis canária.*

waved over the eyes of a Canary while it is lying on its back has the effect of depriving it of all power, so that it will lie quite motionless until taken up. The powder is then shaken carefully out of the feathers, together with the dead and dying mites that may have remained in the plumage, the bird restored to the cage, and the powder and mites burned.

As a few of these pests will escape observation in spite of all precautions, I wait until night has fairly set in, and suddenly taking a very bright lamp into the cage where the birds live, throw its light upon them, and cause the mites to leave the feathers and hurry towards their hiding-places. As at that hour they are always distended with blood, they are easily visible by the light of the lamp, and can at once be killed by being touched with a little oil. Prevention, moreover, is better than cure, and as the mites are chiefly bred in the so-called " nests" which are sold in the shops, no building substances should be given to the birds without having been previously plunged into boiling water. By carefully taking these precautions, the mites will be effectually destroyed in the course of a fortnight or so, and the owner of the birds will find his reward in the recovered sprightliness of his feathered pets, and their speedy restoration to health. The best insect powder that I have yet seen can be obtained at No. 33, Newgate-street, City, at the cost of tenpence per packet. It is instantaneously effectual towards the insects, which it kills or paralyzes, and does no harm to the birds even if it should get into their eyes or mouths. When rammed into a paper tube and burned, it smoulders slowly away, and the smoke will destroy many of the insect pests that infest our houses or injure our plants. In such cases the smoke should be confined as much as possible by coverings of some sort, as its potency is of course according to its concentration.

THE noisy, familiar, impatient SPARROW is one of those creatures that has attached itself to man, and follows him wherever he goes.

Nothing seems to daunt this bold little bird, which is equally at home in the fresh air of the country farm, in the midst of a crowded city, or among the strange sights and sounds of a large railway-station ; treating with equal indifference the slow-paced waggon horses, as they deliberately drag their load over the country roads, the noisy cabs and omnibuses as they rattle over the city pavements, and the snorting, puffing engines, as they dash through the stations with a velocity that makes the earth tremble beneath their terrible rush.

Although its ordinary food consists of insects and grain, both of which articles it can only obtain in the open country, it accommodates itself to a town life with perfect ease, and picks up a plentiful subsistence upon the various refuse that is thrown daily out of town houses, and which, before it is handed over to the dustman, is made by the Sparrow to yield many a meal. Indeed, the appetite of this bird is so accommodating, that there is hardly any article of human diet on which the Sparrow will not feed; and it may often be seen sharing with its family a dry crust of bread, some fragments of potatoes, any of the refuse of a greengrocer's shop, or even sitting upon a bone and picking it with great contentment. In market-places, especially in those where vegetables are sold, as Covent Garden and Farringdon Market, the Sparrow appears in great force, and, in no way daunted by the multitudes of busy human beings that traverse the locality, flutters about their very feet, and feeds away without displaying the least alarm.

In the Zoological Gardens, and indeed in all large aviaries, the Sparrow is quite in its element, pushing its way through the meshes of the wire roofs and fronts, pecking at the food supplied to the birds within, and retreating through the wires if attacked by the rightful owners of the plundered food. Even the majestic eagle is not free from the depredations of the Sparrow, who hops through the bars of the cage with great impudence, feeds quite at its leisure on the scraps of meat that are left by the royal bird, and, within a yard of the terrible beak and claws, splashes about merrily in the eagle's bath. The large animals are also favoured by constant visits from the Sparrows, which hop about the rhinoceros, the elephant, the hippopotamus, or the wild swine with utter indifference, skipping about close to their feet, and picking up grain as if they were the owners of the whole establishment.

When in the country the Sparrow feeds almost wholly on insects and grain, the former being procured in the spring and early summer, and the latter in autumn and winter. As these birds assemble in large flocks and are always very plentiful, they devour great quantities of grain, and are consequently much persecuted by the farmer, and their numbers thinned by guns, traps, nets, and all kinds of devices. Yet their services in insect-killing are so great as to render them most useful birds to the agriculturist. A single pair of these birds have been watched during a whole day, and were seen to convey to their young no less than forty grubs per hour, making an average exceeding three thousand in the course of the week. In every case where the Sparrows have been extirpated, there has been a proportional decrease in the crops from the ravages of insects. At Maine, for example, the total destruction of the Sparrows was ordered by Government, and the consequence was that in the succeeding year even the green trees were killed by caterpillars, and a similar occurrence took place near Auxerre.

Moreover, even in the autumn the Sparrow does not confine itself to grain, but feeds on various seeds, such as the dandelion, the sow-thistle, and the groundsel, all of which plants are placed by the agriculturist in the category of weeds. It has also been observed to chase and kill the common white butterfly, whose caterpillars make such terrible destruction among the cabbage and other garden plants. While feeding, the Sparrows like to be in company, so that they may be seen in bands of variable numbers, all fluttering, and chirping, and pecking, and scolding, and occasionally fighting with amusing pertness. So closely do they congregate, that when I was a boy I used often to shoot the Sparrows with sixpenny toy cannons, by pushing their muzzles through holes bored in the stable-door, pointing them at a little heap of oats thrown there for the occasion, and firing them as soon as a flock of Sparrows had descended to feed.

The Sparrow is not one of the earliest risers among the feathered tribes, but it is quite as wakeful as any of them, beginning to chatter almost with the dawn, and keeping up an animated conversation for nearly an hour before it leaves its roost. The same conduct is observable at night, the birds congregating together before roosting, settling on the tops of houses or in trees, and chattering in a very quarrelsome and noisy fashion before they

SPARROW.—*Passer domésticus.*

retire to their several domiciles. One of my pupils used to catch numbers of Sparrows by hiding in a haystack, watching them home, and drawing them out of their holes after dark.

The nest of the Sparrow is a very inartificial structure, composed of hay, straw, leaves, and various similar substances, and always filled with a prodigious lining of feathers. For, although the Sparrow is as hardy a bird as can be seen, and appears to care little for snow or frost, it likes a warm bed to which it may retire after the toils of the day, and always stuffs its resting-place full of feathers, which it gets from all kinds of sources. Even their roosting-places are often crammed with feathers.

Generally the nest is built in some convenient crevice, such as a hole in an old wall, especially if it be covered with ivy; but the bird is by no means particular in the choice of a locality, and will build in many other situations. Every one who has watched a rookery will have observed the numbers of Sparrows' nests that have been built under the nests of the larger birds, so as to obtain a shelter from rain; and many country house-keepers have learned to their sorrow how fond the Sparrow is of building in water-spouts, thereby choking up the passage, and causing the house to be overflowed. There are generally five eggs, though they sometimes reach the number of six, and their colour is greyish white, profusely covered with spots and dashes of grey-brown. They are, however, extremely variable, and even in the same nest it is not uncommon to find some eggs that are almost black with the mottlings, while others have hardly a spot or stripe about them. The Sparrow is a very prolific bird, bringing up several broods in the course of a season, and has been known to rear no less than fourteen young in a single breeding season.

Sometimes, but very rarely, the Sparrow takes to the trees, and builds a big clumsy nest among the branches, employing a profusion of hay for the exterior, which in that case is always domed, with an entrance in the side, and a great mass of warm feathers for the lining. The reader must remember in this place that he is not to confound this nest with that of the tree Sparrow, the bird next upon our list. The Sparrow is a very affectionate parent, and may often be seen, even in our crowded streets, busily engaged in feeding its young, which sit on the ground, opening their beaks and flapping their wings with hungry eagerness.

Like other familiar birds, the Sparrow sometimes builds its nest in very unsuspected localities, and there are several examples of their nests being placed on different parts of a ship's rigging. For example, while the *Great Britain* was lying in the Sandown graving dock, some Sparrows built two nests in the "bunts" of the main and mizen topsails, *i.e.* the place where the sail is gathered up into a bundle near the mast. As the sail could not be set without disturbing the birds, the sailors augured a speedy and pleasant voyage. Mr. Thompson gives an instance of the Sparrow building upon the furled sail of the *Aurora* of Belfast, but as the sail was loosened during the second voyage to Glasgow, the nest was destroyed and the eggs broken. Again, a pair of Sparrows built their nest under the slings of the fore-yard of the ship *Ann of Shields* just before leaving port, and when the vessel reached the Tyne the birds went ashore and brought back materials wherewith to complete their home.

The colouring of the Sparrow is really rich and pretty, though not brilliant. The top of the head is slaty grey, the chin and throat are black, and the same hue extends from the base of the head over the eyes in a slightly curved line. The sides of the head, the neck, back, and wings are rich ruddy brown, beautifully mottled with black, each feather having a deep black centre. Some of the smaller wing-coverts are tipped with white, the breast is grey-brown, mottled with black, and the abdomen is greyish white. Varieties of pure white, pied, black and dun are not uncommon. The total length of the bird is about six inches. These beautiful shades of colour are only to be seen in the country Sparrows, as the feathered inhabitants of streets soon become tinged with a uniform dingy brown by the smoky atmosphere in which they live.

The TREE SPARROW may readily be distinguished from the preceding species by the chestnut head, the triangular patch of black on the cheeks, and the browner white of the lower surface of the body.

This bird is not nearly so common as the house Sparrow, and generally places its nest in trees in preference to thatch and walls. Sometimes, however, it follows the common Sparrow in the building of its domicile, and has been known to place its nest in the deserted home of a crow or rook, making a dome like that of the common Sparrow when building in trees. Occasionally it has been observed to build its nest in the hollow of a tree, and to take possession of a hole that had formerly been occupied by the woodpecker. The eggs are different in hue from those of the common Sparrow, being dullish white, covered entirely with very light dots of ashen brown. Their number is generally from four to six.

In the *Field* newspaper there is a short communication from a gentleman residing at Penzance. "A Norwegian brig put into Penzance a few days since, and among other incidents of the voyage between Norway and England, the master of the vessel mentioned that midway between the two countries, thousands of small Sparrows paused and alighted on the ship, covering the deck, and rigging. The birds were exhausted and soon died, and some half-dozen were kept from mere curiosity to show to friends. These were brought for my inspection, a day or two since, by a person who begged them of the captain to show me. The six specimens were all *Passer montanus*, the Tree Sparrow, the Mountain Sparrow of Bewick."

Beside the markings which have already been mentioned, the Tree Sparrow has a streak of white, marking the boundary between the chestnut of the neck and the red hue of the back and wings. The lower wing-coverts are not so broadly tipped with white as in the common species, but are of a deep black, with a very narrow edging of white.

TREE SPARROW.—*Passer montánus.*

Below the eye and over the ear-coverts there is a narrow black streak, and the breast and abdomen are white, with a brown tinge, deepening on the flanks. In size the Tree Sparrow is not so large as the common species, by nearly half an inch of length.

The WHITE-THROATED SPARROW is an inhabitant of America, and is one of the partial migrators, passing to and from the northern and southern portions of that continent, according to the season of the year.

Of this bird Wilson speaks as follows : " This is the largest as well as the handsomest of all our Sparrows. It resides in most of the States south of New England. From Connecticut to Savannah I found these birds numerous, particularly in the neighbourhood of the Roanoke river and among the rice plantations. In summer they retire to the higher inland parts of the country, and also farther south, to breed. According to Pennant, they are also found at that season in Newfoundland. During their residence here in winter they collect together in flocks, always preferring the borders of swampy thickets, creeks, and mill-ponds, skirted with alder bushes and long rank weeds, the seeds of which form their principal food.

Early in the spring, a little before they leave us, they have a few remarkably sweet and clear notes, generally in the morning a little after sunrise. About the twentieth of April they disappear, and we see no more of them until the beginning or second week of October, when they again return, part to pass the winter with us, and part on their return farther south."

The colouring of this bird is very graceful. The upper surface of the body and the lower wing-coverts are rather agreeably mottled with black, ashen brown, bay, and clear

WHITE-THROATED SPARROW.—*Zonotrichia albicollis*

ash, the breast is ash, and the chin and the under portions of the body are pure white. The head is striped with black and white, and another white streak which passes over the eye warms into orange-yellow between the eye and the nostril. The female is easily distinguished by the lighter breast, the drab wash upon the white, and the smaller size of the orange line on the head. The legs are flesh-coloured, and the bill has a bluish tinge. The total length of the White-throated Sparrow is about six and a half inches.

The SHARP-TAILED FINCH derives its popular and appropriate title from the peculiar shape of its tail, which can be readily understood by reference to the illustration.

It is an interesting little bird, remarkably swift of foot, and a very excellent climber of reeds and rushes, two accomplishments which are very seldom combined in the same species. The sea-shore is the favourite haunt of this bird, which seems to depend wholly upon the waves for its subsistence. While feeding, it courses along the edge of the water with wonderful celerity, pecking here and there at the little fish and crustaceans which have been flung ashore by the water, and would make good their escape were not they interrupted by the ready beak of their destroyer. As it trips over the sands it has all the appearance of the sandpipers and other shore-living birds, although its legs are shorter and its dimensions smaller.

The low coral-covered islands that edge the Atlantic coast of America are the favoured resorts of the Sharp-tailed Finch, which seldom quits these places of safety, unless driven by continuous and wild easterly gales, which drive the sea over the islands and render them untenable for the time. The bird then flies over to the main land, but still remains close to the sea, preferring to roost on the ground and run about after dark. On examining the stomach of several of these birds, Wilson found that they contained fragments of shrimps, very small molluscs, and broken limbs of small crabs, no other substances ever being found in their interior. Owing to this diet, the flesh of this species is not at all fitted for the table, being rank and fishy.

The crown of the head is olive-brown divided laterally with a streak of slaty blue or light ash. The head and sides of the face are marked with several streaks of white, one of which becomes orange-yellow near the beak. The whole of the upper parts are brownish olive with a perceptible blue wash, the chin and abdomen are pure white, the breast is ashen grey streaked liberally with buff, and the under tail-coverts are buff streaked with black. All the wing-coverts are tipped with narrow white bands, and the wings are rather richly variegated with yellow. The total length of this bird is rather more than six inches.

THE Buntings are known by their sharp conical bills, with the edges of the upper mandible rounded and slightly turned inwards, and the knob on the palate. They are common in most parts of the world, are gregarious during the winter months, and in some cases become so fat upon the autumn grain that they are considered great dainties.

ONE of the most familiar of all these birds is the YELLOW BUNTING, or YELLOW AMMER as it is often called.

This lively bird frequents our fields and hedgerows, and is remarkable for a curious mixture of wariness and curiosity, the latter feeling impelling it to observe a traveller with great attention, and the former to keep out of reach of any missile. So, in walking along a country lane, the passenger is often preceded by one or more of these birds, which always keeps about seventy or eighty yards in advance, and flutters in and out of the hedges or trees with a peculiar and unmistakeable flirt of the wings and tail. It possesses but little song, and is consequently of no value as a cage-bird, remaining scathless while many a poor goldfinch, lark, or thrush falls a victim to the birdcatcher, and passes the remainder of its life cooped in the narrow precincts of a cage.

The song—if it may so be called—of the bird is set in the minor key, and has a peculiar intonation, which is almost articulate, and is variously rendered in different parts of England. For example, among the southern counties it is well represented by

SHARP-TAILED FINCH.—*Ammodrómus caudaoútus.*

the words, " A little bit of bread and *no* cheese ! " the last syllable but one being strongly accented. In Scotland it assumes a sense quite in accordance with the character of its surroundings, and is supposed to say, " De'il, de'il, de'il *take* ye." So, in revenge for the sentiment by which the bird is supposed to be actuated, the rustics persecute the bright little creature most shamefully, killing the parents, breaking the eggs, and destroying the nests, whenever they can find an opportunity. Mr. Thompson says, that to his ears the cry of the Yellow Bunting is of a mournful character, in which opinion I cannot at all agree with him, having many a time been cheered by the odd little tones that were poured forth close to my ear.

The nest of the Yellow Bunting is generally placed upon or very close to the earth, and the best place to seek for the structure is the bottom of a hedge where the grass has been allowed to grow freely, and the ground has been well drained by the ditch. In rustic parlance, a "rough gripe" is the place wherein to look for the Yellow Ammer's nest. It is a neatly built edifice, composed chiefly of grasses, and lined with hair. The

YELLOW BUNTING, OR YELLOW AMMER.—*Emberiza citrinella.*

eggs are five in number, and their colour is white, with a dash of very pale purple, and dotted and scribbled all over with dark purple-brown. Both dots and lines are most variable, and it also frequently happens that an egg appears with hardly a mark upon it, while others in the same nest are entirely covered with the quaint-looking decorations Generally the nest is built later than that of most small birds, but there are instances when it has been completed and the five eggs laid as early as January or even December.

Both parents are strongly attached to each other and to their young, and during the last few days of incubation the mother bird becomes so fearless, that she will sit in her nest even when she is discovered, and in some instances has even suffered herself to be touched before she would leave her charge.

About the end of autumn, all the young birds have been fully fledged, and instead of haunting the hedgerows, they assemble in considerable flocks, and visit the fields in search of food. In the winter, should the weather be severe, they become very bold, and joining the sparrows, and other little birds, enter the farmyards and cultivated grounds, and endeavour to pick up a subsistence. When food is plentiful, the Yellow Ammer becomes very fat, and in some instances is killed for the table, being thought nearly as good as the celebrated ortolan, to which bird it is closely allied.

The reader may probably have remarked, that I have called the bird Yellow Ammer, and not Yellow Hammer, as is mostly the case. The correction is due to Mr. Yarrell, who well observes that, " I have ventured to restore to this bird what I believe to have been its first English name, Yellow Ammer, although it appears to have been printed Yellow Ham and Yellow Hammer from the days of Drs. William Turner and Merrett to the present time. The word Ammer is a well-known German term for Bunting, in very common use. Thus Bechstein employs the names Schnee-ammer, Grau-ammer, Rohr-ammer, Garten-ammer, and Gold-ammer, for our Snow Bunting, Corn Bunting, Reed Bunting, Ortolan or Garden Bunting, and Yellow Bunting. Prefixing the letter H to the word appears to be unnecessary, and even erroneous, as suggesting a notion which has no reference to any known habit or quality in the bird."

The general colour of this bird is bright yellow, variegated with patches of dark brown, and having a richly mottled brownish yellow on the back, with a decided warm ruddy tinge. The primary feathers of the wing are black, edged with yellow, and the remainder of the feathers throughout, with all the wing-coverts, are deep brown-black, edged with ruddy brown. The chin, throat, and all the under parts of the body are bright pure yellow, sobering into rusty brown on the flanks. The female is similarly marked, but is not so brilliant in her hues. The total length of the bird is about seven inches.

The ORTOLAN or GARDEN BUNTING is widely celebrated for the delicacy of its flesh, or rather for that of its fat; the fat of the Ortolan being somewhat analogous to the green fat of the turtle in the opinion of gourmands.

The Ortolan has occasionally been shot in England, but it is most frequently found on the continent, where its advent is expected with great anxiety, and vast numbers are annually captured for the table. These birds are not killed at once, as they would not be in proper condition, but they are placed in a dark room, so as to prevent them from moving about, and are fed largely with oats and millet, until they become mere lumps of fat, weighing nearly three ounces, and are then killed and sent to table. The net and decoy-bird are the means that are generally employed for their capture.

The nest of the Ortolan is placed on the ground, generally among corn, and upon a sandy soil, where some slight defence helps to conceal the nest, and to afford a partial shelter from the wind. The materials of which it is made are grasses of different degrees of fineness, and a few hairs which are placed in the interior. The number of eggs is five or six, and their colour is pale bluish white, covered with spots of black. The nest is generally begun in the early part of May. The Ortolan has no real song, its voice being limited to a few monotonous chirping notes.

The colouring of this bird is as follows: The head is grey with a green tinge, and the back is ruddy brown beautifully mottled with black. The wings are black, with brown edges to the feathers; the chin, throat, and upper portions of the breast are greenish-yellow; and the abdomen is warm buff. The total length of the Ortolan is rather more than six inches.

ORTOLAN.—*Emberiza hortulána.*

As the COMMON BUNTING is not so brilliant a bird as the Yellow Bunting, it is less noticed, though quite as plentiful.

It is a thick-set and heavily made bird, not being possessed of the elegant shape which is found in its yellow relative. During the spring and summer, the Bunting is generally found in the corn-fields, from which habit it is sometimes termed the Corn Bunting, and is but seldom seen among trees or on open pasture-lands. Its food chiefly consists of various grass seeds, especially those of the stronger species, so that it often does good service to the farmer, by preventing the increase of these very stubborn weeds. The millet is a very favourite article of food, as may be supposed from the specific name of *miliaris,* which has been given to the bird by systematic zoologists, and considerable injury is often done to the millet crops by the attacks of the Bunting.

When hard pressed for food, it is capable of doing no small damage, as may be seen from Mr. Knapp's "Journal of a Naturalist." "I was this day (January. 25) led to reflect upon the extensive injury that might be produced by the agency of a very insignificant instrument, in observing the operations of the common Bunting, a bird that seems to live principally, if not entirely, on seeds, and has its mandibles constructed in a very peculiar manner, to aid this established appointment of its life. In the winter season it will frequent the stacks in the farmyard, in company with others, to feed upon any corn that may be found scattered about; but, little inclined to associate with man, it prefers those situations which are most lonely and distant from the village.

BUNTING.—*Emberiza miliaris.*

It would hardly be supposed that this bird, not larger than a lark, is capable of doing serious injury; yet I this morning witnessed a rick of barley standing in a distant field, entirely stripped of its thatching, which this Bunting effected by seizing the end of the straw, and deliberately drawing it out, to search for any grain the ear might yet contain; the base of the rick being entirely surrounded by the straw, one end resting against the ground, the other against the mow, as it slid down from the summit; and so completely was the thatching pulled off, that the immediate renewal of the cover became necessary. The sparrow and other birds burrow into the stack, and pilfer the corn, but the deliberate operation of unroofing the edifice appears to be the habit of this Bunting alone."

During the autumn and winter, the Buntings assemble in flocks, and as they get very fat about that time, the young birds are thought to be great delicacies, and are killed in great numbers for the table, being often sold under the name of larks, from which, however, they may be easily distinguished, by the short bill with the knob on the palate. In some parts of England these birds are captured alive, and fattened with millet for the table. The roosting-place of the Bunting is generally in thick, low bushes, but it sometimes rests for the night in stubble fields, where it is caught, together with the lark, in the nets used for the purpose.

The Bunting has no song, its voice being a harsh and rather jarring scream, which it is very fond of uttering, especially when it is observed or anxious, so that it frequently betrays the position of its home through its very desire to conceal it. The nest of this bird is a tolerably made edifice, composed of grasses, roots, and hair, the latter substance being used as the lining. It contains about five eggs, of a grey-white, tinged with red or purple, and covered with spots and streaks of very dark purplish brown. The usual locality for the nest is among coarse rank herbage or bramble-bushes, and it is seldom more than a few inches from the ground. The nest is completed about the middle of April, and the young birds are in the habit of leaving their home before they can fly, and ramble among the herbage, where they are fed by their parents until they can get their own living.

The general colour of the Bunting is somewhat like that of the lark, so that in some parts of England it goes by the name of Lark Bunting. The upper part of the body is hair-brown, mottled with a darker hue in the centre of each feather, and taking a yellowish tinge on the wing-coverts and quill-feathers. The chin, throat, breast, and abdomen are whitish brown, covered with longitudinal streaks of dark-brown, of a conical shape on the breast, and linear upon the flanks. The total length of the bird is about seven inches.

The BLACK-THROATED BUNTING is a native of America, and is rather less than the preceding species. Of this bird and its habits, Wilson writes as follows.

"They arrive in Pennsylvania, from the south, about the middle of May, descend in the neighbourhood of Philadelphia, and seem to prefer level fields covered with rye grass, timothy or clover, where they build their nest, fixing it on the ground, and forming it of fine dried grass. The female lays five white eggs, sprinkled with specks and lines of black. Like most part of their genus, they are nowise celebrated for musical powers. Their whole song consists of five notes, or more properly of two notes, the first repeated twice, and slowly, the second thrice and rapidly, resembling 'chip-chip, che-che-che.' In their shape and manner they very much resemble the yellow ammers of Britain; like them, they are fond of mounting to the top of some half-grown tree, and there chirruping for half an hour at a time.

In travelling through different parts of New York and Pennsylvania in spring and summer, whenever I came to level fields of deep grass, I have constantly heard these birds around me. In August they become

BLACK-THROATED BUNTING.—*Euspiza Americâna.*

mute, and soon after, that is towards the beginning of September, leave us altogether."

The top of the head is greenish yellow, the neck is dark ashen grey, and the back rusty red, touched with black, the same colour extending to the wings and tail, but of a darker hue, without the black spots. The chin is white, and the throat is marked with a heart-shaped patch of deep black edged with white. The breast is yellow, and a line of the same hue extends over the eyes and into the lower angle of the bill. The lesser coverts are bay, and the abdomen greyish white. The total length of the bird is about six inches and a half.

The LAPLAND BUNTING, SNOW BUNTING, or SNOW FLECK is one of our winter visitors, and is known by a great variety of names owing to the manner in which its plumage is coloured, according to the time of year or age of the individual. In some places it is called the Tawny Bunting, White Lark, or Pied Finch; in others the Mountain Bunting, because it is usually found upon the hilly ranges of the counties which it frequents.

It is an interesting bird, and has engaged the attention of almost every practical ornithologist. It generally arrives in the northern regions of Great Britain at the end of autumn, and remains during the winter; the oldest birds always leaving last and keeping towards the north, while the young birds arrive first, and go farther southward than their elderly relatives. They generally congregate in little flocks, and may be seen scudding over the snow-clad hills, their black wings and tail contrasting strangely with the pure white surface over which they pass. Colonel Montague once shot more than forty out of the same flock, and found that there were hardly any two specimens whose plumage was precisely alike, the feathers varying from the tawny hue of the young bird to the pure white and black of the adult in full winter dress.

LAPLAND BUNTING.—*Plectróphanes nivális.*

While treating of this bird, Mudie gives the following interesting remarks. "There is another trait in the natural history of birds, which, although it may be observed in them all, resident as well as migrate, is yet so conspicuous in the Snow Bunting that this is the proper place for noticing it. The male is the most sensitive to heat, and the female to cold. That difference appears, whether the result of the action of heat be change of place or change of plumage. The males of all our summer birds arrive earlier than the females, and in all resident birds the change of plumage and voice of the male are among the first indications of the spring, taking precedence of most of the vegetable tribes, for the redbreast and the wren sing before the snowdrop flowers appear.

It seems, too, that the song and the attractions of the male are accessories in aid of the warmth of the season, to produce the influence of the season upon the female; and even as the season advances, the female remains a skulking and hideling bird throughout the season, at least until the young have broken the shell and require her labour to feed, and her courage (which she sometimes requires to a wonderful degree at this time) to protect them. Whether it be that instinct leads the female to husband her heat for the purpose of hatching her eggs, or simply that the thinning of the under plumage, which takes place at that time, is the more conspicuous the more closely that the bird sits, it is certain that the females of most birds avoid the sun, and that all cover their eggs from the light during the period of incubation."

Wilson says of this species that it makes its appearance in the northern states early in December, coming in flocks of different sizes, and flying closely together at some little elevation from the ground. They seem to be restless in their disposition, seldom staying long in one spot, and resuming their flight after a short repose. The nest of the Snow Bunting is made in the most retired mountainous districts, and is placed in the cleft of a rock at some distance from other habitations of the same species. It is built of grass and feathers, and is lined with down or the fur of different quadrupeds; the fox and the hare being the most usual. The number of eggs is five, and the colour is white spotted with brown.

The song of the Lapland Bunting is feeble but pleasing, and is continually uttered while the bird is sitting near its nest. There are, besides, several notes peculiar to this bird; one, a sweet, short call, and the other a harsh ringing scream of alarm. In several countries this bird is valued for its flesh, which when it is fat is thought to be very delicate, and in Greenland it is captured in great quantities and dried; the Laplanders have an idea that it fattens on the flowing of the tide and grows lean on the ebb.

The food of this bird is rather various, but greatly consists of seeds. According to Wilson, it "derives a considerable part of its food from the seeds of certain aquatic plants, which may be one reason for its preferring those remote northern countries, so generally intersected with streams, ponds, lakes, and sheltered arms of the sea, that probably abound with such plants. In passing down the Seneca river towards Lake Ontario, late in the month of October, I was surprised by the appearance of a large flock of these birds feeding on the surface of the water, supported on the top of a growth of weeds that rise from the bottom, growing so close together that our boat could with great difficulty make its way through them. They were running about with great activity; and those I shot and examined were filled, not only with the seeds of this plant, but with a minute kind of shell-fish that adheres to the leaves. In this kind of aquatic excursion they are, doubtless, greatly assisted by the length of their hind heel and claws. I also observed a few on Table Rock, above the falls of Niagara, seemingly in search of the same kind of food."

As has already been noticed, the plumage of the Lapland Bunting varies greatly in its colouring, passing through every imaginable stage between the winter and summer dress. The winter plumage of this bird is briefly as follows: The back and part of the wings are dark black-brown, and the whole of the remaining feathers are pure snowy white. In all cases the amount of black is very variable, and in some instances the entire plumage has been white. In the summer, the colour is a tawny brown, speckled with white, and the back is black, mottled with brown. The quill-feathers of the wing and tail are black, variegated with bay and white, and the under surface dull white, deepening into tawny on the flanks. The lenght of the bird is about seven inches.

Besides the examples already given, there are very many other species of Bunting, some inhabiting Great Britain, and others scattered over the surface of the globe, whose history is equally interesting, but cannot be given in a work of the present dimensions. The species, however, which have already been mentioned are good examples of the group, and will serve as types by which the character of the sub-families may be known.

THE Larks may be readily recognised by the very great length of the claw of the hind toe, the short and conical bill, and the great length of the tertiary quill-feathers of the wing, which are often as long as the primaries.

The first example of these birds is the well-known SKYLARK, so deservedly famous for its song and its aspiring character.

This most interesting bird is happily a native of our land, and has cheered many a sad heart by its blithe jubilant notes as it wings skyward on strong pinions, or flutters between cloud and earth, pouring out its very soul in its rich wild melody. Early in the spring the Lark begins its song, and continues its musical effort for nearly eight months, so that on almost every warm day of the year on which a country walk is practicable, the skylark's happy notes may be heard ringing throughout the air, long after the bird which utters them has dwindled to a mere speck, hardly distinguishable from a midge floating in the sunbeams.

The natural impulse of the bird to hurl himself aloft while singing is so powerful, that when kept in confinement it flings itself against the top of the cage, and would damage itself severely were not a piece of green baize strained tightly as a roof, so as to take away the shock of the upward spring. In a state of nature, the Skylark sometimes sings while on the ground, and has been seen to sit on the top of a post, and from that point of vantage to pour forth its light sparkling melody.

Although it is by no means a familiar bird, nor does it seek the society of human beings, it is marvellously indifferent to their presence, and exhibits no discomposure at the close vicinity of the labourer, springing from the ground close to his feet, and singing merrily as it passes by his face. When pressed by danger, it has even been known to place itself under human protection. A gentleman was once riding along a road in Northamptonshire, when a Skylark suddenly dropped on the pummel of his saddle, where it lay with outspread wings, as if wounded to death. When the rider tried to take it up, it shifted round the horse, and finally dropped under the legs of the horse, where it lay cowering, evidently smitten with terror. On looking up, the rider saw a hawk hovering above, evidently waiting to make its stoop, as soon as the Lark left her place of refuge. The Lark presently remounted the saddle, and taking advantage of a moment when the hawk shifted its position, sprang from the saddle, and shot into the hedge, where it was safe.

The following curious instance of a Lark's intelligence I had from the lady who was an eye-witness of the scene.

A pair of Larks had built their nest in a grass field, where they hatched a brood of young. Very soon after the young birds were out of the eggs, the owner of the field was forced to set the mowers to work, the state of the weather forcing him to cut his grass sooner than usual. As the labourers approached the nest, the parent birds seemed to take alarm, and at last the mother bird laid herself flat upon the ground, with outspread wings and tail, while the male bird took one of the young out of the nest, and by dint of pushing and pulling, got it on its mother's back. She then flew away with her young one over the fields, and soon returned for another. This time, the father took his turn to carry one of the offspring, being assisted by the mother in getting it firmly on his back ; and in this manner they carried off the whole brood before the mowers had reached their nest. This is not a solitary instance, as I am acquainted with one more example of this ingenious mode of shifting the young, when the parent-birds feared that their nest was discovered, and carried the brood into some standing wheat.

Mr. Yarrell, moreover, mentions that the Lark has been seen in the act of carrying away her young in her claws, but not on her back as in the previous instance. Perhaps the bird would learn the art of carriage by experience, for the poor little bird was dropped from the claws of its parent, and falling from a height of nearly thirty feet, was killed by the shock. It was a bird some eight or ten days old. The Lark has also been known to carry away its eggs when threatened by danger, grasping them with both feet.

The nest of the Skylark is always placed on the ground, and generally in some little depression, such as the imprint of a horse's hoof, the side of a mole hill, or the old furrow of a plough. It is very well concealed, the top of the nest being only just on a level with the surface of the ground, and sometimes below it. I have known several instances where the young Larks would suffer themselves to be fed by hand as they sat in their nests, but the parent birds always seemed rather distressed at the intrusion into their premises. The materials of which it is made are dry grasses, bents, leaves and hair, the hair being generally used in the lining. It will be seen that the sober colouring of those substances renders the nest so uniform in tint with the surrounding soil, that to discover it is no easy matter. The eggs are four or five in number, and their colour is grey-yellow washed with light brown, and speckled with brown of a darker hue. They are laid in May and are hatched in about a fortnight.

The young birds are rather precocious, and leave the nest long before they are fully

SKYLARK.—*Alauda arvensis.*

fledged. Even when young, the sexes can be distinguished by the deep yellow of the breast and the more upright carriage. Dealers say that the most certain mode of ascertaining the sex of the Skylark is to lay it flat on its back, when, if it be a male, it will spread its tail like a fan.

The flesh of the Lark is very excellent, and thousands of these birds are annually captured and sent to market. Although it may seem a pity to eat a bird of such musical capacities, the Lark multiplies so rapidly that their numbers seem to suffer no perceptible diminution, and possibly their quick death at the hands of the bird-catcher may be a merciful mode of terminating their existence. The food of the Lark consists of grasshoppers, beetles, and other insects, worms, spiders, and various grubs, all of which it finds upon the ground. In the spring and autumn it varies its diet with vegetable food, eating young grass shoots in the spring, and seeds of the wheat in the summer.

The upward flight of this bird is rather remarkable, as it does not consist of a diagonal shoot like that of the pigeon, nor a succession of leaps like that of the eagle and hawk, but is a continual fluttering ascent, taking a spiral course, widening as the

WOODLARK.—*Alauda arbórea.*

bird rises into the air. The form of the spiral has been well described by comparing it to a spiral line wound around the exterior of an ascending column of smoke. Mudie suggests that the bird extends the diameter of the spiral in exact proportion to the sustaining power of the atmosphere, and remarks that while descending the Lark follows the same line which it had taken in its ascent.

During the spring and summer the Skylark lives in pairs, and is assiduously employed in attending to the wants of its family, of which it generally produces two broods in each season. Towards the end of autumn and throughout the winter the Larks become very gregarious, "packing" in flocks of thousands in number, and becoming very fat when snow should cover the ground, in which case they speedily lose their condition. These flocks are often augmented by the arrival of numerous little flocks from the continent, that come flying over the sea about the end of autumn, so that the bird-catchers generally reap a rich harvest in a sharp winter.

The colour of the Skylark is brown of different shades, mingled with a very little white and an occasional tinge of yellow. The feathers on the top of the head form a crest, and are dark brown with paler edges. The whole of the upper parts are brown mottled with a darker hue in the middle of each of them, the throat and upper part of the breast are greyish brown spotted with dark brown, and the abdomen is yellowish white deepening into pale brown on the flanks. The greater part of the tail is brown, dark in the centre of the feathers and lighter upon the edges, the two exterior feathers are white streaked with brown on the inner web, and the two next feathers are dark brown streaked with white on the outer web. The total length of this bird is rather more than seven inches.

ANOTHER species of British Lark is often mistaken for the preceding species, from which, however, it may be distinguished by its inferior dimensions, its shorter tail, and the light streak over the eye. This is the WOODLARK, so called on account of its arboreal tendencies and its capability to perch upon the branches of trees, a power which seems to be denied to the skylark. I have, however, seen one or two letters from persons who

PENCILLED LARK.—*Otócoris pencillátus.*

assert that they have seen the skylark singing in trees, and proved the truth of their assertion by shooting the songster.

The Woodlark is a very sweet singer, not so powerful in tune nor so various in phrase as the skylark, but so pleasing and gentle that by many persons it is preferred to its more powerful relative, and even thought to rival the nightingale. Like that bird, it will often sing long after the sun has set, perching upon the low branch of a tree, or circling in the air and pouring forth its soft plaintive strains during the warm summer evenings. This bird has the curious habit of fluttering to some tree-branch, singing fitfully the while, and after uttering a few low notes it darts from the branch, launches itself into the air, and rises nearly as high as the skylark, its song becoming louder in proportion to the height of its ascent, and sinking as it floats downward towards the ground. On account of the sweetness of the song, the Woodlark is in quite as great demand as the skylark, and is caged and fed after the same fashion.

The nest of the Woodlark is placed on the ground, and is composed of grasses, moss, and hair. The eggs are generally four or five in number, and their colour is pale reddish white covered with little red-brown spots. The bird builds very early, the nest being begun in March and the young birds hatched in May. The Woodlark is not so gregarious a bird as the skylark, being seldom seen in large flocks, the ordinary number being about eight to twelve, and generally being composed of the parents and their young. It is not so common as the skylark, but is by no means a rare bird in the locality which it best loves. It is seldom found in the open country, preferring cultivated lands in the more immediate neighbourhood of copses, woods, plantations, and thick hedges.

The general tint of the Woodlark is rather yellower than that of the skylark, and there is more red about the breast. The tail is differently coloured, the two exterior feathers being light brown, with a deeper patch on the inner web, the two central feathers uniform light brown, and all the others rather dark brown with a white spot at the extremity. The whole of the under surface is pale yellowish brown diversified on the breast by dark spots. The total length of the bird is a little more than six inches.

THERE is a curious genus of Larks called by the name of Otócoris, or Eared Larks, on account of the double pencil, or tuft of feathers, which they bear upon their heads, and which project on each side of the face like the pen of a lawyer's clerk from behind his ear.   Two species of this genus are now well known to ornithologists, the one being the PENCILLED LARK, and the other the Shore Lark.

The Pencilled Lark is a very rare bird, and has comparatively recently been introduced to science.   It is found in Persia, especially about Erzeroum, and is worthy of notice on account of the greatly developed pencils of dark feathers from which it derives its name. It is a prettily, though not brightly, coloured bird.   The upper part of the body is darkish ash, the wings and quill-feathers being of a brownish cast, with the exception of the external primaries, which are white.   The forehead, the chin, ear-coverts, breast, and abdomen are white, and the two projecting pencils are jetty black.   The top of the head and the nape of the neck are also ashen, but with a purple wash.   The tail is dark brown, with the exception of the two central feathers, which are dusky grey.

A CLOSELY allied species is sometimes seen in this country.   This is the SHORE LARK, a bird which has occasionally been seen, and of course killed, on our coasts, although its ordinary dwelling-place is in North America.   Of this bird Wilson speaks as follows :—

"It is one of our winter birds of passage, arriving from the north in the fall; usually staying with us the whole winter, frequenting sandy plains and open downs, and is numerous in the Southern States, as far as Georgia, during that season.   They fly high in loose, scattered flocks, and at these times have a singular cry, almost exactly like the skylark of Britain.

They are very numerous in many tracts of New Jersey, and are frequently brought to Philadelphia market.   They are then generally very fat, and are considered excellent eating.   Their food seems principally to consist of small round compressed seeds, buckwheat, oats, &c. with a large proportion of gravel.   On the flat commons, within the boundaries of the city of Philadelphia, flocks of them are regularly seen during the whole winter.   In the stomachs of them I have found, in numerous instances, quantities of the eggs or larvæ of certain insects, mixed with a kind of slimy earth.   About the middle of March they generally disappear, on their route to the north."

Forster informs us that they visit the environs of Albany first in the beginning of May, but go farther north to breed; that they feed on grass seeds and buds of the spring birch, and run into small holes, keeping close to the ground; from whence the natives call them "chi-chup-pi-sue."   The pencils which decorate the head of this bird are movable, and are raised or depressed at the will of their owner, thereby producing a very grotesque appearance.   It is a remarkable fact that when the bird is dead, they lie so closely among the other feathers, that they can with difficulty be distinguished.

THE well-known BULLFINCH is, perhaps, rather more familiar as a cage bird than as a denizen of the wood, for it is so remarkably shy and retiring in its habits that it keeps itself sedulously out of sight, and though bold enough in the pursuit of food, invading the gardens and orchards with considerable audacity, it yet has a careful eye to its own safety and seldom comes within reach of gunshot.

It cares little for open country, preferring cultivated grounds, woods, and copses, and is very fond of orchards and fruit-gardens, finding there its greatest supply of food.   This bird seems to feed almost wholly on buds during their season, and is consequently shot without mercy by the owners of fruit-gardens.   The Bullfinch has a curious propensity for selecting those buds which would produce fruit, so that the leafage of the tree is not at all diminished.   Although the general verdict of the garden-keeping public goes against the Bullfinch, there are, nevertheless, some owners of gardens who are willing to say a kind word for Bully, and who assert that its mischievous propensities have been much overrated.

It is true that the bird will oftentimes set hard to work upon a fruit-tree, and ruthlessly strip off every single flower-bud, thereby destroying to all appearance the prospects of the crop for that season.   Yet there are cases when a gooseberry-bush has thus been

BULLFINCH.—*Pyrrhula rubicilla.*

completely disbudded, and yet borne a heavy crop of fruit. The reason of this curious phenomenon may probably be, that some of the buds were attacked by insects, and that the kind of pruning process achieved by the Bullfinch was beneficial rather than hurtful to the plant.

The Bullfinch affords a curious instance of the change wrought by domestication.

In its natural state its notes are by no means remarkable, but its memory is so good, and its powers of imitation so singular, that it can be taught to pipe tunes with a sweet and flute-like intonation, having some of that peculiar "woody" quality that is observable in the clarionet. It is always captured very young for this purpose, and from the moment of its capture its instruction begins. The teacher keeps his birds separate, and always plays the tune to be learned upon some instrument, such as a bird-organ or a flageolet, as soon as he has given them their food. The latter instrument always turns out the best birds, as those which are taught with the bird-organ acquire that mechanical precision of note and total absence of feeling which renders the notes of a grinding organ so obnoxious to musical ears.

The birds are always apt to forget their lesson during the moulting season, and if they are permitted at that time to hear other birds, they pick up notes that are entirely foreign to the air which they are meant to perform, and so make a sad jumble. I once knew a piping Bullfinch, a very amusing bird, who had forgotten the first two or three bars of "Cherry-ripe," and always used to commence in a most absurd manner in the very middle of a phrase. He always finished with a long whistle, as of surprise, and then began to chuckle and hop about the table as if greatly charmed with his own performance. He had a great wish to teach me to pipe, and used to give me lessons every time I saw him. Sometimes I would purposely go wrong in the tune, when he would break off his piping, scold harshly, and begin afresh.

The Bullfinch is a remarkably tameable and loving bird, and is easily affected by predilection or dislike for different persons, generally holding fast by its first impulse. The bird which I have just mentioned was most absurd in the violence of his feelings. He was fond of scudding about on a bare mahogany table, and liked to lift up knitting-needles and let them fall, merely for the pleasure of hearing them rattle against the wood. But towards the lady to whom the said needles belonged he had an unappeasable enmity, and so jealous was he, that when she was working at the same table, she dared not touch her thread or scissors without looking to see whether Bully were near, for if he could do so he always dashed across the table and pecked her fingers, hissing loudly with anger, and all his feathers ruffled up.

Mrs. Webber gives a very interesting account of a Bullfinch which was in her possession, and which went the way of most pets, and perhaps a happier way than that which would have been travelled in their neglected state.

"THE loss of our pet, General Bem, was deeply felt. There was a sad vacancy in our house again which we did not soon expect to have filled. However, one morning, while I yet wept for General Bem, W—— came in with a small cage in his hand, containing an English Bullfinch.

'See!' said he, 'I have brought a fine Bullfinch to cheer you; he sings very sweetly several German airs, and it will fill Bem's place a little for you.'

'No, no, I cannot let him stay, no bird can take Bem's place, I do not want another bird to love; take him away.'

Poor little Bobby, I found him in the room of a rough fellow who did not care for him, and who gladly exchanged the sullen bird, as he called him, for some trinket. A little girl I saw there told me how sweetly he sang, and I determined to have him at any rate. 'Must I take the poor bird away? He will be so startled among my clamourers that he will not sing to me.'

'Well, let the fellow stay, though I assure you I cannot love him.'

So he hung the birdcage on a nail in my room, and I tried to turn my back upon him. I could not help observing, however, that he seemed to relish the glow of my wood fire and the warmth of the room greatly, and was commencing to dress his feathers, and to jump about in his little cage with quite a cheerful air.

I thought him at all events a sensible bird, and determined to give him a larger cage during the day. I then discovered that he had been so unfortunate as to lose three of his toes, perhaps in the struggles he had made when he had been taken prisoner, by means of the deceitful bird-limed twig, so that he was almost incapable of resistance, if one chose to catch him while in the cage, and then he would only crouch in a corner, and with his bright black eye and beseeching chirp pray to be left at peace.

For a week or more I took but little notice of him, only admiring his irresistible song, for he became so cheerful as to sing to us once or twice during the twenty-four hours.

One afternoon, however, I caught myself mimicking the droll whistle with which he would break his song, and which had precisely the sound we express by the whew-o-o-o when we make what we know to be some ludicrous mistake.

He instantly repeated it more slowly. I tried again and again till he seemed satisfied, and commenced the first bar of a strain of German music and then paused. I looked up, 'What, do you mean to teach me your song?'

He repeated the notes, and I essayed to reproduce them; my effort, however, seemed to amuse the young master, for he drew out to its fullest extent his whew-o-o-o-o, but instantly commenced the bar again. By this time I had become thoroughly interested, and not liking to be laughed at, made a more successful effort. This time Bob seemed more satisfied, and added a few more notes. When I had achieved these, he repeated all and put me to the test, and so on through his whole song; every few moments, however, evidently enjoying the fantastic mistakes which I made, and uttering his whistle in the most provokingly sarcastic tone. I was greatly amused, and related the story with great gusto on Mr. Webber's return.

The next morning when I came near the cage, the bird came as near me as he could and commenced a pleasant chirping, which evidently meant 'Good morning to you.' This I returned in tones resembling his as nearly as I could, and it finally ended by my taking the young gentleman into my hand and feeding him. He took his seeds from my fingers from that time every morning, for two or three weeks. Then we were to leave C—— for some time, and I sent him back to W——, congratulating myself that I was yet heart-whole as far as Bobby was concerned.

In about a month we returned and we called to see the birds; what was my surprise when Master Bullfinch instantly descended from his perch to the corner of the cage nearest to my face, and after the first chirp of greeting commenced singing in a sweet undertone, hovering and turning, his feathers lifted, his eye gleaming, and his whole expression one of the most profound admiration for little me. I was quite heartless, only shrugging my shoulders and turning away.

But I do not know exactly how it came about; in a few weeks I had the painted finch and the Bullfinch quite domesticated in my room, and though I still said I did not love him, yet I talked a great deal to the bird, and as the little fellow grew more and more cheerful and sang louder and oftener each day, and was getting so handsome, I found plenty of reasons for increasing my attention to him, and then above all things he seemed to need my presence quite as much as sunshine, for if I went away, if only to my breakfast, he would utter the most piteous and incessant cries until I returned to him, when in a breath his tones were changed and he sang his most enchanting airs. He made himself most fascinating by his polite adoration, he never considered himself sufficiently well dressed, he was most devoted in his efforts to enchain me by his melodies. Art and nature both were called to his aid, until, finally, I could no longer refrain from expressing in no measured terms my admiration. He was then satisfied, not to cease his attention, but to take a step further; he presented me with a straw, and even with increased appearance of adulation.

From that time he claimed me wholly; no one else could approach the cage; he would fight most desperately if any one dared to approach, and if they laid a finger on me his fury was unbounded, he would dash himself against the bars of his cage, and bite the wires as if he would obtain his liberty at all hazards, and thus be enabled to punish the offender.

If I went away now he would first mourn, then endeavour to win me back by sweet songs. In the morning I was awakened by his cries, and if I but moved my hand, his moans were changed to glad greetings. If I sat too quietly at my drawing, he would become weary seemingly, and call me to him; if I would not come, he would say in gentle tone, 'Come here! come here!' so distinctly that all my friends recognised the meaning of the accents at once, and then he would sing to me. All the day he would watch me; if I was cheerful, he sang and was so gay; if I was sad, he would sit by the hour watching every movement, and if I arose from my seat I was called, 'Come-e-here;' and whenever he could manage it, if the wind blew my hair within his cage he would cut it off, calling me to help him, as if he thought I had no right to wear anything else than feathers, and if I would have hair it was only suitable to nest-building! If I let him fly about the room with the painted finch, he would follow so close on my footsteps that I was in constant terror that he would be stepped upon or lost in following me from the room. At last he came to the conclusion that I could not build a nest; I never seemed to understand what to do with the nice materials he gave me, and when I offered to return them, he threw his body to one side and looked at me so drolly from one eye that I was quite abashed. From that time he seemed to think I must be a very young creature, and more assiduously fed me at stated periods during the day, throwing up from his own stomach the half-digested food for my benefit, precisely in the manner of feeding young birds.

But I did not like this sort of relationship very much, and determined to keep it down, and forthwith commenced by coldly refusing to be fed, and as fast as I could bring my hard heart to do it, breaking down all the gentle bonds between us.

The result was sad enough; the poor fellow could not bear it. He sat in wondering grief; he would not eat. At night I took him in my hand and held him to my cheek; he nestled closely and seemed more happy, although his little heart was too full to let him speak. In the morning I scarcely answered his tender low call, 'Come-e-here!' but I sat down to my drawing, thinking if I could be so cold much longer to so gentle and uncomplaining a creature.

I presently arose and went to the cage. Oh, my poor, poor bird; he lay struggling on the floor; I took him out, I tried to call him back to life in every way that I knew, but it was useless; I saw he was dying, his little frame was even then growing cold within my warm palm. I uttered the call he knew so well; he threw back his head with its yet undimmed eye and tried to answer—the effort was made with his last breath. His eye glazed as I gazed, and his attitude was never changed; his little heart was broken. I can never forgive myself for my cruelty."

THOSE who desire to find the nest of the Bullfinch must search in the thickets and most retired parts of woods or copses, and they may, perhaps, find the nest hidden very carefully away in some leafy branch at no great height from the ground. A thick bush is a very favourite spot for the nest; but I have more than once found them in hazel branches, so slender that their weight has bent them aside. The eggs are very prettily marked with deep violet and purple-brown streaks and mottlings upon a greenish white ground, and are easily recognisable by the more or less perfect ring which they form round the larger end of the egg. The eggs are generally five in number.

The parents are very fond of their young, and retain them through the autumn and winter, not casting them off until the next breeding season. The families assemble together in little flocks only five or six in number, and may be seen flying about in company, but never associating with birds of any other species.

In confinement it is a very jealous and withal a most combative bird, not easily daunted, and fighting with its fellow-prisoners till one or the other is vanquished or even killed. These birds have been known to fight continually with other inhabitants of the same cage, and even to kill the goldfinch in spite of his long pointed bill and high spirit. Many persons who keep Bullfinches find their plumage getting gradually darker until at last it assumes a black hue. This change of colour is mostly produced by two causes—one the confinement in a smoky atmosphere, and the other the presence of hemp-seed in the food. Hemp-seed when too liberally given has often this effect upon the cage-birds, and even the light colours of the goldfinch will darken into dingy black and brown under its influence. The reason of so curious a phenomenon is not known, but it is virtually a problem which when solved may be of considerable value.

The colour of the adult male bird is as follows:—

The base of the neck and the back are beautiful slaty grey, which has been known to take a roseate hue. The top of the head, the greater wing-coverts, the upper tail-coverts, and the chin are jetty black, and the tips of the wing-coverts are snowy white, so that they form a bold white bar across the wing. The quill-feathers of the wing and tail are deep black with a perceptible violet lustre, and the sides of the head, the throat, breast, and abdomen are light and rather peculiar red with a slight chestnut tinge. As is the case with most birds, varieties are not uncommon. The bill is deep shining black.

The female is not so brilliantly coloured as her mate, the grey of the back being of a rather dingy cast, and the red of the under portions being of a purplish brown hue. Young birds are coloured like the female, except that the head is not black. The total length of the bird rather exceeds six inches.

The CROSSBILLS, of which three species are known to inhabit England, are most remarkable birds, and have long been celebrated on account of the singular form of beak from which they derive their name.

In all these birds, the two mandibles completely cross one another, so that at first sight the structure appears to be a malformation, and to prohibit the bird from picking up seeds or feeding itself in any way. But when the Crossbill is seen feeding, it speedily proves itself to be favoured with all the ordinary faculties of birds, and to be as capable of obtaining its food as any of the straight-beaked birds.

The food of the Crossbill consists almost, if not wholly, of seeds, which it obtains in a very curious manner. It is very fond of apple-pips, and settling on the tree where ripe apples are to be found, attacks the fruit with its beak, and in a very few moments cuts a hole fairly into the "core," from which it picks out the seeds daintily and eats them, rejecting the ripe pulpy fruit in which they had been enveloped. As the Crossbill

CROSSBILL —*Loxia curvirostris*

is rather a voracious bird, the havoc which it will make in an orchard may be imagined.

Some persons say that the bird is able to cut an apple in two with a single bite; but I should fancy that in such cases the apple must be of the smallest and the bird of the largest, for it is hardly larger than the bullfinch, and the head is not at all disproportionate in length to the rest of the body.

This bird is also very fond of the seeds of cone-bearing trees, and haunts the pine-forests in great numbers. While engaged in eating, it breaks the cones from branches, and holding them firmly in its feet after the fashion of the parrots, inserts its beak below the scales, wrenches them away, and with its bone-tipped tongue scoops out the seed. They get their beaks under the scales by partially opening their mouths so as to bring the extremities of the bill immediately over each other, thus forming a kind of wedge. The points of the beak are then easily inserted like a wedge under the scales, and by suddenly drawing the lower mandible sideways, the scale is detached from the cone.

The power of the beak is quite extraordinary, as the bird evinces no difficulty in breaking open almonds while in their shell, and getting at the kernel. This feat is achieved by pecking a hole in the shell, pushing the point of the beak into the aperture, and then wrenching the shell asunder by a sudden turn of the bill. The apparently clumsy beak is thus shown to be an apparatus adapted in the most perfect manner to the wants of its owner, and to be capable, not only of exerting great force on occasions, but of picking up little seeds as well as could be done by a sparrow or a canary. Indeed, the bird can shell hemp and canary seed with perfect ease and readiness.

As might be gathered from the description of the habits of the Crossbill, the beak and all its attendants are of very great strength, the muscles on each side of the face being very conspicuous for their size and development. The position of the two mandibles is not at all uniform, nor does it depend, according to some persons, on the sex of the bird. Sometimes the upper mandible is turned to the right and the lower to the left, while in other individuals the reverse arrangement is followed. In either case, the lower mandible is that which is used for the wrenching asunder of the coverings which hide its food.

The Crossbill is not common in this country, although when it does make its appearance it generally comes over in flocks. Usually it consorts in little assemblies consisting of the parents and their young, but it has often been known to associate in considerable numbers. It is a very shy bird, and has a peculiar knack of concealing itself at a moment's notice, pressing itself closely upon the branches at the least alarm, and remaining without a movement or a sound to indicate its position until the danger has departed.

Mr. Yarrell mentions that on one occasion he had succeeded in shooting seven of these birds upon a tree, and as they still hung upon the boughs, one of the party volunteered to climb the tree in search of them. When he had got among the branches, a flock of eighteen or twenty Crossbills suddenly flew out, uttering a shrill sharp cry of alarm. Sometimes flocks of great extent have been noticed in England, upwards of a hundred individuals having been seen in a single flock.

In Sweden and Norway the Crossbill is a very common bird; and the north of Europe seems to be their proper breeding-place.

The nests are always placed in rather close proximity, so that if one nest is found, others are sure to be at no great distance. The nest is made of little fir-twigs, mosses, and wool, and is of rather a loose texture. It is always found upon the part of the branch that is nearest to the stem. The fir is the tree that is almost always if not invariably employed by this bird as the nesting-place. The eggs are generally three, but sometimes four in number, and are something like those of the greenfinch, but rather larger.

The nest is generally built at the end of February or the beginning of March, and the young are remarkable from the fact that their beaks are not crossed like those of the parents, but made much like those of any other young bird, the crossing not taking place until they are attaining an age and development which will enable them to shift for themselves. On one or two occasions the Crossbill has been known to make its nest in the British islands.

The colour of this bird is variable in the extreme, seeming to depend on external circumstances for its difference of tint and depth of hue.

The adult male assumes several varieties of tint, the plumage being coloured with red, yellow, or orange, which latter hue, as Mr. Yarrell well observes, is partly covered by the mixture of the other two. His description of the different kinds of plumage is very interesting.

" A red male now before me, that had completed his moult during his first autumn, has the back dull reddish brown, darkest in colour towards the tip of the upper mandible; the head, rump, throat, breast and belly tile-red; the feathers on the back mixed with some brown, producing a chestnut brown; wing-coverts, and quills, and tail-feathers nearly uniform dark brown.

A second male bird, killed at the same time as the red bird just described, has the head, rump, and under surface of the body pale yellow tinged with green, the back olive-brown; wings and tail-feathers like those of the red bird.

A third male, killed at the same time, has the top of the head and the back a mixture of reddish brown and dark orange: the rump reddish orange; the upper tail-coverts light orange; the chin, throat, and upper part of the breast red, passing on the lower part of the breast, belly, and sides to orange.

Red males that have moulted in confinement have changed during the moult to greenish yellow, and others to light yellow; thus apparently indicating that the yellow colour is that of the older livery; but young males, as before observed, certainly sometimes change at once to yellow, without going through either the red or the orange coloured stages. The lightest colours, whether green, yellow, red, or orange, pervade the feathers of the rump and the upper tail-coverts.

In captivity I have known several instances of red and yellow coloured specimens changing back to dull brown, as dark or even darker than their early plumage. This

CHILIAN PLANT-CUTTER.—*Phytótoma rara.*

might be the effect of particular food, which is known to exercise such an influence on other birds ; but whether having once assumed light tints, they ever in a wild and healthy state go back to olive brown or more dull colours, has not, I believe, been ascertained."

The young birds are dark green covered with horizontal dashes of black. They afterwards assume their yearling plumage, which is a general dull brown, greyish white on the head, and with the under surface of the body liberally streaked with a darker tint. The female is of a green-yellow, with a dash of brown on the top of the head and the upper surface of the body, changing into a purer yellow on the upper tail-coverts.

The total length of the male bird is rather more than six inches, and the female frequently reaches seven inches in length.

THE PLANT-CUTTERS derive their name from their habit of seizing the plants on which they feed, and nipping their stems asunder with their sharp bills as neatly as if they had been cut with shears. They are all of moderate size, about equalling the bullfinch in dimensions. In order to enable them to obtain their food, their beaks are very sharp and slightly notched.

The CHILIAN PLANT-CUTTER is rather a large species, being equal to a thrush in dimensions. It is a common bird in its native country, and is most destructive to the crops. It is very fond of sprouting corn, and, not content with eating the green blades, it seems to find such pleasure in the exercise of its bill, that it cuts down hundreds of stalks as if in mere wantonness, and leaves the green stems lying strewed about the ground. On account of these destructive propensities, it is greatly persecuted by the agriculturists, who shoot it and trap it, and further aid in its extermination by setting a price on its head, and giving a certain sum to every one who will bring in a dead bird.

The nest is made on the summit of a lofty tree in some very retired situation, so that in spite of all the persecution with which it meets, it still holds its ground against the farmers. In colour it is sober ; the usual tints being grey, with a bronze tinge on the back, and somewhat of a slaty hue upon the breast and abdomen. The quill-feathers of the wing and tail are black. Its voice is rather harsh, and consists of a series of rough broken notes.

2.   K K

SENEGAL COLY.—*Cólius macrocercus.*

THE COLIES form a small family of birds, whose exact place among the feathered tribes seems to be rather uncertain. They are inhabitants of Africa and India; and as their plumage is of a soft and silken character, and generally of sober tints, they often go by the name of Mouse-birds, a title which is also due to their mouse-like manner of creeping among the boughs of trees.

The SENEGAL or LONG-TAILED COLY is found in Africa, in the country from which it derives its name.

It is a pretty bird, and as it traverses the branches has a peculiarly elegant appearance; its long tail seeming to balance it in the extraordinary and varied attitudes which it assumes, and its highly movable crest being continually raised or depressed, giving it a very spirited aspect.

It is gregarious, living in little companies of four or five in number, and is continually jumping and running about the branches in search of its food, which consists of fruit and buds. The grasp of its feet is very powerful, as much so indeed as that of the parrot; and while traversing the boughs, it may often be seen hanging by its feet with its head downward, and occasionally remaining for some time suspended by a single foot. Le Vaillant says that this bird, in common with other members of the same family, is fond of sleeping in this singular attitude, and that in the early morning it may often be found so benumbed with cold, that it can be taken by hand before it can loosen its hold from the bough which it grasps so firmly.

Owing to the formation of its feet, which are almost wholly formed for grasping, it is seldom seen on the ground, and when it has alighted, is awkward in its movements. Among the boughs, however, it is all life and energy, leaping about with a quick vivacity that reminds the observer of our common long-tailed titmouse. In climbing from one branch to another, as in lowering themselves, the Colies frequently use their beaks to aid them, after the well-known practice of the parrots.

The nests of the Colies are all large and rounded, and are generally placed in close proximity to each other, five or six being often found on the same branch. The materials of which they are made are slender twigs externally, lined with mosses and soft feathers.

VIOLET PLANTAIN-EATER.—*Musóphaga violácea.*

The number of the eggs is from four to six. When fat and in good condition, the flesh is said to be delicate and tender. In size it is about equal to a blackbird.

The general colour of this species is a rather light chestnut-grey, brightening into ruddy fawn on the forehead. The crest is composed of fine and slender feathers. The nape of the neck takes a blue tint, and the back is grey, changing to slaty blue on the upper tail-coverts. The chin and the abdomen are pearl-grey, and the chest is of the same light ruddy fawn as the forehead. The beak is thick and sturdy, and is black at the tips, and brown towards the base.

ALLIED to the colies we find another curious and interesting group of birds called the PLANTAIN-EATERS.

These birds are natives of Africa, where they are not at all uncommon, and in the forests which they frequent may be seen flitting among the branches of the lofty trees, gliding among the boughs with great adroitness, and displaying their shining silken plumage to the best advantage. They are wary birds, and seem to have tolerably accurate ideas respecting the range of shot, for they mostly keep to the highest parts of the tree, and can but seldom be approached sufficiently near to be killed by the gun. Their food is almost wholly composed of fruits, and for feeding on such substances they are well suited by their large and peculiarly formed beaks.

They are all handsome birds, their dimensions averaging those of the European jay, and their plumage glancing with violet, green, purple, and red of different shades. One of the finest of the species is the Violet Plantain-eater, a bird which is found about Senegal and

WHITE-CRESTED TOURACO.—*Turacus albocristátus.*

the Gold Coast. It is remarkable for the extraordinary shape and dimensions of the beak, which is everywhere large and prominent, but is especially swollen towards the base, where it expands into a large shield-like mass of horny substance, which spreads over the forehead as far as the crown, where it terminates in a semicircular thickened line. The ridge of the beak is greatly arched, and its sides are much compressed. Its colour is equally singular with its shape, for it is of a fine golden yellow, passing into rich crimson on the upper part of the base.

The top of the head is crimson, not unlike that of the beak, and the feathers are very soft and fine, bearing a velvety or plush-like aspect. The general colour of the plumage is very deep violet, appearing black in the shade, and glossed with rich green in many lights. Part of the primary quills of all the secondaries are carmine, softening into delicate lilac, and tipped with deep violet. The large and powerful legs are black.

ANOTHER beautiful example of this group is the WHITE-CRESTED TOURACO.

This bird is remarkable not only for its handsome plumage, but for its peculiar customs. It is even more suspicious and wary than the previous species, and has a peculiar talent for concealing itself. Let a White-crested Touraco only take the alarm, and in a second of time it will be so well hidden that even a practised eye can scarcely obtain a clue to its whereabouts.

It is generally to be found among the branches of trees, and if it should be alarmed, and fly from one tree to another, it will vanish from sight so rapidly that the only way to get a shot at it is by sending some one up the tree to beat each bough in succession.

BLUE PLANTAIN-EATER —*Schizorhis gigantea.*

While traversing the branches, it runs along them, always keeping its body in the same line with the bough, so that if it fears any danger, it has only to crawl closely to the upper part of the bough to be quite imperceptible from below. Like the European creepers, or the squirrel, it often avails itself of the thick trunk of a tree to hide itself from a supposed enemy, slipping quietly round the trunk, and always keeping on the opposite side.

Some of these birds are extremely inquisitive, and, in spite of their native caution, will follow a traveller for miles ; keeping just out of gunshot, and screaming loudly the while.   The general colour of this species is olive-green above, except on the crest, which is also green, but of a lighter hue, and is edged with a delicate line of white.   The wings take a bluish purple tint, especially upon the primary quill-feathers, and there is a hori-zontal streak of pure white beneath each eye.   It is about as large as a common jackdaw.

The BLUE PLANTAIN-EATER, whose colour may be known by its popular title, is generally to be found on the lofty trees that skirt the edges of streams, either perched demurely on the boughs, or flitting rapidly through them in search of the fruits and insects on which it feeds.

HOATZIN.—*Opisthócomus cristátus.*

The wings of this species are but weak, and are unable to endure a lengthened flight. It is rather remarkable that this species should have two distinct modes of flight: the one —which is its most usual method—is by a succession of rapid and apparently laborious flappings; while the other is a graceful soar, in which the bird floats softly through the air with wings extended and motionless. It never employs its wings if it can avoid doing so, and even in making a short flight it avails itself of every opportunity of alighting, thinking, like the unfortunate people who live in the courts of royalty, that to sit whenever it gets a chance is the wisest course of conduct.

This duplicate kind of movement extends to its feet as well as its wings. Sometimes it will take a lazy fit, and will sit in a lumpish drowsy position, as if it were one of the slowest birds among the feathered tribes, its body all huddled up, and its head sunk between its shoulders. But when roused, it leaps in a single instant from this apathetic condition into graceful vivacity, every movement full of life and sparkling energy, traversing the boughs with wonderful speed, its head and neck being darted in every direction, like that of a snake, its crest rapidly raised and depressed, its eyes full of light, and its voice uttering loud and animated cries.

The voice of this and other Plantain-eaters is always of a loud character. It is quite as shy as its comrades, concealing itself in the same effective manner, and displaying more than ordinary precaution when in the vicinity of human habitations. The nest of this bird is made in the hollow of some decaying tree. The general colour of this bird is dark blue, marked with verditer green. The crest is almost black, the abdomen is greenish, and the thighs chestnut.

THE remarkable bird known by the name of HOATZIN, or CRESTED TOURACO, is the sole example of the family or sub-family, as the case may be, to which it belongs. Its exact place in the catalogue of birds is rather unsettled, some authors considering it to belong to the poultry, or the Gallinaceous birds, and others looking upon it as one of the true Passerines.

It is a very fine bird, being nearly as large as a peacock, and having somewhat of the same gait and mode of carriage. The peculiar construction of the foot, the outer toe of which cannot be turned backward, has induced zoologists of the present day to separate it from the plantain-eaters, and to consider it as a unique representative of a sub-family.

This bird is a native of tropical America, being found in Guiana and the Brazils, where it leads a gregarious life, assembling together in large flocks, on the banks of creeks and rivers. Although so closely resembling the Gallinaceous birds in general appearance and habits, its flesh is, fortunately for itself, quite uneatable, being impregnated with a strong and peculiar odour that deters any but a starving man from making a meal upon it. Perhaps this odour may be caused by its food, which consists almost wholly of the leaves of the arum.

The nest of the Hoatzin is made in the lower part of a tree, and is composed exteriorly of slender twigs, and interiorly of mosses and other soft substances. The eggs are about three or four in number, and their colour is greyish white, besprinkled with red spots. The head of this species is adorned with a tuft of elongated and narrow feathers. Its colour is brown above, striped with white, and the breast and throat are light brown washed with grey. The abdomen is deep chestnut, and the tail tipped with white. The bill is short, thick, very convex, and bent downwards at the tip.

THERE are many strange and wonderful forms among the feathered tribes; but there are, perhaps, none which more astonish the beholder who sees them for the first time, than the group of birds known by the name of HORNBILLS.

They are all distinguished by a very large beak, to which is added a singular helmet-like appendage, equalling the beak itself in some species, while in others it is so small as to attract but little notice. On account of the enormous size of the beak and the helmet, which in some species recede to the crown of the head, the bird appears to be overweighted by the mass of horny substance which it has to carry; but on a closer investigation, the whole structure is found to be singularly light, and yet very strong.

On cutting asunder the beak and helmet of a Hornbill, we shall find that the outer shell of horny substance is very thin indeed, scarcely thicker than the paper on which this description is printed, and that the whole interior is composed of numerous honey-combed cells, with very thin walls and very wide spaces, the walls of the cells being so arranged as to give very great strength when the bill is used for biting, and with a very slight expenditure of material. The whole structure, indeed, reminds us greatly of that beautiful bony network which gives to the skull of the elephant its enormous size and lightness, and which is fully described in the volume on Mammalia, page 733. The general appearance of the dried head of a Hornbill, with its delicate cellular arrangements, and its thin polished bony shell, is not unlike the well-known shell of the paper nautilus, and crumbles in the grasp almost as easily.

Five species of Hornbills are shown in the engraving on page 504. The upper figure is the common Rhinoceros Hornbill (*Búceros Rhinóceros*); the handsome, but smaller, bird on its left is the White-crested Hornbill (*Buceros albocristátus*). Of the two figures that occupy the middle of the drawing, the left bird represents the Crested

GROUP OF HORNBILLS.

Hornbill (*Buceros cristatus*), and that on the right is the Two-horned Hornbill (*Buceros bicornis*). The smaller species at the bottom is the Woodpecker Hornbill (*Buceros Pica*).

PERHAPS the greatest development of beak and helmet is found in the RHINOCEROS HORNBILL, although there are many others which have these appendages of great size, as may be seen by referring to the group of Hornbills.

RHINOCEROS HORNBILL.—*Búceros Rhinoceros.*

As is the case with all the Hornbills, the beak varies greatly in proportion to the age of the individual, the helmet being almost imperceptible when it is first hatched, and the bill not very striking in its dimensions. But as the bird gains in strength, so does the beak gain in size, and when it is adult the helmet and beak attain their full proportions. It is said that the age of the Hornbill may be known by inspecting the beak, for that in every year a wrinkle is added to the number of the furrows that are found on the bill.

The object of the huge helmet-like appendage is very obscure, but the probability is, that it may aid the bird in producing the loud roaring cry for which it is so celebrated. When at liberty in its native forests, the Hornbill is lively and active, leaping from bough to bough with great lightness, and appearing not to be in the least incommoded by its large beak. It ascends the tree by a succession of easy jumps, each of which brings it to a higher branch, and when it has attained the very summit of the tree, it stops and pours forth a succession of loud roaring sounds, which can be heard at a considerable distance.

The flight of the Hornbill is rather laborious, and performed by rapid flappings of the wings. While in the air the bird has a habit of clattering its great mandibles together,

WHITE-CRESTED HORNBILL.—*Búceros albocristátus.*

which, with the noise of the wings, produces a most weird-like sound in the forest depths, which is a fertile source of alarm to the timid traveller.

The food of the Hornbill seems to consist both of animal and vegetable matters, and Lesson remarks that those species which inhabit Africa live on carrion, while those that are found in Asia feed on fruits, and that their flesh acquires thereby an agreeable and peculiar flavour—something, we may presume, like that of the famous lamb fed upon pistachio nuts. Perhaps this statement may be too sweeping, and the birds of both continents may in all probability be able to eat both animal and vegetable food.

At all events, the enormous beak of the Rhinoceros Hornbill, which is one of the Asiatic species, appears to be made for the express purpose of destroying animal life, as is now known to be the case with the corresponding member of the toucan. It is hard to think that so formidable a weapon should be given to the Hornbill merely for the purpose of eating fruits ; and when we remember that many of the species are acknowledged to be carnivorous, and that the toucan employs its huge and similarly formed beak in the destruction of small quadrupeds and birds, it is but rational to suppose that the Hornbill acts often in a similar fashion.

One individual, a Concave Hornbill (*Buceros cavátus*), which was kept in captivity, was much more attached to animal than vegetable food, and, like the toucan, would seize with avidity a dead mouse, and swallow it entire, after squeezing it once or twice between the saw-shaped edges of its beak. The Rhinoceros Hornbill is said to be oftentimes extremely carnivorous in its habits, and to follow the hunters for the purpose of feeding upon the offal of the deer and other game which they may have killed.

While on the ground, the movements of the Hornbill are rather peculiar, for instead of walking soberly along, as might be expected from a bird of its size, it hops along by a succession of jumps. It is but seldom seen on the ground, preferring the trunks of trees, which its powerful feet are well calculated to clasp firmly.

The colour of the Rhinoceros Hornbill is as follows :- The general tint of the body is dusky black, changing to greyish white below. The feathers of the head and neck are long and loose, and more like hairs than feathers. The tail is of a greyish white, with a bold black band running across it near the extremity. The enormous bill is generally of a yellowish white colour. the upper mandible being of a beautiful red at its base, and the lower mandible black. The helmet is coloured with black and white. The length of the bill is about ten inches.

ANOTHER species of this curious group is the WHITE-CRESTED HORNBILL, a bird which is remarkable for the peculiarity from which it derives its name.

Although not nearly so large as the preceding species, it is a truly handsome bird, and, except by an ornithologist, would hardly be recognised as belonging to the same group as the rhinoceros Hornbill. Its beak, although very large in proportion to the rest of the bird, is not so prominent a feature as in the other Hornbills, and its beautiful white fan-shaped crest takes off much of the grotesque aspect which would otherwise be caused by the large bill. Very little of the helmet is visible in this species, as it is of comparatively small dimensions, and is hidden by the plumy crown which decorates the head. The tail is very long, and, as will be seen by the engraving, is graduated and coloured in a very bold manner, each feather being black except at the extreme tips, which are snowy white. The general colour of this bird is deep, dull black, through which a few very small white feathers protrude at distant intervals, the tail is black, each feather being tipped with white, and the crest is white, with the exception of the black shaft and black tip of each feather.

SCANSÓRES, OR CLIMBING BIRDS.

A LARGE group of birds is arranged by naturalists under the title of SCANSORES, or CLIMBING BIRDS, and may be recognised by the structure of their feet. Two toes are directed forward and the other two backward, so that the bird is able to take a very powerful hold of the substance on which it is sitting, and enables some species, as the woodpeckers, to run nimbly up tree-trunks and to hold themselves tightly on the bark while they hammer away with their beaks, and other species, of which the Parrots are familiar examples, to clasp the bough as with a hand. There is some little difficulty in settling the exact limits of this group, so I have preferred to accept the arrangement which has been sanctioned by the authorities of the British Museum.

THE very curious birds that go by the name of TOUCANS are not one whit less remarkable than the hornbills, their beak being often as extravagantly large, and their colours by far superior. They are inhabitants of America, the greater number of species being found in the tropical regions of that country.

Of these birds there are many species, of which no less than five were living in the Zoological Gardens in a single year. Mr. Gould, in his magnificent work, the "Monograph of the Rhamphastidæ," figures fifty-one species, and ranks them under six genera.

The most extraordinary part of these birds is the enormous beak, which in some species, such as the TOCO TOUCAN, is of gigantic dimensions, seeming big enough to give its owner a perpetual headache, while in others, such as the Toucanets, it is not so large as to attract much attention.

As in the case of the hornbills, their beak is very thin and is strengthened by a vast number of honeycomb-cells, so that it is very light and does not incommode the bird in the least. In performing the usual duties of a beak, such as picking up food and pluming the feathers, this apparently unwieldy beak is used with perfect address, and even in flight its weight does not incommode its owner.

COMMON TOUCAN.—*Rhamphastos Ariel.*

The beak partakes of the brilliant colouring which decorates the plumage, but its beautiful hues are sadly evanescent, often disappearing or changing so thoroughly as to give no intimation of their former beauty. The prevailing colour seems to be yellow, and the next in order is red, but there is hardly a hue that is not found on the beak of one or other of the species. As examples of the colouring of the beaks, we will mention the following species. In the Toco Toucan it is bright ruddy orange, with a large black oval spot near the extremity; in the Short-billed Toucan it is light green, edged and tipped with red; in the Tocard Toucan it is orange above and chocolate below; in the Red-billed Toucan it is light scarlet and yellow; in Cuvier's Toucan it is bright yellow and black, with a lilac base; in the Curl-crested Araçari it is orange, blue, chocolate, and white; in the Yellow-billed Toucan it is wholly of a creamy yellow, while in Azara's Araçari it is cream-white with a broad blood-red stripe along the middle. Perhaps the most remarkable bill of all the species is found in the Laminated Hill Toucan (*Andígena laminátus*), where the bill is black, with a blood-red base, and has a large buff-coloured shield of horny substance at each side of the upper mandible, the end next the base being fused into the beak, and the other end free. The use of this singular, and I believe unique, appendage is not known.

The flight of the Toucan is quick, and the mode of carrying the head seems to vary in different species, some holding their heads rather high, while others suffer them to droop. Writers on this subject, and indeed on every point in the history of these birds, are rather contradictory; and we may assume that each bird may vary its mode of flight or carriage in order to suit its convenience at the time. On the ground they get along with a rather awkward hopping movement, their legs being kept widely apart. In ascending a tree the Toucan does not climb, but ascends by a series of jumps from one branch to another, and has a great predilection for the very tops of the loftiest trees, where no missile except a rifle ball can reach him.

The voice of the Toucan is hoarse and rather disagreeable, and is in many cases rather articulate. In one species the cry resembles the word "Tucano," which has given origin to the peculiar name by which the whole group is designated. They have a habit of sitting on the branches in flocks, having a sentinel to guard them, and are fond of lifting up their beaks, clattering them together, and shouting hoarsely, from which custom the natives term them Preacher-birds. Sometimes the whole party, including the sentinel, set up a simultaneous yell, which is so deafeningly loud that it can be heard at the distance of a mile. They are very loquacious birds, and are often discovered through their perpetual chattering.

Grotesque as is their appearance, they have a great hatred of birds who they think to be uglier than themselves, and will surround and "mob" an unfortunate owl that by chance has got into the daylight with as much zest as is displayed by our crows and magpies at home under similar circumstances. While engaged in this amusement, they get round the poor bird in a circle, and shout at him so, that wherever he turns he sees nothing but great snapping bills, a number of tails bobbing regularly up and down, and threatening gestures in every direction.

In their wild state their food seems to be mostly of a vegetable nature, except in the breeding season, when they repair to the nests of the white ant which have been softened by the rain, break down the walls with their strong beaks, and devour the insects wholesale. One writer says that during the breeding season they live exclusively on this diet. They are very fond of oranges and guavas, and often make such havoc among the fruit-trees, that they are shot by the owner, who revenges himself by eating them, as their flesh is very delicate. In the cool time of the year they are killed in great numbers merely for the purposes of the table.

In domestication they feed on almost any substance, whether animal or vegetable, and are very fond of mice and young birds, which they kill by a sharp grip of the tremendous beak, and pull it to pieces as daintily as a jackdaw or magpie. One Toucan, belonging to a friend, killed himself by eating too many ball-cartridges on board a man-of-war. As the habits of most of these birds are very similar, only one species has been figured, for the description of other species would necessarily have been limited to a mere detail of colouring.

Mr. Broderip has given a very interesting account of an Ariel Toucan and its habits, which has been frequently quoted, but is so graphic a description that any work of this nature would be incomplete without it :—

"After looking at the bird which was the object of my visit, and which was apparently in the highest state of health, I asked the proprietor to bring up a little bird, that I might see how the Toucan would be affected by its appearance. He soon returned, bringing with him a goldfinch, a last year's bird; the instant he introduced his hand with the goldfinch into the cage of the Toucan, the latter, which was on a perch, snatched it with his bill. The poor little bird had only time to utter a short weak cry, for within a second it was dead; killed by compression on the sternum and abdomen, and that so powerful, that the bowels were protruded after a very few squeezes of the Toucan's bill.

As soon as the goldfinch was dead, the Toucan hopped with it still in his bill to another perch, and placing it with his bill between his right foot and the perch, began to strip off the feathers with his bill. When he had plucked away most of them, he broke the bones of the wings and legs (still holding the little bird in the same position) with his bill, taking the limbs therein, and giving at the same time a strong lateral wrench. He continued this work with great dexterity till he had almost reduced the bird to a shapeless mass; and ever and anon he would take his prey from the perch in his bill, and hop from perch to perch, making at the same time a peculiar hollow clattering noise; at which times I observed that his bill and wings were affected with a vibratory or shivering motion, though the latter were not expanded.

He would then return the bird to the perch with his bill, and set his foot on it; he first ate the viscera, and continued pulling off and swallowing piece after piece, till the head, neck, and part of the sternum, with their soft parts, were alone left. These, after a little more wrenching while they were held on the perch, and mastication, as it were, while they were held in the bill, he at last swallowed, not even leaving the beak or legs of his prey. The last part gave him most trouble; but it was clear that he felt great enjoyment, for whenever he raised his prey from the perch he appeared to exult, now masticating the morsel with his toothed bill, and applying his tongue to it, now attempting to gorge it, and now making the peculiar clattering noise, accompanied by the shivering motion above mentioned. The whole operation, from the time of seizing his prey to that of devouring the last morsel, lasted about a quarter of an hour; he then cleansed his bill from the feathers, by rubbing it against the perches and bars of his cage.

While on this part of the subject, it may be as well to mention another fact which appears to me not unworthy of notice. I have more than once seen him return his food some time after he had taken it into his crop, and after masticating the morsel for awhile in his bill, again swallow it; the whole operation, particularly the return of the food to the bill, bearing a strong resemblance to the analogous action in ruminating animals. The food on which I saw him so employed was a piece of beef, which had evidently been macerated some time in his crop. While masticating it, he made the same hollow clattering noise as he made over the remains of the goldfinch.

Previous to this operation he had examined his feeding trough, in which there was nothing but bread, which I saw him take up and reject, and it appeared to me that he was thus reduced from necessity to the above mode of solacing his palate with animal food. His food consists of bread, boiled vegetables, eggs, and flesh, to which a little bird is now added about every second or third day. He shows a decided preference for animal food, picking out all morsels of that description, and not resorting to the vegetable diet till all the former is exhausted."

When settling itself to sleep, the Toucan packs itself up in a very systematic manner, supporting its huge beak by resting it on its back, and tucking it completely among the feathers, while it doubles its tail across its back, just as if it moved on a spring hinge. So completely is the bill hidden among the feathers, that hardly a trace of it is visible in spite of its great size and bright colour, and the bird when sleeping looks like a great ball of loose feathers.

In the Toco Toucan the beak is of enormous size, being eight inches and a half long, forming rather more than one third of the entire length. Its colour is rich glowing orange, with a large oval patch near the tip, and a black line round the base. There are also a number of darker red bars upon the sides. The head and body are deep black, and the throat and cheeks are white, changing into brimstone-yellow on the breast, edged with a line of blood red. The upper tail-coverts are greyish white, and the under tail-coverts deep crimson. Around the eye is a large orange circle, within which is a second circle of cobalt-blue. The eye is rather curious, a green ring encircling the pupil, and a narrow yellow ring encircling the green.

In one species, the Curl-crested Araçair, the feathers of the head assume a most unique and somewhat grotesque form, reminding the observer of a coachman's wig dyed black. On the top of the head the shafts of the feathers, instead of spreading out into webs, become flattened, and are rolled into a profusion of bright shining curls, so that the bird really appears to have been under the tongs of the hairdresser. Indeed, it appears almost impossible that this singular arrangement of the feathers should not be the work of art.

PARROTS.

THE general form of the PARROTS is too well known to need description. All birds belonging to this large and splendid group can be recognised by the shape of their beaks, which are large, and have the upper mandible extensively curved and hanging far over the lower; in some species the upper mandible is of extraordinary length. The tongue is short, thick, and fleshy, and the structure of this member aids the bird in no slight degree in its singular powers of articulation. The wings and tail are generally long, and in some species, such as the Macaws, the tail is of very great length, while in most of the Parrakeets it is longer than the body.

THE first sub-family of this group is composed of those birds which are called by the title of Ground Parrakeets. In the generality of the Parrot tribe, the legs are short, but in these birds they are of greater length in order to enable them to run freely on the ground. One of the most striking examples of this little group is the PARRAKEET COCKATOO of Australia.

Although not clothed with the brilliant plumage that decorates so many of the Parrot tribe, this bird is a remarkably pretty one, and is worthy of notice not only for the curious crest with which its head is adorned, but for the grace and elegance of its form. With the exception of the head, on which a little crimson and yellow are seen, the plumage of the Parrakeet Cockatoo is simply tinted with brown, grey, and white; but these colours are so pure, and their arrangement so harmonious, that the eye does not at all look for brighter colouring.

It is mostly seen upon the ground, where it runs with great swiftness, and is very accomplished at winding its way among the grass stems, upon the seeds of which it subsists. It is by no means a shy bird, and will permit of a close approach, so that its habits can be readily watched. When alarmed, it leaves the ground and flies off to the nearest tree, perching upon the branches and crouching down upon them lengthwise so as to be invisible from below. There is no great difficulty in shooting it, which is a matter of some consequence to the hunter, as its flesh is notable for its tenderness and delicate flavour.

The eggs of this species are pure white, which is the case with Parrot eggs generally, and their number is from four to six.

Mr. Gould gives the following description of the Parrakeet Cockatoo :—

" The interior portion of the vast continent of Australia may be said to possess a fauna almost peculiar to itself, but of which our present knowledge is extremely limited. New forms therefore of great interest may be expected when the difficulties which the explorer has to encounter in his journey towards the centre shall be overcome. This

PARRAKEET COCKATOO.—*Nymphicus Novæ Hollandiæ.*

beautiful and elegant bird is one of its denizens. I have, it is true, seen it cross the great mountain ranges and breed on the flats between them and the sea ; still, this is an unusual occurrence, and the few thus found, compared to the thousands observed on the plains stretching from the interior side of the mountains, proves that they have, as it were, overstepped their natural boundary.

Its range is extended over the whole of the southern portion of Australia, and being strictly a migratory bird, it makes a simultaneous movement southward to within one hundred miles of the coast in September, arriving in the York district near Swan River in Western Australia precisely at the same time that it appears in the Liverpool plains in the eastern portion of the country. After breeding and rearing a numerous progeny, the whole again retire northwards in February and March, but to what degree of latitude towards the tropics they wend their way I have not been able satisfactorily to ascertain. I have never received it from Port Essington or any other port in the same latitude, which, however, is no proof that it does not visit that part of the continent, since it is merely the country near the coast that has yet been traversed. In all probability it will be found at a little distance in the interior wherever there are situations suitable to its habits, but doubtless at approximate periods to those in which it occurs in New South Wales.

It would appear to be more numerous in the eastern divisions of Australia than in the western. During the summer of 1839, it was breeding in all the apple-tree (*Angophora*) flats on the Upper Hunter as well as in similar districts on the Peel and other rivers which flow northward.

After the breeding season is over, it congregates in numerous flocks before taking its departure. I have seen the ground quite covered by them while engaged in procuring food; and it was not an unusual circumstance to see hundreds together in the dead branches of the gum-trees in the neighbourhood of water, a plentiful supply of which would appear to be essential to its existence; hence we may reasonably suppose that the interior of the country is not so sterile and inhospitable as is ordinarily imagined, and that it yet may be made available for the uses of man. The Harlequin Bronzewing and the Warbling Grass Parrakeet are also denizens of that part of the country, and equally unable to exist without water."

The head and throat of this species are yellow, and there is a patch of crimson on the ears. Upon the head there is a long, slender, painted crest, yellow at the base and grey at the tip, giving the bird so curious an aspect that at first sight it appears either to be a cockatoo or a Parrakeet as the eye is directed to the crest or the general form. The back and under portions of the body are brown, and a large part of the wings is white. The central tail-feathers are brown, and the rest grey. The female is distinguished from her mate by a green tinge which pervades the yellow of the head and throat, and the numerous bars of yellow and dark blackish brown which cross the tail.

YELLOW-BELLIED PARRAKEET.—*Platycercus Caledonicus.*

THE genus Platycercus, or Wide-tailed Parrakeets, to which the YELLOW-BELLIED PARRAKEET belongs, is a very extensive one, and numbers among its members some of the loveliest of the Parrot tribe. They all glow with the purest azure, gold, carmine, and green, and are almost immediately recognisable by the bold lancet-shaped feathers of the back, and the manner in which each feather is defined by its light edging and dark centre.

The Yellow-bellied Parrakeet inhabits the whole of Van Diemen's Land and the islands of Bass Straits, where it is very plentiful, and often so completely familiar as to cause extreme wonder in the mind of an Englishman who for the first time traverses the roads of

2. L L

this strange land, and finds the Parrakeets taking the place of the sparrows of his native country, quite as familiar and almost as pert, perching on the trees or fences, and regarding him with great indifference. But the novelty soon wears off, and before long his only emotions at the sight of a Parrot are hatred at its thieving propensities, and a great longing to eat it. As to this particular species, its flesh is cultivated for its delicacy and peculiar flavour, and Mr. Gould is so appreciative of its merits, that he waxes quite eloquent when speaking of Parrakeet pie.

These birds are gregarious, assembling in little companies, probably composed of the parents and their young, and haunting almost every kind of locality; trees, rocks, grass, fields, or gully, being equally in favour. They are excellent runners, getting over the ground with surprising ease and celerity; and there are few prettier sights than to behold a flock of these gorgeous birds, decked in all the varied beauty of their feathery garments, scudding over the ground in search of food, their whole movements instinct with vivacity, and assuming those graceful attitudes which are best suited for displaying the beauty of the colouring.

The food of these birds consists mostly of grass seeds, but they also feed upon the flowers of the gum-trees, upon grubs and different insects. Whenever there is a scarcity of food, the Yellow-bellied Parrakeets betake themselves to human habitations, and crowd around the farm-doors with as much confidence as if they formed part of the regular establishment. There is, however, not very much need for this intrusion into the farm-yard, as its natural food is simple and varied, and the powers of wing are sufficiently great to carry the bird over a large extent of country. The flight of this species is powerful, and is achieved by means of a series of very wide undulations. Yet on some occasions the mendicant Parrakeets may be counted by hundreds, as they press around the barn-door, disputing every chance grain of corn with the poultry, and behaving with perfect self-reliance.

In captivity, the Yellow-bellied Parrakeet is a hardy bird, and is well adapted for a caged life.

The nest of this bird is made in the bark of a gum-tree, and the eggs are in colour a pure white, and in number average from six to eight. The season for nest-building is from September to January. When the young are hatched, they are covered with a coating of soft white cottony down.

The colouring of this species is very magnificent. The forehead is rich crimson, and the back is a peculiar mottled green, each feather being of a deep black-green, edged with the same hue, but of a much lighter character. The throat and the middle of the wings are blue, the breast and abdomen are bright golden yellow, and the under tail-coverts are marked with a few red dashes. The two middle feathers of the tail are green, and the remainder are blue, dark at the base, but becoming lighter towards the tip. The female is similarly coloured, but not so brilliantly.

ANOTHER most beautiful example of this genus is found in the ROSE-HILL PARRAKEET, popularly known to dealers by the name of the Rosella Parrot.

This most lovely bird is found in New South Wales and Van Diemen's Land, and although very plentiful in places which it frequents, it is a very local bird, haunting one spot in hundreds, and then becoming invisible for a range of many miles. In the open country it lives in little companies like the preceding species, and is even more familiar, being exceedingly inquisitive, as is the nature of all the Parrot tribe. Plentiful as it is, there are few birds which are likely to suffer more from the gun, as its plumage is so magnificent and its form so elegant that it is in great request among the dealers, who are always sure of a sale when the beautiful skin is properly stuffed and put into a glass case.

The wings of the Rose-hill Parrakeet are not very powerful, and do not seem capable of enduring a journey of very great extent, for the bird always takes opportunities of settling as often as it can do so, and then, after running along the ground for awhile, starts afresh. The flight is composed of a succession of undulations. The voice of this species is not so harsh as that of many Parrots, being a pleasing and not very loud whistle, which is often uttered. As the bird is a hardy one, and can bear confinement well, it is coming much into fashion as an inhabitant of the aviary, and will probably be brought over to England in great numbers. The natural food of the Rose-hill Parrakeet consists of seeds, a diet which it varies by eating many kinds of insects, a food which every Australian bird can have in the greatest variety, and without the slightest fear of stint.

The eggs of this bird are rather numerous, being from seven to ten as a general average, and they are laid in the bark of some decaying gum-tree. Their colour is pure white.

The plumage of the adult Rose-hill Parrakeet is very beautiful, and is coloured as follows: The head, sides of the face, back of the neck, and the breast, are glowing scarlet, connected with each other by a band that passes over the shoulders. The chin and upper part of the throat are pure white. The feathers of the back are very dark black-green, broadly edged with light green of that exquisite hue which is only seen in the early spring leaves of the hawthorn. The upper tail-coverts are wholly of this beautiful leaf-green. The shoulder of the wing is shining lilac mixed with black, which by degrees settles upon the centre of the feathers, so that many of the tertiaries are nearly of the same colour as the feathers of the back. Many of the tertiaries have their centre black-green, their edges bright golden yellow, and a very little bright green spot just on the tip. The primaries are dull blackish brown, with a tinge of purple on their inner webs. The central feathers of the tail are dull green, and the others are lilac-blue, deeper towards their base and becoming nearly white at their tips. They are regularly graduated, the central being the longest, and on their under surfaces are a few scattered dashes of black. The lower part of the breast is yellow, which changes gradually into very light green

ROSE-HILL PARRAKEET.—*Platyœ cus eximius.*

on the abdomen, and the under tail-coverts are light scarlet. In total length it measures about thirteen inches.

ALTHOUGH not endowed with the glowing hues of the preceding species, the GROUND PARRAKEET is a remarkably pretty and interesting bird.

This species derives its name from its ground-loving habits. Mr. Gould says that it never perches on trees; but the author of "Bush Wanderings in Australia" remarks that he has seen it perching upon the tea-tree scrub. From its peculiarly pheasant-like shape and habits, it is sometimes called the pheasant by the colonists.

GROUND PARRAKEET.—*Pezophorus formosus*

It is a very common bird, and is found spread over the whole of Southern Australia and Van Diemen's Land.

It is remarkable that this bird, which has much of the outline of the pheasant, should have many of the habits of our game birds, and a very strong game odour. It runs very rapidly on the ground, and is especially excellent at getting through grass-stems, among which it winds its way with such wonderful celerity, that it can baffle almost any dog. Flight seems to be its last resource; and even when it does take to wing, it remains in the air but a very short time, and then pitches and takes to its feet.

The flight is very low, very quick, as bewilderingly irregular as that of the snipe, but is not maintained for more than a hundred yards. When the dogs come near the place where it is concealed, it crouches closely to the ground, hoping thereby to escape detection; but if this stratagem should prove of no avail, it leaps suddenly into the air, dashes forward for a few yards, and then settles again. This bird makes no nest, and does not even make its home in the hollow of a tree, but lays its white eggs upon the bare ground.

The flesh of the Ground Parrakeet bears some resemblance to that of our British game birds, and is said to be somewhat of the same character as that of the snipe or the quail.

The general colour of this pretty bird is dark green above, mottled with yellow and variegated with a multitude of black semilunar markings. The under surface is yellow, changing to a greener tint upon the throat, and also mottled with a darker hue. The tail is long and slender, the two central feathers are green barred with yellow, and the rest are marked in just the reverse fashion, being yellow barred with dark green.

THE genus Palæornis, of which the RINGED PARRAKEET is an excellent example, is a very extensive one, and has representatives in almost every hot portion of the world, even including Australia.

The Ringed Parrakeet is found both in Africa and Asia, the only difference perceptible between the individuals brought from the two continents being that the Asiatic species is rather larger than its African relative. It has long been the favourite of man as a caged bird, and is one of the species to which such frequent reference is made by the ancient writers, the other species being the Alexandrine Ringed Parrakeet (*Palæornis Alexandri*).

The individual from which the illustration is taken belongs to one of my friends, and is a very great favourite in the house, being looked upon more in the light of a human being than a bird. Her birthday is scrupulously kept, and on that auspicious morning she is always presented with a sponge cake, which she eats daintily while sitting on the mantelpiece, chuckling to herself at intervals. She is a most affectionate little creature, and cannot bear that any of her especial friends should leave the room without bidding farewell; and I once saw her set up such a screech because her mistress happened to go away without speaking to her, that she had to be taken out of her cage and comforted before she would settle quietly.

Her owner, by whose permission the portrait was taken, has kindly presented to me the following account of the bird :—

"You ask me to tell you something about my little Polly; perhaps the simplest plan will be to give a sketch of her history, premising that although I believe my little pet to be a male, still, as I love her so tenderly, I always use the feminine pronoun in speaking of and to her.

Polly's birthplace was Trincomalee, and she was brought over to this country by one of my wife's sons, an officer in the Royal Navy, being accompanied hither by a vast retinue of Parrakeets, almost all of whom fell victims to the rough cold weather which they had to encounter, together with the change of climate.

RINGED PARRAKEET.—*Palæornis torquatus.*

The poor birds literally laid them down and died, the deck being strewn with their elegant forms. Polly, I am thankful to say, was blessed with an excellent constitution, and her nurse, a kind-hearted weather-beaten sailor, loved her, and she lay in his bosom and was so kept warm and comfortable through the cold.

On Polly's arrival at Plymouth, her nurse, being obliged to attend to other matters, left her to her own resources in an old cage in which she usually slept, when her horizon was suddenly darkened by a cloud of bum-boat women from the shore, one of whom, seeing her defenceless situation, seized upon her, like Glumdalclitch upon Gulliver, and conveyed the delicate little creature to her coarse bosom. Fortunately for Polly, she uttered a little sound, which was heard by her nurse, who, seizing the woman by the shoulders, rescued Polly from the vile embrace.

After this *contretemps*, Polly was put into a rickety old cage, with two buns for her nourishment, and sent all by herself in the train to London. On her arrival there she was forwarded to a person who had formerly been confidential servant to my wife. One morning, this good person, hearing a great chattering downstairs, looked in at her back-parlour door, and there, to her infinite surprise, she saw Polly seated upon the cat's back, chattering away at no allowance, while pussy was majestically marching round the room.

Soon after this we came to London, and then saw for the first time our little pet, which soon began to know and love me. Her favourite place is on my shoulder, where at lunch-time she delights to sit and digest after having pecked from my plate whatever she most fancies. If the weather be cold and her feet chilly, she pulls herself up by my whiskers, placing herself on the top of my head, which being partially bald is warm to her little *pattes*. Her favourite resort is generally on my shoulder, and whilst sitting there, her manner of attracting attention is by giving my ear a little peck.

Whenever I come home, and wherever Polly may be, no sooner do I put my key in the lock, or sometimes before I have quite reached the door, than Polly gives a peculiar shrill call, and then it is known for certain that I am in the house. Even when I go to bed, though it may be at one or two in the morning, on my entering the room, however gently, Polly knows I am there, and although apparently asleep and with two thick shawls wrapped round her cage, excluding all light, she immediately utters one little note of welcome.

She has a peculiar way of contracting her eye when preparing to do or actually doing anything mischievous : when so contracted, the pupil of the eye appears as it were a mere speck of jet. I believe that her fondness for and her sympathetic attachment to me was something more than mere instinct, for if I think strangely of her at any time, even in the middle of the night, she is sure to answer me with her own little note, her eyes remaining shut and her head tucked in her shoulder as though she were fast asleep."

I have noticed the peculiar movement of the eye referred to in this narrative, and must add that the entire eyelid partakes of this curious contraction, the bird possessing the power of circularly contracting the lid, at first quite smoothly, but afterwards with a multitude of tiny radiating wrinkles or puckers, until at last the aperture is reduced to the size of a small pinhole. It looks, to use a familiar illustration, as if the eyelid were made of India-rubber, and could be contracted or relaxed at will.

Perhaps this power of reducing the aperture of vision may be given to the bird for the purpose of enabling it to see the better, and may have some connexion with the united microscopic and telescopic vision which all birds possess in a greater or less degree.

This species of Parrakeet is not very good at talking, though it can learn to repeat a few words and is very apt at communicating its own ideas by a language of gesture and information especially its own. It is, however, very docile, and will soon learn any lesson that may be imposed, even that most difficult task to a Parrot—remaining silent while any one is speaking. One of my pupils had one of these birds, of which he was exceedingly fond ; and finding that although his body was in the schoolroom below, his mind was with his Polly in the room above, I allowed her to stay in the room on condition that the lesson should be properly learned. At first, however, Polly used to screech so continually that all lessons were stopped for the time, and I was fearful that Polly must be banished.   However, I soon overcame the difficulty, for every time that Polly screamed I used to put her into a dark cupboard and not release her for some time.   She soon found out my meaning, and it was very amusing to see her push out her head ready for a scream, and then check herself suddenly.

She was a very nice Polly, and became a great favourite.   Her great treat was a half walnut, which she held tightly in one claw while she delicately prized out the kernel with her hooked bill and horny tongue.   The end of the poor bird was very tragic ; she got out of window, flew to a tree, and was there shot by a stupid farmer.   The history of this bird is given more at length in "My Feathered Friends."

The general colour of this species is grass-green, variegated in the adult male as follows :   The feathers of the forehead are light green, which take a bluish tinge as they approach the crown and nape of the neck, where they are of a lovely purple blue. Just below the purple runs a narrow band of rose colour, and immediately below the rosy line is a streak of black, which is narrow towards the back of the neck, but soon becomes broader, and envelops the cheek and chin.   It does not go quite round the neck, as there is an interval of nearly half an inch on the back of the neck.   The quill-feathers of both wings and tail are darkish green ; the wings are black beneath, and the tail yellowish.   The two central feathers of the tail are always much longer than the others, sometimes projecting nearly four inches.   The female is wholly green, and may thereby be distinguished from her mate.   Owing to the variable development of the central feathers of the tail, the length of this bird cannot be accurately given, but may be set down from sixteen to eighteen inches.   The upper mandible is coral-red, and the lower is blackish ; the feet are flesh-coloured.

ONE of the very prettiest and most interesting of the Parrot tribe is the GRASS or ZEBRA PARRAKEET ; deriving its names from its habits and the markings of its plumage.

It is a native of Australia, and may be found in almost all the central portions of that land, whence it has been imported in such great numbers as an inhabitant of our aviaries, that when Dr. Bennett was last in England, he found that he could purchase the birds at a cheaper rate in England than in New South Wales.   This graceful little creature derives its name of Grass Parrakeet from its fondness for the grass lands, where it may be seen in great numbers, running amid the thick grass blades, clinging to their stems, or feeding on their seeds.   It is always an inland bird, being very seldom seen between the mountain ranges and the coasts.

Of the habits of this bird Mr. Gould writes as follows :   "I found myself surrounded by numbers, breeding in all the hollow spouts of the large Eucalypti bordering the Mokai ; and on crossing the plains between that river and the Peel, in the direction of the Turi mountains, I saw them in flocks of many hundreds, feeding upon the grass seeds that are there abundant.   So numerous were they, that I determined to encamp upon the spot, in order to observe their habits and to procure specimens.   The nature of their food and the excessive heat of these plains compel them frequently to seek the water ; hence my camp, which was pitched near some small fords, was constantly surrounded by large numbers, arriving in flocks varying from twenty to a hundred or more.

The hours at which they were most numerous were early in the morning, and some time before dark in the evening.   Before going down to drink, they alight on the neighbouring trees, settling together in clusters, sometimes on the dead branches, and at others on the drooping boughs of the Eucalypti.   Their flight is remarkably straight and rapid, and is generally accompanied by a screeching noise.   During the heat of the day, when sitting motionless among the leaves of the gum-trees, they so closely assimilate in colour, particularly on the breast, that they are with difficulty detected."

The voice of this bird is quite unlike the rough screeching sounds in which Parrots seem to delight, and is a gentle, soft, warbling kind of song, which seems to be contained within the body, and is not poured out with that decision which is usually found in birds that can sing, however small their efforts may be.   This song, if it may be so called, belongs only to the male bird, who seems to have an idea that his voice must be very agreeable to his mate, for in light warm weather he will warble nearly all day long, and often pushes his beak almost into the ear of his mate, so as to give her the full benefit of his song. The lady, however, does not seem to appreciate his condescension as he wishes, and sometimes pecks him sharply in return. Dr. Bennett observes that the bird has some ventriloquial powers, as he has noticed a Grass Parrakeet engaged in the amusement of imitating two birds, one warbling ard the other chirping.

WARBLING GRASS PARRAKEET.—*Melopsittacus undulatus.*

The food of this Parrakeet consists almost chiefly of seeds, those of the grass plant being their constant food in their native country. In England they take well to canary seed, and it is somewhat remarkable that they do not pick up food with their feet, but always with their beaks. It is a great mistake to confine these lively little birds in a small cage, as their wild habits are peculiarly lively and active, and require much space. The difference between a Grass Parrakeet when in a little cage and after it has been removed into a large house, where it has plenty of space to move about, is really wonderful.

This species has frequently bred in England, and nest-making is of very common occurrence, though it often happens that the female deserts her eggs before they are hatched. A correspondent of the *Field* newspaper, Mr. Moore, of Fareham, writes as follows: "Having been very successful in breeding most of our British birds in cages, I was induced to try the Australian Parrakeet, commonly known as the Grass Warbling Parrakeet, and I now have the pleasure of making known to you what I consider my most extraordinary success. Between the 24th of December last and the present month, I have reared eleven from one pair, and having watched their habits very carefully, I venture to make a few remarks upon them.

They do not build a nest as most birds do, but must have a piece of wood with a rough hole in the middle, and this they will finish to their liking. Let it be kept private, and let them pass through a hole to the nesting-place. When the hen has laid, take the egg out, putting a false one in its place till four have been laid. This should be attended to, as she only lays on alternate days, and the young would be so far apart in hatching. By so doing I have ascertained the exact time of incubation, and have found it to be seventeen days. I mention this, as persons might otherwise be led astray. These birds feed their young in the same manner as pigeons; the young never gape, but the old ones take the beak in their mouths, and by a peculiar process disgorge the food, which the young take at the same moment. They begin to breed in December, that being their summer. The young are so tame that they will fly after me anywhere."

In another instance, mentioned in the same journal, the birds laid their eggs upon some sawdust and there hatched two young, the number of eggs having been three. This Parrakeet will breed more than once in the season. The young birds get on very fast after hatching, provided that the room be kept warm and the parent well supplied with food. At thirty days of age the young Parrakeet has been observed to feed itself from the seed-drawer of its cage. Groundsel seems to be a favourite diet with them, but it seems that lettuce does not agree with their constitution. With this exception, the Grass Parrakeet may be fed precisely in the same manner as the canary.

In its native land it is a migratory bird, assembling after the breeding season in enormous flocks as a preparation for their intended journey. The general number of the eggs is three or four, and they are merely laid in the holes of the gum-tree without requiring a nest.

The general colour of this pretty bird is dark mottled green, variegated with other colours. The forehead is yellow, and the head, the nape of the neck, the upper part of the back, the scapularies and the wing-coverts are light yellowish green, each feather being marked with a crescent-shaped spot of brown near the tip, so as to produce the peculiar mottling so characteristic of the species. These markings are very small on the head, and increase in size on the back, and from their shape the bird is sometimes called the Shell or Scallop Parrot. On each cheek there is a patch of deep blue, below which are three circular spots of the same rich hue. The wings are brown, having their outer webs deep green, roped with a yellower tint. The throat is yellow, and the abdomen and whole under surface light grass-green. The two central tail-feathers are blue, and the remainder green, each with an oblique band of yellow in the middle.

The young birds have the scallopings all over the head, and the females are coloured almost exactly like their mate, who may be distinguished by the cere of the upper part of the beak being of a deep purple.

A VERY beautiful species of Parrakeet, and closely allied to the preceding bird, is the BLUE-BANDED GRASS PARRAKEET, also a native of Australia.

This pretty little Parrakeet is a pleasing and interesting creature, not at all uncommon in its favourite localities.

It is a summer visitor to Van Diemen's Land, where it remains from September to February or March. Thickly wooded places are its usual haunts, as it feeds almost wholly on seeds and grasses, and it is generally seen on the ground unless it has been alarmed. It congregates in flocks, and appears to have but little fear of danger, and but very confused notions of placing itself in safety; for as soon as a flock is alarmed, they all rise screaming feebly, and after flying for a hundred yards or so, again alight. During the short time that they are on the wing, their flight is rapid and very irregular, reminding the European sportsman of the snipe, and being not unlike that of the ground Parrakeet already mentioned.

It is a very quick runner, and displays great address in threading its way among the grass stems. Sometimes when frightened it will fly to some neighbouring tree and there perch for awhile ; but it soon leaves the uncongenial branches and returns to the ground. As it is not at all shy, a careful observer can easily approach the flocks within a short distance by moving very slowly and quietly, and can inspect them quite at his ease through a pocket telescope, that invaluable aid to practical ornithologists. As it is a hardy bird and bears confinement well, it is rapidly coming into favour as a cage bird, and will probably earn great popularity, as it is very easily tamed and of a very affectionate nature.

The eggs of this species are six or seven in number, and are generally laid in a convenient hole of a gum-tree, although the bird sometimes prefers the hollow trunk of a prostrate tree for the purpose.

The colour of this bird is green with a slight brown wash ; the wings, the tail, and a

BLUE-BANDED GRASS PARRAKEET.—*Euphéma chrysóstoma.*

band over the forehead are beautiful azure, and around the eyes and on the centre of the abdomen the colour is yellow.

THE pretty bird to which so extravagantly long a name has been given is also a native of Australia, and is found only in New South Wales, being, though plentiful, very local.

The SCALY-BREASTED LORRIKEET is a good example of a very large genus; and as the habits of all the species are very similar, more than a single example is not necessary. The name Trichoglossus signifies " hairy tongue," and is given to these birds in consequence of the structure of that member, which is furnished with bristly hairs like the tongue of the honey-eaters, and is employed for the same purpose. This species may generally be found in those bush ranges which are interspersed with lofty gum-trees, from the blossoms of which it extracts the sweet juices on which it feeds. While employed in feeding, it clings so tightly to the blossoms, that if shot dead its feet still retain their hold. The amount of honey consumed by these birds is really surprising, a teaspoonful of honey having been taken from the crop of a single bird. Whenever the natives kill one of these birds, they always put its head in their mouths and suck the honey out of its crop. Young birds are always very well supplied with this sweet food, and are consequently in great favour with the native epicures.

When captured it is readily tamed, and is sufficiently hardy to live in a cage, provided that it be well supplied with sugar as well as seeds.

It assembles in large flocks of a thousand or more in number; and when one of the vast assemblies is seen perched on a tree, the effect is most magnificent. They are so

SCALY-BREASTED LORRIKEET.—*Trichoglossus chlorolepidótus.*

heartily intent on their food, that they cannot be induced to leave the tree even by the report of a gun or the rattling of shot among them, and at the best will only scream and go to another branch. This species will associate with others very harmoniously, and Mr. Gould has shot at a single discharge four species of Lorrikeet, all feeding in the most friendly manner upon the same tree.

The Lorrikeets are very conversational birds, and discourse in loud and excited tones, so that the noise of a large flock is quite deafening. When the whole flock rises simultaneously, as is generally the case, and moves to another tree, the effect of all the wings beating the air together is extraordinary, and is said to resemble a thunderstorm mixed with wind.

The colour of this species is as follows: The upper surface is rich grass-green, and the under surface, together with a few feathers on the back of the neck, is light yellow with green edges. The under side of the shoulders and the base of the wings are deep scarlet, and the rest of the under surface of the wings is jetty black.

THE MACAWS are mostly inhabitants of Southern America, in which country so many magnificent birds find their home.

They are all very splendid birds, and are remarkable for their great size, their very long tails, and the splendid hues of their plumage. The beak is also very large and powerful, and in some species the ring round the eyes and part of the face are devoid of covering. Three species are well known in our menageries; but as their habits are all very similar, only one example has been figured. This is the great BLUE AND YELLOW MACAW, a bird which is mostly found in Demerara. It is a wood-loving bird, particularly haunting those places where the ground is wet and swampy, and where grows a certain palm on the fruit of which it chiefly feeds.

The wings of this species are strong, and the long tail is so firmly set that considerable powers of flight are manifested. The Macaws often fly at a very high elevation, in large flocks, and are fond of executing sundry aërial evolutions before they alight. With one or two exceptions they care little for the ground, and are generally seen on the summit of the highest trees.

BLUE AND YELLOW MACAW.—*Ara Ararauna.*

Waterton writes as follows of the RED AND BLUE MACAW :—

" Superior in size and beauty to any Parrot of South America, the Ara will force you to take your eyes from the rest of animated nature and gaze at him ; his commanding strength, the flaming scarlet of his body, the lovely variety of red, yellow, blue, and green in his wings, the extraordinary length of his scarlet and blue tail, seem all to join and demand from him the title of emperor of all the parrots. He is scarce in Demerara until you reach the confines of the Macoushi country ; there he is in vast abundance ; he mostly feeds on trees of the palm species.

When the coucourite trees have ripe fruit on them, they are covered with this magnificent Parrot. He is not shy or wary ; you may take your blowpipe and a quiver of poisoned arrows, and kill more than you are able to carry back to your hut. They are very vociferous, and like the common Parrots, rise up in bodies towards sunset and fly two and two to their places of rest. It is a grand sight in ornithology to see thousands of Aras flying over your head, low enough to let you have a full view of their flaming mantle. The Indians find the flesh very good, and the feathers serve for ornaments in their headdresses."

The Blue and Yellow Macaw generally keeps in pairs, though, like the other species, it will sometimes assemble in flocks of considerable size. When thus congregated the Macaws become very conversational, and their united cries are most deafening, and can be heard at a great distance, as any one can understand who has visited the Parrot-house of the Zoological Gardens. In common with the other Macaws, this species is easily tamed, and possesses some powers of imitation, being able to learn and repeat several words or even phrases. It is not, however, gifted with the extraordinary powers of speech which are so wonderfully developed in the true Parrots, and on account of its deafening cries is not an agreeable inhabitant of a house.

The Macaws lay their eggs in the hollows of decaying trees, and are said to alter the size and form of the hole to their taste by means of their powerful beaks, a feat which they certainly have the ability to perform. The eggs are never more than two, and there are generally two broods in the season. Both parents assist in the duties of incubation.

The GREAT GREEN MACAW, a very splendid species, with green body, scarlet and blue head, blue-tipped wings, and red and blue tail, is not so exclusively an inhabitant of the forest nor so wary as the preceding species. Taking advantage of the labours of mankind, it makes raids on the maize and corn fields, and does very great damage in a very short time, for its appetite is voracious, and its beak powerful. Like most birds of similar character, it never ventures upon one of these predatory excursions without placing a sentinel on some elevated post where he can see the whole of the surrounding country, and give the alarm to his comrades whenever he fears the approach of danger. So great is the destruction wrought by these birds, that the agriculturists are forced to protect their property by keeping a watch day and night over their corn fields from the time when the grain begins to ripen to the day when it is cut and carried.

During the rainy season these Macaws leave the country, and do not return until January or February.

The plumage of the Blue and Yellow Macaw is rather roughly set on the body, and is thus coloured : The forehead is green, and the whole of the upper surface ; the wings and tail are bright rich blue of a verditer cast. The cheeks are white and nearly naked, and below the eye are three delicate semilunar streaks of black. Below the chin is a broad black band, which sweeps round towards the ears, and runs round nearly the whole of the white space. The throat, head, and abdomen are rich golden yellow, and the under surfaces of the wings and tail are also yellow, but of a more ochreous cast. The bill is deep black, the eye yellowish white, and the legs and feet blackish grey.

The entire length of this bird is about forty inches, of which the tail alone occupies nearly two feet. It is not, however, the largest species of Macaw, as the Red and Blue Macaw equals it in size.

ANOTHER species of Macaw is found in the more northern portions of America, though it is popularly called a Parrot, and not a Macaw. This is the well-known CAROLINA PARROT, of which so much has been written by Wilson, Audubon, and other American ornithologists.

This bird is much more hardy than the generality of the Parrot tribe, and has been noticed by Wilson in the month of February flying along the banks of the Ohio in the midst of a snow storm, and in full cry. It inhabits, according to Wilson, " the interior of Louisiana, and the shores of Mississippi and Ohio and their tributary waters, even beyond the Illinois river, to the neighbourhood of Lake Michigan in latitude 42° N., and contrary to the generally received opinion, is chiefly resident in all these places. Eastward, however, of the great range of the Alleghany, it is seldom seen farther north than the state of Maryland ; though straggling parties have been occasionally observed among the valleys of the Juniata, and according to some, even twenty-five miles to the north-west of Albany in the state of New York." These accidental visits are, however, rightly regarded by our author as of little value.

The Carolina Parrot is chiefly found in those parts of the country which abound most in rich alluvial soils on which grow the cockle-burs, so dear to the Parrot and so hated by the farmer. In the destruction of this plant the Carolina Parrot does good service to the sheepowner, for the prickly fruit is apt to come off upon the wool of the sheep, and in some places so abundantly as to cover it with one dense mass of burs through which the wool is hardly perceptible. The prickly hooks of the burs also break away from the fruit, and intermingle themselves so thoroughly with the fleece that it is often rendered worthless, the trouble of cleansing it costing more than the value of the wool.

Besides the cockle-burs, the beech-nut and the seeds of the cypress and other trees

CAROLINA PARROT.—*Conúrus Carolinensis.*

are favourite food of the Carolina Parrot, which is said to eat apples, but probably only bites them off their stems for wantonness, as it drops them to the ground and there lets them lie undisturbed.

An idea was and may be still prevalent in its native country, that the brains and intestines of the Carolina Parrot were fatal to cats; and Wilson after some trouble succeeded in getting a cat and her kittens to feed upon this supposed poisonous diet. The three ate everything excepting the hard bill, and were none the worse for their meal. As, however, the Parrot was in this case a tame one, and had been fed upon Indian corn, he conjectured that the wild Parrot which had lived on cockle-burs might be injurious to the cat, although that which had eaten the comparatively harmless diet might do no injury.

The nest of this bird is made in hollow trees.

One of these Parrots was tamed by Wilson, who gave the following animated description of his favourite and her actions:—

" Anxious to try the effects of education on one of those which I procured at the Big Bone Lick, and which was but slightly wounded in the wing, I fixed up a place for it in the stern of my boat, and presented it with some cockle-burs, which it freely fed on, in less than an hour after it had been on board. The intermediate time between eating and sleeping was occupied in gnawing the sticks that formed its place of confinement, in order to make a practicable breach, which it repeatedly effected.

When I abandoned the river and travelled by land, I wrapped it up closely in a silk handkerchief, tying it tightly around, and carried it in my pocket. When I stopped for refreshment I unbound my prisoner and gave it its allowance, which it generally despatched with great dexterity, unhusking the seeds from the bur in a twinkling; in doing which it always employed its left foot to hold the bur, as did several others that I kept for some time. I began to think that this might be peculiar to the whole tribe, and that they all were, if I may use the expression, left-footed; but by shooting a number afterwards while engaged in eating mulberries, I found sometimes the left and sometimes the right foot stained with the fruit, the other always clean; from which, and the constant practice of those I kept, it appears that, like the human species in the use of their hands, they do not prefer one or the other indiscriminately, but are either left or right footed.

But to return to my prisoner. In recommitting it to 'durance vile' we generally had a quarrel, during which it frequently paid me in kind for the wound I had inflicted and for depriving it of liberty, by cutting and almost disabling several of my fingers with its sharp and powerful bill.

The path between Nashville and Natchez is in some places bad beyond description. There are dangerous creeks to swim, miles of morass to struggle through, rendered almost as gloomy as night by a prodigious growth of timber, and an underwood of canes, and other evergreens, while the descent into these sluggish streams is often ten or fifteen feet perpendicular into a bed of deep clay. In some of the worst of these places, where I had as it were to fight my way through, the Paroquet frequently escaped from my pocket, obliging me to dismount and pursue it through the worst of the morass before I could regain it. On these occasions I was several times tempted to abandon it, but I persisted in bringing it along. When at night I encamped in the woods, I placed it on the baggage beside me, where it usually sat with great composure, dozing and gazing at the fire till morning. In this manner I carried it upwards of a thousand miles in my pocket, where it was exposed all day to the jolting of the horse, but regularly liberated at meal times and in the evening, at which it always expressed great satisfaction.

In passing through the Chickasaw and Chactaw nations, the Indians, whenever I stopped to feed, collected around me—men, women, and children—laughing, and seemingly wonderfully amused with the novelty of my companion. The Chickasaws called it in their language 'Kelinky,' but when they heard me call it Poll, they soon repeated the name; and whenever I chanced to stop amongst these people, we soon became familiar with each other through the medium of Poll.

On arriving at Mr. Dunbar's, below Natchez, I procured a cage, and placed it under the piazza, where, by its call, it soon attracted the passing flocks, such is the attachment they have for each other. Numerous parties frequently alighted on the trees immediately above, keeping up a constant conversation with the prisoner. One of these I wounded slightly in the wing, and the pleasure Poll expressed on meeting with this new companion was really amusing. She crept close up to it as it hung on the side of the cage, chattering to it in a low tone of voice as if sympathising in its misfortune, scratched about its head and neck with her bill, and both at night nestled as close as possible to each other, sometimes Poll's head being thrust among the plumage of the other.

On the death of this companion she appeared restless and inconsolable for several days. On reaching New Orleans I placed a looking-glass beside the place where she usually sat, and the instant she perceived her image, all her former fondness seemed to return, so that she could scarcely absent herself from it a moment. It was evident she was completely deceived. Always when evening drew on, and often during the day, she laid her head close to that of the image in the glass, and began to doze with great composure and satisfaction.

In this short space she had learned to know her name, to answer when called on, to climb up my clothes, sit on my shoulder, and eat from my mouth. I took her with me to sea, determined to persevere in her education, but, destined to another fate, poor Poll having one morning about daybreak wrought her way through the cage while I was asleep, instantly flew overboard and perished in the Gulf of Mexico."

The result of this and other experiments was, that Wilson delivered his verdict in favour of the Carolina Parrot, saying that it is a docile and sociable bird, soon becomes perfectly familiar, and is probably capable of imitating the accents of man. Towards its own kind it displays the strongest affection, and if its companions be in danger, it hovers about the spot in loving sympathy. It is very fond of salt, and will frequent the saline marshes in great numbers, covering the whole ground and neighbouring trees to such an extent, that nothing is visible but their bright and glossy plumage.

While thus assembled together Wilson shot a great number of the birds, and was much struck with their affectionate conduct. "Having shot down a number, some of which were only wounded, the whole flock swept repeatedly round their prostrate companions, and again settled on a low tree within twenty yards of the spot where I stood. At each successive discharge, though showers of them fell, yet the affection of the survivors seemed rather to increase, for after a few circuits round the place they again alighted near me, looking down on their slaughtered companions with such manifest symptoms of sympathy and concern as entirely disarmed me."

The same graceful writer then proceeds to observe, with that accuracy of detail for which his works are so valuable, " I could not but take notice of the remarkable contrast between their elegant manner of flight, and their lame, crawling gait, among the branches. They fly very much like the wild pigeon—in close compact bodies, and with great rapidity, making a loud and outrageous screaming, not unlike that of the red-headed woodpecker. Their flight is sometimes in a direct line, but most usually circuitous, making a great variety of elegant and easy serpentine meanders as if for pleasure.

They are particularly attached to the large sycamores, in the hollows of the trunks and branches of which they generally roost ; thirty or forty, and sometimes more, entering at the same hole. Here they cling close to the sides of the tree, holding fast by the claws and also by the bill. They appear to be fond of sleep, and often retire to their holes during the day, probably to take a regular siesta. They are extremely sociable with and fond of each other, often scratching each other's heads and necks, and always at night nestling as close as possible to each other, preferring at that time a perpendicular position, supported by their bill and claws."

The general colour of this bird is green, washed with blue, and diversified with other tints as follows : The forehead and cheeks are reddish orange, the same tint is seen on the shoulders and head and wings, and the neck and back of the head are pure golden yellow. The upper parts of the body are soft green, and the under portions are of the same hue, but with a yellowish cast. The greater wing-coverts are yellow, tinged with green, the primary feathers of the wing are deep purplish black, and the long wedge-shaped tail has the central feathers streaked with blue along their central line. The female is coloured after the same fashion, but not so brightly, and the young of both sexes are green on the neck instead of yellow. The total length of this species is about twenty-one inches.

In the Lories the bill is weaker than in the preceding species, and of smaller size, and the plumage is very beautiful, scarlet being the predominating tint.

The Papuan Lory is, as its name denotes, a native of Papua and other parts of New Guinea, and has always attracted great attention on account of its beautiful form and rich colouring. In its general shape it is not unlike the ring Parrakeet, the contour of the body being very similar and the tail boldly graduated, with the two central feathers projecting far beyond the rest. This elongated form of the tail-feathers is so unusual in the Lories, which mostly have rather short and stumpy tails, that it has induced systematic naturalists to place the bird in a genus distinct from the other Lories. Many specimens of this lovely bird have been sent to England, but, like the birds of paradise, they are often destitute of legs, and in some cases even the long tail-feathers have been abstracted, thus entirely altering the appearance of the bird.

PAPUAN LORY.—*Charmosýna Pápua.*

The colours with which this species is decorated are remarkably rich and intense. The general colour is deep scarlet, relieved by patches of azure, golden yellow, and grass-green. The head, neck, the upper part of the back, and all the lower parts of the body are brilliant scarlet, with the exception of two patches of azure-blue across the top of the head, edged with deep purple. There are also some patches of yellow on the sides of the breast and the thighs. The lower parts of the back, the upper tail-coverts, and the lower part of the legs are deep azure, and the wings are green. The two long feathers of the tail are light grass-green for the greater part of their length, and are tipped with golden yellow. The remaining feathers of the tail have their basal halves deep green, and the remainder golden yellow. The total length of the Papuan Lory is about seventeen inches, of which measurement the two long tail-feathers occupy no less than eleven inches. The bill is orange-red, and the upper mandible is much longer than the lower, but is not very sharply curved.

Another beautiful example of these birds is given in the Purple-Capped Lory a native of the Moluccas and other islands.

The reader will not fail to observe the great difference in form between this and the preceding species, caused chiefly by the shortness and shape of the tail. It is often sent to this country as a cage-bird, and as it is readily tamed, is of an affectionate nature, and can be taught to speak very creditably, is somewhat of a favourite among bird-fanciers. It is a lively and active creature, ever in motion, and is very fond of attracting the notice of strangers and receiving the caresses of those whom it likes.

PURPLE-CAPPED LORY.—*Lórius domicellus.*

Like the Papuan Lory, the principal tint of the plumage is rich scarlet, which is in even greater abundance than in that bird. The top of its head is very deep purple, being nearly black on the forehead, and passing into violet on the hinder part of the head. Upon the upper part of the breast there is a collar of yellow, and with this exception, the whole of the face, neck, back, breast, and abdomen are rich scarlet. The wings are green above, changing to violet on the edges and on the under wing-coverts. The feathers of the tail are rich scarlet at their base, and each feather is banded near its extremity with black, and tipped with yellow. The feathers of the thigh are azure. The bill is yellow, with a tinge of orange, and is rather narrow towards the tip. In spite of its short tail, this bird measures about eleven inches in length, so that it is very much larger than the preceding species.

The true Parrots constitute a group which are easily recognised by their short squared tails, the absence of any crest upon the head, and the toothed edges of the upper mandible. Many species belong to this group, of which we shall find three examples in the following pages.

GREY PARROT. — *Psittacus erythacus.*

The GREY PARROT has long been celebrated for its wonderful powers of imitation and its excellent memory.

It is a native of Western Africa, and is one of the commonest inhabitants of our aviaries, being brought over in great numbers by sailors, and always finding a ready sale as soon as the vessel arrives in port. Unfortunately the nautical vocabulary is none of the most refined, and the sailors have a malicious pleasure in teaching the birds to repeat some of the most startling of their phrases. The worst of the matter is, that the Parrot's memory is wonderfully tenacious, and even after the lapse of years, and in spite of the most moral training, the bird is apt to break out suddenly with a string of very reprehensible observations affecting the eyes, limbs, and general persons of his hearers.

There is no doubt that the Parrot learns in course of time to attach some amount of meaning to the words which it repeats, for the instances of its apposite answers are too numerous and convincing not to prove that the bird knows the general sense of the phrase, if not the exact force of each word.

I am unwilling to reproduce narratives which I have already published, and therefore restrict myself to one or two original anecdotes.

MM 2

There was a Parrot belonging to a friend of our family, a Portuguese gentleman who had married an English wife and resided in England. This Parrot was a great favourite in the house, and being accustomed equally to the company of its owner and the rest of the household, was familiar with Portuguese as well as English words and phrases. The bird evidently had the power of appreciating the distinction between the two languages, for if it were addressed, its reply would always be in the language employed.

The bird learned a Portuguese song about itself and its manifold perfections, the words of which I cannot remember. But it would not sing this song if asked to do so in the English language. Saluted in Portuguese, it would answer in the same language, but was never known to confuse the two tongues together. Towards dinner-time it always became very excited, and used to call to the servant whenever she was late, " Sarah, lay the cloth, —want my dinner !" which sentence it would repeat with great volubility, and at the top of its voice.

But as soon as its master's step was heard outside the house, its tone changed, for the loud voice was disagreeable to its owner, who used to punish it for screaming by flipping its beak. So Polly would get off the perch, very humbly sit on the bottom of the cage, put its head to the floor, and instead of shouting for its dinner in the former imperious tone, would whisper in the lowest of voices, " Want my dinner ! Sarah, make haste, want my dinner !"

In the well-known autobiography of Lord Dundonald, there is an amusing anecdote of a Parrot which had picked up some nautical phrases, and had learned to use them to good effect.

Some ladies were paying a visit to the vessel, and were hoisted on deck as usual by means of a " whip," i.e. a rope passing through a block on the yardarm, and attached to the chair on which the lady sits. Two or three had been safely brought on deck, and the chair had just been hoisted out of the boat with its fair freight, when an unlucky Parrot on board suddenly shouted out, " Let go !" The sailors who were hauling up the rope instantly obeyed the supposed order of the boatswain, and away went the poor lady, chair and all, into the sea.

Its power of imitating all kinds of sounds is really astonishing. I have heard the same Parrot imitate, or rather reproduce, in rapid succession the most dissimilar of sounds, without the least effort and with the most astonishing truthfulness. He could whistle lazily like a street idler, cry prawns and shrimps as well as any costermonger, creak like an ungreased " sheave" in the pulley that is set in the blocks through which ropes run for sundry nautical purposes, or keep up a quiet and gentle monologue about his own accomplishments with a simplicity of attitude that was most absurd.

Even in the imitation of louder noises he was equally expert, and could sound the danger whistle or blow off steam with astonishing accuracy. Until I came to understand the bird, I used to wonder why some invisible person was always turning an imperceptible capstan in my close vicinity, for the Parrot had also learned to imitate the grinding of the capstan bars and the metallic clink of the catch as it falls rapidly upon the cogs.

As for the ordinary accomplishments of Parrots, he possessed them in perfection, but in my mind his most perfect performance was the imitation of a dog having his foot run over by a cart-wheel. First there came the sudden half-frightened bark, as the beast found itself in unexpected danger, and then the loud shriek of pain, followed by the series of howls that is popularly termed " pen and ink." Lastly, the howls grew fainter, as the dog was supposed to be limping away, and you really seemed to hear him turn the corner and retreat into the distance. The memory of the bird must have been most tenacious, and its powers of observation far beyond the common order ; for he could not have been witness to such canine accidents more than once.

The food of this, as well as the green Parrot, consists chiefly of seeds of various kinds, and in captivity may be varied to some extent. Hemp-seed, grain, canary-seed, and the cones of fir-trees are very favourite articles of diet with this bird. Of the cones it is especially fond, nibbling them to pieces when they are young and tender ; but when they are old and ripe, breaking away the hard scales and scooping out the seeds with its very useful tongue. Hawthorn berries are very good for the Parrot, as are several vegetables.

These, however, should be given with great caution, as several, such as parsley and chick-weed, are very hurtful to the bird.

There are few things which a Parrot likes better than nuts and the stones of various fruits. I once succeeded in obtaining the affections of a Parisian Parrot, solely through the medium of peach-stones, which I always used to save for the bird, and for which he regularly began to gabble as soon as he saw me coming along the street. When taken freshly from the peach the stones are very acceptable to the Parrot, who turns them over and over, chuckling all the while to show his satisfaction, and picking all the soft parts from the deep indentations in the stone. As a great favour I sometimes used to crack the stone before giving it to him, and his delight then knew no bounds. Walnuts when quite ripe are in great favour with Parrots ; and it is very curious to see how well the bird sets to work at picking out their contents, holding the nut firmly with its foot, and hooking out its kernel with the bill and tongue. A split walnut will give a Parrot employment for more than an hour.

Woody fibre is generally beneficial to these birds, who often try to gratify their natural longing for this substance by pulling their perches to pieces. The Parrot owner will find the health of his pet improved and its happiness promoted by giving it, every now and then, a small log or branch, on which the mosses and lichens are still growing. Some persons are in the habit of giving their Parrots pieces of meat, fish, and other similar articles of diet, but generally with evil effects. The diet is too stimulating, and keeps up a continual irritation in the system, which induces the bird to be always pecking out its feathers. Many Parrots have almost stripped themselves of their plumage by this constant restlessness, and I knew of an individual that had contrived to pluck himself completely bare in every part of the body which his bill could reach, so that he presented the ludicrous sight of a bare body and a full-plumaged head. The soaked bread and milk which is so often given to these birds is, also, too heating a diet, and their bread should only be steeped in water.

The Parrot has the true tropical love for hot condiments, and is very fond of cayenne pepper or the capsicum pod from which it is supposed to be made. If the bird be ailing, a capsicum will often set it right again. It is rather curious that my cat has a similar taste, having, I presume, caught it from her master. Some months ago, a careless cook made a " curry " with a dessert-spoonful of cayenne pepper instead of curry powder, to the very great detriment of the throats of the intended consumers. " Pret," as usual, pushed her nose against my hand to ask for some of my dinner, so in joke I gave her a very red piece of the meat. To my profound astonishment, she ate the burning morsel with great zest, and became so clamorous for more that I could hardly satisfy her fast enough.

The Parrot should be able to change its position, as it does not like to sit perpetually on a round perch, and is much relieved by a little walking exercise. If possible, it should have some arrangement to enable it to climb ; a matter easily accomplished by means of a little wire cord and a small modicum of ingenuity. There should always be some spot where the Parrot can find a warm perch ; as all these birds are singularly plagued with cold feet, and often catch sundry disorders in consequence. If it is kept in a cage, the Parrot should never be confined in a brass prison ; for the bird is always climbing about the wires by means of its beak, and is likely to receive some hurt from the poisonous verdigris that is sure to make its appearance sooner or later on brass wire. An occasional bath is very beneficial to the Parrot's health ; and if the bird refuses to bathe, tepid water may be thrown over him with very good effect.

When proper precautions are taken, the Parrot is one of our hardiest cage-birds, and will live to a great age even in captivity. Some of these birds have been known to attain an age of sixty or seventy years, and one which was seen by Le Vaillant had attained the patriarchal age of ninety-three. At sixty its memory began to fail ; and at sixty-five the moult became very irregular, and the tail changed to yellow. At ninety it was a very decrepid creature, almost blind and quite silent, having forgotten its former abundant stock of words.

A Grey Parrot belonging to one of my friends was, during the former part of its life, remarkable only for its large vocabulary of highly discreditable language, which it would insist upon using exactly when it ought to have been silent, but suddenly changed its nature and subsided into a tender and gentle foster-mother.

In the garden of its owner there were a number of standard rose-trees, around all of which was a circular wire fence covered with convolvuluses and honeysuckle. Within one of these fences a pair of goldfinches had made their nest, and were constantly fed by the inhabitants of the house, who all had a great love for beasts and birds, and took a delight in helping the little creatures under their charge; and, indeed, were deeply interested in animated nature generally. Polly soon remarked the constant visits to the rose-tree, and the donations of crumbs and seeds that were regularly given, and must follow so good an example. So she set off to the spot; and after looking at the birds for a little while, went to her cage, brought a beakful of her sopped bread, and put it into the nest.

At last the young birds were hatched, much to Polly's delight; but she became so energetic in her demonstrations of attachment that she pushed herself fairly through the wire meshes, and terrified the parents so much that they flew away. Polly, seeing them deserted, took on herself the task of foster-mother, and was so attentive to her little charge that she refused to go back to her cage, but remained with the little birds by night as well as by day, feeding them carefully, and forcing them to open their beaks if they refused her attentions. When they were able to hop about they were very fond of getting on her back, where four of them would gravely sit, while the fifth, which was the youngest, or at all events the smallest, always preferred to perch on Polly's head.

With all these little ones on her back, Polly would very deliberately walk up and down the lawn as if to give them exercise; and would sometimes vary her performance by rising into the air, thus setting the ten little wings in violent motion, and giving the birds a hard task to remain on her back. By degrees they became less fearful, and when she rose from the ground, they would leave her back and fly down. They were but ungrateful little creatures after all; for when they were fully fledged they flew away, and never came back again to their foster-mother.

Poor Polly was for some time in great trouble about the desertion of her foster-children, but soon consoled herself by taking care of another little brood. These belonged to a pair of hedge-sparrows, whose home had been broken up by the descent of some large bird, which was supposed to have been a hawk by the effects produced. Polly found the little birds in dire distress; and contrived in some ingenious manner to get them, one by one, on her back, and to fly with them to her cage. Here she established the little family; never entering the cage except for the purpose of attending to her young charge.

The oddest part of the matter was, that one of the parents survived, and Polly was seen to talk to her in the most absurd manner; mixing up her acquired vocabulary with that universal bird-language that seems to be common to all the feathered tribes, and plentifully interlarding her discourse with sundry profane expressions. At last the instinctive language conquered the human, and the two birds seemed to understand each other perfectly well. At that time Polly was supposed to be about eight or nine years old.

There is a rather general belief that only the male Parrot can talk, but this is merely a popular error. The female Parrot has often been known to be an excellent talker, and at the same time has proved her sex by the deposition of a solitary egg. As might be supposed, such eggs produce no young; but there are accredited instances where the Grey Parrot has bred in Europe. In Buffon's well-known work may be seen a notice of a pair of Parrots that bred regularly for five or six years, and brought up their young successfully. The place chosen for their incubation was a tub, partially filled with sawdust, and was probably selected because it bore some resemblance to the hollow trunk of a tree, which is the usual nesting-place of the Parrots.

The general colour of this bird is a very pure ashen grey, except the tail, which is deep scarlet.

Two species of Green Parrot are tolerably common, the one being the FESTIVE GREEN PARROT, and the other the Amazon Green Parrot.

The former bird is a much larger and altogether finer species than the latter, often measuring sixteen inches in length. It is found in various parts of South America, such

GREEN PARROT.—*Chrysótis festivus.*

as Guiana, Cayenne, and the Brazils, and is very plentiful along the banks of the Amazon. It is a forest-loving bird, frequenting the depths of the vast wooded tracts which cover that country with their wonderful luxuriance, and being seldom seen beyond their outskirts. Being of an affectionate nature and easily tamed, it is in great favour as a cage-bird, and can readily be taught to pronounce words or even sentences.

The general colour of this Parrot is bright green. On the top of the head and behind the eyes the feathers are rather pale cobalt-blue, and a deeper tint of blue is also seen on the outer webs of the primary and secondary feathers of the wings, their interior webs being dark greenish black. The lower part of the back and the upper tail-coverts are deep crimson-red, and the short square tail is green, except the outermost feathers, which are edged with blue. On all the tail-feathers, except the central, there is a spot of pale red near the base. The bill is large and flesh-coloured.

The AMAZON GREEN PARROT is the species most commonly seen in England. It is a handsome bird, and is even a better conversationalist than the last-mentioned species. Like the Festive Parrot, it is a native of Southern America, and especially frequents the banks of the Amazon. It is not, however, so retiring in its habits as that bird, and will often leave the woods for the sake of preying upon the orange plantations, among which it works great havoc. Its nest is made in the decayed trunks of trees.

As a general fact, it is not so apt at learning and repeating phrases as the Grey Parrot, but I have known more than one instance where its powers of speech could hardly be exceeded, and very seldom rivalled. One of these birds which used to live in a little garden into which my window looked, was, on our first entrance into the house, the cause of much perplexity to ourselves and the servants. The nurserymaid's name was Sarah, and the unfortunate girl was continually running up and down stairs, fancying herself called by one of the children in distress. The voice of the Parrot was just that of a child, and it would call Sarah in every imaginable tone, varying from a mere enunciation of the name as if in conversation, to angry remonstrances, petulant peevishness, or sudden terror.

Even after we had been well accustomed to the bird, we were often startled by the sharp cry of "Sarah! Sa-rah, Sa . . . ráh!" Presently it would cry, "Sarah, lay the cloth;" and after a while, "Sarah, why *don't* you lay the cloth?" always contriving to get the name of that domestic into its sentences.

The end of the poor bird was rather tragic. It was the property of a very irritable master, from whom the angry cries for Sarah were probably learned. He was very fond of his Parrot, but one day in playing with her, he teased her so far beyond her patience, that she bit his finger, whereupon, in a fit of passion, he seized her by the neck and dashed her on the ground so hard, that she died on the spot.

SWINDERN'S LOVE-BIRD.—*Psittácula Swinderniána.*

From the Festive Parrot it may easily be distinguished, not only for its lesser size, it being barely twelve inches in length, but by the different arrangement of the colouring. The whole of the cheeks, chin, and the angles at the base of the bill are yellow, the forehead is deep blue-purple, and the feathers of the back of the head and nape of the neck are green, edged with black. When the bird is angry, it raises these feathers like a crest. The plumage of the body both above and below is rich green. The tail-feathers are beautifully marked with green, yellow, and red, and the primary feathers of the wings are tinged with green of various qualities, azure, deep brownish red and black.

GROUP OF COCKATOOS.

The Love-birds derive their name from the great fondness which they display for others of their own species, and the manner in which they always sit close to each other while perched, each trying to snuggle as closely as possible among the soft feathers of its neighbour. They are all little birds, and among the smallest of these is the SWINDERN'S LOVE-BIRD, which measures barely six inches in length.

It is a rather scarce bird, but deserves notice on account of its very small dimensions, and its beautiful plumage.   Like others of its kind, it is very fond of society, and unless furnished with a companion is very apt to droop, refuse nourishment, and die.   Its habits in a wild state are not precisely known, as it is a bird of rare occurrence, and not easily to be watched.

The head of this species is light grass-green ; round the back of the neck runs a black collar, and the chest, together with a band round the neck, just below the black collar, is yellow with a greenish cast.   The general colour of the body is the same grass-green as that of the head, except the upper tail-coverts, which are deep rich azure.   The short and rounded tail is beautifully and richly coloured, the two central feathers being green, and the others bright scarlet for the first half of their length, then banded with a warm bar of black, and the tips green.   The bill is black, and of a stronger make than is usually the case with the Love-birds.   The legs and feet are greyish black.

The Cockatoos are very familiar birds, as several species are common inhabitants of our aviaries, where they create much amusement by their grotesque movements, their exceeding love of approbation, and their repeated mention of their own name.   Wherever two or three of these birds are found in the same apartment, however silent they may be when left alone, the presence of a visitor excites them to immediate conversation, and the air resounds with "Cockatoo !" "Pretty Cocky !" in all directions, diversified with an occasional yell, if the utterer be not immediately noticed.

They are confined to the Eastern Archipelago and Australia, in which latter country a considerable number of large and splendid species are found.   The nesting-place of the Cockatoos is always in the holes of decaying trees, and by means of their very powerful beaks, they tear away the wood until they have augered the hollow to their liking.   Their food consists almost wholly of fruit and seeds, and they are often very great pests to the agriculturist, settling in large flocks upon the fields of maize and corn, and devouring the ripened ears or disinterring the newly sown seeds with hearty goodwill.   The wrath of the farmer is naturally aroused by these frequent raids, and the Cockatoos perish annually in great numbers from the constant persecution to which they are subjected, their nests being destroyed, and themselves shot and trapped.

To those, however, who own no land, and are anxious about no crops, a flock of Cockatoos is a most beautiful and welcome sight, as they flit among the heavy-leaved trees of the Australian forest, their pinky-white plumage relieved against the dark masses of umbrageous shade, as they appear and vanish among the branches like the bright visions of a dream.

The first of the Cockatoos which will be noticed in these pages is the GOLIATH ARATOO, a striking and very remarkable bird.

The generic name "microglossum" which is given to this creature is of Greek origin, and signifies "little-tongue," that member being very curiously formed.   In the generality of the Parrot tribe the tongue is thick and fleshy, but in the Aratoo it is long, tubular, and extensile.   The powerful bill is also of a rather unusual form, the upper mandible being very large, sharply curved, and having its cutting edges two-toothed, while the lower mandible is comparatively small, and only furnished with a single tooth.

It is a native of New Guinea and the neighbouring islands, and is not a very common bird, although specimens may be found in several museums.   The peculiar formation of the tongue and beak would lead the observer to suppose that its habits must be different from those of ordinary Cockatoos ; but as little or nothing is known of its mode of life in a wild state, the precise use of these organs is rather problematical.

In size this bird is one of the largest of the Parrot tribe, being equal to and in some cases exceeding that of the great macaws, although the absence of the long tail renders it a less conspicuous bird.   The general colour of this species is deep black with a greenish gloss, caused chiefly by the large amount of whitish powder which is secreted in certain imperfect quills, and thence scattered among the feathers, giving them a kind of "bloom," like that of the plum or grape.

GOLIATH ARATOO.—*Microglossum atérrimum.*

This substance is found very largely in most of the Parrot tribe, and I well remember getting my coat powdered like that of a miller from playing with the great white Cockatoo in the Zoological Gardens of Dublin. Many other birds, such as the vultures, possess this curious powdery substance, whose office is rather doubtful. The powder is produced from the formative substance of the quill, which, instead of being developed into shaft and web, as in the case of the perfect quills, dries up and is then thrown off in a dusty form. The imperfect quill-feathers can generally be seen intermixed with the rest of the plumage when the Cockatoo bends down its head or plumes itself, and the white substance may be seen in the open ends of the imperfect quills, or lying thickly about them. In the case of the vultures it is thought to be given for the purpose of keeping their skin and plumage undefiled by the putrid animal substances on which those unclean and useful birds feed, but as it is found in equal plenty on the Cockatoos, than who no cleaner feeding or more fastidious birds exist, it is evident that it must serve some purpose that is common to these two dissimilar species. Very little structure is found in this dust when placed under the microscope, but with the aid of the polarizer I have made out several well-marked hexagonal cells.

GREAT WHITE COCKATOO.—*Cacatúa cristátus.*

The green-black hue extends over the whole of the plumage, but around the eye is a large naked space of skin, red in colour, and covered with wrinkles. The head is ornamented with a large and curiously formed crest, which is composed of a number of single feathers, each being long, narrow, and the web rather scanty. The colour of the crest is rather greyer than the remainder of the plumage, probably on account of its less massive construction, and its freedom from the white powdery dust which has just been described. In general the crest lies along the top of the head, and merely exhibits the tips of its feathers projecting over the neck; but when the bird is excited by anger or pleasure, it can erect the crest as well as the common Cockatoo. Some naturalists think that there are two species of Aratoo, the larger being distinguished by the title of M. Goliath, and the smaller called by the name of M. atérrimum, but the general opinion leans in favour of a single species and two varieties.

Two species of Cockatoo are tolerably familiar in England, differing from each other in the colour of their crests.

The first of these is the GREAT WHITE COCKATOO, a remarkably handsome bird, especially when excited. In size it is rather a large bird, equalling a common fowl in dimensions, and assuming a much larger form when it ruffles up its feathers when under the influence of anger. Many of these birds are admirable talkers, and their voice is peculiarly full and loud.

A Great White Cockatoo which I lately saw, was rather celebrated for his powers of conversation; but as he was moulting, his vocabulary was silenced for the time, and he sat in a very disconsolate manner on his perch, looking as if he had fallen into a puddle and not had time to arrange his plumage. All the breast and fore-parts of the body were quite bare of feathers, and even the beautiful crest had a sodden and woe-begone look. By dint, however, of talking to the bird, and rubbing his head, I induced him to favour us with a few words which were given in a voice as full and rounded as that of a strong-voiced man accustomed to talk to deaf people.

Presently we were startled with a deafening laugh, not unlike that of the hyæna, but even louder and more weird-like. On turning round, I saw the Cockatoo suddenly transformed into a totally different bird, his whole frame literally blazing with excitement, his crest flung forward to the fullest extent, and repeatedly spread and closed like the fan of an angry Spanish lady, every feather standing on end and his eyes sparkling with fury while he volleyed forth the sounds which had so startled us. The cause of this excitement was to be found in the persons of two children who had come to look at the bird, and who by some means had excited his ire. He always objected to children, probably with good reason, and being naturally irritable from the effect of moulting, his temper was aroused by the presence of the objects of his dislike.

The plumage of this species is white with a very slight roseate tinge, and the crest is white.

The species of Cockatoo which is most common in England is the SULPHUR-CRESTED COCKATOO, well depicted in the illustration. It may readily be distinguished from the preceding bird by the bright yellow colour of its crest and its more pointed form.

This bird is an inhabitant of different parts of Australia, and is especially common in Van Diemen's Land, where it may be found in flocks of a thousand in number. Owing to the ease with which it is obtained, it is frequently brought to England and is held in much estimation as a pet.

A Cockatoo which I have lately seen, a young bird, displays admirably many peculiarities of the Cockatoo nature.

As yet it is not a very accomplished linguist, although it can repeat many words with much fidelity. It certainly has some notion of the meaning attached to certain words, as it can distinguish between the various members of the family, and when they enter the room will frequently utter their name. Sometimes it will act in the same manner when they leave the room. It can laugh merrily, but in rather too loud a tone for sensitive ears, and promises well for further accomplishments. Like others of the parrot tribe, it rejoices greatly in exercising its sharp beak, and is very fond of biting to pieces every bit of wood that may come in its way.

Empty cotton-reels are favourite toys, and it watches the gradual diminution of the thread with great interest, knowing that it is sure to have the wooden reel after the thread has been used. When the reel is placed on the outside of the cage the bird descends from its perch, pushes one of its feet through the wires and with extended toes feels in every direction for its toy. When the position of the coveted article is found, the bird grasps it with its feet, draws it through the wires, and bites it to pieces. Many times it has been known to split a reel with a single bite. Sometimes its owners give it one of those flat wooden discs on which silk-ribbon has been wound, and in such cases it always takes care to turn the disc edgeways before attempting to bring it through the wires.

So powerful is its beak that it can break up the shell of a periwinkle or even a whelk, and with its curved beak peck out the inhabitant. In a similar manner it will crack nuts to pieces, and extract the kernel; but seems to do so merely for the pleasure of exercising its beak, as it generally allows the kernel to fall on the floor and contents itself with breaking the shell into many little pieces.

SULPHUR-CRESTED COCKATOO.—*Cacatúa galerita.*

When I saw it, the plumage was in very fine order, and the crest with its double fan of bright yellow feathers had a remarkably fine effect as the bird ruffled up its plumage, erected the crest, and began bowing and crying "Pretty Cocky!" in a very excited state of mind.

Although its beak is so powerful, it can climb up the hands or face of any one whom it knows without doing any damage, whereas another Cockatoo of my acquaintance once inflicted unwitting but painful damage on my finger, as it lowered itself from my hand to its perch. I suppose that the bird found the substance of the finger yielding under the pressure of its beak, and fearful lest it should fall, gripped the finger in hope of saving itself, thereby inflicting a rather severe wound, and bruising the surrounding parts to such an extent that the whole finger swelled greatly, and for nearly a week could not be used.

The Cockatoo seems to court notice even more than the parrot, and will employ various ingenious manœuvres in order to attract attention to its perfections. They are mostly good-tempered birds, seldom trying to bite unless they have been teased, and even in that case they generally give fair notice of their belligerent intentions by yelling loudly with anger, and spreading their yellow crests in defiance of their enemy.

LEADBEATER'S COCKATOO.—*Cacatua Leadbeateri.*

The Cockatoo evidently possesses some sense of humour, particularly of that kind which is popularly known as practical joking. A lady had once shown some timidity in approaching a tame Cockatoo, and was evidently afraid of its beak. The bird thought that it was a great joke to frighten any one so much bigger than itself, and whenever the lady came near its perch, it would set up its feathers, yell and make believe to attack her, merely for the pleasure of hearing her scream and seeing her run away.

In its own country the Cockatoo is anything but a favourite on account of its devastation among the crops. In treating of this bird, Mr. Gould writes as follows: "As may be readily imagined, this bird is not upon favourable terms with the agriculturist, upon whose fields of newly sown grain and ripening maize it commits the greatest devastation. It is consequently hunted and shot down wherever it is found, a circumstance which tends much to lessen its numbers. It is still, however, very abundant, moving about in flocks varying from a hundred to a thousand in number, and evinces a decided preference for the open plains and cleared lands, rather than for the dense bushes near the coast.

Except when feeding or reposing on the trees after a repast, the presence of a flock, if not seen, is sure to be indicated by their horrid screaming notes, the discordance of which may be slightly conceived by those who have heard the peculiarly loud, piercing, and grating scream of the bird in captivity; always remembering the immense increase of the din occasioned by the large number of the birds uttering their disagreeable notes at the same moment."

The colour of this Cockatoo is white, with the exception of the crest, which is of a bright sulphur-yellow, and the under surface of the wings and the basal portions of the inner webs of the tail-feathers, which are of the same colour, but much paler in hue. The total length of this species is about eighteen inches.

THE remarkably handsome bird which is represented on page 543 is a native of Australia. It is called by several names, such as the TRICOLOR CRESTED COCKATOO, and the PINK COCKATOO, by which latter name it is known to the colonists. The title of LEADBEATER'S COCKATOO was given to the bird in honour of the well-known naturalist, who possessed the first specimen brought to England.

It is not so noisy as the common species, and may possibly prove a favourite inhabitant of our aviaries, its soft blush-white plumage and splendid crest well meriting the attention of bird-fanciers. The crest is remarkable for its great development, and for the manner in which the bird can raise it like a fan over its head, or depress it upon the back of its neck at will. In either case it has a very fine effect, and especially so when it is elevated, and the bird is excited with anger or pleasure.

The general colour of this bird is white, with a slight pinkish flush. Round the base of the beak runs a very narrow crimson line, and the feathers of the crest are long and pointed, each feather being crimson at the base, then broadly barred with golden yellow, then with crimson, and the remainder is white. The neck, breast, flanks, and under tail-coverts are deeply stained with crimson, and the under surface of the wing is deep crimson red. The beak is pale greyish white, the eyes brown, and the feet and legs dark grey, each scale being edged with a lighter tint. In size it is rather superior to the common white Cockatoo.

A VERY singular form of Cockatoo is that which is known as the PHILIP ISLAND, or the LONG-BILLED PARROT.

This bird is only found in the little island from which it derives its name. It may probably become extinct at no distant period, as its singularly shaped beak renders it an object of attraction to those who get their living by supplying the dealers with skins and various objects of natural history; and its disposition is so gentle and docile, that it readily accommodates itself to captivity. Philip's Island is only five miles in extent; and it is a very remarkable fact, that this Long-billed Parrot is never found even in Norfolk Island, though hardly four miles distant.

Its favourite resorts are among rocky ground interspersed with tall trees, and its food consists mostly of long and succulent vegetable substances. The blossoms of the white Hibiscus afford it a plentiful supply of food, and in order to enable it to obtain the sweet juices of the flowers, the tongue is furnished with a long, narrow, horny scoop at the under side of the extremity, not very unlike the human nail. As earth has often been found upon the long upper mandible, the bird is believed to seek some portion of its food in the ground, and to dig up with its pickaxe of a bill the ground nuts and other subterraneous vegetation. This opinion is strengthened by the fact that another species of Parrot belonging to the same country is known to seek its food by digging.

The hard and strong fruits which are so favoured by other Parrots, are rejected by this species, whose long bill does not possess the great power needed for cracking the shells. In captivity it has been known to feed upon various soft leaves, such as lettuce and cabbage, and displays a decided predilection for ripe fruits, cream, and butter.

While on the ground its mode of progression is not the ungainly waddle generally employed by the Parrot tribe, but is accomplished by hopping something after the fashion of the rooks, the wings aiding in each hop. One species of this genus has been known

PHILIP ISLAND PARROT.—*Nestor productus.*

to imitate the human voice with much accuracy. This is the Southern Nestor, or the Kaka of the natives (*Nestor hypopólius*). The voice of the Long-billed Parrot is harsh, loud, and very disagreeable, and is said to resemble the continual barking of a hoarse-voiced, ill-tempered cur. While ranging among the trees, these birds fill the woods with their dissonant quacking barks. The eggs of this species are white, and, as is generally the case with the Parrots, are laid in the hollow of a decaying tree.

The birds which belong to the genus Nestor may at once be known by their extra-ordinarily long upper mandibles, which curve far over the lower, and remind the observer of the overgrown tooth so common in the rat, rabbit, and other rodent animals. This remarkable structure is very probably for the purpose of enabling the bird to scoop roots and other vegetable substances out of the earth. The length, curve, and shape of the upper mandible differ in the various species. Another peculiarity is that the tips of the tail-feathers are partially denuded of their webs, leaving the shaft to project slightly beyond the feathered portion. Some persons suppose the Long-billed Parrots to form a link between the Parrots and the Cockatoos.

2.   N N

Neither of these birds are remarkable for brilliancy of plumage, the prevailing tints being brown and grey, with a little red and yellow here and there. The Philip's Island Parrot is dark brown on the upper surface of the body, but takes a greyish hue on the head and back of the neck. Each feather of the upper surface is edged with a deeper tinge, so that the otherwise uniform grey and brown is agreeably mottled. The cheeks, throat, and breast are yellow, warming into orange on the face. The inner surface of the shoulders is olive-yellow, and the abdomen and both tail-coverts are deep orange-red. The tail is moderately long, and squared at the extremity. The feathers are crossed at their base by bands of orange-yellow and brown, and the under surfaces of the inner webs are brown, mingled with dusky red. The feet are dark blackish brown, and the long bill is uniformly of a brownish tint. The total length of the adult bird is about fifteen inches.

The BANKSIAN COCKATOO is a good representative of a very curious genus of Cockatoos resident in Australia.

The plumage of these birds, instead of being white or roseate as in the two previous Cockatoos, is always of a dark colour, and frequently dyed with the richest hues. About six species belong to this genus, and they all seem to be wild and fierce birds, capable of using their tremendously powerful beaks with great effect. Their crests are not formed like those of the common Cockatoo, and the tails are larger and more rounded.

The Banksian Cockatoo is only found in New South Wales, inhabiting the vast brush district of that land. Its food is mostly of a vegetable nature, consisting chiefly of the seeds of the Banksia; but the bird will also eat the large and fat grubs of different insects, mostly of a coleopterous nature, which it digs out of the trunks of trees with its strong bill.

It is not seen in such large flocks as the white Cockatoo, being generally in pairs, although little companies of six or eight in number are occasionally met in the bushes. Being a particularly wild and cautious bird, it is not easily approached by a European, except when feeding, at which time it is so occupied that a cautious sportsman may creep within gunshot. The native, however, unencumbered with raiment, and caring nothing for his time, can glide through the bushes noiselessly, and bring down the bird with a well-aimed stick.

The flight of this handsome bird is rather heavy, the wings flapping laboriously, and the progress being rather slow. It seldom mounts to any great height, and as a general fact only flies from the top of one tree to another. The eggs are generally two and sometimes three in number, and are laid in the hollow "spout" of a green tree, without any particular nest.

The chin of the adult male is deep rich black with a green gloss. A broad vermilion band crosses the whole of the tail, with the exception of the two central feathers, and the external webs of the outside feathers. The female is also greenish black, but her plumage is variegated with numerous spots and bars of pale yellow.

THERE are many other species of Australian Cockatoos, which cannot be mentioned in these pages. The native mode of hunting Cockatoos is so curious, and displays so well the character of the birds, that it must be given in the words of the writer, Captain Grey.

"Perhaps as fine a sight as can be seen in the whole circle of native sports, is the killing Cockatoos with the kiley, or boomerang. A native perceives a large flight of Cockatoos in a forest which encircles a lagoon; the expanse of water affords an open clear space above it, unencumbered with trees, but which raise their gigantic forms all around, more vigorous in their growth from the damp soil in which they flourish. In their leafy summits sit a countless number of Cockatoos, screaming and flying from tree to tree, as they make their arrangements for a night's sound sleep.

The native throws aside his cloak, so that he may not have even this slight covering to impede his motions, draws his kiley from his belt, and with a noiseless, elastic step, approaches the lagoon, creeping from tree to tree, and from bush to bush, and disturbing the birds as little as possible. Their sentinels, however, take the alarm, the Cockatoos farthest from the water fly to the trees near its edge, and thus they keep concentrating

BANKSIAN COCKATOO.—*Calyptorhynchus Banksii.*

their force as the native advances; they are aware that danger is at hand, but are ignorant of its nature. At length the pursuer almost reaches the edge of the water, and the scared Cockatoos, with wild cries, spring into the air; at the same instant the native raises his right hand high over his shoulder, and, bounding forward with his utmost speed, to give impetus to his blow, the kiley quits his hand as if it would strike the water; but when it has almost touched the unruffled surface of the lake, it spins upwards with inconceivable velocity, and with the strangest contortions.

In vain the terrified Cockatoos strive to avoid it; it sweeps wildly and uncertainly through the air—and so eccentric are its motions, that it requires but a slight stretch of the imagination to fancy it endowed with life—and with fell swoops in rapid pursuit of the devoted birds, some of whom are almost certain to be brought screaming to the earth. But the wily savage has not yet done with them. He avails himself of the extraordinary attachment which these birds have for one another, and fastening a wounded one to a tree, so that its cries may induce its companions to return, he watches his opportunity, by throwing his kiley or spear, to add another bird or two to the booty he has already obtained."

OWL PARROT.—*Strigops habroptilus.*

THE name given to the curious bird now before us is a very appropriate one, as the creature seems to partake equally of the natures of the Owl and the Parrot.

Even in its habits it has much of the Owl nature, being as strictly nocturnal as any of those birds. During the daytime it conceals itself in holes, under the stumps of trees, and similar localities, and seldom being seen except after sunset. The natives of New Zealand, where it is found, say that during the winter months the Owl Parrots assemble together in great numbers, collecting themselves into certain large caverns, and that while arranging for their winter-quarters, and before dispersing for the summer, they become very noisy, and raise a deafening clamour.

The Owl Parrot is weak of wing and seldom trusts itself to the air, taking but a very short flight whenever it rises from the ground. Neither is it seen much in trees, preferring to inhabit the ground, and making regular paths to and from its nest, by means of which its habitation may be discovered by one who knows the habits of the bird. These tracks are about a foot in width, and so closely resemble the paths worn by the footsteps of human beings that they have been mistaken for such by travellers.

The food of this bird is mostly obtained on the ground, and consists of tender twigs, leaves, and roots, which it digs up with its curved bill, covering that useful organ with earth and mud. The eggs of the Owl Parrot are merely laid upon some decaying wood in the same hollows wherein the bird sleeps during the day. Their number is two, although three are sometimes found. The breeding season commences in February. The natives distinguish this bird by the name of Kakapo.

It is a very large bird, nearly equalling the eagle owl in dimensions ; and, like that bird, standing very upright on its legs. The general colour of the plumage is darkish green profusely mottled with black, and sparingly dashed with yellow. Under the eye is a patch of yellow green. The beak is long and curved, very like that of an owl, and it is nearly concealed by the stiff bristles with which it is surrounded, and many of which cross each other at the tips over the bill. The abdomen is green of a yellower hue than the upper parts of the body, crossed with a few very faint bars of a darker hue. The tail is also green, but marked with brown.

ACCORDING to some authors, the Aratoo, already described on page 538, is closely connected with the very remarkable bird represented in the accompanying illustration.

As in the case of the previous species we find an example of a Parrot following the owl type in its form and many of its habits, we have here an instance of another Parrot bearing a close resemblance to the diurnal predaceous birds. Indeed, from examining the Parrots and their habits, it is impossible not to perceive the analogy that exists between themselves and the birds of prey, many of whom are far less formidably armed than the vegetable-feeding Parrots. Perhaps we may call the Parrots vegetarian raptores.

The rather long generic name of Dasyptilus which has been given to this bird is of Greek origin, signifying " Hairy-plumage," and is appropriated to the bird

PESQUET'S DASYPTILUS.—*Dasyptilus Pesquétii.*

on account of the bristle-like feathers, which cover the head and neck, and the generally bristly character of the plumage. The beak is long, straight for a considerable portion of its length, and then curved suddenly downwards at the tip, just after the manner of the eagles. Indeed, if the head were removed from the body, nine persons out of ten would attribute it to one of the eagles. The lower mandible is, however, more like that of the Parrots, short, thick, and keeled. Around the eye there is a large patch of bare skin, and the bristly feathers of the head and neck very scantily protect those portions. The nostrils are round, and situated in the "cere" at the base of the beak.

The colouring of this bird is very simple. The general tint of the whole upper surface is black-green, like that of the Aratoo, excepting the greater wing-coverts, and the upper tail-coverts, which are of a rich crimson. The abdomen and thighs are also crimson, but with a perceptible vermilion tint. The upper part of the breast and the neck are black, and a very slight white edging appears on some of the feathers. The tail is moderately long, rounded, and very firmly made. The total length of this bird is about twenty inches.

HAIRY-BREASTED BARBET.—*Laimodon hirsutus.*

WE now take our leave of the Parrots, and come to a very interesting family of scansorial birds, known popularly as Woodpeckers, and scientifically as Picidæ.

There are many members of this large family, differing exceedingly in size, colour, and form, but yet possessing a kind of family resemblance not easy to be described, but readily recognisable. For convenience of description modern zoologists have grouped the Woodpeckers into several sub-families, all of which will be represented in the following pages, and which are termed the Capitoninæ or Barbets, the Picumninæ or Piculets, the Picinæ or true Woodpeckers, the Gecinæ or Green Woodpeckers, the Melanerpinæ or Black Woodpeckers, and the Colaptinæ or Ground Woodpeckers.

OUR example of the first sub-family is the HAIRY-BREASTED BARBET.

This is, perhaps, the most curious of all the Barbets, on account of the peculiarity from which it derives its name. The feathers of the breast are much stiffer than the others, and more sharply pointed, and the shafts of the lower breast-feathers are devoid of web, and project to the distance of nearly an inch from the rest of the plumage, looking as if a number of long curved bristles had been inserted among the plumage. All the Barbets possess strong and conical beaks, surrounded with bristles at the base, and their stiff tail-feathers enable them to support their bodies while they are perched upon the upright trunk of the tree on which they are seeking their insect food. They are all found in tropical climates, and the greater number, among which the present species may be included, are natives of Western Africa. In their habits they are said to be rather slow and sluggish birds, not possessed of the fiery vivacity which distinguishes the true Woodpeckers, and their food is not so wholly of an insect nature. The wings and tail are short, and all the species are of small dimensions.

The general colour of this bird is brown on the upper parts of the body, spotted with sulphur-yellow, a round mark of that tint being found on the end of each feather. The head, chin, and part of the throat are black, and there is one white stripe behind the eye, and another running from the angle of the mouth down the neck. The quill-feathers of the wings are deep brown, edged with sulphur-yellow. The whole of the under surface is yellow with a green tinge, and is profusely spotted with black. The total length of this species rather exceeds seven inches.

THE Piculets seem to bear the same proportion to the Woodpeckers as the merlin to the eagle, being about the size of sparrows, and more slenderly framed. Their bills are shorter in proportion than those of the true woodpeckers, and are rather deeper than wide at the base. Their wings are short and rounded, and their tails are also short.

The PIGMY PICULET is a very pretty example of this little sub-family. It is a native of Southern America, and is generally found in the vast forests of that fertile land. It is a lively little creature, running quickly up the trunks of trees after the manner of the English creeper, but seldom appearing to use its tail in aid of its progress, or to seek its food on the tree-trunks in the usual Woodpecker fashion. In general it is seen among the branches, where it sits across the boughs when at rest, and hops quickly from one branch to another while searching after its food.

It is not a gregarious bird, being generally found either singly or in pairs. The nest of this species is made in hollow trees, and its eggs are only two in number.

PIGMY PICULET.—*Picumnus pygmæus.*

This species is a remarkably pretty one, elegant in shape and delicately coloured. The general colour of the back and upper portions of the body is a very soft hair-brown, and the wings are also brown, but of a deeper hue. Over the back are scattered a few oval spots of a much lighter brown, each having a nearly black spot towards one end, and contrasting in a very pleasing manner with the delicate brown of the back. The tail is of the same dark brown as the wings, with the exception of the two central feathers, which are of a light fawn. The most striking portion of this bird is the top of the head, which is decorated with a bright scarlet crest-like crown, covered with velvety-black dots. The rest of the head and the back of the neck are jetty black, interspersed with white dots. The under surface of the body is pale brown variegated with the same curious spots as those of the back. In size this bird hardly exceeds a wren.

WE now arrive at the true Woodpeckers, several species of which bird are familiar from their frequent occurrence in this country.

As is well known, the name of Woodpecker is given to these birds from their habit of pecking among the decaying wood of trees in order to feed upon the insects that are found within. They also chip away the wood for the purpose of making the holes or tunnels wherein their eggs are deposited. In order to enable them to perform these duties, the structure of the Woodpecker is very curiously modified. The feet are made extremely powerful, and the claws are strong and sharply hooked, so that the bird can retain a firm hold of the tree to which it is clinging, while it works away at the bark or wood with its bill. The tail, too, is furnished with very stiff and pointed feathers, which are pressed against the bark, and form a kind of support on which the bird can rest a large proportion of its weight. The breast-bone is not so prominent as in the generality of flying birds, in order to enable the Woodpecker to press its breast closely to the tree, and the beak is long, strong, and sharp.

These modifications aid the bird in cutting away the wood, but there is yet a provision needful to render the Woodpecker capable of seizing the little insects on which it feeds, and which lurk in small holes and crannies into which the beak of the Woodpecker could not penetrate. This structure is shown by the accompanying sketch of a Woodpecker's

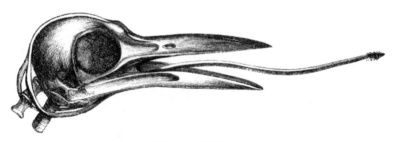

HEAD OF WOODPECKER.

head dissected. The tongue-bones or " hyoid" bones are greatly lengthened, and pass over the top of the head, being fastened in the skull just above the right nostril. These long tendinous-looking bones are accompanied by a narrow strip of muscle by which they are moved.

The tongue is furnished at the tip with a long horny appendage covered with barbs and sharply pointed at the extremity, so that the bird is enabled to project this instrument to a considerable distance from the bill, transfix an insect, and draw it into the mouth. Those insects that are too small to be thus treated are captured by means of a glutinous liquid poured upon the tongue from certain glands within the mouth, and which cause the little insects to adhere to the weapon suddenly projected among them. This whole arrangement is clearly analogous to the tongue of the ant-eater, described in the volume on Mammalia, page 771. Some authors deny the transfixion.

The GREAT SPOTTED WOODPECKER is one of the five British species, and is also known by the names of Frenchpie and Woodpie.

It is found in many parts of England, and, like the other Woodpeckers, must be sought in the forests and woods rather than in orchards and gardens. Like other shy birds, however, it soon finds out where it may take up its abode unmolested, and will occasionally make its nest in some cultivated ground, where it has the instinctive assurance of safety, rather than entrust itself to the uncertain security of the forest.

In the woods frequented by these birds, which are often more plentiful than is generally known, the careful observer may watch their movements without difficulty, by taking a few preliminary precautions.

The rapid series of strokes on the bark, something like the sound of a watchman's rattle, will indicate the direction in which the bird is working; and when the intruding observer has drawn near the tree on which he suspects the Woodpecker to have settled, he should quietly sit or lie down, without moving. At first the bird will not be visible, for the Woodpeckers, like the squirrels, have a natural tact for keeping the tree-trunk or branch between themselves and the supposed enemy, and will not show themselves until they think that the danger has passed away.

Presently the Woodpecker may be seen coming very cautiously round the tree, peering here and there, to assure itself that the coast is clear, and then, after a few preliminary taps, will set vigorously to work. So rapidly do the blows follow each other, that the head of the bird seems to be vibrating on a spring, and the sound can only be described by the comparison already made, namely, a watchman's rattle. Chips and bark fly in every direction, and should the tree be an old one, whole heaps of bark will be discovered at the foot. By the aid of a small telescope, the tongue can be seen darted out occasionally, but the movement is so quick, that unless the attention of the observer be especially directed towards it, he will fail to notice it.

The Woodpecker has several modes of tapping the trees, which can be readily distinguished by a practised ear. First there is the preliminary tap and the rapid whirring strokes already described, when the bird is engaged in seeking its food. Then there is a curious kind of sound made by pushing its beak into a crack, and rattling it in such a manner against the wood, that the insects think their house is falling, and run out to escape the impending danger, just as worms come to the surface when the ground is agitated by a spade or fork. Lastly, there is a kind of drumming sound made by striking the bill against some hollow tree, and used together with the peculiar cry for the purpose of calling its mate.

Although the Woodpeckers were formerly much persecuted, under the idea that they killed the trees by pecking holes in them, they are most useful birds, cutting away the decaying wood, as a surgeon removes a gangrened spot, and eating the hosts of insects which encamp in dead or dying wood,

GREAT SPOTTED WOODPECKER.—*Picus major.*

and would soon bring the whole tree to the ground. They do not confine themselves to trees, but seek their food wherever they can find it, searching old posts and rails, and especially delighting in those trees that are much infested with the green fly, or aphis, as the wood-ants swarm in such trees for the purpose of obtaining the "honey-dew," as it distils from the aphides, and then the Woodpeckers eat the ants. Those destructive creatures generally called wood-lice, and known to boys as "monkey-peas," are a favourite article of diet with the Woodpeckers, to whom our best thanks are therefore due.

But the Woodpeckers, although living mostly on insects, do not confine themselves wholly to that diet, but are very fond of fruits, always choosing the ripest. In this country the forest-land forms so small a portion of the kingdom, that the Woodpeckers are comparatively few, and can do little appreciable mischief to the gardens; but in some lands, such as many parts of America, they do very great damage, stripping the trees of their fruit, and the fields of their crops, to such an extent that they are annually shot by hundreds.

As is the case with all its congeners, the Great Spotted Woodpecker lays its eggs in the hollow of a tree.

The locality chosen for this purpose is carefully selected, and is a tunnel excavated, or at all events altered, by the bird for the special purpose of nidification. Before commencing the operation, the Woodpeckers always find out whether the tree is sound or rotten, and they can ascertain the latter fact, even through several layers of sound wood. When they have fixed upon a site for their domicile, they set determinately to work, and speedily cut out a circular tunnel just large enough to admit their bodies, but no larger. Sometimes this tunnel is tolerably straight, but it generally turns off in another direction.

At the bottom of the hole the female bird collects the little chips of decayed wood that have been cut off during the boring process, and deposits her eggs upon them without any attempt at nest-making. The eggs are generally five in number, but six have been taken from the nest of this species. The young are able to run about the tree some time before they can fly, and traverse the bark quite fearlessly, retiring to the hole and calling their parents whenever they want food.

Generally the nests of birds are kept scrupulously clean; but that of the Woodpecker is a sad exception to the rule, the amount of filth and potency of stench being quite beyond human endurance. The colour of the eggs is white, and their surface glossy, and they are remarkable, when fresh, for some very faint and very narrow lines, which run longitudinally down the shell towards the small end.

The general colour of this species is black and white, curiously disposed, with the exception of the back of the head, which is light scarlet, and contrasts strongly with the sober hues of the body. Taking the black to be the ground colour, the white is thus arranged: The forehead and ear-coverts, a patch on each side of the neck, the scapularies, and part of the wing-coverts, several little squared spots on the wings, and large patches on the tail, are pure white. The throat and the whole of the under surface are also white, but with a greyish cast, and the under tail-coverts are red. The total length of the adult male is rather more than nine inches. The female has no red on the head, and the young birds of the first year are remarkable for having the back of the head black and the top of the head red, often mixed with a few little black feathers.

The DOWNY WOODPECKER derives its name from the strip of loose downy feathers which passes along its back. It is a native of America, and very plentiful in various parts of that country. Its habits are so well described by Wilson, that his own words will be the best comment on this pretty little bird.

" About the middle of May the male and female look out for a suitable place for the reception of their eggs and young. An apple, pear, or cherry tree, often in the near neighbourhood of the farmhouse, is generally pitched upon for this purpose. The tree is minutely reconnoitred for several days previous to the operation, and the work is first begun by the male, who cuts out a hole in the solid wood as circular as if described with a pair of compasses. He is occasionally relieved by the female, both parties working with the most indefatigable diligence. The direction of the hole, if made in the body of the tree, is generally downwards by an angle of thirty or forty degrees for the distance of six or eight inches, and then straight down for ten or twelve more; within roomy, capacious, and as smooth as if polished by the cabinet-maker; but the entrance is judiciously left just so large as to admit the bodies of the owners.

During this labour they regularly carry out the chips, often strewing them at a distance to prevent suspicion. This operation sometimes occupies the chief part of a week. Before she begins to lay, the female often visits the place, passes in and out, examines

DOWNY WOODPECKER.—*Picus pubescens.*

every part of the exterior and interior with great attention, as every prudent tenant of a new house ought to do, and at length takes complete possession. The eggs are generally six, pure white, and laid in the smooth bottom of the cavity. The male occasionally supplies the female with food while she is sitting, and about the last week in June the young are perceived making their way up the tree, climbing with considerable dexterity."

The same writer then proceeds to remark that the process of nest-making is not always permitted to go on without hindrance, for the impertinent little house-wren, who likes to build her nest in hollows, but who is not strong or large enough to scoop a habitation for herself, will often allow the Woodpeckers to make a nice deep hole just fit for a wren's nest, and then drives them off and takes possession of the deserted domicile. One pair of Woodpeckers met with very hard treatment, being twice turned out of their house in one season, and the second time they were even forced to abandon one egg that had been laid.

The holes made by this Woodpecker in trees are very numerous, and have often led more observant orchard-owners to think the bird an enemy to their trees, and to kill it accordingly. Wilson has, however, completely exonerated the bird from the charge, and proved it to be a useful ally to man instead of a noxious foe. " Of all our Woodpeckers, none rid the apple-trees of so many vermin as this; digging off the moss which the negligence of the proprietor had suffered to accumulate, and probing every crevice. In fact, the orchard is his favourite resort in all seasons; and his industry is unequalled and almost incessant, which is more than can be said of any other species we have.

In fall he is particularly fond of boring the apple-trees for insects, digging a circular hole through the bark just sufficient to admit his bill; after that a second, third, &c. in pretty regular horizontal circles round the body of the tree. These parallel circles of holes are often not more than an inch or an inch and a half apart, and sometimes so close together, that I have covered eight or ten of them at once with a dollar. From nearly the surface of the ground up to the first fork, and sometimes far beyond it, the whole bark of many apple-trees is perforated in this manner, so as to appear as if made by successive discharges of buckshot; and our little Woodpecker, the subject of the present account, is the principal perpetrator of this supposed mischief. I say supposed, for, so far from these perforations of the bark being ruinous, they are not only harmless, but, I have good reason to believe, really beneficial to the health and fertility of the tree.

In more than fifty orchards which I have myself carefully examined, those trees which were marked by the Woodpecker (for some trees they never touch, perhaps because not penetrated by insects) were uniformly the most thriving, and seemingly the most productive. Many of them were upwards of sixty years old, their trunks completely covered with holes, while the branches were broad, luxuriant, and loaded with fruit. Of decayed trees, more than three-fourths were untouched by the Woodpecker."

Although a little bird—less than seven inches in length—it is a truly handsome one. The crown of the head is velvety black, its back deep scarlet, and there is a white streak over the eye. The back is black, but is divided by a lateral stripe of puffy or downy white feathers. The wings are black spotted with white, and the tail is also variegated with the same tints. From the base of the beak a black streak runs down the neck. The sides of the neck, the throat, and the whole of the under parts of the body are white. The nostrils are thickly covered with small bristly feathers, probably to protect them from the chips of wood struck off by the beak. The female is known by the greyish white of the abdomen, and the absence of red upon its head.

ALTHOUGH not the largest of the Woodpecker tribe, the IVORY-BILLED WOODPECKER, of North America, is perhaps the handsomest and most striking in appearance.

This splendid bird is armed with a tremendous beak, long, powerful, sharp, and white as ivory, which can be used equally as an instrument for obtaining its food, or as a weapon for repelling the attacks of its enemies, and, in the latter point of view, is a truly formidable arm, as terrible to its enemies as the British bayonet, to which it bears no little resemblance in general shape.

Few birds are more useful than the Ivory-billed Woodpecker, which wages continual war upon the myriad insects which undermine the bark of forest-trees, and saves the forest giants from falling a prey to their diminutive adversaries. In one season several thousand acres of huge pine-trees, from two to three feet in diameter, and many of them measuring one hundred and fifty feet in height, were destroyed by the larvæ of a little insect not bigger than a grain of rice. Besides this creature, there are large grubs and caterpillars that bore their way into the interior of trees, and are the pioneers of the destruction that afterwards follows.

When the Ivory-billed Woodpecker has been hard at work upon a tree, he leaves ample traces of his progress in the heaps of bark and wood chips which surround the tree, and which look, according to Wilson, as if a dozen axe-men had been working at the trunk. Strips of bark seven or eight inches in length are often struck off by a single blow, and the body of the tree is covered with great excavations that seem more like the work of steel tools than of a bird's beak. Yet these apparent damages are really useful to the tree, as the sound wood is allowed to remain in its place performing its proper functions, while the decaying substances are scooped out in order that the bird may get at the grubs and beetles that make their home therein.

As in the case of all Woodpeckers, the beak is also employed in excavating the holes

IVORY-BILLED WOODPECKER.—*Campéphilus principális.*

in which the eggs are laid. The following account of the nesting of this bird is given by Audubon :—

"The Ivory-billed Woodpecker nestles earlier in spring than any other species of its tribe. I have observed it boring a hole for that purpose in the beginning of March. The hole is, I believe, always made in the trunk of a live tree, generally of an ash or a hagberry, and is at a great height.

The birds pay great regard to the particular situation of the tree, and the inclination of its trunk, first, because they prefer retirement, and again, because they are anxious to secure the aperture against the access of water during beating rains. To prevent such a calamity, the hole is generally dug immediately under the juncture of a large branch with the trunk. It is first bored horizontally for a few inches, then directly downwards, and not in a spiral manner, as some people have imagined. According to circumstances, this cavity is more or less deep, being sometimes not more than ten inches, whilst at other times it reaches nearly three feet downwards into the core of the tree. I have been led to think these differences result from the more or less necessity under which the female may be of depositing her eggs, and again have thought that the older the Woodpecker is, the deeper does it make its hole. The average diameter of the different nests which I have examined was about seven inches within, although the entrance, which is perfectly round, is only just large enough to admit the bird.

Both birds work most assiduously at this excavation, one waiting outside to encourage the other whilst it is engaged in digging, and when the latter is fatigued, taking its place. I have approached trees whilst these Woodpeckers were thus busily employed in forming their nest, and by resting my head against the bark could easily distinguish every blow given by the bird. I observed that in two instances, when the Woodpeckers saw me thus at the foot of the tree in which they were digging their nest, they abandoned it for ever. For the first brood there are generally six eggs. They are deposited in a few chips at the bottom of the hole, and are of a pure white colour. The young are seen creeping out of the hole about a fortnight before they venture to fly to any other tree. The second brood makes its appearance about the 15th of August."

The courage and determination of the Ivory-billed Woodpecker is very great, and it will fight with its opponent in a most desperate manner. When wounded, it endeavours to reach the nearest tree, and to run up its trunk, and if intercepted will peck as fiercely at the hand of its pursuer as at the wood and bark, and is able to inflict severe injury with its sharp powerful bill. On account of this bold and fiery disposition, the American Indians pay much honour to the bird, and are in the habit of carrying its head and bill among the numerous charms or "medicines" in which they delight, and which are supposed to transmit to the wearer the good qualities of the slain creature.

The voice of this Woodpecker is seldom uttered while the bird is on the wing, but is frequently heard as soon as the bird has alighted. It is a rather shrill and very loud tone, and can be heard at a great distance.

The cry of the wounded bird is, according to Wilson, just like that of a hurt child. "The first place I observed this bird at, when on my way to the south, was about twelve miles north of Wilmington, in North Carolina. Having wounded it slightly in the wing, on being caught, it uttered a loudly reiterated and most piteous note, exactly resembling the violent crying of a young child, which terrified my horse so, as nearly to have cost me my life.

It was distressing to hear it. I carried it with me in the chair, under cover, to Wilmington. In passing through the streets, its affecting cries surprised every one within hearing, particularly the females, who hurried to the doors and windows with looks of alarm and anxiety. I drove on, and on arriving at the piazza of the hotel where I intended to put up, the landlord came forward, and a number of other persons who happened to be there, all equally alarmed at what they heard; this was greatly increased by my asking, whether he could furnish me with accommodations for myself and my baby. The man looked blank and foolish, while the others stared with still greater astonishment. After diverting myself for a minute or two at their expense, I drew my Woodpecker from under the cover, and a general laugh took place. I took him upstairs, and locked him up in my room, while I went to see my horse taken care of.

In less than an hour I returned, and on opening the door, he set up the same distressing shout, which now appeared to proceed from grief that he had been discovered in his efforts at escape. He had mounted along the side of the window, nearly as high as the ceiling, a little below which he had begun to break through. The bed was covered with large pieces of plaster; the lath was exposed for at least fifteen inches square, and a hole large enough to admit the fist, opened to the weather boards; so that in less than another hour he would certainly have succeeded in making his way through. I now tied a string round his leg, and fastening it to the table, again left him. I wished to preserve his life, and had gone off in search of suitable food for him. As I reascended the stairs, I heard him again hard at work, and on entering had the mortification to perceive that he had almost entirely ruined the mahogany table to which he was fastened, and on which he had wreaked his whole vengeance. While engaged in taking a drawing, he cut me severely in several places, and on the whole displayed such a noble and unconquerable spirit, that I was frequently tempted to restore him to his native woods. He lived with me nearly three days, but refused all sustenance, and I witnessed his death with regret."

The general colour of this bird is black, glossed with green. The fore part of the head is black, and the remainder is covered with a beautiful scarlet crest, each feather being spotted towards the bottom with white, and taking a greyish ashen hue at the base. Of course these colours can only be seen when the crest is erected. From below the eye a white streak runs down the neck, and along the back, nearly to the insertion of the tail, and the secondaries, together with their coverts and the tips of some of the primaries, are also white, so that when the bird shuts its wings, its back appears wholly white. The tapering tail is black above, yellowish white below, and each feather is singularly concave. The wings are also lined with yellowish white. The bill is white as ivory, strong, fluted along its length, and nearly an inch broad at the base. The female is plumaged like the male, with the exception of the head, which is wholly black, without the beautiful scarlet crest. The total length of the Ivory-billed Woodpecker is about twenty inches.

THE commonest of the British Woodpeckers is that which is generally known by the name of the GREEN WOODPECKER. It has, however, many popular titles, such as Rain-bird, Wood-spite, Hew-hole, and Wood-wall. This bird is our representative of the Gecinæ, or Green Woodpeckers.

Although the Green Woodpecker is a haunter of woods and forests, it will sometimes leave those favoured localities, and visit the neighbourhood of man. The grounds between the Isis and Merton College, Oxford, are rather favourite resorts of this pretty bird, and I once performed something of a cruel feat by flinging a brickbat at a Green Woodpecker, without the least idea of hitting it, and crushing its legs with the edge of the brick. I do not think I ever threw a stone at a bird afterwards, and though the event happened some years ago, I have never forgiven myself for it.

GREEN WOODPECKER.—*Gecinus viridis.*

The name of Rain-bird has been given to this species because it becomes very vociferous at the approach of wet weather, and is, as Mr. Yarrell well observes, "a living barometer to good observers." Most birds, however, will answer the same purpose to those who know how and where to look for them. The other titles are equally appropriate, Wood-spite being clearly a corruption of the German term "specht." Hew-hole speaks for itself; and Wood-wall is an ancient name for the bird, occurring in the old English poets.

This species, although mostly found on trees, is a frequent visitor to the ground, where it finds abundance of food. Ants' nests are said to form a great attraction to the Green Woodpecker, which feasts merrily at the expense of the insect community. During the autumn, it also lives on vegetable food, being especially fond of nuts, which it can crack without any difficulty by repeated strokes from its bill. The nest of this Woodpecker is, like that of the other species, a mere heap of soft decaying wood at the bottom of a tunnel dug by the birds, or adapted to their use from an already existing cavity.

The colouring of this species is very pretty. The top of the head is bright scarlet, and from the base of the beak starts a kind of moustache, black, with a scarlet centre. The whole of the upper surface is dark green, mixed with yellow, changing to sulphur-yellow on the upper tail-coverts. The primaries are greyish black spotted with white, and the secondaries and tertials are green on their outer webs, and grey-black spotted with white on the inner. The stiff tail-feathers are greyish black, variegated with some bars of a lighter hue ; and the throat, chest, and all the under surface are ashen green. The colour of the beak is dark horny black. The female may be known from her mate by the wholly black moustache, and the smaller ornament of scarlet on the head. In the young birds of both sexes, the scarlet of the head is mottled with black and yellow, the green feathers of the back are yellow at their tips, and the under surface is dull brownish white, with streaks and bars of greyish black. The total length of this bird rather exceeds one foot. The other British species are the Great Black Woodpecker (*Dryócopus Mártius*), the Northern Three-toed Woodpecker (*Picóïdes tridáctylus*), and the Lesser Spotted Woodpecker (*Picus minor*).

AMERICA possesses many species of these birds, among which the RED-HEADED WOOD-PECKER deserves a short notice, as being a good representative of the Black Woodpeckers. It is one of the commonest of American birds, bold, fearless of man, and even venturing within the precincts of towns. The habits of this bird are well told by Audubon and Wilson. The former author remarks of this bird : " When alighted on a fence stake by the road, or in a field, and one approaches them, they gradually move sideways out of sight, peeping now and then to discover your intention, and when you are quite close and opposite, lie still until you have passed, when they hop to the top of the stake, and rattle upon it with their bill, as if to congratulate themselves on the success of their cunning. Should you approach within arm's length, which may frequently be done, the Woodpecker flies to the first stake or the second from you, bends his head to peep, and rattles again, as if to provoke you to continuance of what seems to him excellent sport. He alights on the roof of the house, hops along it, beats the shingles, utters a cry, and dives into your garden to pick the finest strawberries he can discover."

Every one who has had practical experience of this bird agrees that it is very mischievous in a garden ; and even Wilson, whose kind heart would hardly permit him to see that any feathered creature could be hurtful to man, is forced to admit that its robberies are very extensive, but ought to be conceded as a tribute of thankfulness to the bird for eating so many grubs. " Wherever there is a tree or trees of the wild cherry," writes Wilson, " covered with ripe fruit, there you see them busy among the branches, and in passing orchards you may easily know where to find the earliest and sweetest apples, by observing those trees on or near which the Red-headed Woodpecker is skulking. For he is so excellent a connoisseur in fruit, that wherever an apple or pear tree is found broached by him, it is sure to be among the ripest and best-flavoured ; when alarmed, he seizes a capital one by striking his open bill deep into it, and bears it off to the woods."

When the Indian corn is in its rich, succulent, milky state, he attacks it with great eagerness, opening a passage through the numerous folds of the husk, and feeding on it with voracity. The girdled or deadened timber, so common among cornfields in the back settlements, are his favourite retreats, whence he sallies out to make his depredations. He is fond of the ripe berries of the sour gum, and pays pretty regular visits to the cherry-trees when loaded with fruit. Towards fall he often approaches the barn or farmhouse, and raps on the shingles and weather-boards. He is of a gay and frolicsome disposition, and half a dozen of the fraternity are frequently seen diving and vociferating around the

RED-HEADED WOODPECKER.—*Melanerpes erythrocéphalus.*

high dead limbs of some large tree, pursuing and playing with each other, and amusing the passenger with their gambols.

Their note or cry is shrill and lively, and so much resembles that of a species of tree-frog which inhabits the same tree, that it is sometimes difficult to distinguish the one from the other."

On account of the garden-robbing propensities of this bird, it is held in much odium, and trapped whenever occasion offers itself. In some places the feeling against it was so strong, that a reward was offered for its destruction. It is probable, however, that the services which it renders by the destruction of acknowledgedly noxious insects may more than compensate for its autumnal ravages in the fields and orchards.

Unlike the previous species, which is a permanent inhabitant, the Red-headed Woodpecker is a bird of passage, appearing in Pennsylvania about the beginning of May, and leaving that country towards the end of October. The eggs of this bird are pure white, speckled with reddish brown, mostly towards the larger end, and generally six in number.

The adult male is a really beautiful bird, its plumage glowing with steely black, snowy white, and brilliant scarlet, disposed as follows: The head and neck are deep scarlet, and the upper parts of the body are black, with a steel-blue gloss. The upper tail-coverts, the secondaries, the breast, and abdomen, are pure white. The beak is light blue, deepening into black towards the tip; the legs and feet are blue-green, the claws blue, and round the eye there is a patch of bare skin of a dusky colour. The female is coloured like her mate, except that her tints are not so brilliant. The young of the first year have the head and neck blackish grey, and the white on the wings is variegated with black. The total length of the bird is between nine and ten inches.

The Ground Woodpeckers are represented by the GOLD-WINGED WOODPECKER of America.

This bird may lay claim to the title of the feathered ant-eater, for it feeds very largely on those insects, and has its beak shaped in a somewhat pickaxe-like form, in order to enable it to dig up their nests from the ground and the decaying stumps of trees. In the stomach of one of these birds Wilson found a mass of ants nearly as large as a plum. It also feeds much on woodlice, those destructive creatures which eat the bitterest and the

2.

O O

GOLD-WINGED WOODPECKER.—*Colaptes aurátus.*

toughest substances with the best of appetites, and have been known to render a boat unsafe for sea, in spite of the strong flavour of salt water, pitch, and tar, with which sea-faring boats are so liberally imbued.

It is a brisk, lively, and playful creature, skipping about the trunks of trees with great activity, and "hopping not only upwards and downwards, but spirally, pursuing and playing with its fellow in this manner round the body of the tree." I may here mention that I never yet saw an English Woodpecker hop down the tree's trunk. Like others of its race, it is fond of varying its insect diet with a little vegetable food, eating various fruits, the Indian corn, the wild cherries, and the sour gum and cedar berries.

The Gold-winged Woodpecker seems to be readily tamed, as may be seen from the following account by Wilson.

"In rambling through the woods one day, I happened to shoot one of these birds and wounded him slightly in the wing. Finding him in full feather, and seemingly but little hurt, I took him home and put him into a large cage made of willows, intending to keep him in my own room, that we might become better acquainted.

As soon as he found himself inclosed on all sides, he lost no time in idle fluttering, but throwing himself against the bars of the cage, began instantly to demolish the willows, battering them with great vehemence, and uttering a loud piteous kind of cackling, similar to that of a hen when she is alarmed and takes to wing. Poor Baron Trenck never laboured with more eager diligence at the walls of his prison than this son of the forest in his exertions for liberty; and he exercised his powerful bill with such force, digging into the sticks, seizing and shaking them from side to side, that he soon opened for himself a passage, and though I repeatedly repaired the breach, and barricaded every opening in the best manner I could, yet, on my return into the room, I always found him at large, climbing up the chairs, or running about the floor, where, from the dexterity of his motions, moving backwards, forwards, and sideways with the same facility, it became difficult to get hold of him again.

Having placed him in a strong wire cage, he seemed to give up all hopes of making his escape, and soon became very tame; fed on young ears of Indian corn, refused apples, but ate the berries of the sour gum greedily, small winter grapes, and several other kinds of berries, exercised himself frequently in climbing, or rather hopping perpendicularly along the sides of the cage, and as evening drew on fixed himself in a high hanging or perpendicular position, and slept with his head in his wing.

As soon as dawn appeared, even before it was light enough to perceive him distinctly across the room, he descended to the bottom of the cage and began his attack on the ears of Indian corn, rapping so loud as to be heard from every room in the house. After this he would sometimes resume his former position and take another nap. He was beginning to become very amusing and even sociable, when, after a lapse of several weeks, he became drooping and died, as I conceived from the effects of his wound."

The colouring of the Gold-winged Woodpecker is very complicated. The top of the head is grey, the cheeks are cinnamon, and the back and wings are umber, marked with transverse bars of black. On the back of the head is a semilunar spot of blood-red, the two horns pointing towards the eyes, and a streak of black passes from the base of the beak down the throat. The sides of the neck are grey. The breast, throat, and chin are cinnamon, and a broad crescentic patch of black crosses the chest. The abdomen is yellowish-white, profusely spotted with black. The upper tail-coverts are white, serrated with black. The inner sides of the wings and tail, and the shafts of nearly all the feathers, are of a beautiful golden yellow ; " the upper sides of the tail and the tip below are black, edged with light loose filaments of a cream colour, the two exterior feathers serrated with whitish." The bill is dusky brown colour and slightly bent. The female is coloured, but does not possess the black feathers on each side of the throat. The total length of this bird is about one foot.

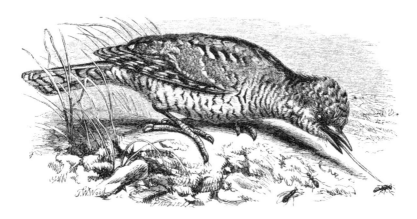

WRYNECK.—*Yunx torquilla.*

THIS curious bird, known under the popular and appropriate name of the WRYNECK, is by some authors considered to be closely allied to the woodpeckers.

The Wryneck is a summer visitant to this country, appearing just before the cuckoo, and therefore known in some parts of England as the cuckoo's footman. There is a Welsh name for this bird, signifying " Cuckoo's knave," " Gwas-y-gôg," the pronunciation of which I must leave to Welsh throats.

The tongue of this bird is long, slender, and capable of being projected to the distance of an inch or so from the extremity of the beak, and its construction is almost exactly the same as that of the woodpecker. As might be supposed, it is employed for the same purpose, being used in capturing little insects, of which ants form its favourite diet. So fond, indeed, is the Wryneck of these insects, that in some parts of England it is popularly known by the name of Emmet-hunter. In pursuit of ants it trips nimbly about the trunks and branches of trees, picking them off neatly with its tongue as they run their untiring course. It also frequents ant-hills, especially when the insects are bringing out their pupæ to lie in the sun, and swallows ants and pupæ at a great pace. When, as in damp or cold weather, the ants remain within their fortress, the Wryneck pecks briskly at the hillock until it breaks its way through the fragile walls of the nest, and as the warlike insects come rushing out to attack the intruder of their home and to repair damages, it makes an excellent meal of them in spite of their anger and their stings.

When ants are scarce and scantily spread over the ground, the Wryneck runs after them in a very agile fashion; but when it comes upon a well-stocked spot, it stands motionless, with the exception of the head, which is darted rapidly in every direction, the neck and central line of the back twisting in a manner that reminds the observer of a snake. When captured or wounded, it will lie on its back, ruffle up its feathers, erect its neck, and hiss so like an angry serpent that it is in some places known by the name of the snake-bird. It is a bird of retiring habits, keeping itself mostly to the wooded parts of the country, and especially favouring fir-woods where the ants most congregate, the dead leaves of the fir-trees forming excellent material for their nests without the trouble of cutting them to a proper length.

As the food of this bird is so dependent on the ants, it only comes to this country when the weather is warm enough to induce the ants to leave their winter quarters; and as soon as they begin to retire into their hiding-places, it takes its departure for warmer lands. During the early part of the season they are rather sociable, and may be captured by a good imitation of their call-note.

Mr. Yarrell, however, seems to hold rather an opposite opinion, and says that "the Wryneck is rather solitary in its habits, being very seldom seen associating with, or even near, any other bird than its own single partner, and that too but for a very limited portion of the year."

In captivity, the Wryneck is tolerably docile; and when taken young can be perfectly tamed. In some countries it is the fashion to tie a string to the leg of a tame Wryneck and take it out for daily exercise for food, letting it run up the trees or on the ground in search of insects. The little bird soon becomes accustomed to this kind of life, and when the string is pulled returns to its owner, and runs about his clothes until he gives it permission to take another excursion.

The nest of the Wryneck is hardly deserving of that name, being merely composed of chips of decaying wood. The eggs are laid in the hollow of a tree, not wholly excavated by the bird, as is the case of the woodpeckers, its beak not being sufficiently strong for such a task, but adapted to the purpose from some already existing hole.

From a letter of a correspondent to Mr. Yarrell, it seems that although the Wryneck makes no nest, it does not hesitate in appropriating the deserted home of any other bird which it may find in the hollow which it selects for nidification. The bird had chosen a hole in an old apple-tree for that purpose, and the eggs were laid upon a mass of hair, moss, and fibrous roots, evidently a deserted nest of a redstart. The pertinacity with which the Wryneck adhered to the tree was really extraordinary, for she suffered her nest to be disturbed and replaced five times, and to be robbed four times of her eggs before she would finally leave the spot. The number of eggs laid by the Wryneck is rather great, as many as ten having often been found in a single nest. In the instance just mentioned, no less than twenty-two eggs were taken at the four intervals. Their colour is beautiful white with a pinky tinge, not unlike those of the kingfisher; and as this pink colour is produced by the yolk showing itself through the delicate shell, it is, of course, lost when the egg is emptied of its contents. The plumage of this little bird, although devoid of brilliant hues, and decked only with brown, black, and grey, is really handsome, from the manner in which those apparently sombre tints are disposed. In Yarrell's "British Birds" the markings of the Wryneck are given so concisely that they cannot be altered without damage. "The top of the head greyish brown, barred across with streaks of darker brown and white: neck, back, rump, and upper tail-coverts grey, speckled with brown. From the occiput (i.e. back of the head) down the middle line of the back of the neck and

GREAT HONEY GUIDE.—*Indicator major.*

between the scapularies, is a streak of dark brown mixed with black; the wings brown, speckled with lighter yellow brown, and a few white spots; the primary quill-feathers barred alternately with pale yellow, brown, and black; the tertials on the upper surface marked with a descending line of black; upper surface of the tail-feathers mottled with grey and brown, and marked with four irregularly transverse bars of black; chin, throat, ear-coverts, and neck, in front, pale yellow-brown with narrow transverse black lines; breast, belly, sides, and under tail-coverts, dull white tinged with yellow-brown, and spotted with black; under surface of tail-feathers pale greyish brown, speckled and barred with black; legs, toes, and claws brown." The total length of the adult male bird is about seven inches, and the female is a little smaller than her mate.

THE Cuckoos constitute a large family, containing several smaller groups, and many species. Representatives of the groups will be found in the following pages. All these birds have a rather long, slender, and somewhat curved beak, which in some species takes a curve so decided, that it gives quite a predaceous air to its owner. Examples of the Cuckoo tribe are to be found in almost every portion of the globe, and are most plentiful about the tropics.

The first group is that of which the celebrated GREAT HONEY GUIDE is our typical example. The Honey Guides derive their name from the fact that they are extremely fond of wild bees and their honey, and by their eager cries attract keen-eared and sharp-eyed hunters to the spoil. It has been said that the birds intentionally ask the aid of mankind to dig out the nests when the combs are placed in too secure a spot, and that they utter their peculiar cry of "Cherr! cherr!" to call attention, and then precede their human assistants to the nest, fluttering their wings, and keeping a few yards in advance. That they do lead travellers to the bees' nests is true enough, but that they should seek out human beings, and intentionally bring them to the sweet stores, seems doubtful, though it has been affirmed by many travellers.

At all events, even up to the present time, whenever the Honey Guide does succeed in leading the Hottentot to a store of honey, the men are grateful to it for the service, and do not eat the whole of the honey, leaving some for their confederate. Neither will they

RAIN-BIRD.—*Sauróthera vétula.*

kill the bird, and they are offended if they see any one else do so. Sparrman remarks that the present species is seldom seen near Cape Town, as it cannot find a supply of its food so near the habitations of man, and that he never saw any except on the farm of a single colonist, who had succeeded in living some wild swarms by fixing convenient boxes on his grounds.

One thing is certain, that the Honey Guide is by no means a safe conductor, as it will sometimes lead its follower to the couching-place of a lion or tiger, or the retreat of a poisonous snake. Gordon Cumming, as well as other travellers, testifies to this curious mode of conduct.

The feathers of the Honey Guide are thick, and the skin is tougher than is usually the case with birds, so that if the irritated bees should attack them, little harm is done unless a sting should penetrate the eye or the bare skin around it.

Honey Guides are found in various parts of Africa, India, and Borneo, and in all cases their habits seem to be very similar. Two species are very common in Southern Africa, namely, the bird figured in the engraving, and a smaller species (*Indicator minor*). The nesting of both these birds is very similar, their homes being pendent from the branches of trees, and beautifully woven into a bottle-like form, the entrance being downward. The material of which they are composed is bark torn into filaments. The eggs are from three to four in number, and their colour is a brownish white. Both parents assist in the duties of incubation.

These birds are very soberly clad, the Great Honey-eater being brown above, darker on the wings and tail, and greyish white on the under surface of the body.

We now arrive at the Ground Cuckoos, all of which are inhabitants of tropical America and the neighbouring islands, and are represented by the RAIN-BIRD.

This curious Cuckoo, which is popularly known in Jamaica by the name of RAIN-BIRD, is tolerably common in the West Indian Islands.

According to Mr. Gosse, who has given a very interesting account of this species in his "Birds of Jamaica," the Rain-Bird is so inquisitive at the sight of any new object, and so reckless of danger while gratifying its curiosity, that it is often called by the name of Tom Fool. Indeed, the first Rain-Bird which he saw lost its life by a stone, while sitting on a bush only a few feet distant, so occupied with the two featherless bipeds that were approaching, that it suffered itself to be struck from its perch by a missile that might have been avoided with the least precaution.

The wings of this bird are rather short and weak, so that it does not fly to any great distance when alarmed, but merely flits to a branch a few yards in advance, and then turns round and contemplates the intruder. It has a curious habit of sitting across a branch with its head lower than its feet, and balanced by the long tail, which hangs nearly perpendicularly. The voice is a harsh cackle, something like the words "ticky-ticky," pronounced with very great rapidity. It feeds on animal substances, preferring insects and spiders to any other kind of food, but not disdaining to prey upon the smaller reptiles and mammalia. The nest seems to be made in the fork of a branch. The colour of this bird is soft brown-grey upon the back, dullish yellow on the under parts of the body, and rusty red upon the wings. The long tail is beautifully barred with black and white.

Of the Coccyginæ, or Lark-heeled Cuckoos, so called from their long hind toe, we shall select two examples; the one being an Australasian bird, and the other an inhabitant of America.

The PHEASANT CUCKOO derives its popular appropriate name from the great length of its tail, which gives to the bird an outline bearing some resemblance to that of the pheasant, a similitude which is further carried out by the bold markings of its plumage. This handsome bird is a native of New South Wales, where it is not uncommon, although rather a local bird, seldom wandering to any great distance from the spot which it loves. It frequents low-lying and swampy lands; living almost entirely among the rank herbage of such localities, and keeping itself concealed among the brushes. When alarmed it flies to the nearest tree, alights on the lowest branches, rapidly makes its way through the boughs to the very summit, and then takes to wing.

The nest of this bird is placed on the

PHEASANT CUCKOO.—*Centropus phasiánus.*

COW-BIRD. —*Coccygus Americanus.*

ground, shaded by a convenient tuft of grass. It is a large and rather clumsily constructed edifice; having two apertures, through one of which the hen, while sitting, thrusts her head, and through the other she pokes her tail. The eggs are generally from three to five in number, and are more spherical than is generally the case among birds. Their colour is greyish white, sometimes blotched with brown, and they are remarkable for the roughness of their shells.

The colours of this bird are not brilliant, but are rich and warm in their tone and disposed so as to form very bold markings. The upper surface of the body is black devoid of gloss, with the exception of the shafts of the feathers; which are highly polished and glittering. The wing-coverts are brown mottled richly with black. The wings are ruddy chestnut barred with black, and the tail is dark brown glossed with green, freckled with brown, barred with white, and tipped with the same colour. The young birds are much lighter in colour than their parents, are more liberally streaked, and have more white about them.

In a former part of this work, the reader was warned not to confound the cow-troopial with the Cow-Bird which is here presented to his notice.

This bird is a native of America, and is one of the most familiar of the feathered tribes which inhabit that country. The name of Cow-Bird is derived from the cry which sounds like the word "Cow, cow!" constantly repeated. From the colour of its beak, it is sometimes known by the name of the Yellow-billed Cuckoo. It is one of the migrators, arriving in Pennsylvania about the end of April, and returning to the south towards the middle of September. Respecting this bird and its habits, Wilson has the following interesting remarks :—

"The singular, I will not say unnatural, conduct of the European Cuckoo (*Cuculus canorus*), which never constructs a nest for itself, but drops its eggs in those of other birds, and abandons them to their mercy and management, is so universally known, and so proverbial, that the whole tribe of Cuckoos have, by some inconsiderate people, been stigmatised as destitute of all parental care and affection. Without attempting to account for this remarkable habit of the European species, far less to consider as an error what the wisdom of Heaven has imposed as a duty upon the species, I will only remark, that the bird now before us builds its own nest, hatches its own eggs, and rears its own young; and, in conjugal and parental affection, seems nowise behind any of its neighbours of the grove.

Early in May they begin to pair, when obstinate battles take place among the males. About the tenth of that month they commence building. The nest is usually fixed among the horizontal branches of an apple-tree; sometimes in a solitary thorn, crab, or cedar, in some retired part of the woods. It is constructed, with little art, and scarcely any concavity, of small sticks and twigs, intermixed with green weeds and blossoms of the common maple. On this almost flat bed, the eggs, usually three or four in number, are placed; these are of a uniform greenish blue colour, and of a size proportionable to that of the bird. While the female is sitting, the male is generally not far distant, and gives the alarm, by his notes, when any person is approaching. The female sits so close, that you may almost reach her with your hand, and then precipitates herself to the ground, feigning lameness, to draw you away from the spot, fluttering, trailing her wings, and tumbling over, in the manner of the partridge, woodcock, and many other species. Both parents unite in providing food for the young. This consists, for the most part, of caterpillars, particularly such as infest apple-trees. The same insects constitute the chief part of their own sustenance.

They are accused, and with some justice, of sucking the eggs of other birds, like the crow, the blue jay, and other pillagers. They also occasionally eat various kinds of berries. But, from the circumstance of destroying such numbers of very noxious larvæ, they prove themselves the friends of the farmer, and are highly deserving of his protection."

The general colour of this bird is dark drab with a silken greenish gloss. The quill-feathers of the wings are ruddy cinnamon, and the tail is composed of black feathers tipped with white, with the exception of the two outer feathers, which are of the same green-glossed drab as the back. The whole under surface is pure white. The bill is rather long and curved, and is greyish black above and yellow beneath. The female may be known by the fact that the four central tail-feathers are drab, and the white takes a greyish tinge. Its total length is rather more than a foot.

THE Anis are all inhabitants of tropical climates, and are found chiefly in forest-lands, being most common in the dense woods of South America. They are by no means large birds, seldom exceeding the dimensions of the common English Blackbird. These birds are known by their compressed and arched beaks, and the decided keel or ridge upon the upper mandible.

The SAVANNAH BLACKBIRD is acknowledged to be the typical species of these birds, as it exhibits the peculiar form of the beak in a very marked manner. As it is rather a conspicuous bird, it is known by several other names, among which are Razor-billed Blackbird and Great Blackbird. In some places it is called the Black Parrakeet, and in Mexico its native title is Cacalototl.

The food of the Savannah Blackbird is mostly of an animal nature, and consists chiefly of grasshoppers, locusts, and similar insects, although the bird is very fond of lizards and other small vertebrates, a prey which its peculiar beak is well calculated to secure. Seeds are also said to be eaten by this bird.

In some cases their insect-loving nature is directed in a manner very useful to the cattle-owners. In those regions, the cows are greatly troubled with ticks and other parasitic insects, which fasten upon their backs where the poor beasts cannot reach them. The Anis are fortunately very fond of these noxious insects, and perching upon the cow's back, soon rid them of their unpleasant companions. The cows are so well aware of the services rendered to them by these birds, that when they find themselves much annoyed by ticks, they lie down in order to permit the Anis to pursue their avocation without disturbance. Sometimes, according to Brown, in his History of Jamaica, the Anis remind the cows of their reciprocal duties, and if the great quadruped

SAVANNAH BLACKBIRD.—*Crotóphaga Ani*

forgets to lie down for their mutual benefit, they hop about just in front of its nose as it grazes, and give it no peace until it complies with their request.

It is gregarious in its habits, associating in large flocks, and is a very fearless creature, caring little for the report or the effects of a gun. Whether this insensibility to danger be due to love of its comrades and to real courage, or only to that spurious bravery which fears nothing because it knows nothing, is not certain; but it is well known that if a flock of Anis be fired at, and many killed by the discharge, the survivors will only rise and fly to a short distance, and there settle as composedly as if no danger were at hand.

The Anis are very noisy, as is often the case with gregarious birds, and the combined loquacity of a large flock of Anis is almost deafening. They do not seem to use their wings to any great extent, their flight being low and short. They are easily tamed, soon become amusing inhabitants of the house, and can be taught to utter several words. Fortunately for itself, the flesh of the Savannah Blackbird is thought to be very disagreeable, so that it is not killed for the table.

The nesting of this bird is rather peculiar. It haunts bushes, the skirts of woods, and similar localities, and builds its nest on the branches of trees. The nest is extremely large, and is said to be in common to several pairs of birds, which live amicably under the same roof like the sociable weaver birds of Africa.

In size the Savannah Blackbird rather exceeds the generality of its kind, equalling a pigeon in dimensions, the long tail adding to the apparent length. Its colour is black, glossed with green.

THE very remarkable bird known by the name of CHANNEL-BILL inhabits part of Australia and some of the Eastern Islands. Its large and curiously formed beak gives it so singular an aspect, that on a hasty glance it might almost be taken for a species of toucan or hornbill.

It is most common in New South Wales, and is migratory in its habits, arriving in October and departing in June. It is a gregarious bird, being seen in little flocks or companies varying in number from three to eight, and sometimes living in pairs. The voice of the Channel-Bill is by no means pleasing, and is exercised at the approach of

CHANNEL-BILL.—*Scythrops Novæ Hollandiæ.*

rainy weather or the presence of a hawk. In either instance, the bird utters a series of vigorous yells, which are well understood by those who have studied its habits.

Although one of the migrators, it is slow and heavy of wing. Apparently, it is not easily tamed, for Mr. Gould mentions an instance where one of these birds was wounded and kept alive for two days, during the whole of which time it refused to be reconciled to captivity, screaming and pecking fiercely at its cage and captor. Its food consists of the seeds of the red gum and peppermint, and it also feeds upon beetles, phasmidæ, and other large insects of the land which it frequents.

It is a very handsome and elegantly coloured bird. The head and breast are grey, and the spaces around the eyes and nostrils are scarlet. The back is a deep greyish green, each feather being tipped with black, so as to give that portion of the bird a boldly mottled aspect. The under parts are white tinged with buff, and faintly barred with greyish brown. The long tail has the two central feathers black to the very tip, and the others are barred with black and tipped with white. Both sexes are alike in their colouring; the chief difference being that the female is smaller than her mate. In dimensions the Channel-Bill is about equal to the common crow, but owing to the long and broad tail, which causes the bird to measure more than two feet in total length, it appears much larger than is really the case.

THERE are few birds which are more widely known by good and evil report than the common CUCKOO.

As the harbinger of spring, it is always welcome to the ears of those who have just passed through the severities of winter; and as a heartless mother, an abandoner of its offspring, and an occupier of other homes it has been subjected to general reprobation. As is usual in such cases, both opinions are too sweeping; for the continual cry of "Cuck-oo! cuck-oo!" however agreeable it may be on the first hearing, soon becomes monotonous and fatiguing to the ear; and the mother Cuckoo is not so far lost to all feelings of maternity as to take no thought for her young, but ever remains near the place where it has deposited her egg and seems to keep watch over the foster-parents.

It is well known that the female Cuckoo does not make any nest, but places her egg in the nest of some small bird, and leaves it to the care of its unwitting foster-parents. Various birds are burdened with this charge, such as the hedge-warbler, the pied-wagtail, the meadow-pipit, the red-backed shrike, the blackbird, and various finches. Generally, however, the three first are those preferred. Considering the size of the mother-bird, the egg of the Cuckoo is remarkably small, being about the same size as that of the skylark, although the latter bird has barely one-fourth the dimensions of the former. The little birds, therefore, which are always careless about the colour or form of an egg, provided that it be nearly the size of their own productions, and will be perfectly contented with an egg-shaped pebble or a scraped marble, do not detect the imposition, and hatch the interloper together with their own young.

The general colour of the Cuckoo's egg is mottled reddish grey, but the tint is very variable in different individuals, as I can testify from personal experience. It has also been noted that the colour of the egg varies with the species in whose nest it is to be placed, so that the egg which is intended to be hatched by the hedge-warbler is not precisely of the same colour as that which is destined for the nest of the pipit.

Several experienced naturalists now lean to the opinion that the female Cuckoo really feels a mother's anxiety about her young; and this theory—a somewhat recent one—is corroborated by an account kindly sent to me by a lady, at that time unknown to me. A young Cuckoo had been hatched in the nest of some small bird, and after it was able to leave the nest for a short time, was taken under the protection of a female Cuckoo, who had been hovering about the place, and which at once assumed a mother's authority over the young bird, feeding it and calling it just like any other bird.

On inquiring whether the old Cuckoo ever helped the young one back into the nest, nothing could be ascertained. The children of the family, who were naturally interested in the affair, used sometimes to pick up the young bird and put it back into the nest, but it was often found in its warm home without human intervention, and as it was too helpless and timid to perform such a feat unaided, the natural assumption was that the old bird had given her assistance.

The mode by which the Cuckoo contrives to deposit her eggs in the nest of sundry birds was extremely dubious, until a key was found to the problem by a chance discovery made by Le Vaillant. He had shot a female Cuckoo, and on opening its mouth in order to stuff it with tow, he found an egg lodged very snugly within the throat.

When hatched, the proceedings of the young Cuckoo are very strange. As in process of time it would be a comparatively large bird, the nest would soon be far too small to contain the whole family; so the young bird, almost as soon as it can scramble about the nest, sets deliberately to work to turn out all the other eggs or nestlings. This it accomplishes by getting its tail under each egg or young bird in succession, wriggling them on to its back, and then cleverly pitching them over the side of the nest. It is rather curious that in its earlier days it only throws the eggs over, its more murderous propensities not being developed until a more advanced age.

There seems to be some peculiarity in the nature of the Cuckoo which forces other birds to cater for its benefit, as even in the case of a tame and wing-clipped Cuckoo, which was allowed to wander about a lawn, the little birds used to assemble about it with food in their mouths, and feed it as long as it chose to demand their aid.

Sometimes two Cuckoo's eggs have been laid in the same nest; when they are hatched there is a mutual struggle for the sole possession of the nest. Dr. Jenner, in his well-known and most valuable paper on this bird, gives the following account of such a strife.

CUCKOO.—*Cucúlus canórus.*

"Two Cuckoos and a hedge-sparrow were hatched in the same nest this morning ; one hedge-sparrow's egg remained unhatched. In a few hours after, a combat began between the Cuckoos for the possession of the nest, which continued undetermined until the next afternoon, when one of them, which was somewhat superior in size, turned out the other, together with the young hedge-sparrow and the unhatched egg. This contest was very remarkable. The combatants alternately appeared to have the advantage, as each carried the other several times nearly to the top of the nest, and then sank down again oppressed by the weight of its burden, till at length, after various efforts, the strongest prevailed, and was afterwards brought up by the hedge-sparrows."

In order to enable the young Cuckoo to perform this curious feat, its back is very different in shape from that of ordinary birds, being very broad from the shoulder downwards, leaving a well-marked depression in the middle, on which the egg or young bird rests while it is being carried to the edge of the nest. In about a fortnight this cavity is filled up, and the young bird has nothing extraordinary in its appearance.

From its peculiar mode of foisting off its young upon other birds, its character would seem to be of a solitary nature. Such, however, is not the case, for at some periods of the year these birds may be seen in considerable numbers, playing with each other or feeding in close proximity. Upwards of twenty have been observed in a single field, feeding on the caterpillars of the burnet moth, and several communications have been addressed to the *Field* and other journals in which the subject of natural history is discussed, relating similar occurrences. One of these correspondents records a large assembly of Cuckoos seen by herself in the month of August, the locality being near Leicester, and another relates a curious anecdote of a number of Cuckoos, which he saw on the wing, playing over and near a large grey stone in Eskdale. It seems that these birds are very partial to prominent objects, such as bushes, tree-stumps, large stones, &c., and that they are fond of congregating in their vicinity.

The peculiar note of the Cuckoo is so well known as to need no particular description, but the public is not quite so familiar with the fact that the note changes according to the time of year. When the bird first begins to sing, the notes are full and clear ; but towards the end of the season, they become hesitating, hoarse, and broken, like the breaking voice of a young lad. This peculiarity was noticed long ago by observant persons, and many are the country rhymes which bear allusion to the voice and the sojourn of the Cuckoo. For example :—

> " In April
> Come he will.
> In May
> He sings all day.
> In June
> He alters his tune.
> In July
> He prepares to fly.
> In August
> Go he must."

About Derbyshire and the north of England, this rhyme is slightly varied, and is given as follows :—

> " In April Cuckoo sings her lay ;
> In May she sings both night and day
> In June she loses her sweet strain ;
> In July she is off again."

An old writer, John Haywood, who " flourished," according to Mangnall, about 1580, has the following quaint and very graphic rhyme upon the voice of the Cuckoo at different periods of the year :—

> " In April the Coocoo can sing her song by rote.
> In June oft time she cannot sing a note.
> At first, koo, koo ; koo, koo ; sings till can she do
> At last, kooke, kooke, kooke ; six kookes to one koo."

The voice of the female bird is quite distinct from that of the male, and has been compared to the sound made by pouring water out of a narrow-necked bottle, and to the quacking clutter of the dabchick.

Sometimes the Cuckoo has been known to sing at night, having been seen to perch in a tree and then to commence its song. Many such instances are recorded, as also of the Cuckoo's song heard very early in the season ; but in all such instances where the bird was not actually seen, great caution must be used in accepting evidence. For the note of the Cuckoo is so peculiar, and so easily imitated, that boys are often in the habit of hiding in the copses and behind hedges for the purpose of deluding people into the idea that the Cuckoo has arrived. There have even been instances where such delinquents have confessed their bad practices when they attained to mature years, and wrote on natural history themselves.

When the stomach of the Cuckoo is opened, it is found to be lined with brown hairs, which on investigation with the microscope have been found to be those of the long-haired caterpillars, such as the " woolly-bear," *i.e.* the larva of the tiger-moth (Arctia caja), on which the Cuckoo loves to feed.

In captivity it feeds on many substances, always preferring caterpillars and raw beef chopped fine. It also likes worms, hard-boiled eggs, flies, wasp-grubs, and similar food. According to some persons, the young Cuckoo is a very easy bird to rear ; while according to others it gives the greatest trouble. One writer goes so far as to say that he would sooner rear a baby single-handed than a Cuckoo. However this may be, the first winter is always a trying season to the young bird, and there are very few which get well through it.

In general appearance the Cuckoo bears some resemblance to a bird of prey, but it has little of the predaceous nature. It is rather curious that small birds have a tendency to treat the Cuckoo much as they treat the hawks and owls, following it wherever it flies in the open country, and attending it through the air.

The colour of the plumage is bluish grey above, with the exception of the wings and tail, which are black, and barred with white on the exterior feathers. The chin, neck, and breast are ashen grey, and the abdomen and under wing-coverts are white, barred with slaty grey.

Sometimes the colour varies from these tints, and a white specimen may occasionally be found. Yearling birds of both sexes are hair-brown above, barred profusely with brownish red ; the quill-feathers of the wing are reddish brown, barred with white, while those of the tail are of the same dark tinge, but without the white bars, and spotted with white along the centre of the feathers. The whole of the under portions of the body are grey-white, barred with brown, and the short tail is tipped with white. A little white also appears on the tips of some of the feathers on the upper surface of the body. The total length of the adult bird is about fourteen inches. The female is rather smaller than her mate, and on her first arrival in England may be distinguished from the opposite sex by the brown bars upon her neck, and the brown tinge upon the back and wings.

COLUMBÆ; OR, DOVES AND PIGEONS.

THE large order of COLUMBÆ, or the Pigeon tribe, comes now under our notice. It contains very many beautiful and interesting birds ; but as its members are so extremely numerous, only a few typical examples can be mentioned in these pages.

All the Pigeons may be distinguished from the poultry, and the gallinaceous birds in general, by the form of the bill, which is arched towards the tip, and has a convex swelling at the base, caused by a gristly kind of plate which covers the nasal cavities, and which in some species is very curiously developed. In order to enable the parent birds to feed their young, the gullet swells into a double crop, furnished with certain large glands during the breeding season, which mingle their secretions with the food, and soften it, so that when the bird throws up the food after its fashion, to feed its young, the whole mass has acquired a soft and pulpy consistence, suitable to the delicate digestive powers of the tender young. Other peculiarities of form will be found in the appendix to this volume.

In their habits, the Pigeons greatly resemble each other, mostly haunting trees, but sometimes preferring the soil as a hunting-ground. Generally, the family likeness between the Pigeons is sufficiently strong to enable even a novice to know a Pigeon when he sees it ; but there are one or two remarkable exceptions to this rule, such as the Dodo and the Tooth-billed Pigeon, birds which need careful examination to be recognised as belonging to the present order.

The powers of wing are generally very great, the Pigeons being proverbially swift and enduring ; but even this rule has its exceptions. They are found in almost all parts of the globe, being most plentiful in the warmer regions. In this country the colours of the Pigeons, although soft and pleasing, and in some portions of the bird, such as the neck, glowing with a changeful beauty, are not particularly striking for depth or brilliancy. But in the hotter regions of the world, especially towards the tropics, the Pigeons are among the most magnificent of the feathered tribes, their plumage being imbued with the richest colours, and often assuming very elegant forms.

Our first example of this order is the OCEANIC FRUIT PIGEON.

The whole of the birds belonging to the genus Carpóphaga are notable for the curious knob that is found upon the base of the upper mandible, and which only makes its appearance during the breeding season. During the rest of the year, the base of the beak is more flattened than is generally the case with the Pigeons ; but as soon as the breeding season approaches, a little swelling is observable in this part, which rapidly grows larger, until it assumes the aspect shown in the engraving. Towards the end of the breeding season, the knob becomes smaller, and is gradually absorbed, leaving the bill in its former flattened condition.

This species is found in the Pelew and neighbouring islands, and is a forest-loving bird, taking up its residence in the woods, where it finds abundance of food. The diet which this bird most favours is the soft covering of the nutmeg, popularly known as "mace," and the flavour which this aromatic food imparts to the flesh is so peculiarly delicate, that the Oceanic Fruit Pigeon is in great request for the table, and is shot by hundreds. During the nutmeg season, these Pigeons find such an abundance of food that they become inordinately fat, and are sometimes so extremely plump, that when they are shot, and fall to the ground, they burst asunder.

Setting aside the gastronomical properties of this bird, it is a most useful creature, being the means of disseminating far and wide the remarkable nutmeg-tree. The Pigeon, being a bird of large appetite, swallows the nutmeg together with the mace, but only the latter substance is subject to digestion, the nutmeg itself passing through the system with its reproductive powers not only uninjured, but even improved. The sojourn within the body of the bird seems to be almost necessary in order to induce the nutmeg to grow; and when planted by human hands, it must be chemically treated with some preparation before it will strike root.

OCEANIC FRUIT PIGEON.—*Carpóphaga oceanica.*

The colour of this species is as follows: The forehead, cheeks, and throat are greyish white, and the rest of the head and the back of the neck are grey with a slaty blue wash. The back and upper portions of the body are light metallic green. The lower part of the throat and the breast are rusty grey, and the thighs and abdomen are deep brownish red. The under surface of the tail is also green, but with a reddish gloss. The total length of the bird is about fourteen or fifteen inches.

AMONG the most extraordinary of birds, the PASSENGER PIGEON may take very high rank, not on account of its size or beauty, but on account of the extraordinary multitudes in which it sometimes migrates from one place to another. The scenes which take place during these migrations are so strange, so wonderful, and so entirely unlike any events on this side of the Atlantic, that they could not be believed but for the trustworthy testimony by which they are corroborated. To abridge or to condense the spirited narrations of Wilson and Audubon would be impossible, without losing, at the same time, the word-painting which makes their descriptions so exceedingly valuable; and accordingly, these well-known naturalists shall speak for themselves.

PASSENGER PIGEON.—*Ectopistes migratorius.*

After professing his belief that the chief object of the migration is the search after food; and that the birds having devoured all the nutriment in one part of the country take wing in order to feed on the beech-mast of another region, Wilson proceeds to describe a breeding-place seen by himself in Kentucky, which was several miles in breadth, was said to be nearly forty miles in length, and in which every tree was absolutely loaded with nests. All the smaller branches were destroyed by the birds, many of the large limbs were broken off and thrown on the ground, while no few of the grand forest-trees themselves were killed as surely as if the axe had been employed for their destruction. The Pigeons had arrived about the tenth of April, and left it by the end of May.

"As soon as the young were fully grown, and before they left the nests, numerous parties of the inhabitants, from all parts of the adjacent country, came with waggons, oxen, beds, cooking utensils, many of them accompanied by the greater part of their families, and encamped for several days at this immense nursery. Several of them informed me that the noise in the woods was so great as to terrify their horses, and that it was difficult for one person to hear another speak without bawling in his ear.

2.   P P

The ground was strewed with broken limbs of trees, eggs and young squab pigeons which had been precipitated from above, and on which herds of hogs were fattening. Hawks, buzzards, and eagles were sailing about in great numbers, and seizing the squabs from their nests at pleasure ; while from twenty feet upwards to the top of the trees, the view through the woods presented a perpetual tumult of crowding and fluttering multitudes of pigeons, their wings roaring like thunder, mingled with the frequent crash of falling timber. For now the axe-men were at work cutting down those trees which seemed to be most crowded with nests, and contriving to fell them in such a manner that in their descent they might bring down several others, by which means the falling of one large tree sometimes produced two hundred squabs, little inferior in size to the old ones, and almost one mass of fat.

On some single trees upwards of one hundred nests were found, each containing *one* young only, a circumstance in the history of this bird not generally known to naturalists. It was dangerous to walk under these flying and fluttering millions, from the frequent fall of large branches, broken down by the weight of the multitudes above, and which, in their descent, often destroyed numbers of the birds themselves.

I had left the public road to visit the remains of the breeding place, near Shelbyville, and was traversing the woods with my gun, on my way to Frankfort, when, about one o'clock, the pigeons which I had observed flying the greater part of the morning northerly began to return in such immense numbers as I never before had witnessed. Coming to an opening by the side of a creek called the Benson, I was astonished at their appearance.

They were flying with great steadiness and rapidity, at a height beyond gunshot, in several strata deep, and so close together that could shot have reached them, one discharge would not have failed of bringing down several individuals. From right to left, as far as the eye could reach, the breadth of this vast procession extended, seeming everywhere equally crowded.

Curious to determine how long this appearance would continue, I took out my watch to note the time, and sat down to observe them. It was then half-past one. I sat for more than an hour, but instead of a diminution of this prodigious procession, it seemed rather to increase both in numbers and rapidity ; and anxious to reach Frankfort before night, I rose and went on. About four o'clock in the afternoon, I crossed the Kentucky river, at the town of Frankfort, at which time the living torrent above my head seemed as numerous and as extensive as ever. The great breadth of front which this mighty multitude preserved would seem to intimate a corresponding breadth of their breeding place, which by several gentlemen who had lately passed through part of it, was stated to me at several miles."

A few observations on the mode of flight of these birds must not be omitted.

" The appearance of large detached bodies of them in the air, and the various evolutions they display, are strikingly picturesque and interesting. In descending the Ohio by myself in the month of February, I often rested on my oars to contemplate their aerial manœuvres.

A column, eight or ten miles in length, would appear from Kentucky, high in air, steering over to Indiana. The leaders of this great body would sometimes gradually vary their course, until it formed a large bend of more than a mile in diameter, those behind tracing the exact route of their predecessors. This would continue sometimes long after both extremities were beyond the reach of sight; so that the whole, with its glittering undulations, marked a space on the face of the heavens resembling the windings of a vast and majestic river. When this bend became very great, the birds, as if sensible of the unnecessarily circuitous course they were taking, suddenly changed their direction, so that what was in column before became an immense front, straightening all its indentures until it swept the heavens in one vast and infinitely extended line.

Other lesser bodies united with each other as they happened to approach, with such ease and elegance of evolutions, forming new figures, and varying them as they united or separated, that I was never tired of contemplating them. Sometimes a hawk would make a sweep on a particular part of the column, when, almost as quick as lightning, that part shot downwards out of the common track ; but soon rising again, continued advancing at the same rate as before. This reflection was continued by those behind, who on arriving at this point dived down almost perpendicularly to a great depth, and rising, followed the exact path of those before them."

Let us now see what Audubon has to say on this subject. The reader will remark the brilliant account given by Wilson, of the effects produced by the attack of a hawk on a flock. Audubon has also remarked the same circumstance, and says : " But I cannot describe to you the extreme beauty of their aerial evolutions when a hawk chanced to press upon the rear of a flock. At once, like a torrent, and with a noise like thunder, they rushed into a compact mass, pressing upon each other towards the centre. In these almost solid masses, they darted forward in undulating and angular lines, descended and swept close over the earth with inconceivable velocity, mounted perpendicularly so as to resemble a vast column, and when high, were seen wheeling and twisting within their continued lines, which then resembled the coils of a gigantic serpent."

Writing of the breeding places of these birds, the same author proceeds as follows :—
" One of these curious roosting-places on the banks of the Green River in Kentucky I repeatedly visited. It was, as is always the case, a portion of the forest where the trees are of great magnitude, and where there was little underwood. I rode through it upwards of forty miles, and found its average breadth to be rather more than three miles. My first view of it was about a fortnight subsequent to the period when they had made choice of it, and I arrived there nearly two hours before sunset.

Few pigeons were then to be seen, but a great number of persons with horses and waggons, guns and ammunition, had already established encampments on the borders. Two farmers from the vicinity of Russelsville, distant more than a hundred miles, had driven upwards of three hundred hogs to be fattened on the pigeons that were to be slaughtered. Here and there the people employed in plucking and salting what had already been procured were seen sitting in the midst of large piles of these birds. Many trees two feet in diameter I observed were broken off at no great distance from the ground ; and the branches of many of the largest and tallest had given way, as if the forest had been swept by a tornado. Everything proved to me that the number of birds resorting to this part of the forest must be immense beyond conception. As the period of their arrival approached, their foes anxiously prepared to receive them ; some were furnished with iron pots containing sulphur—others with torches of pine-knots,—many with poles, and the rest with guns. The sun was lost to our view, yet not a pigeon had arrived. Everything was ready, and all eyes were gazing on the clear sky which appeared in glimpses amidst the tall trees.

Suddenly there burst forth a general cry of 'Here they come.' The noise which they made, though yet distant, reminded me of a hard gale at sea, passing through the rigging of a close reefed vessel. As the birds arrived and passed over me I felt a current of air that surprised me. Thousands were soon knocked down by the pole-men ; the birds continued to pour in ; the fires were lighted, and a most magnificent as well as wonderful and almost terrifying sight presented itself. The pigeons arriving by thousands alighted everywhere, one above another, until solid masses as large as hogsheads were formed on the branches all round. Here and there the perches gave way with a crash, and falling on the ground destroyed hundreds of the birds beneath, forcing down the dense groups with which every stick was loaded.

It was a scene of uproar and confusion : no one dared venture within the line of devastation : the hogs had been penned up in due time, the picking up of the dead and wounded being left for next morning's employment. The pigeons were constantly coming, and it was past midnight before I perceived a decrease in the number of those that arrived. Towards the approach of day the noise in some measure subsided ; long before objects were distinguishable the pigeons began to move off in a direction quite different from that in which they had arrived the evening before, and at sunrise all that were able to fly had disappeared. The howlings of the wolves now reached our ears, and the foxes, lynxes, cougars, bears, racoons, and opossums were seen sneaking off, whilst eagles and hawks of different species, accompanied by a crowd of vultures, came to supplant them, and enjoy their share of the spoil."

The chief food of the Passenger Pigeon is beech-mast, but the bird feeds on numerous other grains and fruits, such as acorns, buckwheat, hempseed, maize, holly-berries, huckle-berries, and chestnuts.   Rice is also a favourite article of food, and pigeons have been killed with rice still undigested in their stomachs, though the nearest rice plantation was distant several hundred miles.   The amount of food consumed by these birds is almost incredible.   Wilson calculates that taking the breadth of the great column of pigeons mentioned above, to be only one mile, its length to be two hundred and forty miles, and to contain only three Pigeons in each square yard (taking no account of the several strata of birds, one above the other,) and that each bird consumes half-a-pint of food daily; all which assumptions are below the actual amount, the quantity of food consumed in each day would be seventeen million bushels.   Audubon makes a similar calculation, allowing only two birds to the square yard.

Although these birds are found in such multitudes, there is only a single young one each time of hatching, though there are probably two or even three breeds in a season.   The young birds are extremely fat, and their flesh is very delicious, only, as during their stay every one eats pigeons all day and every day, they soon pall upon the taste.   So plump are these birds, that it is often the custom to melt them down for the sake of their fat alone.

When they begin to shift for themselves they pass through the forest in search of their food, hunting among the leaves for mast, and appear like a prodigious torrent rolling along through the woods, every one striving to be in the front.   "Vast numbers of them are shot while in this situation.   A person told me that he once rode furiously into one of these rolling multitudes and picked up thirteen pigeons, which had been trampled to death by his horse's feet.   In a few minutes they will beat the whole nuts from a tree with their wings, while all is a scramble, both above and below, for the same."   The young, the males and females, have a curious habit of dividing into separate flocks.

One or two specimens of this bird have been taken in Europe, and one individual was shot in Fifeshire in 1825.   This species has bred in the Zoological Gardens, and it is rather remarkable that the female made the nest while her mate performed the duties of hodman by bringing materials.   The nest is very slight, being only composed of a few twigs rudely woven into a platform, and so loosely made that the eggs and young can be seen from below.   In this instance the nest was begun and finished in the same day. The young bird was hatched after sixteen days.

The colour of the Passenger Pigeon is as follows : The head, part of the neck and the chin are slate-blue, and the lower part and sides of the neck are also deep slate "shot" with gold, green, and purplish crimson, changing at every movement of the bird. The throat, breast, and ribs are reddish hazel ; the back and upper tail-coverts dark slaty blue, slightly spotted with black upon the shoulders.   The primary and secondary quill-feathers of the wings are black, the primary being edged and tipped with dirty white. The lower part of the breast is a pale purplish red, and the abdomen is white.   The long and pointed tail has the two central feathers deep black, and the rest white, taking a bluish tint near their bases, and being marked with one black spot and another of rusty red on the inner webs.   The beak is black, the eye fiery orange, and a naked space around it is purplish red.   The female is known by her smaller size, her oaken-brown breast and ashen neck, and the slaty hue of the space round the eyes.   The total length of the adult male is about sixteen inches.

The STOCK-DOVE derives its name from its habit of building its nest in the stocks or stumps of trees.   It is one of our British Pigeons, and is tolerably common in many parts of England.

It is seldom found far northward, and even when it does visit such localities, it is only as a summer resident, making its nest in warmer districts.   As has already been mentioned, the nest of this species is made in the stocks or stumps of trees, the birds finding

RING-DOVE.—*Columba palumbus.*   STOCK-DOVE.—*Columba œnas.*

out some convenient hollow, and placing their eggs within. Other localities are, however selected for the purpose of incubation, among which a deserted rabbit-burrow is among the most common. The nest is hardly worthy of the name, being a mere collection of dry fibrous roots, laid about three or four feet within the entrance, just thick enough to keep the eggs from the ground, but not sufficiently woven to constitute a true nest. In some places when the keepers discover a brood they make a network of sticks at the mouth of the hole, so that the young cannot escape, although they can be fed by the parents from without, and when they are sufficiently large and plump they are taken for the table.

Now and then the Stock-Dove takes up its residence under thick furze-bushes, especially those which have grown close to the ground, and into which little openings have been made by the rabbits. The voice of the Stock-Dove is rather curious, being a hollow rumbling or grunting kind of note, quite unlike the well-known cooing of the Ring-Dove.

The head, neck, and back and wing-coverts are bluish grey, the primary quill-feathers of the wing taking a deeper hue, the secondaries being pearl-grey deepening at the tips, and the tertials being blue-grey with two or three spots. The chin is blue-grey, the sides of the neck slaty grey glossed with green, and the breast purplish red. The specific name of "œnas," or wine-coloured, is given to the bird on account of the peculiar hue of the throat. The whole of the under surface is grey, and the tail-feathers are coloured with grey of several tones, the outside feathers having the basal portion of the outer web white. The beak is deep orange, the eyes scarlet, and the legs and toes red. The total length is about fourteen inches, the female being a little smaller.

THE bird which now comes before our notice is familiar to all residents in the country under the titles of RING-DOVE, WOOD-PIGEON, WOOD-GUEST, and CUSHAT.

This pretty Dove is one of the commonest of our British birds, breeding in almost every little copse or tuft of trees, and inhabiting the forest grounds in great abundance. Towards, and during the breeding season, its soft complacent cooing—coo-goo-roo-o-o-o ! coo-goo-roo-o-o-o !—is heard in every direction, and with a very slight search its nest may be found. It is a strange nest, and hardly deserving that name, being nothing more than a mere platform of sticks resting upon the fork of a bough, and placed so loosely across each other, that when the maternal bird is away, the light may sometimes be seen through the interstices of the nest, and the outline of the eggs made out. Generally the Ring-Dove chooses a rather lofty branch for its resting-place, but it occasionally builds at a very low elevation. I have found the nest of this bird in a hedge only a few feet from the ground, so low indeed, that I could look down upon the eggs while standing by the hedge, and more like the work of the turtle-dove than of the Ring-Dove.

The eggs are never more than two in number, and perfectly white, looking something like hen's eggs on a small scale, save that the ends are more equally rounded. The young are plentifully fed from the crops of their parents, and soon become very fat. Just before they are able to fly they are held in great estimation for the table, and in some places ingenious boys are in the habit of going round to the Ring-Dove's nest while the young are still in their infantile plumage, tying a piece of string to their legs, passing it through the interstices of the nest, and fastening it to the branch. The young birds are thereby prevented from escaping, and are sure to be at hand when wanted. Even when adult, the Ring-Dove is a favourite article of food, and is shot by hundreds when they flock together in the cold weather. They also exhibit a decided partiality for certain roosting-places, and can be shot by waiting under the trees to which they have taken a liking.

The food of this Dove consists of grain and seeds of various kinds, together with the green blades of newly sprung corn and the leaves of turnips, clover, and other vegetables. Quiet and harmless as it may look, the Ring-Dove is a wonderful gormandizer, and can consume great quantities of food. The crop is capacious to suit the appetite, and can contain a singular amount of solid food, as indeed seems to be the case with most of the Pigeon tribe, so that when the birds assemble together in the autumn, the flocks will do great damage to the farmer.

The Ring-Dove may be easily known by the peculiarity from which it derives its name, the feathers upon the side of the neck being tipped with white so as to form portions of rings set obliquely on the neck. The head, chin, and part of the neck are blue-grey ; the remainder of the neck and the breast are purple-red, and the bare skin about the base of the beak is nearly white. The upper parts of the body are also blue-grey, but of a more slaty hue than the head and neck. The wings are also of the same dark hue, the primary quill-feathers having black shafts and a narrow band of white extending along the edges of their outer webs. The wing coverts are mostly blue-grey, but some of the feathers are more or less white, so that when the bird spreads its wings they form a very bold white patch, but when the wings are closed the white feathers of the coverts only form a line along the top of the wing. The tail is marked with several shades of grey, and the abdomen is soft pearly grey ; the beak is warm orange, and the eyes topaz yellow. It is a larger bird than the preceding species, being about seventeen inches in length.

THE many varieties of size, form, and colour which may be seen in the accompanying illustration afford an excellent example of the wonderful variations of which animals are susceptible under certain circumstances. Different as are the DOMESTIC PIGEONS, some of which are most ably figured on the next page by a practical pigeon-fancier as well as an accomplished artist, they all are modifications of the common BLUE ROCK PIGEON, and if permitted to mix freely with each other, display an inveterate tendency to return to the original form, with its simple plumage of black bars across the wing, just as the finest breeds of lop-eared rabbits will now and then produce upright-eared young.

BARBS. GROUP—DOMESTIC PIGEONS.—ROCK-DOVE.—*Columba Livia.* BALD-HEAD.
CARRIERS. POUTER. NUNS. TRUMPETERS. TURBITS.
OWLS. FANTAILS. JACOBINS. TUMBLERS.

The Rock-Dove derives its popular name from its habit of frequenting rocks rather than trees, an idiosyncracy which is so inherent in its progeny, that even the domestic Pigeons, which have not seen anything except their wooden cotes for a long series of generations, will, if they escape, take to rocks or buildings, and never trouble themselves about trees, though they should be at hand. Some years ago, one of my friends, living opposite Merton College, lost all his pigeons, by their gradual desertion of the loft in which they and their progenitors had been born, in favour of Merton tower, where they finally took up their residence in amiable proximity to multitudinous jackdaws and several owls, and may be seen hovering about the towers, but always remaining near its summit.

This species seems to have a very considerable geographical range, for it is common over most parts of Europe, Northern Africa, the coasts of the Mediterranean, and has even been found in Japan.

As a general rule, anyone who wants Pigeons about his house, and is not particular about the breed, can obtain them without the least trouble, by getting a good cote put up on his premises, and painting it white. The Pigeons are sure to be attracted by the glittering object, and will take possession of it spontaneously. I think that in many cases the cotes are deserted by the birds because they are left so long uncleansed, and are made on too small a scale. Among rocks or ruins, cleanliness is no such great matter, because there is plenty of air, and the birds can change their places freely ; but in the case of the wooden cotes, the space is very limited, and the ventilation almost reduced to a nullity. Vermin, too, swarm in such places, and the birds show their good sense in getting away from so unhealthy a situation. The cotes should always be well cleaned at intervals, and the owner will be repaid by the health and rapidly increasing number of his birds.

In a domesticated state, although it is better to feed them at home and so keep them from straying, they will always forage for themselves and young without any assistance, a flight of ten miles or so being a mere nothing to these strong-winged birds. Indeed, the Pigeons that inhabit the Hague, are known to cross the sea as far as the coast of Norfolk for the sake of feeding on the vetches.

The colour of the Rock-Dove is as follows : The head is grey, and the neck of the same colour, but "shot" with purple and green. The chin is blue-grey, and the throat changeable green and purple. The upper surface of the body is also grey, but of a different tone ; the greater coverts are barred with black at their tip, forming a decided band across the wing; the tertials are also tipped with black, and another black band crosses the wing a little below the first-mentioned bar. These conspicuous black bars are difficult to eradicate from the domestic breeds, and are always apt to make their appearance most unexpectedly, and annoy the fancier greatly. The lower part of the back is pure white, the upper tail-coverts are pearl grey, and the breast and abdomen of the same hue. The total length of this bird is not quite a foot.

From this stock, the varieties that have been reared by careful management are almost innumerable, and are so different in appearance that if they were seen for the first time, almost any systematic naturalist would set them down as belonging not only to different species, but to different genera. Such, for example, as the pouter, the jacobin, the trumpeter, and the fantail, on page 583, the last-mentioned bird having a greater number of feathers in its tail than any of the others.

As this work is not intended to be of a sporting or "fancy" character, a description of the various fancy Pigeons cannot be given. But the "homing" faculty of this bird, and the use to which it has been put, is too important to be passed over without a notice.

It has long been known that Pigeons have a wonderful power of finding their home, even if taken to great distances, and the mode by which the birds are enabled to reach their domiciles has long been the object of discussion, one party arguing that it is an instinctive operation, and the other, that it is entirely by sight. In my opinion the latter party have the better of the argument, though perhaps the element of instinct ought not wholly to be omitted. I have been told by those who have hunted on vast plains, where no object serves as a guide, that the only way to get safely back is to set off on the homeward track without thinking about it, for that when a man begins to exercise his reason, his instinct fails him in proportion, and unless he should be furnished with a compass, he will probably be lost.

TOP-KNOT PIGEON.—*Lopholaïmus antárcticus.*

Still, that the sense of sight is the principal element, cannot, I think, be denied. For in training a bird, the instructors always take it by degrees to various distances, beginning with half-a-mile or so, and ending with sixty or seventy miles in the case of really good birds, which will travel from London to Manchester in four hours and a half. In foggy weather the birds are often lost, even though they have to pass over short distances, and when a heavy fall of snow has obliterated their landmarks and given the country an uniform white coating, they are sadly troubled in finding their way home. The fancy Carrier Pigeon, with the large wattles on the beak, is said to be no very good messenger, the trainers preferring the Belgian bird, with its short beak, round head, and broad shoulders.

It is a curious, but a well ascertained fact, that the accuracy of Pigeon flight depends much on the points of the compass, although each individual bird may have a different idiosyncracy in this respect. Some birds, for example, always fly best in a line nearly north and south, while others prefer east and west as their line of flight. This remarkable propensity seems to indicate that the birds are much influenced by the electric or magnetic currents continually traversing the earth. When starting from a distance to reach their home, these Pigeons rise to a great height, generally hover about for a while in an undecided manner, and then, as if they had got their line, dart off with an arrowy flight. Missives written on very thin paper and rolled up tightly, are secured to the bird in such a way that they will not be shaken off by the flapping of the wings, or encumber the bird in its flight; and before the introduction of the electric telegraph, this mode of correspondence was greatly in use, mostly in political or sporting circles.

The splendid Top-Knot Pigeon is one of the handsomest of the tribe, and in any collection of birds would be one of the most conspicuous species.

It is a native of Southern and Eastern Australia, and, according to Mr. Gould, is most plentifully found in the bushes of the Illawarra and Hunter rivers. The powerful feet and general structure point it out as an arboreal bird, and it is so exclusively found in the trees that it will not even perch among the underwood, but must needs take its place on the branches of lofty trees. When perched it sits boldly and uprightly, having an almost hawk-like air about it.

It is a gregarious bird, assembling in large flocks, and being very fond of constant proximity to its neighbours, whether it be swiftly flying through the air, or quietly perched upon a branch. When a flock of Top-Knot Pigeons directs its flight towards a tree, the rushing sound of wings can be heard at a considerable distance, and when the birds perch simultaneously upon the boughs, bending them down with their weight, or fluttering their wings and displaying their beautiful crests, they present a very animated scene. Their wings are proportionately powerful to their feet, and they have a custom of ascending high into the air and taking very long flights, packed so closely together that the spectator involuntarily wonders how they can move their wings without striking their companions.

The food of this bird consists mostly of fruits, and it is very fond of the wild fig, and the berries of the cabbage palm. Its throat is wonderfully capacious, and Mr. Gould says it could swallow a walnut without inconvenience. Fortunately for itself, it is not good eating, the flesh being dry and coarse.

The crest of the forehead and top of the head, together with the hackle-like feathers of the throat and breast, are silver grey, showing the darker hues on the breast. On the back of the head the crest is of a ruddy rust colour. From the eye to the back of the head runs a dark streak shaded by the crest. The upper surface of the body is dark slaty grey, and the primaries and secondaries, together with the edge of the wing, are black. The tail is grey of two shades, having a broad band of black across the centre, and the extremity deeply tipped with the same dark hue. The under surface is silver grey like the breast. The eye is fiery orange, surrounded with a narrow crimson line; the base of the bill is blue and the remainder red, and the feet are purplish red. The length of this fine bird is about seventeen inches.

The world-famed Turtle-Dove is, although a regular visitor of this country, better known by fame and tradition than by actual observation. This bird has, from classic time until the present day, been conventionally accepted as the type of matrimonial perfection, loving but its mate and caring for no other until death steps in to part the wedded couple. Yet it is by no means the only instance of such conjugal affection among the feathered tribes, for there are hundreds of birds which can lay claim to the same excellent qualities, the fierce eagle and the ill-omened raven being among their number.

The Turtle-Dove seems to divide its attention pretty equally between Africa and England, pausing for some little time in southern Italy as a kind of half-way house. It arrives here about the beginning of May, or perhaps a little earlier in case the weather be warm, and after resting for a little while, sets about making its very simple nest and laying its white eggs. The nest of this bird is built lower than is generally the case with the Wood-Pigeon, and is usually placed on a forked branch of some convenient tree, about ten feet or so from the ground. Both parents aid in the duties of incubation as they ought to do, and both are equally industrious in the maintenance of their small family. The eggs are laid rather late in the season, so that there is seldom more than a single brood of two young in the course of the year.

TURTLE-DOVE.—*Turtur auritus.*

The Turtle-Dove is far more common in the southern than in the northern counties, and I have reason to believe that in Derbyshire, where I was greatly fond of bird-nesting for some years, it is not of very frequent occurrence, at least as far as personal experience goes, which, however, is only of a negative character in this instance. The white eggs are rather more sharply pointed than those of the Wood-Pigeon, but all the English Pigeons' eggs are much alike and can with difficulty be distinguished from each other.

The food of the Turtle-Dove mostly consists of seeds, such as corn, peas, rape, and similar seeds.

It is a bird of strong flight, and on its migrating journeys prefers to travel in company, associating in little flocks of ten or twelve. The end of August and September are the periods most in favour for the annual emigration.

The Turtle-Dove may be readily known by the four rows of black feathers tipped with white, which are found on the sides of the neck. The top of the head is ashen slate, deepening into a browner hue on the back of the neck. The chin and neck are pale brown, tinged with purple upon the breast. The upper surface of the body is pale brown mottled with a darker hue, and the wing-coverts are another shade of brown edged with warm ruddy chestnut. The quill-feathers of the wing are brown, and the upper tail-coverts are also brown with a slight ruddy tinge. The two central tail feathers are of the same colour, and the remaining feathers are dark brown tipped with white. Both edges of the tail are also white. The abdomen and under tail-coverts are white. The eye is chestnut, and under it there is a little patch of bare pink skin; the legs and toes are brownish yellow, and the beak is brown. The young birds of the year are differently shaded with brown, the head is wholly of that colour, the wing-coverts are tipped with yellowish white, and the quill-feathers of the wing are edged with a rusty hue.   The tail, too, is without the white that distinguishes the adult bird. The total length of this species is rather more than eleven inches.

THE little CRESTED PIGEON, although not so conspicuous as some of its relations, is one of the most elegant in form and pleasing in colour among this tribe.

It is a native of central Australia, and, according to Mr. Gould, is fond of haunting the marshy ground by the side of rivers and lagoons, and there assembling in large flocks. The gregarious propensities of this bird are indulged to an extent that seems almost ridiculous, for a large flock of Crested Pigeons will fly to the same tree, sit closely packed upon the same branch, and at the same moment descend in a mass to drink, returning in a similar manner to their perch.   The flight of this bird is strong, and rather curiously managed.   When it starts from the tree on which it is sitting, it gives a few quick strokes with its wings, and then darts off on steady pinion with an arrowy flight.   When it settles, it flings up its head, erects its crest, and jerks its tail over its back, so that the crest and tail nearly touch each other.   Its nest is, like that of most pigeons, made of little twigs, and placed on the low forking branch of some convenient tree.   While sitting on the nest, or perching quietly on the bough, the crest lies almost upon the back, and from below is hardly distinguishable from the rest of the plumage.

The head, face, and most of the under portions are pearl grey, the long slender crest being jetty black, and the sides of the neck tinged slightly with pink.   The back of the neck, the back, flanks, and both tail-coverts are light brown ; the feathers at the insertion of the wing are buff, crossed with black nearer their tips, and the great coverts are shining bronze green edged with white.   The primary feathers of the wing are brown, some partially edged with brownish white, and the rest with pure white.   The secondaries are brown in their inner webs, and their outer webs are bronzy purple at the base, tipped with brown, and edged with white.   The two central feathers of the tail are brown, the rest are blackish brown, with a green gloss on their outer webs and tipped with white. The bill is olive black, deepening at the tip, the feet are pink, and the eye orange set in a pink orbit.

The BRONZEWING PIGEON is also an Australian bird, and with the exception of the Wonga-Wonga Pigeon, hereafter to be described, is the most celebrated for the delicacy of its flesh.

It is a plump, and readily fattening bird, weighing about a pound when in good condition.   The breast is particularly large, as may be supposed from the great force of its wings, and when the bird is fat, is the most esteemed portion.   To the Australian traveller the Bronzewing is invaluable, as it is a great water drinker, and its flight will direct the thirsty wanderer to the stream or spring.   Mr. Gould, who has had long experience of this as well as of many other birds, gives the following interesting account of its habits :—

"Its amazing powers of flight enable it to pass in an incredibly short space of time over a great expanse of country, and just before sunset it may be observed swiftly winging its way over the plains or down the gullies to its drinking place.

During the long drought of 1839–40, when I was encamped at the northern extremity of the Brezi range, I had daily opportunities of observing the arrival of this bird to drink ; the only water for miles, as I was assured by the natives, being that in the immediate vicinity of my tent, and that was merely the scanty supply left in a few natural basins in the rocks, which had been filled by the rains of many months before. This peculiar situation afforded me an excellent opportunity for observing not only the Bronzewing, but many other birds inhabiting the neighbourhood.   Few, if any, of the true insectivorous or fissirostral birds came to the water holes, but on the other hand, those species that live upon grain and seeds, particularly the parrots and honey-eaters (*Trichoglossi* and *Meliphagi*), were continually rushing down to the edges of the pools, utterly regardless of my presence, their thirst for water quite overcoming their sense of danger ; seldom, if ever, however, did the Bronzewing make its appearance during the

CRESTED PIGEON.—*'Ocyphaps lophótes.*   BRONZEWING PIGEON.—*Phaps chalcóptera.*

heat of the day, but at sundown, on the contrary, it arrived with arrow-like swiftness, either singly or in pairs.

It did not descend at once to the edge of the pool, but dashed down to the ground at about ten yards' distance, remained quiet for a short time, then walked leisurely to the water, and after taking libations deep and frequent, winged its way to its roosting-place for the night. With a knowledge, therefore, of the habits of this bird, the weary traveller may always perceive when he is in the vicinity of water; and however arid the appearance of the country may be, if he observes the Bronzewing wending its way from all quarters to a given point, he may be certain to procure a supply of food and water. When rain has fallen in abundance, and the rivers and lagoons are filled not only to the brim, but the water has spread over the surface of the surrounding country, the case is materially altered; then the Bronzewing and many other birds are not so easily procured, the abundant supply of the element so requisite to their existence, rendering it no longer necessary that they should brave every danger in procuring it."

This Pigeon does not assemble in flocks, but in many parts of the country is so plentiful and is so attached to certain localities that forty or fifty may be killed in a day after the breeding season, when it is in best condition. It feeds almost invariably on the ground, its diet consisting chiefly of leguminous seeds. The nest is a frail structure of twigs, rather more hollowed than is usually the case with the houses of Pigeons, and is placed on the low forking branch of a gum tree near water. The bird is presumed to undergo a partial migration.

WONGA-WONGA PIGEON.—*Leucosárcia picáta.*

In colour, the forehead is buff, the head is dark brown changing to deep plum colour at the sides, the sides of the neck are grey, and there is a white waved line under the eye, and running partly down the chin. The upper surface of the body is dark brown. The coverts are marked with bronze-green spots, and the tertiaries have a large oblong shining green spot, edged with buff. The two central feathers of the tail are brown and the rest grey, banded with black near the tip. The breast is purple-brown, fading into grey on the abdomen. The eyes are reddish brown, and the legs and feet crimson.

OF all this group of birds, the WONGA-WONGA PIGEON is the most celebrated for the whiteness, plumpness, and delicacy of its flesh, which, when eaten with bread sauce, is of such remarkable excellence, that the remembrance always excites the liveliest reminiscences in those who have partaken of so great a dainty.

The Wonga-Wonga Pigeon is a native of Australia, but is not spread generally over the country, being found mostly, if not wholly, among the bushes along the coast of New South Wales, or the sides of the hills of the interior. According to Mr. Gould, it inhabits the same district as the bush turkey, the satin bower bird, and the lyre bird. It lives mostly on the ground, feeding upon the stones and seeds of fallen fruit. When disturbed, it suddenly rises from the ground with a loud whirring rush like that of the pheasant, and like that bird, rather startles the novice with the noise. It does not maintain a long flight; but either directs its course to a neighbouring tree, or again settles upon the earth.

In colour it is a very conspicuous bird. The forehead and chin are white, and a jetty black line passes from the eye to the base of the bill. The sides of the head are grey, the back and upper surface are slate-grey, and the chest is deep blackish grey, with a very broad white band crossing the chest and running up the sides of the neck. The abdomen is white, the under coverts dark brown tipped with buff, and the flanks are also white, but agreeably diversified with a bold black spot near the tip of each feather. The beak is red tipped with black, the eyes are dark brown with pink orbits, and the legs are bright pink.

NICOBAR PIGEON. — *Calœnas Nicobórico.*

The Nicobar Pigeon may fairly be reckoned among the more magnificent species belonging to the Pigeon tribe; the long pointed feathers of the neck and shoulders glowing with resplendent green, bronze, and steely blue, and having a peculiarly attractive effect as they droop towards the ground, their loose points waving in the wind, and their hues changing with every movement. Like others of the sub-family to which it belongs, it is mostly a terrestrial bird. As its name imports, it is most commonly found in Nicobar; but it also inhabits Java, Sumatra, and many neighbouring islands.

The head of this Pigeon is slaty blue, with a purplish cast, which is more conspicuous in certain lights. The beautiful long-pointed feathers of the neck are greatly like the hackles of the game cock, except that they hang lower on the neck. Their colour is rich refulgent green, deepening into a warm copper when the light falls obliquely upon them, and the wing-coverts are of the same hue, and pointed after a similar fashion. The back and whole of the upper surface is glowing green, with bronze and steel-blue reflections, and the under surface partakes of the same colouring, but without its peculiar resplendence. The short, square tail is pure white. It is rather remarkable that in the breeding season a rounded fleshy knob makes its appearance upon the upper mandible, similar to that which has already been noticed in the Fruit Pigeon, on page **576**. The total length of this bird is about fourteen inches.

The splendid Crowned Pigeon is indisputably the most conspicuous of all its tribe; its great size and splendid crest rendering it a most striking object, even at a considerable distance.

So large and so un-pigeon like is this bird, that few on first seeing it would be likely to determine its real relations to the rest of the feathered race, and would be more likely to class it among the poultry than the pigeons. If, however, the reader will lay a card upon the crest so as to expose only the head, he will see that the general outline of the head and beak is clearly that of a pigeon. It is a native of Java, New Guinea, and the Moluccas.

In the Zoological Gardens are several specimens of this splendid bird, whose manners are very curious and interesting. Their walk is quite of a royal character, stately and

CROWNED PIGEON.—*Goura coronata.*

majestic, and well according with the beautiful feathered crown which they bear upon their heads. The crest seems to be always held expanded. They have a quaint habit of sunning themselves upon the hot pavement of their prison by lying on one side, laying the head flat on the ground, tucking the lower wing under them, and spreading the other over their bodies so as to form a very shallow tent, each quill-feather being separated from its neighbour, and radiating around the body. Sometimes the bird varies this attitude by stretching the other wing to its full length, and holding it from the ground, at an angle of twenty degrees or so, as if to take advantage of every sunbeam and every waft of air.

While lying in this unique attitude, it might easily pass at a little distance for a moss-covered stone, a heap of withered leaves, or a rugged tree-stump, with one broken branch projecting to the side. No one would think of taking it for a bird. Unfortunately, it is a difficult matter to take a sketch of the bird while thus reposing, for there are so few salient points, that a very careful outline is needed, and its companions are sure to come and peck it up before the sketch can be concluded.

TOOTH-BILLED PIGEON.—*Didunculus strigirostris.*

The cry of this bird is loud and sonorous, and not very easy of description. Some authors compare it to the gobbling of a turkey cock, but I can perceive no resemblance to that sound. It is more of a loud, hollow boom, than anything else, a kind of mixture between a trombone and a drum, and every time that the bird utters this note, it bows its head so low that the crest sweeps the ground.

The nest of the Crowned Pigeon is said to be made in trees, the eggs being two in number, as is generally the case with this group of birds. Its flesh is spoken highly of by those who have eaten it. The general colour of this bird is a deep and nearly uniform slate-blue; the quill-feathers of the wing and tail being very blackish ash, and a patch of pure white and warm marroon being found on the wing.

In the Samoan islands of the Pacific, is found a bird of extreme rarity of form, which is, as far as is known, unique among the feathered tribes that now inhabit the earth. I say, now inhabit, because in former days, when the Dodo was still in existence, that remarkable and ungainly bird presented a form and structure greatly similar to those of the TOOTH-BILLED PIGEON.

On account of its close relationship with the Dodo, it has received from some systematic zoologists the generic name of Didunculus, or Little Dodo, while others have given it the title of Gnathodon, or Toothed-jaw, in allusion to the structure of its beak. The food of this bird consists largely of the soft bulbous roots of several plants. The whole contour of the Tooth-bill is remarkable, and decidedly quaint; its rounded body seeming hardly in accordance with the large beak, which is nearly as long as the head, and is greatly arched on the upper mandible. The lower mandible is deeply cleft into three distinct teeth near its tip.

In colour it is rather a brilliant bird. The head, neck, breast, and abdomen are glossy greenish black, and upon the shoulders and the upper part of the back the feathers are velvety black, each having a crescent-shaped mark of shining green near its extremity. The rest of the back, the wings, tail, and under tail-coverts, are deep chestnut. The primary and secondary quill-feathers of the wing are greyish black, and the large arched bill is orange. The total length of this bird is about fourteen inches.

2.   Q Q

THE position held by the celebrated DODO among birds was long doubtful, and was only settled in comparatively late years by careful examination of the few relics which are our sole and scanty records of this very remarkable bird.

For many years the accounts given by the early voyagers of the Dodar, or Walgh Vögel, found in the Mauritius and other islands, were thought to be merely fabulous narratives, a mental reaction having set in from the too comprehensive credulity of the previous times; and the various portraits of the Dodo to be found in the books of travel were set down as examples, not of the Dodo, but of the inventive faculties possessed by the authors. Truth, however, stood its own ground, as it always will do, and steadily withstood the batteries of negative reasonings that were brought to bear on the subject. An entire bird was quietly lodged in the Ashmolean Museum at Oxford; portions of other specimens made their way to Europe among the curiosities which sailors are so fond of bringing home, and there is every reason to believe that a living example of this bird was exhibited in Holland, if not in England.

It is curious that, but for a code of far-seeing regulations, providing that when the stuffed skin of a bird was so far decayed as to be useless as a specimen, the head and feet should be preserved, our best and most perfect relics of the Dodo would have been burned as useless rubbish. The specimen at Oxford was suffered to fall into decay, no one seeming to be aware of its priceless value, and when the skin was destroyed, the head and feet were laid aside and put away with other objects, among which they were afterwards discovered to the great joy of the finder. These were sufficiently perfect to prove the real existence of the bird, and the correctness with which it had been depicted by many draughtsmen; some portraits being of the rudest description, while others were the work of eminent artists, and most valuable for their high finish and accuracy of detail. The position of the bird among the feathered tribes was long doubtful, and it was provisionally placed between the ostriches and bustards, until, after a careful examination of the relics, it was found to belong to the pigeon tribe. This decision received a valuable confirmation in the discovery of the tooth-billed pigeon, just described.

For further information respecting the anatomical and scientific details of this bird, the reader is referred to Strickland and Melville's instructive and interesting work on the subject.

Many of the earlier travellers have spoken of the Dodo—a name, by the way, corrupted from the Dutch term Dod-aers—and their accounts are as quaint as the bird which they describe. For example, Bontius writes as follows: "The Dronte, or Dod-aers, is for bigness of mean size between an ostrich and a turkey, from which it partly differs in shape and partly agrees with them, especially with the African ostriches, if you consider the rump, quills, and feathers; so that it was like a pigmy among them if you regard the shortness of its legs.

It hath a great ill-favoured head, covered with a kind of membrane, resembling a hood; great black eyes; a bending, prominent, fat neck; an extraordinary long, strong, bluish-white bill, only the ends of each mandible are a different colour, that of the upper black, that of the nether yellowish, both sharp-pointed and crooked. Its gape, huge wide, as being naturally very voracious. Its body is fat and round, covered with soft grey feathers after the manner of an ostrich's; in each side, instead of hard wing-feathers or quills, it is furnished with small soft-feathered wings, of a yellowish ash colour; and behind, the rump, instead of a tail, is adorned with five small curled feathers of the same colour. It hath yellow legs, thick, but very short; four toes in each foot; solid, long, as it were scaly, armed with strong black claws.

It is a slow-paced and stupid bird, and which easily becomes a prey to the fowlers. The flesh, especially of the breast, is fat, esculent, and so copious that three or four Dodos will sometimes suffice to fill one hundred seamen's bellies. If they be old, or not well boiled, they are of difficult concoction, and are salted and stored up for provision of victual. There are found in their stomachs stones of an ash colour, of divers figures and magnitudes, yet not bred there, as the common people and seamen fancy, but swallowed by the bird; as though by this mark also nature would manifest that these fowls are of the ostrich kind, in that they swallow any hard things though they do not digest them."

DODO.—*Didus ineptus.*

Other travellers, such as Leguat and De Bry, agree with Bontius in his description of the bird, and coincide in his opinion of the excellence of its flesh; but one writer, Sir T. Herbert, who visited the Mauritius about 1625, differs greatly in his estimation of the value of the Dodo as an article of food. In his book of travels, which is perhaps the quaintest and raciest to be found among such literature, he speaks as follows of this bird :—

"The Dodo, a bird the Dutch call Walghvogel, or Dod Eersen; her body is round and fat, which occasions the slow pace, or that her corpulencie, and so great as few of them weigh less than fifty pound: meat it is with some, but better to the eye than stomach, such as only a strong appetite can vanquish. . . . It is of a melancholy visage, as sensible of nature's injury in framing so massie a body to be directed by complimental wings, such, indeed, as are unable to hoise her from the ground, serving only to rank her among birds. Her traine, three small plumes, short and improportionable, her legs suiting to her body, her pounces sharpe, her appetite strong and greedy. Stones and iron are digested; which description will better be conceived in her representation."

So plentiful were the Dodos at one time, and so easily were they killed, that the sailors were in the habit of slaying the birds merely for the sake of the stones in their stomachs, these being found very efficacious in sharpening their clasp-knives. The nest of the Dodo was a mere heap of fallen leaves gathered together on the ground, and the bird laid but one large egg. The weight of one full-grown Dodo was said to be between forty and fifty pounds. The colour of the plumage was a greyish brown in the adult males, not unlike that of the ostrich, while the plumage of the females was of a paler hue.

CRESTED CURASSOW.—*Crax Alector.*

LEAVING the pigeons, we now come to the large and important order of birds, termed scientifically the Gallinæ, and, more popularly, the Poultry. Sometimes they are termed Rasores, or scrapers, from their habit of scraping up the ground in search of food. To this order belong our domestic poultry, the grouse, partridges, and quails, the turkeys, pheasants, and many other useful and interesting birds. In almost every instance the Gallinæ are handsome birds, and interesting in their habits, but as their number is legion, and our space is rapidly diminishing, we must content ourselves with such species as afford the best types of the order to which they belong.

OUR first example of these birds is the CRESTED CURASSOW, the representative of the genus Crax, in which are to be found a number of truly splendid birds. All the Curassows are natives of tropical America, and are found almost wholly in the forests.

The Crested Curassow inhabits the thickly wooded districts of Guiana, Mexico, and Brazil, and is very plentifully found in those countries. It is a really handsome bird, nearly as large as the turkey, and more imposing in form and colour. It is gregarious in its habits, and assembles together in large troops, mostly perched on the branches of trees. It is susceptible of domestication, and, to all appearances, may be acclimatized to this country as well as the turkey or the pheasant.

There is special reason that the Curassows should be added to our list of domesticated poultry, for their flesh is peculiarly white and well flavoured, surpassing even that of the turkey, and they are of a pleasant temper, and readily tamed by kindness. A dry soil is absolutely necessary for their well-being, as they suffer greatly from damp, which produces a disease of the foot and toes, often causing the toes to mortify and fall off.

Trees are also needful, as these birds are fond of perching at some height from the ground and the situation must be sheltered from wind or rain.

In their native country the Curassows build among the trees, making a large and rather clumsy-looking nest of sticks, grass stems, leaves, and grass blades. There are generally six or seven eggs, not unlike those of the fowl, but larger and thicker shelled. The voice of the Crested Curassow is a short croak, but the various species differ slightly in this respect. The male Globose Curassow, for example, has a voice that sounds like a short hoarse cough, and every time that it utters the cry it jerks up its tail and partially spreads the feathers. The voice of the female is unlike that of her mate, being a gentle whining sound. While perambulating the ground or traversing the branches, the Curassow continually raises and depresses its crest, giving itself a very animated aspect.

The colour of the Crested Curassow is very dark violet, with a purplish green gloss above and on the breast, and the abdomen is the purest snowy white, contrasting beautifully with the dark velvety plumage of the upper parts. The bright golden yellow of the crest adds in no small degree to the beauty of the bird.

The Guans also belong to the same family as the Curassows. They are also inhabitants of the forests of tropical America; and are easily to be recognised by the naked and dilatable skin of the throat. They are not gregarious, like the curassow, but are mostly solitary in their habits, feeding chiefly on fruits and remaining on the branches. They are not so susceptible of domestication as the curassow, nor are they so large, being of a more delicate and slender shape. The flesh of these birds is very excellent.

Several very singular birds are found in Australia and New Guinea, called by the name of Megapodinæ or Great-footed birds, on account of the very large size of their feet; a provision of nature which is necessary for their very peculiar mode of laying their eggs and hatching their young.

The first of these birds is the Australian Jungle Fowl, which is found in several parts of Australia, but especially about Port Essington. In that country great numbers of high and large mounds of earth exist, which were formerly thought to be the tombs of departed natives, and, indeed, have been more than once figured as such. The natives, however, disclaimed the sepulchral character, saying that they were origins of life rather than emblems of death; for that they were the artificial ovens in which the eggs of the Jungle Fowl were laid, and which, by the heat that is always disengaged from decaying vegetable substances, preserved sufficient warmth to hatch the eggs.

The size of these tumuli is sometimes quite marvellous; in one instance, where measurements were taken, it was fifteen feet in perpendicular height, and sixty feet in circumference at its base. The whole of this enormous mound was made by the industrious Jungle Fowl, by gathering up the earth, fallen leaves, and grasses with its feet, and throwing them backwards while it stands on the other leg. If the hand be inserted into the heap, the interior will always be found to be quite hot. In almost every case the mound is placed under the shelter of densely leaved trees, so as to prevent the sun from shining upon any part of it. This precaution is probably taken in order to prevent the rays of the sun from evaporating the moisture. The aspect of the heap depends much on the surrounding objects; and in one instance it was placed close to the sea, just above high-water mark, and was composed of sand, shells, and black mould. It was situated in the midst of a large yellow-blossomed hibiscus, by which it was enveloped.

The bird seems to deposit its eggs by digging holes from the top of the mound, laying the egg at the bottom, and then making its way out again, throwing back the earth that it had scooped away. The direction, however, of the holes is by no means uniform, some running towards the centre and others radiating towards the sides. They do not seem to be dug quite perpendicularly; so that although the holes in which the eggs are found may be some six or seven feet in depth, the eggs themselves may be only two or three feet from the surface.

A further detailed account of these tumuli and the manner in which the bird lays its eggs is given by Mr. Gilbert, whose researches are quoted in Gould's Birds of Australia.

AUSTRALIAN JUNGLE FOWL.—*Megapódius túmulus.*

"The birds are said to lay but a single egg in each hole, and after the egg is deposited, the earth is immediately thrown down lightly until the hole is filled up ; the upper part of the mound is then smoothed and rounded over. It is easily known where a Jungle Fowl has been recently excavating, from the distinct impression of its feet on the top and sides of the mound ; and the earth being so lightly thrown over, that with a slender stick the direction of the hole is readily detected, the ease or difficulty of thrusting the stick down indicating the length of time that may have elapsed since the bird's operations.

Thus far it is easy enough, but to reach the eggs requires no little exertion and perseverance. The natives dig them up with their hands alone, and only make sufficient room to admit their bodies and to throw out the earth between their legs. By grubbing with their fingers alone, they are enabled to feel the direction of the hole with greater certainty, which will sometimes, at a depth of several feet, turn off abruptly at right angles, its direct course being obstructed by a clump of wood or some other impediment.

Their patience is, however, often put to severe trials. In the present instance the native dug down six times to a depth of, at least, six or seven feet, without finding an egg, and at the last attempt came up in such a state of exhaustion that he refused to try again. But my interest was now too much excited to relinquish the opportunity of verifying the native's statement, and by the offer of an additional reward I induced him to try again. This seventh trial proved successful, and my gratification was complete when the native, with equal pride and satisfaction, held up an egg, and after two or three more attempts, produced a second ; thus proving how cautious Europeans should be in disregarding the narratives of these poor children of nature, because they happen to sound extraordinary or different from anything with which they were previously acquainted."

On one occasion, Mr. Gilbert caught a young Jungle Fowl in a hole, about two feet in depth, and the little creature, which appeared to be only a few days old, was lying upon some dry leaves. It was a wild and intractable bird despite its tender age, and though it was treated well and ate largely of the food with which it was supplied, it continued to be restless and uneasy, and in two or three days contrived to escape. Even at that age it possessed the earth-heaping propensities of its kind, and used to be continually flinging about the sand which filled the box in which it was placed. Although so small a bird, not larger than a young quail, it could grasp a quantity of sand, and throw it from one end of the box to the other, without apparently exerting itself, and was so constantly engaged in that occupation that it deprived its owner of sleep during the few nights that it remained in his possession.

The same patient and acute observer gives the following account of the general habits of this species :—

"The Jungle Fowl is almost exclusively confined to the dense thickets immediately adjacent to the sea-beach ; it appears never to go far inland except along the banks of creeks. It is always met with in pairs or quite solitary, and feeds on the ground ; its food consisting of roots, which its powerful claws enable it to scratch up with the utmost facility, and also of seeds, berries, and insects, particularly the larger kind of coleoptera.

It is at all times a very difficult bird to procure ; for although the rustling noise produced by its stiff pinions when flying away be frequently heard, the bird itself is seldom to be seen. Its flight is heavy and unsustained in the extreme. When first disturbed, it invariably flies to a tree, and on alighting, stretches out its head and neck in a straight line with its body, remaining in this position as stationary and motionless as the branch upon which it is perched ; if, however, it becomes fairly alarmed, it takes a horizontal but laborious flight for about a hundred yards, with its legs hanging down as if broken. I did not myself detect any note or cry, but from the natives' description and imitation of it, it much resembles the clucking of the domestic fowl, ending with a scream like that of the peacock.

I observed that the birds continued to lay from the latter part of August to March, when I left that part of the country ; and, according to the testimony of the natives, there is only an interval of about four or five months, the driest and the hottest part of the year, between their season of incubation."

The colouring of this bird is simple, but the tints are soft and pleasing. The head is rich ruddy brown, the back of the neck blackish grey, and the back and wings brownish cinnamon, deepening into dark chestnut on the tail-coverts. The whole of the under surface is blackish grey. The legs are orange, and the bill rusty brown.

The LEIPOA or NATIVE PHEASANT of the colonists, so called on account of the pheasant-like aspect of its head and neck, and the general outline of the body, is also an Australian bird, inhabiting the north-western parts of that country, and the sandy plains of the interior.

Like the preceding species, it lays its eggs in a mound of earth and leaves, but the mound is not nearly so large, seldom exceeding three feet in height and eight or nine in diameter, so that it bears some resemblance to a large ant-heap, a similitude which is greatly strengthened by the large number of ants which are always found in the mounds, and by the indurated substance of its lower portion, which is sometimes so hard that the eggs cannot be got at without the aid of a chisel. These nests are generally well hidden away from observation, being placed in the driest and sandiest spots, in which a thick dense bush grows so plentifully that a human being can hardly force his way through them, though the bird is able to traverse their intricacies with great celerity.

The mound is composed of sand and soil, containing a mass of leaves and grass, in the midst of which the eggs are laid, each egg being carefully placed separately from the others. There are many eggs, often more than a dozen, and one of these mounds is quite a

NATIVE PHEASANT.—*Leipoa ocellata.*

little property to the person who is fortunate enough to find it, as the bird will suffer her nest to be robbed repeatedly, and will lay over and over again, thus affording a bountiful supply of eggs to the discoverer. The colour of the eggs is white with a very slight tinge of red.

The Leipoa is an active bird, chiefly depending on its legs, like the pheasant of our own country, and never seeking to escape by flight unless absolutely driven to such a course. When startled, its usual plan is to take to its legs, and run off at full speed, threading the bushes with great rapidity and being very likely to escape if the bush be thick, But if it be surprised when the ground is tolerably open, it may be run down and captured without much difficulty, as it possesses a stupid habit which was formerly attributed to the ostrich. Looking naturally upon the bushes as its home, it makes at once for the nearest bush, dashes into it, and there remains until the pursuer comes up and drags it from its fancied refuge.

The head of the Leipoa is decorated with a well-defined crest, which, like the remainder of the head, is blackish brown. The neck and shoulders are dark ashen grey, and the front of the neck and the upper part of the breast are covered with long black pointed feathers, each having a white stripe along its centre. The primary feathers of the wings are dark brown, having some sharply toothed lines near the tip, and the feathers of the back and remainder of the wings are marked near their extremities with three bands of greyish white, brown, and black, forming a series of "eyes" upon the feathers. The under surface is buff, the flanks being barred with black. The tail is deep blackish brown with a broad buff tip, the bill is black, and the legs blackish brown. In size the Leipoa is about equal to a very small turkey.

BRUSH TURKEY.—*Tallegalla Lathami.*

ANOTHER very remarkable bird possesses many of the same habits as the two preceding species. This is the BRUSH TURKEY, sometimes called the WATTLED TALLEGALLA or the NEW HOLLAND VULTURE, the latter extraordinary title having been given to it on account of its head and neck, which in some parts are devoid of feathers, in others are covered only with short hair, and in others are decorated with naked fleshy wattles. The native name is Weelah.

This bird is far from uncommon in many parts of New South Wales, and inhabits the densest bushes of that country. Like the Leipoa, when pursued it endeavours to effect its escape by running through the tangled brush, a feat which it can perform very adroitly, but it is not so silly as to allow itself to be taken by hand as in the case of the preceding species. When very close pursued, and unable to escape by speed, it jumps into the lowest branch of some tree, leaps from bough to bough until it has reached the top, and either perches there or flies off to another part of the brush.

The Brush Turkey is a gregarious bird, living in small companies, and, like the true turkey, is very wary and suspicious. The great enemy of this bird is the dingo or native dog, which persecutes the flocks sadly, and often hunts them down. From this foe they are safe by flying into a tree; but this elevated position only makes them the more subject to the colonist's gun, and as the birds seem stunned or bewildered by the report, they will suffer several rounds to be fired before they will fly away. Moreover, they have a habit of resorting to the trees at midday, and sheltering themselves from the sun under the spreading foliage, so that any one who has a knowledge of the customs of this bird may be sure of good sport and a heavy bag.

The food of the Brush Turkey mostly consists of seeds and vegetable substances, though insects of various kinds have been found in its stomach, which is exceedingly muscular. Like other gallinaceous birds, it is fond of dusting itself, and as it loves to resort to the same spot, it scrapes considerable depressions in the earth, which lead the practised hunter to its residence. The voice of the Brush Turkey is a rather loud clucking sound. Its flesh is particularly excellent, and there are hopes that this fine bird may also be in time added to our list of domesticated poultry.

The egg mound—for it cannot rightly be called a nest—of this bird is extremely large, containing, according to Mr. Gould, several cartloads of materials, and being formed into a conical or somewhat pyramidal shape. It is not made by a single pair of birds, but is the result of united labour, and is used from year to year, fresh materials being supplied each season in order to make up the deficiency caused by the decomposition of the vegetable matter below. Mr. Gould, to whom we are indebted for the greatest part of our knowledge respecting these curious birds, gives the following account of the nidification of the Brush Turkey :—

"The mode in which the materials composing these mounds are accumulated is very singular, the bird never using its bill, but always grasping a quantity in its foot, throwing it backwards to one common centre, and thus clearing the surface of the ground for a considerable distance so completely that scarcely a leaf or a blade of grass is left. The heap being accumulated, and time allowed for a sufficient heat to be engendered, the eggs are deposited, not side by side as is usually the case, but *planted* at the distance of nine or twelve inches from each other, and buried at nearly an arm's depth, perfectly upright, with the large end upwards; they are covered up as they are laid, and allowed to remain until hatched. I have been credibly informed, both by natives and settlers living near their haunts, that it is not an unusual event to obtain nearly a bushel of eggs at one time from a single heap; and as they are delicious eating, they are eagerly sought after.

Some of the natives state that the females are constantly in the neighbourhood of the heap about the time the young are likely to be hatched, and frequently uncover and cover them up again, apparently for the purpose of assisting those that may have appeared; while others have informed me that the eggs are merely deposited, and the young allowed to force their way unassisted. In all probability, as Nature has adopted this mode of reproduction, she has also furnished the tender birds with the power of sustaining themselves from the earliest period; and the great size of the egg would equally lead to this conclusion, since in so large a space it is reasonable to suppose that the bird would be much more developed than is usually found in eggs of smaller dimensions. In further confirmation of this point, I may add that in searching for eggs in one of the mounds, I discovered the remains of a young bird, apparently just excluded from the shell, and which was clothed with feathers, not with down, as is usually the case."

In the "Guide to the Gardens of the Zoological Society," by Mr. P. L. Sclater, the Secretary to the Society, is the following most valuable account of the habits of this bird in a state of captivity. The date of the notice is May, 1861.

"Since the year 1854, the singular phenomenon of the mound-raising faculty of the Tallegalla, which had been well ascertained in Australia by Mr. Gould, has been annually displayed in this country.

On being removed into an inclosure, with an abundance of vegetable material within reach, the male begins to throw it up into a heap behind him, by a scratching kind of motion of his powerful feet, which project each footful as he grasps it for a considerable distance in the rear. As he always begins to work at the outer margin of the inclosure, the material is thrown inwards in concentric circles, until sufficiently near the spot selected for the mound to be jerked upon it. As soon as the mound is risen to a height of about four feet, both birds work in reducing it to an even surface, and then begin to excavate a depression in the centre. In this, in due time, the eggs are deposited as they are laid, and arranged in a circle, about fifteen inches below the summit of the mound, at regular intervals, with the smaller end of the egg pointing downwards. The male bird watches the temperature of the mound very carefully: the eggs are generally covered, a cylindrical opening being always maintained in the centre of the circle for the purpose of giving air to them, and probably to prevent the danger of a sudden increase of heat from the action of the sun or accelerated fermentation in the mound itself. In hot days the eggs are nearly uncovered two or three times between morning and evening.

On the young bird chipping out of the egg, it remains in the mound for at least twelve hours without making any effort to emerge from it, being at that time almost as deeply covered up by the male as the rest of the eggs.

On the second day it comes out, with each of its wing-feathers well developed in a sheath which soon bursts, but apparently without inclination to use them, its powerful feet giving it ample means of locomotion at once. Early in the afternoon, the young bird retires to the mound again, and is partially covered up for the night by the assiduous father, but at a diminished depth as compared with the circle of eggs from which it emerged in the morning. On the third day, the nestling is capable of strong flight, and on one occasion one of them, being accidentally alarmed, actually forced itself, while on the wing, through the strong netting which covered the inclosure. The accounts of the habits of the Tallegalla, given by Mr. Gould in his Birds of Australia, in 1842, strange as it appeared at the time, are thus perfectly verified in every respect."

The general colour of the adult male Talegalla is blackish brown above, and the same on the under surface with a silver grey gloss produced by the grey tips of the back feathers. The cheeks are naked, the head and neck covered with short hair-like feathers of a dark blackish hue, and the front of the neck is furnished with a large naked fleshy wattle, something like that of a turkey, and being of a bright yellow warming into orange-red at its junction with the neck. The bill is black; the eyes brown chestnut, and the legs and feet dark brown. The male bird is about the size of an ordinary turkey, and the female is about one-fourth less. Her plumage is like that of the male, from which she may be readily distinguished by the smaller size of the wattle.

THE large family of the Peacocks, or Pavonidæ, now claims our attention. For convenience of description, these birds have been separated into several sub-families, which are defined with tolerable certainty. Of the Pavoninæ, we shall find two examples in the following pages.

The PEACOCK may safely be termed one of the most magnificent of the feathered tribe, and may even lay a well-founded claim to the chief rank among birds in splendour of plumage and effulgence of colouring. We are so familiar with the Peacock that we think little of its real splendour; but if one of these birds had been brought to Europe for the first time, it would create a greater sensation than even the hippopotamus or the gorilla.

The Peacock is an Asiatic bird, the ordinary species being found chiefly in India, and the Javanese Peacock in the country from which it derives its name. In some parts of India the Peacock is extremely common, flocking together in bands of thirty or forty in number, covering the trees with their splendid plumage, and filling the air with their horridly dissonant voices. Captain Williamson, in his "Oriental Field Sports," mentions that he has seen at least twelve or fifteen hundred peacocks within sight of the spot where he stood.

These birds are great objects of sport, and are mostly killed by the gun, though a good rider may sometimes run them down by fair chase. The Peacock takes some little preparation to get on the wing, and if hard pressed is not able to rise into the air. The horseman then strikes at the bird with his long lashed whip, so as to get the lash round its neck, and soon masters the beautiful quarry. "When upon the wing," says Captain Williamson, "they fly very heavy and strong, generally within an easy shot. It may reasonably be supposed that they fall very heavy, but if only winged they soon recover, and if not closely pursued will, nine times in ten, disappear. When the peepul berries, or figs, are in season, their flesh is rather bitter; but when they have fed

PEACOCK.—*Pavo cristátus.*

a while among the corn-fields, they become remarkably sweet and juicy. This is to be understood of the young birds, which make excellent roasters. The older birds are sometimes put to the spit, but are by no means so good as when the breasts are made into cutlets, and the residue boiled down into a rich soup. I have always thought such Peacocks as frequented the mustard-fields after the pods were formed to be very superior.

They abound chiefly in close wooded forests, particularly where there is an extent of long grass for them to range in. They are very thirsty birds, and will only remain where they can have access to water. Rhur plantations are their favourite shelter, being close above so as to keep off the solar rays, and open at the bottom sufficiently to admit a free passage for the air. If there be trees near such spots, the Peacocks may be seen mounting into them every evening towards dark to roost; and in which they generally continue till the sun rises, when they descend to feed, and pass the midday in the heavy coverts.

They are very jealous of all quadrupeds, especially of dogs; no doubt from finding the jackal, and probably the tiger, to be such inveterate enemies. When Peacocks are discovered in a tree, situated on a plain, if a dog be loose and hunt near it, the bird will rarely move from it, though it will probably show extreme uneasiness.

But the most certain mode of killing one or two birds is by stealing under the trees at night; if there be a clear moon, so much the better. In this way, by looking up among the foliage, the Peacocks may be readily distinguished. When they are very numerous, and only one bird is wanted, as certain a mode as any is to lie in wait behind a bush near their feeding haunts; but without the most perfect silence this will not succeed.

Though Pea-fowls invariably roost in trees, yet they make their nests on the ground, and ordinarily on a bank raised above the common level, where in some sufficient bush they collect leaves, small sticks, &c. and sit very close. I have on several occasions seen them in their nests, but as I refrained from disturbing them, they did not offer to move, though they could not fail to know that they were discovered. They usually sit on about a dozen or fifteen eggs. They are generally hatched about the beginning of November; and from January to the end of March, when the corn is standing, are remarkably juicy and tender. When the dry season comes on, they feed on the seeds of weeds and insects, and their flesh becomes dry and muscular."

Peacock-shooting, although an exciting sport, is a dangerous one, the tiger feeling himself suited by the rhur and other vegetation in which the peacock delights, so that an inexperienced sportsman may suddenly find himself face to face with a tiger, and run a strong chance of being himself the object of pursuit. Old hunters, however, who know the habits of the Peacock, find that bird extremely useful in denoting the presence of tigers. When the Peacock find itself in close proximity to a tiger or even a wild cat, it raises the sound of alarm, which is a loud hoarse cry, answered by those within hearing. The bird then utters a series of sharp quick grating notes, and gets higher into the trees so as to be out of reach of the tiger's claws.

In this country the Peacock is very common, and forms a magnificent adjunct to the lawn, the park, the garden, and the farmyard. The evident admiration and self-consciousness with which a Peacock regards himself are truly amusing, the bird always looking out for spectators before it spreads its train, and turning itself round and round so as to display its beauties to the best advantage. At night it always roosts in some elevated spot; and invariably sits with its head facing the wind. Several Peacocks, whom I used to see daily, always roosted upon the thatch of a corn-rick, their long trains lying along the thatch so closely that towards dark they could hardly be seen. In character, the Peacock is as variable as other creatures, some individuals being mild and good-tempered, while others are morose and jealous to the extreme.

One of these birds, living in the north of Ireland, was a curious mixture of cruelty and fun. He had four wives, but he killed them all successively by pecking them to death, for what cause no one could find out. Even its own children shared the same fate, until its owner put the Pea-fowl eggs under a sitting hen, and forced her to hatch the eggs and tend the young far out of his sight.

His great amusement was to frighten the chickens. There were two iron troughs in which the food for the chickens was placed daily, and to which they always resorted as soon as their food was poured into their troughs. No sooner had they all assembled than the Peacock would erect his train, rattle his quills together with that peculiar rustling sound that is so characteristic of these birds, and march slowly towards the chickens. The poor little birds would slowly back away from the trough as the Peacock advanced, not liking to lose sight of their food, and not daring to remain in defiance of their persecutor. By degrees he got them all into a corner, crouching together and trembling, when he would overshadow them with his train, place the ends of the feathers against the wall so as to cover them completely, rattle the quills heartily so as to frighten them extremely, and then would walk off, looking quite exultant at the trick he had just played. He did not care for eating their food, but left the trough untouched.

The train of the male Peacock, although popularly called its tail, is in reality composed of the upper tail-coverts, which are enormously lengthened, and finished at their extremities with broad rounded webs, or with spear-shaped ends. The shafts of these feathers are almost bare of web for some fourteen or fifteen inches of their length, and then throw out a number of long loose vanes of a light coppery green. These are very brittle and apt to snap off at different lengths. In the central feathers the extremity is modified into a wide flattened battledore-shaped form, each barbule being coloured with refulgent emerald green, deep violet-purple, greenish bronze, gold and blue, in such a manner as to form a distinct "eye," the centre being violet of two shades, surrounded with emerald, and the other tints being arranged concentrically around it. In the feathers that edge the train there is no "eye," the feathers coming to a point at the extremity, and having rather wide but loose emerald green barbules on its outer web, and a few scattered coppery barbules in the place of the inner web. The tail-feathers are only seven or eight inches in length, are of a greyish brown colour, and can be seen when the train is erected, that being their appointed task.

On the head is a tuft or aigrette of twenty-four upright feathers, blackish upon their almost naked shafts and rich golden green shot with blue on their expanded tips. The top of the head, the throat and neck, are the most refulgent blue, changing in different lights to gold and green. On the back the feathers are golden green, edged with velvety black, giving a peculiar richness of effect. The wings are darker than the rest of the plumage, the quill-feathers being marked with black, and having some red aboutt hem The abdomen is blackish with a green gloss, and the feathers of the thighs are fawn. The female is much smaller than her mate, and not nearly so beautiful, the train being almost wanting, and the colour ashy brown with the exception of the throat and neck, which are green. A white or albino variety of this bird is not at all uncommon, and in this case the characteristic "eyes" are faintly indicated in neutral tint.

THE generic term Polyplectron signifies "many-spurred," and is given to a genus of gallinaceous birds because they have two or sometimes three spurs on each leg. There are several species, all very handsome birds, and one of the most conspicuous is the CRESTED PEACOCK PHEASANT. As is the case with all the species, the tail is greatly enlarged, so as to be spread into a flat, wide, fan-like form, with two ranges of feathers placed one above each other, and decorated with a double row of large lightly coloured spots. It probably inhabits Soudan and the Moluccas, but there is little known of its habits.

The beautiful crest which adorns the head is very deep shining violet-blue, and the head, neck, and breast are of the same colour. Over the eye runs a white streak, and a white patch is placed just below and behind the eye, contrasting very boldly with the deep violet of the surrounding plumage. The back is brown, covered with irregular wavy lines of a paler hue, and the wing-coverts and secondaries are bright azure tipped with velvety black. The tail is brown, covered with innumerable little spots of yellowish white, and each feather is marked near the tip with a large oval spot of shining metallic green, surrounded first with a waved line of black and then with a broader line of pale brown. Close to the tip each feather is bordered with black, and the extremity is pale fawn. The abdomen is dull black. In total length, this bird measures about twenty inches.

THE Pheasants come next in order, and the grandest and most imposing of this group, although there are many others that surpass its brilliant colouring, is the ARGUS PHEASANT,

PEACOCK PHEASANT.—*Polyplectron Napoleónis.*

so called in remembrance of the ill-fated Argus of mythology, whose hundred eyes never slept simultaneously until charmed by the magic lyre of Mercury.

This magnificent bird is remarkable for the very great length of its tail-feathers and the extraordinary development of the secondary feathers of the wings. While walking on the ground, or sitting on a bough, the singular length of the feathers is not very striking, but when the bird spreads its wings, as shown in the smaller figure in the background, they come out in all their beauty. As might be supposed from the general arrangement of the plumage, the bird is by no means a good flier, and when it takes to the air, only flies for a short distance. In running, its wings are said to be efficient aids.

Although the Argus is hardly larger than an ordinary fowl, the plumage is so greatly developed that its total length measures more than five feet. The head and back of the neck are covered with short brown feathers, and the neck and upper part of the breast are warm chestnut-brown covered with spots of yellow and black, and similar tints are formed on the back. The tail is deep chestnut, covered with white spots, each spot being surrounded with a black ring. When the bird chooses, it can raise the tail, so that it stands boldly in the air between the wings and is partially spread. The secondaries of the wings are most wonderful examples of plumage, and would require many pages to describe them fully. Suffice it to say that the gradations of jetty black, deep rich brown, orange, fawn, olive and white are so justly and boldly arranged as to form admirable studies for the artist, and totally to baffle description.

In one feather now before me there are seventeen large "eyes" on the outer web, each being surrounded with a ring of jetty black, then with a dash of chocolate within the

ARGUS PHEASANT.—*Argus gigantéus.*

ring, then olive with the least possible tinge of purple, and lastly with a spot of pure white near the tip, fading imperceptibly into the olive on one side and the chocolate on the other. Between these "eyes" some leopard-like mottlings diversify the rich fawn of the ground colour, and outside them four wavy bands of dark brown run along the feather towards the edge, breaking up into spots about an inch before they reach the edge. The inner web is pale fawn covered with black spots, surrounded with buff, and the tip of the whole feather is deep brown, spotted profusely with white. The shaft is black at its base, and yellow towards its termination.

In another feather both webs are marked just like a leopard, with dark spots on a fawn ground, only the spots are arranged in diagonal rows. But along the shaft runs a band, about three-quarters of an inch wide, of rich chocolate, profusely speckled with the tiniest white spots, also arranged in rows. This band does not quite extend to the end of the feather, which at its tip is pale fawn very sparingly studded with deep brown rosettes, surrounded with chestnut. These are but two feathers, and I might take twenty as wonderful. In the female the secondary feathers, instead of measuring nearly a yard in length, are little more than a foot, and the eyes are much more obscure. The Argus Pheasant inhabits Sumatra and neighbouring localities.

THE well-known PHEASANT affords a triumphant instance of the success with which a bird of a strange country may be acclimatized to this island with some little assistance from its owners.

Originally the pheasant was an inhabitant of Asia Minor, and has been by degrees introduced into many European countries, where its beauty of form and plumage and the delicacy of its flesh made it a welcome visitor. In this country it is probably dependent to a great extent on "preserves" for its existence, as, even putting aside the marauding attacks of poachers, whether biped or quadruped, the bird requires much shelter and plenty of food. Even with the precautions that are taken by the owners of preserves, the breed is to some

PHEASANT.—*Phasiánus Cólchicus.*

degree artificially kept up by the hatching of Pheasant's eggs under domestic hens, and feeding them in the coop like ordinary chickens, until they are old and strong enough to get their own living.

The food of this bird is extremely varied. When young it is generally fed on ants' eggs, maggots, grits, and similar food, but when it is fully grown it is possessed of an accommodating appetite, and will eat many kinds of seeds, roots, and leaves. The tubers of the common buttercup form a considerable item in its diet, and the bird will also eat beans, peas, acorns, berries of various kinds, and has even been known to eat the ivy leaf as well as the berry.

The Pheasant is a ground-loving bird, running with great speed, and always preferring to trust to its legs rather than its wings. It is a crafty creature, and when alarmed, instead of rising on the wing, it slips quietly out of sight behind a bush or through a hedge, and then runs away with astonishing rapidity, always remaining under cover until it reaches some spot where it deems itself to be safe. The male Pheasant is not in the least given to the domestic affections, passing a kind of independent existence during part of the year, and associating with others of its own sex during the rest of the season. It is a very combative bird, and can maintain a stout fight even with a barn-door cock. When the two fight, an event of no very unfrequent occurrence, the Pheasant often gets the better of the combat by his irregular mode of proceeding. After making two or three strokes, up goes the Pheasant into a tree to breathe awhile, leaving the cock looking about for his antagonist. Presently, while his opponent is still bewildered, down comes the Pheasant again, makes another stroke and retires to his branch. The cock gets so puzzled at this mode of fighting that he often yields the point.

It is rather curious that the Pheasant should display so great a tendency to mate with birds of other species. Hybrids between the Pheasant and common hen are by no means uncommon, and the peculiar form and colour of the plumage, together with the wild and suspicious mien are handed down through several generations. The grouse is also apt to mate with the Pheasant, and even the turkey and the guinea fowl are mentioned among the members of these curious alliances.

As these pages are not intended for sporting purposes, the art and mystery of Pheasant shooting will be left unnoticed. The ingenious mode employed by Mr. Waterton for the deception of poachers, is however, too amusing to be omitted. Those nocturnal marauders were accustomed to haunt the fir plantations at night, and by looking upwards could easily see the Pheasants as they sat asleep across the branches, and bring them down with the gun, or even a noose on a long rod. So, thinking that prevention was better than prosecution, he first planted a number of thick holly clumps, dark as night in the interior, and quite impervious to human beings unless cased in plate armour. The Pheasants soon resorted to these fortresses, but their places were filled with a few hundred rough wooden Pheasants, which were nailed upon the fir branches, and at night looked so exactly like the birds that the most practised eye could not discover the difference. After these precautions had been taken, the astute inventor was able to rest quietly at home and chuckle to himself at the nocturnal reports in the direction of the fir-wood.

The nest of the Pheasant is a very rude attempt at building, being merely a heap of leaves and grasses collected together upon the ground, and with a very slight depression, caused apparently quite as much by the weight of the eggs as by the art of the bird. The eggs are numerous, generally about eleven or twelve, and their colour is an uniform olive brown. Their surface is very smooth. When I was a boy I well remember finding a Pheasant's nest in a copse, taking the whole clutch and blowing them on the spot with perfect openness, being happily ignorant of the penalties attached to such a proceeding, and not in the least acquainted with the risk until I exhibited my prize to some friends, and saw their horrified looks.

The adult male Pheasant is a truly beautiful bird. The head and neck are deep steely blue, "shot" with greenish purple and brown ; and the sparkling hazel eye is surrounded with a patch of bare scarlet skin, speckled profusely with blue-black. Over the ears there is a patch of brown. The upper part of the back is beautifully adorned with light golden red feathers, each being tipped with deep black ; and the remainder of the back is of the same golden red, but marked with brown and a lighter tint of yellow without any admixture of red. The quill-feathers of the wing are brown of several shades, and the long quills of the tail are oaken brown changing to purple on the edge of the outer web, and barred with jetty black on the outer web and brown on the inner. The breast and front of the abdomen are golden red with purple reflections, and diversified by the black edge of each feather ; the rest of the abdomen and under tail-coverts are blackish brown. In total length the full-grown male Pheasant is about three feet. The female is much more sober in her colours and less in size than her mate, her body being of a pale yellow-brown, and her length only some two feet.

THE gorgeous bird which is now known by the name of REEVES' PHEASANT, but which has undergone so many changes of title, is a native of Surinagur and Northern China.

It is a truly remarkable bird, for although its body does not surpass the ordinary Pheasant in size, the total length of a full-grown male will often exceed eight feet, owing to the very great development of the two central tail-feathers, which alone will measure six and seven feet in length, and are very wide at the base. This species has been brought alive to England and placed in the Zoological Gardens, where it throve tolerably well ; and was sufficiently hardy to warrant a hope that it might be acclimatized to this country. Its habits in a wild state are little known, but those specimens which have been kept in captivity behaved much like the ordinary Pheasant. Although so splendid and highly coloured a bird, it inhabits very cold regions, the mountains of Surinagur being covered with snow. In that country it is known by the appropriate name of Doom-durour or Long-tail.

No amount of artificial colouring could give the full effect of the gorgeous and ever-changing beauty which adorns the plumage of this magnificent bird; while the simple black and white of an engraving gives but a very faint notion of its real magnificence. The absence of colours must therefore be faintly supplied with a brief description in words.

The head is white, except a patch of light scarlet naked skin around the eyes, edged by a band of black which runs over the forehead, under the chin, and is rather broader over the ear-coverts. The neck is also broadly collared with white. The back of the neck, and the back itself are covered with shining scale-like feathers, each being a light golden yellow and edged at the extremity by a band of deep velvety black, thus producing an extremely rich appearance. The feathers of the breast and abdomen are snowy white, banded and tipped with the same velvety black as those of the upper parts with the exception of the middle of the breast and abdomen, which are deep black, and the under tail-coverts, which are also black covered with golden yellow spots. The two central feathers of the tail are delicate grey, covered with numerous transverse and rather curved bands of rich dark brown, edged with a lighter tint of the same colour. In one of these feathers only four feet in length, Mr. Temminck counted forty-seven bands. The remaining feathers of the tail are greyish white, also profusely barred with deep brown, and passing into chestnut at their edges. They can be folded over each other, and they appear very narrow.

Two very lovely birds are shown in the illustration on page 612, one glowing like the sun in the full radiance of gold and crimson, and the other shining like the moon with a soft silvery lustre, not so splendid, but even more pleasing.

The GOLDEN PHEASANT is a native of China, where it is a great favourite, not only for its splendid plumage and elegant form, but for the excellence of its flesh, which is said to surpass in delicacy even that of the common Pheasant.

For the purposes of the table, however, it is hardly likely to come into general use, as there are great difficulties in the way of breeding it in sufficient

REEVES' PHEASANT.—*Phasianus Reevesii.*

SILVER PHEASANT.—*Gallophásis Nycthémerus.*   GOLDEN PHEASANT.—*Thaumália picta.*

number, and one feels a natural sensation of repugnance to the killing of so beautiful a bird merely for the sake of eating it. As it is a tolerably hardy bird, bearing confinement well, and breeding freely, especially in the southern parts of England, it has been turned out into preserves with the common Pheasant, but as yet without sufficient success to warrant the continuation of the experiments.

This bird, together with another which will be briefly mentioned, is remarkable for the large ruff of broad squared feathers which folds round its neck, as well as for the finely developed crest. This crest is of rich golden yellow with a tinge of carmine. The feathers of the ruff are squared, and disposed in a scale-like fashion; their colour is rich orange edged with velvety black. The whole ruff can be raised or depressed at will. Fly-fishers hold the crest and ruff of this bird in great value, as many of their best artificial baits owe their chief beauty to the Golden Pheasant. Just below the ruff comes a patch of scale-like rounded feathers of dark glossy green, over which the ends of the ruff feathers play as the bird moves its head, and below them the back is wholly of a bright golden yellow, enriched on the upper tail-coverts by a crimson edging. The primary and secondary feathers of the wings are rich brown barred with chestnut, and their bases are deep blue. The breast and abdomen are brightest scarlet, and the tail is rich chestnut mottled with black. The eye is bright, glancing, and of a whitish yellow.

These magnificent colours only belong to the male bird, the female being reddish brown, spotted and marked with a darker hue, and the tail is short.

The second ruffed Pheasant is that which is known by the name of AMHERST'S Pheasant (*Thaumálea Amhérstiœ*), also a native of China. This magnificent bird has a wonderfully long and broad tail, quite as remarkable as that of Reeves' Pheasant. The crest of this beautiful bird is scarlet, the tippet is snowy white, each feather being tipped with velvety black, the shoulders are rich shining green, the abdomen pure white, and the tail is white, barred with dark green, and strikingly varied with the scarlet tips of the upper tail-coverts, which are much elongated.

The SILVER PHEASANT is another inhabitant of China, and is found chiefly in the northern portions of that country.

It is one of the largest and most powerful of the tribe to which it belongs, and is said to be a match for a gamecock in fair combat. It is a hardy bird, and like the Golden Pheasant, has been turned loose into British preserves, but with even less success. The weight of the bird is generally too great in proportion to its strength of wing, so that it does not readily raise itself from the ground, and thereby runs a risk of being devoured by the carnivorous quadrupeds that infest every preserve. Moreover, it is so large, so strong, and so combative, that it fights the common Pheasants and drives them out of the coverts, so that at present we have to content ourselves with rearing it under the safe protection of brick and wire.

The crest on the top of the head is deep purple-black, and the naked skin round the eyes, which forms a kind of wattle over the nostrils and below the chin, is a bright scarlet. The upper surface of the body is pure silver-white, delicately pencilled with wavy black lines. The tail is also white, pencilled boldly with black, except the two central feathers, which are wholly white, long and curved. The breast and abdomen are of the same deep purple black as the crest. The colours of the female are quite dissimilar, so that the bird would hardly be recognised as belonging to the same species. She is much smaller in size, has a smaller crest, and a shorter tail, of a brown colour, streaked on the outer feathers with black and white. Instead of the silvery white of the male, her back is greyish brown, irregularly marked and waved with narrow black bars. The breast and abdomen are greyish white, marked with brown and barred with black.

THE very handsome FIREBACK is an Asiatic bird, inhabiting Sumatra and in all probability several other neighbouring localities.

The popular name of Fireback is very appropriate, being given to the bird on account of the fiery red feathers which decorate a considerable portion of the back. It is remarkable for the great size of the naked skin about the eyes, which nearly covers the whole head, running over the ears and forehead, and descending well below the chin. The colour is of a bluish purple during the life of the bird, but after its death the colour darkens into dark brown, as is generally the case with bare skin both in beasts and birds, and in the stuffed species it shrinks, like wetted leather, and entirely loses its former fulness and shape.

The head is decorated with an elegant crest of upright feathers, their shafts being nearly devoid of web, and expanding at the extremities into a number of delicate barbs. The general colour of the bird is rich deep satiny violet, appearing black except in certain lights, and the feathers of the lower part of the back are flaming orange-red, the depth of hue being changeable according to the light. The tail is smaller than that of the domestic cock, and the central feathers are snowy white, the others being deep green glossed with purple. The total length of the adult male is about two feet. The female is smaller, and her plumage is warm cinnamon-brown above and greyish white below.

WE now arrive at the typical genus of the Gallinæ, to which our ordinary barn-door poultry, with all its multitudinous varieties belongs. Our first example of this genus is the beautiful SONNERAT'S JUNGLE FOWL.

This fine bird is a native of India, and is found chiefly in the wooded districts. Although smaller than the common domestic fowl, it is a wonderfully powerful bird in

VEILLOT'S FIREBACK.—*Gallophasis Veillotii.*

proportion to its size, and so fierce and determined a combatant that the native sportsmen, who set great store upon fighting cocks, always prefer a Jungle Cock as their champion. As in general appearance it is something like the domestic fowl, some persons have supposed that it is the stock from which our poultry were derived. The Bankiva Fowl, however, is thought with more reason to be the original progenitor of these useful birds. The very peculiar formation of the hackles affords a good reason for believing that the domestic fowl is not the offspring of Sonnerat's Jungle Fowl. The webs of the hackles and upper tail-coverts are dark grey, but their shafts are bright orange, dilating in the centre and at the tip into flat, shining horny plates of a brilliant orange hue, which give a peculiar splendour to the plumage, and are discernible at a considerable distance, their tips being rounded instead of lancet-shape.

The voice of this bird is rather startling, for at first sight it looks so like a game-cock, that its crow strikes the ear in a very absurd manner. Every one knows the ludicrous attempts made by a young cock to crow like his elders; how he breaks down just when he thinks he is doing best, like a young lad with a cracked voice, trying to talk with a manly intonation, and going unexpectedly from hoarse bass to sharpest treble. Give the young cock a sharp attack of whooping-cough, and that will afford a tolerably good notion of the crowing of this Jungle Fowl.

The head of this bird is adorned with well-developed wattles, deeply notched at the tip. The beautiful hackles have already been described, with their flattened ends shining like the gold coins gleaming on the dark tresses of Oriental beauties. The back and lower portions of the body are deep grey, and the tail is long, arched, and beautifully

SONNERAT'S JUNGLE FOWL.—*Gallus Sonneratii.*

coloured with changing hues of purple, green, and gold. The female is a smaller and very sober-looking bird, without comb or wattles, and devoid of the curious horny hackles that decorate her mate.

The BANKIVA JUNGLE FOWL is now supposed to be the original stock of the domesticated poultry.

It is a native of Java, and the male very closely resembles the game-cock of England. It is a splendid creature, with its light scarlet comb and wattles, its drooping hackles, its long arched tail, and its flashing eye. The comb and wattles are of the brightest scarlet, the long hackles of the neck and lower part of the back are fine orange-red, the upper part of the back is deep blue-black, and the shoulders are ruddy chestnut. The secondaries and greater coverts are deep steely blue, and the quill-feathers of the wing are blackish brown edged with rusty yellow. The long, arched and drooping tail is blue-black glossed with green, and the breast and under parts black, so that in general aspect it is very like the black-breasted red game-cock.

The domesticated bird is of all the feathered tribe the most directly useful to man, and is the subject of so many valuable treatises that the reader is referred to them for the best mode of breeding, rearing, and general management of poultry. On the accompanying illustration are shown some of the most useful or remarkable of the varieties of this bird.

Towards the top, and on the left hand, may be seen some examples of the famous Cochin Fowl, whose enormous size and ungainly appearance took England so completely by storm some few years ago. Nothing was talked of but Cochin China Fowls, and the

BANKIVA FOWL.—*Gallus Bankiva.*—DOMESTIC POULTRY.

| COCHINS. | GAME. | BANTAMS. |
| SPANISH. | DORKINGS. | POLISH. |

sums given for these birds almost rose to the fabulous. A first-rate hunter, or three or four valuable cows, or a tolerable flock of sheep might have been purchased for the money that was freely given for a single Cochin China Fowl.

Occupying the centre of the plate are a pair of the Game Fowls, certainly the finest of all the varieties. The time has now passed away, when these splendid birds were openly trained for combat, and cock-fights were held in every village and town throughout the kingdom. The law has rightly prohibited this savage amusement, and cock-fighting like dog-fighting is now confined to a small and continually decreasing knot of sporting men. For this purpose, the birds are trained in the most regular and scientific manner, as great pains being taken about them as about a race-horse on the eve of the Derby. In order to deprive the antagonist of the advantage which it would gain by pecking the comb, which is very tender and bleeds freely, the comb was cut off and the horny spurs were replaced with steel weapons, long, sharp-edged and pointed. These precautions were, after all, not so barbarous as they seem on a first view, for the comb was " dubbed " at so early an age that its growth was prevented rather than its substance mangled, and the substitution of metal for horny spurs served to set the combatants on more equal terms, just as a sword sets a small man on an equality with a large one. Irrespective of these advantages, the Game-cock is an hereditary gladiator, delighting in combat and instinctively practising the art of defence as well as that of assault. So superior is it to the ordinary breeds in these respects, that I have seen a little old one-eyed Game-cock cut down, as if with a sword, a great swaggering barn-door cock that looked as if it could have killed its puny antagonist with a blow and eaten him afterwards.

There seems to be no limits to the courage of the Game-cock, which will attack not only his own kind, but any other creature that may offend it. One of these birds has been known to fly at a fox that was carrying off one of his wives, and to drive his spur deep into the offender's eyes. There are instances innumerable of similar rescues from cats, rats, and other marauders. Sometimes, however, the Game-cock takes upon himself to defend certain localities, and then often becomes dangerous. One such bird, of whose ferocity I have often had personal proof, was accustomed to parade, with the air of an emperor, the yard in which he was necessarily confined, and would fly at every living being that came within the prohibited precinct. A besom was kept by the door and always used by every one who passed through the yard, for the purpose of repelling the attacks of this savage bird. Many a time have I tried to tire him out, knocking him over with the broom, or pushing him back against the wall, but I was always tired first, and had to vacate the premises, leaving him to get on a water-butt and crow forth his triumph. Sometimes he would slip past the broom, and then the stroke of his spur was no trifle, feeling like the blow of a stone thrown by a strong arm, and leaving a black-and-blue mark for days afterwards.

The flesh of the Game breed is very excellent, but they are troublesome birds to keep, the males always fighting among each other, and having to be separated before they are fully grown. Crosses with the Game breed are common.

Just below the game bird is seen an odd-looking fowl, with a head so covered with a monstrous plume of drooping feathers that its features are not more discernible than those of a skye-terrier under his thick hair. This wealth of cranial plumage seems, however, to impoverish the brain, for the large-crested Polish fowls are generally stupid birds, and apt to meet with accidents which might easily be avoided.

On the opposite side of the plate is the Spanish fowl, a very fine variety, glossy black, with a very large comb, and notable for the white naked skin below the ear. It is a very large breed, coming next in size to the Cochin China, and very far surpasses that large but uncouth bird in the symmetry of its form. The flesh of this breed is excellent, and as the hens are regular layers, these birds are deservedly favourites among poultry owners.

On the foreground of the plate are some examples of those birds whose many excellencies have rendered a town famous. These are the Dorking Fowls, short-legged, round-bodied, plump-fleshed, and remarkable for having at least one, and sometimes two supplementary toes. These useful birds are mostly to be found in Surrey and Kent, the northern and marshy districts not suiting them. The Dorking Fowls are excellent for the table, their flesh being peculiarly plump and white, and the hens are remarkably prolific layers.

Lastly comes the odd, quaint, opiniated little Bantam, with its feathered legs, full breast, and bold fearless carriage. This minikin member of the poultry tribe is, despite

HORNED TRAGOPAN.—*Ceriornis Lathami.*

his small dimensions, as bold as any of them, and if he thinks himself aggrieved will attack a great Cochin China or Spanish cock with such spirited audacity that he will not unfrequently come off victor in the contest. The Bantam is of little use to the poultry keeper, and may be classed among the fancy fowls, of which there are so many and ever-varying breeds. Two examples of these birds may be seen perched on the paling just above the game-cock's tail.

The common Barn-Door Fowl is of no particular breed, no pains being taken to prevent crossing, but is a kind of compound of all the preceding, except perhaps the bantam, which ought to be kept away from them as tending to diminish the size of the birds and their eggs. The regular egg trade is a very complicated and curious affair, giving a livelihood to thousands, and possessing a national importance of which few would dream whose only notion of eggs is connected with the breakfast table or the salad bowl.

A MOST singular group of birds now comes before our notice, of which the HORNED TRAGOPAN affords an example. The males are remarkable for the loose pendent skin which hangs from the base of the lower mandibles, and can be inflated at the pleasure of the bird, and for the two lengthened protuberances behind the eyes which generally hang listlessly down the cheeks, but can be erected at will and then look as shown in the illustration. In all these birds the plumage is ample and the tail short. As far as is at present known they are found in the higher and more mountainous districts of Asia, having been taken in Thibet, Nepâl, and the Himalayas.

TURKEY.—*Meleagris gallopavo.*

They all are beautifully coloured, and the present example may challenge competition with any of the species, if not for absolute brilliance of plumage, yet for delicacy of tint, and pleasing marking of its feathers. The bare skin around the eyes, together with the wattles and horns are bluish purple, and the feathers of the crest, together with the chin and back of the neck are deep black. The upper part of the breast, the neck and shoulders are light cinnamon with a dash of carmine and purple, and variegated by the white eye-like tips of the feathers. The wings and part of the back are rich amber mottled with brown, and also decorated with white spots. The spots are largest and most conspicuous upon the flanks. The tail-coverts are also amber-brown, spotted with white, and extend to such a length as nearly to conceal their short rounded tail. In size the Tragopan about equals a common Spanish fowl.

THE now well-known Turkey is another example of the success with which foreign birds can be acclimatized in this country, and is one of the creatures that affords great encouragement to the members of the Acclimatization Society to persevere in their valuable efforts. Indeed, if so wild a bird as the Turkey, and one so delicate in its youth, can be thus transferred from America to England, there seems every reason that the numerous birds and beasts mentioned by Mr. Buckland in his well-known lecture on this subject, may find a suitable home in this country.

As to its qualities as a poultry bird, there is little to be said, as every reader will have had practical experience thereof, and the mode of breeding and rearing it belongs to the regular treatises on poultry, and does not come within the province of this work.

Admirable descriptions of the Turkey when wild are given by Audubon and other writers, and their narratives must be condensed very briefly in consequence of our rapidly decreasing space.

The Turkey is spread over many parts of America, such as the wooded parts of Arkansas, Louisiana, Alabama, Indiana, &c. but does not seem to extend beyond the Rocky Mountains. It begins to mate about the middle of February, and the males then acquire those ludicrous gobbling sounds which have caused the bird to be called Gobbler, or Bubbling-Jock by the whites, and Oocoocoo by the Cherokees. In Persia, a pair of these birds, who had wandered there in some strange manner, were thought to speak very good Arabic, though the particular dialect was beyond the comprehension of the hearers.

The female makes her nest in some secluded spot, and is very guarded in her approaches, seldom employing the same path twice in succession; and if discovered using various wiles by which to draw the intruder from the spot. As soon as the young are hatched she takes them under her charge, and the whole family go wandering about to great distances, at first returning to the nest for the night, but afterwards crouching in any suitable spot. Marshy places are avoided by the Turkey, as wet is fatal to the young birds until they have attained their second suit of clothes, and wear feathers instead of down. When they are about a fortnight old they are able to get up into trees, and roost in the branches, safe from most of the numerous enemies which beset their path through life.

The great horned owl is, however, still able and willing to snatch them from the branches, and would succeed oftener in its attempts, were it not baffled by the instinctive movements of the Turkey. Even the slight rustling of the owl's wings sets the watchful Turkeys on the alert, and with anxious eyes they note his movements as he sails dark and lethal over them in the moonbeams, his large lambent eyeballs glowing with opalescent light; a feathered Azrael impending over them, and with fearful deliberation selecting his victim. Suddenly the stoop is made, but the intended victim is ready for the assault; ducks down its head, flattens its tail over its back, and the owl, striking upon this improvised shield, finds no hold for his claws, and slides off his prey like water from a duck's back. The whole flock drop from the boughs, and are safely hidden among the dark underwood before their enemy has recovered himself and renewed the attack.

The lynx is a terrible foe to the Turkeys, bounding suddenly among them, and as they hastily rise into the air to seek the shelter of the branches, the lynx leaps upwards and strikes them down with his ready paw just as a cat knocks down sparrows on the wing. Various other animals and birds persecute the inoffensive Turkey throughout its existence, but its worst enemy is the featherless biped. Snares of wonderful construction, traps, and " pens," are constantly employed for the capture of this valuable bird; the " pen " being so simple and withal so ingenious, that it merits a short description.

A little square hut is made of logs, without window or door. A trench is cut in the ground, some ten or twelve feet in length, passing under the wall of the hut and terminating in its centre. A kind of bridge of flattened logs or sticks is then laid across the trench in the interior of the hut, close to the wall. The roof is then laid, and the pen is complete. Its mode of action is as follows. A quantity of corn is strewn in the pen and along the trench, and is sparingly scattered at intervals so as to lead the Turkeys to the trench. When they see the corn they follow it up, feeding as they go, and finding that the trench is so well supplied, they traverse its length and pass into the pen. There is no trap-door to prevent them from escaping, neither is there need of it. As is the custom of trapped birds in general, they walk round the walls of their prison, trying to find a hole at which to escape, and peering anxiously through the interstices between the logs. When they come to the trench, they never think of going out by the way that they entered, but keeping close against the wall, they walk over the little bridge and recommence their tour. In this way great numbers of Turkeys are taken annually.

The Turkey is a very migratory bird, passing over great distances, and retaining the habit in its tamed state, giving no small amount of trouble to the poultry owner. In describing one of these migrations, Audubon speaks as follows:—

"About the beginning of October, when scarcely any of the seeds and fruits have fallen from the trees, these birds assemble in flocks, and gradually move towards the rich bottom-lands of the Ohio and Mississippi. The males, or as they are more commonly called the *gobblers*, associate in parties from ten to a hundred, and search for food apart from the females, while the latter are seen either advancing singly, each with its brood of young, then about two-thirds grown, or in union with other families, forming parties often amounting to seventy or eighty individuals, all intent on shunning the old cocks, who, when the young birds have attained this size, will fight with and often destroy them by repeated blows on the head. Old and young, however, all move in the same course, and on foot, unless their progress be intercepted by a river, or the hunter's dog force them to take wing.

When they come upon a river, they betake themselves to the highest eminences, and there often remain a whole day, and sometimes two, as if for the purpose of consultation. During this time the males are heard gobbling, calling, and making much ado, and are seen strutting about as if to raise their courage to a pitch befitting the emergency. Even the females and young assume something of the same pompous demeanour, spread out their tails, and run round each other, purring loudly and performing extravagant leaps.

At length when the weather appears settled, and all around is quiet, the whole party mount to the tops of the highest trees, whence at a signal, consisting of a single *cluck* given by a leader, the flock takes flight for the opposite shore. The old and fat birds get easily over, even should the river be a mile in breadth, but the younger and less robust frequently fall into the water, not to be drowned, however, as might be imagined; they bring their wings close to their body, spread out their tail as a support, and striking out their legs with great vigour, proceed rapidly towards the shore; on approaching which, should they find it too steep for landing, they cease their exertions for a few moments, float down the stream until they come to an accessible part, and by a violent effort generally extricate themselves from the water. It is remarkable that immediately after crossing a large stream, they ramble about for some time as if bewildered. In this state they fall an easy prey to the hunter."

The colouring of the wild male Turkey is briefly as follows. The small head and half of the neck are covered with a warty, naked, bluish skin, hanging in wattles from the base of the bill and forming a long fleshy protuberance, hanging from the base of the bill and having a tuft of hairs at its tip. This excrescence is capable of elongation under excitement. There is also a long tuft of strong black hairs hanging from the junction of the neck and breast. The general colour of the plumage is very beautiful, gleaming with golden-bronze, banded with black and "shot" with violet, green, and blue. In total length this bird measures about four feet.

THE splendid HONDURAS TURKEY is even a more magnificent bird than the preceding species. It is found, as its name imports, in the wooded districts of Honduras and Yucatan.

Two specimens of this splendid bird, a male and female, were brought to the Zoological Gardens; and I am indebted to the kindness of Mr. T. W. Wood, the artist of the accompanying illustration, for the following short account of its habits in a state of captivity, I being at the time unable through ill health to visit the gardens. "In the spring, the male became highly excited, and stalked about with his tail spread, wings drooping, and all his feathers puffed up, looking as if he would burst with pride. At such a time his head was thrown back so far and his breast-feathers projecting so far that he could not observe the ground beneath him, and consequently he often stepped into the water, much to his annoyance and the visitor's amusement."

The colouring of this bird is peculiarly fine. The naked skin of the head and neck is delicate violet-blue, covered with a number of round pea-like knobs, arranged in a cluster upon the crown and of a pale buff-orange, a row over the eye, and others scattered about the neck without any particular arrangement. The wattle hanging from the base of the neck is light orange at its tip. The skin round the eyes and the knobs on

HONDURAS TURKEY.—*Meleágris ocelláta.*

the neck are carmine. The hairy tuft on the breast is not seen in this species. The feathers are finely webbed, rounded, and scale-like, and their colours are truly splendid. On the lower part of the neck and upper part of the back they are bronze-green banded with black and gold; and towards the tail the green assumes a flashing emerald hue, and the gold band becomes wider and darker with fiery-red, like the throat of the ruby-throated humming-bird. The tail-coverts are furnished with bold "eyes" at their tips, and the lower parts of the body are also bronze-green and black, but without the lustre of the upper parts. The primary feathers of the wings are black edged with white, and the secondaries have the outer webs wholly white. The greater coverts are rich chestnut, and the legs and feet are lake. In size this bird is rather smaller than the common turkey.

THE prettily spotted GUINEA FOWL or PINTADO is, although now domesticated in England, a native of Africa, and with some exceptions, has much of the habits and propensities of the turkey, which bird it evidently represents.

Like the turkey, it is a confirmed wanderer, travelling continually during the day, and perching on the branches to roost at night It differs from the turkey however in its choice of locality, for whereas the turkey always keeps itself to the driest spots, shunning the low-lying lands as fatal to its young, the Guinea Fowl has a special liking for the marshes, and may generally be found among the most humid spots or upon the banks of rivers. It is a gregarious bird, assembling in large bands which traverse the country in company. The flight of the Pintado is seldom extended to any great distance, as the body is heavy in proportion to the power of wing, and the bird is forced to take short and hasty flights with much flapping of the wings, and to trust mostly to its legs for loco-

GUINEA FOWL.—*Númida meleagris.*

motion. On the ground the Guinea Fowl is a very swift bird, as is well known to those who have tried to catch it in an open field.

Both in the wild and the captive state the Guinea Fowl is wary and suspicious, and particularly careful not to betray the position of its nest, thus often giving great trouble to the farmer. Sometimes when the breeding season approaches, the female Pintado will hide herself and nest so effectually that the only indication of her proceedings is her subsequent appearance with a brood of young round her. The number of eggs is rather large, being seldom below ten and often double that number. Their colour is yellowish-red, covered with very little dark spots, and their size is less than that of the common fowl. Their shells are extremely hard and thick, and when boiled for the table require some little exertion to open properly.

Every one knows the curious, almost articulate cry of the Guinea Fowl, its " Come-back ! come-back !" being continually uttered wherever the bird is kept, and often affording a clue to its presence. This bird has been imported into America and several of the West Indian islands, where it has entirely acclimatized itself, and has increased so much in numbers as to be reckoned among the game birds and shot accordingly. In the poultry-yard it is not always a desirable inmate, partly on account of its wandering habits, sometimes extending over a mile or two of the surrounding country, and partly because it is so pugnacious, quarrelling with the fowls and pecking them sharply with its hard beak. Still, as its flesh when young is very good, and the cost of its keep very trifling, it is a profitable bird if well watched.

The forehead of the Guinea Fowl is surmounted with a horny casque, and the naked

IMPEYAN PHEASANT.—*Lophóphorus Impeyánus.*

skin round the eyes falls in wattles below the throat. In the male the wattles are purplish red, and in the female they are red without any mixture of blue, and are of smaller size. The legs are without spurs. The pretty spotted plumage of this bird is too well known to need description. Another species of the same genus, the CRESTED GUINEA FOWL, is remarkable for a large crest of arched feathers upon its head, taking the place of the casque of the common species. The colour of the Crested Guinea Fowl is blue-black, each feather having from four to six greyish spots. The primary feathers of the wings are oaken brown, and the edges of the secondaries snowy white, forming a bold contrast with the extremely dark plumage of the body.

Although less in size than the peacock, and without the wonderful train of that bird, the IMPEYAN PHEASANT or MONAL is nearly as splendid a creature, and but for the absence of the train, would even surpass it in the glory of its hues.

On looking at a living or well-stuffed male Monal it strongly reminds the observer of the humming-birds, and looks as if one of those glittering little beings had been suddenly magnified to a thousand times its size. The plumage of the Impeyan Pheasant has the

PARTRIDGE.—*Perdix cinereus.*

appearance of having been cut out of thin flakes of nacre or mother-of-pearl, their shining polished surface, their deep changing hues of azure, metallic green, amethystine purple, and fiery orange, being just like the effect produced by the finest nacre when rightly cut.

Although possessed of such flashing hues, which are mostly the offspring of a tropical sun, the Impeyan Pheasant inhabits the cold, snowy regions of the Himalayas, a climate not unlike that of our Scotch mountains. There is, therefore, some hope that we may be able to have this wondrously magnificent bird fairly settled down in the British Islands, for it is hardy enough in this country, breeds without difficulty, and endures our severest frosts with impunity. As far as is known, it remains entirely in the higher regions of its native land, and never descends to the plains. The food of this bird consists mostly of bulbous roots, which it digs out of the ground with its peculiarly curved and sharp beak. Even in captivity the Impeyan Pheasant will often indulge in many quaint and grotesque actions, especially towards the pairing time, when all birds like to show themselves off to the best advantage.

The colouring of this gorgeous bird may be briefly described as follows : The head and throat are of a metallic golden green, and the feathers of the crest are bare shafted for the greater part of their length, and spread at their tips into flattened spatula-shaped ends. The lower part of the neck and top of the back are rich shining purple with green and red reflections, and the feathers are all lancet-shaped. Across the lower part of the back there is a broad band of pure snowy white, and the tail is reddish brown, barred irregularly with a darker hue. The rest of the plumage is deep steely blue. The legs are spurred, and the general form is strong and robust. The female is a very sober-plumaged bird, without the lofty crest or gorgeous colours of her mate. Her feathers are mostly dull brown, mottled with grey and ochry yellow, and there is a broad white patch under the chin and throat. She is also smaller than her mate.

OF the many members of the Perdicine group, we shall take only five examples, the first of which is the well-known English PARTRIDGE.

2.   S S

This bird, so dear to British sportsmen, is found spread over the greater part of Europe, always being found most plentifully near cultivated ground. It feeds upon various substances, such as grain and seeds in the autumn, and green leaves and insects in the spring and early summer. In all probability this bird, although it may do some damage to the corn-fields, may still be very useful to the farmer by its unceasing war upon the smaller "vermin," that devastate the fields and injure the crops. Small slugs are a favourite diet with the Partridge, which has a special faculty for discovering them in the recesses where they hide themselves during the day, and can even hunt successfully after the eggs of these destructive creatures. Caterpillars are also eaten by this bird, and the terrible black grub of the turnip is consumed in great numbers by the Partridges. Even the white cabbage butterfly, whose numerous offspring are so hurtful in the kitchen garden, falls a victim to the quick-eyed Partridge, which leaps into the air and seizes it in its beak as the white-winged pest comes fluttering unsuspectingly over the bird's head.

The Partridge begins to lay about the end of April, gathering together a bundle of dried grasses into some shallow depression in the ground, and depositing therein a clutch of eggs, generally from twelve to twenty in number. Sometimes a still greater number have been found, but in these cases it is tolerably evident from many observations that several birds have laid in the same nest. Now and then a number of pheasants' eggs are found in the nest of a Partridge, and *vice versâ*, the pheasant seeming, however, to be the usurper in most instances. The Partridge is singularly careless of the position of her nest, placing it in the most exposed situations, and sitting upon the eggs with perfect contentment, although within a yard or two of a footpath. Indeed, I have found the nest of this bird, with six or seven eggs, so close to a frequented pathway running through a little copse, that a careless step to one side might have broken the eggs. In colour the eggs are not unlike those of the pheasant, being of a smooth olive-brown.

The mother-bird sits very closely, and is not easily frightened from her charge; and during the last day or two of incubation she is so fearless that she will not suffer herself to be disturbed, and will allow the scythe of the mower to kill her on her nest rather than desert her home. Sitting Partridges have sometimes allowed themselves to be taken by hand. When imminent danger threatens the nest, the mother-bird has been known to carry off the eggs and convey them to a place of safety, executing the task in a wonderfully short space of time. Mr. Jesse mentions one such instance, where there were twenty-one eggs, the whole of which were removed to a distance of forty yards in about twenty minutes. It is probable that the cock bird assisted his mate in her labours.

When the young are hatched they are strong on their legs at once, running about with ease, and mostly leaving the nest on the same day. The mother takes her little new-born brood to their feeding-places, generally ant-hills or caterpillar-haunted spots, and aids them in their search after food by scratching away the soil with her feet. The nests of the wood-ant, which are mostly found in fir plantations or hilly ground, being very full of inhabitants, very easily torn to pieces, and the ants and their larvæ and pupæ being very large, are favourite feeding-places of the Partridge, which in such localities is said to acquire a better flavour than among the lower pasture lands.

The young brood, technically called a "covey," associate together, and have a very strong local tendency, adhering with great pertinacity to the same field or patch of land. When together they are mostly rather wild, and dart off at the least alarm with their well-known whirring flight, just topping a hedge or wall and settling on the other side till again put up; but when the members of the covey are separated they seem to dread the air, and crouch closely to the ground, so that it is the object of the sportsman to scatter the covey and to pick them up singly. They are always alarmed at a soaring bird, whether of prey or not, and squat closely to the ground. When they are very wild and shy, the sportsmen take advantage of this propensity, and fly a kite shaped like a hawk over them, thus inducing them to lie frightened on the ground until the dog can point them in the proper fashion. Even a common, long-tailed, round-shouldered boy's kite will answer the purpose well enough. Some punctilious sportsmen, however, denounce the kite as a trick only worthy of a poacher, and would rather walk after the birds all day without getting a shot than secure a full bag by the use of such a device.

About the middle or end of February, according to the mildness or inclemency of the season, the Partridge begins to pair; and as the male birds are very numerous, they fight desperate battles for the object of their love. While engaged in combat, they are so deeply absorbed in battle, that they may be approached quite closely, as they whirl round and round, grasping each other by the beak, and have even been taken by hand. So strong, however, is the warlike instinct, that, when released, the furious birds recommenced the quarrel.

The females take no part in these battles; waiting quietly, like the strong-minded heroines of romance, to abide the issue of the combat, and to reward the victors with their love. Not that they are devoid of courage, but they reserve its display for a better purpose, namely, the defence of their young. Should a hen Partridge be disturbed while in charge of her little brood, she will endeavour to put them out of danger, and to draw the intruder aside by the exertion of many a crafty wile. But should the enemy come upon them too suddenly to be deceived by cunning, she will boldly dash at the foe, and, with self-sacrificing courage, attack with beak, foot, and wing, until the enemy has left the ground or herself is killed, knowing that her young charge are taking advantage of the time to place themselves in safety. Small though the bird may be, it can strike with considerable force, and has been known to inflict some painful wounds on the faces of human beings who have suddenly disturbed a brood of young.

Though strong and rapid of flight for a short distance, the Partridge loves not to trust itself over much to the air, and cannot fly to any great distance without alighting. When these birds are forced to pass over wide rivers or arms of the sea, they are often so wearied that they fall into the water, and these are mostly drowned, having but little idea of swimming, beyond the idea that they are to sit still and trust to their fortune. A bird thus fallen into the sea will sometimes be washed to shore, should the tide be favourable, but in fresh water it is generally drowned, or snapped up by a hawk from above, or a big pike from below, should such fresh-water sharks feed in that locality.

The plumage of the Partridge is brown of several shades above, mingled with grey. The breast is grey, with a horseshoe-like patch of rich chestnut on its lower portion, and the sides and flanks are barred with chestnut. The total length of the male bird is rather more than a foot; the female is smaller than her mate, and the chestnut bars on the flanks are broader than those of the male.

The RED-LEGGED PARTRIDGE is a larger and stronger bird than the common species, from which it may at once be distinguished by the black bar over the forehead, behind the eye and round the breast, as well as by the black streaks that pass from the neck towards the tail, and the conspicuous grey, fawn, and black bars on the flanks.

This bird is common in France and Italy, and also is a denizen of the Channel Islands, whence it has been introduced into our country, and thrives so well that, like the Norwegian rat, it has in some places fairly driven away the original breed, and usurped their territory. It is much stronger on the wing than the common Partridge, and yet is so swift and active of foot that it cannot easily be induced to rise, but runs away from the dogs with such speed that it often baffles their best efforts to start it within shot range. According to Yarrell, they are difficult of capture even when wounded, as they have a habit of running into rabbit-holes or similar sanctuaries, whence they cannot be dislodged without costing too much of the sportsman's time. These birds seem to prefer heaths and commons to the turnip and corn fields as frequented by the common Partridge.

The eggs of this species are very numerous, averaging sixteen or seventeen in each nest; and their colour is unlike those of the ordinary species, being yellowish white with a dash of yellow, and covered with spots of reddish brown. The food is the same as that of the ordinary breed.

The plumage of this bird is altogether smoother than that of the last-mentioned species. The upper parts of the body are soft brown. Before and behind the eye there is a line of white, and a bold stripe of black runs over the forehead to the eye, then starts from behind the eye, and runs along the sides of the neck over the breast, where it is very broad.

RED-LEGGED PARTRIDGE.—*Cáccabis rufa.*

A number of black dotted streaks extend from the black stripe so as to form an interrupted band of black over the shoulders. The breast is grey, the abdomen is fawn, and the feathers of the flanks and sides are marked with curved bands of grey, white, black, and fawn. The legs and beak are red. The total length of this bird is between thirteen and fourteen inches. The female is like the male, but smaller and not quite so brightly coloured.

The SANGUINE FRANCOLIN may fairly be reckoned as the finest of its group.

This splendid bird inhabits the great Himalayan range, and is thought to be peculiar to that region. Very little is known of its habits, the fullest account being that given by Dr. Hooker, and quoted by Mr. Gould in his "Birds of Asia."

"This, the boldest of the Alpine birds of its kind, frequents the mountain ranges of Eastern Nepâl and Sikkim, at an elevation varying from 10,000 to 14,000 feet, and is very abundant in many of the valleys among the forests of pine (*abies Webbiana*) and juniper. It seldom or never crows, but emits a weak cackling noise. When put up, it takes a very short flight, and then runs to shelter. During winter it appears to burrow under the hills among the snow, for I have snared it in January, in regions thickly covered with snow, at an altitude of 12,000 feet. I have seen the young in May.

The principal food of this bird consisting of the tops of the pine and juniper in spring, and the berries of the latter in autumn and winter, its flesh has always a very strong flavour, and is, moreover, uncommonly tough; it was, however, the only bird I obtained at these great elevations in tolerable abundance for food, and that not very frequently.

The Bhoteas say that it acquires a distinct spur every year; certain it is, that they are more numerous than in any other bird, and that they are not alike on both legs. I could not discover the cause of this difference, neither could I learn if they were produced at different times. I believe that five on one leg and four on the other is the greatest number I have observed."

The colouring and arrangement of their plumage are very complete, and entirely different in the two sexes. In the male, the forehead and a line round the eyes are black, and the crest is grey with buff streaks. The chin and throat are deep blood-red, and the upper part of the breast is white streaked with black. The feathers of the back and whole of the upper surface are slaty grey, each having a streak of white crossed with

SANGUINE FRANCOLIN.—*Ithaginis cruentus.*

black down the centre; and the breast and upper part of the abdomen are light green, streaked with blood-red and white. The lower part of the abdomen is brown-grey. The upper tail-coverts are blood-red, with a long narrow streak of yellow down the centre of each feather; and the tail is white at the tip, and each feather is broadly crossed with blood-red at the base. The bill is black at the tip and red at the base, and the legs and feet are deep pinky red. The female is a bird of very sober plumage, being reddish brown, lighter on the head and neck, and freckled with black on the back. The under surface is rather redder than the upper. In size the Sanguine Francolin about equals an ordinary fowl.

THE odd, short-legged, round-bodied, quick-footed QUAIL is closely allied to the partridge in form and many of its habits. Of these birds there are many species; but as all are much alike, there is no need of many examples.

The common Quail is found spread over the greater part of Europe, and portions of Asia and Africa, coming to our island in the summer, though not in very great numbers. In England the bird is not sufficiently plentiful to be of any commercial value; but in Italy and some of the warmer lands which the Quails traverse during their periodical migrations, the inhabitants look forward to the arrival of the Quail with the greatest anxiety. In those countries they are shot, snared, and netted by thousands; and it is chiefly from the foreign markets that our game shops are supplied with these birds. When fat, the flesh of the Quail is very delicious; and the most approved way of cooking the bird is to envelop it in a very thin slice of bacon, tie it up in a large vine-leaf, and then roast it.

In their migrations the Quails fly by night, a peculiarity which has been noted in the Scriptural record of the Exodus, where it is mentioned, that " at even the Quails came up and covered the camp." Mr. Yarrell suggests, that the object of this nocturnal journeying may be to save the defenceless birds from the attacks of the numerous birds of prey, which would probably assail them were they to travel during the daytime. There are, however, larger and more powerful birds, which need no such safeguard, and yet are in the habit of travelling by night, as well as the Quail.

QUAIL.—*Coturnix commŭnis.*

It is rather curious, that the males precede the females by several days, and are consequently more persecuted by the professional fowlers.

The male bird does not pair like the partridge, but takes to himself a plurality of wives, and, as is generally the case with such polygamists, has to fight many desperate battles with others of its own sex. Although ill provided with weapons of offence, the Quail is as fiery and courageous a bird as the gamecock; and in Eastern countries is largely kept and trained for the purpose of fighting prize-battles, on the result of which the owners stake large sums. The note of the male is a kind of shrill whistle, which is only heard during the breeding season.

The nest of the Quail is of no better construction than that of the partridge, being merely a few bits of hay and dried herbage gathered into some little depression in the bare ground, and generally entrusted under the protection of corn-stalks, clover, or a tuft of rank grass. The number of eggs is generally about fourteen or fifteen, and their colour is buffy white, marked with patches or speckles of brown. The young are able to run about almost immediately after they leave the eggs, and are led by their parent to their food. However wild they may be, many of these birds are killed by a very simple device. The sportsman having marked down a covey of Quails, walks round them in circles sufficiently large not to alarm them, and as he returns towards the spot whence he started, he strikes off for another circle of less diameter. By describing a gradually lessening spiral, he drives all the Quails together in the middle, where they pack closely and suffer themselves to be killed in numbers.

The colouring of the Quail is simple, but pleasing. The head is dark brown, except a streak of pale brown over the eyes, and another on the top of the head passing towards the nape of the neck. The whole upper surface is brown streaked with yellower brown, and the feathers with lighter shafts. The chin and throat are white, and around the throat run two semicircular bands of dark brown, their points reaching as high as the ear-coverts, and having a black patch in front. The breast is rather pale but warm brown, variegated by the polished straw colour of the shafts, and the remainder of the under surface is ochry white deepening into chestnut on the flanks. The female may be known by the absence of the two dark semicircles on the throat, which even in the male are not acquired

VIRGINIAN QUAIL.—*Ortyx Virginiàna.*

until the second year, and the little dark spots on the feathers of the breast. The total length of the Quail is about seven inches.

An allied species is found in many parts of North America, and is known by the name of the VIRGINIAN QUAIL. In popular parlance, however, it is generally called the Partridge, greatly to the confusion of young ornithologists. On account of its peculiar cry, it is also called "Bob-White," its clear call-note bearing considerable resemblance to those words.

The Virginian Quail generally keeps itself to the open ground, preferring those spots where grain is plentiful. Sometimes, however, it shelters itself among the trees or brushwood, but even then seems to pass but little of its time in such retreats. During the winter it gains courage by hunger, approaching human habitations in search of food, and boldly fighting with the poultry for the grain thrown to them. Oftentimes the eggs are placed under the domestic hen, and in that case the young birds are very tame, provided that the foster-mother is of a quiet stay-at-home temper, and not given to roam. Wilson informs us that two young Quails, which had been hatched by a hen, attached themselves to the cows, accompanying them regularly to the field; standing by them when they were milked, retiring with them in the evening and roosting in the stable. These interesting little birds unfortunately disappeared in the spring.

As the flesh of the Quail is particularly excellent, it is greatly persecuted in the winter time, when it is easily attracted by baits. Ten or fifteen at a time are often caught in a contrivance that much resembles the common sieve-trap of our own country, saving that a kind of coop supplies the place of the sieve.

In the wild state the Quail makes its rude nest under the shelter of corn or grass-tufts, and then lays from fifteen to twenty-four pure white eggs. As is the case with the European Quail, the young are able to run about as soon as they are fairly free of the shell, and are guided by their mother to the best feeding-places. The old bird is peculiarly watchful of her charge, and if she should be suddenly surprised, she endeavours to draw off the attention of the intruder by feigning lameness, flapping along the ground as if with a broken wing, in order to gain time for the helpless young to conceal themselves. At night the Quails prefer to roost on some elevated spot in the middle of a field, and it appears that they sit in a circle with their heads radiating outwards and their tails almost touching each other.

The top of the head and the upper part of the breast are warm reddish brown, the chin is pure white, and a streak of white runs from behind the eye along the neck. The sides of the neck are also reddish brown spotted with black and white. The upper surface of the body is reddish brown sprinkled with ashy grey and black. The wings are grey-brown, and the tertials edged with yellowish white. The abdomen and lower parts of the breast are yellowish white, marked with spear-head dashes of black. The female is known by the yellowish brown of the chin and sides of the head. It is a larger bird than the European Quail, being about nine inches long.

ALTHOUGH once a common inhabitant of the highland districts of Great Britain, the CAPERCAILLIE has now been almost wholly extinct for some years, a straggling specimen being occasionally seen in Scotland, and shot "for the benefit of science." This bird is also known by the following names: Cock of the Woods, Mountain Cock, Auerhahn, and Capercailzie.

It is now most frequently found in the northern parts of Europe, Norway and Sweden being very favourite homes. From those countries it is largely imported into England by the game-dealers.

The Capercaillie is celebrated, not only for its great size and the excellence of its flesh, but for its singular habits just previous to and during the breeding season. Mr. Lloyd has given so excellent an account of these curious proceedings, that they must be told in his own words:—

" At this period, and often when the ground is still deeply covered with snow, the cock stations himself on a pine and commences his love song, or *play* as it is termed in Sweden, to attract the hens about him. This is usually from the first dawn of day to sunrise, or from a little after sunset until it is quite dark. The time, however, more or less depends upon the mildness of the weather and the advanced state of the season.

During his 'play,' the neck of the Capercaillie is stretched out, his tail is raised and spread like a fan, his wings droop, his feathers are ruffled up, and, in short, he much resembles an angry turkey-cock. He begins his play with a call something resembling *Peller! peller! peller!* These sounds he repeats at first at some little intervals; but as he proceeds, they increase in rapidity, until at last, and after perhaps the lapse of a minute or so, he makes a sort of gulp in his throat and finishes with sucking in, as it were, his breath.

During the continuance of this latter process, which only lasts a few seconds, the head of the Capercaillie is thrown up, his eyes are partially closed, and his whole appearance would denote that he is worked up into an agony of passion. At this time his faculties are much absorbed, and it is not difficult to approach him. . . . The play of the Capercaillie is not loud, and should there be any wind stirring in the trees at the time, it cannot be heard at any considerable distance. Indeed, during the calmest and most favourable weather, it is not audible at more than two or three hundred paces.

On hearing the call of the cock, the hens, whose cry in some degree resembles the croak of the raven, or rather perhaps the sounds, *Gock! Gock! Gock!* assemble from all parts of the surrounding forest. The male bird now descends from the eminence on which he was perched, to the ground, where he and his female friends join company.

The Capercaillie does not play indiscriminately over the forest, but he has his certain stations (Tjader-lek, which may perhaps be rendered, his playing-grounds). These, however, are often of some little extent. Here, unless very much persecuted, the song of these birds may be heard in the spring for years together. The Capercaillie does not, during his play, confine himself to any particular tree, for, on the contrary, it is seldom he is to be met with exactly on the same spot for two days in succession.

On these *lek*, several Capercaillie may occasionally be heard playing at the same time; Mr. Grieff, in his quaint way, observes, 'It then goes gloriously.' So long, however, as the old male birds are alive, they will not, it is said, permit the young ones, or those of the preceding season, to play. Should the old birds, however, be killed, the

CAPERCAILLIE.—*Tétrao urogallus*

young ones, in the course of a day or two, usually open their pipes. Combats, as it may be supposed, not unfrequently take place on these occasions; though I do not recollect having heard of more than two of these birds being engaged at the same time.

Though altogether contrary to law, it is now that the greatest slaughter is committed among the Capercaillie; for any lump of a fellow who has strength to draw a trigger may, with a little instruction, manage to knock them down. As the plan, however, of shooting these noble birds during their play is something curious, I shall do my best to describe it.

It being first ascertained where the *lek* is situated, the sportsman proceeds to the spot and listens in profound silence until he hears the call of the cock. So long, however, as the bird only repeats his commencing sound, he must, if he be at all near to him, remain stationary; but the instant the Capercaillie comes to the wind-up, the gulp, &c., during which, as I have just now said, its faculties of both seeing and hearing are in a degree absorbed, then he may advance a little. This note, however, lasts so short a time, that the sportsman is seldom able to take more than three or four steps before it ceases; for the instant that is the case, he must again halt, and if in an exposed situation remain fixed like a

COCK OF THE PLAINS.—*Tétrao urophasiánus.*

statue. This is absolutely necessary; for during his play, excepting during the gulp, &c., the Capercaillie is exceedingly watchful, and easily takes the alarm. If all remain quiet, however, the bird usually goes on again immediately with his first strain, and when he once more comes to the final note, the sportsman advances as before.

.To become a proficient at this sport requires a good deal of practice. In the first place, a person must know how to take advantage of the ground when advancing upon the Capercaillie; for if in full daylight, this is hardly practicable in exposed situations; and in the next, that he may not move forward excepting upon the note which is so fatal to that bird. This is likely enough to happen if it be an old cock that has been previously exposed to shots, for he often runs on with *Peller, peller, peller*, until one supposes that he is just coming to the gulp, when he suddenly makes a stop. If, therefore, a person were then incautiously to advance, he would, in all probability, instantly take to flight."

The nest of the Capercaillie is made upon the ground, and contains eight to ten eggs; when hatched, the young are fed upon insects, more especially ants and their pupæ. The adult birds feed mostly on vegetable substances, such as juniper, cranberry, and bilberries, and the leaves and buds of several trees.

The colour of the adult male bird is chestnut-brown covered with a number of black lines irregularly dispersed, the breast is black with a gloss of green, and the abdomen is simply black, as are the lengthened feathers of the throat and the tail. The female is easily known by the bars of red and black which traverse the head and neck, and the reddish yellow barred with black of the under surface. In size, the Capercaillie is nearly equal to a turkey.

The COCK OF THE PLAINS is closely allied to the preceding species.

It is an American bird, being found in the dry plains in the interior of southern California. Like the cock of the woods, this bird is accustomed during the breeding season to disport himself after a peculiar and grotesque manner, drooping his wings, spreading his tail like a fan, puffing out his crop until the bare yellow skin stands prominently forward, somewhat after the fashion of the pouter pigeon, and erecting the long silken plumes of the neck. Thus accoutred, he parades the ground with much dignity, turning himself about so as to display his shape to the best advantage, assuming a variety of rather ludicrous attitudes, and uttering a loud booming cry that is compared to the sound made by blowing strongly into a large hollow reed.

The nest of this bird is made of dried grasses and small twigs, and is placed on the ground under the shelter of bushes or rank herbage. It is rather carefully made, and generally contains from thirteen to seventeen brown eggs blotched with chocolate on the large end. The Cock of the Plains is a gregarious bird, assembling in little troops in the summer and autumn, and in large flocks of several hundred in number during the winter and spring. The flesh of this bird is eatable, but dark in colour and not of a very good flavour.

The male is a very handsome bird, brown on the upper surface and mottled with very dark brown and yellowish white. The skin of the crop is deep orange-yellow, and on each side of it is a tuft of long and very slender feathers, having the shafts nearly naked, and dotted at the tip with a pencil of black bands. The throat and head are white profusely variegated with black, and the white feathers of the sides are firm, rounded, and of a scale-like form. The shafts of the breast-feathers are black and stiff. In total length this bird measures about twenty-two inches. The female is less in size, is without the feather-tufts on the neck and the scale-like plumage of the sides.

THE well-known BLACK GROUSE, or BLACK COCK, is a native of the more southern countries of Europe, and still survives in many portions of the British Isles, especially those localities where the pine-woods and heaths afford it shelter, and it is not dislodged by the presence of human habitations.

Like the two preceding species, the male bird resorts at the beginning of the breeding season to some open spot where he utters his love-calls, and displays his new clothes to the greatest advantage, for the purpose of attracting to his harem as many wives as possible. The note of the Black Cock when thus engaged is loud and resonant, and can be heard at a considerable distance. This crowing sound is accompanied by a harsh, grating, stridulous kind of cry, which has been likened to the noise produced by whetting a scythe. The Black Cock does not pair, but leaves his numerous mates to the duties of maturity and incubation, and follows his own desires while they prepare their nests, lay their eggs, hatch them, and bring up the young. The mother-bird is a fond and watchful parent, and when she has been alarmed by man or beasts of prey, has been known to remove the eggs to some other locality, where she thinks they will not be discovered.

The nest is a careless kind of structure, of grasses and stout herbage, and is placed on the ground under the shelter of grass or bushes. The female lays about six or ten eggs of a yellowish grey diversified with spots of light brown. The young are fed first upon insects and their larvæ, and afterwards on berries, grain, the buds and young shoots of trees.

It is a wild and wary bird, requiring much care on the part of the sportsman to get within fair gunshot. The old male which has survived a season or two is particularly shy and crafty, distrusting both man and dog, and running away as fast as his legs can carry him as soon as he is made aware of the approaching danger.

In the autumn the young males separate themselves from the other sex, and form a number of little bachelor establishments of their own, living together in harmony until the next breeding season, when they all begin to fall in love; the apple of discord is thrown among them by the charms of the hitherto repudiated sex, and their rivalries lead them into determined and continual battles, which do not cease until the end of the season restores them to peace and sobriety, and they need fear no foes save the beasts and birds of prey, and their worst enemy, the autumnal British statesman.

BLACK GROUSE.—*Tétrao tetrix.*

The general colour of the adult male bird is black glossed with blue and purple, except a white band across each wing. The under tail-coverts are white. The remarkable form of the tail is caused by the peculiar development of the exterior feathers, three, four, or even five of which are laterally curved, the outermost being the longest and having the most decided curve. Their ends are somewhat squared. The colouring of the female is quite different. Her general colour is brown, with a tinge of orange, barred with black and speckled with the same hue, the spots and bars being larger on the breast, back, and wings, and the feathers on the breast more or less edged with white. The under tail-coverts are greyish white. The total length of the adult male is about twenty-two inches, and that of the female from seventeen to eighteen inches. She also weighs nearly one-third less than her mate, and is popularly termed the Heath Hen.

ANOTHER fine species of this group is the PINNATED GROUSE of North America. This bird is found almost wholly in open dry plains on which are a few trees or tufts of brushwood, pines and shrub-oaks being the most favoured shelter. Like the greater part of the group, the males "play" at the breeding season, ruffling their feathers, erecting their neck-tufts, swelling out their wattles, and uttering their strange love-cries. At this time the Pinnated Grouse is particularly remarkable for the large size and bright orange colour of the naked sacculated appendages which hang at each side of the neck, and which can be filled with air until they are nearly of the same size and colour as a Seville orange, or can be permitted to hang loosely along the neck. The males are great fighters on these occasions, and dash fiercely against each other, though to all appearance these combats are more notable for display than for effect, little or no damage seeming to be done or suffered by either party.

Mr. Webber gives the following interesting account of some of the habits of this species :—

" The most extraordinary phenomenon produced by the necessities of the climate, and as a protection against the terrible winds which sweep over that apparently illimitable beach at the approach of winter, consists in the assembling of these birds, from a distance of many miles around, to roost upon the same spot, something after the manner of the wild pigeon. This fact seems also to have escaped M. Audubon's notice.

At the opening of winter a spot is selected on the open prairies, in the upper part of the Missouri country, which is more sheltered than the surrounding regions, by the character of the ground, from the biting force of the north-east winds. Here the prairie-hens begin to assemble early in the evening ; and by the time dusk comes, an immense number are collected. They approach the scene in small flocks, in a leisurely manner, by short flights. They approach the place of gathering silently, with nothing of that whirr of wings for which they are noted when they are suddenly put up, but they make ample amends when they arrive ; as in the pigeon-roost, there is a continued roar, caused by the restless shifting of the birds and sounds of impatient struggle emitted by them, which can be heard distinctly for several miles. The numbers collected are incalculably immense, since the space covered sometimes extends for over a mile in length with a breadth determined by the character of the ground.

This is a most astonishing scene when approached in the early part of the night on horseback ; the hubbub is strangely discordant and overwhelmingly deafening, They will permit themselves to be killed in great numbers, with sticks or any convenient weapon, without the necessity of using guns. They, however, when frequently disturbed in the first of the season, will easily change their roosting-place ; and when the heavy snows have fallen, by melting which by the heat of their bodies, and by trampling it down, they have formed a sort of sheltered yard, the outside walls of which defend them against the winds, they are not easily driven away by any degree of persecution. Indeed, at this time they become so emaciated as to afford but little inducement to any human persecutors, by whom they are seldom troubled, indeed, on account of the remoteness of these locations ; from foxes, wolves, hawks, and owls, &c., their natural enemies, they·have, of course, to expect no mercy at any time.

The noise of their restless cluckings, flutterings, and shiftings, begins to subside a few hours after dark. The birds have now arranged themselves for the night, nestled as close as they can be wedged, every bird with his breast turned to the quarter in which the wind may be prevailing. This scene is one of the most curious that can be imagined, especially when they have the moonlight on the snow to contrast with their dark backs. At this time they may be killed by cartloads, as only those in the immediate neighbour-hood of the aggressors are disturbed apparently. They rise to the height of a few feet with a stupified and aimless fluttering, and plunge into the snow within a short distance, where they are easily taken by the hand. In these helpless conditions such immense numbers are destroyed, that the family would be in danger of rapid extermination but that the fecundity of the survivors nearly keeps pace with the many fatalities to which they are liable.

These birds are distributed over an immense northern territory ; and though they are everywhere in the more sheltered regions found to exhibit the propensity to collect in numbers greater or smaller, during the extreme cold weather, in low spots where they will have some shelter from the accidental peculiarities of the locality, yet nowhere else except just upon these wide plains are they to be found in such astonishing congregations as we have here described. The universal habit of all this family of Gallinaceæ is rather to run and roost in little squads or flocks. Whence this difference in the habits of the same bird, who knows? Ah! whence the difference ? That is the question."

The nest of the Pinnated Grouse is a rude structure of grasses and leaves, and placed under the shadow of a bush or a tuft of thick grass. The eggs are brownish white, and about fifteen in number.

PINNATED GROUSE.—*Tétrao Cupido.*

The colour of the Pinnated Grouse is mottled with black, white, and chestnut-brown, the male having two wing-like appendages on the neck, composed of eighteen feathers, five long and black, and thirteen shorter, streaked with black and brown. The male is also known by the slight crest on the head, a semicircular comb of orange-coloured skin over each eye, and the naked appendages to the neck already described. He is also larger than his mate. The under parts are brown marked with white in broken transverse bars, and the throat is white with mottlings of reddish brown and black. The length is about nineteen inches.

The RUFFED GROUSE is spread over the greater portion of the United States, where it is known either as Partridge or Pheasant, according to the locality.

Its habits are described at length by Wilson, in his " American Ornithology," to which work the reader is referred for fuller information. His account of the " play " of the male Ruffed Grouse must, however, be given in his own words: " In walking through the solitary woods frequented by these birds, a stranger is surprised by suddenly hearing a kind of thumping very similar to that produced by striking two full-blown bladders together, but much louder. The strokes at first are slow and distinct, but gradually increase in rapidity, till they run into each other, resembling the rumbling sound of very distant thunder, dying away gradually on the ear. After a few minutes' pause, this is again repeated; and in a calm day may be heard nearly a mile off. This drumming is most common in spring, and is the call of the cock to his favourite female. It is produced in the following manner :—

The bird, standing on an old prostrate log, generally in a retired and sheltered situation, lowers his wings, erects his expanded tail, contracts his throat, elevates the tufts of feathers on the neck, and inflates his whole body, something in the manner of the turkey-cock, strutting and wheeling about with great stateliness. After a few manœuvres of this kind, he begins to strike with his stiffened wings in short and rapid quick strokes, which become more and more rapid till they run into each other, as has already been described. This is most common in the morning and evening, though I have heard them drumming at all hours of the day."

RUFFED GROUSE.—*Tétrao umbellus.*

This bird pairs in April, and builds its nest in May, placing it on the ground at the root of a bush or a stump, and laying from nine to fifteen brownish-white eggs. The mother-bird has a system of decoying intruders from her nest, very similar to that which has already been mentioned in connexion with several other birds. The best time for shooting the Ruffed Grouse is September and October, when it is very fat, having fed on whortle-berries and other fruits which give its flesh a delicate and somewhat aromatic flavour. In winter these birds feed much on the buds of alder and laurel, and are then thought to be poisonous.

In general colour the male is rich chestnut-brown, variegated with abundant mottling of dark brown and grey. The curious tufts on the shoulder are rich velvety black glossed with green, and just below them the skin is bare. The tail is grey, barred with blackish brown. The length of the male bird is about eighteen inches. The female is smaller, and is known by the brown colour of the neck-tufts and the bar on the tail.

The RED GROUSE seems to be exclusively confined to the British Islands, and is found in the north of England, Wales, Ireland, and Scotland, and some of the Channel Islands. The birds of this genus are separated from the remainder of the group in consequence of the feathered toes, which are thickly clothed with short plumage, earning thereby the name of Lágopus, or Hare-footed,

It inhabits the moors, where heather is in abundance, as it feeds chiefly on the tender leaves of that plant together with whortle-berries, grain, and similar substances. The bird pairs early in the spring, and makes its nest of grass and ling stems, occasionally

RED GROUSE.—*Lágopus Scóticus.*

interspersed with feathers, and places it on the ground under the shelter of a heather-tuft. As soon as hatched, the young are able to run about, and are led to feed by both parents. These birds are greatly persecuted by sportsmen; but, in spite of their annual losses, they increase rather than diminish in number, except in seasons like the present (1861), when they are suffering greatly from internal parasites.

The colour of the Red Grouse is extremely variable, differing according to the locality or the season of year; and cream-coloured and speckled varieties are most uncommon. The ordinary plumage is as follows: In winter the adult male is chestnut-brown upon the upper surface, barred and speckled with black, and diversified by a few feathers of light yellowish brown. The head and neck are also chestnut-brown, but of a warmer tint than the back. Over the eye is a crescent-shaped patch of light scarlet bare skin, slightly fringed above. The tail is brown, with a tinge of red on the central feathers. The breast is brown, and the remainder of the under surface and flanks is of the same hue, each feather being tipped with white. The short plumage of the legs and toes is greyish white. In summer the red is lighter, and the body is sprinkled with yellow. The female is smaller and lighter than her mate, with more yellow and less red. In total measurement the male bird is about sixteen inches in length. This bird is also called the Red Ptarmigan and the Brown Ptarmigan.

The COMMON PTARMIGAN (*Lágopus vulgáris*) belongs to the same genus. This is the smallest of the British Grouse, and is not restricted to Great Britain, being found in Northern and mountainous Europe, especially in Norway and Sweden, and is also an inhabitant of North America.

This bird has a habit of resorting to stones and broken ground covered with lichens, which so exactly harmonise with the colours of its plumage that it is hardly distinguishable from the ground on which it is sitting, and under such circumstances it squats very closely. A person may walk through a flock without seeing a single bird. Mr. McGillivray says: " When squatted, they utter no sound, their object being to conceal themselves; and if you discover the one from which a cry has proceeded, you generally find him on the top

SAND GROUSE.—*Ptérocles bicinctus.*

of a stone, ready to spring off the moment you show an indication of hostility. If you throw a stone at him, he rises, utters his call, and is immediately joined by all the individuals around, which to your surprise, if it be your first rencontre, you see spring up one by one from the bare ground." A flock of these birds flitting along the sides of a mountain has a very curious effect, their speckled bodies being hardly visible as they sweep along, and when they alight they vanish from view as if by magic. In the winter, too, when the snow lies thickly on the ground, the Ptarmigan assumes a white coat, hardly distinguishable from the snow. When perceived by a hawk, the Ptarmigan has been seen to dash boldly into the deep snow, and to find a refuge under the white covering until its enemy had left the spot.

In the winter, the plumage of the male Ptarmigan is almost wholly white, the exceptions being a small patch behind the eye, the shafts of the primaries, and the bases of the fourteen exterior tail feathers, which are black. There is also a patch of red, bare skin round the eye. In the summer, the black retains its position, but the white is mottled and barred with black and grey. The length of the adult male is rather more than fifteen inches.

The SAND GROUSE are mostly found in the sandy deserts of Africa and Asia, though one or two species are inhabitants of Europe. The wings of all these birds are long and pointed, denoting considerable powers of flight, and in many species the two central feathers of the tail are much elongated and project beyond the others.

These birds are mostly gregarious, assembling in large flocks, but still retaining a division into pairs. One species, the Pin-tailed Sand Grouse, is found in such vast multitudes that they are killed by boys, who arm themselves with sticks and fling these rude missiles at the winged armies. It has been suggested by some writers that this bird is the quail of Scripture. The Sand Grouse runs with considerable rapidity; and as the legs are very short, and the body consequently carried close to the ground, the effect produced very much resembles the toy mice which are wound up like watches, and run about the floor. When coming directly towards the observer, the bird has a very comical aspect, the feet being hardly visible beneath the broad body, and the steps being very short, quick and tripping.

2.   T T

WHITE SHEATH-BILL.—*Chionis alba.*

The female Sand Grouse makes no nest, but lays her eggs, generally about two or three in number, on the bare ground. The young birds are very strong of foot, and as soon as their plumage has dried after their exit from the shell, they run about with their mother, and can afterwards lead a vagrant life.

The male bird has its forehead whitish, then a black patch and then white. The upper part of the plumage is dusky brown, mottled with buff; and its tail is buff, barred profusely with blackish brown, the tip being buff, and the last bar very broad and black. The breast is pale buff, and between the breast and abdomen runs a semilunar white band, reaching up to the shoulders. Just below the white there is an equally conspicuous black band, also running up under the wings to the shoulders. The abdomen and flanks are pale buff, mottled transversely with black-brown. In the female the plumage is of a more yellow cast, the black patch on the forehead and black band round the chest are wanting, and the white band has a grey tinge.

ANOTHER curious group of birds is known by the title of Sheath-bills, on account of the remarkable sheath of horny substance, which is situated on the base of the bill, and under which lie the nostrils. The use of this appendage is rather obscure. The whole of the bill is short and stout, and it is considerably arched towards the tip.

One of the commonest species of this group is the WHITE SHEATH-BILL, a native of Australia, New Zealand, and neighbouring islands.

This bird is almost exclusively found upon the coasts, finding its food among the molluscs, small crustaceæ, and bestranded fish and other similar substances. Perhaps, under some circumstances, it may subsist on carrion, and thereby give an evil flavour to its flesh, as there are very contradictory reports as to its value for the table, some specimens

ELEGANT TINAMOU.—*Tinamótis élegans.*

having been of so vile an odour that even the sailors, with their proverbial appetites for fresh meat, could not touch the ill-savoured flesh, while in other cases the bird is reported to be of excellent quality and equal to duck in tenderness and flavour. The legs of the Sheath-bill are rather long in proportion to the size of the body; and as it always frequents the sea-side, running in and out of the water in search of its food, and possesses many of the habits of the waders, it has been classed by some naturalists among these birds.

The White Sheath-bill is a pretty bird, its whole plumage being pure white, and the legs reddish black. The generical name, Chionis, is derived from a Greek word, signifying snow, and is given to this bird in allusion to its pure snow-white plumage. In total length, the adult male measures about fifteen inches.

The ELEGANT TINAMOU is one of the handsomest, though not the largest, of the family to which it belongs.

The Tinamous are only found in South America, where they are tolerably common. The word Tinamou is the native name for these birds; and as they bear some resemblance to bustards, the generic title of Tinamotis, although rather a barbarous combination of languages, is sufficiently appropriate.

The Tinamous are found in the open fields, preferring those which lie on the borders of woods. They are very seldom known to perch on branches, and are not very willing to use their wings, trusting rather to the swiftness of their legs. Of one species of Tinamou, Mr. Darwin writes as follows: "These birds do not go in coveys, nor do they conceal themselves like the English kind. It appears a very silly bird. A man on horseback, by riding round and round in a circle, or rather in a spire, so as to approach closer each time, may knock on the head as many as he pleases. The more common method is, to catch them with a running noose or little lasso, made of the stem of an ostrich's feather, fastened to the end by a long stick. A boy on a quiet old horse will frequently thus catch thirty or forty in a day."

The food of the Tinamous consists mostly of grain ; and after the fields of corn and maize are sown, these birds do considerable damage by running over the ground, and picking out all the seeds which have not been entirely covered by the soil. The eggs of these birds are about seven or eight in number, and are laid in the centre of some convenient tuft of herbage.

The Elegant Tinamou is a native of Chili, and is rather larger than the generality of its kind, as it slightly exceeds a grouse in dimensions, and has a much longer neck. The head and neck are light greyish buff with short delicate longitudinal streaks, and upon the head there is a long curved crest, each feather being brown with a dark streak along its centre. The back is spotted and barred with buff and blackish brown, and on the breast and general under-surface the feathers are irregularly barred with the same hue, the bars being wider and darker on the flanks.

CURSÓRES.

With the Ostrich commences a most important group of birds, containing the largest and most powerful members of the feathered tribe, and termed Cursores, or Running-Birds, on account of their great speed of foot and total impotence of wing. All the birds belonging to this order have the legs developed to an extraordinary degree, the bones being long, stout, and nearly as solid as those of a horse, and almost devoid of the air-cells which give such lightness to the bones of most birds. By the aid of the micro-scope, the peculiar character of the bone is clearly shown, though the bone of an Ostrich or cassowary is very different from the same bone in a fowl or a pigeon. The wings are almost wanting externally, their bones, although retaining the same number and form as in ordinary birds, being very small, as if suddenly checked in their growth. The huge wing muscles which give such prominence to the breast of flying birds, are therefore not required, and the breast-bone is consequently devoid of the projecting keel, and is quite smooth and rounded.

The common Ostrich is so well known that little need be said of its habits, its use to mankind, and the mode of hunting it, a very brief description being all that is necessary.

This magnificent creature, the largest of all existing birds, inhabits the hot sandy deserts of Africa, for which mode of life it is wonderfully fitted. In height it measures from six to eight feet, the males being larger than their mates, and of a blacker tint. The food of the Ostrich consists mostly of the wild melons which are so beneficently scattered over the sandy wastes, absorbing and retaining every drop of moisture condensed in the comparatively cool temperature of night, or fallen in the brief but severe rainstorms which serve to give new vigour to the scanty desert vegetation and to replenish the rare water springs.

Besides these melons, which the Ostrich, in common with the lion and other inhabitants of the desert, eats as much for drink as for food, the bird feeds on grasses and hard grain, which it is able to crush in its powerful gizzard, the action of which internal mill is aided by stones and other hard substances, which the Ostrich picks up and swallows just as ordinary grain-eating birds swallow sand and small pebbles. In captivity the Ostrich will swallow almost anything that comes in its way, such as brickbats, knives, old shoes, scraps of wood, feathers, and tenpenny nails, in addition to the legitimate stones. It has even been seen to swallow in succession a brood of ducklings ; but whether in that case the bird was impelled by normal hunger, whether it was afflicted by a morbid appetite, or whether it was merely eating the young birds for sheer mischief, are questions open for consideration.

The Ostrich is a gregarious bird, associating in flocks, and being frequently found mixed up with the vast herds of quaggas, zebras, giraffes, and antelopes which inhabit the

OSTRICH.—*Strúthio Camélus.*

same desert plains. It is also polygamous, each male bird having from two to seven wives. The nest of the Ostrich is a mere shallow hole scooped in the sand, in which are placed a large number of eggs, all set upright, and with a number of supplementary eggs laid round the margin.

The eggs are hatched mostly by the heat of the sun; but, contrary to the popular belief, the parent birds are very watchful over their nest, and aid in hatching the eggs by sitting upon them during the night. Both parents give their assistance in this task. The eggs which are laid around the margin of the nest are not sat upon, and consequently are not hatched, so that when the eggs within the nest are quite hard, and the young bird is nearly developed, those around are quite fit for food. Their object is supposed to be to give nourishment to the young birds before they are strong enough to follow their parents and forage for themselves.

Each egg will weigh on the average about three pounds, being equal to two dozen ordinary fowl's eggs. Yet one of them is not thought too much for a single man to eat at a meal, and in one instance two men finished five in the course of an afternoon. The approved method of dressing Ostrich eggs is to set the egg upright on the fire, break a round hole at the top, squeeze a forked stick into the aperture, leaving the stem protruding, and then to twist the stick rapidly between the hands so as to beat up the contents of the egg while it is being cooked. Within each egg there are generally some little smooth bean-shaped stones, which are composed of the same substance that forms the shell.

These eggs are put to various useful purposes. Not only are they eaten, but the shell is carefully preserved and chipped into spoons and ladles, or the entire shell employed as a water vessel, the aperture at the top being stuffed with grass. The mode of filling these shells from sandy pools is ingenious and simple. The business of procuring water is entrusted to the women, each of whom is furnished with a hollow reed, a bunch of grass, and her egg-shells. She makes a hole in the bed of the water-pool as deep as her arms will reach, ties the bunch of grass at the end of the reed, pushes it to the bottom of the hole, and rams the wet sand tightly round it. After waiting a little for the water to accumulate, she applies her mouth to the upper end of the reed, drawing the water through the tuft of grass at the bottom and so filtering it. Having filled her mouth with water, she puts another reed into the egg-shell, and pours the water from her mouth into the shell. In this manner a whole village is supplied with water, the shells being carefully buried to prevent evaporation.

The Bushmen make terrible use of these water shells. When they have determined on a raid, they send successive parties on the line, loaded with Ostrich egg-shells full of water, which they bury in spots known to themselves alone. The tiny but resolute little warriors start off on their expedition, get among the dwellings of their foes, carry off as many cattle as they can manage, shoot the rest with poisoned arrows, and then retiring over the burning desert are able to subsist upon their concealed water stores, while their enemies are totally unable to follow them.

After removing the eggs from the nest, the approved method of carrying them is to take off the "crackers" or leather trousers, tie up the ankles firmly, fill the garment with eggs, and set it astride the shoulders if the captor be a pedestrian, or in front of the saddle should he be on horseback. The shells are so strong that they are able to bear this rather curious mode of conveyance without damage, provided that no extreme jolting take place. A frisky horse will, however, sometimes smash the whole cargo, with disastrous consequences to himself and the vessel in which they were carried.

Among the Fellatahs, an Ostrich egg on the top of a pole fixed to the roof of the hut is the emblem of royalty. The Copts call it the emblem of watchfulness, and carry out the idea by making the empty shell defend their church lamps from the rats, which crawl down the cords by which the lamps are suspended, and drink the oil. Their plan is to run the cord through an Ostrich shell, which is placed at some little distance above the lamp, and, by its smooth polished surface, forms an impassable barrier even to rats.

The feathers are too well known to need description. On an average, each feather is worth about a shilling. The best time for obtaining them is in the months of March and April. The greater number are furnished by means of the poisoned arrow, the native hunter scraping a hole in the sand near the nest, and lying concealed there until the birds come to their eggs, when a few rapid discharges will kill as many birds. Sometimes the hunter envelops himself in the skin of an Ostrich, his natural legs doing duty for those of the bird, and his arm managing the head and neck in such a way as to simulate the movements of the bird when feeding—an imitation so admirably managed that at a short distance it is impossible to distinguish the sham bird from the true. The enterprising little hunter is thus enabled to get among a flock of Ostriches, and to shoot one after the other with great ease, the birds not being able to understand the reason why their comrades should suddenly run away and then lie down, and permitting their enemy to follow them up until they share the same fate.

In some tribes each Ostrich feather worn on the head is an emblem of an enemy slain in battle.

The flesh of the Ostrich is tolerably good, and is said to resemble that of the zebra. It is, however, only the young Ostrich that furnishes a good entertainment, for the flesh of the old bird is rank and tough. The fat is highly valued, and when melted is of a bright orange colour. It is mostly eaten with millet flour, and is also stirred into the eggs while roasting, so as to make a rude but well-flavoured omelet.

Those who are fond of hunting employ a more sportsman-like though less profitable mode of procuring this bird. Mounted on swift horses they give fair chase to the nimble-footed bird, and generally manage to secure it by sending one of their number to head it on its course, and shooting it as it dashes by. The speed of the Ostrich is very great, though hardly so considerable as has been supposed. Some writers set it down as running sixty miles per hour, while others only give it half that rate. When going at full speed, its legs move so rapidly that they hardly seem to touch the ground; and as the pace of a running adult Ostrich is from ten to fourteen feet in length, its exceeding swiftness may be imagined.

For a short distance, the speed of the Ostrich is perhaps quite as great as the higher of the above statements; but it seldom keeps up that astonishing rate of going for more than half a mile, and then settles down into a more steady rate of progress. Being a long-winded bird, it would tire out most horses, did not it always run in curves, so that the horse-man by taking a direct course saves much ground, and is able to get a shot as the huge bird comes dashing by him. By the accompanying sketch of the Ostrich skeleton, the reader will be better enabled to understand the great powers of the bird and the curious modifications of its structure better than by many pages of description. The long and powerful legs with their two toes at their extremity are firmly yet flexibly jointed into their sockets, and their form is wonderfully adopted for the attachment of the stalwart muscles

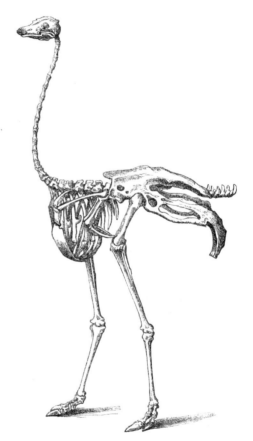

SKELETON OF OSTRICH.

which move them. Not only are the legs employed for progression, but they can be used with tremendous effect as offensive weapons, knocking over a hyæna with a stroke, and deterring even the agile leopard from coming within their reach. The Ostrich always kicks forward, and when hunted with dogs it is sure to inflict severe injuries on young and inexperienced hounds before it is pulled down. The strong sharp claw with which its toe is armed gives dreadful effect to the blow, and, like the claw of the kangaroo, has been known to rip up an antagonist at a stroke. When driven to bay it will turn and fight desperately even with man, and, unless due precautions are taken, will strike him down and trample upon him. In captivity, the bird has been frequently known thus to assault intruders or strangers, and to be very formidable to them, although to its keeper it soon becomes affectionate.

The voice of the Ostrich is a deep, hollow, rumbling sound, so like the roar of the lion that even practised ears have been deceived by it, and taken the harmless Ostrich for a prowling lion. In its wild state the Ostrich is thought to live from twenty to thirty years.

In the male bird, the lower part of the neck and the body are deep glossy black, with a few white feathers, which are barely visible except when the plumage is ruffled. The plumes of the wings and tail are white. The female is ashen brown, sprinkled with white, and her tail and wing-plumes are white, like those of the male. The weight of a fine adult male seems to be between two and three hundred pounds.

The EMEU inhabits the plains and open forest country of Central Australia, where it was in former days very common, but now seems to be decreasing so rapidly in numbers that Dr. Bennett, who has had much personal experience of this fine bird, fears that it will, ere many years, be numbered with the Dodo and other extinct birds.

The Emeu is not unlike the ostrich, which it resembles in many of its habits as well as in its form and general aspect. It is very swift of foot, but can be run down by horses and dogs without much difficulty. The dogs are trained to reserve the attack until the bird is thoroughly tired out, and then spring upon the throat in such a manner as to escape the violent kicks which the Emeu deals fiercely around, and which are sufficiently powerful to disable an assailant. The Emeu does not kick forwards like the ostrich, but delivers the blow sideways and backwards like a cow.

The flesh of the Emeu is thought to be very good, especially if the bird be young. The legs are always the coarsest and worst-flavoured portions, the flesh of the back being thought equal to fowl. The natives will not permit women or boys to eat the flesh of the Emeu, reserving that diet for warriors and counsellors. A rather valuable oil is obtained from this bird, as much as six or seven quarts being secured from a fine specimen. It chiefly resides in the skin, but also collects in great quantities about the rump, and between the scapularies and the sternum. It is obtained easily enough by plucking the feathers, cutting the skin into pieces, and boiling them in a common cooking-pot. A still simpler plan, though not so productive, is to toast the skin before the fire, and catch the oil in a vessel as it drips from the heated skin. This oil is of a light yellow colour, and is considered very valuable, being largely used as an embrocation to bruises or strains, either by itself or mixed with turpentine. As it does not readily congeal, or become glutinous, it is useful for oiling the locks of fire-arms. The natives prefer to roast the Emeu with the skin still upon it, thinking that the oil makes the flesh more luscious. When quite fresh, it is almost free from taste or smell, and is quite transparent.

The food of the Emeu consists of grass and various fruits. Its voice is a curious, hollow, booming, or drumming kind of note, produced by the peculiar construction of the windpipe. The legs of this bird are shorter and stouter in proportion than those of the ostrich, and the wings are very short, and so small that when they lie closely against the body they can hardly be distinguished from the general plumage.

The nest of the Emeu is made by scooping a shallow hole in the ground in some scrubby spot, and in this depression a variable number of eggs are laid. Dr. Bennett remarks that "there is always an odd number, some nests having been discovered with nine, others with eleven, and others, again, with thirteen." The colour of the eggs is, while fresh, a rich green, of varying quality, but after the shells are emptied and exposed to the light, the beautiful green hue fades into an unwholesome greenish brown. The parent birds sit upon their eggs, as has been related of the ostrich. The Emeu is not polygamous, one male being apportioned to a single female.

In captivity, the Emeu soon accommodates itself to circumstances, and even in

EMEU.—*Dromaius Novæ Hollandiæ.*

England breeds freely, and seems as much at ease as if it were in its proper climate. It is a most inquisitive bird, inspecting every novelty with great attention. "I once," writes Dr. Bennett, "saw a fine pair of full-grown specimens in a paddock near Sydney. Stopping to observe one which was at a short distance from the fence, he immediately came down to have a look at me. The second bird was some distance off, but, with their usual keenness of vision, on perceiving me viewing his companion, he came stalking down rapidly, and they both stared at me most attentively, stretching out their necks for the sake of making a nearer acquaintance, when, finding no result from our interview, and their curiosity being satisfied, they quietly stalked away.

In the Domain, near the Government House, some tame Emeus may be seen walking about, and often, near the Grand House, marching with measured pace, as if keeping guard with the soldiers on duty. One day, during the levée, when the Domain was crowded with people to see the arrivals and listen to the band, the Emeus mingled with the crowd, apparently enjoying the gay scene around them, when some strangers, who were afraid of these birds, ran away. On seeing this, the Emeus, enjoying a chase, pursued, and overtaking one of the gentlemen, took off his hat, to his great surprise. The above circumstance demonstrates their fearless nature, and how readily these noble birds might be domesticated."

The colour of the adult bird is lightish brown and grey, but when it is young, its plumage is decorated with four broad, black, longitudinal stripes down the back, and four on each side, and four more down the neck and breast. These stripes run in pairs, the two streaks of each pair being divided by a narrow line of white. Towards the head, the stripes are broken into spots and dashes. The feathers are very loose, and hairy in their appearance, and, as is the case with all the Struthiones, will repay a close examination, on account of the great development of the accessory plumes, springing from the shafts of the feathers. The height of a fine male Emeu is from six to seven feet.

Another species belonging to the same genus, the SPOTTED EMEU (*Dromaius irrorátus*), is found in the same country, and can be distinguished by its black head and neck, and the dashes of brownish black and grey upon its plumage.

AMERICA is not without representatives of this fine group of birds, three distinct species being even now (1861) in the gardens of the Zoological Society.

The RHEA is a native of South America, and is especially plentiful along the river Plata. It is generally seen in pairs, though it sometimes associates together in flocks of twenty or thirty in number. Like all the members of this group, it is a swift-footed and wary bird, but possesses so little presence of mind that it becomes confused when threatened with danger, runs aimlessly first in one direction, and then in another, thus giving time for the hunter to come up and shoot it, or bring it to the ground with his "bolas"—a terrible weapon, consisting of a cord with a heavy ball at each end, which is flung at the bird, and winds its coils round its neck and legs, so as to entangle it, and bring it to the ground.

The food of the Rhea consists mainly of grasses, roots, and other vegetable substances, but it will occasionally eat animal food, being known to come down to the mud banks of the river for the purpose of eating the little fish that have been stranded in the shallows.

Our knowledge of the Rhea and its habits is almost wholly derived from Mr. Darwin's writings, and, as an original narrative is mostly superior to a second-hand description, part of his account will be given in his own words. The reader must remember that the Rhea is popularly called the Ostrich in South America.

"This bird is well known to abound on the plains of La Plata. To the north it is found, according to Azara, in Paraguay, where, however, it is not common; to the south, its limit appears to have been from 42° to 43°. It has not crossed the Cordilleras, but I have seen it within the first range of mountains in the Uspallata plain, elevated between six and seven thousand feet. They generally prefer running against the wind, yet, at the instant, they expand their wings, and, like a vessel, make all sail. On one fine hot day I saw several Ostriches enter a bed of tall rocks, where they squatted concealed till nearly approached.

It is not generally known that Ostriches readily take to the water. Mr. King informs me that at Patagonia, in the Bay of St. Blas, and at Port Valdez, he saw these birds swimming several times from island to island. They ran into the water both when driven down to a point, and likewise of their own accord, when not frightened. The distance crossed was about two hundred yards. When swimming, very little of their bodies appears above water, and their necks are stretched a little forward; their progress is slow. On two occasions I saw some Ostriches swimming across the Santa Cruz River, where it was about four hundred yards wide, and the stream rapid.

The inhabitants who live in the country readily distinguish, even at a distance, the male bird from the female. The former is larger and darker coloured, and has a larger head. The Ostrich, I believe the cock, emits a singular deep-toned hissing note. When

RHEA.—*Rhea Americána*

first I heard it, while standing in the midst of some sand hillocks, I thought it was made by some wild beast, for it is such a sound that one cannot tell from whence it comes, or from how far distant.

When we were at Bahia Blanca, in the months of September and October, the eggs were found in extraordinary numbers all over the country. They either lie scattered singly, in which case they are never hatched, and are called by the Spaniards "huachos," or they are collected together into a hollow excavation, which forms the nest. Out of the four nests which I saw, three contained twenty-two eggs each, and the fourth twenty-seven. In one day's hunting on horseback sixty-four eggs were found; forty-four of these were in two nests, and the remaining twenty, scattered huachos. The Gauchos unanimously affirm, and there is no reason to doubt their statement, that the male bird alone hatches the eggs, and that he, for some time afterwards, accompanies the young. The cock, while in the nest, lies very close; I have myself almost ridden over one. It is asserted that at such times they are occasionally fierce, and even dangerous, and that they have been known to attack a man on horseback, trying to kick and leap on him. My informant pointed out to me an old man whom he had seen much terrified by one of these birds chasing him."

In captivity it is rather an amusing bird, and easily domesticated. Sometimes it seems to be taken with a fit, and runs up and down its inclosure as if it were being chased, holding its wings from the body and appearing in the most desperate state of alarm. This is only a sham after all, a mere outburst of frolic, for the bird immediately subsides into quietude, and resumes its leisurely walk as if nothing had happened. If startled or vexed ; it utters a kind of grunt as a warning, and if the offence be repeated, hisses sharply, draws back its head, and seems poising itself for a stroke. The grunt is a hollow sound, something like the noise produced by striking a tin can with a wooden mallet, and every time that it is produced the throat swells and sinks convulsively. The young are pretty little birds, pert, brisk, and lively, and are coloured rather prettily, their general hue being grey, striped with black, each stripe having a cream-coloured line along its centre. In the Zoological Gardens there are several of these pretty creatures, which have been hatched in the Society's incubator.

The Rhea is darkish grey, taking a blackish hue above, and being rather lighter below. The plumes of the wings are white, and a black band runs round the neck, and passes into a semilunar patch on the breast. The neck is completely feathered. The average height of the Rhea is about five feet.

Three species of Rhea are, however, all inhabitants of South America, namely, the common Rhea just described, DARWIN'S RHEA (*Rhea Darwinii*), and the LARGE-BILLED RHEA (*Rhea macrorhyncha*).

THE well-known CASSOWARY, long thought to be the only example of the genus, is found in the Malaccas.

This fine bird is notable for the glossy black hair-like plumage, the helmet-like protuberance upon the head, and the light azure, purple, and scarlet of the upper part of the neck. The "helmet" is a truly remarkable apparatus, being composed of a honey-combed cellular bony substance, made on a principle that much resembles the structure of the elephant's skull, mentioned in the previous volume of this work treating of the Mammalia. It yields readily to a sharp knife or a fine saw, and may be cut through by a steady hand without leaving ragged edges. This helmet is barely perceptible in the young bird when newly hatched, and increases in proportion with its growth, not reaching its full development until the bird has obtained adult age. A similar phenomenon may be observed in the common Guinea fowl. The beak is high in proportion to its width, and is therefore unlike the flattened and comparatively weak bills of the ostrich.

The plumage of the body is very hair-like, being composed of long and almost naked shafts, two springing from the same tube, and one always being longer then the other. At the roots of the shafts there is a small tuft of delicate down, sufficiently thick to supply a warm and soft inner garment, but yet so small as to be hidden by the long hair-like plumage. Even the tail is furnished with the same curious covering, and the wings are clothed after a similar manner, with the exception of five black, stiff, strong, pointed quills, very like the large quills of the porcupine, and being of different lengths, the largest not exceeding one foot, and generally being much battered about the point. When stripped of its feathers, the whole covering only extends some three inches in length, and is evidently a mere indication of the limb.

The eye of the Cassowary is fierce and resolute, and its expression is carried out by the character of the bird, which is tetchy of disposition, and apt to take offence without apparent provocation. Like the bull, it is excited to unreasoning ire at the sight of a scarlet cloth, and, like the dog or the cat, has a great antipathy towards ragged or unclean persons, attacking such individuals with some acerbity merely because their garments or general aspect do not please its refined taste. It is a determined and rather formidable antagonist, turning rapidly about and launching a shower of kicks which can do no small damage, their effect being considerably heightened by the sharp claws with which the toes are armed. In the countries which it inhabits, the native warriors are accustomed to use the innermost claw of the Cassowary's foot as the head of their spears.

The food of this bird in a wild state consists of herbage and various fruits, and in captivity it is fed on bran, apples, carrots, and similar substances, and is said to drink

CASSOWARY.—*Casuárius Emu.*

nearly half a gallon of water per diem. The eggs are somewhat like those of the rhea, save that their surface is more tubercular, and the shades of green more varied. The colour of the plumage is black, glossy above, as if made of shining black horsehair, and rather duller below. At the lower part of the neck there are two wattles, and the upper part of the neck is coloured with beautiful blue, purple, and scarlet. The legs are feathered. An adult male is about five feet in height.

THE other species of Cassowary was discovered by Captain Devlin, and, having been taken to Sydney in 1857, was there purchased and sent to England by Dr. Bennett, after whom it has been very appropriately named. Its native title is MOORUK, and its home is in the island of New Britain.

Dr. Bennett's description of the Mooruk is as follows: "The height of the bird is three feet to the top of the back, and five feet when standing erect. Its colour is rufous, mixed with black on the back and hinder portions of the body, and raven-black about the neck and breast; the loose wavy skin of the neck is beautifully coloured with iridescent tints of bluish purple, pink, and an occasional shade of green, quite different from the red and purple caruncles of the Cassowary; the feet and legs, which are very large and strong, are of a pale ash colour, and exhibit a remarkable peculiarity in the extreme length of the claw of the inner toe of each foot, it being nearly three times the length which it attains in the claws of the other toes. This bird also differs from the cassowary in having a horny plate instead of a helmet-like protuberance on the top of the head, which callous plate resembles mother-of-pearl darkened with black lead."

The voice of the Mooruk is a kind of whistling chirp. It is a very cleanly bird, keeping its plumage free from stain, and being very fond of washing, lying down to have repeated bucketfuls of water poured over its body, and squatting on the ground in heavy rain. In the month of May, 1861, three examples of the Mooruk were in the Zoological Gardens, all presented by Dr. Bennett. Their proceedings when in captivity are most amusingly told by their owner in his valuable "Gatherings of a Naturalist in Australasia," and although too long to be entirely inserted, are so interesting and so indicative of the Mooruk's character, that a portion must find a place in these pages.

"On the 29th of October I succeeded in purchasing the birds. When placed in the yard, they walked about as tame as turkeys. They approached any one who came in, as if desirous of being fed, and were very docile. They began pecking at a bone they found lying about (probably not having tasted any meat for some time), and would not, while engaged upon it, touch some boiled potatoes which were thrown to them; indeed, it was found afterwards that they fed better out of a dish than from the ground, having no doubt been early accustomed to be fed in that manner. They seemed also fond of scraping about the dunghill, and appeared to pick up food from it, probably insects or grubs. They were as familiar as if born and bred among us for years, and did not require time to reconcile them to their new situation, but were sociable and quite at home at once.

We found them on the following day rather too tame, or, like spoilt pets, too often in the way. One or both of them would walk into the kitchen, and while one was dodging under the tables and chairs, the other would leap up on the table, keeping the cook in a state of excitement; or they would be heard in the hall or in the library, in search of food or information; or they would walk upstairs, and then quickly descend again, making their peculiar chirping, whistling noise; not a door could be left open, but in they walked. They kept the servants constantly on the alert: if one went to open the door, on turning round she found a Mooruk behind her; for they seldom went together, generally wandering apart from each other.

If any attempt was made to turn them out by force, they would dart rapidly about the room, dodging about under the tables, chairs, and sofas, and then end by squatting down under a sofa or in a corner; indeed, it was impossible to remove the bird, except by carrying it away. On attempting this, the long muscular legs would begin kicking and struggling, when it would soon get released, and politely walk out of its own accord. I found the best method was to entice them out as if you had something eatable in your hand, when they would follow the direction in which you wished to lead them. On the housemaid attempting to turn the bird out of one of the rooms, it kicked her and tore her dress. They walk into the stables among the horses, poking their bills into the manger. When writing in my study, a chirping, whistling noise is heard; the door, which is ajar, is pushed open, and in walk the Mooruks, who quietly pace round the room inspecting everything, and then as peaceably go out again.

Even in the very tame state of these birds, I have seen sufficient of them to know that, if they were loose in a wood, it would be impossible to catch them, and almost as difficult to shoot them. One day, when apparently frightened at something that occurred, I saw one of them scour round the yard at a swift pace, and disappear under the archway so rapidly that the eye could hardly follow it, upsetting all the poultry in its progress, as they could not get out of the way. The lower half of the stable door, about four feet high, was kept shut, to prevent them going in; but this proved no obstacle, as it was easily leaped over by these birds.

They never appeared to take any notice of or be frightened at the jabiru, or gigantic crane, which was in the same yard, although that sedate, stately bird was not pleased at their intrusion. One day I observed the jabiru spreading his long wings, and clattering his beak, opposite one of the Mooruks, as if in ridicule of their wingless condition. The Mooruk, on the other hand, was preening its feathers, and spreading out its funny little apology for wings, as if proud of displaying the stiff horny shafts with which they were adorned. The Mooruks often throw up all their feathers, ruffling them, and then they suddenly fall flat as before. Their wings aid them in running, but are never used for

MOORUK.—*Casuárius Bennéttii.*

defence. Captain Devlin says, that the natives consider them to a certain degree sacred, and rear them as pets. He does not know whether they are used as food, but if so, not generally; indeed, their shy disposition, and power of rapid running, darting through brake and bush, would almost preclude their capture.

The natives carry them in their arms, and entertain a great affection for them, which will account for their domesticated state with us. The noise of these birds, when in the yard, resembled that of the female turkey; at other times, the peculiar chirping noise was accompanied by a whistling sound, which often reminded me of the chirp of the Guinea fowl. The contrast of these birds with the jabiru, or gigantic crane (*Myctéria Austrális*), was very great. The Mooruks were sometimes seen moving about like the female turkey, but were more often in a state of rapid motion or excitement; when walking quietly, they were very inquisitive, poking their beaks into everything, and familiar with every person. The jabiru, on the other hand, was a perfect picture of sedate quietness, looking upon all play as injurious to his constitution or derogatory to his dignity, remaining stiff in his gait and serious in his demeanour. The Mooruks, by their activity and noise, would let every one know they were in the yard, whereas no one would be aware of the presence of the jabiru except by sight; and when he moves away, it is with a quiet sedate gait.

The Mooruk has, when seen in full face, a fine eagle-like expression of countenance, having the same vivid, piercing eye and curved beak. The instant the Mooruk saw an egg laid by a hen, he darted upon it, and breaking the shell, devoured it immediately, as if he had been accustomed to eggs all his life. A servant was opening a cask of ale: as soon as the birds heard the hammering, they both ran down to it, and remained there while it was unpacked, squatting down on each side, most intently watching the process, and occasionally pecking at the straw and contents.

When the carpenter was in the yard, making some alteration in the cage of these birds, previous to their voyage to England, it was very amusing to see them squat down upon their tarsi, like dogs, watching the man, with the greatest apparent interest in all his actions, enjoying the hammering noise, and occasionally picking up a nail, which was not in this instance swallowed, but again dropped; one of them swallowed his "oilstone," which so alarmed the man that he considered the bird had committed suicide, and hurried to inform me of the circumstance, when, to his surprise, I told him if he did not take care they would swallow his hammer, nails, and chisel. The birds kept close to the man until he left for dinner, when they went about the yard as usual, resuming their position near him as soon as he returned to his work, and not leaving until he had finished.

These birds invariably retire to roost at dusk, and nothing more is seen or heard of them until daylight, as they never leave their usual roosting-place after retiring; indeed, their usual time of roosting is as soon as the sun is on the verge of setting, even before the poultry depart; and on looking at them about this time in their retirement, they utter their usual greeting chirps, and one may be observed reposing upon the breast, the other upon the tarsi. The door may be safely left open during the night, as they will not move, nor leave their sleeping-place, until the dawn of day. If, during any hour of the night, I approached their resting-place, they immediately greeted me with their peculiar chirping noise, being evidently, like geese, very watchful, or, according to the common saying, 'sleeping with one eye open;' when gazed at, they not only chirped, but, if I continued too long, I was saluted by a loud growl.

One morning the male Mooruk was missing, and was found in the bedroom upstairs, drinking water out of the water-jug. There were some silkworms in the room at the time, but they were fortunately covered; otherwise, I have no doubt, he would have made a meal of them. The same bird swallowed a bung-cork which measured one and a half inch in diameter; indeed, they both seem to swallow anything, from butter and eggs to iron, in the form of small bolts or nails, and stones. The bird did not appear well; he was sulky and heavy all day; and when, in this sickly state, any one approached him, instead of being greeted with a cheerful chirping, he uttered a loud sulky growl: we were afraid he was dying. On the following day he was as lively as ever, having passed the cork in a perfectly undigested state.

To show how dangerous it was to leave any object capable of being swallowed, I will relate the following occurrence: The servant was starching some muslin cuffs, and having completed one and hung it up to dry, she was about to finish the other, when, hearing the bell ring, she squeezed up the cuff, threw it into the starch, and attended to the summons. On her return the cuff was gone, and she could not imagine who had taken it during her brief absence, when she discovered that the Mooruk was the thief, its beak and head being covered with starch: he had without doubt swallowed it. This occurred at eleven A.M., and at half-past five P.M. the cuff was passed, quite undigested and uninjured, and with a little washing was as good as ever.

They could not digest unboiled potato. Maize, or any unboiled grain, was likewise indigestible. When a piece of bread was offered them at a height beyond their reach, they would first stretch up the body and neck as much as possible, and then, finding they could not get it, they would jump up for it like a dog. They were frequently seen running and tumbling about the yard together in high spirits. It is well to warn persons, inclined to keep these birds as pets, of their insatiable propensities. When about the house, they displayed extraordinary delight in a variety of diet; for, as I have previously related, one day they satisfied their appetites with bones, whetstones, corks, nails, and raw potatoes, most of which passed perfectly undigested; one dived into thick starch and devoured a muslin cuff, whilst the other evinced a great partiality for nails and pebbles; then they stole the jabiru's meat from the water. If eggs and butter were left upon the kitchen table, they were soon devoured by these marauders; and when the servants were at their dinner in the kitchen, they had to be very watchful, for the long necks of the birds appeared between their arms, devouring everything off the plates; or, if the dinner table was left for a moment, they would mount upon it and clear all before them.

At other times they stood at the table, waiting for food to be given to them, although they did not hesitate to remove anything within their reach. I have often seen them stand at the window of our dining-room, with keen eye, watching for any morsel of food that might be thrown to them. The day previous to the departure of the pair for England, in February, 1859, the male bird walked into the dining-room, and remained by my side during the dessert. I regaled him with pine-apple and other fruits, and he behaved very decorously and with great forbearance. Having had these birds for a considerable time

APTERYX, OR KIWI-KIWI.—*'Apteryx austrális.*

in my possession, I had ample opportunity of hearing all the notes uttered by them. I never heard them utter a sound like 'Mooruk.' I am inclined to consider the name signifies, in the native language, 'swift'—resembling closely the Malay term 'a muck,' or mad career."

In the same work is much more curious and valuable information respecting this bird, and to its pages the reader is referred for further information concerning this and many other objects of natural history.

The Mooruk is not devoid of offensive weapons, for it can kick very sharply, delivering the stroke forward like the ostrich, and deriving much aid from the long-pointed claw which has already been mentioned. Its attitudes are much more various, and its form more flexible, than would be supposed by persons who have not seen the bird in a living state. Sometimes it squats down with the legs bent under it, and so sits upright like a dog that has been taught to "beg;" sometimes it lies on its side, stretching the legs straight behind it; sometimes it flattens itself against the ground, its legs tucked under its body, and its head and neck stretched at full length on the ground. This latter position is a favourite one. Like the emeu, it is often taken with an ebullition of joyousness, and then dashes about its inclosure as if half mad, jumps against a tree or post, trying to kick it at a great height from the ground, and tumbling flat on its back when it misses its aim. Then it will suddenly cease its vagaries, and walk about very composedly, but panting for breath with open bill.

This bird may be distinguished from the cassowary by the four (instead of five) spines of the wings, and the shape of the helmet.

PERHAPS the very strangest and most weird-like of all living birds is the APTERYX, or Kiwi-Kiwi.

2.

U U

This singular bird is a native of New Zealand, where it was once very common, but, like the dinornis, is in a fair way of becoming extinct, a fate from which it has probably been hitherto preserved by its nocturnal and retiring habits.

Not many years ago, the Apteryx was thought to be a fabulous bird, its veritable existence being denied by scientific men as energetically as that of the giraffe in yet older days, or the duck-bill in more modern times. A skin brought from New Zealand was given to a taxidermist to " set up," and the man, taking it for one of the penguins on account of its very short wings and the total absence of a tail, stuffed it in a sitting posture, such as is assumed by the penguine tribe, and arranged the head and neck after the same model.

In this bird there is scarcely the slightest trace of wings, a peculiarity which has gained for it the title of Apteryx, or wingless. The plumage is composed of rather curiously shaped flat feathers, each being wide and furnished with a soft, shining, silken down for the basal third of its length, and then narrowing rapidly towards the extremity, which is a single shaft with hair-like webs at each side. The quill portion of the feathers is remarkably small and short, being even overlapped by the down when the feather is removed from the bird.

The skin is very tough and yet flexible, and the chiefs set great value upon it for the manufacture of their state mantles, permitting no inferior person to wear them, and being extremely unwilling to part with them even for a valuable consideration. The bird lives mostly among the fern ; and as it always remains concealed during the day in deep recesses of rocks, ground or tree roots, and is remarkably fleet of foot, diving among the heavy fern-leaves with singular adroitness, it is not very easy of capture. It feeds upon insects of various kinds, more especially on worms, which it is said to attract to the surface by jumping and striking on the ground with its powerful feet. The natives always hunt the Kiwi-kiwi at night, taking with them torches and spears. The speed of this bird is very considerable, and when running it sets its head rather back, raises its neck, and plies its legs with a vigour little inferior to that of the ostrich.

The fine specimen in the Zoological Gardens has already proved a very valuable bird, as she has laid several eggs, thereby setting at rest some disputed questions on the subject, and well illustrates the natural habits of the species. During the day she remains hidden behind the straw, which is piled up in a corner of her box, and declines to come forth unless removed by force. When brought to the light, she looks sadly puzzled for a short time, and when placed on the ground, she turns her back—not her tail, as she has no such appendage—and runs off to her box in the most absurd style, looking as if she were going to topple over every moment. I noticed that she always goes round her box and slips in between the box and the wall, insinuating herself behind the straw without even showing a feather. Before hiding herself, she lingered a few moments to eat some worms from her keeper's hand, taking them daintily with the end of the bill, and disposing of them at a rapid rate.

Upon her box is placed, under a glass shade, the shell of one of her eggs. These eggs are indeed wonderful, for the bird weighs just a little more than four pounds, and each egg weighs between fourteen and fifteen ounces, its length being four inches and three-quarters, and its width rather more than two inches, thus being very nearly one-fourth of the weight of the parent bird. There have been six eggs laid between 1852, when it was first introduced to the Gardens, and 1861, when I last saw the bird, and each egg has varied between thirteen and fourteen and a half ounces in weight.

The long curved beak of the Apteryx has the nostrils very narrow, very small, and set on at each side of the tip, so that the bird is enabled to pry out the worms and other nocturnal creatures on which it feeds, without trusting only to the eyes. The general colour of the Apteryx is chestnut-brown, each feather being tipped with a darker hue, and the under parts are lighter than the upper. The height is about two feet.

Three species of Apteryx are known—namely, the one already described, OWEN'S APTERYX (*Apteryx Owenii*), remarkable for the puffy downiness of its plumage, and MANTELL'S APTERYX (*Apteryx Mantellii*), and it is very probable that there are still other species at present unknown.

GREAT BUSTARD.—*Otis tarda.*

ALTHOUGH the progress of civilization has conferred many benefits on this country, it has deprived it of many of its aboriginal inhabitants, whether furred or feathered, the GREAT BUSTARD being in the latter category.

This splendid bird, although in former days quite a usual tenant of plains and commons, and having been an ordinary object of chase on Newmarket Heath, is now so very rare, that an occasional specimen only makes its appearance at very rare intervals, and is then generally found—and shot—on Salisbury Plain. In the countries which it still inhabits, it is a most wary bird, and very difficult of approach, being generally shot with rifles after a careful and lengthened chase that rivals deer-stalking in the watchfulness and perseverance that are requisite before the sportsmen can get within shot. They are carried in carts, covered with ordinary farm produce, and having an aperture through which they can aim; they put on various disguises; they enact the part of agricultural labourers, plying their work, and gradually slipping towards the wary birds; they walk behind cows, and, in fine, put into practice every device which their ingenuity, sharpened by experience, can suggest.

The Great Bustard is not fond of flying, its wings having but a slow and deliberate movement; but on foot it is very swift, and tests the speed of dog and horse before it can be captured.

The nest—if a hole in the ground may be called a nest—of this bird is generally made among corn, rye, &c., although it is sometimes situated in rather unexpected localities. The eggs are two or three in number, and of an olive-brown colour, splashed with light brown, in which a green tinge is perceptible. The food of the bird is almost wholly of a vegetable nature, though it is said to feed occasionally upon mice, lizards, and other small vertebrates. The flesh of the Bustard is very excellent, but the extreme rarity of these birds prevents it from being often seen upon English tables. When caught young, the Bustard can be readily tamed, and soon becomes quite familiar with those who treat it kindly.

The head and upper part of the neck are greyish white, and upon the side of the neck there is a small patch of slaty blue bare skin, almost concealed by the curious feather tuft which hangs over it. The upper part of the body is pale chestnut, barred with black, and the tail is of similar tints with a white tip, and a very broad black band next to the white extremity. The wing-coverts, together with the tertials, are white, and the primaries black. The under surface of the body is white. The total length of an adult male is about forty-five inches.

The LITTLE BUSTARD is also an occasional visitor to this country; and whenever it does make its appearance, it almost invariably chooses the winter time.

It is by no means uncommon in several parts of Europe, and in Russia, assembles in little flocks. Towards the shores of the Caspian Sea it is found in greater numbers, the flocks being of considerable size, and all appearing (in the month of December) to consist of birds which have not put on, or which have already put off, their nuptial plumage. This bird feeds upon insects, herbs, grasses, and seeds, and its flesh is very good, having been compared to that of a young pheasant. The eggs are placed on the ground among a tuft of rank herbage in which the bird can lie concealed; their number is about four, and their colour olive-brown.

The male, when in full plumage, is a decidedly handsome bird. The top of the head is fawn and black, and the sides of the face and neck are slaty grey. Around the neck runs a broad gorget of black, cut by two white bands, one narrow and forming a ring round the neck, and the other, broader and of semilunar shape, just across the top of the breast. The upper parts of the body are fawn, mottled profusely with black, and the wings are beautifully marked with black and white. The under surface of the body is white. The female is without the beautiful black and white stripes on the neck and chest, and her breast, sides, and flanks are barred with black. Except during the breeding season, the male has the same plumage. The total length of this bird is about seventeen inches.

THERE are many other Bustards scattered over the world, some being well known in India under the title of Florikans, and others being distributed over Africa. The HOUBARA, or RUFFLED BUSTARD, is well known, on account of its curious-plumed ruffles and the sport which it affords to Algerian falconers. There are also two South African species, the Pauw and the Koran, which are often casually mentioned in the works of African travellers. Both these birds belong to the genus Eupodotis, and of them Captain Drayson, R.A., has kindly given me the following account:—

"The PAUW bird is more sought for by the pot-hunter than any other in South Africa. Its size is about that of a turkey, and its flesh delicious. On the breast of this bird there are two coloured meats. First, there is a dark brown, similar to that of the grouse; but beneath this there is white meat, which is similar in appearance to chicken's flesh.

The Pauw is usually found on the plains, which it prefers to bushy country; for as it is a very crafty bird, it does not like to give the sportsman an opportunity to stalk it. When the long grass of the plains has been burnt, and the young grass began to shoot up, then would numbers of Pauws assemble on the ground, and search for the worms and slugs which became visible. There was little chance, however, of approaching within two

LITTLE BUSTARD.—*Otis tetrax.*

hundred yards of the bird at these times, as the whole flock would take flight immediately they believed themselves in danger, and they had formed a very fair estimate of the distance at which a smooth bore would be dangerous. The flight of the Pauw was something like the heron's, except that, when it purposed settling, it would skim for a considerable distance with its wings quite rigid.

The bird being rather heavy, with the appearance of a full habit of body, it could not take flight very readily. When it was possessed of a good feeding locality it seemed disinclined to fly away, although its sense of danger was apparent. The sportsman might then probably reach to within one hundred yards of the bird, particularly if there happened to be only one near him, and if he did not look attentively in the direction of the Pauw. It was still necessary, however, to ride round the circumference of a circle of which the Pauw was the centre, and, by decreasing the radius, to approach nearer and nearer. If the Pauw crouched, then it usually depended upon the accuracy of the shooting whether or not the bird was killed; for the sportsman might then gradually narrow the radius of his circle, until he was within seventy or eighty yards, when he might dismount, if on horseback, and run in towards the bird, discharging the dose of buck-shot just as the Pauw opened wide his wings. These birds are not confined to any particular locality, but seem to range over any country within a radius of a hundred miles or so.

The CORAN is much smaller than the pauw, is longer, in proportion, in the leg, and is rarely seen in flocks. It is quite as much esteemed for the table as the larger bustard, and possesses also the two coloured meats. The Coran may be expected where the grass is long, near rivers or ponds, and where there are some portions of marshy ground; but it

GREAT PLOVER, OR THICK-KNEE.—*Œdicnémus crépitans.*

avoids showing itself much in the open. The poet has very appropriately designated this bird as the 'listless Coran,' for its flight is slow and short, and, if possible, will be avoided altogether.

In consequence of these characteristics this bird, if once seen, is almost certain to be 'bagged.' It will allow the sportsman to almost ride over it before it will rise; then a slow, lazy, owl-like flight of about two hundred yards will satisfy its organ of caution. Upon being pursued, it will again lie close, and has to be almost kicked before it will leave the ground; after which its slow flight affords even an indifferent shot an excellent chance of killing, for the Coran can carry off very little shot."

The Wading Birds are well furnished with legs and feet formed for walking, and in many species the legs are greatly elongated, so as to enable them to walk in the water while they pick their food out of the waves.

In the British Museum the Plovers head the list of Waders, of which our first example is the GREAT PLOVER, or THICK-KNEE, of England.

It is not an uncommon visitor to this country, spending the months between April and September within our coasts, and being found in various parts of England, where it is known under the names of STONE CURLEW and NORFOLK PLOVER. As it comes from the south, it is more common in the southern than in the northern counties. It moves about chiefly in the dark, its large full eyes enabling it to take advantage of the waning light, and to pounce upon the slugs, worms, and insects that come forth by night. The bird is also thought to kill and devour lizards, frogs, and mice; and the remains of the large hard-shelled beetles have been found within its stomach.

The note of this Plover is almost human in its intonation, sounding like that strange whistle produced by putting the fingers in the mouth and blowing shrilly through them. The Thick-knee frequents open country and plains, disliking inclosures, and being very fond of downs where sheep are fed in large flocks. It is a cautious and very shy bird, so that the sportsman cannot, without great trouble, come within shot range. More-over, it is singularly tenacious of life, and will carry away a large charge of shot without seeming much the worse at the time.

PRATINCOLE.—*Glaréöla pratincola.*

The eggs of this bird are laid upon the bare ground, and are two in number. Their colour is rather light dingy brown, covered with splashes and streaks of slaty blue and dark brown. The male bird is supposed to aid in the duties of incubation. When hatched, the young birds are covered with a soft spotty down, so like the stones and soil in which they repose, that they can hardly be discovered even within a yard or two. For the same reason, the eggs are very safe from unpractised eyes. About October the birds take their departure, assembling together in flocks before they start on their travels.

The general colour of the Thick-knee is mottled brown and black. The head is brown streaked with black; there is a light-coloured stripe from the forehead to the ear-coverts, and the chin and throat are white. The back is brown streaked with black, and the quill-feathers of the wing are nearly black, with a few patches of white. The neck and breast are extremely pale brown, streaked with a darker hue, and the abdomen is nearly white, with a few long and very narrow longitudinal streaks. In total length the bird measures about seventeen inches.

THE close compact plumage of the PRATINCOLE, its long pointed wings, its deeply forked tail, and swallow-like form, point it out as a bird of swift wing and enduring flight.

The Pratincole is by no means plentifully found in the British Isles, making its usual residence in the east of Europe and Central Asia. Like the swallows, to which it is so similar in form and habits that even modern zoologists have doubted whether it ought not to find a place among those birds rather than with the Waders, the Pratincole feeds much upon the wing, snapping up the insects as they come across its path, and especially delighting in picking the aquatic insects out of their native element without even staying its aerial course. Its endurance is equal to its speed, and a flight of two or three hundred miles is but an easy journey to this bird, which can thus pass over a very great extent of country in a few days.

The nest of the Pratincole is made among thick aquatic herbage, and the eggs are generally about five or six in number. The general colour of the Pratincole is shining yellowish brown above. The chin is whitish, and the front of the throat reddish white. A

CREAM-COLOURED COURSER.—*Cursórius Gállicus.*

narrow black streak runs from the eyes over the ear-coverts, and round the throat, forming the "collar," by which the bird is so readily known. The breast is light brown, and the abdomen, as well as the upper tail-coverts, is white. The quill-feathers of the wings are dark blackish brown, and the deeply forked tail is white at its basal half, and dark blackish brown to the tip.

THE very rare bird which, on account of its speed of foot and the colour of its plumage, is termed the CREAM-COLOURED COURSER, is found even less frequently than the preceding species.

It seems to live chiefly in Barbary or Abyssinia, though specimens have been obtained from almost every country in Europe. One of these birds, shot in Kent, was remarkable for its boldness. When the gun that was aimed at it missed fire, the bird only flew away for a short distance, and then alighted within a hundred yards of the gunner. It ran with great velocity, picking up objects from the ground in its course, and it was with difficulty raised from the ground so as to afford a fair shot. The note of this species is very peculiar, and is uttered on the wing.

The crown of the head is fawn, fading into' grey behind, and the chin is white. From the eye over the ear-coverts is a black curved streak, and immediately above it is a similar white streak. The whole upper parts of the body are pale reddish brown, the primary feathers of the wing are jetty black, and there is a curious black spot near the end of each tail-feather. The whole under surface is cream-white, becoming white on the under tail-coverts. Both sexes are similarly coloured, and the total length is rather more than ten inches.

THE well-known LAPWING, or PEEWIT, is celebrated for many reasons. Its wheeling, flapping flight is so peculiar as to attract the notice of every one who has visited the localities in which it resides, and its strange, almost articulate, cry is equally familiar. When it fears danger, it rises from the nest, or rather from the eggs, into the air, and continually wheels around the intruder, its black and white plumage flashing out as it inclines itself in its flight, and its mournful cry almost fatiguing the ear with its piercing

LAPWING.—*Vanellus cristátus.*

frequency. "Wee-whit! wee-e whit!" fills the air, as the birds endeavour to draw away attention from their home; and the look and cry are so weird-like that the observer ceases to wonder at the superstitious dread in which these birds were formerly held. The French call the Lapwing "Dix-huit," from its cry.

It is the male bird which thus soars above and around the intruder, the female sitting closely on her eggs until disturbed, when she runs away, tumbling and flapping about as if she had broken her wing, in hopes that the foe may give chase and so miss her eggs. It is certainly very tempting, for she imitates the movements of a wounded bird with marvellous fidelity.

The eggs of the Lapwing are laid in a little depression in the earth, in which a few grass stalks are loosely pressed. The full number of eggs is four, very large at one end and very sharply pointed at the other, and the bird always arranges them with their small end inwards, so that they present a somewhat cross-like shape as they lie in the nest. Their colour is olive, blotched and spotted irregularly with dark blackish brown, and they harmonize so well with the ground on which they are laid that they can hardly be discerned from the surrounding earth at a few yards' distance. Under the title of "Plover's eggs" they are in great request for the table, and are sought by persons who make a trade of them, and who attain a wonderful expertness at the business. The eggs are generally laid in marshy grounds, heaths, and commons, where they are sometimes found by dogs trained for the purpose. They are, however, often placed in cultivated grounds, and I have found numbers in ploughed fields in the months of April and May. At first, the novice may pass over the ground three or four times without finding an egg, and may have the mortification of seeing a more experienced egg-hunter go over the very same ground and fill his bag. After a while, however, the eye becomes accustomed to the business, and the speckled eggs stand out boldly enough against the ground. Even the protruding ends of the bents and grass stems on which they are laid take the eye, and there are very few eggs that can escape.

The food of the Lapwing consists almost wholly of grubs, slugs, worms, and insects. It is easily tamed, and is often kept in gardens for the purpose of ridding them of these destructive creatures. In the garden next our own a Lapwing was kept, and lived for some years, tripping featly over the grass and thoroughly at home.

In its colouring the Lapwing is rather a handsome bird. The top of the head is black, as is the long-pointed crest, which can be raised or depressed at will. The sides of the face and neck are white, speckled with black ; the chin, throat, and breast are jetty black, and from the chin a black streak runs under the eye. The upper part of the body is shining coppery green, glazed with purple, and the primary feathers of the wing are black, with some greyish white at their tips. The upper tail-coverts are chestnut, and the tail is half white and half black, the exterior feather on each side being almost wholly white. The under parts are white, changing to fawn on the under tail-coverts. In winter the chin and throat are white. The yearling birds are mottled with buff on the back. The total length of the bird rather exceeds one foot.

THE three Plovers represented in the engraving are common throughout England.

The GOLDEN PLOVER, sometimes called the YELLOW PLOVER from its prettily coloured plumage, is common in many parts of England, being found mostly in the more northern districts of Great Britain, moving southward in the autumn. The spots which it selects for its breeding-places are generally situated on open moors, where the vegetation is but scanty, and water is at hand, although well below the level of the nest, rather high ridges, with a dell slope, being its most favoured spots. It makes its simple preparations in the beginning of April or the end of March, according to the season, choosing some little depression in the soil, scratching it tolerably level, and laying in it a few bents and grass stems. The eggs are usually four in number, and their colour is yellowish olive, blotched with dusky brown. Like the eggs of the lapwing, they are arranged with their small ends inwards. The Golden Plover also puts in practice sundry devices to draw an intruder away from the nest, rising into the air when it has succeeded in its object, and uttering an exultant whistling cry as it wheels off in safety. The female is very careful about her eggs. While sitting, she crouches so low upon them that her speckled plumage can hardly be distinguished from the earth ; and when she leaves her nest, she runs to some little distance along the ground before she rises into the air, and returns after the same cautious fashion.

The young birds are active on foot, and are able to follow their parent within a very short time after their escape from the egg-shell. They are pretty little creatures, covered with thick dusky mottled down, and not easily to be discovered.

The plumage of the Golden Plover varies generally according to age and the season of year. In the summer, the top of the head and whole of the upper surface are greyish black, mottled with triangular spots of golden yellow. The face, chin, throat, and under surface of the body are jetty black, a white streak passing over the eyes and forehead, and separating the mottlings of the head from the black of the face. The primaries are nearly black, and the tail is barred with whitish grey and blackish brown. Below the wing there is a band of white, and the under tail-coverts are white. In the winter the chin is white, and the breast also dusky white, spotted with yellow ; and in late autumn and early spring the changing plumage is curiously mottled with black, yellow, and white. The yearling birds are more grey on the breast and lower parts than when they have attained their second year's plumage. In total length this bird measures not quite one foot.

The DOTTEREL has long been held as the type of stupidity, and to call a man a Dotterel is considered as great an insult as to term him a goose or a donkey.

Certainly, the Dotterel is not a very wise bird in some things, having but little of the general wary habits of the Plovers, and allowing itself to be approached without displaying much uneasiness. It was once thought to be so very inquisitive and so foolish as to imitate all the actions of the fowler, holding out a wing if he held out an arm, lying flat if he did the same, and so permitting the net to be thrown over it before it was aware of any danger. It is not now so plentiful as it used to be, its numbers having been much thinned by guns and nets. Its flesh is thought very good, and the bird finds a ready sale in the poulterer's shop. The specific title, Morinellus, signifies a little fool. The cry of the Dotterel is a kind of piping whistle.

GOLDEN PLOVER.—*Charádrius pluviális.*　　　DOTTEREL.—*Charádrius morinellus.*
KENTISH PLOVER.—*Charádrius cantiánus.*

The breeding-places are selected on high grounds, and the eggs, mostly three in number, are placed on a few grass stems laid carelessly in a depression in the soil, sheltered in most cases by a large stone or fragment of rock. The colour of the eggs is like that of the golden Plover.

The top of the head and back of the neck are dark brown; above the eye a rather broad white streak runs towards the nape of the neck, and the chin and sides of the face are white, speckled with darker tints. The back is ashen brown, and the scapularies and wing-coverts are edged with buff. The primaries are ashen grey mixed with white. The throat is ashen grey, and the breast is rich dark fawn, crossed by a bold white streak, extending completely across the breast and terminating at the shoulders. The abdomen is black, and the under tail-coverts buffy white. In the summer the breast is buffy white. The total length of this bird is not quite ten inches.

THE pretty little KENTISH PLOVER may be seen on some of our shores, running along the edge of the waves with surprising celerity, pecking here and there as the waves retreat, and uttering its happy whistling little notes as it runs.

It bears a considerable resemblance to the ringed Plover (*Charadrius hiaticula*), but may be distinguished from that bird by the smaller size and the broken black collar on the neck, which does not extend completely across the breast. The best mode of observing this bird, or, indeed, the many species that haunt the shores, is to get on the cliffs, lie down among the high grass and herbage, and make use of a good double field-glass. With an ordinary telescope the birds get out of the field too rapidly, and they are liable to be alarmed by the movements of the tube.

The eggs of this bird are laid in a hollow scraped in the sand or the fine shelly shingle. There is no nest excepting the sand. The colour of the eggs is yellowish olive, with streaks and spots of black.

The top of the head is rich chestnut, the forehead white, with a black patch immediately above the white, and a slight streak of white passes near the eye. The ear-coverts are black, and the edge of the neck is greyish white. The chin, sides of the throat, breast, and under parts are white, except a black collar which very nearly crosses the breast, but leaves a white space in front. The back and upper parts are ashen brown, and the primaries dull black. The length of the adult bird is not quite seven inches.

OYSTER-CATCHER.—*Hæmátopus ostrálegus.*

THE handsome OYSTER-CATCHER is another of our coast birds, and is tolerably plentiful upon the shore. From the black and white hues of its plumage, it is sometimes called the Sea-Pie.

It generally keeps to the shore, haunting sandy bays, interspersed with partially submersed rocks, and picking up its subsistence with great animation. It feeds mostly on molluscs, mussels and limpets being ordinary articles of its food. It is able to detach the firmly-clinging limpet from the rock by striking a sharp blow with its wedge-like beak, and detaching the mollusc before it has had time to take the alarm and draw itself firmly against its support. It is swift of foot, and a good swimmer, frequently taking to the water in search of food, and being able to dive when alarmed. Diving, however, does not seem to be a favourite accomplishment, and is seldom resorted to unless under peculiar circumstances.

In some parts of England, the Oyster-Catcher makes short inland migrations during the summer, but even in such cases it displays its aquatic propensities by keeping near the river banks, and feeding on the worms, slugs, and similar creatures.

The nest of the Oyster-Catcher is merely a hole scraped in the ground, wherein lie three or four eggs of a yellowish olive, spotted with grey and brown. They are generally placed on the beach well above high-water mark, but the bird sometimes makes its home at some distance from the sea. The flat sandy coasts of Lincolnshire seem to be the localities most favoured by the Oyster-Catcher. The young are covered with soft down of a greyish brown colour.

The head, neck, upper part of the breast, scapularies, quill-feathers, and latter half of the tail-feathers are deep shining black, and the rest of the plumage is pure white. The curious beak is three inches in length, very much compressed—*i. e.* flattened sideways—and towards the point is thinned off into a kind of wedge or chisel-shaped termination. The rich ruddy colour is deepest at the base. During some of the winter months there is no white collar round the throat, and in the yearling bird the back and wings are mottled with brown. The total length of the Oyster-Catcher is about sixteen inches.

THE handsomely plumed TURNSTONE is, though a little bird, so boldly decorated with black, white, and ruddy orange, that it is more conspicuous upon the coast than birds of double its size.

The name is derived from its movements when feeding, at which times it runs along the beach, picking up sandhoppers, marine worms, and other creatures, and turning over the stones in its course for the purpose of getting at the small crustaceæ that are generally found in such situations. This bird is spread over a considerable portion of the world, and is found even in Northern America, where it retains the same habits which distinguish it in Europe. According to Wilson, it feeds almost wholly, during May and June, on the spawn of the king-crab, and is known by the name of the Horse-foot Snipe, the king-crab being popularly called the horse-foot crab. It runs with some speed, but not the rapidity that characterises many shore-loving birds, and spends some time in examining any spot of ground to which it has taken a fancy, tossing the pebbles from side to side, and picking up the unfortunate being that may have lain under their shelter.

TURNSTONE.—*Cinclus intérpres.*

The nest of this bird is situated upon the coast, and the bird is very valiant in its attacks upon the gulls which approach too near its home. A nest found by Mr. Hewitson " was placed against a ledge of rock, and consisted of nothing more than the drooping leaves of the juniper-bush, under a creeping branch, by which the eggs, four in number, · were snugly concealed, and admirably sheltered from the many storms by which these bleak and exposed rocks are visited, allowing just sufficient room for the bird to cover them. The several nests that we examined were placed in the same situation as the one described, with the exception of two, one of which was under a slanting stone, the other on the bare rock ; all the nests contained four eggs each. Their time of breeding is about the middle of June. The eggs measure one inch seven lines in length, by one inch two lines in breadth, of an olive-green colour spotted and streaked with ash-blue and two shades of reddish brown." These nests were found on the coast of Norway.

The top of the head is white streaked with black, and a black band crosses the forehead and passes over the eyes. The chin, face, and sides of the neck are white, and the breast is jetty black, throwing out black branches shaped like the gnarled boughs of the oak, which run to the base of the bill, the lower eyelid, the back of the neck, and the shoulders. The upper part of the back is also black, with a band of bright rust-red, and the lower part white, with a broad band of black just above the tail-coverts. The under parts are pure white, and the legs and toes are scarlet-orange. The length of the bird is rather more than nine inches.

THE two birds represented on the following engraving are natives of Tropical America.

GOLDEN-BREASTED TRUMPETER.—*Psophia crepitans.*   CARIÁMA.—*Cariáma cristáta.*

The GOLDEN-BREASTED TRUMPETER is a handsome bird, remarkable for the short velvety feathers of the head and neck, and their beautiful golden green lustre on the breast. The body of this bird is hardly larger than that of a fowl, but its legs and neck are so long as to give it the aspect of being much larger than it really is. Like most birds of similar structure, it trusts more to its legs than its wings, and is able to run with great speed and activity. It is generally found in the forests.

As it is very easily tamed, it is a favourite inmate of the house, where it soon constitutes itself the self-chosen guardian, watching the premises as jealously as any dog, and permitting no other bird or beast to share its owner's favours at the same time. Dogs and cats it dislikes, and turns them out of the room when meal-times approach. The dog sometimes fights for its privileges, but mostly in vain, for the Trumpeter has a way of rising into the air, coming down on the dog's back and striking him with bill and feet, that effectually puzzles the four-footed foe and forces him to vacate the field of battle. It is said to learn to drive sheep, and to perform this arduous duty as well as any dog.

The name of Trumpeter is derived from the strange hollow cry which it utters without seeming to open the beak. This cry is evidently produced by means of the curiously formed windpipe, which is furnished with two membranous expansions, and, during the utterance of the cry, puffs out the neck very forcibly, just as the rhea does when grunting. The nest of the Trumpeter is said to be a hole scratched in the ground at the foot of a tree, and to contain about ten or twelve light green eggs. The head and neck are velvety black, and on the breast the feathers become large, rounded, and more scale-like, and their edges are beautifully bedecked with rich shining green with a purplish gloss in some lights and a lustrous golden hue in others. The back is grey, the feathers being long and silken and hanging over the wings. The wings, under surface, and tail are black, and the feathers of the tail are soft and short.

The ÇARIAMA is rather larger than the trumpeter, and has many of the same habits. It is chiefly remarkable for the feathery crest on the crown and forehead.

The Çariama is an admirable runner, getting over the ground with astonishing speed, and turning and twisting with such adroit rapidity that even the admirable horsemen of its native land find it put their skill to the sharpest test. Not until it is quite wearied out, and crouches under a bush or other shelter, does the hunter endeavour to use either rifle or lasso, the two deadly weapons of his land. The walk of this bird is peculiarly bold and easy, its paces are long, its lithe neck moves with every step as it continually turns its little sharp-looking head from side to side, and its full intelligent eyes gleam through their heavy lashes as they survey every object within their ken. The eyes are truly beautiful, large, round, and translucent, of a clear pearly grey, with many little dark changing spots, much like the eye of a living dragon-fly.

It is easily tamed, and soon becomes so attached to its new home that it is accustomed to roam about at will, and to return to its owner like the common fowl. The nest of this bird is placed upon the branches of a rather low tree, it is made of sticks, and generally contains two white eggs.

The general colour of the Çariama is pale brown, with numerous irregular splashes of dark brown. The crest is always held erect, and the feathers of the forehead project slightly over the beak. The wing is blacker brown than the rest of the body, and is covered with narrow white streaks, dotted with black. The under parts are greyish white, the bill is red, and the legs orange. In total length it measures about thirty-two inches.

ALTHOUGH in former days tolerably common in England, the CRANE has now, with the bustard, almost disappeared from this land, a single specimen being seen at very long and increasing intervals. In some parts of England and Ireland the popular name of the heron is the Crane, so that the occasional reports which sometimes find admission into local newspapers respecting the Crane have often reference, not to that bird, but to the heron.

The Crane is found in various parts of the continent of Europe, migrating from place to place, and flying in large flocks at a great elevation in the air. They continue their aerial journeys for great distances, and seldom descend but for the purpose of feeding. When they alight, it is generally on marshy ground, the banks of rivers, or the coasts of the sea, where they can find a bountiful supply of marine and aquatic animals; and sometimes they are attracted by a field of newly-sown corn, among which they make sad havoc, stocking up the seed with their long bills, or eating the newly sprouted blades. The food of the Crane is various, mostly consisting of worms, slugs, frogs, lizards, newts, and similar creatures; but the bird will often feed upon grain and the leaves of different plants.

The voice of the Crane is loud, resonant, and trumpet-like, and has a singular effect when heard from the great elevation at which the bird prefers to fly. The peculiar resonance of the note is caused by a remarkable structure of the windpipe, which is elongated, and instead of running straight down the neck, passes into the breast-bone, lodges between the two plates of bone which form the keel, and, after making some contortions which vary according to the age of the bird, leaves the breast-bone and proceeds as usual to the lungs.

The Crane makes its nest mostly on marshy ground, placing it among osiers, reeds, or the heavy vegetation which generally flourishes in such localities. Sometimes, however, it prefers more elevated situations, and will make its nest on the summit of an old deserted ruin. The eggs are two in number, and their colour is light olive, covered with dashes of a deeper hue and brown. The well-known plumes of the Crane are the elongated tertials, with their long drooping loose webs, which, when on the wings of the bird, reach beyond the primaries.

The forehead, top of the head, and neck are rather dark slaty ash, and a patch of

CRANE.—*Grus cinerea.*

greyish white extends from behind the eyes, partially down the neck on each side. The general surface of the body is soft ashen grey, and the primaries are black. The long plumy tertials form two crest-like ornaments, which can be raised or depressed at will. The eyes are red, and the beak is yellow, with a green tinge. The total length of the adult Crane is about four feet, but it is rather variable in point of size, and the males are rather larger than the females.

THE two following birds are remarkable, not only for their beauty of form and plumage, but for the extraordinary antics in which they occasionally indulge.

The DEMOISELLE, or NUMIDIAN CRANE, is common in many parts of Africa, and has been seen in some portions of Asia and occasionally in Eastern Europe. The movements of this beautiful bird are generally slow and graceful, with a certain air of delicate daintiness about them which has earned for it the title of Demoiselle. But on occasions it is seized with a fit of eccentricity, and puts itself through a series of most absurd gambols, dancing about on the tips of its toes, flapping its wings, and bowing its head in the most grotesque fashion. It may sometimes be seen performing these antics in the Zoological Gardens, but it is very capricious in its habits, and, like the parrot, will seldom perform its tricks when it is most desired to do so.

It is a very pretty bird, the soft texture of the flowing plumage, and the delicate greys of the feathers, harmonizing with each other in a very agreeable manner. The general tint of the plumage is blue-grey, taking a more leaden tone on the head and neck, and offering a beautiful contrast to the snowy-white ear-tufts, issuing from velvety black, which decorate the head. There is also a tuft of long flowing plumes of a deep black grey,

DEMOISELLE CRANE.—*Scops Virgo.*   CROWNED CRANE.—*Baleárica pavonina.*

hanging from the breast. Its secondaries are much elongated, and hang over the primaries and tail-feathers. In height the Demoiselle Crane is about three feet six inches.

The Crowned Crane is even more striking than the demoiselle, its coronet of golden plumes and the scarlet cheeks making it a very conspicuous bird.

This species is a native of Northern and Western Africa, where it is usually found in swampy and marshy localities, which it frequents for the purpose of feeding on the insects, molluscs, reptiles, and fishes which are to be caught abundantly in such places. Like the demoiselle, the Crowned Crane occasionally indulges in fantastic gambols, and on account of the conspicuous crest and general aspect of the bird, they have an effect even more ludicrous.

In captivity the Crowned Crane thrives well, and its habits can be readily watched. At the Zoological Gardens there are some fine specimens of these birds, and an hour may be pleasantly spent in watching their proceedings. Sometimes they rest still and stately, one leg tucked under them quite out of sight, and the body balanced on the other. Sometimes they like to sit on their bent legs, their feet projecting far in front of them, and their knees, or rather their ankles, sustaining the weight of the body. At another time they will walk majestically about their inclosure, or begin their absurd dances, while a very favourite amusement is to run races at opposite sides of the wire fence, and then come to a halt, each bird trying which can yell the loudest. The voice is very loud, and has something of a trumpet in its hollow ringing resonance.

The forehead is black, the feathers being short and velvety. From the top of the head rises a tuft of long straight filamentary plumes, of a golden hue, fringed with very delicate black barbules. The skin of the cheek is bare, and the greater part of it is bright scarlet, the upper part being white, and running into a small wattle on the throat. The general colour of the plumage is slaty grey, and the primaries and quill-feathers of the tail are black, the long secondaries are brown and the wing-coverts snowy white. The height of this species is about four feet.

THE three birds represented on the accompanying illustration afford examples of the Herons. Two of them are natives of England, though yearly becoming more scarce owing to the rapid spread of cultivation, and the third is a European species, but not as yet known on our British Islands.

The EGRET is a native of several parts of America, having its principal residence in the southern portions of that continent, and visiting the more northern districts during several months of the year, arriving generally about February or March. As it finds its food among inundated and swampy grounds, it is generally seen haunting the rice-fields, the marshy river-shores, and similar localities, and seldom if ever visits the high inclosed regions. The food of the Egret consists of the smaller mammalia, little fish, frogs, lizards, snakes, and insects. It is a handsome and elegant bird, and is conspicuous among the low marshy grounds which it frequents, on account of its large size and snowy plumage.

The beautiful loose feathers of the train, which fall from the shoulders over the back, are not fully developed until the third year, and are then greatly in request for feather brushes, either to dust delicate furniture or to flap away the flies. The old birds are, however, so very wary that they cannot be approached without great difficulty, and these brushes are therefore sold at a high price. The train-feathers are also employed in the decoration of head-dresses. The Egret breeds chiefly in extensive cedar-swamps, placing its nest on the branches of trees, and laying three or four large pale blue eggs. The young are usually hatched about the end of June; and when they are strong enough to walk about, they associate in little flocks of twenty or thirty in number.

The colour of the Egret is pure snowy white, with the exception of the train, which has a creamy yellow tinge. The feathers of the train are so long that when they are fully developed they hang over the tail and quite conceal it. The long sharp bill is nearly six inches in length, and its colour is rich golden orange, darkening into black at the tip. The long legs are black and the eye is rather pale orange. In total length the adult bird is about four feet, if the measurement be taken to the end of the train. Both sexes have the same plumage.

THE well-known HERON was once one of our commonest English birds, but on account of the draining of swamps and their conversion into fertilized and habitable ground, is now seldom to be seen except in certain localities which still retain the conditions that render them so acceptable to this bird. There are some places where Herons are yet plentiful, especially those localities where the owner of the land has established or protected the nests, or where a wide expanse of wild uncultivated ground affords them a retreat. Only a few days ago I came suddenly on three of these beautiful birds fishing quietly in the Avon, and permitting my approach within a few yards before they spread their wide wings for flight.

The food of the Heron consists mostly of fish and reptiles, but it will eat small mammalia, such as mice, or even water-rats. In the stomach of one of these birds were found seven small trout, a mouse, and a thrush. Eels also are a favourite food of the Heron, but on account of their lithe bodies and active wrigglings are not so easy to despatch as ordinary fish, and are accordingly taken on shore and banged against the ground until disabled. Dr. Neill, quoted by Yarrell, mentions a curious instance of the Heron feeding on young water-hens. "A large old willow-tree had fallen down into the pond, and at the extremity, which is partly sunk in the sludge and continues to vegetate, water-hens breed. The old cock Heron swims out to the nest and takes the young if he can. He has to swim ten or twelve feet, where the water is between two and three feet deep. His motion through the water is slow, but his carriage stately. I have seen him

EGRET.—'*Ardea egretta.*   HERON.—'*Ardea cinerea.*   BITTERN.—*Botaurus stelláris.*

fell a rat at one blow on the back of the head, when the rat was munching at his dish of fish."

Like many other birds, the Heron is able to disgorge the food which it has swallowed, and resorts to this measure when it is chased by birds of prey while going home after a day's fishing.

While engaged in its search for food, the Heron stands on the water's edge mostly with its feet or foot immersed, and there remains still as if carved out of wood, with its neck retracted, and its head resting between the shoulders. In this attitude its sober plumage and total stillness render it very inconspicuous, and as it mostly prefers to stand under the shadow of a tree, bush, or bank, it cannot be seen except by a practised eye, in spite of its large size. The back view of the bird while thus standing partakes largely of the ludicrous, and reminds the observer of a large jargonelle pear with a long stalk stuck in the ground. Sometimes it likes to squat on its bent legs, the feet being pushed out in front, and the knees, or rather ankles, bent under its body. It generally suns itself in this position, partially spreading the wings and slightly shaking them. Usually it sits with the head resting on the shoulders; but if alarmed at any unexpected sound, it shuts its wings, stretches its neck to its utmost extent, and then presents a most singular aspect.

The flight of the Heron is grand and stately, for the wings are long and wide, and in spite of the long neck and counterbalancing legs, the bird moves through the air with a-noble and rapid flight. It is curious to see a Heron pass directly overhead. The head, body, and legs, are held in a line, stiff and immovable, and the gently waving wings carry the bird through the air with a rapidity that seems the effect of magic.

The long beak of the Heron is very sharp and dagger-like, and can be used with terrible force as an offensive weapon. The bird instinctively aims its blow at the eye of its adversary, and if incautiously handled is sure to deliver a stroke quick as lightning at the captor's eye. There seems to be some attraction in the eye, for a gentleman who turned a tame Heron into an aviary where five owls were kept, found next day that the Heron had totally blinded four owls and only left the fifth with a single eye. Even the game-cock can make nothing of the Heron, as has been seen in a short battle that raged between those birds. The cock made his first fly very boldly, but not being used to such long-legged foes, missed his stroke. Returning to the attack, he was met by a blow from the Heron which astonished him to such a degree that he declined further combat and ever afterwards avoided so unpleasant an antagonist. The beak of a species of Heron set upon a stick is used by some savage tribes as a spear.

The nest of the Heron is almost invariably built upon some elevated spot, mostly the top of a large tree, but sometimes on rocks near the coast. It is a large and rather clumsy looking edifice, made of sticks and lined with wool. The eggs are from four to five in number, and their colour is pale green.

The general colour of the Heron is delicate grey on the upper surface of the body, with the exception of the primaries, which are black, and the tail, which is deep slaty grey. The head is very light grey, and the beautiful long plume is dark slaty blue. The throat and neck are white, covered along the front with dashes of dark blue-grey, and at the junction of the neck with the breast the white feathers are much elongated, forming a pendent tuft. The breast and abdomen are greyish white covered with black streaks. The total length of the bird is about three feet. On the inside of the middle claw of each foot the horny substance is developed into a sort of shallow-toothed comb, the use of which is very problematical. This peculiarity runs through the genus, and several objects have been assigned to it, combing the plumage being the favourite theory, but clearly untenable on account of the shortness of the teeth.

The BITTERN is now seldom seen in this country, partly because it is a rare bird and becoming scarcer almost yearly, and partly because its habits are nocturnal, and it sits all day in the thickest reeds or other aquatic vegetation. The marshy grounds of Essex seem to be the spots most favoured by this bird at the present day, although specimens are annually killed in various parts of the country.

In habits and food, the Bittern resembles the heron, except that it feeds by night instead of by day. Like that bird it uses its long sharp beak as a weapon of offence, and chooses the eye of its adversary as the point at which to aim. The feet and legs are also powerful weapons, and when disabled from flight, the Bittern will fling itself on its back, and fight desperately with foot and bill.

NANKEEN NIGHT HERON.—*Nycticorax Caledónicus.*

The nest of the Bittern is placed on the ground near water, and concealed among the rank vegetation that is found in such localities. It is made of sticks and reeds, and generally contains about four or five pale brown eggs. The voice of the Bittern varies with the season of year. Usually it is a sharp harsh cry uttered on rising, but in the breeding season the bird utters a loud booming cry that can be heard at a great distance.

The general colour of this fine bird is rich brownish buff, covered with irregular streaks and mottlings of black, dark brown, grey, and chestnut. The top of the head is black with a gloss of bronze, the cheeks are buff, and the chin white tinged with buff. Down the front of the neck the feathers are marked with bold longitudinal dashes of blackish and reddish brown, and the feathers of the breast are dark brown broadly edged with buff. The under surface of the body is buff streaked with brown, the beak is greenish yellow, and the feet and legs are green. In total length the Bittern measures about thirty inches. Several species of herons have been seen in England, nine being mentioned by Yarrell, including one species of Egret, two Bitterns, and a Night Heron.

The Night Herons derive their name from their nocturnal habits. One of these, the Common Night Heron (*Nycticorax Europœus*), has several times been found in England; though its usual residence is on the continents of Europe, Asia, Africa, and North America. In North America it is common, and an admirable description of its habits may be found in the works of Wilson and Audubon.

The NANKEEN NIGHT HERON is a native of Australia, and is thus described by Mr. Gould.

" This beautiful species is universally dispersed over the continent of Australia, but is far less abundant over the western than over the eastern coast. In the summer latitudes it is only a summer visitant, arriving in New South Wales and South Australia in August and September, and retiring again in February. As its name implies, it is nocturnal in its habits, and from its frequenting swamps, inlets of the sea, the sedgy banks of rivers, and other secluded situations, it is seldom seen. On the approach of morning it retires to the forests, and perches among the branches of large trees, where, shrouded from the heat of the sun, it sleeps the whole day, and when once discovered is easily taken, as it seldom moves unless shot at, or driven from its perch by some other means, and when forced to quit its perch, it merely flies a short distance and again alights. Its flight is slow and flapping, and during its passage through the air the head is drawn back between the shoulders and the legs are stretched out backwards after the manner of true Herons. When perched upon the trees or resting on the ground, it exhibits none of the grace and elegance of those birds, its short neck resting on the shoulders.

When impelled by hunger to search for a supply of food it naturally becomes more animated, and its actions more active and prying ; the varied nature of its food in fact demands some degree of activity—fishes, water-lizards, crabs, frogs, leeches, and insects being all partaken of with equal avidity. It breeds in the months of November and December, and generally in companies like the true Herons; the favourite localities being the neighbourhood of swampy districts, where an abundant supply of food is to be procured ; the branches of large trees, points of shelving rocks and caverns, are equally chosen as a site for the nest, which is rather large and flat, and generally composed of crooked sticks loosely interwoven.

The eggs, which are usually three in number, are of a pale green colour, and average two inches and five-eighths in length by one inch and a half in breadth. So little difference exists in the colouring of the sexes, that it is extremely difficult to distinguish the male from the female, and never with certainty unless dissection be resorted to ; both have the three beautiful elongated occipital plumes, the use of which except for ornament is not easily imagined. The young on the contrary differ so greatly from the adult, that they might readily be regarded as a distinct species."

The general colour of the adult bird is a rich cinnamon-brown, the top of the head and nape of the neck are black, and the head-plumes, cheeks, a stripe over the eye, and whole of the lower surface are pure white, melting softly into cinnamon-brown on the sides of the neck. The bare skin round the eye is greenish yellow, and the eyes orange. The bill is black, with a little yellow at the tip or on the lower mandible, and the legs and feet are rich yellow. As is frequently the case among the feathered tribes, the plumage of the young bird, instead of being adorned with broad uniform tints, is richly mottled and streaked, the upper surface being buff streaked with deep brown, and the under surface ochry white diversified with a dark stripe down the centre of each feather. The primaries of the wings and quill-feathers of the tail are very dark chestnut at their base, deepening into black near their extremities, which are buff-white.

THE very remarkable BOAT-BILL HERON inhabits Southern America, and is tolerably plentiful in Guiana and Brazil.

It derives its popular name from the singular form of its beak, which, although it really preserves the characteristics of the Heron's bill, is modified after a rather strange fashion, probably for the purpose of aiding it in its search after food. Generally the beak is straight; slender, and sharp, but in this case, although it retains the same amount of substance, its shape is materially altered. Both mandibles are much shortened, rather flattened, and greatly hollowed, so as to assume the aspect of a pair of boats laid upon each other, gunwale to gunwale, the keel being well represented by the corresponding portion of the upper mandible.

This bird is generally found near water, haunting the rivers, marshes, and swamps, where it finds ample supplies of food. Sometimes it traverses the sea-coast, picking up the various crustacea that are to be found at low water, but its usual places of resort are rivers and inland swamps. Its mode of angling is not unlike that of the kingfisher, as the Boat-bill perches upon some branch that overhangs the water, and thence pounces

BOAT-BILL.—*Cáncroma cochleária.*

upon the prey below. It is not a large bird, the body being hardly bigger than that of a common duck, and the legs are rather short in proportion to the size of the body.

The adult male bird has the top of the head decorated with a long and full plume of jetty black feathers, pointed and drooping over the back. In the female the elongated feathers are wanting. The tuft or plume of the neck and breast is greyish white. The feathers of the back are elongated, and their colour is grey with occasionally a wash of rusty red; there is also a patch of the same hue, but of a deeper tone, upon the middle of the under surface. The tail is white and the sides black. The bill is blackish brown, and the legs nearly of the same colour, but not quite so dark. Specimens of this bird have been kept in England, and were fed principally upon fish.

THE well-known SPOONBILL affords another instance of the endless variety of forms assumed by the same organ under different conditions; both the beak and the windpipe being modified in a very remarkable manner.

The Spoonbill has a very wide range of country, being spread over the greater part of Europe and Asia, and inhabiting a portion of Africa. Like the bird to which it is closely allied, this species is one of the waders, frequenting the waters, and obtaining a subsistence from the fish, reptiles, and smaller aquatic inhabitants, which it captures in the broad spoon-like extremity of its beak. It is also fond of frequenting the sea-shore, where it finds a bountiful supply of food along the edge of the waves and in the little pools that are left by the retiring waters, where shrimps, crabs, sand-hoppers, and similar animals are crowded closely together as the water sinks through the sand. The bird also eats some vegetable substances, such as the roots of aquatic herbage, and when in confinement will

SPOONBILL.—*Platalea leucoródia.*

feed upon almost any kind of animal or vegetable matter, providing it be soft and moist. The beak of an adult Spoonbill is about eight inches in length, very much flattened, and is channelled and grooved at the base. In some countries the beak is taken from the bird, scraped very thin, and polished, and is then used as a spoon, and is thought a valuable article, being sometimes set in silver.

It has often been found in this country, but is now very scarce, owing to the increasing drainage of marshy soil. The breeding-places of the Spoonbill are usually open trees, the banks of rivers, or in little islands and tufts of aquatic herbage. In the latter cases the nest is rather large, and is made of reeds piled loosely together, and set on a foundation of water-weeds heaped sufficiently high to keep the eggs from the wet. There is no lining to the nest. The eggs are generally four in number, and their colour is greyish white, spotted with rather pale rusty brown.

The Spoonbill seems to have no power of modulating its voice, a peculiarity which is explained by the structure of the windpipe. Upon dissecting one of these birds, the windpipe is seen to be bent into a kind of 8-like shape, the coils not crossing, but just applied to each other, and held in their place by a thin membrane. At the junction of the windpipe with the bronchial tubes that communicate with the lungs, there is none of the bony structure nor the muscular development by which the modulations of the voice are effected, and which are found so strongly developed in the singing and talking birds. This curious formation does not exist in the very young bird, and only assumes its perfect form when the Spoonbill has arrived at full age.

The colour of the adult bird is pure white, with the slightest imaginable tinge of soft pink. At the junction of the neck with the breast there is a band of buffy yellow. The

STORK.—*Ciconia alba.*

naked skin on the throat is yellow, the eyes are red, the legs and feet black, and the bill yellow at the expanded portion, and black for the remainder of its length. The total length of the male bird is about thirty-two inches, but the female is not quite so large, and her crest is smaller than that of the other sex. There are six or seven known species of these curious birds.

The STORK is another of the birds which, in the olden days, were tolerably frequent visitors to the British Islands, but which now seldom make their appearance in such inhospitable regions, where food is scarce and guns are many.

It is sufficiently common in many parts of Europe, whither it migrates yearly from its winter quarters in Africa, makes its nest and rears its young. In most countries it is rigidly protected by common consent; partly on account of the service which it renders in the destruction of noisome reptiles and unpleasant offal, and partly because it is surrounded with a kind of halo of romantic traditions handed down from time immemorial to successive generations.

The Stork is not slow in taking advantage of its position, and attaches itself to man and his habitations, building its huge nest on the top of his house, and walking about in his streets as familiarly as if it had made them. It especially parades about the fish-markets, where it finds no lack of subsistence in the offal ; and in Holland, where it is very common, it does good service by destroying the frogs and other reptiles which would be likely to become a public nuisance unless kept down by the powerful aid of this bird.

The habits of the Stork are well told by Colonel Montague in his account of a Black Stork (*Ciconia nigra*) domesticated by him.

" Like the white Stork, it frequently rests upon one leg, and if alarmed, especially by the approach of a dog, it makes a considerable noise by reiterated snappings of the bill, similar to that species. It soon became docile, and would follow its feeder for its favourite morsel, an eel. When very hungry, it crouches, resting the whole length of the legs upon the ground, and suppliantly seems to solicit food by nodding the head, flapping its un-wieldy pinions, and forcibly blowing the air from the lungs with audible expirations. Whenever it is approached, the expulsion of air, accompanied by repeated noddings of the head, is provoked.

The bird is of a mild and peaceful disposition, very unlike many of its congeners, for it never makes use of its formidable bill offensively against any of the companions of its prison, and even submits peaceably to be taken up without much struggle. From the manner in which it is observed to search the grass with its bill, there can be no doubt that reptiles form part of its natural food ; even mice, worms, and the larger insects probably add to its usual repast. When searching in thick grass or in the mud for its prey, the bill is kept partly open ; by this means I have observed it take eels in a pond with great dexterity ; no spear in common use for taking that fish can more effectually secure it between its fangs than the grasp of the Stork's mandibles. A small eel has no chance of escaping when once roused from its lurking-place.

But the Stork does not gorge its prey instantly, like the cormorant ; on the contrary, it reitres to the margin of the pond, and there disables its prey by shaking and beating it with its bill before it ventures to swallow it. I never observed this bird attempt to swim, but it will wade up to the belly and occasionally thrust the whole head and neck under water after its prey. It prefers an elevated spot on which to repose ; an old ivy-bound weeping willow that lies prostrate over the pond is usually resorted to for that purpose. In this quiescent state the neck is much shortened by resting the hinder part of the head on the back, and the bill rests on the fore part of the neck, over which the feathers flow partly so as to conceal it, making a very singular appearance."

The Stork is fond of making its nest upon some elevated spot, such as the top of a house, a chimney, or a church spire ; and in the ruined cities of the East, almost every solitary pillar has its Stork's nest upon the summit. The nest is little more than a heterogeneous bundle of sticks, reeds, and similar substances heaped together, and with a slight depression for the eggs. These are usually three or four in number, and their colour is white with a tinge of buff. The young are puffy, big-beaked, long-necked, ungainly little things, and remain in their lofty cradle until they are well fledged and able to achieve the downward flight. The mother-bird is exceedingly devoted to her young, and there are many well-known tales of this parental affection. On account, probably, of this trait of character, the Stork is looked upon with a feeling of reverence in many countries, and is encouraged to build its nest on the houses, the inhabitant thinking that the bird will bring him good fortune.

The flight of the Stork is extremely high, and the birds fly in large flocks, in some instances numbering many thousand individuals. So great an aerial assembly of such large birds necessarily causes a loud and peculiar rushing sound of huge wings ; but except an occasional sharp clattering of the beaks, the flocks make no noise. Like many of the long-legged birds, the Stork, when resting, stands on one leg, its neck doubled back, and its head resting on its shoulder.

The colour of the adult Stork is pure white, with the exception of the quill-feathers of the wings, the scapularies and greater wing-coverts, which are black. The skin round the eye is black, the eyes are brown, and the beak, legs, and toes red. The length of the full-grown bird is about three feet six inches, and when erect, its head is about four feet from the ground.

ADJUTANT. - *Leptoptilos Argala.*

SOME remarkable members of this group now come before our notice. The first is the well-known ADJUTANT or ARGALA of India, the former name being derived from its habit of frequenting the parade-grounds.

This fine bird is notable for the enormous size of the beak, which is capable of seizing and swallowing objects of considerable size, a full-grown cat, a fowl, or a leg of mutton being engulfed without any apparent difficulty. The Adjutant is a most useful bird in the countries which it inhabits, and is protected with the utmost care, as it thoroughly cleans the streets and public places of the various offal which is flung carelessly in the way, and would be left to putrefy but for the constant services of the Adjutant and creatures of similar habits. The vulture is valuable in devouring dead animals of a large size, as its beak is capable of tearing the hide and flesh from the bones, which are in their turn the prey of the hyæna; but the Adjutant is chiefly important in swallowing the refuse of slaughtered animals, and killing snakes and other unpleasant reptiles. It is remarkable that the bird, though very far removed from the vulture, should have a decidedly vulturine aspect; its nearly naked head and neck adding greatly to the semblance.

The attitudes assumed by the Adjutant are varied, and generally partake of the grotesque. It has a curious habit of airing itself on a hot day, by standing still with the huge beak drooping towards the ground and nearly touching the earth, and its wings stuck out straight from the body. In this odd attitude it will remain for a considerable time, immovable as if carved in stone, and has about as grotesque an appearance as can well be imagined. Sometimes it squats on the ground with its legs tucked under its body, and sits looking about it with a superb air of dignity as of an enthroned monarch. Sometimes it stalks menacingly along, its neck stretched to the utmost, its head thrust forward and its huge bill open, looking a most formidable creature.

It is, however, a cowardly kind of bird, and its assumption of valour is of the most flimsy description, for it will run away from a child if boldly faced, and would as soon face a bantam cock as a tiger. Some enemies, however, from which man would flee, are attacked and killed by the Adjutant, who thus redeems himself from a wholly pusillanimous character. Serpents fall an easy prey to this bird, which has a fashion of knocking them over before they can strike, and after battering them to death swallows them whole. During the inundations the Adjutants are invaluable, as they follow the course of the rising waters, and make prey of the reptiles that are driven from their holes by the floods.

The capacity of the Adjutant's stomach seems to be almost unlimited, and its digestion is so rapid that it can consume a very large amount of food daily. It will swallow a whole joint of meat, or even so impracticable a subject as a tortoise, its stomach being endowed with the power of dissolving all the soft and digestible parts, and ejecting the indigestible, such as the shell and bones.

It is easily tamed, and soon attaches itself to a kind owner; sometimes, indeed, becoming absolutely troublesome in its familiarity. Mr. Smeathman mentions an instance where one of these birds was domesticated, and was accustomed to stand behind its master's chair at dinner-time and take its share of the meal. It was, however, an incorrigible thief, and was always looking for some opportunity of stealing the provisions, so that the servants were forced to keep watch with sticks over the table. In spite of their vigilance it was often too quick for them; and once it snatched a boiled fowl off the dish and swallowed it on the spot.

The exquisitely fine and flowing plumes, termed "Marabou feathers," are obtained from the Adjutant and a kindred species, the Marabou of Africa (*Leptoptilos Marabou*).

The general colour of the Adjutant is delicate ashen grey above and white beneath. The great head and proportionately large neck are almost bare of covering, having only a scanty supply of down instead of feathers. From the lower part of the neck hangs a kind of dewlap, which can be inflated at the will of the bird, but generally hangs loose and flabby.

The JABIRUS rank among the giants of the feathered race. They are very similar in general form to the marabous, but may be distinguished from them by the form of bill, which slightly turns up towards the extremity. The head and part of the neck are also nearly destitute of feathers. There are very few species known, and they all seem to have similar habits; haunting the borders of lakes, marshy grounds, and the banks of rivers, where they find abundance of the fish and aquatic reptiles on which they feed. Of one species, the AUSTRALIAN JABIRU, Dr. Bennett has treated so fully and with such graphic powers of narration, that a condensation of his interesting account must be transferred to these pages. The whole narrative may be found in his "Gatherings of a Naturalist in Australia." One of these birds was taken at Port Macquarie and brought safely to Dr. Bennett's home.

"The first evening it was at my house, it walked into the hall, gazed at the gas-lamps which had just been lighted, and then proceeded to walk upstairs, seeking for a roosting-

AUSTRALIAN JABIRU.—*Myctéria Austrális.*

place; but not liking the ascent came quickly down again, returned into the yard, and afterwards went to roost in the coach-house between the carriages, to which place it now retires regularly every evening soon after dark. It may always be found in that part of the yard where the sun is shining, and with its face invariably directed towards it. When hungry it seeks for the cook, who usually feeds it; and if she has neglected its food, looks into the kitchen as if to remind her of her neglect, and waits quietly, but with a searching eye, during the time the meat is cutting up, until it is fed.

It is amusing to observe this bird catch flies; it remains very quiet as if asleep, and on a fly passing, it is snapped up in an instant. The only time I observed any manifestation of anger in it, was when the mooruks were introduced into the yard where it was parading about. These rapid, fussy, noisy birds, running about its range, excited its indignation; for on their coming near, it slightly elevated the brilliant feathers of the head, its eyes became very bright, it ruffled its feathers, and chattered its mandibles, as if about to try their sword-like edge upon the intruding mooruks, but the anger subsided without further demonstration than an occasional flapping of its powerful wings. One day, however, on one of the mooruks approaching too near him, he seized it by the neck with his mandibles, on which the mooruk ran away and did not appear in any way injured.

When he was first placed in the yard where some poultry were kept, he stared at the fowls, and they ran away on his approach, although he did not make the least attempt to molest them; and when striding round the yard, all the poultry fled before him, although it did not appear to be an intentional chase on his part.

There happened to be a pugnacious, fussy little bantam-cock in the yard, who would not permit the intrusion of any stranger, and on seeing the Jabiru, he strutted up with expanded and fluttering wings and ruffled feathers in a violent state of excitement, cackling and screaming most vehemently, and making efforts as energetic as so diminutive a bird was capable of, to frighten and drive him out of the yard. The Jabiru with his keen bright eyes regarded the little fluttering object with cool contempt, and walked about as before; the bantam followed. At last the Jabiru turned and strode after the consequential little urchin as if to crush him under his feet; when the bantam, seeing matters take this serious turn, made off as fast as possible, like all little bullies, and did not again venture to attack so formidable an opponent. In a few days the Jabiru became quite domesticated among the poultry, and they evinced no fear; even the little bantam tolerated his presence, but whether from fear or affection I know not.

This bird is as tame as my Native Companion when in captivity, but it will not follow any one about as that bird will, nor has it uttered any sound; it seems to be voiceless.

The bird appears timid when any one is looking at him from a short distance, and he then watches acutely all the actions of the intruder; but when startled by any one coming suddenly upon him, he appears frightened, and spreads his wings as if preparing for flight; it is then possible, by a little activity, to capture him by his long bill and wings. When the mooruks came too close to him, he looked at them with flashing eyes, and flapped his wings as if to express his contempt towards them on account of their wingless condition, and at the same time the mooruks spread their rudimentary wings, as if to show that they have some stumps resembling wings, and appeared proud of their appendages also.

When the Jabiru was sunning himself as usual, and any of the mooruks came between him and the sun, he manifested great indignation at their intrusion by clattering his beak, ruffling his feathers, and flapping his wings at them; if these hints were disregarded, he gave them a blow with his beak, which soon made them walk away.

The Jabiru was occasionally observed lying upon its breast, with its legs doubled up underneath so as to resemble a large goose with a most disproportionate size of bill. I have noticed him watch the ground very attentively under the trees, and then dart his bill into the ground and bring up larvæ, which I found to be those of locusts (*Tettigoniæ*, or Treehoppers). When the bird observed a slight motion of the soil, he darted his beak down and devoured the insect as it was emerging from the soil. On any of these insects falling from the trees upon the ground, he would rapidly pick them up and devour them. On giving him one, he first crunched it between his mandibles, and throwing it up caught and devoured it. He appeared to relish these insects very much, and was eager to procure them.

He became latterly so familiar and domesticated that he would permit the person who was in the habit of feeding him to touch and examine his plumage and wings. When called to be fed, he ran from any part of the yard, and so regular was he in his habits, that when not called at the usual hour, he would stand at the place where he was accustomed to be fed, until his meat was given to him. When the person who fed him called him, he clapped his mandibles and ran up. He seemed to delight in standing in the rain, and did not appear in the least uncomfortable when his feathers were dripping wet. He frequently slept in the open air all night, preferring it to the shelter of the coach-house.

He strutted about the yard a long time after dark. When caught by the wings or otherwise annoyed, he displayed his anger by no other sound than a loud and violent clattering of the mandibles, nor did he attempt any act of aggression upon his captors with his powerful beak. He would often run about the yard, spreading and fluttering his wings, merely for exercise."

The Australian Jabiru appears to be a very rare bird; and as it is extremely wary, and haunts wide expanses where but little cover can be found, it can with difficulty be approached. The natives, with their eagle eyes, their snake-like movements, and the exhaustless patience of men to whom time is of no value, manage to creep within range of their weapons; but even to them the task is a difficult one, and to Europeans almost impracticable. One good sportsman, who succeeded at last in killing a Jabiru, followed it for several days before he could get within long range of the suspicious bird.

The food of this species mostly consists of fish, and eels seem to be the favourite diet. Ordinary fish it swallows at once, but eels and gar-fish are battered about until dead before the bird attempts to devour them. Nearly two pounds of eels and small fish have been found in the stomach of a shot Jabiru.

In its colouring the Australian Jabiru is a very handsome bird, and its movements are quiet, majestic, easy and graceful. The large head and neck are rich shining green, changing to rainbow tints of violet and purple upon the back of the head, the feathers gleaming in the sun with a light metallic radiance. "The greater wing-coverts, scapularies, lower part of the back and tail are dark brown mixed with rich bluish green, which changes in the adult to a rich glossy green tinged with a golden lustre. The smaller wing-coverts, lower part of the neck and back, and upper part of the breast are white speckled with ashy brown, but become pure white in the adult: lower part of the breast, thighs, and inner part of the wings, white. Eyes brilliant and hazel in colour. The legs are blackish with a dark tinge of red, becoming of a bright red colour in the adult; and when the bird flies with the legs stretched out, looking like a long red tail. . . . My specimen measures three feet ten inches to the top of the head, and is not yet full grown; they are said to attain four or five feet in height." The specimen belonging to Dr. Bennett died after a captivity of about seven months, nearly four of which were passed in Dr. Bennett's residence. The cause of his death was not known—probably the diet might have been injurious.

THE singular WHALE-HEADED STORK is the most striking of its tribe.

This bird lives in Northern Africa, near the Nile, but is seldom seen on the banks of that river, preferring the swampy districts to the running water. Mr. Petherick, who first brought this bird to England, found it in the Rhol district, about latitude 5° to 8°, in a large tract of country about a hundred and fifty miles in extent, where the ground is continually swelled by rains, and has by degrees modified into a huge morass, some parts flooded with water, others blooming with vegetation, and the whole surrounded by thick bush. "This spot," writes Mr. Petherick in his "Egypt, the Soudan, and Central Africa," "is the favourite home of the Balæniceps."

These birds are seen in clusters of from a pair to perhaps one hundred together, mostly wading in the water; and when disturbed, will fly low over its surface and settle at no great distance. But if frightened and fired at, they rise in flocks high in the air, and after hovering and wheeling around settle on the highest trees, and as long as their disturbers are near, will not return to the water. Their roosting-place at night is, to the best of my belief, on the ground.

Their food is principally fish and water-snakes, which they have been seen by my men to kill and devour. They will also feed on the intestines of dead animals, the carcases of which they easily rip open with the strong hook of their upper bill.

Their breeding-time is in the rainy season, during the months of July and August, and the spot chosen is in the reeds or light grass immediately on the water's edge or on some small elevated and dry spot entirely surrounded by water. The bird before laying scrapes a hole in the earth, in which, without any lining of grass or feathers, the female deposits her eggs. Numbers of these nests have been robbed by my men of both eggs

WHALE-HEADED STORK.—*Balæniceps Rex.*

and young, but the young birds so taken have invariably died. After repeated unsuccessful attempts to rear them, continued for two years, the eggs were eventually hatched under hens, which were procured at a considerable distance from the Raik negroes.

As soon as the hens began to lay, and in due time to sit, a part of their eggs were replaced with half the number of those of the Balæniceps, as fresh as possible from the nest, the locality of which was previously known, and several birds were successfully hatched. These young birds ran about the premises of the camp, and, to the great discomfort of the hens, would persist in performing all sorts of unchickenable manœuvres, with their large beaks and extended wings, in a small artificial pool constantly supplied with water by several negresses retained for their especial benefit. Negro boys were also employed to supply their little pond with live fish, upon which, and occasionally the intestines of animals killed for our use, chopped into small pieces, they were reared."

The chief point in this fine bird is the huge bill, which, from its resemblance in size and shape to a shoe, has gained for its owner the title of Shoe-bird. It is enormously expanded at each side of the beak, the edges of the upper mandible overhang those of the lower, and its tip is furnished with a large hook, curved and sharp as that of an eagle, and well suited for tearing to pieces the substances on which the bird feeds. Its colour is brown, mottled profusely with a deep mahogany tinge. The general colour of the plumage is dark slaty grey above, each feather being edged with a narrow band of greyish white. The feathers of the front of the neck are pointed, very dark in the centre, and broadly edged with grey. The under surface is grey. In the British Museum a skeleton of this bird is placed near the stuffed specimen, and gives an excellent idea of the singular formation of the beak and head.

SACRED IBIS.—*Ibis religiósa.*   GLOSSY IBIS.—*Ibis falcinellus.*

The SACRED IBIS is one of a rather curious group of birds. With one exception they are not possessed of brilliant colouring, the feathers being mostly white and deep purplish black. The Scarlet Ibis, however, is a most magnificent, though not very large bird, its plumage being of a glowing scarlet, relieved by a few patches of black.

The Sacred Ibis is so called because it figures largely in an evidently sacred character on the hieroglyphs of ancient Egypt. It is a migratory bird, arriving in Egypt as soon as the waters of the Nile begin to rise, and remaining in that land until the waters have subsided, and therefore deprived it of its daily supplies of food. The bird probably owes its sacred character to the fact that its appearance denotes the rising of the Nile, an annual phenomenon on which depends the prosperity of the whole country.

Sometimes the Ibis stalks in solitary state along the banks of the river or the many watercourses that intersect the low country, but sometimes associates in little flocks of eight or ten in number. Its food consists mostly of molluscs, both terrestrial and aquatic, but it will eat worms, insects, and probably the smaller reptiles. The Ibis was at one time thought to kill and eat snakes, and this idea was strengthened by the fact that Cuvier detected the scales and bones of snakes within a mummied corpse of an Ibis which was found in the tombs of Egypt and which is known to be identical with the present species. Recent specimens, however, seldom contain anything but molluscs and insects.

The walk of the Ibis is quiet and deliberate, though it can get over the ground with considerable speed whenever it chooses. Its flight is lofty and strong, and the bird has a habit of uttering a loud and peculiar cry as it passes through the air. By the natives of Egypt it is called the Abou Hannes, *i.e.* Father John, or Abou Menzel, *i.e.* Father Sickle Bill, the former name being in use in Upper and the other in Lower Egypt.

The colour of the adult bird is mostly pure silvery white, the feathers being glossy and closely set, with the exception of some of the secondaries, which are elongated and hang gracefully over the wings and tail. These, together with the tips of the primaries, are deep glossy black, and the head and neck are also black, but being devoid of feathers have a slight brownish tinge, like that of an ill-blacked boot, or an old crumpled black kid glove. While young, the head and neck are clothed with a blackish down, but when the bird reaches maturity, even this slender covering is shed, and the whole skin is left bare. The body is little larger than that of a common fowl.

ANOTHER species, the GLOSSY IBIS, is also an inhabitant of Northern Africa, but is sometimes found in this country, where the fishermen know it by the name of Black Curlew. It is probably the Black Ibis mentioned by Herodotus.

The Glossy Ibis is sometimes found in different parts of America, rarely in the northern States, but of more frequent occurrence in the centre or south. Audubon remarks that he has seen great numbers of these birds in Mexico, where it is a summer resident only. In England a stray specimen or two alight on our shores in the course of the migration, and in the ornithological annals of this country there are few years without the mention of a Glossy Ibis being seen or killed in some part of the British Isles.

The habits and food of the Glossy Ibis are much the same as those of the last-mentioned species, and, like that bird, it was invested while living with sacerdotal honours by the ancient Egyptians, embalmed and honoured after death with a consecrated tomb, in common with the bull, the cat, and the sacred Ibis.

The plumage of the Glossy Ibis varies somewhat according to the age of the bird, so that according to Yarrell the same species has been termed the Glossy Ibis, the Green Ibis, and the Bay Ibis, by various authors, the difference of colour being due to the more or less advanced age of the individual. Both sexes have similar plumage, but the female is smaller than her mate.

In the full-grown bird, the head, neck, and part of the back between the shoulders are dark chocolate, and the wing-coverts and tertials are a still darker brown glossed with purple and green. The quill-feathers of the wings are dark blackish brown glossed with green, and the tail is of a similar hue, but glossed with purple. The breast and under surface of the body are chocolate brown, changing to a duller hue under the wings and upon the under tail-coverts. The beak is dark brown with a tinge of purple, the naked skin round the eyes is greyish green, the eyes are hazel, and the legs and toes green. In total length this species measures not quite two feet. The young bird is more mottled than the adult, and has little of the bright glossiness of the plumage. The head and neck are dull brown streaked with grey, the whole of the upper surface, together with the wings and tail, are dark reddish brown, and there are a few irregular patches of white upon the breast.

The STRAW-NECKED IBIS derives its name from the tuft of stiff naked feather-shafts which hang from the front of the neck and breast, and greatly resemble small yellow straws. These curious feathers, with their light polished, golden surface, afford a pretty contrast to the glossy green black of the chest and wings, and the pure white of the neck and abdomen. The following description of the bird and its habits is written by Mr. Gould, in the " Birds of Australia."

" This beautiful Ibis has never yet been discovered out of Australia, over the whole of which immense country it is probably distributed, as it is more abundant in certain localities at one season than at another ; its presence, in fact, appears to depend upon whether the season be or be not favourable to the increase of the lower animals upon which the vast hordes of this bird feed. After the severe drought of 1839, it was in such abundance on the Liverpool plains, that to compute the number in a single flock was impossible. It was also very numerous on the seaside of the great Liverpool range, inhabiting the open downs and flats, particularly such as were studded with shallow lagoons, through which it would wade knee-high in search of shelled molluscs, frogs, newts, and insects ; independently of the food I have mentioned, it feeds on grasshoppers and insects generally. The natives informed me that sometimes many seasons elapsed without the bird being seen. Where then does it go ? To what country does it pass ?

Does there not exist a vast oasis in the centre of Australia, to which the bird migrates when it is not found in the located parts of the country? We may reasonably suppose such to be the case.

The Straw-necked Ibis walks over the surface of the ground in a very stately manner; it perches readily on trees, and its flight is both singular and striking, particularly when large flocks are passing over the plains, at one moment showing their white breasts, and at the next, by a change in their position, exhibiting their dark-coloured backs and snow-white tails. During the large semicircular sweeps they take over the plains, and when performing a long flight, they rise tolerably high in the air; the whole flock then arrange themselves in the form of a figure or letter similar to that so frequently observed in flights of geese and ducks.

The note is a loud, hoarse, croaking sound, which may be heard at a considerable distance. When feeding in flocks they are closely packed, and from the constant movement of their bills and tails, the whole mass seems to be in perpetual motion. In disposition this bird is rather shy than otherwise; still, with a very little care, numerous successful shots may be made with an ordinary fowling-piece.

The sexes, when fully adult, exhibit the same beautiful metallic colouring of the plumage. The female is, however, smaller, and has the straw-like appendages on the neck less prolonged and less stout than in the male. Mature birds only have the whole of the head and back of the neck quite bare of feathers."

The colouring of the Straw-necked Ibis is very conspicuous, and the lines of demarcation between the different tints are sharply drawn. The head and part of the neck are deep sooty black, which suddenly changes into a beautiful white downy plumage, clothing the remainder of the neck. From the fore part of the neck and throat hang a number of delicate fringe-like feathers.

STRAW-NECKED IBIS.—*Geronticus spinicollis*

The whole of the upper surface is coloured of a deep and glistening green-black, "shot" with purple, and changing its tints at every variation of light. Irregular bars of the same colour as the head are drawn across the back, and the entire under surface is pure white. During the life of this bird the thighs are slightly coloured with crimson, but this tinting soon vanishes after death.

THE two birds which have been chosen to represent the large genus Numénius are well-known inhabitants of this country.

The CURLEW, or WHAUP, is mostly found upon the sea-shore and open moorlands, and partly on account of its wild, shy habits, and partly because its flesh is very delicate and well flavoured, is greatly pursued by sportsmen. These birds are most annoying to a gunner who does not understand their ways, having a fashion of keeping just out of gun range, rising from the ground with a wild mournful cry which has the effect of alarming every other bird within hearing, and flying off to a distance, where they alight only to play the same trick again. Moreover, they are strong on the wing and well feathered, so that they require a sharp blow to bring them down, and necessitate the use of large shot. When thus alarmed they generally skim along at a low elevation, averaging about four or six feet from the ground, and consequently afford little mark.

Sir W. Jardine writes as follows concerning the habits of the Curlew. "They retired regularly inland after their favourite feeding-places were covered. A long and narrow ledge of rocks runs into the Frith (Solway) behind which we used to lie concealed for the purpose of getting shots at various sea-fowl returning at ebb. None were so regular as the Curlew. The more aquatic were near the sea and could perceive the gradual reflux; the Curlews were far inland, but as soon as we could perceive the top of a sharp rock standing above water, we were sure to perceive the first flocks leave the land, thus keeping pace regularly with the change of tides. They fly in a direct line to their feeding-grounds, and often in a wedge shape; on alarm a simultaneous cry is uttered, and the next coming flock turns from its course, uttering in repetition the same alarm-note. In a few days they become so wary as not to fly over the concealed station."

The breeding-grounds of the Curlew are more inland, the locality varying according to the character of the district, wild heath and high hilly grounds being chosen in some places, while marshy and boggy soils are favoured in others. The nest of this bird is very slight, being only a small heap of dry leaves or grasses scraped together under the shelter of a tuft of heather or a bunch of rank grass. There are usually four eggs, placed, as is customary with such birds, with their small ends together, and being much larger at one end than at the other. Their colour is brownish green with some blotches and splashes of dark brown and a darker green. The young are curious little birds, long-legged, short-billed, covered with puffy down, and with very little indications of either wings or tail.

The general colouring of the Curlew is brown, lighter upon the head and neck, and darker upon the back, each feather being darker in the centre than on the edges. The upper tail-coverts are white streaked with brown, the smaller wing-coverts are edged with greyish white, and the tail is grey-white barred with brown. The wings are black, and some of the quills have white shafts. The chin is white, and the under parts are also white, but with a tinge of grey and streaked with short marks of dark brown. The under tail-coverts are white. Both sexes are coloured alike, and the average length is rather more than twenty inches.

AT first sight the WHIMBREL looks something like a diminutive curlew, save that the bill is not so long, so thick, nor so sharply curved as in the preceding species. On account of this resemblance it is in some places known by the name of Half-Curlew, and in others it is called the Jack Curlew. In the Shetland Isles it is known by the popular name of Tang-Whaap.

The habits of the Whimbrel much resemble those of the curlew. Mr. Thompson makes the following remarks respecting this bird. "In spring, Whimbrels, probably for want of good company—no godwits &c. being here—keep generally 'by themselves,' and as such are literally a host; but in autumn they exhibit a very sociable disposition, and are frequently to be found in company with curlews: with godwits too they not uncommonly associate; I have obtained both species at the same shot. At this time they accompany the curlew, and under a safer guardian the most trustworthy of friends could not place them. Never while under the surveillance of the curlew, but only when trusting to themselves, have Whimbrels fallen to my gun. Like that bird, they fly much

WHIMBREL.—*Numénius phœopus.* CURLEW.—*Numénius arquáta.*

about during the autumnal nights, be they dark or moonlight, but they prefer the latter They may always be distinguished from other species by the cry, resembling in sound the word *titterel*, the provincial name applied to them in Sussex. They fly from the sea inland as well as in the opposite direction, and take both courses during every state of the tide ; hence we may consider that they are night-feeding birds."

The general colouring of the Whimbrel is much like that of the curlew, but it may be easily known from that bird by the browner tinge of the light-coloured feathers, the pale brown streak at the top of the head, the dark line from the base of the bill to the eye, and the light streak over the eye and ear-coverts. The quill-feathers of the wing are greyish black and the secondaries are barred with white. The length of the Whimbrel is various according to the individual, but the usual average is about eighteen inches.

THE two species of Godwits are known in England, the Common, or Bar-tailed, and the Black-tailed Godwit. These birds may be known from each other by the peculiarity from which they derive their name, the one species being distinguished by the uniform black hue of the latter two-thirds of the tail, and the other by the brown and grey bars which cross the tail-feathers.

The Godwit is a noisy bird, continually uttering its odd cry, which has been well compared to the word *grutto* rapidly repeated. The flesh of this bird is held in high estimation ; and in some parts of England where these Godwits are found, it is the custom to catch them alive and to keep them for some time before killing them, fattening them well with bread and milk. The Godwit generally haunts marsh lands, where it finds its food by probing the soft oozy soil with its bill. The fens of Lincolnshire are favourite

BLACK-TAILED GODWIT.—*Limósa œgocéphala.*

resorts of this bird, but it may be found occasionally in most localities where the ground is wet and soft and where it can obtain cover.

The flight of the Godwit is strong, and when an intruder comes near the nest, the bird rises into the air and wheels uneasily over the spot after the fashion of the lapwing, uttering the while its screaming cry. The nest is placed on the ground sheltered by herbage, and is made of leaves and grass inartificially put together. The eggs are generally four in number, and their colour is light brown with a greenish tinge, covered with spots and blotches of a darker hue.

The shape of the Godwit is light and graceful, and the tints of its plumage, although not at all brilliant, are rich and soft of their kind and arranged after a pleasing fashion. The general colour of the head, neck, and upper surface is ashen brown, a whitish patch is seen over and in front of the eye, and the feathers of the coverts and tertials are edged with grey-brown. The primaries are dull black with white shafts, and nearly all of them are white towards the base, so that an undefined bar is formed across the wings. The tail is white for the first third, and the remainder is black; the two exterior feathers have, however, more white than the others. The whole under surface of the body is ashen grey, becoming white upon the under tail-coverts. This is the winter plumage; the summer clothing is rather different, and requires a separate description.

In summer the adult male has the head reddish brown streaked with black, diversified by a light brown streak round the eyes and a dark spotted stripe extending from the base of the beak to the eye. The neck is simple ruddy fawn. The back is curiously mottled with rich tints of black and warm chestnut, each feather having the darker hue on the centre and the lighter on the edge. The primaries retain their dull black hue, but the white band is of a purer tone and better defined, and the tail undergoes no alteration. The breast is nearly white, barred with brown of several tints. The young birds have a ruddy colouring on the neck, but are to be distinguished by the ashen brown of the neck and upper part of the breast, the grey-white of the lower portion of the breast, and their smaller size. The average length of the male is sixteen inches, and the female is about one inch longer than her mate.

GREEN SANDPIPER.—*Tótanus óchropus.*　　　　COMMON SANDPIPER.—*Tringóides hypoleuca.*

THE two birds on the accompanying illustration are good examples of the Totaninæ or Sandpipers.

The GREEN SANDPIPER is, like the whole of its tribe, a frequenter of wet and marshy lands, and seems not to be so fond of the sea-shore as many allied species. Salt-water marshes are, however, favourite spots with these birds, and whenever the brackish water spreads from the sea-coast over the adjoining country, as is the case along part of the southern coast of England, there the Green Sandpiper may generally be found. It is a quick and active bird, running about with much agility, and flirting its short tail up and down as it moves along. It is rather noisy, its cry being a shrill whistle remarkably loud in proportion to the size of the bird, and very constantly repeated. When flushed it begins to scream, and flies rapidly away at a low elevation, keeping as much as possible over the water.

The food of the Green Sandpiper consists of worms, and the myriad insects and other small creatures that swarm in moist and watery situations. Generally, this Sandpiper moves off to Northern Europe for the breeding season, but it not unfrequently remains with us for the purpose of rearing its young. The nest is mostly placed on the bank by the side of a stream, hidden by overhanging grass or some such protection. The eggs are four in number and are greyish white with a tinge of green, spotted rather profusely with grey and ruddy brown. In colouring, however, the eggs are variable, the number and colour of the spots differing in individual specimens. The flesh of this Sandpiper is highly esteemed, and in the countries where the bird is plentiful it is largely captured by limed twigs and other snares.

The general colour of the upper surface of this species is brownish green with waved markings of deeper brown. Over the eye runs a rather broad line of white. The primaries are black, and the scapularies and tertials greenish brown profusely spotted with a lighter tint. The tail-feathers are white for the first half of their length, and are then banded with blackish brown. The chin, breast, and abdomen are white, and the front and sides of the neck are also white, but are streaked with narrow lines of grey-brown. In total length the Green Sandpiper is not quite ten inches.

The COMMON SANDPIPER or SUMMER SNIPE is a well-known visitor to this country, and has derived its name of Summer Snipe from its habit of remaining in England only during the summer months, arriving here about April or May and leaving before October.

It is a pretty and lively little bird, running about the edges of rivers, lakes, or ponds, poking in all directions after food, and occasionally swimming or diving when alarmed. Even the young birds, before they can fly, will take instinctively to the water when frightened, and plunging beneath the waves will dive to some distance before they again rise to the surface. I have seen three or four of them follow each other into a river like so many sheep through a gate, disappearing below the water and not emerging within sight. The river was a winding one, and the banks were thickly studded with trees, so that I believe the birds to have emerged just round a neighbouring bend of the stream. Owing to the dark shadows of the trees I could not see the birds when under water, but it is said to employ its wings in urging itself along.

The nest of the Sandpiper is made on the ground, mostly on a bank of a river, or sometimes in a field in the vicinity of water. The mother-bird is careful of her eggs and young; and has a habit of feigning lameness, after the well-known custom of the lapwing, when any intruder comes near her nest. A correspondent of the "Magazine of Natural History" gives an interesting account of the Sandpiper and her little family. "I this year started an old one from her nest at the root of a fir-tree. She screamed out, and rolled about in such a manner, and seemed so completely disabled, that, although perfectly aware that her intention was to allure me from her nest, I could not resist my inclination to pursue her; and in consequence I had great difficulty in finding the nest again.

It was built of a few dried leaves of the Weymouth pine, and contained three young ones, just hatched, and an egg, through the shell of which the bill of the young chick was just making its way. Yet, young as they were, on my taking out the egg to examine it, the little things, which could not have been out of their shells more than an hour or two, set off out of the nest with as much celerity as if they had been running about for a fortnight. As I thought the old one would abandon the egg if the young ones left the nest, I caught them again, and covering them up with my hand for some time, they settled down again. Next day all four had disappeared."

The top of the head, back of the neck, back, upper tail-coverts, and central feathers of the tail are greenish brown mottled with black, each feather being darker on the centre than on the edges. The secondaries are tipped with white, and all the black primaries, except the first, have a patch of greyish white on the inner web. From the base of the beak to the eye runs a dark streak, and a light coloured stripe passes over the eye. The tail is barred with greenish ·black; most of the feathers are tipped and patched with white. The chin is white, the upper part of the breast ashen grey streaked with brown, and the whole of the under surface of the body is snowy white. The average length of the Sandpiper is between seven and eight inches.

Several other members of this sub-family are inhabitants of England; among which may be named the Common and Spotted Redshank, the Wood Sandpiper, the Greenshank, the Spotted and Buff-breasted Sandpipers.

The AVOCET is one of the most remarkable among English birds, and is easily recognisable by its long, curiously curved beak, and its boldly pied plumage.

The Avocet is not a common bird in England, and is now but seldom seen, though in former days it used to be tolerably plentiful on the sea-coasts and in marshy lands. The long and oddly curved beak is very slender and pointed, and from its peculiar shape has earned for its owner the name of Cobbler's Awl Bird. While obtaining its food the Avocet scoops the mud with its beak, leaving sundry unmistakable marks behind; and is called in some countries the Scooper. The food of the Avocet consists almost wholly of worms, insects, and little crustaceans; and while the bird is engaged in the search after these creatures, it paddles over the oozy mud with its webbed feet and traverses the soft surface with much ease and some celerity. The cry of the Avocet is a sharp,

STILT PLOVER.—*Himántopus cándidus.*   AVOCET.—*Recurvirostra avocetta.*

shrill kind of yelp, and is uttered whenever the bird is alarmed. The flight is strong and rapid.

The nest of the Avocet is placed on the ground in some convenient hollow, and the eggs are yellowish brown with black marks. The mother will feign lameness when observed, like the preceding species.

The greater part of the plumage of this bird is pure white, but the top of the head, the back of the neck, the scapularies, lesser wing-coverts, and the primaries are jetty black. It is a rather large bird, measuring about eighteen inches in total length. The beak is extremely thin, and has been well compared by Yarrell to "two thin pieces of whalebone coming to a point and curving upwards."

The STILT PLOVER is nearly as conspicuous for its long legs as the Avocet for its curved bill.

This bird, which really looks if the legs were intended for a body at least twice its size, is sometimes, but very rarely, found in England, and whenever it is found within these islands, generally prefers the swampy or marshy ground. Owing to the great scarcity of this species, and its speedy fate from powder and shot, very little is known of its habits ; but if we may judge by the Black-necked Stilt of America, it employs its long legs in wading through the water in search of food, and picks up the various aquatic inhabitants which come in its path.

Wilson remarks of the Black-necked species, that when these birds alight on the ground " they drop their wings, stand with their legs half bent and trembling as if unable to support the weight of their bodies. In this ridiculous position they will sometimes stand for several minutes, uttering a curring sound, while from the corresponding quiverings of their wings and long legs they seem to balance themselves with great difficulty. This singular manœuvre is no doubt intended to induce a belief that they may easily be caught, and so turn the attention of the person from the pursuit of their eggs and young to themselves."

The Stilt is able to swim, but generally contents itself with wading up to its belly in water. The flight of this bird is strong, and the long legs are trailed far behind the tail, looking at a little distance as if it had carried off a piece of string fastened to its toes. Five or six species of Stilt are known to science. The eggs of the Stilt are of a bluish hue covered with streaks and blotches of dusky green and dark brown.

The greater part of the plumage of this bird is white, but the back and wings are of a deep black with a gloss of green. In the female the black takes a brownish tone. The beak is black, the eggs red, and the legs and toes pink. The total length of this bird is about thirteen inches.

Like many other birds which depend for their existence upon marshy and uncultivated grounds, the Ruff is gradually being turned out of England, and may in time be nothing more than a rare and occasional visitor.

It is one of the migratory species, arriving in this country in April and leaving by the end of September. Formerly it was so common in the fenny districts that six dozen have been taken by one birdcatcher in a single day. The flesh of these birds is remarkably excellent, and they fatten fast, so that the trade of catching and fattening Ruffs was at one time a very lucrative occupation, though it now hardly repays the trouble, time, and expense. So readily can these birds be fattened, that a Ruff weighing only six ounces when first placed in the cage, will weigh ten when removed for the table. Generally the young birds of the first year are chosen for slaughter, as they are more tender and bear captivity better than the older birds. As soon as captured the Ruffs will begin to eat, and if a basin of food be placed among a number of these birds they will fight so eagerly for it that each bird would starve rather than allow any but itself to partake of the provisions. The feeders, therefore, humour their selfish disposition by placing several dishes of food in the cages and filling them all.

The Ruff is a most pugnacious bird, rivalling if not exceeding the gamecock in irritability of temper and reckless courage. Their attitude in fighting is not unlike that of the cock, but as they have no spurs, they cannot inflict severe wounds, and after a fierce contest neither party will be much the worse. Prolonged and obstinate combats are waged among the Ruffs for the possession of the females, popularly called Reeves, and as the birds make a great noise about their affairs, and in their eager combat trample down the grass on the little hills where they love to resort, the fowler knows well where to lay his nets. Many birds can be taken at one sweep of the net by an experienced fowler who knows his business.

The Ruff is chiefly remarkably for the peculiarity from which it derives its name, the projecting ruff of long, closely set feathers, which surrounds the neck, and can be raised or lowered at pleasure. This ruff only belongs to the adult males, and is assumed by them during the short breeding season, being in greatest perfection about the beginning of June and falling off by degrees from July to August and September, after which time the plumage of the male assumes the ruff-less and sober tints of its mate.

It is a remarkable fact that the male Ruffs are never coloured exactly alike, and if a hundred individuals be compared together, each will be found to present some diversity of tinting. Captivity also has a great effect in altering the form and colour of the plumage. The general colouring of the male Ruff is briefly as follows : The head, ruff, and shoulders are black glossed with purple, and barred with chestnut. The back is chestnut spotted with black, except the greater wing-coverts, which are ashen brown. The wings are brownish black, and each feather has a white shaft. The tail is brown of various hues, and mottled with black. The breast is chestnut mottled with black, and the abdomen is

RUFF.—*Philomachus pugnax.*

white sparely spotted with brown. The length of the adult bird is between twelve and thirteen inches.

The female is ashen brown upon the head and neck, spotted with dark brown, the back is nearly black, each feather being edged with ashen brown, and the wing-coverts are barred with chestnut. The tail is also ashen brown barred with chestnut and black. The chin is greyish white, and the neck and breast are grey spotted profusely with black. The abdomen is white. The female is smaller than her mate, being rather under eleven inches in length.

The KNOT, so called in honour of King Knut, or Canute as the name is generally spelled, is one of the English members of the interesting genus Tringa.

This pretty bird is found upon our coast in varying numbers, at one season flying and settling on the shore in flocks of a thousand or more in number, and at another being so scarce that hardly one bird can be seen where a hundred had formerly made their appearance. Mr. Thompson mentions that in Belfast Bay he has seen them in such profusion, that upwards one hundred and seventy were killed at a single discharge from a swivel-gun. Sometimes they are silent while on the ground, but at others they utter a peculiar chucking kind of note, which seems to indicate their position to the expectant female.

The Knot loves to feed on the large expanses of sea-grass (*Zostera marina*) which are left bare by the receding tide, and is often found with a mixed assembly of godwits, dunlins, and redshanks. As far as is known, the Knot does not breed in England, preferring more southern countries for that business. The eggs are five in number, and are merely laid on a tuft of grass.

When attired in its full summer plumage the male Knot is a really handsome bird. The sides of the head are bright chestnut with a few dark spots, and the top of the head is a deeper chestnut with dark brown streaks. The upper part of the back is richly mottled, the centre of each feather being black, and the edges warm chestnut and white. The greater wing-coverts are ashen grey, the primaries black with white shafts, the secondaries edged with white, and the upper tail-coverts rusty white, edged with white and barred with black. The tail is dark ash edged with white, and the under surface is warm ruddy chestnut, fading into white on the under tail-coverts. After the breeding season all the rich warm tints are lost, and the bird assumes a sober dress of ashen grey above, black wings, and the under surface white streaked with grey. The length of the Knot is about ten inches.

The PIGMY CURLEW, or CURLEW SANDPIPER, is so called on account of the form of its beak, which bears some resemblance to that of the Curlew, although it is much smaller and not so sharply curved.

This bird is a visitor to our shores, but is not plentiful. Mr. Thompson remarks that "as it appears on the shore it is a graceful, pretty bird, and particularly interesting from presenting so pleasing a miniature of the great Curlew. I have often known the Pigmy Curlew to be killed in company with dunlins, occasionally with them and ring dottrells, once with those two species and godwits, in a single instance with redshanks and knots." In some years these birds are more plentiful, and may be seen in little flocks of thirty or forty in number. Sixty were killed at a single shot in Cork harbour in the year 1847, in the month of October.

The voice of this bird is a kind of chattering noise. The eggs are yellowish white, spotted with dark brown.

The summer plumage of the male is warm chestnut, slightly streaked with black and white upon the head and neck. The back is beautifully mottled with black and ruddy chestnut, each feather being of the darker hue in the centre, and edged with the lighter. The wings are black with white shafts, the upper tail-coverts are white with dark spots, and the tail is ashen grey. The breast and abdomen are warm chestnut with faint bars of black, and the under tail-coverts are white spotted with black. In winter the bird assumes quite a different aspect. The head and back of the neck are ashen brown, and a stripe of white runs over the eye. The back is also ashen brown mottled with white ; the wings are black, and the whole of the under surface of the body is white. The upper tail-coverts retain their white hue, and by these alone the Pigmy Curlew may be distinguished from the dunlin. In length this bird rather exceeds eight inches.

TEMMINCK'S STINT is remarkable for being the smallest of the British Sandpipers, the average length being about five inches and a half.

This little bird is rarely found on our coast, preferring inland rivers and sheets of water, where it feeds upon worms and aquatic insects. It is said by Nilsson to breed on the shores of the seas of northern Europe.

The colouring is briefly as follows : The head is black with a little rusty red, and a light streak passes over the eye. The back is dull black, mottled with greyish white and rust colour. The wings are blackish brown with whitish edges, and the tail has the two middle feathers dusky, the next pair ashen edged with ruddy chestnut, and the remainder white. The breast is ashen yellow streaked with white, and the under parts of the body are white. This is the summer plumage. In winter the feathers of the back are brown edged with grey, and the breast becomes white streaked sparingly with brown.

The DUNLIN is known under a variety of names, such as the Stint, the Ox-bird, the Sea-snipe, and the Purre, the last of which is the most common.

This bird is the commonest of the sea-loving Sandpipers, and comes to our shores in large flocks, keeping close to the edge of the waves, running along the sands and pecking eagerly at the molluscs, worms, and smaller crustacea, which are so plentiful on the margin of the retiring waves. They are nimble-limbed birds, always on the move, and

PIGMY CURLEW.—*Tringa subarquata.*   TEMMINCK'S STINT.—*Tringa Temminckii.*
KNOT.—*Tringa canútus.*   DUNLIN.—*Tringa cinclus.*

are sure to be either engaged in running about after food or flying from one feeding-place to another. While flying they present rather a curious aspect, as they seem to change from white to black alternately, according to the point of view in which they are seen; their dark backs and white under surfaces contrasting boldly with each other.

The nest of the Dunlin is placed rather inland, and is mostly so well concealed by the heather and long grass among which it is situated, that it is not discovered without some difficulty. The mother-bird sits very closely, and towards the latter end of incubation will permit herself to be taken by hand rather than relinquish her sacred charge. During the first period of incubation, however, she will slip quietly away from her nest, run to some distance under cover of the herbage, and then begin a series of tricks, which are intended to allure the intruder from her nest. The full number of eggs is four, and their colour is greenish white covered with numerous spots and splashes of dark rusty brown.

The peculiar flight of the Dunlin is well described by Mr. Thompson in his diary, from which the following extract is condensed. "When immense flocks divide, fly right and left and shoot into single strings, they strike upon the eye while the sun shines upon them and the dark banks of the bay serve as background, like silver lines, occasionally of great length. A flock flying for a great distance just above the margin of the flowing tide has strongly resembled, from their white plumage being displayed, a single wave sweeping rapidly onwards.

I was particularly attracted by the beauty of a large flock, one moment shooting out in the form of a cornucopia, the next gathered into a circle; one instant almost dazzling by their extreme brightness, the next dark in hue, and again on the turn of the wing

exhibiting both light and darkness. . . . Descending from on wing, they all swept down
in the same direction and covered an extent of bank in such a manner as to remind one
of grain thrown from the hands of the sower. Every bird on alighting moved at the
same moderate pace, between walking and running, about equidistant from each other;
and their heads being all similarly elevated, they had a most formal and singular
appearance.

A friend being out shooting early in the morning saw a flock of several thousands.
He described their appearance as the sun rose to have been one of the most beautiful
sights he ever witnessed. The great body first appeared glancing in the sun, then it broke
up into a dozen flocks, which rose and fell in the air like molten silver, or, as his companion
observed, like showers of new shillings—a most apt image. One of the finest effects is
when the background is so dark that the birds are only seen in silvery whiteness, flashing
their under plumage upon us. The uncertainty as to when they may next appear, like
that of lightning from an extensive mass of thunder-cloud, adds much to the effect."

The Dunlin is a very loquacious bird, and the noise made by one of these vast flocks
is very great. The bird is able to swim on occasions, and when wounded will make its
way through the water with ease.

In the summer, the adult bird has the top of the head black mottled with a rusty hue,
the neck greyish white with black streaks, and the back black with reddish edges to each
feather. The primaries are dull black with white shafts, and the secondaries are of a
similar black, but edged with white. The upper tail-coverts are mottled with black, rusty
red and ash colour, and the tail is ashen grey except the two central feathers, which are
rather long and of a dark brown with light edges. The breast is black mottled with
white, and the under tail-coverts are wholly white.

In the winter the upper part of the body, together with the head and neck, is uniform
ashen grey, the centre of each feather being rather darker than the edge. The wings
retain their summer tinting, and the breast and under parts are white. In this state this
bird is called the Purre. The average length of the Dunlin is about eight inches, the
females being rather larger than the males.

THE accompanying illustrations represent three examples of British Snipes.

The GREAT SNIPE is rather a rare bird in England, but may occasionally be seen in
favourable localities, where even on the wing it may be distinguished from the common
species by the peculiar fan-like shape of the tail. While flying it hardly looks larger
than the common Snipe. It is not readily roused from the ground, but will permit itself
to be almost trodden on before it will rise, trusting to its brown mottled plumage, which
harmonizes so well with the ground that the bird is not readily perceived. When flushed
it only flies to a little distance and then settles among heather or rank grass. The flesh
of this species is very good, as the bird becomes exceedingly fat when it finds a good
feeding-place, so much so indeed that it can hardly fly, and according to Mr. Grieff is in
autumn so fat that it almost bursts its skin.

The nest of the Great Snipe is merely a small quantity of grass and leaves scraped
into a hollow and containing four olive-brown eggs spotted with reddish brown. The food
of this bird appears to consist chiefly of the larvæ or grubs of the common "daddy
longlegs," a fly which does infinite harm to pasture-land, the grubs destroying the turf as
effectually as those of the cockchaffer. As is the case with all the species, the Great
Snipe haunts watery places and heaths.

The plumage of the Great Snipe is not unlike that of the common Snipe, presently to
be described, but the bird may be distinguished by the pale hue of the cheeks, the deeper
mottling of the breast and flanks, and the conspicuous straight dark line which is drawn
like a continuation of the bill from the base of the beak to the eye. The length of the
Great Snipe is about one foot.

The COMMON SNIPE is too well known to need much description. Its habits, however, are interesting and deserve some notice.

This bird may be seen all over England wherever damp and swampy places are found.
When first flushed it shoots off in a straight line for a few yards and then begins to twist

GREAT SNIPE.—*Gallinago major.*

and turn in a strangely zigzag fashion, and at last darts away, thereby puzzling juvenile sportsmen greatly, and often escaping before its enemy has got his aim.

The male bird has a curious habit of rising to a great height in the air, circling repeatedly over the same ground, and uttering continually a peculiar cry like the words, " chic! chic! chic-a, chic-a, chic-a," constantly repeated. Every now and then the bird makes a downward stoop, and then emits a very singular sound, something between the bleating of a goat and the humming of a slack harp-string. How this sound is produced has long been a controversy, but I am convinced that it is produced by the wings—at all events that it is not from the mouth.

During a recent stay in the New Forest, I set myself to the elucidation of this problem, and in company with two friends went towards sunset to an excellent cover near a large marsh, in which Snipes were almost as plentiful as sparrows. From this post we could watch the Snipes to great advantage, and the birds would come circling over our heads, piping and drumming vigorously. On several occasions, when a Snipe was passing over us at so low an elevation that his long drooping beak was distinctly visible, he stooped directly over our heads and uttered his " chic-a, chic-a!" simultaneously with the drumming, both sounds being distinctly heard at the same time. The first time that we clearly heard the double sound was on June 27, but we repeatedly witnessed it on subsequent occasions. The Snipe remains a long time upon the wing while thus engaged, contrary to its usual habit, which is to fly for a short distance, and then to pitch again.

Sometimes the Snipe clings very closely to the ground, flies very short distances, and will almost suffer itself to be trodden upon before it will rise. Indeed after a few flushes it will often permit itself to be approached so closely that it can be knocked down by a stick, as I have often seen.

The Snipe can be tamed successfully, as may be seen by the following communication addressed to the *Field* newspaper by Mr. W. B. Scott, of Chudleigh, Devonshire :—

" John Constantius Upham, Esq. of Starcross, Devon, has a common Snipe, which is extremely tame and familiar, and answers to the name of Jenny. In December last she was caught by some boys near the warren, and was brought to Mr. Upham in a starving state. She was recovered by forcing her to eat some very minute pieces of raw mutton. Worms having been procured, she soon commenced feeding herself, and eventually would follow Mr. Upham round the room for a worm. Her bath is a good-sized pie-dish, her *salle à manger* an eight-inch flower-pot, and her amusement probing a large damp sod of rushes placed for her fresh every day on a good thick piece of brown paper.

We were three of us who went to see her on Wednesday, the 27th ult. On our entering the parlour where she is allowed to run about, she evinced no alarm, and presently commenced feeding. The upper mandible of a Snipe's bill being a little longer than the under one, it was with some perseverance and some difficulty that she picked up from the carpet a worm which was thrown to her. Except when she is very hungry, she generally washes the worms before eating them. The flower-pot is half full of earth and worms; it is placed on its side. The Snipe, when she feeds, probes the earth for a worm; having caught one, she carries it to the pie-dish. After carefully washing it, she disables the worm by pinching it all over with the tip part of her bill; then she takes it by the middle and throws it back to swallow, in doing which the head of the worm is on one side of the bill and the tail on the other. The head and tail soon disappear, and the worm goes down double, even if it be as thick as a goose-quill.

The Snipe constantly goes in and out of the pie-dish, and probes round at its bottom with her bill. She frequently washes herself, throwing the water over her back and flapping and splashing it with her wings, after which she comes out of the dish and preens her feathers, spreading her tail like a fan, bending it round with great flexibility in a curious manner, and keeping it in constant motion. She is very fond of the fire, and stands before it on one leg for hours together. She has on two or three occasions exhibited symptoms of impatience at confinement by flying against the window; on the last occasion she flew against the ceiling of the room with some violence, and came down much hurt, so that the feathers of one of her wings have been cut.

Mr. Upham is getting a place made to collect and store worms; her consumption of them is almost incredible, for she consumes in twelve hours nearly double her own weight. Three sorts of worms she takes, the dew-worm, and two other small red sorts, the names of which are unknown to me; the brandling, the lobb, the gilt-tail, or indeed any worm from a dung heap, she will not touch. She is also very fond of snails' eggs, very small young snails, woodlice, or small *planorbis*, and several other fresh-water shells, eating shell and all; she also picks up gravel like other birds. I watched the bird for more than an hour, and saw her eat more than twenty worms. The pie-dish is a blue one, and as it was thought to be not quite deep enough for her, a larger one was searched for; but Starcross could not furnish a larger blue dish, so a yellow one was bought, but she would not go near it; it was even banked up with turf, but it would not do, so the old blue pie-dish was brought back to her again. Mr. Upham is keeping a diary, and notes down the habits and peculiarities he observes in his pet Snipe; he much fears she will not survive the ensuing summer. I was so much interested that I hope to pay the Snipe another visit very soon."

The nest of the Snipe is a simple heap of leaves placed under the shelter of a tuft of furze, heath, or grass, and the eggs are four in number of an olive-white, spotted and dashed with brown of different tones towards and upon the large end. The mother-bird has been known to carry away her young when threatened by danger.

The colouring of the Common Snipe is briefly as follows : The top of the head is dark brown, a light brown streak runs along the centre, the cheeks are pale brown with a dark streak from the bill to the eye, and over the dark streak is another of a paler hue. The back is beautifully mottled with two shades of brown, and four bold lines of warm buff run along the upper surface of the body. The wings are black, some of the feathers being tipped with white. The chin is very pale brown, the neck is also light brown, but spotted with a darker hue, the breast and abdomen are white and the flanks grey-white with dull black bars. The under tail-coverts are cream-coloured with a brown tinge and barred with grey-black. The average length of the Snipe is between ten and eleven inches.

COMMON SNIPE.—*Gallinágo média.*   JACK SNIPE.—*Gallinágo gallinula.*

The little JACK SNIPE is seldom seen in this country except in the winter, and is remarkable for its tenacity in clinging to the ground even on the near approach of an enemy. Terror seems to have some part in this propensity, for Mr. Yarrell remarks that a Jack Snipe has allowed itself to be picked up by hand before the nose of a pointer. It has also a strong attachment to localities, adhering closely to one spot, and always returning to the same place after a while. It is not an easy bird to shoot unless taken at the rise, when there is danger that so diminutive a bird may be blown to pieces by the first discharge, for it dodges about and skims just over the heather, in a vastly perplexing manner. There is a story told of a gentleman not remarkable for his skill in shooting, who was found lamenting over the corpse of a Jack Snipe which he had succeeded in killing, and whose death had deprived him of the amusement which he had enjoyed for many weeks, as the bird could always be found in the same place every morning, and be hunted up and down all day without going out of the grounds.

The plumage of the Jack Snipe is very like that of the common species, but may be at once distinguished by the absence of the pale brown streak over the top of the head.

Three more species of Snipe are known in England, but are very scarce. One is the SABINE'S SNIPE (*Gallinágo Sabinii*), notable for the total absence of white upon its plumage; the second is the BROWN, or RED-BREASTED SNIPE (*Gallinágo grísea*), properly an American bird, and distinguishable by the ruddy breast and the streak of white from the bill to the eye; and the third is rather a dubious species, known by the name of BREHM'S SNIPE (*Gallinágo Brehmii*).

The WOODCOCK is nearly as well known, though not so plentiful as the snipe, to which bird it bears a considerable resemblance in form, plumage, and many habits.

Generally it is only a winter visitor, arriving about October, and leaving England in March or April. Sometimes, however, it will breed within the British isles, and there remain throughout the summer. During their migration the Woodcocks fly at a great altitude, and descend almost perpendicularly upon the spot where they intend to rest. They fly in companies of varying numbers, and prefer hazy and calm weather for their journey.

It is rather peculiar in some of its habits; so that an experienced Woodcock-shooter will find plenty of birds, and fill his bag, in places where a novice will hardly get a single shot. It is not to be seen until long after noon, and prefers the earliest dawn and the hours of dusk for feeding and going abroad. At daybreak it rises from its covert to some height, uttering its peculiar call, wheels about at a considerable elevation, and then starts off to its feeding-grounds, which are like those of the snipe, wet and marshy, though the bird always chooses a dry and elevated spot for its couch. After satisfying its hunger, it returns to its domicile, and remains quietly hidden until about three or four in the afternoon, when the short day is at its close. After spending some time in feeding, it returns to its couch after some preliminary wheeling and twisting in the air, and lies quiet until the dawn of the next day calls it again to activity.

The food of the Woodcock consists mostly of worms, which it obtains with extraordinary skill, thrusting its beak as far as the nostrils into the soft moist earth, and hitting upon the hidden worms with unerring skill. A tame Woodcock has been seen to probe large turfs with its bill, and to draw out a worm at every thrust of the long slender beak. It is thought that the sense of smell enables the bird to discover the worms beneath the surface. It moves about chiefly on misty days, and is said by experienced Woodcock-shooters to prefer the northern side of a hill to the southern.

It is a very silent bird, seldom uttering its cry except when first starting for its feeding-places, and hardly even crying when flushed. The flight of the Woodcock is wonderfully swift, although the wings do not appear to move very fast; and the bird has a custom of jerking and dodging about so quickly when it sees the sportsman, that it often escapes his shot. One bird, mentioned by Mr. Thompson, used to baffle an experienced sportsman by always feeding near an archway, and slipping through it before the gun could be brought to bear. Not that the frequent escapes of the bird are owing solely to the quickness and irregularity of its flight, because a good eye and ready hand soon become accustomed to such movements; but the individual birds are so variable in their mode of flight that the movements of one form hardly any criterion to those of another.

Yet, when it is hit, it falls a very easy prey, as it falls to a very slight blow, so that the smaller sizes of shot can be employed; and when wounded it has not much idea of running and hiding itself, and can be easily found by a good spaniel. The first day of a thaw is said to be the best for Woodcock shooting, but the birds are apt to desert the spot within a day or two.

It has been remarked, that when a certain number of Woodcocks inhabit a copse, the same number will always be found in the same spot, in spite of those that are killed. Anglers have remarked a similar custom with trout, especially the larger fish, for if they have taken a fine trout out of a certain spot, the deserted tenement is sure to be occupied immediately by another large trout.

The nest of the Woodcock is made of leaves—those of the fern being favourites—closely laid together, but without any particular skill in arrangement, and without lining. The full number of eggs is four, and their colour is buffy white with rusty brown blotches. The mother bird has been known to carry away her young when threatened by danger; and from reliable accounts, she places them upon her spread feet, pressing them between the toes and the breast. According to Mr. St. John, "regularly as the evening comes on, many Woodcocks carry their young ones down to the soft feeding-grounds and bring them back again to the shelter of the woods before daylight. I have often seen them going down to the swamps in the evening, carrying their young with them.

WOODCOCK.—*Scólopax rusticola.*

Indeed, it is quite evident that they must in most instances transport the newly hatched birds in this manner, as their nests are generally placed in dry heathery woods, where the young would inevitably perish unless the old ones managed to carry them to some more favourable feeding-ground. Nor is the food of the Woodcock of such a nature that it could be taken to the young from the swamps in any sufficient quantity; neither could the old birds bring with them the moisture which is necessary for the subsistence of all birds of this kind. In fact, they have no means of feeding their young, except by carrying them to their food, for they cannot carry their food to them."

Per contra, it is said that the drooping body and tail give the female an appearance, especially if viewed from behind, of having a young bird in her embrace, although she really is unburdened. Still, however, the numerous well-accredited instances of this custom set the matter beyond a doubt.

The general colour of the Woodcock is brown of several shades, pale wood-brown upon the cheeks, rich dark brown upon the back, mottled with a lighter hue; throat, breast, and abdomen, wood-brown barred with dark brown. The tail is black above tipped with grey. The average length is about fourteen inches, but the weight is extremely variable. An ordinarily good bird weighs about thirteen ounces, but a very fine specimen will weigh fourteen or fifteen ounces, and there are examples of Woodcocks weighing twenty-six and twenty-seven ounces.

The GREY PHALAROPE is one of the rare British birds, belonging rightly to the limits of the Arctic circle, and coming southward in the autumn.

It is a light and active bird, flying and swimming with great address, and braving the raging sea with easy courage. The body of this bird is singularly buoyant, so that the Phalarope rides on the waters like a cork, bidding defiance to the waves, and circling about the surface with an ease and rapidity that reminds the observer of the whirligig beetles that urge their ceaseless wheels on the surface of our shady streams. It is a bold bird, caring little for the presence of human beings, and suffering itself to be approached without displaying fear. Moreover, when made prisoner it becomes familiar with its

GREY PHALAROPE.—*Phaláropus fulicárius.*

captor in a very short time, and in a few hours will eat out of his hand. The general food of the Grey Phalarope consists of marine creatures, such as molluscs and little crustaceans, but when it comes to the shore the bird will feed on larvæ and various insects. While swimming it has a habit of nodding its head at each stroke. The flight is said to resemble that of the terns.

The plumage differs greatly according to the season of year. In the winter the head, back of the neck, and back are dark blackish brown and these feathers are surrounded by an orange border. There is a white band on the wing, and the front of the neck and lower parts of the body are dull red. In the winter the upper parts of the body are pearly ash, the centres of the feathers being darker than their edges. The white band on the wing is retained, and the breast and lower parts are pure white. The bill is brown at the point, and orange-red at the base. The Grey Phalarope is but a small bird, its length being about eight inches. The Red Necked Phalarope (*Phaláropus hypobóreus*) is also an inhabitant of England.

THE two curious birds which are depicted in the accompanying illustration are examples of the Jacanas. All these birds are remarkable for the extraordinary development of their toes, which are so long and so slender that they seem to have been drawn out like wire, and to impede the progress of their owner. These elongated toes are, however, of the greatest use, as they enable the bird to walk upon the floating leaves which overspread the surface of many rivers, and to pick its food from and between the leaves on which it walks. As the bird marches upon the leaves, the long toes dividing the pressure upon several leaves at each step, they are slightly sunk below the surface by the weight, so that the bird appears to be really walking upon water.

The COMMON JACANA is a native of Southern America, and there are other species scattered over Africa, Asia, and Australia. Mr. Gould tells us that the Australian species is a good diver, but a bad flier. "Their powers of diving and of remaining under water are equal to those of any bird I have ever met with; on the other hand, the powers of flight are very weak. They will, however, mount up fifteen or twenty yards and **fly** from one end of the lake to the other, a distance of half or three-quarters of a mile ; but generally they merely rise above the surface of the water and fly off for about a hundred yards. During flight their long legs are thrown out horizontally to their full length. While feeding, they utter a slowly repeated ' cluck, cluck.' The stomach is extremely muscular, and the food consists of aquatic insects and some kind of vegetable matter."

The general colour of the Common Jacana is black, with a slight greenish gloss, taking a rusty red tinting on the back and wing-coverts. The primary quill-feathers of the wing are green, and the wings are furnished at the bend with long and sharp claws. In the African species these spurs are hardly perceptible. At the base of the beak is a curious leathery appendage, rising upon the forehead above and depending towards the chin below. The claws are all very long, especially that of the hind toe, which is nearly straight, and longer than the toe from which it proceeds.

The pretty CHINESE JACANA well deserves the title of Hydrophasiánus, or Water Pheasant, a name which has been given to it on account of the two long tailfeathers which droop gracefully in a gentle curve. The quill-feathers of the wings are also remarkable for certain little appendages, like hairy plumes, which proceed from the tip of each shaft.

The Chinese Jacana, or MEEWA, is not confined to the country from which it derives its popular name, but is found in various parts of Asia, and has been obtained from the Himalayas and the Philippines.

It is a very active bird on foot or in the water, swimming with easy grace, and traversing the floating herbage in search of its food, which resembles that of most aquatic birds. A tamed specimen fed readily on shrimps. It is not very fond of using its wings, but when flying, extends

JACANA.—*Parra Jacana.*

the legs backwards after the same fashion as the heron. The flesh of the Chinese Jacana is very excellent, and has been likened to that of the snipe, so that the bird is in some request among sportsmen. Shooting it is, however, no very easy task, as a wounded bird is seldom if ever recovered, diving at once and remaining submerged until the foe has left the spot, or death has released it from its sufferings.

The nest of the Chinese Jacana is made of reeds and grasses, is flat in form, and is supported upon the woven stems of aquatic plants. The eggs are about six or seven in

CHINESE JACANA.—*Hydrophasiànus Sinensis.*

number, and their colour is olive-brown. These birds breed during the rains, and choose those spots where the lotus is plentiful. The voice resembles the mewing of a distressed kitten, to which fact is owing its native name of Meewa.

The colours of the male bird are bold and striking. The back and under parts are deep chocolate-brown, the elongated tail is a still darker brown; and the wings, top of the head, throat, and part of the neck are white. The back of the neck is orange, and a narrow black line separates it from the white of the throat. The legs, toes, and beak are grass-green. At the end of the primary feathers are certain filamentous and somewhat lancet-shaped appendages, which, according to some writers, hinder the bird in its flight. The female is quite sober in her plumage. The upper part of the body and head is pale brown, warming to red on the forehead. From the bill a dark streak passes through the eye and down the side of the neck, and above that is another streak of buffy orange. The throat and under parts are white, and a broad collar of dark brown encircles the junction of the neck and breast.

THE sub-family of the Screamers is here represented by two very curious birds, both of them being large birds, having their wings armed with formidable claws capable of being used with much effect as weapons of offence.

The HORNED SCREAMER, or KAMICHI, is a native of Central America, and is found in the vast swamps and morasses of that hot and moist country, where the vegetation springs up in gigantic luxuriance and the miasmatic morasses give birth to reptiles and creeping things innumerable. The large spurs on the wings are valuable to the bird in repelling the attacks of the numerous snakes, and guarding itself and young from their

HORNED SCREAMER.—*Palamedéa cornúta.*   CRESTED SCREAMER.—*Chauna chavária.*

rapacity. In size the Horned Screamer nearly equals a common turkey, so that a blow from its armed wing can be struck with considerable force. The bird is not, however, fond of using its weapons, and, unless attacked, is quiet and harmless.

The food of the Horned Screamer consists chiefly of vegetable substances, such as the leaves and seeds of aquatic plants, in search of which it wades through the reptile-haunted morasses. Its flight is strong and easy, its walk is erect and bold, and its mien lofty like that of the eagle. Upon the head of the present species is a curious horn-like appendage, from three to four inches in length, and about as large as a goose-quill. The use of this horn is quite unknown. The voice of the Horned Screamer is loud and shrill, and is uttered suddenly and with such vehemence that it has a very startling effect.

The general colour of this bird is blackish brown above; the head and upper part of the neck are covered with downy feathers of blackish brown sprinkled with white.

ANOTHER well-known example, the CRESTED SCREAMER, or CHAJA, is a finer-looking bird than the preceding species, though its head is without the singular appendage that gives the horned Screamer so unique an aspect. The name of Chaja is given to this bird on account of its cry, that of the male bird being "chaja" and of the female "chajali."

It is a native of .Brazil and Paraguay, and is generally found near the banks of rivers. It is a shy and generally solitary bird, being mostly seen singly, sometimes in pairs, and now and then in small flocks. Like the horned Screamer, the Chaja is armed with two spurs on each wing, and can employ them to such purpose, that it can drive away even a vulture. Unless attacked, however, the bird is quiet; and as it is easily tamed, it may be often seen domesticated in the houses.

The walk of this bird is bold and dignified; the body is held rather horizontal and the head and neck erect. The flight is strong and sweeping, and the bird rises on circling wings somewhat after the manner of the eagle, after obtaining so great an elevation as to be hardly discernible against the sky. The food of the Chaja consists mostly of aquatic plants, which it obtains by wading. The nest of this bird is a rather large edifice of sticks and leaves, and is placed near water. The number of eggs is two, and the young are able to follow their parents almost as soon as hatched.

The general hue of the Chaja is a leaden-blue colour diversified with black. The bend of the wing is white, and there is a large spot of the same hue at the base of the primaries. Round the neck is a black collar, the small head is furnished with a crest, the upper part of the neck is clothed with down, and the space round the eye is naked and blood-red in colour.

WATER RAIL.—*Rallus aquáticus.*

We now come to the large family of the Rails, a curious group of birds, formed for rapid movement either on the ground or through the water, but not particularly adapted for long flights. Many species inhabit England.

The WATER RAIL is one of the British examples of this family, and is but seldom seen, partly because it really is not very plentiful, and partly on account of its shy and retiring habits, and its powers of concealment. It frequents ponds, lakes, and similar localities, haunting those places where luxuriant reed-beds afford it shelter and covert. On the least alarm it sets off for the place of refuge, diving to a considerable distance and always pressing towards the reeds, through which it glides with wonderful address, and is immediately out of danger. Even a trained dog can hardly flush a Water Rail when once it has reached its reedy refuge, as the bird can thread the reeds faster than the dog can break its way through them, and has always some deep hole or other convenient hiding-place where a dog cannot reach it.

The food of the Water Rail consists mostly of insects, worms, leeches, molluscs, and similar creatures, all of which can be found either upon the aquatic herbage or in the muddy banks. Mr. Thompson mentions a curious instance of the readily domesticative and insect-hunting propensities of this bird. "On the 15th of September, 1832, I saw in a gunsmith's shop, in Belfast, one of these birds, which had been taken alive a day or two before. It was very expert in catching flies in the shop window, running a tilt at them quite regardless of the presence of the stumbling-blocks which beset its path in the form of pistols, turn-screws, &c. When approached, this bird struck wickedly with its bill and feet, but never with its spurred wings." In the stomach of these birds the same writer found the remains of aquatic molluscs, worms, and a few seeds, and portions of leaves. In captivity it will thrive on raw meat chopped small.

While walking, the bird has a habit of flirting up its odd little tail, so as to show the white under tail-coverts. The nest of the Water Rail is sheltered by the thickest herbage of the covert, and is made of coarse grass. There are about seven or eight eggs, and their colour is buffy white spotted with brown. The young are odd little creatures, round, and covered with soft thick down. Almost immediately after their emancipation from the egg-shell, these little puffy balls of down tumble into the water, and swim about as merrily as if they had been accustomed to the exercise for years.

The general colour of the Water Rail is buffy brown above, richly mottled with velvety black. The throat is grey ; the sides of the neck, the breast, and abdomen are slaty grey changing on the flanks into greyish black barred with white and buff, and to cream-white on the under tail-coverts. The bill is brown at the tip, and light orange at the base. The length of the Water Rail is about one foot.

The VIRGINIAN RAIL is an inhabitant of America, reaching the Northern States about May, and retiring to warmer regions in November. Wilson writes of this bird as follows:—" It is frequently seen along the borders of our salt-marshes, and also breeds there, as well as among the meadows that border on large rivers. It spreads over the interior as far west as the Ohio, having myself shot it in the barrens of Kentucky early in May. The people there observe them in wet places, in the groves, only in spring. It feeds less on vegetable and more on animal food than the common rail. During the months of September and October, when the reeds and wild oats swarm with the latter species, feeding on their nutritious seeds, a few of the present kind are occasionally found, but not one for five hundred of the others.

The food of the present species consists of small snail shells,

VIRGINIAN RAIL.—*Rallus Virginianus.*

worms, and the larvæ of insects, which it extracts from the mud : hence the cause of its greater length of bill, to enable it the more readily to reach its food. On this account also, its flesh is much inferior to that of the others. In most of its habits, its thin, compressed form of body, its aversion to take wing, and the dexterity with which it runs or conceals itself among the grass and sedge, are exactly similar to those of the common rail."

In some parts of America it is known under the name of the Fresh-Water Mud-Hen, because it frequents those parts of the marshes where fresh-water springs rise through the morass. " In these places it generally constructs its nest, one of which we had the good fortune to discover. It was built in the bottom of a tuft of grass in the midst of an almost impenetrable quagmire, and was composed altogether of old wet grass and rushes. The eggs had been flooded out of the nest by the extraordinary rise of the tide in a violent north-east storm, and lay scattered about the drift weed. The usual number of eggs is from six to ten. They are of a dirty white or pale cream colour, sprinkled with specks of reddish and pale purple, most numerous near the great end."

The top of the head and the upper surface of the body are black streaked with brown ; the cheeks and a streak over the eye are ashen grey ; and by the lower eyelid there is a white mark. The wing-coverts are light chestnut, the quills are dusky black ; there is a white streak on the bend of the wings ; the chin is white, and the whole lower surface is orange-brown. The female may be distinguished from the male by the pale breast and the greater amount of white on the chin and throat. The average length of the adult male is ten inches, the female being about half an inch shorter.

The well-known CORNCRAKE, or LANDRAIL, is common in almost every part of the British Islands, its rough grating call being heard wherever the hay-grass is long enough to hide the utterer.

The bird runs with wonderful speed through the tall grass, and its cry may be heard now close at hand, now in the distance, now right, and now left, without any other indication of the bird's whereabouts ; for so deftly does it thread the grass stems that not a shaken blade indicates its presence, and it is so wary that it keeps itself well hidden among the thick herbage. The cry of the Corncrake may be exactly imitated by drawing a quill or a piece of stick smartly over the large teeth of a comb, or by rubbing together two jagged strips of bone. In either case the bird may be decoyed within sight by this simple procedure.

The Corncrake is not fond of its wings, and very seldom takes to the air, even preferring to be caught by the dog than to escape by flight. When captured it has a habit of simulating death, and often contrives to escape when the eye of its captor is otherwise engaged. One of these birds which had been picked up by a pointer allowed itself to be placed in the game-bag, carried home, and laid on the table without exhibiting any indications of life. When it thought itself unwatched, it sprung up and dashed at the window, which being closed frustrated this poor bird in its bold attempt.

The Corncrake can be readily tamed, as will be seen from the following account by Mr. Thompson:—" It became quite tame and partook of food very various in kind, such as groats (few, however, of these), raw meat, bread and milk, stirabout and milk, yolk of boiled eggs and butter, which last was especially relished. It also ate worms, snails, slugs, &c., and has been seen to take small sticklebacks that happened to be in the water. This bird was very cleanly, and washed every morning in a basin of water set apart for the purpose. It was accustomed to be taken upstairs at night and brought down in the morning, and of its own accord went habitually out of the cage into a basket containing moss, where the night was passed, and in the morning likewise left the basket and entered the cage in which it was carried downstairs.

When allowed to go about the house, the persons to whom it was attached were sought for and followed everywhere. On becoming unwell, the poor bird took possession of the lap of a member of the family, and looked up to her apparently for relief ; though when in health it resisted all attempts at being handled, flying up at the intruder and snapping its mandibles together. Every spring it called with the usual *crake*, beginning very early in the morning ; this was usually commenced in March, but on one occasion was uttered as early as the 3d of February. As was remarked of the bird after this period, ' it would crake quite impudently in the parlour when orought there to be shown off.'

Moulting took place in the month of August, but no symptoms of uneasiness appeared then or at any particular season. At pairing time this bird was very comical, coming up with its wings spread and neck stretched out after the manner of a turkey cock, and uttering a peculiar croaking note. It would then make a sort of nest in the cage, croaking all the while, and carry a worm or piece of meat about in its bill. So great a favourite was this Corncrake, that its death was duly chronicled as taking place on the 14th of January, 1830, after having been kept for above six years."

The nest of the Corncrake is placed on the ground, and is made of dried grass arranged in a suitable depression. It generally contains from eight to twelve eggs, of a buffy white covered with rusty brown spots. The shell is rather thick, and the size of the egg large in proportion to the dimensions of the bird. The position of the nest and the lateness of the hatching season expose both mother and young to great danger, as the nest is often laid low and the mother killed by a sweep of the mower's scythe. The parent is very fearless when engaged in incubation; and on one occasion when a female Corncrake had been severely wounded by a scythe and taken into the farm-house for two hours, she returned to her nest in spite of its shelterless condition and her own wounded state, and was rewarded by the successful rearing of the brood.

The flesh of the Corncrake is very delicate and well-flavoured.

CORNCRAKE, OR LANDRAIL.—*Ortygometra crex.*

The upper parts of the body are elegantly mottled with dark blackish brown, ashen, and warm chestnut; the first tint occupying the centre of each feather, the second the edges, and the third the tips. The wing-coverts are rusty red. The throat and abdomen are white, and the breast is greenish ash, warming into reddish rust striped with white on the sides. In total length the Corncrake is not quite ten inches.

The HYACINTHINE GALLINULE is a rather curious example of the next sub-family of the Crakes, being remarkable for the large size of its beak and the length of its toes. All the species belonging to the genus Porphyrio are fond of the water, although they are oftener seen on land than is the case with the water-hens. They feed upon seeds and other hard substances, which they crack easily with their powerful bills. Their very long toes enable them to walk upon the floating herbage nearly as well as the Jacanas, and upon land they are very quick of foot. They use their long toes for carrying food to their beak, a habit which has often been observed in the common coot.

The Hyacinthine Gallinule is spread over a large extent of range, being found in many parts of Africa, Asia, and Europe. It is graceful and quick in its movements, but is said to be rather a stupid bird. Perhaps future observers may give a better account of its intellect. The nest of this species is made in the sedgy parts of the morasses which it frequents, and contains a rather small number of nearly white eggs.

The colour of the Hyacinthine Gallinule is rich blue, taking a dark indigo tone upon the back, and assuming a beautiful turquoise hue upon the head, neck, throat, and breast.

HYACINTHINE GALLINULE.—*Porphyrio véterum.*

The under tail-coverts are white, the bill is light red, and the legs and feet are pinky red. The length of this bird is about eighteen inches.

Our most familiar example of the Gallinules is the WATER HEN, sometimes called the MOOR HEN.

This bird may be seen in plenty in every river in England, and mostly on every pond or sheet of water where the reedy or rushy banks offer it a refuge. It is a bold bird, though sufficiently wary on occasions ; and while it will slip quickly out of sight of a dog or a man with a gun, will swim about with perfect self-possession in a pond by the side of a railway, quite undisturbed by the sound and sight of the rushing train. When startled it flies rapidly across the water with quick beating wings and dangling legs, leaving a long track behind it, which will remain for some little time, like the wake of a ship. As it nears its reedy refuge it sinks nearer the surface of the water, so that at the last yard or two of its progress it drives the water before it, and seems equally to run or to fly.

When startled it often dives on the instant, and, emerging under floating weeds or rubbish, just pokes its bill above the surface, so that the nostrils are uncovered by the water, and remains submerged until the danger is past, holding itself in the proper position by the grasp of its strong toes upon the weeds. If wounded, it will often escape by diving, so that unless the sportsman kills his birds on the spot he may lose bird after bird unless he has a good dog with him. Sometimes it pretends to be wounded, and drops into the shelter of reeds or bushes in so death-like a manner that the gunner is deluded into the idea that he has killed his bird very neatly, and while he is reloading away goes the Water Hen to some secure retreat.

I once took a snap-shot from a boat at a fine male specimen, in a little pond at the end of an inlet, and to my astonishment, after backing to the mouth of the little stream, saw him swimming and nodding his head as coolly as if nothing had happened. I was going to give him the second barrel, but, being short of ammunition, determined to paddle quietly up the inlet in which the bird was swimming, and to knock it over with an oar. The bird took not the least notice of the boat, so I pushed the blade of the oar under it, lifted it out of the water, and brought it into the boat. On examination I found that it had been struck through the head with a shot ; I believe that in such cases the powers of volition are suddenly extinguished, and that the bird continues to act according to the last impression upon the brain. Many birds, as every sportsman knows, will tower when shot, and I have found that in such instances they exhibit a singular tenacity of animal life.

When free from persecution, the Water Hen soon becomes familiar with man, and will mix familiarly with domestic poultry, traversing the garden or farmyard with easy confidence. It is apt, however, to be rather mischievous, eating fruit and vegetables of various kinds. The Reverend Mr. Atkinson writes : " The Moor Hens having been much encouraged, were very numerous, both about the moat and in two or three flaggy ponds in the adjoining pastures. I have seen as many as fourteen or fifteen at once upon one bed of cabbage plants. They picked the peas, the strawberries, the currants, the gooseberries, all in early stages of their growth, and they stripped the leaves of the newly planted young cabbages and greens, until nothing was left but ragged fragments of the midrib and stalks."

The nesting of this bird is very peculiar. The Water Hen builds a large edifice of sedges, sticks, and leaves, either on the bank close to the water's edge, upon little reedy islands, or on low banks overhanging the water, and generally very conspicuous. The mother-bird has a habit of scraping leaves and rushes over her eggs when she leaves the nest, not, as some persons fancy. to keep the eggs warm, but to hide them from the prying eyes of crows and magpies, jays, and other egg-devouring birds.

WATER HEN.—*Gallinula chlóropus.*

The Moor Hen is by no means niggardly in her labour, but will build one or more extra nests, or rather rafts, for the accommodation of her young brood ; and in some cases will, without apparent reason, discard the nest in which the eggs have been hatched, build a new one, and transfer to it her little family. When thoroughly pleased with a locality the Moor Hen evinces a strong attachment to it, and returns to the same spot through several successive seasons.

Should the water rise beyond its ordinary level, this bird is equal to the emergency, and rapidly elevates the nest by adding sticks and inserting them into the fabric. One bird generally remains by the nest and acts as builder, while the other searches for materials and brings them to its mate. Mr. Selby mentions an instance where the bird removed the eggs during the process of elevation, and replaced them after the completion of its labours.

The young are able to swim almost as soon as hatched, and for some time remain close to their parents. I once, to my great regret, shot by mistake several young Moor Hens, still in their first suit of black puffy down, and paddling about among the water lilies and other aquatic herbage where I could not see them. Pike are rather apt to carry off the little creatures by coming quickly under the weeds and jerking them under water before they take the alarm.

The male bird is dark olive green above ; the head, neck, and under parts are blackish grey. The under tail-coverts and edges of the wings are white. The bill is green towards the tip and red at the base, the latter hue being brightest at the breeding season. The legs and toes are green, and the naked part of the thigh is red. The female has not so much of the olive as her mate.

The COMMON COOT, or BALD COOT, as it is sometimes called, is another of our familiar British water-birds, being seen chiefly in lakes, large ponds, and the quiet banks of wide rivers.

The habits of the Coot much resemble those of the water hen, and it feeds after a similar fashion upon molluscs, insects, and similar creatures, which it finds either in the water or upon land. It is an admirable swimmer, swift and strong, and can grasp the branches firmly when perching, owing to the contraction of the foot, which is furnished with a wide flattened membrane on the edges of each toe, thus presenting a broad surface to the water, and at the same time permitting the foot to be used in grasping. The Coot may be seen either swimming or traversing the floating weeds in search of food, or wandering

COOT.—*Fúlica atra.*

over the fields with quick but rather eccentric gait, pecking here and there at the herbage, and devouring a great quantity of destructive insects, snails, and slugs. When a very severe winter has frozen the ponds and lakes, the Coot will make off to the nearest coast, and along its unlocked shores obtain a living until the warm breezes of spring have loosened the icy body of its more congenial haunts.

The nest of the Coot is a huge edifice of reeds and rank-water herbage, sometimes placed at the edge of the water, and sometimes on little islands at some distance from shore. I have often been obliged to wade for thirty or forty yards to these nests, which have been founded upon the tops of little hillocks almost covered with water. The whole nest is strongly though rudely made; and if the water should suddenly rise and set the nest floating, the Coot is very little troubled at the change, but sits quietly on her eggs waiting for the nest to be stranded. Several instances are known where the nest and bird have been swept into a rapid current, and carried to a considerable distance. The eggs are generally about eight or ten in number, and their colour is olive-white sprinkled profusely with brown. The shell is rather thick in proportion to the size of the egg, so that Coots' eggs can be carried away in a handkerchief without much danger of being broken.

The head and neck of the Coot are greyish black, the upper parts are deep blue-black, and the under parts are blackish grey with a tinge of blue. The bill is rather pale orange-red, and the horny plate on the forehead is rosy red in the breeding season, fading into white at other times of the year, from which circumstance the Coot derives its *sobriquet* of "bald." The legs are yellow-green, the naked part of the thigh orange-red, and the eye bright red. The length of the Coot is about seventeen or eighteen inches.

FLAMINGO.—*Phœnicópterus ruber.*

The well-known FLAMINGO brings us to the large and important order of Anseres, or the goose tribe.

The common Flamingo is plentiful in many parts of the Old World, and may be seen in great numbers on the sea-shore, or the banks of large and pestilential marshes, the evil atmosphere of which has no effect on these birds, though to many animals it is most injurious, and to man a certain death. When feeding the Flamingo bends its neck, and placing the upper mandible of the curiously bent beak on the ground or under the water, separates the nutritive portions with a kind of spattering sound, like that of a duck when feeding. The tongue of the Flamingo is very thick, and of a soft oily consistence, covered with curved spines pointing backwards, and not muscular.

A flock of these birds feeding along the sea-shore have a curious appearance, bending their long necks in regular succession as the waves dash upon the shore, and raising them as the ripple passes away along the strand. At each wing is always placed a sentinel bird which makes no attempt to feed, but remains with neck erect and head turning constantly about to detect the least indication of danger. When a flock of Flamingos is passing overhead, they have a wonderfully fine effect, their plumage changing from pure white to flashing rose as they wave their broad wings.

When at rest and lying on the ground, with the legs doubled under the body, the Flamingo is still graceful, bending its neck into snaky coils, and preening every part of its plumage with an ease almost incredible. Its long and apparently clumsy legs are equally under command, for the bird can scratch its cheeks with its toes as easily as a sparrow or a canary.

SPUR-WINGED GOOSE.—*Plectróphanes Gambensis.*

When flying the Flamingo still associates itself with its comrades, and the flock form themselves into regular shapes, each band evidently acting under the command of a leader. The nest of the Flamingo is rather curious, and consists of mud and earth scraped together so as to form a tall hillock with a cavity at the summit. In this cavity the eggs are laid, and the bird sits easily upon them, its limbs hanging down at each side of the nest like a long-legged man sitting on a milestone. The eggs are white, their number is two or three, and the young birds are all able to run at an early age. Like many other long-legged birds, the Flamingo has a habit of standing on one leg, the other being drawn up and hidden among the plumage.

The curious beak of this bird is orange-yellow at the base and black at the extremity, and the cere is flesh-coloured. When in full plumage the colour is brilliant scarlet, with the exception of the quill-feathers, which are jetty black. A full-grown bird will measure from five to six feet in height.

THE curious bird represented in the engraving brings us nearer to the true Geese.

The SPUR-WINGED GOOSE inhabits Gambia and Senegal, and is remarkable for the peculiarity from which it derives its name. The reader will remember that several birds, such as the jacana and the screamers, are armed with horny claws or spurs upon the bend of the wing; and it is rather remarkable that the same formation is found in one genus of the Goose tribe, the wings of the Spur-winged Goose being supplied with two of these appendages. The head, too, is notable for a bold elevated crest, which starts from the base of the bill, and which during the life of the bird is of a light red colour. This protuberance is really part of the skull, and has a very curious aspect when the skeleton is prepared. One or two specimens of this bird have been taken in England. Mr. Yarrell mentions two such examples, and in the "Annals of Sporting" there is a notice of a third specimen having been killed in 1827 at Donnington Grove on a large piece of water. Several swans were in the same locality, but the bird always avoided them.

The colouring of this species is bold and simple. The general tint of the plumage is deep black glossed with purple, but the throat, front of the breast, and abdomen are white. In size it rather exceeds the domestic goose.

CAPE BARRON GOOSE.—*Cereopsis Novæ Hollandiæ.*

WE now arrive at the true Geese, our first representative being the CAPE BARRON GOOSE or CEREOPSIS, so called from the cere which covers a large portion of the beak. This fine bird is a native of New Holland, and is found, as its name implies, at Cape Barron Island in Bass's Straits. It is of large size, fattens easily, its flesh is good, and it breeds without difficulty when in confinement, so that it possesses many of the qualifications for domestication. It has, however, one drawback, for it is very quarrelsome, and its powerful beak and large dimensions make it a dreaded foe in the poultry-yard. It feeds on grass like the common Goose, and requires but little care on the part of the owner, and if it could only be induced to lay aside its quarrelsome habits would be quite an acquisition to our limited list of domestic poultry.

For some time after its first discovery it was so fearless of man that it would suffer itself to be approached and knocked down with sticks, but it has now learned caution through bitter experience, and at the sight of a human being seeks safety in flight. Although one of the true Geese it cares little for the water, and in this respect, as well as in others, resembles the wading birds. In England it has bred freely, the specimens in the Zoological Gardens having yearly increased in number. The eggs of the Cereopsis are cream-coloured, and the voice of the bird is loud, hoarse, and has a decided trumpet-like tone that can be heard at a considerable distance.

The general colour of this bird is brownish grey, mottled on the back with a lighter hue, and spotted with black on the wing-coverts and scapularies. On the head the grey fades nearly into white. The bill is short, sharp, and hard, and can be used with great force as an offensive weapon. Its colour is black, and it is covered with a very large greenish yellow cere. The legs are pinkish and the eyes bright red.

To the first of the two birds on page 272 we are indebted for the Domestic Goose, with its few and unimportant varieties.

The GREY-LAG GOOSE is found in many parts of the world, and in a wild state makes occasional visits to this country, and it is probable that the Domestic Geese may derive some of their blood from the other species of the same genus. The white colour of the

2. 3 A

GREY-LAG GOOSE.—*Anser ferus.*   BEAN GOOSE.—*Anser ségetum.*

adult Domestic Gander seems to be the result of careful breeding, probably because white feathers sell at a higher price than the dark and grey plumes. In a state of domestication the Goose lives to a great age, and when treated kindly becomes strongly attached to its friends, and assumes quite an eccentric character. Of the breeding and management of the Goose nothing can be said in these pages, the reader being referred to the numerous extant works on domestic poultry. When wild its flavour is not so delicate as after it has been domesticated and properly fed, and when a wild Goose is shot in the northern climates the sportsman always buries it in the earth some hours before cooking it, a process which removes the rank savour of the flesh. Even the fishy-flavoured sea-birds can be rendered eatable by this curious process.

The Grey-Lag Goose may be known from its congeners by the pinky bill, with its white horny nail at the tip of the mandible. The head, nape, and upper part of the back are ashen brown, and the lower part of the back bluish grey. The quill-feathers are leaden grey; the chin, neck, and breast are grey; and the abdomen white. The average length of the adult male is not quite three feet.

The BEAN GOOSE is another of our English examples of this genus, but is only a visitant of our shores, having its chief residence in the Arctic circle and high northern latitudes, and coming southward about October.

Mr. Selby mentions that the Bean Goose breeds annually upon several of the Sutherland lakes, and in some places it becomes nearly as tame as the common species, but refuses to associate with them. These birds fly in flocks, varying in form according to their size, a little band always flying in Indian file, while a large flock assumes a

BERNICLE GOOSE.—*Bernicla leucopsis.*

V-like form, the sharp angle being always forward. These flocks alight on fields and cultivated grounds, and often commit sad ravages before they again take to wing. On account of this habit the bird is called the Harvest Goose in France.

The beak of this species is rather slender and pointed, and its colour is black with an orange centre. The head and upper parts are brownish grey, the primaries are of a darker hue, both tail-coverts are white, the throat and breast are greyish white, and the abdomen is pure white. The length of the bird is about thirty-four inches.

The BERNICLE GOOSE is also found on our shores, and seems to prefer the western to the eastern coasts.

The name of Bernicle Goose is given to this bird because the olden voyagers thought that it was produced from the common barnacle shell, and this notion had taken so strong a hold of their minds that they published several engravings representing the bird in various stages of its transformation. The positive manner in which they put forth their declaration is very amusing. " What our eyes have seen, and hands have touched," writes Gerard in his " Herbalist," " we shall declare. There is a small island in Lancashire, called the Pile of Foulders, wherein are found the broken pieces of old and bruised ships, some whereof have been cast thither by shipwracke, and also the trunks and bodies with the branches of old and rotten trees, cast up there likewise ; wherein is found a certain spume or froth, that in time breedeth into certain shels, in shape like those of the muskle, but sharper pointed and of a whitish colour, wherein is contained a thing, in form like a lace of silk finely woven as it were together, of a whitish colour ; one end whereof is fastened into the inside of the shel, even as the fish of oisters and muskles are ; the other end is made fast unto the belly of a rude masse or lumpe, which in time commeth to the shape and form of a bird : when it is perfectly formed, the shel gapeth open, and the first thing that appeareth is the foresaid lace or string ; next come the legs of the bird hanging out, and as it groweth greater it openeth the shel by degrees, till at length it is all come forth and hangeth only by the bill : in short space after it commeth to full maturitie, and falleth into the sea, where it gathereth feathers, and groweth to a fowle, bigger than a mallard and less than a goose. . . . For the truth hereof if any doubt, may it please them to repaire unto me, and I shall satisfie them by the testimonie of good witnesses."

The Bernicle Goose generally assembles in large flocks and haunts large salt-marshes near the coast, and feeds on grasses and various algæ. It is a very wary bird and not easily approached. The eggs of this species are large and white. The flesh is considered good. The bill of the Bernicle Goose is black, with a reddish streak on each side. The cheeks and throat are white, a black streak runs from the beak to the eye, the upper parts are boldly marked with black and white, and the lower parts are white. It is rather a small bird, the total length barely exceeding two feet.

There are many other species of Geese which visit our shores in more or less abundance, among which may be mentioned the Egyptian Goose (*Chenalopex Ægyptiaca*), the Brent Goose (*Bernicla Brenta*), the Red-breasted Goose (*Bernicla ruficollis*), the Canada Goose (*Bernicla Canadensis*), the Pink-footed Goose (*Anser brachyrhynchus*), remarkable for its pinky feet and short and narrow beak, the White-fronted Goose (*Anser erythropus*), and the Chinese Goose (*Anser cygnoides*).

THE beautiful Swans now come before our notice. There are nine or ten species of these fine birds, which are well represented in the British Isles, four species being acknowledged as English birds.

OUR most familiar species is the TAME or MUTE SWAN, so called from its silent habits. This elegant and graceful bird has long been partially domesticated throughout England, and enjoys legal protection to a great extent; heavy penalties being proclaimed against any one who kills a Swan without a legal right. The Swan is presumed to be a royal bird, *i.e.* the property of the Crown, and only to be possessed by a subject under a special grant. To each licence thus granted was attached a " swan mark," which was cut on the upper mandible of the birds, in order to show the right of the owner. Swans of a certain age, not marked, become Crown property, except in some instances where a grant conveys the right to seize and keep any adult Swan which has not been marked. Such birds are termed "clear-billed." The "marks" are of endless variety, partly heraldic, and contrived so as to pain the bird as little as possible. One of the most peculiar marks is the double chevron employed by the Vintners' Company, which has given rise to the well-known sign of the Swan with two necks, *i.e.* two nicks. The present royal mark consists of five diamonds, with rounded angles, two cut longitudinally at the base of the beak, and the other three transversely towards the tip. The mark granted to the University of Oxford is a cross with equal arms, each arm being again crossed near its extremity, and that of Cambridge is three buckles, one large, in the middle of the beak, with the point towards the head, and the other two smaller at the tip, with their tongues pointing in different directions.

The process of marking the Swans is termed Swan-upping, a name which has been corrupted into Swan-hopping, and is conducted with much ceremony. The technical term of the Swan-mark is *cigninota*. Swan-upping of the Thames takes place in the month of August, the first Monday in the month being set aside for the purpose, when the markers of the Crown and the Dyers' and Vintners' Companies take count of all Swans in the river, and mark the clear-billed birds which have reached maturity. The fishermen who protect the birds and aid them in nesting are entitled to a fee for each young bird. The mark of the Dyers' Company is a notch on one side of the beak.

The food of the Swan consists mostly of vegetable substances, and the bird can be readily fattened on barley, like ordinary poultry. The young birds, called cygnets, ought not to be killed after November, as they then lose their fat, and the flesh becomes dark and tough. Sometimes the Swan will feed upon animal food, and has been seen to catch and swallow small fish, such as bleak and roach. In the spawning season the Swan is a terrible enemy to the fish, haunting all the spawning-grounds, and swallowing the eggs till it can eat no longer. The Swan will find out the spawn as it hangs on the submerged branches, and strip them of their valuable load. They will follow the carp to their breeding-grounds, and swallow their eggs by the quart, and in many cases

BEWICK'S SWAN.—*Cygnus minor.*   MUTE SWAN.—*Cygnus olor*
WHISTLING SWAN.—*Cygnus ferus.*

they have almost entirely destroyed the fish which inhabited the pond or stream in which they live.

A good idea of the damage done to anglers by the Swan may be formed from the forcible though unrefined description given by one of the piscatorial fraternity, quoted in the

Field newspaper. " There never was no manner of doubt about the dreadful mischief the Swans do. They eats up the spawn of every kind of fish till they have filled out their bags, and then on to shore they goes, to sleep off their tuck out, and then at it again." At such times the birds are so greedy after their feast that they can hardly be driven away, and will often show fight rather than leave the spot.

The nest of the Swan is a very large mass of reeds, rushes, and grasses set upon the bank, close to the water, in some sheltered spot. Generally the bird prefers the shore of a little island as a resting-place for its nest. Like other water-birds, the Swan will raise the nest by adding fresh material before the rising of the water near which it is placed. There are generally six or seven eggs ; large, and of a dull greenish white. The young are of a light bluish grey colour, and do not assume the beautiful white plumage until maturity. The mother is very watchful over her nest and young, and in company with her mate assaults any intruder upon the premises. During the first period of their life the young Swans mount on their mother's back, and are thus carried from one place to another. If in the water, the Swan is able to sink herself so low that the young can scramble upon her back out of the water, and if on land she helps them up by means of one leg.

The HOOPER, ELK SWAN, or WHISTLING SWAN, may at once be distinguished from the preceding species by the shape and colour of the beak, which is slender, without the black tubercle, and is black at the tip and yellow at the base, the latter colour stretching as far as the eye.

The name of Hooper is given to this bird because its cry resembles the word " hoop " very loudly uttered, and repeated many times successively. The bird arrives in England in the winter, mostly coming over in little bands. At the Orkney Islands a few Hoopers remain throughout the year, and large flocks make their appearance about October, departing for the north in April. On the wing these birds generally fly in the form of a wedge, and cry loudly as they go. The curious sound is produced by means of the formation of the windpipe, which is very long, doubled upon itself, and traverses nearly the entire length of the breastbone, which is hollowed to receive it. The length of wind-pipe depends on sex and age, the adult males exhibiting this curious structure in the greatest perfection. In the Mute Swan the windpipe is short, and does not enter the breastbone at all.

The nest of the Hooper is like that of the mute Swan, and the eggs are pale brownish white. The length of the Hooper is about the same as that of the mute species, *i.e.* five feet.

BEWICK'S SWAN, another British species, resembles the hooper in many respects, but may be distinguished from that bird by its smaller size, the large patch of orange at the base of the beak, and the structure of the windpipe and breastbone, which are found in the same place as those of the hooper, but with considerable modification. This is not nearly so graceful a bird as either of the preceding species, sitting on the water more like a goose than a Swan, and having been frequently mistaken for the wild goose, especially when on the wing. When flying, they generally go in a line. The length of this bird is only four feet.

ANOTHER species, the POLISH or IMMUTABLE SWAN, is occasionally found in England. This bird derives its name of immutable from the fact that the young are white like their parents, and do not pass through the grey stage of plumage. It may be readily distinguished by the orange colour, which covers almost the whole of the beak, and the shape and position of the nostrils, which are entirely surrounded by the orange hue. There is a slight tubercle at the base of the beak.

HOWEVER emblematical of ornithological fiction a BLACK SWAN might have been in ancient times, it is now almost as familiar to English eyes as any of the white species.

This fine bird comes from Australia, where it was first discovered in 1698. It is a striking and handsome bird, the blood-red bill and the white primaries contrasting

BLACK SWAN.—*Cygnus atrátus.*

beautifully with the deep black of the plumage. It is not so elegant in its movements as the white Swan, and holds its neck stiffly, without the easy serpentine grace to which we are so well accustomed in our British Swans. The young are not unlike those of the white Swan, and are covered with a blackish grey down. Dr. Bennett mentions that in the Australian Museum is preserved a white or albino specimen with pink eyes.

It is a very prolific bird, producing two and sometimes three broods in a season, commencing to breed about October, and ceasing at the middle of January. The nest is like that of the Swan, and the eggs are from five to eight in number, of a pale green washed with brown. These birds are found in the southern district of Australia and Jamaica, and are sometimes so abundant that Dr. Bennett recollects "a drove of Black Swans being driven up George Street (Sydney) like a flock of geese."

THE beautiful MANDARIN DUCK is worthy of heading the true Ducks, for a more magnificently clothed bird can hardly be found when the male is in health and in his full nuptial plumage.

These bird are natives of China, and are held in such esteem that they can hardly be obtained at any price, the natives having a singular dislike to seeing their birds pass into the possession of Europeans. "A gentleman," writes Dr. Bennett, "very recently wrote from Sydney to China, requesting some of these birds to be sent to him. The reply was, that from the present disturbed state of China, it would be easier to send him a pair of mandarins than a pair of Mandarin Ducks." This bird has the power of perching, and it is a curious sight to watch them perched on the branches of trees overhanging the

MANDARIN DUCK, OR CHINESE TEAL. —*Aïx galericulata.*

pond in which they live, the male and female being always close together, the one gorgeous in purple, green, white, and chestnut, and the other soberly apparelled in brown and grey.

This handsome plumage of the male is lost during four months of the year, *i.e.* from May to August, when the bird throws off his fine crest, his wing-fans, all his brilliant colours, and assumes a sober tinted dress resembling that of his mate. The Summer Duck of America (*Aix sponsa*) bears a close resemblance to the Mandarin Duck, both in plumage and manners ; and at certain times of the year is hardly to be distinguished from that bird. The Mandarin Duck has been successfully reared in the Zoological Gardens, some being hatched under the parent bird, and others under a domestic hen, the latter hatching the eggs two days in advance of the former. The eggs are of a creamy brown colour.

The crest of this beautiful Duck is varied green and purple upon the top of the head, the long crest-like feathers being chestnut and green. From the eye to the beak the colour is warm fawn, and a stripe from the eye to the back of the neck is soft cream. The sides of the neck are clothed with long painted feathers of bright russet, and the front of the neck and breast are rich shining purple. The curious wing-fans, that stand erect like the wings of a butterfly, are chestnut, edged with the deepest green, and the shoulders are banded with four stripes, two black and two white. The under surface is white. The female is simply mottled brown, and the young are pretty little birds, covered with downy plumage of a soft brown above, mottled with grey, and creamy white below.

The SHIELDRAKES, of which there are two British species, namely, the common Shieldrake (*Tadorna vulpanser*) and the Ruddy Shieldrake (*Casarka rutila*), are handsome birds, and remarkable for the singular construction of the windpipe, which is expanded just at the junction of the two bronchial tubes into two very thin horny globes, one being nearly twice the size of the other. They are sometimes called Burrow Ducks, because they lay their eggs in rabbit-burrows made in sandy soil, and are often discovered by the impression of their feet at the entrance of the holes. The nests are made of grass, lined with down plucked from the breast of the parent, and the eggs are generally from ten to twelve in number.

WIGEON.—*Móreca Penélope*.

THE well-known WIGEON is very plentiful in this country, arriving about the end of September or the beginning of October, and assembling in large flocks.

These birds, although wary on some occasions, are little afraid of the proximity of man and his habitations, feeding boldly by day, instead of postponing their feeding-time to the night, as is often the case with water-fowl. The food of the Wigeon mostly consists of grass, which it eats after the fashion of the common goose. About March or April the Wigeons leave us to pass to their northern breeding-places, although a few pairs remain in the north of Scotland, and there breed their young. The nest of the Wigeon is made of decayed reeds and rushes, and is lined with the soft down torn from the parent's body. The eggs are rather small and of a creamy white colour. The number of eggs is from five to eight. The flesh of this bird is very delicate, and it is largely sold in our markets.

The forehead and top of the head of the adult male are creamy white, and the cheeks and back of the neck rich chocolate, and there is a dark green streak from the eye backwards. The back is greyish white pencilled with irregular lines of black, the wing-coverts are white with black tips, the primaries are dark brown, the secondaries green on their outer webs edged with black, and the outer webs of the tertials are black edged with white. The chin and throat brownish black, the breast chestnut, the flanks white with irregular black lines, the abdomen white, and under tail-coverts deep black. The tail is long, pointed, and nearly black.. The female is ruddy brown on the head and neck, with dark specks, the back is brown, and the under surface white. The quill-feathers are of the same tints as in the male. After the breeding season the male loses his bright apparel, and is not unlike the female in the sobriety of his dress. The windpipe swells at its junction into one large globular sac.. The length of the Wigeon is about eighteen inches.

The PINTAIL DUCK (*Dafila acuta*), so called on account of its long and sharply pointed tail, is one of our winter visitors, arriving in October, and departing in the spring. The male is a very handsome bird, its head and neck being rich dark brown, its back beautifully pencilled with black on a grey ground, and the throat, breast, and abdomen snowy white, and a line of the same hue running up the sides of the neck as far as the head. The length of a male bird is about twenty-six inches, the female is shorter, because her tail-feathers are not so well developed.

THE common MALLARD or WILD DUCK now comes before our notice.

This is by no means one of the least handsome of its tribe, the rich glossy green of the head and neck, the snowy-white collar, and the velvet black of the odd little curly feathers of the tail, giving it a bold and striking appearance, which, but for its familiarity, would receive greater admiration than it at present obtains. It is the stock from which descended our well-known domestic Duck, to which we are so much indebted for its flesh and its eggs.

In its wild state the Mallard arrives in this country about October, assembling in large flocks, and is immediately persecuted in every way that the ingenuity of man can devise. Sportsmen go out to shoot it, armed with huge guns that no man can hold to his shoulder, and have to be mounted on gimbals in a boat, thus bringing down whole clouds of birds at a discharge.

Nets and snares of various kinds are in great request, the principal of which is the Decoy, a very complicated piece of apparatus, requiring several acres of water, shaped in a peculiar manner, surrounded by reeds and bushes, and furnished with an elaborate system of tunnels made of net strained over wooden arches, very large towards the mouth, and tapering to a point at the extremity. Along these tunnels are set a number of reed screens, so made that a person standing behind them is hidden from birds on the lake, while he can be seen by those in the tunnel. Various methods are employed to induce the Ducks to enter the mouth of the tunnels, scattered grain being a kind of bait, and certain trained decoy Ducks acting as lures which beguile the wild birds to their death.

As soon as a sufficient number of Ducks have entered a tunnel, the fowler, standing behind the first screen, waves his hat so as to frighten the birds, and makes them swim away from him, *i.e.* along the tunnel. He then runs forwards, frightens them again, and so drives them to the end, where they find their further progress impeded by the narrow funnel-like termination, and their egress cut off by the structure of the net. The fowler then takes off the pocket at the end of the tunnel, which is movable, kills the Ducks quite at his leisure, replaces the pocket, and returns to his work. Generally, each decoy has about six of these tunnels radiating from the centre, and the fowler is assisted by his dog, which is trained to play about the tunnels in such a fashion that the Ducks become inquisitive, and going to investigate the phenomenon, come upon the grain and the decoys, and are lured to the deadly pocket. A small piece of water about three or four acres in extent is better for a decoy than a large expanse, as in the latter case the Mallards might not see the dog or the decoy Ducks. Flat fenny soils afford the best localities for the decoys.

The nest of the Mallard is made of grass lined and mixed with down, and is almost always placed on the ground near water, and sheltered by reeds, osiers, or other aquatic plants. Sometimes, however, the nest is placed in a more inland spot, and it now and then happens that a Duck of more than usual eccentricity builds her nest in a tree at some elevation from the ground, so that when her young are hatched, she is driven to exert all her ingenuity in conveying them safely from their lofty cradle to the ground or the water. Such a nest has been observed in an oak-tree twenty-five feet from the ground, and at Heath Wood, near Chesterfield, one of these birds usurped possession of a deserted crow's nest in an oak-tree. Many similar instances are on record.

The eggs of the Mallard are numerous, but variable according to the individual which lays them, some being far more prolific than others. The eggs are rather large and of a greenish white colour.

MALLARD.—*Anas boschas.*

Like many other birds, the male Mallard or Drake lays aside his best clothes during certain months of the year, and is then hardly to be distinguished from his mate. Mr. Waterton has given an admirable description of the process. "About the 24th of May, the breast and back of the Drake exhibit the first appearance of a change of colour. In a few days after this, the curled feathers above the tail drop out, and grey feathers begin to appear among the lovely green plumage which surrounds the eyes. Every succeeding day now brings marks of rapid change. By the 23d of June scarcely one single green feather is to be seen on the head and neck of the bird. By the 6th of July every feather of the former brilliant plumage has disappeared, and the male has received a garb like that of the female, though of a somewhat darker tint. In the early part of August this new plumage begins to drop off gradually, and by the 10th of October the Drake will appear again in all his rich magnificence of dress, than which scarcely anything throughout the whole wide field of nature can be seen more lovely or better arranged to charm the eye of man."

The head and neck of the adult male Mallard are rich shining green, with a collar of pure white at the lower part of the neck. The back is chestnut-brown deepening into black on the upper tail-coverts. The four central tail-feathers are velvety black and curled over as if with a barber's tongs, and the rest are ashen grey edged with white. The greater wing-coverts have a bold white bar and are tipped with velvety black, and the wings are beautifully coloured with shining purple, snowy white, and velvet-black. The upper part of the breast is dark chestnut, and the remainder of the breast and the whole of the under parts are greyish white, pencilled under the wings with dark grey lines. The total length is about two feet. The female is rather less, and her whole plumage is brown of various shades.

The pretty little TEAL is the smallest and one of the most valuable of the British Ducks, its flesh being peculiarly delicate and its numbers plentiful.

It arrives on our shores about September, coming over in large flocks, and remains with us until the commencement of its breeding season. Some few birds, however, remain in the British Isles throughout the year. Like other Ducks, it is found on lakes, ponds,

TEAL.—*Querquédula Crecca.*

and in marshy places, choosing the last-mentioned localities for its home. The nest of the Teal is made of a large heap of leaves, grasses, and sedges, lined with down and feathers. The number of eggs is about eight or ten, and their colour is buffy white. The Teal is caught in decoys together with the Mallard and other Ducks.

The colour of this little bird is rather complicated. The forehead and top of the head are chestnut-brown, the sides of the face are dark shining green on the upper half, and rich chestnut on the lower half. Above and below the eye run two narrow streaks of buff, sharply dividing the green and the chestnut from each other. The chin is black, the nape of the neck and back are grey-white covered with a multitude of narrow pencillings; the wings are brown, velvet-black and purple, with a bold white bar formed by the white tips of the secondaries, and the tail is blackish brown. The breast is white tinged with purple, and covered with circular black spots, and the abdomen is white. In total length the Teal is not quite fifteen inches. The female is almost wholly brown.

Another species of the same genus, the BIMACULATED DUCK (*Querquedula bimaculata*), is found in England, as is a closely allied species, the GARGANEY or SUMMER TEAL (*Pterocyanea circia*).

THE curious SHOVELLER DUCK may be at once known by the form of the beak, which is much widened on each side near the tip, and bears some resemblance to the beak of the Spoonbill.

The Shoveller is a winter visitant of this country, and, as is often the case with others of its tribe, a few pairs remain in the British Isles throughout the year. It is found on lakes, ponds, and on marshy grounds, feeding upon worms, insects, and various vegetable substances. Snails and the small fry of fishes have been found in the stomach of the Shoveller Duck. The flesh of this bird is singularly good, and is thought by some practical judges to equal that of the far-famed canvas-backed Duck.

The nest of the Shoveller is placed near water, but on some dry spot, and is made of long slender grass-blades covered with down. The average number of eggs in each nest is eight or nine, and their colour is cream-white with a wash of green. The young do not possess the curiously dilated beak.

The head and upper part of the neck of the adult male are rich green, and the lower part of the neck white, this tint extending to the scapularies, and some of the tertials. The back is brown, the tip of the wing, the lesser wing-coverts, and part of the tertials are pale blue, the primaries being blackish brown. The upper tail-coverts and tail are

SHOVELLER DUCK.—*Spatula clypeata.*

black, and the breast and abdomen are chestnut-brown. The female is brown of various tones above, and pale brown below. The total length of the bird is about twenty inches.

Before taking leave of the Anatinæ, or true Ducks, we must cast a casual glance at the GADWALL (*Chaulelasmus strepera*), a noisy and soberly plumaged species, which by right belongs to North America, but sometimes comes over and visits the British Islands. The well-known MUSCOVEY or MUSK DUCK also belongs to this sub-family.

WE now arrive at another sub-family of Ducks, termed Fuligulinæ, of which we have many representatives in England.

The POCHARD DUN-BIRD or REDHEADED POKER is one of our winter visitors, appearing, as is usual with such birds, in October and departing in the spring. It is a wary and timid bird, and being an excellent diver is often able to escape from the decoys by submerging itself at the first alarm, and making its way under water to the mouth of the tunnel. It is, however, possessed of little presence of mind, and according to Montagu can be taken plentifully by a very rude kind of process.

"Poles were erected at the avenues to the decoys, and after a great number of these birds had collected for some time on the pond to which the wild fowl resort only by day, and go to the neighbouring fens to feed by night, a net was at a given time erected by pulleys to these poles, beneath which a deep pit had previously been dug ; and as these birds, like the woodcocks, go to feed just as it is dark, and are said to always rise against the wind, a whole flock was sometimes taken in this manner. For if once they strike against the net they never attempt to return, but flutter down the net until they are received into the pit, whence they cannot rise, and thus we are told twenty dozen have been taken at one catch."

The voice of the Pochard is a kind of harsh croaking sound when the bird is alarmed, but at other times is a low whistle. The bird swims rather deep in the water, and though it dives well, is a bad and awkward walker.

The nest of the Pochard is placed upon the edges of lakes and ponds, and is protected by the sedges, rushes, or similar coarse vegetation. The eggs are ten or twelve in number, and their colour is buffy white.

In the adult male the head and upper part of the neck are chestnut-red, and the lower part of the neck deep velvet-black. The back is grey profusely sprinkled with a darker tint. The secondaries are tipped with white, the remaining quill-feathers are dark grey, some of them being tipped with brown. The tail is greyish brown. The breast and

POCHARD.—*Nyróca ferina.*

abdomen are white finely pencilled with grey, and the under tail-coverts are black. The bill is black at the point and base, and pale blue in the middle. The female is without the deep black of the lower neck. The total length of this bird is about twenty inches.

Several curious species of Duck are closely allied to the Pochard, among which may be noticed the TUFTED DUCK (*Fuligula cristata*), the SCAUP DUCK (*Fuligula marila*), both British birds, and the celebrated CANVAS-BACK DUCK of America (*Fuligula Vallisneria*). The GOLDEN-EYE DUCK (*Clangula glaucion*), the curiously mottled HARLEQUIN DUCK (*Clangula histrionica*), and the light and active SPIRIT DUCK, or BUFFEL-HEADED DUCK (*Clangula albeola*), are also allied, as is the LONG-TAILED DUCK (*Harelda glacialis*), all of which are found in England.

IN the southern parts of England the EIDER DUCK is only a winter visitant, but remains throughout the year in the more northern portions of our island, and in the north of Scotland.

This bird is widely celebrated on account of the exquisitely soft and bright down which the parent plucks from its breast and lays over the eggs during the process of incubation. Taking these nests is with some a regular business, not devoid of risk on account of the precipitous localities in which the Eider Duck often breeds. The nest is made of fine seaweeds, and after the mother-bird has laid her complement of eggs she covers them with the soft down, adding to the heap daily until she completely hides them from view.

The plan usually adopted is to remove both eggs and down, when the female lays another set of eggs and covers them with fresh down. These are again taken, and then the male is obliged to give his help by taking down from his own breast, and supplying the place of that which was stolen. The down of the male bird is pale coloured, and as soon as it is seen in the nest, the eggs and down are left untouched in order to keep up the breed. Mr. Yarrell mentions that on one of the northern coasts of England, the Eiders had been nearly exterminated by foolish persons, who robbed the nests, and sold the eggs for consumption. A gentleman, however, who was employed in building a light-house on the rocks prohibited any such robberies while he was in authority, and in consequence of his judicious management the Eiders became plentiful again.

The Eider is a shy retiring bird, placing its nest on islands and rocks projecting well into the sea. It is an admirable diver, its legs being set very far back, and obtains much of its food by gathering it under water. The same structure, however, which gives the

EIDER DUCK.—*Somatéria mollissima.*

bird its facility in diving, renders it a bad and awkward walker. The number of eggs laid by the Eider is five or six, and their colour is pale green. There are generally two broods in the year.

In the male bird the top of the head is velvety black, and the cheeks are white. The ear-coverts and back of the head are pale green. The back is white, the primaries and secondaries dull black, and the tertials white, long, and drooping, having a very striking effect as they hang over the dark wings. The neck and upper parts of the breast are white, the lower parts of the neck pale buff, and the breast and abdomen black, relieved by a patch of white on the flanks. The bill and legs are green. The female is reddish brown, mottled with darker brown. The total length of this bird rather exceeds two feet.

An allied species, the KING DUCK (*Somateria spectabilis*), is sometimes found in England, though it is a rare bird. It is about the size of the eider, and is notable for its black and white body, and the light red beak and legs.

The RACEHORSE, or STEAMER DUCK (*Micropterus brachypterus*), is a very remarkable species, having very short wings, so that it cannot fly, and legs so formed that it scuds over and drives through the water with amazing speed. While passing over the surface it sends the water flying behind it like a paddle-steamer, and has thereby earned its name. It is a very large species, forty inches in length, and its colour is leaden grey above, and white below. It is found in the Falkland Islands and Patagonia.

WANT of space compels us to omit the SCOTER DUCKS (*Oidemia*), and to pass at once to the next sub-family, represented by two species.

The GOOSANDER is one of our winter visitors, making its appearance about November, and departing in March. Some few birds, however, remain throughout the year. It is generally to be found on the northern coasts, where it may be observed diving after fish with great address, and bringing them to the surface in its long, slender, deeply notched beak, which is so plentifully provided with so-called teeth, that the bird in some places goes by the name of Jacksaw.

GOOSANDER.—*Mergus Castor.*   SMEW.—*Mergellus albellus.*

The nest of this species is placed near water, under the shelter of long grass or the hollow of a decayed tree, and is made of grasses and roots, and lined with down. The eggs are generally six or seven in number, and their colour is very pale buff. In Lapland the eggs are regularly taken by the natives, who always leave a few in order to keep up the breed.

The Goosander is a handsome bird; the head and upper parts of the neck are deep shining green, with a sort of tuft on the back of the neck. The back is black, fading to dark grey near the tail. The wing-coverts are white, as is the front of the wing, the primaries being black. The lower part of the neck, breast, and abdomen are soft warm buff, the bill is scarlet, and the legs reddish orange. In the female the head and neck are reddish brown, and the back grey. The length of the bird rather exceeds two feet.

THE pretty little SMEW is another of our winter visitors, and is a very common bird, being found not only upon the seashore, but frequenting inland lakes and ponds.

It is rather a shy bird, and not very easily approached, especially as it is a swift and active diver, vanishing below the surface at the least alarm, and emerging at some distance.

GREAT NORTHERN DIVER.—*Colymbus glaciális.*

It flies well, but, like most diving birds, walks badly. The food of the Smew consists of fish, small crustaceans, molluscs, and insects, which it obtains under the surface as easily as above it. The eggs of this species are warm buff in colour, and they are generally eight or ten in number.

The head, chin, and neck of the adult male are white. At the base of the bill at each side there is a black patch which surrounds the eye, and over the back of the head runs a green streak forming a kind of crest with some white elongated feathers. The back is black, and the tail grey, the wings are black and white, and the under surface pure white, pencilled with grey on the flanks. The female has her plumage mostly reddish brown and grey. The length of the Smew is about seventeen inches, the female being three inches shorter.

ANOTHER species of the same genus is occasionally found in England, the HOODED MERGANSER (*Mergus cuculládus*), remarkable for the pure white crest on the jetty black head. It is a most active bird on the wing and in the water, diving with wonderful rapidity, and is apt to rise suddenly in the air, and shoot off as if fired from a gun.

The RED-BREASTED MERGANSER (*Mergus serrator*) is not an uncommon species, and sometimes comes far southwards, having been killed on the Thames, and even in Devonshire. It is popularly known by the title of Sawbill.

WE now come to the family of Colymbidæ, or Divers, several examples of which are British birds.

The GREAT NORTHERN DIVER is common on the northern coasts of the British Islands, where it may be seen pursuing its arrowy course through and over the water, occasionally dashing through the air on strong pinions, but very seldom taking to the shore, where it is quite at a disadvantage.

Perhaps there is no bird which excels the Northern Diver in its subaqueous powers, although the penguins and cormorants are equally notable in that respect. Its broad webbed feet are set so very far back that the bird cannot walk properly, but tumbles and scrambles along much after the fashion of a seal, pushing itself with its feet, and scraping its breast on the ground. In the water, however, it is quite at its ease, and, like the seal, no sooner reaches the familiar element, than it dives away at full speed, twisting and turning under the surface as if in the exuberance of happy spirits. So swiftly can it glide through the water that it can chase and capture the agile fish in their own element, thus exhibiting another curious link in the interchanging capacities of various beings; the bats, for example, surpassing many birds in airy flight, the cursorial birds running faster than most quadrupeds, the seals and others equalling the fish in their own watery domain, and some of the fish, again, being able to pass for a considerable distance through the air.

To shoot this bird is a matter of no ordinary difficulty, and is best achieved by the gunner concealing himself behind rocks near the water's edge, where the bird will often come near enough for an effectual shot. As to fair chase, the capture of the Diver is impossible, and to shoot it is almost impracticable. Sir W. Jardine mentions that he chased a Northern Diver for a considerable time in the Frith of Forth, and though rowed by four strong men, he could not get near the active bird. The Rev. J. C. Atkinson corroborates this and other similar narratives. "I have myself pursued the Great Northern Diver (which was shot through the neck at the first discharge, and seriously weakened by the wound, and the consequent loss of blood) for a lengthened space of time ; and though I was assisted by a friend, armed like myself with a heavy flight gun, the bird dived so quickly, and re-emerged at such a distance from the point of his disappearance, that it was not until after we had fired twelve or fifteen shots at him, that we were able at last to take him at such a disadvantage as to disable him for further exertion."

Like many other diving birds, it is able to sink itself in the water, the head disappearing after the body and neck. Young birds are more frequently found on our coasts than the adult. The eggs of the Northern Diver are generally two in number, and of a dark olive-brown, spotted sparingly with brown of another tone. They are laid upon the bare ground, or on a rude nest of flattened herbage near water, and the mother-bird does not sit, but lies flat on the eggs. If disturbed, she scrambles into the water and dives away, cautiously keeping herself out of gunshot, and waiting until the danger is past. Should she be driven to fight, her long beak is a dangerous weapon, and is darted at the foe with great force and rapidity.

The head of the adult Northern Diver is black, glossed with green and purple, and the cheeks and back of the neck are black without the green gloss. The back is black, variegated with short white streaks, lengthening towards the breast, and the neck and upper part of the breast are white spotted with black, and cinctured with two collars of deep black. The breast and abdomen are white. The total length of the bird is not quite three feet. The immature bird is greyish black above, each feather being edged with a lighter hue, and the under parts of the body are dull white. In some places this bird is called the Loon.

Two other species of Divers belong to the British Islands ; namely, the BLACK-THROATED DIVER (*Colymbus arcticus*), notable for the bold alternate bars of black and white which decorate the back and tertials, and the RED-THROATED DIVER (*Colymbus septentrionalis*), known by its smaller size and the red throat. On account of the havoc which this Diver makes among the shoals of sprats, it is sometimes termed the Sprat Loon.

THE sub-family of the Grebes is represented in England by several well-known species. All these birds may be readily distinguished by the peculiar form of the foot, in which each toe is furnished with a flattened web, the whole foot looking something like a horse-chestnut leaf with three lobes.

GREAT CRESTED GREBE.—*Pódiceps cristátus.*   EARED GREBE.—*Pódiceps aurítus.*

The GREAT CRESTED GREBE is the largest of these birds, and is found throughout the year in several parts of England, preferring the lakes and the fenny districts. Like the divers, the Grebes are very bad walkers, but wonderfully active in the water, and tolerably good fliers. They very seldom attempt to walk, and when sitting they bend their legs under them, and assume a very upright attitude. The Crested Grebe swims low in the water, and dives with great facility, chasing and capturing fish, on which it chiefly feeds. It is a rather curious fact that all these birds have a habit of swallowing feathers, apparently plucked from their own breasts.

On account of the great activity of the bird, Grebe-shooting from a boat is a favourite amusement on the Lake of Geneva, the odds being fearfully against the bird—boats, guns, and telescopes being all employed and all needed. Moreover, unless the boatmen are very skilful, well acquainted with the habits of the bird, and know exactly in which direction to row when the Grebe dives, the great activity and remarkable endurance of the bird will certainly secure its safe retreat.

The nest of the Great Crested Grebe is made of decaying vegetation, and is placed so close to the water that it is seldom dry, the bird feeling no discomfort in a wet bed. There are generally three or four white eggs, and the young are very pretty little birds, covered with soft downy plumage, and boldly striped with grey and black from beak to tail.

In the adult Crested Grebe the top of the head is dark brown, the cheeks white, and the curious tippet reddish chestnut, darkening round the edge. The back of the neck and upper surface is dark brown, the secondaries are white, and the whole under surface is white, with a peculiar satiny lustre. The flanks are fawn, and the legs and toes dark green on their outer surfaces, changing to yellow on the inner. The crest is small in yearling birds, and does not attain its full dimensions until the third year. The length of this species is about twenty-two inches.

The EARED GREBE is the rarest of the British species, and derives its name from a tuft of rich golden feathers which arise behind the eye of the adult bird. It is not nearly so large as the preceding species, being only one foot in length, and it may be distinguished from a young crested Grebe by the slight upward curve of the bill, and the absence of rusty red feathers between the eye and base of the beak. The food of this bird consists of small fish, molluscs, aquatic insects, and some water-plants. The nest is made among thick herbage, and the eggs are creamy white, and three or four in number.

The head and neck of the Eared Grebe are black, and the back very dark brown. A triangular patch of bright warm chestnut feathers is placed behind the eye, and the secondaries are white. The under surface of the body is silken white, deepening to chestnut on the flanks, and the legs are green.

ANOTHER very fine British species is the SCLAVONIAN or HORNED GREBE (*Podiceps cornutus*), remarkable for its splendid ruff of dark brown feathers. This bird has been seen to carry its young upon its back, and to dive with them when alarmed.

DABCHICK.—*Podiceps minor.*

THE well-known DAB-CHICK, or LITTLE GREBE, is the smallest and commonest of the British species, being found in most rivers, lakes, or large ponds, where the weeds and rushes afford it a concealment, and a foundation for its nest.

It is a pretty little bird, quick and alert in its movements, and, like the rest of the Grebes, has a great love for water, and an invincible antipathy to land. When alarmed it dives so instantaneously that the eye can hardly follow its movements, and if at the moment of its emergence it perceives itself still in danger it again dives, not having been on the surface for a single second of time. Like many other aquatic birds, it can sink itself in the water slowly, and often does so when uneasy, rising again if relieved from its anxiety, or disappearing as if jerked under the surface from below. I have often seen them in a little pond only a few yards across, thus diving and popping up again with almost ludicrous rapidity.

This bird can fly moderately well, and can rise from the water without difficulty, when it will circle about the spot whence it rose, and keep some five or six feet above the surface, uttering the while its curious rattling cry.

The nest of this bird is made of water-weeds, and is placed among the rank aquatic herbage. It is scarcely raised above the surface, and is mostly wet. The eggs are five or six in number, and their normal colour is white, though they soon become stained with the decaying vegetable matter on which they rest, and before hatching are of a muddy brown hue. The mother-bird always covers her eggs with leaves and aquatic algæ before leaving them. The bird has a curious habit of building a kind of supplementary nest, in which it sits until it has completed the structure in which the eggs are to be laid. The young soon take to the water, and are, on their first introduction to the waves, nearly as adroit as their parents.

The food of the Dabchick consists of insects, molluscs, little fish, and the smaller crustaceans.

In its summer plumage the head, neck, and upper portions of the body are dark brown, except the secondaries, which are white. The sides of the face are warm chestnut, and the under surface is greyish· white. In the winter the upper part of the body is chocolate-brown, and some of the primaries are white, the chin is white, the front of the neck ashen brown, and the under surface greyish white. The total length of the Dabchick rather exceeds nine inches.

THE sub-family of the Alcinæ or Auks has several British representatives, among which the GREAT AUK is the rarest.

This bird, formerly to be found in several parts of Northern Europe, in Labrador, and very rarely in the British Isles, has not been observed for many years, and is thought to be as completely extinct as the Dodo. Almost the last living specimens known were seen in the Orkneys, and were quite familiar to the inhabitants under the name of the King and Queen of the Auks. So agile is (or was) this bird in the water, that Mr. Bullock chased the male for several hours without being able to get within gunshot, although he was in a boat manned by six rowers. After his departure the bird was shot and sent to the British Museum. The female had been killed just before his arrival.

GREAT AUK.—*Alca impennis.*

The egg of this bird is laid close to the water's edge, and is a very large one, marked after a rather curious fashion, hereafter to be described.

One of these birds was caught by a hook, and the story is told by Audubon. " Mr. Henry Havell, brother of my engraver, when on his passage from New York to England, hooked a Great Auk on the banks of Newfoundland in extremely boisterous weather. On being hauled on board it was left at liberty on the deck. It walked very awkwardly, often tumbling over, bit every one within reach of its powerful bill, and refused food of all kinds. After continuing several days on board, it was restored to its proper element." According to the same writer, the Great Auk is said to have been tolerably plentiful near Newfoundland and about Nahant.

According to Mr. Lloyd, this bird formerly frequented certain parts of Iceland, a certain locality called the Auk-Skär being celebrated for the number of Auks which nested upon it. The Skär, however, is so difficult of approach on account of the heavy surf which beats upon it, that few persons have the daring to land. In 1813 a number of Auks were taken from the Skär, and, horrible to relate, they were all eaten except one.

For more recent accounts of this bird I am indebted to the kindness of Mr. R. Champley, who has brought together a large mass of information on the subject, and most liberally furnished me with many valuable notes. The following passages are portions of letters addressed to Mr. Champley by recent travellers.

Sir L. M'Clintock, of the celebrated little vessel *Fox*, writes as follows :—" The Great Auk has not been met with by any of the modern arctic expeditions. I was told in South Greenland, twenty-five years ago, that a young specimen was obtained, but am not certain of the fact. The resident Europeans are quite aware of the value attached by naturalists to the bird, so have kept a sharp look-out for it." The same correspondent again writes :—" Nothing has come to my knowledge respecting the Great Auk during my late voyage to Iceland, Greenland, and Labrador."

Dr. Rae writes as follows :—" I regret I have little or no information to give you about Alca Impennis, although I questioned many persons in Iceland about this rare, if not extinct bird. An ineffectual search was made for them some time ago on an island N.W. of Iceland, where they had previously been not uncommon." Captain Allan Young, R.N., of the *Fox*, writes :—" To my own knowledge, we never saw the Great Auk, nor has the bird been seen for many years on the south coast of Greenland."

Mr. Proctor, curator of the Durham Museum, writes :—" One bird was bought, 1834, by the Rev. I. Gisborne, for 8*l.* I was in Iceland in 1833, and made every inquiry and sought for it, but never saw a single bird. I went to the northern parts of Iceland in 1837 in search of it again, and travelled through the northern parts, and as far as Gremsey, Iceland—a small island forty miles north of the main land—but could not meet with it. I have never seen the bird alive, neither have I ever had the eggs."

Mr. David Graham mentions, that in 1846 he met a fisherman in Iceland who had two Auks and two eggs which he offered for 20*l.*, an offer which was refused, to the writer's great subsequent regret. The two birds were sent to Copenhagen and the eggs were broken. In 1821 Mr. Maclellan, of Scalpay, obtained a Great Auk alive, but allowed it to escape after a while. Another specimen was taken by Mr. Adams, of Lewis, Hebrides, which had been caught in a net. Nicholson says that in 1804, at the mouth of the Bull River, in Greenland, the Great Auks were caught in great numbers, and served as sustenance to the inhabitants through the months of February and March ; their down being made into outer garments. In 1858 Messrs. J. Wolley and Alfred Newton went to Iceland, but were unable to land on the Auk-Skär, or even see a bird. They obtained bones from the inhabitants.

Owing to the extreme value attached to this bird and its eggs, and the high price which it brings in the market, ingenious attempts have been made to forge copies of the eggs. Not many years ago, several apparently genuine Auk's eggs were offered at a low price, but turned out to be nothing more than forgeries admirably manufactured, and really valuable as copies of the true egg. I am acquainted with the names of the persons connected with this imposition, but need not mention them.

Mr. Champley has lately made a tour through Europe, and collected evidence of every bird and genuine egg in existence, and has kindly forwarded to me a copy of his list. It is too long to be transcribed at length, but may be condensed as follows :—

Altogether, thirty-four birds and forty-two eggs are known, which are distributed as follows :—England possesses fourteen birds and twenty-four eggs. The British Museum has two eggs and two birds ; and very few collections, either private or public, can boast of both egg and bird. Sir W. Milner has one of each, and so have Mr. J. Hancock and Mr. Troughton. Mr. Champley's collection is, however, the most valuable, as it includes one bird and four eggs. A photograph of his bird and an egg is now before me.

America has two eggs, both in Philadelphia; France has three birds and six eggs; Germany and Prussia have six birds and six eggs; Austria has two birds; Switzerland one bird, and Italy three birds; Russia one bird; Denmark two birds and two eggs, and Holland two of each. It is worthy of notice, that in England two birds and one egg are changing hands, and in Germany the same may be said of one bird and two eggs.

The eggs are variable in size, and colour, and markings, some being of a silvery white and others of a yellowish white ground; and the spots and streaks are greatly different in colour and form, some being yellowish brown and purple, others purple and black, and others intense blue and green.

The dimensions and weight of Mr. Champley's eggs are as follows:—First egg, five inches long, two inches ten and a half lines wide, weight thirty-one scruples ten grains; second egg, four inches ten and a half lines long, two inches eleven and a half lines wide, weight forty-one scruples nine grains; third egg, four inches seven lines long, three inches one line wide, weight forty scruples nine grains; fourth egg, five inches one line long, three inches wide, weight thirty-eight scruples fifteen grains.

The upper surface of this bird is black, except a patch of pure white round and in front of the eye, and the ends of the secondaries, which are white. The whole of the under surface is white, and in winter the chin and throat are also white. The young are mottled with black and white. The total length of the bird is thirty-two inches.

ANOTHER species of the same genus, the RAZOR-BILL (*Alca torda*), is tolerably common in the Arctic seas, and is occasionally found in Great Britain. A young bird was obtained from a rock in the Isle of Wight. The eggs of this species are singularly variable. Mr. Champley informs me that he possesses five hundred distinct specimens.

PUFFIN.—*Fratercula árctica.*

THE odd little PUFFIN, so common on our coasts, is remarkable for the singular shape, enormous size, and light colours of its beak, which really looks as if it had been originally made for some much larger bird. Owing to the dimensions of the beak it is often called the SEA-PARROT, or the COULTERNEB.

The Puffin can fly rapidly and walk tolerably, but it dives and swims supremely well, chasing fish in the water, and often bringing out a whole row of sprats at a time ranged along the side of its bill, all the heads being within the mouth and all the tails dangling outside. It breeds upon the rocks and in the rabbit-warrens near the sea, finding the ready-made burrows of the rabbit very convenient for the reception of its eggs, and fighting with the owner for possession of the burrow. Where rabbits do not exist, the Puffin digs ts own burrow, and works hard at its labour. The egg is generally placed several feet within the holes, and the parents defend it vigorously. Even the raven makes little of an attack, for the Puffin gripes his foe as he best can, and tries to tumble into the sea, where the raven is soon drowned, and the little champion returns home in triumph. The egg is white, but soon becomes stained by the earth. The food of this bird consists of fish, crustaceans, and insects.

The top of the head, the back, and a ring round the neck are black, and the cheeks and under surfaces are white. The beak is curiously striped with orange upon bluish grey, and the legs and toes are orange. The length of this bird is about one foot.

KING PENGUIN.—*Aptenodytes Pennantii.*

THE Penguins form a very remarkable sub-family, all its members having their wings modified into paddles, useless for flight, but capable of being employed as fore-legs in terrestrial progression when the bird is in a hurry, and probably as oars or paddles in the water. There are many species of Penguins, but as they are very similar in general habits, we must be content with a single example.

The KING PENGUIN is a native of high southern latitudes, and is very plentiful in the spots which it frequents. It swims and dives wonderfully well, and feeds largely on cuttle-fish. Dr. Bennett has given an admirable description of this bird and its habits, as it appeared on Macquarrie's Island in the South Pacific Ocean.

"The number of Penguins collected together in this spot is immense, but it would be almost impossible to guess at it with any near approach to truth, as during the whole of the day and night thirty or forty thousand of them are continually landing, and an equal number going to sea. They are arranged, when on shore, in as compact a manner and in as regular ranks as a regiment of soldiers, and are classed with the greatest order, the young birds being in one situation, the moulting birds in another, the sitting hens in a third, the clean birds in a fourth, &c.; and so strictly do birds in similar condition congregate, that should a bird that is in moulting intrude itself among those which are clean, it is immediately ejected from among them.

The females hatch their eggs by keeping them close between their thighs; and if approached during the time of incubation, move away, carrying their eggs with them. At this time the male bird goes to sea and collects food for the female, which becomes very fat. After the young is hatched, both parents go to sea and bring back food for it: it soon becomes so fat as scarcely to be able to walk, the old birds getting very thin. They sit quite upright in their roosting-places and walk in the erect position until they arrive at the beach, when they throw themselves on their breasts in order to encounter the very heavy sea met with at their landing-places."

There is only a single egg, and its colour is greyish white. The young are covered with puffy grey wool. As the bird walks or rather shuffles along, the body gives a half turn at every step and the feet cross each other. When, however, the Penguin is hurried, it throws itself on its breast, uses its wings as fore-feet, and runs so quickly that it might be mistaken for a quadruped by any one who was not accustomed to the bird and its habits. Captain Fitzroy gives an amusing account of the manner in which the Penguins feed their young. "The old bird gets on a little eminence and makes a great noise between quacking and braying, holding its head up in the air as if it were haranguing the penguinnery, while the young one stands close to it, but a little lower. The old bird, having continued its chatter for about a minute, puts its head down and opens its mouth widely, into which the young one thrusts its head and then appears to suck from the throat of its mother for a minute or two, after which the chatter is again repeated and the young one is again fed. This continues for about ten minutes."

The colouring of the King Penguin is as follows: The upper part of the head and the throat are black, round which runs a broad band of light golden yellow, becoming narrower as it passes to the middle of the throat, and then taking a turn to each side until it is lost in the silvery white of the under surface. The back is dark bluish grey, and the under surface pure silvery white, the feathers being extremely thick and set upon each other after a peculiar fashion that preserves the bird from cold and wet. During life, the base of the under mandible is rich warm purple and the rest of the beak is black. The front of the throat is in some request on account of its bright colouring, and is occasionally made into waistcoats and slippers. The total length of the King Penguin is about three feet, so that it is a really large bird, the tail being almost absent and consequently adding little to the measurement.

There are many species of Penguins, among which the CRESTED PENGUIN (*Eúdypes chrysocoma*) is perhaps the handsomest, its bright golden crest, which can be erected at pleasure, giving it a very conspicuous appearance. It is also a Patagonian species. The Penguins are very noisy birds, and the loud chorus of their voices can be heard at a great distance. Indeed, there are several instances known where the nocturnal cries of these and other sea birds have warned sailors of the rocks on which they were heedlessly rushing.

THE common GUILLEMOT is an example of the next sub-family.

This bird is found plentifully on our coasts throughout the year, and may be seen swimming and diving with a skill little inferior to that of the divers. It can, however, use its legs and wings tolerably well, and is said to convey its young from the rocks on which it is hatched, by taking it on the back and flying down to the water.

Mr. Yarrell remarks, that "about the middle of May, the common Guillemot, with many other species of birds, frequenting rocks at that season, converge to particular points, where, from the numbers that congregate and the bustle apparent among them, confusion of interests and localities might be expected; but on the contrary, it will be found that the Guillemots occupy one station or line of ledges upon the rock; the razor-bills another, the puffins a third, the kittiwake gulls a fourth; while the most inaccessible pinnacles seem to be left for the use of the lesser black-backed and the herring gulls. Two distinct species scarcely ever breed by the side of each other."

The Guillemot lays one egg, singularly variable in colour. I possess several eggs, all unlike, and Mr. Champley has five hundred, no two of which are similar, the ground colouring being of every shade from pure white to intense red, and from pale stone-colour to light and dark green. The shape of the egg is very like that of a jargonelle pear, and the general colour is bluish green with spots, splashes, and streaks of reddish brown and black. The eggs are obtained by men who are let down by ropes over the precipices, and gather up their spoil from the narrow ledges on which they are placed. So narrow, indeed, are some of the ledges, that the egg seems liable to be blown over the edge with every breeze.

GUILLEMOT.—*Uria Troïle.*

The head and upper surface of this bird, with the exception of the white tips to the secondaries, are dull black, and the under surface is pure white. The bill is black, and the legs and toes dark brown. The length of the adult bird is about eighteen inches.

There are several other British species belonging to this sub-family, among which may be mentioned the BLACK GUILLEMOT (*Uria gryllé*), known by its smaller size and its black plumage with a large white patch on the wing. The LITTLE AUK, or ROTCHE (*Arctica allë*), so well known in northern seas, also belongs to the same group.

THE curious family of the Petrels now comes before us. A well-known British example is the STORMY PETREL, known to sailors as the MOTHER CAREY'S CHICKEN, and hated by them after a most illogical manner because it foretells an approaching storm, and therefore by a curious process of reasoning is taken for its cause. A sailor once told me very frankly, after I had held a short argument with him, that "they mostly takes things wrong-side forrards," and so it is with the Stormy Petrel, the pilot-fish, and many other creatures.

This bird has long been celebrated for the manner in which it passes over the waves, pattering with its webbed feet and flapping its wings so as to keep itself just above the surface. It thus traverses the ocean with wonderful ease, the billows rolling beneath its feet and passing away under the bird without in the least disturbing it. It is mostly on the move in windy weather, because the marine creatures are flung to the surface by the chopping waves and can be easily picked up as the bird pursues its course. It feeds on the little fish, crustaceans, and molluscs which are found in abundance on the surface of the sea, especially on the floating masses of algæ, and will for days keep pace with a ship for the sake of picking up the refuse food thrown overboard. Indeed, to throw the garbage of fish into the sea is a tolerably certain method of attracting these birds, who are sharp-sighted and seldom fail to perceive anything eatable. It is believed that the Petrel does not dive. The word Petrel is given to the bird on account of its powers of walking on the water, as is related of St. Peter.

It does not frequent land except during the breeding-season, and can repose on the surface of the ocean, settling itself just at the mean level of the waves, and rising and falling quietly with the swell. This Petrel breeds on our northern coasts, laying

STORMY PETREL.—*Thalassidroma pelágica.*

a white egg in some convenient recess, a rabbit-burrow being often employed for the purpose.

Mr. Reid, of Kirkwell, Orkneys, has kindly given the following short but graphic description of these birds while breeding. "They land in our islets every breeding-season. I have had them handed to me alive, frequently together with their eggs, and stinking little things they were, as bad I suppose as the fulmar."

This bird possesses a singular amount of oil, and has the power of throwing it from the mouth when terrified. It is said that this oil, which is very pure, is collected largely in St. Kilda by catching the bird on its egg, where it sits very closely, and making it disgorge the oil into a vessel. The bird is then released and another taken. The inhabitants of the Faroe islands make a curious use of this bird when young and very fat, by simply drawing a wick through the body and lighting it at the end which projects from the beak. This unique lamp will burn for a considerable period. Sometimes the Petrel appears in flocks, and has been driven southwards by violent storms, some having been shot on the Thames, others in Oxfordshire, and some near Birmingham.

The general colour of this bird is sooty black, and the outer edges of the tertials and the upper tail-coverts are white. Its length is barely six inches.

A VERY much larger species, the FULMAR PETREL, is also one of our British birds.

This Petrel is very plentiful in the island of St. Kilda, and an excellent account ot the bird and its importance to the inhabitants has been given by Mr. McGillivray, who visited the island in 1860. "This bird exists here in almost incredible numbers. . . . It forms one of the principal means of support to the inhabitants, who daily risk their lives in its pursuit. The Fulmar breeds on the face of the highest precipices, and only on such as are furnished with small grassy shelves, every spot on which, above a few inches in extent, is occupied with one or more of its nests. The nest is formed of herbage, seldom bulky, generally a mere shallow excavation in the turf and the withered tufts of the sea-pink, in which the bird deposits a single egg of a pure white colour when clean, which is seldom the case. . . . On the 30th of June, having partially descended a nearly perpendicular precipice six hundred feet in height, the whole face of which was covered with the nests of the Fulmar, I enjoyed an opportunity of watching the habits of this bird, and describe from personal observation.

The nests had all been robbed about a month before by the natives, who esteem the eggs of this species above all others. Many of the nests contained each a young bird a

FULMAR PETREL.—*Procelláriu glaciális.*

day or two old at the farthest, thickly covered with long white down. The young birds were very clamorous on being handled, and vomited a quantity of clear oil, with which I sometimes observed the parent birds feeding them by disgorging it. The old birds, on being seized, instantly vomit a quantity of clear amber-coloured oil, which imparts to the whole bird, its nest and young, and even to the rock which it frequents, a peculiar and very disagreeable odour. Fulmar oil is among the most valuable productions of St. Kilda. The best is obtained from the old bird. The Fulmar flies with great buoyancy and considerable rapidity, and when at sea is generally seen skimming along the surface of the waves at a slight elevation, though I never observed one to alight or pick up anything from the water."

In the Arctic regions the Fulmar Petrel accompanies the whaler, flocking round the ship as soon as a whale is captured, and eagerly devouring all the stray bits of blubber that are wasted during the process of flensing. It is an amusing sight to watch these birds during the process, for they squabble and fight over their meal in their anxiety to secure the best and largest morsels, and contrive to swallow pieces of blubber that seem large enough to choke them. It is said that the birds will sometimes assemble near a living whale, and so indicate to the practised eye the whereabouts of the huge quarry.

The colour of the Fulmar Petrel is white upon the head and neck, pearl-grey on the upper surface, and pure white on all the lower surface. The length of the adult bird is not quite twenty inches.

THERE are several other British examples of the Petrels, such as the CAPPED PETREL (*Procellaria hœsitata*), notable for the patch of white on the top of the head; the pretty FORK-TAILED PETREL (*Thalassidroma Leachii*) with its slender legs, black plumage, and white patch on the upper tail-coverts; WILSON'S PETREL (*Thalassidroma Wilsonii*), much resembling the last-mentioned species, but differing in its more forked tail; BULWER'S PETREL (*Thalassidroma Bulweri*), which is wholly of a sooty black hue; and the SHEAR-WATERS (*Puffinius*), of which genus three species are known in England, and are remarkable for the proportionate length and slenderness of their bills.

THE well-known WANDERING ALBATROS is an excellent example of the next sub-family, being the largest and finest of all the species.

This fine bird is possessed of wondrous powers of wing, sailing along for days together without requiring rest, and hardly ever flapping its wings, merely swaying itself easily

WANDERING ALBATROS.—*Diomedéa éxulans.*

from side to side with extended pinions. Sometimes the bird does bend the last joint, but apparently merely for the purpose of checking its progress, like a ship backing her topsails. It is found in the Southern Seas, and is very familiar to all those who have voyaged through that portion of the ocean. Like the petrel, it follows the ships for the sake of obtaining food, and so voracious is the bird that it has been observed to dash at a piece of blubber weighing between three and four pounds, and gulp it down entire. After this dainty morsel, the bird was not able to rise from the water, but yet swam vigorously after another piece of blubber on a hook, snapped at it, and was only saved from capture by the hook breaking in its mouth.

Angling for Albatros is quite a favourite amusement, and the bird often gives good sport, sometimes rising into the air and being drawn down on deck like a boy's kite, but generally hanging back with all its might and resisting the pull of the line by means of its wings squared in the water. It is no easy matter to haul in an Albatros under such circumstances, and the bird often escapes by the hook tearing out or the line breaking. Nothing, however, teaches it wisdom, for in a few minutes it is quite as ready to take the bait again. Even those which have been captured, marked by a ribbon tied round their necks and set at liberty, will follow the vessel as soon as they recover themselves.

When an Albatros is hooked, the others become very angry, thinking that their companion is monopolising the tempting food. Down they sweep accordingly, pounce on the spot, and when settled on the water are very much astonished to see their companion towed away and themselves left sitting on the waves with nothing to eat. Should one of these birds be shot, the remainder pounce upon it at once and soon entomb their late companion in their rapacious maws. These birds may under some circumstances be dangerous to human beings, as they have been observed to swoop upon the head of a man who had fallen overboard; and their long powerful beaks are fearful weapons when urged by those huge pinions.

In spite of the strong oily flavour of the Albatros, some portions of it can be eaten if properly prepared, and the long bones of the wings are in great request for pipe-stems. When the skin and flesh are removed from the head of the Albatros, a large depression is found just over the eyes, which in a good skull now before me exactly accommodates my bent thumb. This cavity is filled during life with a gland, the use of which is dubious. Dr. Bennett dissected several of the glands and found them to be formed " of a hard granulated substance, and pale colour, consisting of numerous distinct minute oval bodies, and on being cut, it is found to be abundantly nourished by blood-vessels; the nerves supplying it came from the minute foramina seen on the floor of the cavity, and are distributed in and about the substance of the gland." In a dissection in the College of Surgeons, a duct is demonstrated to lead towards the nasal outlet on the upper mandible.

The Albatros makes its home on the lofty precipices of Tristan d'Acunha, the Crozettes, the Marion Islands, and other similar localities. Mr. Earl, who visited their nesting-places, writes in forcible language of the strange and weird-like scene. "A death-like stillness prevailed in these high regions, and to my ear our voices had a strange unnatural echo, and I fancied our forms appeared gigantic, whilst the air was piercingly cold. The prospect was altogether sublime, and filled the mind with awe. The huge Albatros here appeared to dread no interloper or enemy, for their young were on the ground completely uncovered and the old ones were stalking round them. They lay but one egg on the ground, where they make a kind of nest by scraping the earth round it; the young is entirely white, and covered with a woolly down, which is very beautiful. As we approached, they snapped their bills with a very quick motion, making a great noise; this and the throwing up the contents of the stomach are the only means of offence and defence they seem to possess. I again visited the mountain about five months afterwards, when I found the young Albatroses still sitting in their nests, and they had never moved away from them."

The nests of the SOOTY ALBATROS (*Diomedea fulginosa*) are built in similar localities, but are rather better made, consisting of mud raised some five or six inches, with a slight depression on the top. The YELLOW-BEAKED ALBATROS (*Diomedea chlororhynchus*) makes a still more elaborate nest, the heap of mud being about one foot in height, with a kind of little trench round the base. All these species were quite undismayed at the presence of human beings, having to be kicked out of the way before they would allow their visitors to pass, and then returning to their posts with perfect composure. Some of them, however, retaliated after their own fashion, by squirting oil out of their mouths upon the clothes of their disturbers. With this oily substance the parents feed their young.

Captain Carmichael, who visited these islands, tried a curious experiment with an Albatros, with a very unexpected result. "We had the curiosity to take one of them by the point of the wings and fling it over the rock; yet, though it had several hundred feet of clear fall, it never recovered itself, but dropped down like a stone."

The Wandering Albatros is rather variable in plumage, but in general the head, neck, back, and wings are grey or brown, and the rest of the plumage white.

THE large family of the Gulls is here represented by four species, all of which are among our British birds.

The SKUA is a large, fierce, and powerful bird, tyrannizing in a shameful manner over its weaker relations, and robbing them without mercy. It feeds mostly on fish, but prefers taking advantage of the labours of others to working honestly for its own living. As the lesser Gulls are busily fishing, the Skuas hover about the spot, and as soon as a poor Gull has caught a fish, and is going off to his family, down comes the Skua upon him with threatening beak and rocking wings, and when the victim drops his burden, to escape with greater facility, the Skua darts after the falling fish, and snaps it up before it reaches

COMMON GULL.—*Larus canus.*
SKUA GULL.—*Stercorárius catarrhactes.*

BLACK-BACKED GULL.—*Larus máximus.*
KITTIWAKE GULL.—*Rissa tryddctyla.*

the water. It also eats eggs and the smaller birds, a propensity which is shared by other Gulls than the Skua.

The Skua is no coward, however tyrannously it may behave, and is quite as ready to repel a foe as to attack a victim. Even the golden eagle makes no head against the Skua, and Mr. Dunn mentions that he has seen a pair of these birds beat off an eagle from their nesting-place on Rona's Hill. This species breeds in some of the British islands, but properly belongs to more northern regions. It does not associate in flocks, and it is seldom that more than three or four are seen together. The eggs are two or three in number, and their colour is olive-brown, mottled with a darker hue.

The head and neck of the Skua are brown streaked with chestnut, the back and tail are brown, and the throat and under parts are clove-brown. In length the bird rather exceeds two feet. The name of Skua is said to be derived from the cry of the bird, which somewhat resembles the word "Skui."

Several other species of Skua are included in our list of British birds, among which is the POMARINE SKUA (*Stercorarius pomarinus*), known by the mottled plumage of the back; and BUFFON'S SKUA (*Stercorarius cepphus*), remarkable for the elongated central feathers of the tail; and one or two others.

THE common GULL is too familiar to need much description, as it is well known to all who have visited the seashore, or the mouth of any of our larger rivers. It is a bold bird, caring little for man, and even following a steamer so closely that the gleam of its eyes can be plainly seen. It can easily be tamed, and is a rather useful bird in a garden, destroying vermin of various kinds, and occasionally killing and eating any small bird that may incautiously venture within reach of the strong bill. Cheese seems to be an acknowledged dainty with these birds, which have often been known to contract so great an affection for the place of their captivity as to return to it voluntarily, and even to introduce a mate to the well-remembered hospitalities.

Its ordinary food consists of the refuse matter flung up by the tide, as well as of various marine creatures, and to obtain them the Gulls may be seen covering the shores with their white plumage as soon as the tide retires. A very favourite resort of the Gull is the upper part of some sands, such as the Goodwins, which are submerged during high water, and only uncovered as the water recedes. The bird will, however, make considerable incursions inland, and may be seen very amicably following the plough for the purpose of picking up the worms and grubs that are thrown to the surface by the share.

The head and neck of the Gull are pure white, the upper surface grey, variegated with the white edges and tips of the secondaries and tertials. The primaries are black on the outer web, and the two first quills are black at their tips. The tail-coverts, tail, and whole under surface are pure white, and the legs and feet are ashen green. The total length of an adult male is about eighteen inches; the female is rather smaller.

The GREAT BLACK-BACKED GULL is a very fine bird, not very plentiful on our coasts, but spread over the greater part of the British shores.

This bird prefers low-lying and marshy lands, and is found on the flat shores of Kent and Essex at the mouth of the Thames, where it is popularly known under the name of the Cob. It is very plentiful on the shores of Sweden and Norway, and on some of the islands of Shetland and Orkney it breeds in abundance, the eggs being highly valued on account of their rich flavour and their large size. It is the custom in those localities to lay in a stock of these eggs, and to take two sets in succession, leaving the third for the bird to hatch. Mr. Hewitson mentions that upon an island of barely half an acre in extent, sixty dozen eggs were secured.

The food of the Great Black-backed Gull consists mostly of fish, but it has a very accommodating appetite, and will eat almost any kind of garbage, besides destroying young birds. It has even been known to kill weakly lambs. It is a fierce bird, and when wounded will fight vigorously for its liberty. The nest of this species is of grass, and generally contains three eggs of greenish dun flecked with grey and brown. In the summer plumage the head and neck of the Great Black-backed Gull are white, the upper surface of the body is dark leaden grey, with some white upon the quill-feathers of the wings. The whole of the under surface is pure white, and the legs and feet are pinkish. The length of this bird is about thirty inches. In the winter-plumage the bird is streaked with grey.

SCISSOR-BILL.—*Rhyncops nigra.*

THE pretty KITTIWAKE GULL is tolerably plentiful on many of our shores, and breeds upon the rocky portions of the coast. Owing to the diversity of its plumage according to the age, the Kittiwake has been called by several names; "Tarrock" being the best known and belonging to the bird while young. The name of Kittiwake is given in allusion to its cry, which bears some resemblance to that word rather slowly pronounced. The nest of the Kittiwake is made of seaweed, and placed on narrow ledges of rock at a great elevation. The nests are placed in close proximity to each other, and generally contain three eggs of a brownish olive, covered with spots of grey and brown.

The head and neck of the Kittiwake are white, the upper parts of the body silvery grey, the wings being diversified with a little black and much white. The under surface is pure white. Yearling birds are much flecked with black on the back of the neck, and many of the wing-feathers are liberally tipped and edged with the same hue, and the tail is tipped with black. The length of the Kittiwake is nearly sixteen inches. The specific name of *tridactylus*, or three-toed, is given to the bird because the hinder toe is wanting, its place being represented by a small tubercle.

THERE are many other species of British Gulls, too numerous to be described. Among these we may notice the LITTLE GULL (*Larus minutus*), remarkable for its jetty black head and neck and its small size, its length being little more than ten inches; the LAUGHING GULL (*Larus atricilla*), which derives its name from its curious screaming cry; the HERRING or SILVERY GULL (*Larus argentatus*), a fine species, about two feet in length, with a pure white head and neck, and a soft grey back, and jetty black primaries with a spot or two of white at the tips; and the IVORY GULL (*Pagophila eburnea*), so called on account of the pure white of its summer plumage.

THE name of SCISSOR-BILL is very appropriately given to the species now before us.

This remarkable Gull has a long and much compressed beak, the lower mandible being much longer than the upper, rather flatter, and shutting into the upper like a knife-blade into the handle. This beak is orange-red at the base, deepening into black at the tip. The Scissor-bill skims just over the surface of the sea, its knife-like bill cutting through the water, and picking up the crustaceans, molluscs, and fish that come to the surface. While thus darting along, the bird utters loud and exultant cries. It does not, however, trust solely to the wide seas for its food, for, according to Lesson, who was an eye-witness of the scene, the bird feeds much upon bivalves, adroitly inserting its beak into their shells as they lie open, and then banging the shell against a rock or stone, so as to break the hinge and expose the inhabitant, which is immediately scooped out and swallowed.

The Scissor-bill is found along the coast of America and part of Africa. It breeds on marshes and sandy islands, laying three eggs of a white colour spotted with ash. The colour of this bird is dark brown on the top of the head and upper surface, with a bar of white across the wings, and the under surface white. Its length is about twenty inches.

TERN.—*Sterna hirundo.*

THE common TERN, or SEA SWALLOW, is very plentiful on our coasts, and may be seen flying along on rapid wing, its long forked tail giving it so decidedly a swallow-like air, that its popular name of Sea Swallow is well applied.

It is mostly seen on the wing, not often going to shore, except during the breeding-season, but reposing occasionally on floating logs of wood, buoys, and similar resting-places. Its food consists mostly of small fish, which it obtains by darting down from the air upon them. When seeking for food it does not rise to any great elevation, but hovers at a few feet above the water ready to pounce upon a passing fish that may be unwary enough to come to the surface, and sufficiently small to be eaten. Sand eels are said to be a favourite article of food with the Tern. It is a noisy bird, ever on the move, chasing its companions when it is tired of fishing, and uttering continually its loud jarring cry.

The Tern breeds on the low-lying lands, and makes a very rude nest, being indeed nothing more than a shallow depression in the earth into which are scraped a few sticks, stones, and dry grasses. The Tern reaches this country about May and departs in September. An adult bird in summer plumage has the tip of the head and nape of the neck jetty black, the upper part of the body ashen grey, the under surface white, and the legs, feet, and bill coral-red, the bill deepening into black at the tip. The length of the Tern rather exceeds fourteen inches; much of it is due to the long forked feathers of the tail.

ANOTHER rather celebrated species of Tern, though rare in England, is the NODDY.

This bird is spread over many portions of the world, but is a very rare visitant to our shores. It often alights on vessels by night, and as it does not see well except in daylight, suffers itself to be easily caught. This habit has sometimes had a most providential effect, and saved the lives of sailors adrift in a boat without provisions. Audubon writes of the Noddy : " The flight of this bird greatly resembles that of the night-hawk when passing over meadows or rivers. When about to alight on the water the Noddy keeps its wings extended upwards, and touches it first with its feet. It swims with considerable buoyancy and grace, and at times immerses its head to seize on a fish. When seized in the hand it utters a rough cry. On such occasions it bites severely with quickly repeated movements of the bill, which on missing the object aimed at, closes with a snap." The bird picks up its food as it skims along the surface.

NODDY.—*Anöus stólidus.*

The nesting of the Noddy is rather variable. Sometimes it places its nest upon a ledge of rock overhanging the water, and at other times builds it among bushes or low trees. The nest is always a clumsy-looking affair, and as the bird is in the habit of adhering to the same spot for several successive seasons, and adding fresh material each time, the nests soon acquire a very large size, some being masses of seaweed about two feet in thickness, with only a slight depression at the top. In any case, the nest is very dirty and badly kept, and the odour of a crowded nesting-place is anything but pleasant. Mr. Gilbert mentions that in Western Australia, where the Noddy breeds largely, its numbers are much thinned by a kind of lizard, which eats the young and occasionally kills the old birds. He is of opinion that not one in twenty lives long enough to take wing.

The eggs of the Noddy are usually three in number, of a dark orange colour, splashed and spotted with red and purple. They are very good eating, and sailors collect them largely for the table. The Noddy is a dark-looking bird, the forehead and top of the head being buff, the back of the head dusky grey, and the whole of the remaining plumage chocolate-brown. Even the legs, feet, and bill are black. The average length of the Noddy is about fourteen or fifteen inches.

Several other species of Terns are found on our coasts, among which may be mentioned the SOOTY TERN (*Sterna fuliginosa*), sooty black above and white below ; the LESSER TERN (*Sterna minuta*), a very small species, only eight or nine inches in length ;

the Caspian Tern (*Sterna Caspia*), a fine species, twenty inches long ; the Roseate Tern (*Sterna paradisea*), remarkable for the rosy white hue of its under surface ; and several other species, all resembling each other in habits and general form.

We now arrive at the last family of birds, the Pelicans, a group which includes many species, all remarkable for some peculiarity, and many of them really fine and handsome birds.

As its name implies, the Tropic Bird is seldom to be seen outside the tropics unless driven by storms. It is wonderfully powerful on the wing, being able to soar for a considerable period, and passing whole days in the air without needing to settle. It is a beautiful and delicately graceful bird, and always calls forth the admiration of the beholders, as it hovers above the vessel or darts into the water in pursuit of prey. While on the wing it utters a loud, shrill, and grating cry, which often indicates its presence at

TROPIC BIRD.—*Phaëton æthereus.*

night as well as by day. On account of this shrill cry, the sailors call it the Boatswain Bird. They also call it by the name of Startail, on account of the long projecting tail-feathers.

As a general fact they do not fly to very great distances from land, three hundred miles being about the usual limit ; but Dr. Bennett observed them on one occasion when the nearest land was about one thousand miles distant. The long tail-shafts of the Tropic Bird are much valued in many lands, the natives wearing them as ornaments, or weaving them into various implements. The feathers of a closely allied species, the Roseate Tropic Bird (*Phaeton phœnicurus*), are used in the South Seas for various purposes, and are obtained by visiting the birds during the time of incubation, when they sit closely in their nests, and quietly plucking out the coveted plume. Dr. Bennett observes, " The 'red caps' mentioned by Captain Cook as worn by the natives of the Friendly Islands are formed principally from these red shafts, and I observed the same use of them in the island of Rotúma, South Pacific Ocean, the caps (named *shoul*, or war head-dress of the natives of that island, and worn as a decoration by warriors in battle) being formed from the red tail-feathers of the Roseate Tropic Bird, which the natives procure with some difficulty, and they are consequently very highly valued. The cap is in the form of a semicircle, without any crown, and is tied on the forehead. I have also seen many neat baskets in which the red shafts of this bird had been very ingeniously interwoven ; they were exposed for sale at the Sandwich Islands, and even stated to be brought from some part of the coast of California."

DARTER.—*Plotus anhinga.*

The Tropic Bird breeds in the Mauritius. The young are odd little things, not in the least resembling their parents, round as balls and covered with white down. The flight of the Tropic Bird is of a rather peculiar jerking or shooting character. The total length of the bird is about two feet six inches, of which the tail-feathers occupy about fifteen inches.

THE singular DARTERS are inhabitants of two parts of the world, one species inhabiting Africa, and another being found in America. With their slender heads, their long snake-like necks, and their evidently aquatic bodies, they really look as if they had been formed on the same model as the well-known plesiosaurus.

The common Darter inhabits many parts of America, and is found along the banks of rivers and marshy grounds. Mr. Ord writes as follows of these birds: "The first individual that I saw in Florida was making away to avoid me along the shore of a reedy marsh, which was lined with alligators, and the first impression on my mind was that I beheld a snake, but the recollection of the habits of the bird soon undeceived me. To pursue these birds at such times is useless, as they cannot be induced to rise or even to expose their bodies.

Wherever the limbs of a tree project over and dip into the water, there the Darters are sure to be found, these situations being convenient resting-places for the purpose sunning and preening themselves, and probably giving them a better opportunity of observing their finny prey. They crawl from the water upon the limbs and fix themselves in an upright position, which they maintain in the utmost silence. If there be foliage or long moss, they secrete themselves in it in such a manner that they cannot be perceived unless one be close to them. When approached, they drop into the water with such surprising skill that one is astonished how so large a body can plunge with so little noise, the agitation of the body being apparently not greater than that occasioned by the gliding of an eel."

While in the tree this bird has a habit of rapidly darting its snaky head and neck through the foliage, so that at a first glance it would be taken for a serpent, and even when swimming its body is submerged, and the only part visible is the long neck writhing about just like an aquatic serpent.

The nest of the Darter is a rather large edifice of sticks, placed upon the trees that grow in the marshy lands which the Darters frequent. The eggs are blue. In the adult bird the general colour is very deep green. A strip of brownish white runs from the eye partially down the sides of the neck, and the scapulary feathers are long and slender, with a stripe of white along their centre. The wings are black variegated with silvery white. The total length of this bird is not quite three feet.

ANOTHER species, LEVAILLANT'S DARTER or SNAKE-BIRD (*Plotus Levaillantii*), is a native of Africa, and its habits and general form resemble the preceding species. In the water it is most agile, and can often dive at the flash of a gun and place itself in safety by going under water for a considerable distance in a direction where it was not expected, and then rising into the air and flying off.

The GANNET, SOLAN GOOSE, or SPECTACLED GOOSE, is a well-known resident on our coasts, its chief home being the Bass Rock in the Frith of Forth, on which it congregates in vast numbers.

The Gannet is a large bird, nearly three feet long; and being powerful on the wing, and possessed of a large appetite, it makes great havoc among the fish which it devours. Herrings, pilchards, sprats, and similar fish are the favourite food of the Gannet, and as soon as the shoals of herrings approach the coast, the Gannets assemble in flocks and indicate to the fishermen the presence and position of the fish. The bird is able to catch its prey at some distance below the surface, and accomplishes its object by shooting directly downwards with partially closed wings and seizing the fish before it has had time to take alarm. While engaged in feeding upon the shoals of herrings that are inclosed in the nets, the Gannets are frequently taken by becoming entangled in the meshes. They are also caught by fastening a herring on a board and setting it afloat. The Gannet sees the fish, but takes no notice of the board, and so comes down with a pounce, drives its beak through the herring, and into the board, and is of course killed by the concussion. The Gannet is enabled to guard itself against too deep a submergence and to break the shock of its body falling against the water by means of certain subcutaneous air-cells, which according to Montagu are capable of containing three full breaths from human lungs, and are equal to a capacity of one hundred and eighty entire inches.

The birds that breed on the Bass Rock are very tame in consequence of their immunity from persecution, and will even suffer themselves to be handled when on their nests without displaying any uneasiness, except uttering a little dissatisfied croaking. The nest of the Gannet is a heap of grass, seaweed, and similar substances, on which is laid one very pale blue egg, which, however, does not long retain its purity. The young are clothed with white puffy down, which after a while changes to nearly black feathers, the white plumage not being assumed until the bird has reached full age. The head and neck of the full-grown bird are buff, the primaries black, and the rest of the plumage white. The yearling bird is almost wholly black covered with streaks and triangular marks of greyish white. The total length of this bird is about thirty-four inches.

GANNET, OR SOLAN GOOSE. – *Sula Bassánea.*

THE well-known BOOBY (*Sula fusca*), so called from its stupidity when attacked whether by man or the frigate bird, is closely allied to the gannet. This bird is found in most of the warmer latitudes settled upon the islands and rocky shores, and catching fish all day for the benefit of the frigate birds who attack and rob it. The colour of the Booby is brown above and whitish grey beneath.

THE common CORMORANT is well known for its voracious habits, its capacities of digestion having long become proverbial.

This bird is common on all our rocky coasts, where it may be seen sitting on some projecting ledge, or diving and swimming with great agility, and ever and anon returning to its resting-place on the rock. It is an admirable swimmer and a good diver, and chases fish with equal perseverance and success, both qualities being needful to satisfy the wants of its ever-craving maw. Eels are favourite morsels with the Cormorant, which, if the eel should be small, swallows it alive in spite of the writhings and struggling of its victim, and the many retrogressions which it will make from the interior of its devourer, until it is finally accumulated and digested, the latter being a process of wonderful celerity. If the eel is rather large and powerful, the Cormorant batters it against some hard substance and then swallows it easily. Mr. Fortune gives a ludicrous narrative of a number of tame Cormorants and their behaviour at feeding-time; how they were supplied with eels; how they swallowed them as fast as possible; how after all had disappeared, one of the swallowed eels returned into the air and was immediately fought for by the birds, greatly to the discomfiture of the individual whose property it had been; and how he tried to reimburse himself by means of a similar mishap on the part of some of his companions.

CORMORANT.—*Gráculus Carbo.*   CRESTED CORMORANT.—*Gráculus cristátus.*

The Cormorant can easily be tamed, and in China, where everything living or dead is utilized, the bird is employed for the purpose of catching fish. The Cormorants are regularly trained to the task, and go out with their master in a boat, where they sit quietly on the edge until they receive his orders. They then dash into the water, seize the fish in their beaks, and bring them to their owner. Should one of these birds pounce upon a fish too large for it to carry alone, one of its companions will come to its assistance, and the two together will take the fish and bring it to the boat. Sometimes a Cormorant takes an idle fit, and swims playfully about instead of attending to its business, when it is recalled to a sense of duty by its master, who strikes the water with his oar and shouts at the bird, who accepts the rebuke at once and dives after its prey. When the task is completed, the birds are allowed their share of fish. A detailed and interesting account of these birds may be found in Mr. Fortune's work on China.

The Cormorant, although a web-footed bird, is able to perch, and may often be seen sitting on a post or a rail overhanging a stream, watching the fish as they pass below.

The nest of the Cormorant is made of a large mass of sticks, seaweed, and grass, and the eggs are from four to six in number, rather small in proportion to the dimensions of the parent, and of a curious chalky texture externally, varied with a pale greenish blue. Many nests are placed in close proximity to each other, and, as may be imagined from the habits of the bird, the nesting-places are very ill-savoured and by no means adapted to delicate nostrils. When the young are first hatched they are very odd creatures, quite without feathers and covered only with a hard-looking black skin, upon which the clothing of black woolly down does not appear until several days. They soon take to the water, and, like most aquatic birds, swim long before they can fly.

In the summer time the head and back of the neck are black sprinkled with hair-like feathers of a pure white, and elongated on the back of the head so as to form a kind of crest. The upper part of the body is brown mottled with black, and the front of the throat and the whole of the under surface are velvety bluish black. There is a white patch on the thighs, and the legs and toes are likewise black. The young bird is much lighter in colour, without the black on the head or breast, and the lower parts of the body of a dull greyish white mottled with a little brown. The total length of this bird is about three feet; the female is a few inches shorter. The pouch under the throat is yellow edged with white.

ANOTHER well-known British species of this genus is the CRESTED CORMORANT, GREEN CORMORANT, or SHAG, a bird which can at once be distinguished from the preceding species by the green colour of the plumage and the difference in size, the length of an adult male being only twenty-seven inches. In habits this species resembles the common Cormorant.

The nest of the Crested Cormorant is placed upon a ledge of rock, generally in a spot very difficult of access, and like that of the preceding species gives forth a horrible stench caused by the mixture of decaying seaweed and putrefying fish. The nest is made rather ingeniously of stalks and roots of seaweed, lined with grass in the centre. The colour of this Cormorant is rich dark green, except the wing and tail quill-feathers, which are black. The young birds are greenish brown above, and brown and white below.

WE now arrive at the well-known PELICAN, which is universally accepted as the type of the family.

This bird is found spread over many portions of Africa and Asia, and is also found in some parts of Southern Europe. It is a sociable bird, assembling in large flocks, and often mingling with the flamingoes, its white plumage contrasting finely with the scarlet raiment of its long-necked allies. The wings of the Pelican are very long and powerful, and the flight is singularly bold and graceful. Oftentimes the birds fly in large bodies, sometimes forming themselves into some definite shape, sometimes spreading themselves in any order and then gathering together into a compact mass, and often rising to such a height as to be almost beyond ken in spite of their large size.

The pouch of the Pelican is enormously large, capable of containing two gallons of water, and is employed by the bird as a basket wherein to carry the fish which it has caught. The Pelican is a good fisherman, hovering above the water watching for a shoal of fish near the surface. Down sweeps the bird, scoops up a number of fish in its capacious pouch, and then generally goes off homeward. Sometimes it is interrupted by a large species of hawk, which robs the Pelican in a very ingenious manner. As the industrious bird flies home laden with the finny spoil, the hawk swoops down upon it and ruffles its wings in so threatening a manner that the Pelican screams with terror. The hawk snatches the fish out of the bird's pouch and flies off, leaving the poor Pelican to return and replenish its pouch. It is a rather curious fact, that in the Pelican the tongue, which is so large and curiously constructed in many birds, is almost wanting, and is represented by a little fleshy knob hardly the size of a finger-top. In spite of the large size of the pouch, the Pelican can wrinkle it up in such a manner that it is barely perceptible.

The nest of the Pelican is placed on the ground in some retired spot, usually an island in the sea, or the borders of some inland lake or a river. It is made of grasses, and contains two or three white eggs. The female sits on the eggs, and her mate goes off to fish for her; and when the young are hatched they are fed by the parents, who turn the fish out of their pouches into the mouths of the young. In order to perform this operation, they press the bill against their breast, so that its scarlet tip looks like a blood-spot against the white feathers, and has given rise to the fable that the Pelican feeds her young with her own blood.

In spite of the huge size of the beak and pouch, the Pelican can preen its plumage with perfect ease, and uses its feet to those parts of the neck and head which it cannot

PELICAN.—*Pelecánus onocrótalus.*

reach with the bill. When it is annoyed, it has a habit of slapping the mandibles together with a sound that can best be described by the word " walloping," caused by the flapping of the loose pouch. It assumes a great variety of positions; and when two individuals quarrel, the grotesque attitudes with which they denote their anger are irresistibly ludicrous. When playful it tumbles about in the water, dabbles with the bill as if it were catching fish, dashes the water on every side with its huge wings, and then walking on shore stands still and flaps its pinions with a force and decision that strongly remind the observer of a cabman on a very cold day.

The colour of the Pelican is white with a delicate roseate tinge like that of a blush rose. On the breast the feathers are elongated and of a golden yellow. The quill-feathers are black, but hardly seen until the bird expands its wings; the bill is yellow, tipped with red. The length of the bird is almost six feet, and the expanse of wing about twelve feet.

There are several other species of Pelican, but the habits of all are very similar.

· THE last bird on our list is the well-known FRIGATE BIRD, SEA HAWK, or MAN-OF-WAR BIRD, an inhabitant of the tropical seas. So admirable and comprehensive a biography of this bird has been given by Dr. Bennett in his "Wanderings of a Naturalist," that I cannot do better than transfer his account to these pages.

" The Frigate Bird is often seen frequenting the Austral and other tropical islands. It somewhat resembles the cormorant in its general appearance. . . . It is interesting to observe these birds soaring in the air with a flight widely different from that of the birds around them, and conspicuous from the symmetry of their form and the power and extent

FRIGATE BIRD.—'*Atagen Aquila*.

of their wings. Owing to the enormous comparative size of the pectoral muscles, which are so large as to weigh nearly one-fourth as much as the whole body of the bird, they are capable of sustaining very long flights. From the formation of their feet they are incapable of alighting and resting upon the surface of the water; and as they are seen at great distances from the land, they must possess immense power of wing to enable them to sustain such prolonged exertion without repose.

We find accordingly that the whole structure of this elegant bird is adapted to a rapidity of flight surpassing that of all others. It has the power of soaring to so great an elevation in the air as to appear a mere speck in the regions above; and when seen hovering over the ship, adorned with its beautiful glossy plumage, it attracts attention by its noble bearing and graceful evolutions as it sails in gentle undulations in mid-air, or by its rapidity of motion when darting upon its prey. This bird, being incapable of swimming and diving, may generally be seen on the alert for flying fish when they are started into the air by albicores or bonitos, and when unsuccessful it is compelled to resort to a system of plundering other sea-birds. The quiet and industrious birds, the gannets and sea-swallows, are generally selected as objects of attack, and on returning to their haunts to feed their young brood, after having been out fishing all day, are stopped in mid-air by the marauding Frigate Bird, and compelled to deliver up some of their prey, which being disgorged by them, is most dexterously caught by the plunderer before it reaches the water.

The gannets can well afford to be relieved of some of their booty, as they are often seen so full of fish as to be unable to close their beaks. When the Frigate Bird attacks the gannet, it attains its object by hovering over the victim, and then, darting rapidly down, strikes it upon the back of the head. Their usual mode of fishing, however, is generally more quiet. It is not uncommon to observe a single gannet selected from a flock as the object of attack, while the rest in the meantime continue their heavy flight towards land.

It sometimes happens, however, that a spirit of independence rouses even the dull gannet to a determination of resisting the plunderer. It manœuvres to avoid the blow of the enemy by darting about, dropping from its elevation in the air, raising the beak in a perpendicular direction, using every effort to elude the foe, so that sometimes both fall into the water together. On this occurring the gannet gains its object, for although the Frigate Bird has the advantage over the gannet while hovering in the air, the latter has the best of it in the ocean, and generally escapes, leaving the piratical bird to get out of the water in the manner most agreeable to himself.

. . . They have also been observed to soar over the masthead of a ship and tear away the pieces of coloured cloth appended to the vane; this I have frequently seen. When soaring in mid-air, its wings, spread to their utmost degree of expansion, are apparently motionless, while the long forked tail is seen expanding and closing with a quick alternate motion, until the bird rises by degrees and slowly to so lofty an elevation in the sky as at last to appear a mere speck."

The long black feathers of the tail are in great request among the Society Islanders, being woven as ornaments into the head-dresses of the chiefs. The nest of the Frigate Bird is sometimes built upon trees and bushes where the low shores afford no cliffs, but its usual locality for breeding is on the summit of some rocky cliff. On the rock there is no nest, but when the bird breeds among trees, it makes a rude scaffolding of sticks like the nest of the wood pigeon. There is only one egg, of a peculiar chalky whiteness, and while sitting the bird is very bold and will not stir even if pushed with a stick, snapping and biting at the obnoxious implement. The voice of this bird is rough and harsh, and is likened to the sound produced by turning a winch.

The colour of the adult Frigate Bird is shining black glossed with green, the female being dull black above and white streaked with cinnamon upon the head, breast, and under parts. The pouch on the throat is scarlet, and when distended has a very curious effect against the dark black of the throat and neck. Including the long tail, the male measures three feet in length, but the body is extremely small. The expanse of the wings is about eight feet.

CLASS.—AVES OR BIRDS.

Animals possessed of vertebræ, breathing atmospheric air with lungs; having warm blood, and heart with four chambers—Young hatched from eggs—Mouth modified into a horny beak without true teeth—Fore limbs developed into wings, mostly clothed with feathers and used for flight; hind limbs always used for standing, or progression on earth or in water.

Order. **ACCIPITRES.**—Bill rather compressed (*i.e.* flattened sideways). Upper mandible sharp and hooked at tip; base with a bare skin termed the "cere," in which the nostrils are pierced. Wings long and pointed. Feet strong. Tarsi (*i.e.* joint between toes and knee) covered with scales. Toes three in front and one behind, the soles rough, and the claws strong and curved.

Sub-order. ACCIPITRES DIURNI.—Eyes at sides of head. Cere large and prominent. Tarsi moderate.

Family. **VULTURIDÆ.**—Bill compressed. Tarsi short and with net-like scales. Middle toe longer than tarsus, hinder toe rather elevated, claws blunt. Mostly with down on head and neck, and with a large crop.

Sub-family. **Gypaetinæ.**—Feathered on head and neck. Cere hidden by bristly hairs.

Genus. GYPAETUS.—Bill long, sharply curved at tip, tuft of bristly hairs from lower mandible. Wings, second and third quills longest and nearly equal. Tail long and wedge-shaped. Tarsi feathered, very short. Front toes united by membrane at base. Habitat—Mountain chains of Europe, Asia, and Africa.

Sub-family. **Sarcorhamphinæ.**—Bill long and slender, cere soft, nostrils longitudinal, large and oval. Tarsi long with netted scales; third toe mostly short and weak, two outer front toes short and connected with membrane.

Genus. SARCORHAMPHUS.—Bill moderate, cere about one-third its length, nostrils in middle of cere. Wings, third and fourth quills equal and longest. Tail moderate and even. Tarsi short, feathered below knee, netted. Two exterior toes equal and united with middle by membrane; hind toe weak. Head, neck and breast bare, fleshy caruncle above nostrils of male. Habitat—Most parts of America.

Genus. CATHARISTA. — Bill long, cere two-thirds its length, nostrils in front of cere. Wings long, pointed, third and fourth quills longest. Tarsi short, feathered below the knee and netted. Head and upper part of neck bare.

Genus. NEOPHRON.—Bill very long and slender, cere two-thirds its length, nostrils near middle of bill. Wings, third quill longest. Tail wedge-shaped. Tarsi moderate, feathered below knee, netted. Toes with strong scales above, inner the strongest. Front of head and throat, and the cheeks, bare. Habitat—Africa and India.

Sub-family. **Vulturinæ.**—Bill long, deeper than broad, cere nearly half its length, tip very sharply curved, nostrils in cere, and oblique. Middle toe longer than tarsi, others much shorter.

Genus. VULTUR.—Bill large, much arched from cere to tip. Wings, first quill short, third and fourth longest. Tail, shafts projecting. Middle toe united to outer. Feathered ruff on neck, mostly downy crest on back of head, head covered with down. Habitat—Warm portions of Old World.

Genus. OTOGYPS.—Head and neck quite bare, long wattles beneath the ears. Rest like Vultur.

Genus. GYPS.—Bill rather swollen at sides. Head and neck covered with short down, ruff of long pointed, or long downy feathers. Rest like Vultur. Habitat—Warm portions of Old World.

Family. **FALCONIDÆ.**—Bill with partial cere, compressed, and keel curved from cere to tip, which is much curved and sharp, edges toothed or waved. Wings long and pointed. Tail broad. Tarsi varied and strong, and claws sharp and curved. Head and neck feathered, eyes sunk and overshadowed by brow.

Sub-family. **Aquilinæ.**—Bill straight at base, and then greatly arched at tip, edges waved; nostrils in front of cere. Tail long and large. Tarsi long, inner toe and claw strongest.

Genus. AQUILA.—Bill very much curved at tip and sharp, sides much compressed, nostrils large and oblique. Wings, fourth and fifth quills longest and equal. Tarsi feathered to base of toes. Habitat—Most parts of world.

Genus. SPIZAETUS.—Tarsi long and slender. Toes long and powerful, inner much longer than outer, middle and outer united by membrane, all covered with small scales except a few larger at tip. Rest like Aquila. Habitat—Asia, South America, and Africa.

Genus. HERPETOTHERES (or CACHINNA).—Bill short and arched throughout; nostrils large and round. Wings, third, fourth, and fifth quills longest and equal. Tarsi moderate, and netted. Toes with a few large scales at tips, rest with small scales. Habitat—South America.

Genus. CIRCAETUS.—Bill like Aquila, nostrils large, oblique, and rather oval. Wings like Aquila. Tail long and even. Tarsi long, feathered below the knee and netted. Toes short, scaled, outer united to middle. Habitat—Most parts of world.

Genus. THRASAETUS.—Bill like Aquila, nostrils narrow and set rather crosswise. Wings, fourth, fifth, and sixth quills longest. Tail long and rounded. Tarsi short, stout, large-scaled in front, small-scaled at side. Toes powerful, small-scaled at base, rest large-scaled. Habitat—Tropical America.

Genus. MORPHNUS.—Bill like Aquila. Wings very long, third, fourth, and fifth quills longest. Tail long and even. Tarsi very long, small-scaled on sides, large-scaled on front and back. Habitat—South America.

Genus. PANDION.—Bill short and curved from base. Wings reaching to end of tail, second and third quills equal and longest. Tail moderate. Tarsi, short, strong, netted. Toes long, united at base to middle toe, claws much curved. Habitat—Most parts of world.

Genus. HALIAETUS.—Bill like Aquila, nostrils narrow and oblique. Wings long, third, fourth, and fifth quills longest. Tarsi short and strong, narrow scales in front, rest with small irregular scales. Habitat—Old and New World.

Sub-family. **Polyborinæ.**—Bill slightly hooked, and waved. Wings long. Tarsi long, slender, and irregularly scaled.

Genus. IBYCTER.—Bill much compressed, slightly waved, nostrils large, rounded, in fore-part of cere. Sides of head and throat naked. Tail long, broad, and rounded. Wings, fourth quill longest. Tarsi moderate, feathered below knee, scales large in front, rest small. Toes, inner stronger and shorter than outer. Habitat—South America.

Genus. MILVAGO.—Bill rather stout, nostrils with horny tubercle in centre. Wings, third and fourth quills longest. Tail long, broad, and rounded. Tarsi rather long, feathered below knee. Toes, two exterior nearly equal. Habitat—Southern America and Antarctic Islands.

Genus. POLYBORUS.—Bill short and much arched, slightly waved near tip, a few hairs on cere, nostrils oblique and narrow. Wings, third quill longest. Tail moderate. Tarsi long, scales on front and behind, divided along their middle. Claws of inner and hind toes largest. Habitat—South America.

Sub-family. **Buteoninæ.**—Bill short, broad at base, much curved, cere occupying more than half its length. Wings long. Tarsi long, with broad transverse scales in front and behind. Toes rather short.

Genus. BUTEO. — Bill, nostrils large and oval, hairy feathers from base of bill to eye. Wings very long, third and fourth quills longest. Tail moderate and even. Tarsi long, naked, and scaled. Toes, three front joined at base. Habitat—Most parts of world.

Genus. ARCHIBUTEO.—Tarsi feathered to base of toes. Habitat—Many parts of world.

Sub-family. **Milvinæ.**—Bill short and weak, waved, nostrils at base. Wings long and pointed. Tarsi short and strong. Toes broad and well padded beneath.

Genus. PERNIS.—Bill curved from base to tip, nostrils long, narrow, oblique. Wings, third, fourth, and fifth quills longest. Tail long, broad, and rounded. Tarsi, half feathered, rest netted. Inner toe longest. Habitat—Eastern parts of Old World.

Genus. MILVUS.—Bill straight at base, then curved to tip, margins rather straight; nostrils oval. Wings very long, third and fourth quills longest. Tail long, rounded or forked. Tarsi very short, feathered, and scaled. Outer toe united at base to middle. Habitat—Old World.

Genus. ELANUS.—Bill broad at base, compressed towards tip. Wings very long, second quill longest. Tail long and forked. Outer toe much shorter than inner. Habitat—Most parts of world.

Genus. ELANOÏDES (or NAUCLERUS).—Wings very long, second and third quills longest. Tail long and very deeply forked. Hind toe long. Habitat—South America and Africa.

Genus. CYMINDIS.—Bill much compressed, keel gradually sloping to tip; nostrils narrow and oblique. Tail long and rounded. Tarsi short and strong, plumed and feathered. Inner toe longer than outer. Habitat—Tropical America.

Genus. ICTINIA.—Bill broad at base, keel arched to tip, upper mandible deeply scooped near tip; nostrils rather rounded. Wings very long, pointed, third and fourth quills longest. Tail short and slightly forked. Tarsi with transverse scales in front, small scales behind. Inner toe longer than outer; claws sharp and compressed. Habitat—America.

Sub-family. **Falconinæ.**—Bill short and curved throughout, toothed on tip; cere long and rounded. Wings long and pointed. Tail long and rounded.

Genus. FALCO.—Bill strong and toothed at tip; cere short; nostrils round, with central tubercle. Wings, second and third quills longest, first and second notched near tip. Tarsi, feathers hanging over knee, covered with small irregular scales. Toes long and powerful. Habitat—Throughout world.

Genus. HYPOTRIORCHIS.—Tarsi long and slender, large hexagonal scales in front. Rest like Falco. Habitat—Throughout world.

Genus. IERACIDEA.—Wings moderate, second and third quills longest. Tarsi as Hypotriorchis. Hind toe rather short. Habitat—Australia.

Genus. TINNUNCULUS.—Tarsi long, strong, with transverse hexagonal scales. Rest like Falco. Habitat—Most parts of world.

Genus. HARPAGUS.—Bill compressed, two teeth at tip. Wings moderate, third, fourth, and fifth quills longest. Tail moderate and rounded. Tarsi moderate, slender, transverse scales in front. Two exterior toes unequal. Habitat—South America.

Sub-family. **Accipitrinæ.**—Bill short, much arched, and deeply waved. Wings mostly long. Tail large. Tarsi long, large scales in front.

Genus. ASTUR.—Bill broad at base, generally compressed to tip, festooned in middle; nostrils large and oval, in cere. Wings very long, third, fourth, and fifth quills longest. Tarsi rather long, large scales in front and behind. Inner and hind toes equal. Habitat—Throughout world.

Genus. ACCIPITER.—Bill very short, much compressed; nostrils in fore part of cere, partly hidden by hair-like feathers. Wings moderate, fourth and fifth quills longest. Habitat—Most parts of world.

Genus. MELIERAX.—Cere occupying nearly half of bill. Wings long, third, fourth, and fifth quills longest. Tarsi long. Habitat—Africa.

Sub-family. **Circinæ.**—Bill moderate, short, and curved. Tarsi mostly long and slender, and toes short.

Genus. SERPENTARIUS.—Bill broad and elevated at base, sides compressed at tip. Wings long, blunt spur on shoulder, third, fourth, and fifth quills longest. Tail very long, wedge-shaped, two middle feathers projecting. Tarsi very long, large scales in front. Toes very short, all front toes joined at base, hind toe shortest and rather elevated, all covered above with transverse scales. Round eyes the skin is naked. Habitat—Africa.

Genus. CIRCUS.—Bill festooned on margins; nostrils large, oval, and partly hidden by hairy feathers. Wings long, third and fourth quills longest. Tail long and rounded. Tarsi long and compressed, large scales on outer sides, small on inner. Outer toe longer than inner; claws long and slender. Habitat—Throughout world.

Sub-order. ACCIPITRES NOCTURNI.—Eyes set more or less in front of the head, and surrounded with a radiating circle, more or less perfect, of stiff hairy feathers.

Family. **STRIGIDÆ.**—Bill small, much arched, sides compressed, base hidden under hairy feathers, tip much hooked. Tail broad. All plumage very downy. Tarsi feathered to toes, which are either feathered or haired; claws long and sharp.

Sub-family. **Surninæ.**—Head small, no feather tufts; facial disc imperfect over eyes.

Genus. SURNIA.—Bill nearly hidden by plumes, nostrils entirely concealed. Wings long and pointed, third quill longest. Tail long and wedge-shaped. Tarsi short, plumed. Toes short, plumed. Habitat—Northern Europe and America.

Genus. NYCTEA.—Bill as Surnia. Wings rather long, third quill longest. Tail short and rounded. Tarsi and toes short, strong and densely plumed, claws very long and curved. Habitat—Arctic circle.

Genus. ATHENE.—Bill much as preceding. Wings long and rounded, third and fourth quills longest. Tarsi rather long, plumed; toes short and haired, claws long and sharp. Habitat—Warm portions of both hemispheres.

Sub-family. **Buboninæ.**—Head large, flat at top, two feathery tufts or "ears" on the sides; facial disc imperfect above eyes.

Genus. EPHIALTES.—Bill moderate, base broad; nostrils round, and placed in fore part of cere. Wings long, second, third, and fourth quills longest. Tail short. Tarsi rather

·long, plumed to base of toes. Toes long and scaled. Habitat—Most parts of the World.

Genus. BUBO.—Bill moderate. Wings rather long, second, third, and fourth quills longest. Tail moderate, rounded. Tarsi and toes shortish, thickly plumed. Hind toe short. Habitat—Most parts of world.

Sub-family. **Surninæ.**—Head with two small tufts; facial disc complete.

Genus. NYCTALE.—Bill small; nostrils small and transverse; cere somewhat swollen. Wings moderate, rounded, third and fourth quills longest. Tail long and rounded. Tarsi short, clothed with hairy feathers. Habitat—Europe and North America.

Genus. SURNIUM.—Bill as usual. Wings long and rounded, fourth and fifth quills longest. Tail long and rounded. Tarsi and toes rather short, and densely plumed. Habitat—Many parts of the world.

Genus. OTUS.—Bill as usual. Wings very long, second and third quills longest. Tail moderate. Tarsi long and plumed. Toes moderate, plumed. Habitat—Most parts of both hemispheres.

Sub-family. **Striginæ.**—Head without tufts; facial disc complete.

Genus. STRIX.—Bill long; nostrils large, partially covered by membrane. Wings very long, second quill longest. Tail short. Tarsi rather long, slender, and softly plumed. Toes long, haired; outer toe much shorter than inner, hind toe short and thick.· Habitat—Throughout world.

Order. PASSERES, or PERCHERS.—Toes usually three in front, one behind; sometimes outer toe can¯ be carried backward.

Tribe. FISSIROSTRES.—Bill very wide. Gape beneath eyes. Keel curved to tip. Feet weak.

Sub-tribe. **Fissirostres nocturnæ.**—Eyes very large. Plumage very soft.

Family. **CAPRIMULGIDÆ.**—Bill short, flat, and broad. Tarsi short. Toes long and powerful, hind toe joined to inner at base.

Sub-family. **Steatorninæ.**—Bill, tip hooked, base of bill with feathers and bristles.

Genus. STEATORNIS.—Bill long as head and elevated, much curved and rounded, edges nearly straight and toothed near tip, nostrils large and oblong. Wings long and pointed, third and fourth quills longest and equal. Tail broad, long and graduated. Tarsi with a few hairs. Outer toe larger than inner. Habitat—Guadaloupe, Trinidad, and Bogotá.

Genus. PODARGUS.— Bill, sides suddenly compressed near sharp tip, edge of the upper mandible overlapping the lower, which curves down at tip; nostrils covered with membranous scales, and narrow base of bill covered with bristles. Wings long and pointed, second and third quills longest. Tail long. Tarsi short, with broad scales. Toes with strong scales above. Habitat—Australia and New Guinea.

Genus. BATRACHOSTOMUS.—Wings long and rounded, sixth quill longest. Toes short. Rest like the preceding genus. Habitat—Indian Archipelago.

Genus. ÆGOTHELES.—Bill small and very flat, tip hooked and blunt; nostrils narrower behind, base of bill hidden by plumage and bristles. Wings rounded, third and fourth quills longest. Tarsi rather long and slender, scaled. Hind toe long, outer larger than inner. Habitat—Australia.

Sub-family. **Caprimulginæ.**—Bill short and weak, very broad, and with bristles at base. Wings long. Tail rather long. Tarsi short, scaled or plumed. Middle toe longest, its claw long and toothed like comb.

Genus. CAPRIMULGUS.—Bill very short, very wide; nostrils at sides and tubular, opening partly exposed. Wings long and pointed, second quill longest. Tail long, broad, sometimes rounded, sometimes forked. Tarsi plumed. Hind toe very short. Habitat—Most parts of world.

Genus. CHORDEILES.—Bill very small, a few very short hairs at base, nostrils tubular, opening rounded. Wings long and pointed, second quill longest. Tail long and

broad, rounded or forked. Tarsi short and partly plumed. Hind toe short and slender. Habitat—America.

Sub-family. **Podagerinæ.**—Bill very flat, keel compressed on sides to hooked tip. Wings long and pointed. Tail long and broad. Tarsi usually plumed or with transverse scales in front. Inner toe mostly longer than outer, and both joined to middle toe, which is the longest, and furnished with a long toothed claw.

Genus. SCOTORNIS.—Bill with strong bristles, nostrils with membranous scales over opening. Wings long and pointed, second and third quill longest. Tail extremely long and graduated. Tarsi plumed and scaled. Toes unequal, inner longer than outer. Habitat—Africa.

Genus. MACRODIPTERYX.—Bill with very long bristles. Wings very long and pointed, second or third quill longest, and inner quill extremely lengthened, shaft bare, webbed at tip. Tail long. Tarsi plumed and scaled. Two exterior toes nearly equal. Habitat—Africa.

Genus. PODAGER.—Nostrils partly covered by plumes. Wings very long, first quill longest. Tail short. Tarsi very short and plumed. Inner toe longer than outer. Habitat South America.

Sub-tribe. **Fissirostres diurnæ.**—Eyes moderate, plumage close.

Family. **HIRUNDINIDÆ.**—Bill short and weak, very broad at base, suddenly compressed to tip. Wings long, sharp, and narrow. Tail forked. Tarsi weak.

Sub-family. **Cypselinæ.**—Wings very long and curved like scimetars. Toes short, strong, hinder toe genera'ly directed forwards; claws curved and sharp.

Genus. ACANTHYLIS.—Nostrils in membranous groove. Wings long, first quill longest. Tail with shafts proj cting. Tarsi short and naked. Toes compressed. Habitat—America, India, and Australia.

Genus. MACROPTERYX.—Wings long, first and second quills longest. Tail very long and deeply forked. Tarsi naked. Habitat—India, East India Islands, and Africa.

Genus. CYPSELUS.—Bill with sides gradually compressed to the tip; nostrils with little feathers round opening. Wings, second quill longest. Tail mostly forked. Tarsi very short, plumed to toes. Toes all directed forward. Habitat—Warm and temperate parts of world.

Genus. COLLOCALIA.—Bill very small, nostrils in membranous groove. Wings, second quill longest. Tail moderate. Tarsi slender and naked. Toes short and thick. Habitat—Indian Archipelago.

Sub-family. **Hirundininæ.**—Bill short, sides gradually compressed to tip; nostrils at base of bill and rounded. Wings long, first quill longest. Tail forked. Tarsi mostly scaled. Toes mostly long and slender, claws moderate.

Genus. ATTICORA.—Bill, keel curved to tip. Tail very long and deeply forked. Tarsi short and scaled. Exterior toes unequal. Habitat—South America, New Guinea, Africa, and Australia.

Genus. HIRUNDO.—Nostrils partly covered by membrane. Tail, exterior feathers sometimes much lengthened. Habitat—Most parts of world.

Genus. PROGNE.—Bill strong, edges curved to tip. Tail moderate. Tarsi short. strong, and scaled. Exterior toes equal. Habitat—New World.

Genus. COTTLE.—Bill very flat, nostrils prominent. Ta'l moderate, slightly forked. Tarsi scaled and slender. Toes short and slender, exterior unequal. Habitat—Old and New Worlds.

Genus. CHELIDON.—Bill short and strong. Tail moderate. Tarsi rather long and plumed. Toes also plumed, exterior unequal. Habitat—Old World.

Family. **CORACIADÆ.**—Bill long, broad at base, tip hooked; nostrils at base. Wings long and pointed. Tail mostly short and even. Tarsi short. Toes moderate.

Sub-family. **Coracianæ.**—Bill rather high, tip overhanging lower mandible; nostrils narrow and oblique. Wings and tail moderately long. Tarsi short. Toes moderate and free.

Genus. EURYSTOMUS.—Bill rather short, strong, and flat; nostrils partly covered by plumed membrane. Wings long and pointed, second quill longest. Tail even. Tarsi very short and with transverse scales. Outer toe longer than inner, hind toe long. Habitat—Australia and parts of Old World.

Genus. CORACIAS.—Bill long and straight, nostrils narrow and partly covered by feathered membrane. Wings, second and third quills longest. Tail long and rounded. Tarsi short. Inner toe a very little longer than outer.

Genus. BRACHYPTERACIAS.—Bill long and broad, keel gradually coming to tip. Wings short, third to seventh quills longest. Tail long and rounded. Tarsi long, slender, and scaled transversely. Outer toe longer than inner; claws short, sharp, and curved. Habitat—Madagascar.

Sub-family. Todinæ.—Bill long, tip round or sharp; nostrils exposed. Wings short and round. Tail slightly forked. Tarsi slender. Outer toe longer than inner.

Genus. TODUS.—Bill with edges straight and finely notched, short bristles round base; nostrils in a short groove. Wings, fourth to sixth quills longest and equal. Tarsi with one long scale in front. Outer toe united to second joint, inner to first joint; claws compressed and curved. Habitat—Tropical America.

Sub-family. Eurylaiminæ.—Bill large, very broad at base, keel much arched, nostrils near base. Wings rounded, third and fourth quills longest. Tail moderate. Toes unequal, outer joined to middle.

Genus. EURYLAIMUS.—Bill rather flattened, edges dilated at base. Tail rounded. Tarsi feathered below knee, broad scales in front. Outer toe joined to middle for some distance, hind toe long. Habitat—India and Indian Archipelago.

Genus. CYMBIRHYNCHUS.—Bill with sides gradually compressed, a few long and stout bristles at base. Rest like previous genus. Habitat—India and Archipelago.

Sub-family. Momotinæ.—Bill long, slightly curved, and edges strongly notched. Wings short and round. Two middle feathers of tail generally longest. Outer toe longer than inner, joined to middle as far as second joint, hind toe short.

Genus. MOMOTUS.—Bill, with nostrils at base, small and round. Wings, fourth to sixth quills longest. Tarsi scaled in front. Habitat—Tropical America and islands.

Family. TROGONIDÆ.—Bill short, somewhat triangular, broad at base, mostly toothed; bristles at gape. Wings moderate and rounded. Tail long. Tarsi short and plumed. Toes two in front and two behind.

Genus. TROGON.—Bill, keel much curved to tip, edges notched; nostrils hidden by plumes. Wings, fourth and fifth quills longest. Tarsi hidden by plumes. Two front toes unequal, united to first joint. Habitat—Tropical America.

Genus. APALODERMA.—Bill, edges of upper mandible nearly smooth, of lower deeply notched. Tarsi nearly bare and scaled. Front toes not united. Habitat—South Africa.

Genus. PRIOTELUS.—Bill, with edges of both mandibles notched, nostrils partly hidden by plumes. Tail with tip of each feather diverging. Tarsi short and scaled. Two front toes not united. Habitat—China.

Genus. HARPACTES.—Bill, with edges smooth, nostrils partly hidden by membrane. Tarsi short and half plumed. Two front toes united for half their length. Habitat—India and Archipelago.

Genus. CALURUS.—Beak, edges smooth and curved to tip. Wing-coverts long and curved. Upper tail-coverts very long, tail moderate. Tarsi partly plumed. Two front toes united for a little distance. Habitat—South America.

Family. ALCEDINIDÆ.—Bill long, nostrils straight and sharp. Wings long and rounded. Tail short and rounded. Tarsi short. Toes variously arranged.

Sub-family. Bucconinæ.*—Bill, nostrils hidden by plumes. Tail moderate. Toes, two in front and two behind, outer front toe longest.

Genus. BUCCO.—Bill long, broad at base, suddenly curved at tip, and hooked, edges straight; nostrils in membranous groove. Wings, first quill short, fourth longest. Tail broad and even. Tarsi with transverse scales, claws long, compressed and sharp. Habitat—Tropical America.

Genus. MONASA.—Bill slightly curved and sharp. Wings, fourth and fifth quills longest. Tail long, broad, and round. Tarsi with transverse scales in front; claws short and much compressed. Habitat—Tropical America.

Genus. CHELIDOPTERA.—Bill short and curved throughout. Wings long, third and fourth quills longest. Tail very short. Tarsi with broad transverse scales. Habitat—Tropical America.

Sub-family. Halcyoninæ.—Bill broad at base, compressed towards tip and edges, straight or curved upwards.

Genus. DACELO.—Bill long, very broad at base, keel straight to tip, which is slightly hooked, edges curved upwards at end, lower mandible deepest in middle, and curving upwards to tip; nostrils at sides of base, small and narrow. Wings moderate, first quill long, third and fifth longest. Tail moderate. Tarsi very short and scaled. Inner toe short, united to middle to first joint, outer long and united to third joint, all with broad pads below; claws long and curved. Habitat—Australia and New Guinea.

Genus. HALCYON.—Bill long, keel nearly straight, keel of lower mandible rather more angular than in Dacelo; nostrils in small membranous space, partly hidden by plumes. Wings, first quill long, third longest. Tail moderate. Tarsi short and scaled transversely. Outer toe united to third joint of middle toe, inner to second joint. Habitat—Africa, India, Australia, and South Sea Islands.

Genus. TANYSIPTERA.—Bill, nostrils rounded and exposed. Wings, fourth quill longest. Tail wedge-shaped, central feathers prolonged, with wide tips. Tarsi short and scaled transversely. Inner toe short, united to second joint of middle toe, outer long and joined to third joint. Habitat—New Guinea and Philippines.

Genus. CEYX.—Bill long and slender, keel of lower mandible straight to tip, nostrils narrow, in membranous space. Wings, first quill long, second and third longest. Tail very short and round. Tarsi very short. Inner toe absent, middle and outer toes in front, and united to third joint. Habitat—India and Archipelago.

Sub-family. Alcedininæ.—Bill long and slender, keel of both mandibles sloping to sharp tip.

Genus. CERYLE.—Bill long and strong, keel encroaching on forehead, edges wider at base. Tail long and rounded. Tarsi very short. Habitat—Africa, India, and most of America.

Genus. ALCEDO.—Bill, nostrils oblique, covered with plumed scale. Wings, second and third quills longest, first nearly as long. Tail short. Outer toe united to middle nearly to tip, hind toe short and broad. Habitat—Most parts of Old World.

Genus. ALCYONE.—As Alcedo, but inner toe wanting. Habitat—Australia, Indian Archipelago, and New Guinea.

Sub-family. Galbulinæ.—Bill various, but always long. Wings moderate, fourth quill longest. Tail long and graduated. Tarsi short. Toes, three or four, two front toes united to near end of inner toe.

Genus. GALBULA.—Bill squared and straight, two long slender bristles at base; nostrils in small groove. Tarsi feathered. Inner hinder toe very small or wanting. Habitat—Tropical America and Islands. Includes JACAMARALCYON.

Genus. JACAMEROPS.—Bill curved. Inner hind toe shortest, claws long, compressed, and sharp. Habitat—Tropical America.

Family. MEROPIDÆ.—Bill long, both mandibles curved and very sharp; nostrils on sides of base, partly hidden by

* These are better called Puff-birds than Barbets, for which see page 550.

short bristles. Wings long and pointed. Tail long and broad. Tarsi very short; toes long, two exterior united to middle, hind toe long with broad pad beneath.

Genus. MEROPS.—Wings reaching two-thirds the length of tail, second quill longest. Tail with two middle feathers prolonged. Tarsi strong and scaled. Outer toe longer than inner. Habitat—Most parts of Old World.

Genus. MELITTOPHAGUS.—Wings only reaching to middle of tail and rounded, third quill longest. Tail broad and even. Habitat—Africa.

Genus. NYCTIORNIS.—Bill grooved above, nostrils covered with short plumes. Wings reaching rather beyond base of tail, fourth quill longest. Tail slightly forked. Outer toe longer than inner. Habitat—India and Archipelago.

Tribe. TENUIROSTRES.—Bill slender, upper mandible mostly sharp at tip. Tarsi short; toes long, hind toe mostly lengthened.

Family. UPUPIDÆ.—Bill long, sides compressed to tip; nostrils small and at base. Wings long and rounded. Tail long. Outer toe united to middle, hind toe long.

Sub-family. Epimachinæ.—Bill slender and arched; nostrils in wide groove. Wings moderate, rounded. Tail various. Tarsi moderately long and strong. Hind toe very long and strong, its claw powerful.

Genus. NEOMORPHA.—Wings, fifth to seventh quills longest. Tail long, broad, and round. Tarsi with wide scales in front. Outer toe larger than inner. Habitat—New Zealand.

Genus. EPIMACHUS.—Wings, quills sometimes truncated. Tail long. Tarsi with large thick scales in front. Outer toe longer than inner. Sides of body with flattened and ecomposed feathers. Habitat—Australia and New Guinea.

Sub-family. Upupinæ.—Bill with prominent keel, tip sharp. Claws long, sharp, curved, and strong.

Genus. UPUPA. Bill slender, and curved throughout; nostrils covered with membranous scale. Tail long and even. Tarsi strong, broadly scaled. Outer toe longer than inner, and united to first joint; claw of hind toe long and nearly straight. Habitat—Europe, Asia, and Africa.

Family. PROMEROPIDÆ.—Bill long, slender; nostrils at base, and covered with a scale. Wings moderate. Tarsi moderate, broad scaled; toes moderate, with curved sharp claws.

Sub-family. Promeropinæ.—Bill curved, base broad; nostrils small. Wings moderate. Tail long. Outer toe longer than inner.

Genus. NECTARINIA.—Bill, keel large and round between nostrils, edges of lower jaw very finely notched; nostrils in a short broad groove. Wings, third and fourth or fourth quills only, longest. Two middle feathers of tail elongated. Habitat—Africa, India, and Archipelago.

Genus. DREPANIS.—Wings, first quill wanting, third and fourth equal and longest. Tail moderate and even. Habitat—Sandwich Islands.

Genus. DICÆUM.—Bill curved, short, and broad, edges of tip very finely notched; nostrils in a membranous groove. Wings, first quill wanting; third and fourth longest. Tail short and even. Habitat—Australia, India, and Archipelago.

Sub-family. Cærebinæ.—Bill rather long; nostrils at its base, and covered with a membranous scale. Wings, long and pointed. Tail short. Tarsi rather long; toes short, outer rather longer than inner.

Genus. CÆREBA.—Bill long, arched, and slender, tip sharp. Wings, first quill spurious, third and fourth longest. Tail short and squared. Habitat—South America.

Family. MELIPHAGIDÆ.—Bill long, curved; nostrils in a large groove. Wings moderate. Tongue with tuft of slender fibres at its tip. Tail long and broad. Tarsi short and strong; toes rather long, outer united to middle at its base.

Sub-family. Meliphaginæ.—Wings rounded, fourth to sixth quills usually longest. Toes moderate, hind toe long and strong.

Genus. MELIPHAGA.—Bill slender, broad at base; nostrils covered by membranous scale. Wings, first quill short, and others graduated to fourth and fifth, which are equal and longest. Tail long, very slightly forked. Claws very sharp and long. Habitat—Australia.

Genus. PROSTHEMADERA.—Wings, fifth and sixth quills equal and longest. Tail rounded on sides. Habitat—New Zealand and Auckland Islands.

Genus. ANTHOCHÆRA.—Wings, fifth and sixth quills equal and longest. Tail graduated on sides. Habitat—Australia.

Genus. TROPIDORHYNCHUS (including ENTOMYZA).—Bill elevated at base. Wings, second quill double the length of the first, fourth and fifth equal and longest. Tail long and rounded. Habitat—Australia and New Guinea.

Sub-family. Melithreptinæ.—Bill short and rather conical. Wings, fourth and seventh quills mostly longest.

MYZANTHA (MANORHINA).—Bill, nostrils at base in broad groove, partly covered by plumes and with a membranous scale. Wings, fourth and fifth equal and longest, third, second, and first gradually shorter. Tail long and rounded on sides. Habitat—Australia.

Family. TROCHILIDÆ.—Bill very slender, tip sharp; nostrils covered with a large scale, sometimes feathered. Wings long and sharp. Tail usually long. Tarsi very short; toes long and slender.

Genus. TROCHILUS.—Bill very slight and curved, edges of upper mandible overlapping lower at base; nostrils with partly feathered scale. Wings, first quill shorter than second. Tail forked, two feathers much projecting. Tarsi stout, partly plumed. Inner toe shorter than outer. Habitat—Jamaica.

Genus. THAUMASTURA.—Bill small, delicate, nearly straight. Wings moderate. Throat with scaly metallic feathers. Two central tail feathers very long, rest abruptly graduated, forming a very deep and sudden fork. Habitat—Guatemala.

Genus. DOCIMASTER.—Bill enormously elongated, slightly recurved. Wings large. Tail forked. Habitat—Peru, Santa Fé de Bogotà.

Genus. HELIACTIN.—Bill short and spine-like. Male with long shining crest over eyes and ears. Throat with drooping pointed gorget. Tail graduated, long, narrow, pointed feathers. Habitat—Brazil.

Genus. ERIOCNEMIS.—Bill moderate, straight. Wings moderate. Tail short and broad. Tarsi tufted to toes with downy muff.

Genus. TOPAZA.—Bill slightly curved; nostrils with scale, mostly concealed by plumes. Wings, first quill longest. Tail moderate two exterior feathers on each side much lengthened. Tarsi slender, partly plumed. Outer toe shorter than inner. Habitat—Central America.

Genus. SPATHURA.—Bill straight, slender. Throat shining green. Tail very deeply forked; exterior feathers very long, bare almost to end, where they have a racket-shaped web. Tarsi covered with large muff of soft down. Habitat—Peru, Santa Fé de Bogotà, Bolivia.

Genus. DISCURA.—Resembles Spathura, but is without the tarsal muffs. Habitat—Cayenne, Surinam, Demerara, and Northern Brazil.

Genus. RHAMPHOMICRON.—Bill very small, short, and delicate. Gorget effulgent. Wings small. Tail moderately forked. Habitat—Santa Fé de Bogotà.

Genus. GOULDIA.—Bill short. Wings small. Gorget green. Tail, exterior feathers long, bristly, and pointed, central very short, others regularly graduated. Habitat—Brazil.

Genus. SELASPHORUS.—Bill straight and delicate. Gorget composed of refulgent large scaly feathers. Wings delicate. Tail feathers narrow. Habitat—Veragua.

Genus. PHAETHORNIS.—Bill very long, slightly curved, base broad; nostrils with partly plumed scale. Wings, first quill longest. Tail long, two central feathers elongated. Tarsi slender, and plumed; exterior toe united to middle at base. Habitat—South America.

2.

Genus. FLORISUGA.—Bill arched. Wings moderate. Tail short, forming rounded fork when closed ; tail coverts very large, projecting beyond central feathers of tail. Habitat—Tobago, Brazil.

Genus. OXYPOGON.—Bill short and sharp. Head of male with long pointed crest, black in centre, rest white ; long drooping plumes on chin. Wings moderate. Tail large, slightly forked when closed. Habitat—Cordilleras.

Genus. HELIOMASTER.—Bill long, straight, or nearly so. Gorget very large, extended laterally by long lancet-shaped feathers. Tail ample. Feet small. Habitat—Brazil, Antilles.

Genus. EUTOXERES.—Bill rather long, very much curved. Tail-coverts long. Tail rapidly graduated, composed of lancet-shaped feathers. Habitat—Veragua, Santa Fé de Bogotà.

Genus. HELIANGELUS.—Bill long as head, straight, rather flat at base, rest cylindrical. Throat with scaly lustrous feathers, mostly a white or buff gorget. Wings strong, outer primary well curved. Tail moderate, rounded. Habitat—Venezuela, New Granada, &c.

Genus. LOPHORNIS.—Bill short, slender, straight, and sharp ; head crested. Neck furnished with fan-like plumes. Tail moderate, rounded. Habitat — Guiana, Cayenne, Brazil.

Genus. PETASOPHORA.—Bill long, slender, almost, if not quite, straight. Ear-coverts composed of large, scaly, metallic feathers, standing boldly from neck. Wings large. Tail squared. Habitat—Mexico.

Genus. TRYPHÆNA.—Bill long and awl-shaped ; gorget large. Wings rather small. In male, tail long, forked to upper tail-coverts. Habitat—Mexico.

Genus. COMETES.—Bill long, cylindrical, boldly curved downwards. Tail very long, very deeply forked, feathers long, wide, and squared at ends. Tarsi bare. Habitat—Peru and Bolivia.

Genus. CALOTHORAX.—Bill long and curved, base broad, covered with plumes ; nostrils concealed by plumes. Wings moderate, first quill longest. Tail long. Tarsi slender, partly plumed. Outer toe shorter than inner. Habitat—Central America.

Genus. CYNANTHUS.—Bill straight. Tail very deeply forked, the two exterior feathers being broad, squared, and twice the length of the second pair. The remainder are lancet-shaped, and rapidly graduated to the central, which are very short. Habitat—Peru, Santa Fé de Bogotà.

Genus. AVOCETTA.—Bill much curved upward in front. Tail short. Wings moderate.

Genus. HELIOTHRIX.—Bill very slightly arched, and wedge-shaped ; nostrils hidden by plumes. Wings long, largely feathered. Feet very small. Habitat—Brazil, Guiana.

Genus. ORTHORHYNCHUS.—Bill short and straight. Head with large loose tuft forming a long peak. Includes CEPHALEPIS. Habitat—Brazil, &c.

Genus. OREOTROCHILUS.—Bill slightly curved ; nostrils in long groove, concealed by projecting plumes. Wings, first quill longest. Tail long and rounded. Tarsi slender, partly plumed. Toes free. Habitat—South America.

Genus. MELLISUGA.—Bill straight ; nostrils hidden by plumes. Wings, first quill longest. Tail long. Tarsi slender. Exterior toes united to middle at base ; outer toe longer than inner. Habitat—New World.

Family. CERTHIDÆ.—Bill long, arched ; nostrils small and with membranous scale. Wings, tail, tarsi, and toes extremely varied.

Sub-family. Furnarinæ.—Toes long, outer longer than inner, and united at base to middle toe.

Genus. FURNARIUS.—Nostrils oval, covered with plumes of forehead. Wings, quills graduated to fourth and fifth, which are longest. Tail moderate, rather squared. Tarsi rather long ; toes moderate. Habitat—South America.

Sub-family. Certhinæ.—Toes very long and delicate, outer toe longer than inner, and joined to middle joint beyond first joint ; inner toe united as far as first joint ; hind toe very long and slender ; claws long and curved.

Genus. CERTHIA.—Bill curved, nostrils lunate, partly covered with membranous scale. Wings, first quill short, fourth and fifth equal and longest. Tail long, graduated, tips of feathers sharp and stiff. Tarsi short and slender. Habitat—Europe, Asia, and North America.

Genus. TICHODROMA.—Bill slightly curved, nostrils with membranous scale. Wings long and rounded, quills as in Certhia. Tail short, tips of feathers soft. Habitat—Europe and Asia.

Sub-family. Dendrocolaptinæ.—Bill various ; nostrils small and exposed. Tail, shaft of each feather projecting beyond web.

Genus. DENDROCOLAPTIS (XIPHORHYNCHUS).—Bill much curved, broad at base, then compressed to tip, nostrils oval. Wings, fourth quill longest. Tarsi short and strong, broad scaled. Outer toe longer than inner, and united to middle toe. Habitat—South America.

Sub-family. Sittinæ.—Outer toe longer than inner, and united as far as first joint to the middle toe.

Genus. SITTA.—Bill straight, nostrils in broad groove. Wings, first quill very short, third and fourth longest. Tail short and broad. Tarsi strong ; hind toe longer than middle. Habitat—Europe and North America.

Sub-family. Menurinæ.—Bill long and slender ; nostrils with horny scale or membrane. Wings short and round. Outer toe longer than inner, hind toe long.

Genus. MENURA.—Bill strong and broad ; nostrils in broad triangular groove. Wings, sixth to ninth quills longest. Tail very long, in males lyre-shaped. Tarsi rather long, broadly scaled in front ; claws long, powerful, and blunt. Habitat—Australia.

Genus. TROGLODYTES.—Bill straight ; nostrils in a groove, protected by membrane. Wings, fourth to sixth quills longest. Tail short and round. Tarsi rather short, broadly scaled in front. Toes long and slender. Habitat—Europe, Asia, and America.

Tribe. DENTIROSTRES.—Tip of upper mandible slightly toothed and hooked. Outer toe united to centre toe at base.

Family. LUSCINIDÆ.—Bill sharp, slender, straight, sides rather compressed towards tip. Wings rather long. Tarsi slender.

Sub-family. Malurinæ.—Nostrils open, in membranous groove. Tail long and rounded. Hind toe with strong claw.

Genus. ORTHOTOMUS.—Bill slightly flattened at base ; nostrils with longitudinal opening. Wings, fourth to eighth quills longest. Tail graduated, feathers narrow. Tarsi with single scale in front. Outer toe longer than inner. Habitat—India and Archipelago.

Genus. DRYMOICA.—Bill short, with short bristles at base ; nostrils with semi-lunar opening, partly covered by a scale. Wings, fourth and fifth quills longest. Tail long, broad, and rounded. Tarsi broadly scaled in front. Habitat—Africa, Southern Europe, India, and Australia.

Genus. STIPITURUS.—Bill with long bristles at base ; nostrils with oval opening. Wings, fourth to sixth quills longest. Tail long and graduated, six long slender shafts with filamentous webs. Tarsi with large scale in front. Habitat—Australia.

Sub-family. Luscininæ.—Bill slender and straight ; nostrils in membranous groove. Wings and tail moderate. Tarsi slender and scaled. Toes long and slender, claws sharp.

Genus. SYLVIA.—Bill with few weak bristles at base ; nostrils in broad short groove, with partly semi-lunar opening. Wings, first quill very short, third and fourth longest. Outer toe united at base to middle ; hind toe with strong claw. Habitat—Old World.

Genus. LUSCINIA.—Bill short, no bristles ; nostrils with rounded opening. Wings, third quill longest. Tarsi with one long scale in front. Habitat—Old World.

Genus. CALAMODYTA.—Bill small, sometimes with very weak bristles ; nostrils with oval opening. Wings, first quill very short, third and fourth longest. Tarsi with broad scales in front. Habitat—Old World.

Genus. REGULUS.—Bill small, broad at base ; nostrils semi-lunar, covered with membranous scale. Wings, first quill very short, fourth and fifth longest. Tail very slightly forked. Tarsi with one long scale in front. Habitat—Europe, Asia, and America.

Sub-family. **Erythacinæ.**—Bill with tip entire, bristles at base. Wings and tail various. Tarsi, long, slender, with single scale. Toes moderate.

Genus. SAXICOLA.—Bill rather flat at base, nostrils in membranous groove, opening rounded. Wings long, third and fourth quills longest. Tail even. Outer toe longer than inner. Habitat—Most parts of world.

Genus. PRATINCOLA.—Bill with short bristles, nostrils with rounded opening, nearly hidden by plumes of forehead. Wings long and round, fourth and fifth quills longest. Tail short and broad. Outer toe longer than inner. Habitat—Old World.

Genus. RUTICILLA.—Bill broad at base ; nostrils in membranous groove, opening rounded. Wings long and rounded ; fourth and fifth quills longest. Tail long, broad, and somewhat forked. Toes short and slender. Habitat—Old World.

Genus. ERYTHACUS.—Bill strong ; nostrils in groove, opening elongated. Wings moderate, fourth and fifth quills longest. Tail moderate, broad, and rather forked. Outer longer than inner, united to middle at base, hind toe long and slender. Habitat—Europe.

Genus. CYANECULA.—Bill straight and slender, nostrils at base, rounded. Wings moderate, third and fourth quills longest. Tail moderate, rounded. Tarsi long ; outer toe longer than inner. Habitat—Old World.

Genus. COPSYCHUS.—Bill wide at gape, with short bristles ; nostrils in slight groove. Wings, fifth quill longest, first to fourth graduated. Tail rather long. Tarsi rather long, hind toe long and strong. Habitat—India and Africa.

Genus. SIALIA.—Bill short, broad at base ; nostrils in groove, opening elongated. Wings very long and pointed, third and fourth quills longest. Hind toe moderate. Habitat—America.

Sub-family. **Accentorinæ.**—Bill short, straight, nostrils in groove, opening exposed. Wings long. Tail moderate. Tarsi moderate. Outer toe united to middle at base, hind toe long.

Genus. ACCENTOR.—Bill, nostrils very narrow. Wings, third to fifth quills longest, first very short. Tarsi broadly scaled in front. Claw of hind toe long and compressed. Habitat—Europe and Asia.

Sub-family. **Parinæ.**—Bill straight, short, and strong ; nostrils at base. Wings moderate, first three quills graduated. Tail rounded and even. Tarsi scaled in front. Inner toe shortest ; claws curved and strong.

Genus. PARUS.—Bill, nostrils concealed by plumes of forehead. Wings, fourth and fifth quills longest. Hind toe and claw very long. Habitat—Old World and North America.

Genus. PAROIDES.—Wings, third and fourth quills longest. Tail rather long. Habitat—Europe and Africa.

Sub-family. **Mniotiltinæ.**—Bill conical, nostrils exposed, opening rather large. Wings long. Tail moderate. Tarsi broadly scaled ; toes long and slender.

Genus. ZOSTEROPS.—Bill with few short bristles at base ; nostrils in broad groove, protected by semilunar scale. Wings, fourth and fifth quills longest, first very small. Tail broad. Outer toe longer than inner. Habitat—India, Africa, and Australia.

Sub-family. **Motacillinæ.**—Bill moderate, keel rather curved at end, sides much compressed ; nostrils in small groove. Wings long and pointed. Tail long. Tarsi long, slender, with transverse scales. Toes and claws long.

Genus. MOTACILLA.—Bill, nostrils oval, partially hidden by a membrane. Wings, second and third quills longest. Tail very long, even or forked. Hind toe long. Habitat—Most parts of world.

Genus. EPHTHIANURA.—Bill, nostrils in broad groove, covered with membranous scale. Wings, third to fifth quills longest. Tail rather short. Toes rather short. Habitat—Australia.

Genus. ANTHUS.—Bill straight and slender ; nostrils in short broad groove, rounded, and partly covered by membrane. Wings, first to third quills longest. Claws of front toes short and curved, that of hind toe long and sharp. Habitat—Most parts of world.

Family. **TURDIDÆ.**—Bill rather strong, sides rather compressed. Wings long. Tail moderate. Tarsi and toes various.

Sub-family. **Formicarinæ.**—Bill straight ; nostrils in membranous groove, reached but not covered by plumes. Wings short and rounded. Tarsi with divided scales in front. Outer toe longer than inner.

Genus. PITTA.—Bill broad at base, opening of nostrils oblique and exposed. Wings, third and fourth quills longest. Tail very short and even. Tarsi very long with broad scales. Outer toe longer than inner, and united to middle at base ; hind toe and claw very long. Habitat—India, Australia, and Western Africa.

Genus. HYDROBATA.—Bill moderate, slightly curved upwards ; opening of nostrils semilunar. Wings rounded, third and fourth quills longest, first spurious. Tail very short and even. Outer toe rather longer than inner. Habitat—Europe, Asia, and America.

Sub-family. **Turdinæ.**—Bill as long as head ; nostrils open, in small groove. Wings, first quill very short. Tail long and broad. Tarsi long. Outer toe longer than inner, united to middle at base, hind toe long and strong.

Genus. MIMUS.—Bill with short bristles at base ; nostrils oval. Wings, fourth, fifth, sometimes seventh quills longest, second shorter than third. Tarsi with broad scales in front. Habitat—America, West Indian and Galapagos Islands.

Genus. TURDUS.—Bill with weak bristles at base ; nostrils as in Mimus. Wings, third and fourth quills longest. Tail slightly forked. Tarsi with single long scale in front. Habitat—All parts of the world.

Sub-family. **Timalinæ.**—Bill moderate, keel curved ; nostrils exposed. Wings short and rounded. Tail graduated. Tarsi long and strong. Toes long, strong, and with large scales above ; claws compressed and sharp.

Genus. TIMALIA.—Bill, sides much compressed to tip ; few short bristles at base ; nostrils in small groove, semilunar opening with a small scale. Wings, fifth to seventh quills longest. Tarsi with one long scale in front. Habitat—India and Java.

Genus. ICTERIA.—Bill short, strong, elevated and curved at base, tip slightly scooped ; nostrils in small groove. Wings, third and fourth quills longest. Tarsi long, with single large scale in front. Outer toe longer than inner. Habitat—America.

Genus. CINCLOSOMA.—Bill moderate, rather slender, slightly compressed ; a few very short bristles at base ; nostrils with membranous scale. Wings, third to fifth quills longest. Tarsi with broad scales. Inner toe longer than outer. Habitat—Australia.

Genus. GARRULAX.—Bill, nostrils in short groove, opening rounded and hidden by plumes and bristles. Wings, fifth and sixth quills longest. Tail broad and rounded. Tarsi strong, hind toe very long. Habitat—India and China.

Sub-family. **Oriolinæ.**—Bill long as head, broad at base, compressed on sides, slightly curved above ; nostrils partly closed by membrane. Wings long, third and fourth quills longest. Tarsi short, strong scales.

3 D 2

Genus. ORIOLUS.—Bill, nostrils oval. Wings, first quill only half length of third and fourth. Tail moderate and rounded. Tarsi with broad scales. Toes free, outer longer than inner, claws strong and curved. Habitat—Old World.

Family. MUSCICAPIDÆ.—Bill various, curved above, sides compressed to tip. Tail long. Tarsi short ; toes long.

Sub-family. Alectrurinæ.—Bill broad and rather flattened at base ; nostrils rounded and exposed. Wings long. Tarsi slender.

Genus. GUBERNETES.—Bill short, with bristles at base ; nostrils in short membranous groove, opening rounded. Tail long and deeply forked. Tarsi with transverse scales. Outer toe united at base to middle ; hind toe long. Habitat —South America.

Sub-family. Tyranninæ.—Bill long, broad and flat at base, sides compressed to tip, which is hooked ; nostrils hidden by plumes and bristles. Wings long and pointed. Tail moderate. Tarsi broadly scaled. Outer toe longer than inner, united to middle at base ; claws short and sharp.

Genus. TYRANNUS.—Bill with long bristles at base ; nostrils small and rounded. Wings, second, third, and sometimes to fifth quills equal, all sharply pointed. Tail slightly forked. Tarsi slender. Habitat—Tropical America.

Genus. MILVULUS.—Bill, nostrils on side and rounded. Wings, second quill longest, first and third equal, all waved and pointed. Tail long, very deeply forked. Tarsi slender. Habitat—Tropical America.

Sub-family. Muscicapinæ.—Bill broad at base, narrowing to tip, base with bristles. Wings long and pointed. Outer toe longer than inner.

Genus. RHIPIDURA.—Bill with long bristles at base, nostrils partly covered with plumes and bristles. Wings, fourth and fifth quills longest, first short. Tail broad and graduated. Tarsi broadly scaled ; front toes short, hind toe long. Habitat—India, New Zealand, and islands of Indian Ocean.

Genus. TCHITREA.—Wings, fourth and fifth quills longest, first to third graduated. Tail long and graduated. Tarsi with slightly divided scales. Hind toe long. Habitat— Africa and India.

Genus. MUSCICAPA.—Bill short, nostrils partly hidden by plumes. Wings, third and fourth quills longest, first very short. Tail even. Front toes short, hind toe long. Habitat—Old World.

Genus. MUSCIVORA.—Bill straight, rather flattened above, bristles very long and stiff, nostrils oval and exposed. Tail nearly even. Tarsi with one long scale. Hind toe long. Habitat—South America.

Family. AMPELIDÆ.—Bill moderate, broad at base, rather flattened, sides gradually compressed to tip. Wings long. Tail moderate. Toes moderate, outer united to inner at base.

Sub-family. Pachycephalinæ.—Bill with few slight bristles at base. Wings rounded. Tarsi long and slender ; toes long.

Genus. PARDALOTUS.—Bill very short, nostrils covered by membrane. Wings, first to third quills longest. Hind toe rather long. Habitat—Australia.

Sub-family. Piprinæ.—Bill, base very broad, nostrils at side, nearly hidden by plumes. Wings moderate and pointed. Tail very short and even. Tarsi long and slender ; outer toe united to middle beyond second joint.

Genus. PIPRA.—Bill short, nostrils partly closed by a membrane. Wings, third and fourth quills longest. Outer toe longer than inner. Habitat—Tropical America.

Genus. RUPICOLA.—Bill strong, nostrils large, oval, partly closed by a membrane. Wings, fourth and fifth quills longest. Tarsi partly covered by plumes. Outer toe longer than inner, and united to middle beyond second joint, the inner united at base ; hind toe long and strong. Habitat— South America.

Genus. CALYPTOMENA.—Bill short, very broad at base, nostrils rounded. Wings, third and fifth quills equal, fourth longest. Tarsi broadly scaled in front. Outer toe united to inner beyond second joint. Habitat—Sumatra.

Sub-family. Ampelinæ.—Bill rather long, very wide gape ; nostrils oval. Wings moderate. Tail short. Outer toe slightly united at base.

Genus. AMPELIS.—Bill moderate, gradually curved above; nostrils on side, concealed by plumes. Wings, second quill longest. Claws short, much curved, compressed, and sharp. Habitat—Europe, Asia, and Northern America.

Genus. ARAPUNGA (PROCNIAS).—Bill very short at base ; nostrils in membranous groove, plumes reach the opening. Wings, third and fourth quills longest. Tarsi short, transversely scaled. Outer toe longer than inner. Habitat— Tropical America.

Sub-family. Campephaginæ.—Bill short, few bristles at base ; nostrils rounded. Wings moderate, third to fifth quills longest. Tail long and rounded. Tarsi short, transversely scaled.

Genus. PERICROCOTUS.—Bill, sides much compressed ; nostrils in sunken groove. Tail graduated. Tarsi slender, with narrow scales. Lateral toes equal. Habitat—India and Archipelago.

Sub-family. Dicrurinæ.—Bill well keeled above and curved ; strong bristles at base. Wings long. Tarsi and toes short.

Genus. ARTAMUS.—Bill moderate, keel rounded, few bristles, nostrils small and partly hidden by plumes. Wings very long, pointed, second quill longest, and first only rudimental. Tarsi short, transversely scaled. Toes short, outer united at base to middle. Habitat—India, Australia, and Madagascar.

Genus. DICRURUS.—Bill, keel elevated and curved to tip ; nostrils rounded, and hidden by plumes and bristles. Wings, fourth and fifth quills longest. Tail long and deeply forked, sometimes extensive feathers, webbed only at ends. Outer toe united to middle at second joint, inner at base. Habitat—India and Africa.

Family. LANIIDÆ.—Bill long, strong, straight, mostly hooked, keel curved, and sides compressed to tip. Wings moderate. Tail long. Tarsi strong. Hind toe long, claws long, curved, and very sharp.

Sub-family. Laninæ.—Bill moderate ; nostrils on side and rounded. Wings rather long. Outer toe longer than inner, and united to middle at base. Hind toe long, broadly padded.

Genus. LANIUS.—Bill broad at base, tip strongly hooked ; few short bristles at base ; nostrils partly hidden by bristles. Wings, fourth quill longest. Tail graduated. Habitat— Europe, Asia, and Africa.

Genus. ENNEOCTONUS.—Wings, third quill longest. Tail much rounded. Rest like Lanius. Habitat—Europe and India.

Sub-family. Thamnophilinæ.—Bill long, keel arched, tip hooked, base with bristles. Wings moderate. Tail long. Tarsi broadly scaled. Outer toe united to middle at base.

Genus. THAMNOPHILUS.—Bill, nostrils at side of base, rounded and exposed. Wings rounded, fourth to seventh quills longest. Tarsi with transverse scales before and behind. Habitat—America.

Tribe. CONIROSTRES.—Bill strong, conical. Wings moderate, mostly pointed. Tarsi strongly scaled ; toes moderate.

Family. CORVIDÆ.—Bill, keel arched, sides compressed to tip ; nostrils at base, concealed by plumes. Tail varied. Tarsi and toes moderate.

Sub-family. Phonogaminæ.—Bill long, base broad ; nostrils long and narrow. Wings various. Tail long. Tarsi and toes scaled ; outer toe united to middle at base.

Genus. GYMNORHINA.—Bill longer than head, keel slightly curved and encroaching on forehead ; nostrils like straight slits in beak. Wings very long, pointed, third and fourth quills longest, first very short. Tail even. Toes nearly even. Habitat—Australia.

Genus. STREPERA.—Bill, keel rounded and elevated at base. Wings, fourth and fifth quills longest, first very short. Tail squared. Habitat—Australia.

Sub-family. **Garrulinæ.**—Bill moderate, compressed, keel suddenly curved at tip. Wings moderate, rounded. Tail various. Tarsi moderate, strongly scaled, toes long.

Genus. GARRULUS.—Bill not so long as head; nostrils hidden by plumes. Wings, fourth, fifth, and sixth quills longest. Tail nearly even. Outer toe longer than inner; hind toe long and strong. Habitat—Old World.

Genus. CYANOCORAX.—Bill slightly compressed, nostrils partly hidden by plumes. Wings, fourth to sixth quills longest. Tail long and rounded. Habitat—New World.

Genus. CISSA.—Bill large, keel elevated, tip hooked, sides much compressed; nostrils partly covered with plumes. Wings, fourth and fifth quills longest. Tail long and graduated. Outer toe little longer than inner. Habitat—India and Archipelago.

Sub-family. **Calleatinæ.**—Bill short, much curved above, sides compressed. Wings short and rounded. Tail long and graduated. Tarsi broadly scaled in front.

Genus. TEMNURUS.—Bill, nostrils small, rounded, hidden by plumes and bristles. Wings, fourth to sixth quills longest. Outer toe longer than inner. Habitat—India.

Genus. CRYPSIRHINA.—Bill, tip notched, base and nostrils hidden by plumes. Wings, fifth and sixth quills longest. Outer toe longer than inner. Habitat—India.

Sub-family. **Corvinæ.**—Bill, base broad, rather curved above. Wings long and rounded. Tail various. Tarsi broadly scaled in front. Toes moderate, strong; two exterior equal.

Genus. NUCIFRAGA.—Bill long, nostrils at sides of base, and covered by bristles. Wings, fourth and fifth quills longest. Tail moderate, rounded on sides. Toes long, strong, and broadly scaled. Habitat—Europe and Asia.

Genus. PICA.—Bill long, tip slightly hooked; nostrils rounded and hidden by bristles. Wings, fourth and fifth quills longest, first very short. Tail long and graduated. Tarsi long, toes short and strong. Habitat—Old World and North America.

Genus. CORVUS.—Bill strong and mostly long; nostrils covered by bristly plumes, except when forehead bare. Wings long and pointed, third and fourth quills longest. Habitat—Most parts of world.

Genus. PICATHARTES.—Bill rather flattened; nostrils in large membranous groove and exposed, upper part of head and cheeks bare. Wings, sixth and seventh quills longest. Tail, very long and rounded. Tarsi long and slender. Habitat—Africa.

Sub-family. **Gymnoderinæ.**—Bill straight, rather flattened, sides compressed to tip; nostrils in membranous groove. Wings long and pointed. Tail moderate and rounded. Tarsi long, transversely scaled. Toes long, outer nearly equal to middle; claws long, curved, and sharp.

Genus. GYMNODERUS.—Bill rather short, broad at base, groove of nostrils clothed with down. Wings, third to fifth quills longest. Tail even. Sides of neck bare, head and front of neck scantily clothed with down. Habitat—Guiana, &c.

Genus. CEPHALOPTERUS.—Bill long, opening of nostrils large. Male with large overhanging crest on head, breast bare with pendulous pencil of plumes. Wings, third quill longest. Habitat—Brazil.

Sub-family. **Pyrrhocoracinæ.**—Bill long, slender, curved; nostrils hidden by broad plumes. Wings long and pointed. Tail long.

Genus. CORACIA.—Wings, fourth quill longest. Tail moderate, nearly equal. Tarsi short, broadly scaled in front, exterior toes nearly equal. Habitat—Europe and Asia.

Family. **PARADISEIDÆ.**—Bill long, strong, curved above; nostrils hidden by short plumes. Wings long and rounded. Tarsi with single long scale. Outer toe longer than inner, united to middle at base, hind toe very long, claws long and curved.

Genus. PARADISEA.—Bill, sides compressed to tip. Wings, fourth to seventh quills longest. Tail long and broad. Tarsi rather long. Sides of neck, breast, body, tail, and sometimes head with long drooping feathers. Habitat—New Guinea.

Family. **STURNIDÆ.**—Bill, sides compressed. Wings long and pointed. Tail rather long. Tarsi strong and broadly scaled in front. Toes long and strong, hind toe very long.

Sub-family. **Ptilonorhynchinæ.**—Bill moderate; upper mandible scooped at tip.

Genus. PTILONORHYNCHUS.—Bill, nostrils sunken, large, nearly covered by plumes. Wings, fourth and fifth quills longest, three first graduated. Tail short, even. Outer toe longer than inner. Habitat—Australia.

Genus. CHLAMYDERA.—Bill, nostrils rounded, exposed, in a membrane. Wings, third and fourth quills longest, first two unequal. Tail long and rounded. Habitat—Australia.

Genus. ASTRAPIA.—Bill, sides much compressed; nostrils sunken, partly hidden by plumes. Wings, third and fourth (?) quills longest. Tail very long, and deeply graduated. Habitat—New Guinea.

Sub-family. **Graculinæ.**—Bill broad at base, much compressed, slightly curved above; nostrils exposed. Wings long, first quill short, third and fourth longest. Tail short. Tarsi short. Toes long.

Genus. GRACULA.—Bill long. Wings, first quill rudimental. Tail various. Tarsi strong, hind toe very long. Parts of head bare. Habitat—India, New Guinea.

Sub-family. **Buphaginæ.**—Bill moderate, broad, slightly flattened above; nostrils partly closed by membrane. Tail long, graduated, each feather pointed. Tarsi short and strong. Toes moderate.

Genus. BUPHAGA.—Bill, sides of lower mandible very broad, projecting under eyes. Wings, third quill longest. Habitat—Africa.

Sub-family. **Sturninæ.**—Bill, tip rather blunt and flattened; nostrils in membranous groove. Tail short. Tarsi long. Toes long and strong.

Genus. PASTOR.—Bill, groove of nostrils clothed with short feathers. Wings, second quill longest, first spurious. Tail even. Tarsi transversely scaled. Habitat—Asia, Africa, and temperate Europe.

Genus. STURNUS.—Bill sharp, straight, and conical groove of nostrils feathered. Wings, second quill longest, third nearly as long, first spurious. Habitat—Most parts of world.

Sub-family. **Quiscalinæ.**—Bill long, slightly curved at tip; nostrils in triangular groove. Tail long and graduated. Hind toe long.

Genus. QUISCALUS.—Bill, upper mandible overhanging lower; nostrils oval, partly closed by membrane. Wings, second to fourth quills longest. Tail, sides turned up. Habitat—North America.

Sub-family. **Icterinæ.**—Bill rather long. Tail long. Tarsi moderate and broadly scaled. Toes moderate and strong.

Genus. CACICUS.—Bill advancing in crescent on forehead, sides compressed, tip sharp; nostrils oval, pierced in bill. Wings, third and fourth quills longest. Tail graduated. Habitat—Tropical America.

Genus. XANTHORNIS.—Bill arched, upper mandible advancing in point on forehead; nostrils covered with membrane. Wings, third and fourth quills longest, second nearly as long. Tail wedge-shaped. Habitat—America and West Indies.

Genus. YPHANTES.—Bill, upper mandible advancing in point on forehead; nostrils oval, covered with membrane. Wings, second and third quills longest, first nearly equal. Tail squared. Habitat—North America.

Sub-family. **Agelainæ.**—Bill very conical, flattened above. Tarsi long as middle toe. Toes long, slender, outer united at base to middle.

Genus. MOLOTHRUS.—Bill short, upper mandible slightly advanced on forehead and arched at base ; nostrils partly covered by membrane and plumes. Wings, first and second quills longest. Tail rounded. Habitat—America.

Genus. AGELAIUS.—Bill flattened at base, edges slightly waved. Wings, second and third quills longest. Tail long and rounded. Habitat—America.

Genus. DOLICHONYX.—Bill nearly straight to tip ; nostrils covered with membrane. Wings, first and second quills longest. Tail, tips of feathers sharp and stiff. Tarsi slender. Habitat—America and West Indies.

Family. **FRINGILLIDÆ.**—Small birds.—Bill short, thick, mostly angular at base.

Sub-family. **Ploceinæ.**—Bill, keel projecting on forehead, arched to tip. Wings rounded, first quill very short. Legs and toes strong, scaled, hind toe long and strong.

Genus. HYPHANTORNIS.—Bill, keel pointed on forehead, nostrils oval, pierced in bill. Wings, fourth quill longest, third and fifth quill nearly equal. Tail short and even. Tarsi moderate. Claws much curved. Habitat—Africa.

Genus. PLOCEUS.—Bill compressed ; nostrils partly hidden by plumes. Wings, third to fifth quills longest, first very short ; tertials nearly as long. Tail moderate. Habitat—India and Africa.

Genus. TEXTOR.—Bill broad at base, curved towards tip, edges waved ; nostrils pierced in bill. Wings, third and fourth quills longest, second nearly equal. Toes short, claws curved. Habitat—Africa.

Genus. PHILETÆRUS.—Bill much compressed, rather arched above. Wings, second to fourth quills longest, first rudimentary. Habitat—South Africa.

Genus. PLOCEPASSER.—Bill large, advancing in point on forehead. Wings, second and third quills longest. Tarsi with single scale in front. Claws strong and curved. Habitat—Africa.

Genus. VIDUA.—Bill compressed, advancing in point on forehead ; nostrils hidden by plumes. Wings, third to fifth quills longest, first spurious. Tail-coverts and tail-feathers lengthened variously. Tarsi with divided scales in front. Habitat—Africa.

Sub-family. **Coccothraustinæ.**—Bill large, short, very broad at base, curved to tip. Wings long and pointed. Tail short. Tarsi and toes moderate.

Genus. CARDINALIS.—Bill advanced on forehead, edges waved in middle ; nostrils rounded, hidden by plumes. Wings, fifth quill longest, four first graduated. Tail rather long, rounded. Tarsi strong. Claws short. Habitat—America.

Genus. COCCOTHRAUSTES.—Bill, edges angular at base ; nostrils oval, hidden by plumes. Wings, second and third quills longest. Tail short, rather forked. Outer toe longer than inner. Habitat—Europe, North America, Northern India.

Sub-family. **Tanagrinæ.**—Bill triangular at base, rather arched to tip, which is scooped. Wings moderate, pointed. Tarsi and toes short and slender.

Genus. PYRANGA.—Bill compressed to tip ; nostrils small, partly hidden by plumes. Wings, third quill longest. Tail moderate. Habitat—New World.

Sub-family. **Fringillinæ.**—Bill short, sloping to tip. Wings long and pointed. Tarsi moderate, slender, transversely scaled.

Genus. AMADINA.—Bill very broad at base, flattened above and pointed on forehead ; nostrils sunken, hidden by plumes. Wings, third and fourth quills longest, second nearly equal, first very small. Tail short. Habitat—Africa, Asia, and Australia.

Genus. FRINGILLA.—Bill broad at base ; nostrils in small groove. Wings, second and third quills equal and longest, first nearly as long. Tail moderate, slightly forked. Outer toe united to middle at base. Habitat—Most parts of world.

Genus. PASSER.—Bill broad at base, slightly scooped at tip ; nostrils partly covered by plumes. Wings, second and third quills longest. Tail moderate. Habitat—Europe, Asia, and Africa.

Genus. ZONOTRICHIA.—Bill, tip not scooped, nostrils in small groove. Wings, second to fourth quills longest. Tail rather long and broad. Tarsi short. Outer toe united at base, hind toe long, with very long curved claws. Habitat—America.

Genus. AMMODROMUS.—Bill straight, keel prominent at base ; nostrils in small groove. Wings, second to fourth quills longest, first short. Tail, lateral feathers graduated. Hind toe long, and with long claws. Habitat—America.

Sub-family. **Emberizinæ.**—Bill rather flattened above, straight, edges waved, a hard knob on palate of upper mandible. Wings pendant and pointed. Tarsi moderate, scaled ; hind toe longer than inner.

Genus. EUSPIZA.—Bill slightly scooped at tip ; nostrils oval, partly hidden by plumes. Wings, second to fourth quills longest. Tail slightly forked. Tarsi with long scales. Habitat—Asia, Eastern Europe, and America.

Genus. EMBERIZA.—Bill small, nostrils partly hidden by plumes. Wings, second and third quills longest. Tail rather forked, feathers sharp. Claw of hind toe long. Habitat—Old and New World.

Genus. PLECTROPHANES.—Bill short, advancing on forehead, palatial knob elongated. Wings, second and third quills longest. Claw of hind toe nearly straight. Habitat—Europe and North America.

Sub-family. **Alaudinæ.**—Wings with tertials nearly or quite as long as primaries. Claws long, curved. Head of hind toe very long and nearly straight.

Genus. ALAUDA.—Bill slightly arched above, nostrils oval, hidden by plumes. Wings, third quill generally longest, first sometimes spurious. Tail moderate, slightly forked. Habitat—Old World.

Genus. OTOCORIS.—Bill slender. Wings long, first to third quills longest. Tail long and even. Tarsi and toes short. Habitat—Northern Europe and America.

Sub-family. **Pyrrhulinæ.**—Bill very stout, short, arched above. Wings moderate, rounded. Tail moderate, slightly forked. Tarsi short.

Genus. PYRRHULA.—Bill rather flattened at base, wider than high. Wings, second to fourth quills longest. Tarsi and toes strongly scaled. Habitat—Europe and Northern India.

Sub-family. **Loxianæ.**—Bill broad, compressed towards tips, which sometimes cross, or upper overhangs lower, curved. Wings rather long. Tail moderate. Tarsi short.

Genus. LOXIA.—Bill moderate, mandibles crossing. Wings, first to third quills longest. Tail short, slightly forked. Tarsi scaled, flattened below knee.

Sub-family. **Phytotominæ.**—Bill short, edges notched.

Genus. PHYTOTOMA.—Bill, gradually compressed to tip, keel rather arched, edges finely notched. Wings, third to fifth longest. Hind toe long. Habitat—South America.

Family. **COLIDÆ.**—Bill moderate, keel elevated at base, sides compressed to tip ; nostrils in large membranous grooves. Wings short. Tail long and graduated. Toes long, hind toe directed forwards.

Genus. COLIUS.—Bill, edges waved. Wings, second to fourth quills longest. Tail, feathers narrow. Habitat—Africa.

Family. **MUSOPHAGIDÆ.**—Bill, keel much elevated and curved, sides much flattened, tip deeply scooped. Wings long and round. Tail long and broad. Tarsi moderate, transversely scaled in front.

Sub-family. **Musophaginæ.**—Outer toe capable of being turned back.

Genus. MUSOPHAGA.—Bill large, much advanced on forehead ; nostrils near middle. Wings rather short, fourth and fifth quills longest, tertials broad. Tail rounded. Habitat—Africa.

Genus. TURACUS.—Bill short; nostrils in middle partly covered with plumes. Wings, fourth to seventh quills longest. Habitat—Africa.

Genus. SCHIZORHIS.—Bill short, much arched, nostrils near base, and elongated. Wings, fourth to sixth quills longest. Tail long and equal. Edges of toes with a membrane. Habitat—Africa.

Sub-family. **Opisthocominæ.**— Toes long, outer not capable of being turned back.

Genus. OPISTHOCOMUS.—Bill, upper mandible hooked over base, which is suddenly terminated; nostrils surrounded by a membrane, in middle of bill. Wings, sixth quill longest, first five graduated. Sides of head bare. Habitat—South America.

Family. **BUCEROTIDÆ.**—Bill long, curved, broad at base, compressed to tip; keel mostly with bony helmet; nostrils at base. Wings moderate. Tail broad and graduated.

Genus. BUCEROS.—Bill very large, helmeted, edges notched in adult. Wings, third and fourth quills longest. Tail mostly long. Tarsi strong, broadly scaled in front. Toes united at base, outer to second joint. Face and throat nearly bare. Habitat—Africa and India.

Order. **SCANSORES.**—Toes arranged two in front and two behind.

Family. **RHAMPHASTIDÆ.**—Bill very long and wide, curved above, compressed, edges notched.

Genus. RHAMPHASTOS.—Bill, nostrils at base, nearly hidden by projection of keel. Wings short, rounded, fifth quill longest, four first graduated and narrow at tip. Tail short, even. Toes, outer pair larger than inner; claws strong. Habitat—South America.

Family. **PSITTACIDÆ.**—Bill large and powerful, much arched, tip elongated, base with a cere containing nostrils. Wings and tail usually long.

Sub-family. **Pezoporinæ.**—Bill moderate, tip sometimes toothed. Tarsi short and strong. Tail broad, long, and graduated, feathers narrowed at tip.

Genus. NYMPHICUS.—Bill strongly toothed, compressed; nostrils large. Wings very long, second quill longest. Tail, two central feathers lengthened. Tarsi with small scales. Outer front toe longest. Habitat—Australia.

Genus. PLATYCERCUS.—Bill, sides swollen, tip rather blunt, edges very slightly toothed or entire, cere small and rounded. Wings concave, second and third quills longest. Habitat—Australia, New Zealand, and New Guinea.

Genus. PEZOPORUS.—Bill not toothed, cere broad, rounded. Wings, second and third quills longest. Tail slender, ends of feathers sharpened. Habitat—Australia.

Genus. PALÆORNIS.—Bill large, under mandible small, cere narrow. Wings, second quill longest. Tail, two middle feathers very long and slender. Habitat—India and Australia.

Genus. EUPHEMA.—Bill short, tip toothed, cere short and rounded. Wings, first to third quills longest. Habitat—Australia.

Genus. MELOPSITTACUS.—Bill with several small dentations near tip; cere broad, large and swollen. Wings, second quill longest. Habitat—Australia.

Genus. TRICHOGLOSSUS.—Bill slender, cere narrow. Wings, first to third quills longest. Tarsi partly hidden by feathers of thighs. Tongue with bristly brush at tip. Habitat—Australia and Indian Archipelago.

Sub-family. **Araina.**—Bill large, keel much arched to lengthened tip; lower mandible very deep. Tail long and graduated, tips of feathers narrow.

Genus. ARA.—Bill very large, nostrils hidden by plumes. Wings long and pointed, second and third quills longest. Sides of head rather bare. Habitat—South America.

Genus. CONURUS.—Bill, lower mandible very broad at base. Wings, first and third quills longest. Orbits bare. Habitat—America.

Sub-family. **Lorinæ.**—Bill slender, curved to tip, edges waved or notched.

Genus. CHARMOSYNA.—Wings long, first to third quills longest. Tail, two central feathers elongated. Habitat—New Guinea.

Genus. LORIUS.—Wings moderate, second and third quills longest. Tail moderate and rounded. Habitat—Moluccas, New Guinea, and Borneo.

Sub-family. **Psittacinæ.**—Bill large, sides compressed, arched to lengthened tip, edges toothed or festooned. Wings long and pointed. Tail squared. Tarsi short.

Genus. PSITTACUS.—Bill, upper mandible deeply scooped, lower deeply waved and sharp edged. Wings, second and third quills longest. Habitat—Africa and South America.

Genus. CHRYSOTIS.—Bill, upper mandible scooped, lower waved. Wings, third quill longest. Tarsi very short. Habitat—South America.

Genus. PSITTACULA.—Bill, edges festooned. Wings, first and second quills longest. Tail with ends of feathers squared or pointed. Habitat—America, Africa, and Asia.

Sub-family. **Cacatuinæ.**—Bill large, compressed, arched to tip. Tail long, broad, and even.

Genus. MICROGLOSSUM.—Bill very large, upper mandible with two teeth, lower much scooped near tip; nostrils hidden by feathers on cere. Cheeks and front of throat bare. Habitat—New Guinea, &c.

Genus. CACATUA.—Bill, keel flattened, edges deeply festooned, keel of lower mandible distinct, cere narrow. Wings, second to fourth quills longest. Habitat—Australia and Moluccas.

Genus. NESTOR.—Bill very long, keel of lower mandible flattened. Wings, third and fourth quills longest. Tail, shafts of feathers protrude beyond web. Habitat—Australia and New Zealand.

Genus. CALYPTORCHYNCHUS.—Bill large, lower mandible small, deeply scooped at tip. Wings, second to fourth quills longest. Habitat—Australia.

Genus. STRIGOPS.—Bill grooved on sides, edges toothed in middle, lower mandible deeply grooved, feathers at base with hair-like shafts. Wings, fifth and sixth quills longest. Tail, end of feathers pointed and shafts lengthened. Habitat—South Pacific Islands.

Genus. DASYPTILUS.—Bill long, lower mandible deeply scooped; keel bold. Wings, fourth quill longest.

Family. **PICIDÆ.**—Bill long, straight, sharp, and compressed at tip.

Sub-family. **Capitoninæ.**—Bill broad at base, with bristles. Tail short and nearly even.

Genus. LAIMODON.—Bill rather arched above, edges irregularly notched. Wings, third to sixth quills longest. Tarsi broadly scaled in front. Habitat—Africa.

Sub-family. **Picuminæ.**—Bill rather short. Tail short, tip of each feather broad and rounded.

Genus. PICUMNUS.—Wings rounded, third to fifth quills longest. Two front toes united to first joint. Habitat—South America and India.

Sub-family. **Picinæ.**—Bill truncated at tip; sides of upper mandible with a distinct ridge.

Genus. PICUS.—Bill, height and breadth equal; nostrils hidden by bristles. Wings short and pointed, fourth quill longest. Tail long, rounded. Habitat—Most parts of world.

Genus. CAMPEPHILUS.—Bill wider than high. Wings, third to fifth quills longest. Tarsi broadly scaled in front; inner hinder toe very short.

Sub-family. **Gecininæ.**—Bill, tip sharp and truncated, sides sloping, ridge extends to two-thirds length of upper mandible.

Genus. GECINUS.—Bill, keel slightly curved, edges straight. Wings, fourth quill longest. Tail, tip of feathers sharp and stiff. Tarsi short; claws large. Habitat—Old World.

Sub-family. **Melanerpinæ.**—Bill, lateral ridge half way between keel and edges.

Genus. MELANERPES.—Wings long, fourth quill longest, third nearly equal. Tail long, tips of feathers stiff and pointed. Habitat—America.

Sub-family. **Colaptinæ.**—Bill, lateral ridge very small or wanting.

Genus. COLAPTES.—Wings, fourth and fifth quills longest, first very short. Tail long, graduated, tips of feathers stiff. Habitat—America and South Africa.

Sub-family. **Yuncinæ.**—Bill short, straight, and sharp. Wings pointed. Tail rounded, feathers soft.

Genus. YUNX.—Bill, nostrils partly hidden by membrane. Tarsi scaled, partly covered with feathers. Two front toes united at base. Habitat—Europe, India, Africa.

Family. **CUCULIDÆ.**—Bill rather slender and compressed, curved above, tip scooped ; nostrils in membranous groove. Wings long. Tail rounded.

Sub-family. **Indicatorinæ.**—Bill short, broad at base, slightly curved above. Tail slightly forked. Tarsi very short.

Genus. INDICATOR.—Bill, tip scarcely scooped ; nostrils long and near keel. Wings, third to fifth quills longest. Front pair of toes united at base. Habitat—Africa, India, and Borneo.

Sub-family. **Saurotherinæ.**—Bill long, suddenly curved at tip, much compressed. Wings rounded. Tail long and graduated. Tarsi with transverse scales in front.

Genus. SAUROTHERA.—Bill, nostrils wide, partly covered by membrane. Wings, fourth to sixth quills longest. Outer lateral toes longest. Habitat—South America and West Indies.

Sub-family. **Coccyzinæ.**—Bill elevated at base, nostrils narrow, partly closed by a scale. Wings rounded. Tail long and graduated. Tarsi broadly scaled.

Genus. CENTROPUS.—Bill short, edges much curved, nostrils in short broad groove, opening oblique. Wings, fourth to sixth quills longest. Tail broad, rounded on sides. Outer front toe longest, inner hind toe with a long straight claw. Habitat—Warmer parts of world.

Genus. COCCYZUS.—Bill long, slender, nostrils in short membranous groove, opening oval. Wings, third and fourth quills longest. Habitat—America and parts of Europe.

Sub-family. **Crotophaginæ.**—Bill arched above, sides much compressed ; nostrils in substance of bill. Wings short and rounded. Tail long, broad, and graduated. Tarsi long, with large transverse scales. Two outer toes longest. Claws short.

Genus. CROTOPHAGA.—Bill short, keel deviated, tip hooked ; nostrils partly closed by membrane. Wings, fourth to sixth quills longest. Habitat—South America and West Indies.

Genus. SCYTHROPS.—Bill long and strong, hooked at tip, sides channelled. Wings, third quill longest. Two front toes united at base. Habitat—Australia.

Sub-family. **Cuculinæ.**—Bill broad, flattened at base, sides compressed to tip. Wings long. Tail long. Tarsi short, feathered and scaled.

Genus. CUCULUS.—Bill, nostrils in short, broad membranous groove, opening round. Wings, third quill longest. Tail, two outer feathers shortest. Tarsi feathered below knee. Outer front toe longest, united at base to inner. Habitat—Old World.

Order. **COLUMBÆ.**—Bill short, straight, compressed, front half of mandible vaulted, base slight, and with fleshy membrane, in which the nostrils are pierced. Wings moderate. Tarsi strong, and toes well padded beneath.

Family. **COLUMBIDÆ.**—Included in above.

Sub-family. **Columbinæ.**—Bill, nostrils narrow, longitudinal in fore-part of membrane. Wings pointed. Toes long.

Genus. CARPOPHAGA.—Bill, base flattened. Wings, second to fourth quills longest. Tail long. Tarsi very short, covered with down below the knee. Outer toe longer than

inner. Habitat—India, Moluccas, Australia, and Pacific Islands.

Genus. LOPHOLAIMUS.—Bill much compressed, nostrils oblique, near middle of bill. Wings very long, third quill longest. Tarsi half clothed with down. Habitat—Australia.

Genus. COLUMBA.—Bill, membrane swollen above nostrils. Wings, second quill longest. Tail moderate. Tarsi very short. Habitat—Most parts of world.

Genus. ECTOPISTES. — Bill, nostrils longitudinal, in middle. Wings long and pointed, first and second quills longest. Tail long, four central feathers sharp. Tarsi feathered below the knee. Habitat—North America.

Genus. TURTUR.—Bill slender. Wings long, second and third quills longest. Tail moderate, rounded. Tarsi scaled in front. Outer toe shorter than inner. Habitat—Europe, Africa, and Asia.

Sub-family. **Gourinæ.**—Bill moderate, straight. Tail moderate, rounded. Tarsi strong. Toes long, edged with membrane, hind toe very long, claws short.

Genus. OCYPHAPS.—Tip of both mandibles vaulted, nostrils longitudinal. Wings, third quill longest, and narrowed. Outer toe longer than inner. Habitat—Australia.

Genus. PHAPS.—Bill, nostrils at sides, membranous, narrow. Wings, second and third quills longest. Tarsi very short. Habitat—Australia.

Genus. LEUCOSARCIA (GEOPHAPS). — Bill, nostrils at sides of base, curved. Wings, third to fifth quills longest. Outer toe shorter than inner. Hind toe slender. Habitat —Australia.

Genus. CALÆNAS.—Bill strong, united near middle. Base of upper mandible with a wattle. Wings, third quill longest. Feathers of neck long and drooping. Habitat— Indian Archipelago.

Genus. GOURA.—Bill slender, slightly plumed at base ; nostrils in groove in middle of bill. Head with compressed crest. Wings, fourth to sixth quills longest. Tarsi with rounded scales. Habitat—New Guinea, and Indian Archipelago.

Sub-family. **Didunculinæ.**—Bill long, depressed at base, arched to sharp hooked tip ; lower mandible with three distinct teeth ; tip truncated ; nostrils narrow and oblique. Wings concave.

Genus. DIDUNCULUS.—Bill strong. Wings with blunt tubercle at bend, second to fourth quills longest. Skin round eye and sides of throat bare. Habitat—Samoan Islands.

Sub-family. **Didinæ.**—Bill large, covered with membrane for two-thirds its length, horny and vaulted in front ; nostrils in fore-part of membrane. Wings and tail almost rudimentary. Tarsi short and strong, with small scales.

Genus. DIDUS.—Bill, keel straight for some distance, then suddenly arched, and curving over tip of lower mandible. Outer toe shorter than inner, claws short, strong, and blunt. Habitat—Mauritius. (Extinct.)

Order. **GALLINÆ.**—Tarsi long and strong, often spurred. Toes connected at base ; hinder toe elevated from the ground.

Family. **CRACIDÆ.**—Bill moderate, rather arched, nostrils at sides of base and exposed. Wings short and rounded. Tail long and broad. Hind toe long, and on same plane with others.

Sub-family. **Cracinæ.** — Bill rather long, sides compressed, tip blunt. Nostrils large, partly closed by a membrane.

Genus. CRAX.—Bill much curved. Nostrils placed in a cere, and with their openings crescent-shaped. Wings, sixth to eighth quills equal and longest. Tarsi covered in front with broad scales ; toes also covered with large scales. Habitat—Tropical America.

Family. **MEGAPODIDÆ.**—Bill vaulted on tip. Wings very round. Tarsi covered with scales. Hind toe long and resting on ground. Claws thick, long, and slightly curved.

Sub-family. **Tallegallinæ.**—Bill, sides compressed.

Genus. TALLEGALLUS.—Head and neck very bare. Bill strong, opening of nostrils large, in membranous groove. Wings, fifth and sixth quills equal and longest. Tail long, rounded on sides. Tarsi powerful, feathered below knee, and covered with scales in front ; toes long, claws powerful and sharp. Habitat—Australia and New Guinea.

Sub-family. **Megapodinæ.**—Bill rather weak, sides compressed.

Genus. MEGAPODIUS.—Bill straight, rather vaulted towards tip ; nostrils with oval opening in fore part of membranous groove. Wings large and round. Tail short and round. Tarsi very powerful. Habitat—Australia and part of Asiatic Archipelago.

Genus. LEIPOA.—Bill like Megapodius ; nostrils in short membranous groove, with oblique opening. Tail long, broad, and rounded. Claws long, sharp, and flattened. Habitat—Australia.

Family. **PHASIANIDÆ.**—Bill moderate. Wings moderate and round. Tail long and broad. Tarsi mostly spurred. Toes moderate, hind toe short and elevated.

Sub-family. **Pavoninæ.**—Tail and coverts developed and depressed.

Genus. PAVO.—Bill apical, half arched. Wings short and round, sixth quill longest. Tail long and round, coverts very long, extending far beyond tail. Tarsi with conical spur. Toes moderate. Habitat—India.

Genus. POLYPLECTRON.—Bill, nostrils with longitudinal opening, partially concealed by membrane. Wings, fifth and sixth quills longest. Tail long, broad, and round. Tarsi with two or three spurs in male. Habitat—India.

Sub-family. **Phasianinæ.**—Tail long and wedge-shaped.

Genus. ARGUS.—Head and neck nearly bare. Bill rather long ; nostrils large, with opening partly closed by membrane. Wings short and round, secondaries much larger than quills. Tail, two central feathers very long and slender. Tarsi not spurred. Toes, outer longer than inner. Habitat—Indian Archipelago.

Genus. PHASIANUS.—Bill moderate ; nostrils in groove at base. Wings, fourth and fifth quills longest. Tail long, each feather narrowing at tip. Tarsi spurred in male. Toes, outer longer than inner. Habitat—Asia.

Sub-family. **Gallinæ.**—Bill moderate ; nostrils large, nearly semicircular, in membranous groove, and protected by a scale. Wings, secondaries equalling quills. Tail compressed. Tarsi spurred.

Genus. GALLOPHASIS.—Sides of head bare, skin hanging in wattles. Wings, fourth to seventh quills equal and longest, secondaries broad. Tarsi covered on both sides with large scales, spurred on inner side near hind toe. Claws short and curved. Habitat—India.

Genus. GALLUS.—Fleshy crest on head, and wattled below chin ; cheeks bare. Bill strong, nostrils exposed. Tail compressed, and covered by lengthened coverts. Tarsi covered in front with broad scales, and long curved spur. Hind toe little elevated. Habitat—India.

Genus. CERIORNIS.—Bill short and thick, tip blunt, nostrils oval and naked. Wings very concave. Tail large and round. Tarsi strong, scales in front, and short spur. Long horn over each eye. Two naked wattle-like spaces below chin. Habitat—Central Asia.

Sub-family. **Meleagrinæ.**—Tail short and drooping. Head and neck naked, haired, or wattled ; sometimes base of lower mandible wattled.

Genus. MELEAGRIS.—Bill strong, keel arched to tip, nostrils in groove. Wings short, fifth and sixth quills longest. Tail broad and rounded. Tarsi long and strong, scaled in front, and bluntly spurred. Front toes united at base. Habitat—America.

Genus. NUMIDÆ.—Bill, nostrils large, oval, partly covered by membrane. Wings, fifth quill longest. Tarsi with broad divided scales. Inner toe shorter than outer. Habitat—Africa.

Sub-family. **Lophophorinæ.**—Bill broad at base, upper mandible projecting, nostrils partly covered with feathers and membrane. Wings moderate. Tail large. Tarsi short, strong, with divided scales. Front toes united.

Genus. LOPHOPHORUS.—Bill flattened at base. Wings, fourth and fifth quills longest. Tarsi spurred in male. Habitat—Himalayas.

Family. **TETRAONIDÆ.**—Bill broad at base and compressed, arched above to blunt tip. Wings short and rounded. Tail rounded.

Sub-family. **Perdicinæ.**—Edges of bill entire, nostrils covered with a hard scale. Tarsi long, with divided scales in front.

Genus. ITHAGINIS.—Bill short. Wings, fourth to sixth quills longest. Tarsi with two or three spurs. Outer toe longer than inner ; hind toe long. Habitat—Himalayas.

Genus. PERDIX.—Bill short, nostrils with rounded scale. Wings, third to fifth quills longest. Tail short. Tarsi not spurred. Outer toe longer than inner. Habitat—Old World.

Genus. CACCABIS.—Bill, nasal groove with short plumes, scale bare. Wings, second to fifth quills longest. Tail short. Tarsi with a blunt tubercle. Habitat—Europe, Asia, and Northern Africa.

Genus. COTURNIX.—Bill short. Wings, second to fourth quills longest. Tail very short. Tarsi not spurred. Habitat —Old World and Australia.

Sub-family. **Odontophorinæ.**—Bill, lower mandible with two teeth, nostrils in short groove, covered with a scale. Wings and tail rounded. Tarsi with divided scales.

Genus. ORTYX.—Edges of beak curved and waved. Wings, third to sixth quills longest. Outer toe united to inner at base. Habitat—America, West Indies.

Sub-family. **Tetraoninæ.**—Bill short, broad, gradually compressed to tip, nostrils feathered. Wings rounded. Tarsi feathered, toes long.

Genus. TETRAO.—Wings, third and fourth quills longest. Tarsi feathered to base of toes. Toes with rough scales. Eyebrows bare, with red warty skin. Habitat—Northern—Europe and America.

Genus. BONASA.—Basal half of tarsi haired, remainder scaled. Habitat—Europe and America.

Genus. LAGOPUS.—Tarsi and toes wholly haired. Habitat—Europe and America.

Sub-family. **Pteroclinæ.**—Bill short, curved to tip, sides compressed. Wings and tail long and pointed. Tarsi feathered ; hind toe very small.

Genus. PTEROCLES.—Bill small, nostrils partly concealed by membrane. Wings, first and second quills longest. Tarsi, front and inner sides feathered. Toes short, united at base by membrane extending along sides of toes. Habitat—Southern Europe, Asia, and Africa.

Family. **CHIONIDÆ.**—Bill moderate, nostrils protected by horny sheath. Wings long and pointed. Tarsi short, front toes united at base, hind toe short.

Sub-family. **Chionidinæ.**—Horny sheath extending over the basal half of the bill.

Genus. CHIONIS.—Bill short, base broad ; base and cheeks covered with naked skin. Wings, second quill longest, tubercle at bend. Tarsi roughly scaled. Habitat—Islands of Antarctic Ocean.

Family. **TINAMIDÆ.**—Bill straight, flattened, with membrane at base, nostrils large. Wings short and concave. Toes long. (Sub-family, Tinaminæ, with same characteristics.)

Genus. TINAMUS.—Bill rather short, hooked at tip: sides compressed ; nostrils towards base. Wings, third and fourth quills longest, tips curved. Tail very short, coverts lengthened. Claws short and thick. Habitat—South America.

Order. **STRUTHIONES.**—Very large size. Wings almost rudimental.

Family. **STRUTHIONIDÆ.**—Wings very short. Toes short, broad, and unequal. Tarsi very long and powerful.

Sub-family. **Struthioninæ.**—Bill broad, flattened, rounded in front. Toes, second or third directed forwards.

Genus. STRUTHIO.—Bill, tip overlapping lower mandible, nostrils in broad membranous groove near middle. Wings with long soft plumes. Tail of curved drooping feathers. Claws short and flattened. Habitat—Africa.

Genus. RHEA.—Bill, nostrils in membranous groove at middle ; membrane extends over base of keel. Wings with long soft feathers. Tail not visible. Claws strong and compressed. Habitat—South America.

Genus. DROMAIUS.—Bill sloping to tip, nostrils in front of membranous groove. Wings and tail not visible. Habitat—New Holland.

Genus. CASUARIUS.—Bill long, compressed, curved to tip ; base of keel and head with elevated helmet ; neck naked, and with two wattles. Wings, five rounded shafts, no webs. Tail not visible. Toes long, inner with long powerful claw. Habitat—New Guinea and Moluccas.

Sub-family. **Apteryginæ.**—Bill very long, slender, base with horny cere, tip overhanging lower mandible, nostrils at tip.

Genus. APTERYX.—Bill, sides grooved towards end, base with long hairs. Wings very short, and covered with long feathers. Hind toe very short, with long, strong, and sharp claws. Habitat—New Zealand.

Sub-family. **Otidinæ.**—Bill compressed, tip scooped, nostrils in large membranous groove, opening oval. Wings moderate. Tail broad and rounded. Tarsi long ; toes and claws short.

Genus. OTIS.—Bill, straight at base, then curved to tip ; nostrils partly closed by membrane. Wings, second to fourth quills longest, tertials long as quills. Inner toe shorter than outer. Habitat—Europe and Asia.

Order. **GRALLÆ.**—Tarsi long, rounded, slender ; thighs naked at lower part.

Family. **CHARADRIADÆ.**—Bill short, weak at base, strong at tip ; nostrils in deep longitudinal groove. Hind toe wanting, or small and set high.

Sub-family. **Œdicneminæ.**—Bill long as head, keel of lower mandible angulated. Three toes in front.

Genus. ŒDICNEMUS.—Bill, nasal groove nearly triangular. Wings, second quill longest. Tail wedge-shaped. Tarsi with hexagonal scales. Toes united at base. Habitat—Most parts of world.

Sub-family. **Glareolinæ.**—Bill short, broad at base, compressed to tip. Wings very long, first quill longest, three toes in front, one behind.

Genus. GLAREOLA.—Bill flattened at base, arched to tip. Wings, first quill longest. Tail forked. Legs moderate, hind toe very short, claws of middle toe slightly toothed on one side. Habitat—Old World.

Sub-family. **Cursorinæ.**—Bill moderate, arched beyond nostrils, which are in short triangular groove. Wings long and pointed. Three slender toes in front.

Genus. CURSORIUS.—Bill broad at base, compressed to tip, straight above, and then arched ; nostrils narrow. Wings, first and second quills longest. Tail short and even. Habitat—Asia, Africa, and Europe.

Sub-family. **Charadrinæ.**—Bill slender, flattened at base, vaulted at tip, sides compressed and grooved ; nostrils in groove. Wings long and pointed. Tail broad. Tarsi long and slender. Outer toe longer than inner.

Genus. VANELLUS.—Bill straight, sides grooved two-thirds of length. Wings, second and third quills longest. Thigh sometimes feathered to knee. Four toes, hind toe short. Habitat—Europe, Asia, and America.

Genus. CHARADRIUS.—Bill short, strong ; groove of upper mandible for two-thirds of length. Wings, first quill longest. Hind toe wanting. Habitat—Most parts of world.

Sub-family. **Hæmatopodinæ.**—Bill long, strong, front half very much compressed, tip blunt ; nostrils in membranous groove. Wings long and pointed. Toes three, united at base.

Genus. HÆMATOPUS.—Bill, groove reaching nearly to middle. Wings, first quill longest. Habitat—Old and New World.

Sub-family. **Cinclinæ.**—Bill short, straight, compressed to tip ; nostrils in membranous groove. Wings long, first quill longest. Tarsi short, with small scales. Toes long and free, hind toe slender.

Genus. CINCLUS.—Bill flattened at base, straight above, edges curved upwards to tip ; groove extends half length of upper mandible. Tarsi with broad scales. Hind toe elevated, tip touching ground. Habitat—Most parts of world.

Family. **ARDEIDÆ.**—Bill long, sharp, compressed. Tail rounded. Tarsi long and slender. Outer toe longer than inner, united at base.

Sub-family. **Psophinæ**—Bill vaulted towards tip, which overhangs lower mandible ; nostrils in membranous groove reached by plumes. Hind toe short.

Genus. PSOPHIA.—Bill short, curved. Wings, fourth to sixth quills longest. Tail very short. Tarsi with transverse scales. Outer toe longer than inner ; claws long, compressed, sharp. Habitat—South America.

Genus. CARIAMA. Bill strong, curved ; nostrils in front of plumed groove. Wings, fifth to seventh quills longest, first very short. Tail long. Tarsi very long. Habitat—South America.

Sub-family. **Gruinæ.**—Bill long, mandibles equal ; nostrils in deep groove. Wings and tertials long. Tail short and even. Tarsi very long, toes short.

Genus. GRUS.—Bill slightly flattened at base, and curved to tip ; nostrils large and partly hidden by membrane. Wings, third and fourth quills longest. Tarsi with transverse scales. Outer toe united at base, hind toe very short. Habitat—Many parts of world.

Genus. SCOPS.—Bill, nostrils very narrow. Wings, third and fourth quills longest. Habitat—Europe, Asia, and Africa.

Genus. BALEARICA.—Bill rather short and thick. Wings, third quill longest. Cheeks naked, wattles on base of bill and throat. Habitat—Africa and Islands of Mediterranean.

Sub-family. **Ardeinæ.**—Bill long, sharp, gape extending beneath the eyes ; nostrils in lateral groove. Wings long. Tail short and even. Tarsi and toes long and slender.

Genus. ARDEA.—Bill slender, tip scooped, edges sometimes notched, nostrils guarded by membranous scale. Wings, second and third quills longest. Tarsi with transverse scales. Outer toe longer than inner. Habitat—Most parts of world.

Genus. BOTAURUS.—Bill rounded, tip deeply scooped, nostrils narrow, near base. Wings, first to third quills longest. Habitat—Many parts of world.

Genus. NYCTICORAX.—Bill, nostrils closed by membranous scale. Wings, second and third quills longest. Tarsi irregularly scaled. Habitat— Many parts of world.

Genus. CANCROMA.—Bill long, very broad towards middle, sides compressed to tip. Keel very large and grooved to tip, which is hooked ; nostrils partly closed by membrane. Wings, third and fourth quills longest. Tarsi irregularly scaled ; hind toe long. Habitat—South America.

Genus. PLATALEA.—Bill straight, flattened and widened at tip, upper mandible overhanging lower, groove commencing on forehead. Wings, second quill longest. Tarsi with netted scales. Habitat—Many parts of world.

Sub-family. **Ciconinæ.**—Bill long, compressed to tip ; nostrils narrow, pierced through beak. Wings large. Tail moderate. Tarsi long ; front toes united at base.

Genus. CICONIA.—Bill, keel bold and straight. Wings, third and fourth quills longest. Tarsi with netted scales, hind toe elevated, touching ground. Habitat—Europe, Asia, and Africa.

Genus. LEPTOPTILUS.—Bill very large, high at base, keel straight, nostrils small. Head and neck naked. Habitat—India and Africa.

Genus. MYCTERIA.—Bill very large, tip turned up. Wings, second and third quills longest. Habitat—South America, Africa, and Australia.

Genus. BALÆNICEPS.—Bill, sides enormously expanded, keel widened and flattened, with two grooves, edges overhanging lower mandible, tip deeply scooped and hooked, nostrils at base in groove and oblique, gape beyond eyes. Wings long. Toes very long and straight.

Sub-family. **Tantalinæ.**—Bill long, slender, and curved, sides gradually compressed to tip. Wings long. Tail even. Inner toe shorter than outer, hind toe long.

Genus. IBIS.—Bill, nostrils narrow, in narrow groove extending throughout the bill. Wings, first and second quills longest. Head partly bare. Habitat—Europe, Asia, and America.

Genus. GERONTICUS.—Wings, third and fourth quills longest. Head and neck rather bare. Tarsi with hexagonal scales. Habitat—Asia, Africa, America, and Australia.

Family. **SCOLOPACIDÆ.**—Bill long, slender, compressed and grooved to tip; nostrils in groove at base, closed by membrane. Wings long and pointed. Tarsi long and slender, hind toe short or absent.

Sub-family. **Limosinæ.**—Bill curved or straight from base. Tail mostly short and even. Toes long, united at base.

Genus. NUMENIUS.—Bill curved from base, upper mandible projecting. Wings, first quill longest. Tail short and even. Tarsi with narrow transverse scales in front. Hind toe slender, partly resting on ground. Habitat—Most parts of world.

Genus. LIMOSA.—Bill inclined upwards to tip. Wings, first quill longest. Outer toe united to middle as far as first joint. Habitat—Most parts of world.

Sub-family. **Totaninæ.**—Bill, groove as far as or beyond middle of bill, nostrils very narrow. Hind toe rather long and slender, barely reaching the ground.

Genus. TOTANUS.—Bill slightly curved at tip; groove half length of bill. Wings, first quill longest. Tarsi with very narrow scales in front. Habitat—Both hemispheres.

Genus. TRINGOIDES.—Bill rather straight above, curved at tip; groove extending nearly whole length of bill. Tail rounded, broad. Habitat—Old and New Worlds.

Sub-family. **Recurvirostrinæ.**—Nostrils very narrow, membranous. Tail short and rounded. Tarsi with netted scales in front. Thigh naked above knee.

Genus. RECURVIROSTRA.—Bill, keel flattened at base. Tarsi rather compressed; toes united by indented web; hind toe very short. Habitat—Most parts of world.

Genus. HIMANTOPUS.—Bill long and straight, opening of nostrils long and narrow. Wings, first quill longest. Toes united at base, hind toe wanting. Habitat—Many parts of the world.

Sub-family. **Tringinæ.**—Bill rather long, keel near tip, rather flat and wide; nostrils in groove, extending two-thirds of bill. Toes united at base.

Genus. PHILOMACHUS.—Bill straight. Wings, first and second quills longest. Outer toe united as far as first joint, inner free; hind toe short and elevated. Habitat—Europe and part of Asia.

Genus. TRINGA.—Wings, first quill longest. Toes edged by membrane. Habitat—Many parts of world.

Sub-family. **Scolopacinæ.**—Bill straight, rather flattened and bent downward at tip, which projects over lower mandible. Hind toe short, elevated, reaching ground.

Genus. GALLINAGO.—Nostrils oval. Wings, first and second quills longest. Thigh bare a little above knee. Claw of hind toe long and curved. Habitat—Most parts of world.

Genus. SCOLOPAX.—Nostrils narrow. Wings, first quill longest. Tarsi feathered below knee. Hind toe rather long and elevated. Habitat—Old World.

Sub-family. **Phalaropodinæ.**—Bill straight, but curved at tip. Tarsi short. Toes united at base and lobed on sides. Hind toe elevated, edged with narrow membrane.

Genus. PHALAROPUS.—Bill long. Wings, first and second quills longest. Toes united by membrane edging each toe. Habitat—Northern and temperate regions.

Family. **PALAMEDEIDÆ.**—Bill long and slender, keel rather flat, vaulted at tip, which overhangs lower mandible. Nostrils at sides of bill and longitudinal. Wings long, mostly spurred at shoulder. Tail short and round. Tarsi long and slender. Toes very long.

Sub-family. **Parrinæ.**—Bill, tip not scooped; groove long and narrow, containing nostrils in middle. Toes with long slender claws.

Genus. PARRA.—Bill, sides compressed, nostrils small and oval. Wings, third quill longest. Tail partly hidden by coverts. Claws long, hind toe and claw very long. Base of bill and part of head naked and wattled. Habitat—Asia, Africa, and America.

Genus. HYDROPHASIANUS.—Wings very long, second quill longest; shaft of first three long, fourth to seventh narrow and scooped. Tail narrow, four central feathers very long, rest short and graduated. Base of bill and head fully feathered. Habitat—India.

Sub-family. **Palamedeinæ.**—Bill short, compressed, keel curved to tip; nostrils large, in membranous groove. Wings with two spurs on shoulder. Tail moderate. Tarsi long, strong, small scaled. Front toes united at base, claws long and curved.

Genus. PALAMEDEA. Nostrils oval. Head with cylindrical horn. Wings, third and fourth quills longest. Toes with squared scales above. Habitat—South America.

Genus. CHAUNA.—Bill, tip vaulted and hooked. Wings, third and fourth quills longest. Habitat—Southern and Central America.

Family. **RALLIDÆ.**—Bill long, curved at tip, sides compressed, nostrils in membranous groove. Wings moderate. Tail rounded. Tarsi and toes long and slender.

Sub-family. **Rallinæ.**—Bill long and slender, keel bold, sides compressed. Toes free at base.

Genus. RALLUS.—Bill curved from nostrils to tip, which is slightly scooped; nostrils in groove, extending two-thirds the length of bill, opening narrow. Wings, second and third quills longest. Hind toe short and slender. Habitat—Many parts of world.

Genus. ORTRYGOMETRA.—Bill rather short, nostrils near middle of groove. Wings, second and third quills longest. Outer toe longer than inner, hinder toe very slender and rather short. Habitat—Most parts of world.

Sub-family. **Gallinulinæ.**—Bill short, keel advancing on forehead, sides compressed. Wings short and rounded. Toes very long, slender, edged with membrane, hind toe long.

Genus. PORPHYRIO.—Bill much elevated at base, which is flat and broad on forehead; nostrils in small groove. Wings, second to fourth quills longest. Tarsi with broad scales. Outer toe longer than inner; claws long and slender.

Genus. GALLINULA.—Bill suddenly curved at tip, nostrils near middle of bill in a groove. Wings, second to fourth quills longest. Habitat—Many parts of world.

Genus. FULICA.—Bill deep, keel straight, forming a flattened shield on forehead, curved near tip. Wings, second and third quills longest. Toes much lobed, inner with two lobes, middle with three, and outer with four. Habitat—Most parts of world.

Order. **ANSERES.**—Tarsi short, compressed, set far back. Toes webbed.

Family. **ANATIDÆ.**—Bill flat, broad, laminated on sides.

Sub-family. **Phœnicopterinæ.**—Bill long, rather compressed, suddenly bent downwards in middle. Tarsi very long, thighs also long and naked. Toes short.

Genus. PHŒNICOPTERUS.—Nostrils in groove, narrow, covered by membrane. Wings, first and second quills longest. Tail short. Hind toe nearly touching ground. Habitat—Warmer parts of world.

Sub-family. **Plectropterinæ.**—Bill long, with broad horny tip. Knee and end of thigh bare. Tarsi with squared scales. Hind toe long.

Genus. PLECTROPHANES.—Bill, keel sloping to tip, base large ; nostrils oval. Naked protuberance on base of keel. Wings, second to fourth quills longest, a spur on the bend. Tail short and round. Cheeks and part of neck bare. Habitat—Africa.

Sub-family. **Anserinæ.**—Bill not larger than head, keel elevated at base, sloping to tip, which has a hard horny tip. Knee bare, hind toe short, partially lobed.

Genus. CEREOPSIS.—Bill very short, arched above till near tip, and then flattened ; nostrils large, rounded, in cere. Wings, first quill short. Toes with indented web, hind toe not lobed. Habitat—Australia.

Genus. ANSER.—Bill rather long, laminæ at edge very wide, edge of upper mandible arched at base ; nostrils in middle of bill, longitudinal. Habitat—Europe, Asia, and America.

Genus. BERNICLA.—Bill shorter than head, laminæ not exposed, but wide ; nostrils narrow, in middle of bill. Wings, first and second, or second only, longest. Tail short and rounded. Hind toe very short and elevated. Habitat —Many parts of world.

Sub-family. **Cygninæ.**—Bill long as head, with a soft cere, bill equally broad throughout. Front toes with large web, hind toe not lobed, keel very long.

Genus. CYGNUS.—Cere extending to eye, tip horny. Wings, second and third quills longest. Tail short and rounded. Habitat—Northern Europe, Asia, and America.

Sub-family. **Anatinæ.**—Bill flattened towards tip, which has a horny termination. Hind toe long and slightly lobed.

Genus. AIX. — Bill shorter than head, horny tip very large, edges straight. Wings, second quill longest. Toes with large web. Habitat—America and China.

Genus. MARECA. — Bill equally broad throughout ; laminæ prominent. Wings, first and second quills longest. Habitat—Many parts of world.

Genus. ANAS.—Bill longer than head ; nostrils near base of keel. Wings, first quill longest, tertial long and sharp. Tail short and wedge-shaped. Habitat — Most parts of world.

Genus. QUERQUEDULA.—Bill long as head, horny tip, hooked and narrow. Wings, second quill longest, secondaries long and sharp. Habitat—Many parts of world.

Genus. SPATULA.—Bill longer than head, narrowed at base, having tip small and hooked ; laminæ of upper mandible long and slender ; nostrils oval, near base. Habitat— Most parts of world.

Sub-family. **Fuligulinæ.**—Bill elevated at base, flat and broad towards tip, which has a broad strong horny nail. Tarsi short and compressed. Toes long, and well webbed ; hind toe short, with broad membranous web.

Genus. FULIGULA.—Bill nearly long as head, edges curved upwards. Wings, first quill longest. Habitat—New Zealand, Northern Europe, Asia, and America.

Genus. NYROCA.—Bill long as head, laminæ not prominent, nostrils oval, near base. Wings, first and second quills longest. Habitat—Most northern parts of world.

Genus. SOMATERIA.—Bill divided at base by feathers ; front flattened and narrowed, laminæ wide. Wings, first and second quills longest. Habitat—Northern Europe and America.

Sub-family. **Merginæ.**—Bill straight, much compressed, keel elevated at base, convex towards tip, edges notched. Wings pointed. Tail short and rounded. Front toes well webbed, hind toe edged with web.

Genus. MERGUS.—Bill slender, tip hooked, edges widely notched. Wings, first and second quills longest. Tail graduated. Habitat—Northern regions.

Genus. MERGELLUS.—Bill much shorter than head, tip broad and hooked, edges notched closely, nostrils near middle of bill. Habitat—Northern regions.

Family. **COLYMBIDÆ.**—Bill long, compressed, and straight, nostrils in groove. Tail very short. Tarsi short and much compressed ; three front toes webbed, hind toe short, edged with slight membrane.

Sub-family. **Colymbinæ.**—Bill, tip curved, nostril at base. Wings long and pointed.

Genus. COLYMBUS.—Wings, first and second quills longest. Tail rounded. Habitat—Northern regions.

Sub-family. **Podicepinæ.**—Bill slightly curved above at tip ; nostrils oblong. Wings, first quill longest. Tail not visible. Front toes broadly lobed.

Genus. PODICEPS.—Nostrils in short groove. Wings, first and second quills slightly scooped at tips. Outer toe longer than inner, hind toe short, strongly lobed. Habitat—Most parts of world.

Family. **ALCIDÆ.**—Bill long, mostly curved to tip. Wings short. Tail short and graduated. Hind toe small or absent.

Sub-family. **Alcinæ.**—Bill short, compressed, boldly keeled above and below, top of upper mandible hooked ; nostrils narrow. Wings moderate, first quill longest. Hind toe wanting.

Genus. ALCA.—Beak plumed at base, front half horny, much compressed, keel of lower mandible curved downwards, nostrils at base. Wings pointed. Habitat—Northern regions.

Genus. FRATERCULA.—Bill short, all horny, very deep, very much compressed ; keel deep and sharp. Naked skin at gape. Habitat—Northern regions.

Sub-family. **Spheniscinæ.**—Bill long and straight, sides compressed and grooved, keel rounded at tip, nostrils in groove. Wings short, imperfect, feathers scale-like. Tail very short and stiff. Front toes webbed ; hind toe very small.

Genus. APTENODYTES.—Bill slender, base plumed, lower mandible covered with bare skin. Tarsi very short and plumed. Claws large. Habitat—Antarctic regions.

Sub-family. **Urinæ.**—Bill moderate, tip scooped, nostrils in plumed groove. Wings pointed. Tail short and rounded. Hind toe wanting.

Genus. URIA.—Bill, keel slightly curved. Wings, first quill longest. Tarsi with small scales. Habitat—Arctic regions.

Family. **PROCELLARIDÆ.**—Bill long, straight, compressed, very deeply grooved, tip strong, arched, and suddenly hooked, nostrils tubular and exposed.

Sub-family. **Procellarinæ.**—Nostrils in base of keel.

Genus. THALASSIDROMA.—Bill shorter than head, and slender ; nostrils elevated in front above keel, with a single aperture in front. Wings, second quill longest. Tail forked. Legs long and slender, thighs partially naked. Toes short, hind toe triangular. Habitat—Many parts of globe.

Genus. PROCELLARIA.—Bill not longer than head, tip compressed and sharp, nostrils with a single crescent-shaped opening. Wings, first quill longest. Tip of thigh nearly feathered, hind toe triangular. Habitat—High latitudes of both hemispheres.

Sub-family. **Diomedeinæ.**—Nostrils short, tubular, widest in front, and near base of lateral groove.

Genus. DIOMEDEA.—Bill longer than head, strong, lower mandible weak, compressed, truncated at tip. Wings very long and narrow, second quill longest. Two exterior toes edged on outside with narrow membrane, hind toe wanting. Habitat—Both hemispheres.

Family. **LARIDÆ.**—Bill straight, compressed. Wings long and pointed. Tail long. Tarsi with transverse scales in front. Hind toe mostly short.

Sub-family. **Larinæ.**—Bill long, straight above at first, then curved to tip.

Genus. STERCORARIUS.—Keel of bill covered with membranous or bony cere, tip wattled and strong; nostrils narrow, in fore part of cere. Wings, first quill longest; two central feathers of tail sometimes lengthened. Hind toe very small. Habitat—High latitudes of both hemispheres.

Genus. LARUS.—Bill not longer than head, nostrils at side and near middle, longitudinal. Wings, first quill longest. Tail even. Habitat—Most parts of world.

Genus. RISSA.—Bill longer than head. Hind toe rudimental. Habitat—Northern parts of world.

Sub-family. **Rhyncopinæ.**—Upper mandible much shorter than lower, and grooved to receive edge of lower, much compressed throughout. Wings, long, sharp, curved at tip. Tail forked. Front toes partly united by indented web.

Genus. RHYNCOPS.—Bill broad at base, then suddenly compressed; lower mandible truncated. Wings, first quill longest. Hind toe elevated, touching ground with tip, claws long and curved. Habitat—Tropical parts of world.

Sub-family. **Sterninæ.**—Bill long, slender, straight, and sharp; nostrils narrow at side of base. Wings long and pointed. Tail forked. Toes webbed.

Genus. STERNA.—Nostrils with plumes reaching the opening. Wings, first quill longest. Front toes united by indented web; hind toe very short. Habitat—Most parts of world.

Genus. ANOUS.—Bill longer than head, gradually curved above to tip. Tail with sides rounded or forked. Wings, first quill longest. Toes long and fully webbed, outer larger than inner. Hind toe long and slender. Habitat—Tropical seas.

Family. **PELECANIDÆ.**—Bill long, broad at base, straight, and compressed. Nostrils very narrow. Sometimes a pouch from base of lower mandible. Wings long, first quill longest. Tarsi short and strong. All toes united by membrane. Face and throat partially feathered.

Sub-family. **Phaetoninæ.**—Bill long as head, sharp, gently curved above. Two middle feathers of tail very long and narrow.

Genus. PHAETON.—Edges of bill notched, nostrils partly closed by membrane. Wings, first quill longest. Claws small. Habitat—Tropical seas.

Sub-family. **Plotinæ.**—Bill with edges finely notched, nostrils covered by a shield. Tail long and widening towards end. Front toes broadly webbed. Hind toe united to inner by web.

Genus. PLOTUS.—Bill longer than head, straight, slender, very sharp; nostrils scarcely visible. Wings, second and third quills longest. Outer toe long as middle. Habitat—Warmer parts of Asia, Africa, and America.

Sub-family. **Pelecaninæ.**—Bill long, flattened above, compressed, hooked at tip; nostrils in grooves, hardly visible. Tail short. All toes united by web. Lower mandible and throat with membranous pouch.

Genus. SULA.—Bill straight, slightly curved towards tip, grooved, and edges unequally notched. Wings, first and second quills longest. Tail graduated. Tarsi with keel behind. Claw of middle toe notched, that of hind toe rudimental. Below lower mandible and on breast, a naked expansile space. Habitat—Many parts of world.

Genus. GRACULUS.—Bill straight, slender. Wings, second and third quills longest. Tail rounded. Habitat—Most parts of both hemispheres.

Genus. PELECANUS.—Bill very long, rounded above at base, flat towards tip, which is strongly hooked, lower mandible broader at base than upper, narrowing to tip. Wings, second quill longest, secondaries nearly as long as quills. Tail short and rounded. Lower mandible with large extensile pouch. Habitat—Many parts of world.

Genus. ATAGEN.—Bill longer than head, flattened and concave above, then suddenly hooked and sharp, sides compressed and grooved. Wings very long and narrow, first and second quills longest. Tail very long and deeply forked. Tarsi very short, half feathered. Throat naked, capable of being dilated into a pouch as far as breast. Habitat—Tropics.

INDEX.

END OF BIRDS.